Y0-AAX-974

BLOOD VESSELS
AND
LYMPHATICS

BLOOD VESSELS
AND
LYMPHATICS

Edited by

DAVID I. ABRAMSON

Department of Physical Medicine and Rehabilitation
Department of Medicine
University of Illinois, College of Medicine
Chicago, Illinois

1962

ACADEMIC PRESS
New York and London

ACADEMIC PRESS INC.
111 FIFTH AVENUE
NEW YORK 3, N. Y.

United Kingdom Edition
Published by
ACADEMIC PRESS INC. (LONDON) LTD.
BERKELEY SQUARE HOUSE, BERKELEY SQUARE, LONDON W. 1

Library of Congress Catalog Card Number 62-13040

PRINTED IN THE UNITED STATES OF AMERICA

Contributors

Barry J. Anson, *Department of Anatomy, Northwestern University Medical School, Chicago, Illinois. Pages* 320, 703

Herbert J. Berman, *Department of Biology, Boston University, Boston, Massachuesetts. Page* 170

Robert J. Booher, *Department of Surgery, Cornell University Medical College; Memorial Cancer Center, New York, New York. Page* 633

Hubert R. Catchpole, *Department of Pathology, University of Illinois, College of Medicine, Chicago, Illinois. Page* 164

Kermit Christensen, *Department of Anatomy, St. Louis University School of Medicine, St. Louis, Missouri. Pages* 118, 192, 345, 486, 512

Jay D. Coffman, *Department of Medicine, Boston University School of Medicine; Peripheral Vascular Laboratory, Massachusetts Memorial Hospital, Boston, Massachusetts. Pages* 269, 274

Julius H. Comroe, Jr., *Department of Physiology, Cardiovascular Research Institute, University of California Medical Center, San Francisco, California. Page* 310

Hans Elias, *Department of Anatomy, Chicago Medical School, Chicago, Illinois. Page* 360

Harry C. H. Fang, *Division of Neurology, Birmingham Medical College, University of Alabama, Birmingham, Alabama. Pages* 230, 239, 249

* Eugene M. Farber, *Department of Dermatology, Stanford Medical Center, Palo Alto, California. Page* 489

Edwin R. Fisher, *Department of Pathology, Laboratory Service, University of Pittsburgh and Veterans Administration Hospital, Pittsburgh, Pennsylvania. Page* 654

Björn U. G. Folkow, *Department of Physiology, University of Göteborg, Göteborg, Sweden, Page* 120

George P. Fulton, *Department of Biology, Boston University, Boston, Massachusetts. Pages* 134, 137, 145

Vincent V. Glaviano, *Department of Physiology, Stritch School of Medicine, Loyola University, Chicago, Illinois. Page* 617

A. David M. Greenfield, *Department of Physiology, The Queen's University of Belfast, Belfast, Northern Ireland. Page* 197

Donald E. Gregg, *Department of Cardiorespiratory Diseases, Walter Reed Army Institute of Research, Washington, D. C. Pages* 269, 274

Francis J. Haddy, *Departments of Medicine and Physiology, Northwestern University School of Medicine; Professional Services for Re-*

* Collaborator.

search, Veterans Administration Research Hospital, Chicago, Illinois. Pages 61, 406

* Edgar G. Harrison, Jr., *Department of Surgical Pathology, Mayo Clinic; Department of Pathology, Mayo Foundation, Graduate School, University of Minnesota, Rochester, Minnesota. Page* 719

George M. Hass, *Department of Pathology, University of Illinois School of Medicine; Presbyterian-St. Luke Hospital, Chicago, Illinois. Pages* 566, 587

Robert H. Haynes, *Department of Biophysics, Committee on Biophysics, University of Chicago, Chicago, Illinois. Page* 26

Patrick J. Kelly, *Department of Orthopedic Surgery, Mayo Clinic-Mayo Foundation, Rochester, Minnesota. Page* 531

John E. Kirk, *Department of Medicine, Division of Gerontology, Washington University School of Medicine, St. Louis, Missouri. Pages* 82, 211, 583

William H. Knisely, *Department of Anatomy, University of Kentucky School of Medicine, Lexington, Kentucky. Page* 294

Richard H. Licata, *Congenital Heart Disease Research and Training Center, Hektoen Institute for Medical Research of Cook County Hospital, Chicago, Illinois. Pages* 3, 258, 292, 318, 388

† Brenton R. Lutz, *Department of Biology, Boston University, Boston, Massachusetts. Page* 137

Hymen S. Mayerson, *Department of Physiology, School of Medicine, Tulane University, New Orleans, Louisiana. Pages* 157, 701, 708

Malcolm B. McIlroy, *Department of Medicine, Cardiovascular Research Institute, University of California Medical Center, San Francisco, California. Page* 310

Gardner C. McMillan, *Department of Pathology, Pathological Institute, McGill University, Montreal, Canada. Pages* 648, 670, 680

Kenneth I. Melville, *Department of Pharmacology, Faculty of Medicine, McGill University; Montreal General Hospital; Royal Victoria Hospital, Montreal, Canada. Pages* 77, 209, 340, 380, 411, 523, 670

Nicholas A. Michels, *Department of Anatomy, Daniel Baugh Institute of Anatomy of the Jefferson Medical College, Philadelphia, Pennsylvania. Pages* 9, 324, 349, 390

* Theodore R. Miller, *Department of Surgery, Cornell University School of Medicine; Memorial Hospital for Cancer and Allied Diseases, New York, New York. Page* 640

Anthony P. Moreci, *Stanford Medical Center, Palo Alto, California. Page* 489

* Collaborator.

† Deceased.

GEORGE T. PACK, *Department of Clinical Surgery, New York Medical College; Cornell University School of Medicine; Memorial Hospital for Cancer and Allied Diseases, New York, New York. Page* 640

JAMES C. PATERSON, *Department of Medical Research, University of Western Ontario; Westminster Hospital, London, Ontario, Canada. Page* 688

CARL M. PEARSON, *Department of Medicine, University of California Medical Center, Los Angeles, California. Pages* 512, 515, 526

DANIEL C. PEASE, *Department of Anatomy, School of Medicine, University of California Medical Center, Los Angeles, California. Pages* 12, 233, 395, 537

CLARENCE N. PEISS, *Department of Physiology, Stritch School of Medicine, Loyola University, Chicago, Illinois. Page* 96

* LOWELL F. A. PETERSON, *Department of Orthopedic Surgery, Mayo Clinic-Mayo Foundation, Rochester, Minnesota. Page* 531

CONRAD L. PIRANI, *Department of Pathology, University of Illinois College of Medicine; University of Illinois Research and Educational Hospital, Chicago, Illinois. Pages* 413, 606

ELIZABETH M. RAMSEY, *Department of Embryology, Carnegie Institution of Washington, Baltimore, Maryland. Page* 465

WALTER C. RANDALL, *Department of Physiology, The Stritch School of Medicine, Loyola University, Chicago, Illinois. Page* 105

ELLIOT RAPAPORT, *Department of Medicine, University of California School of Medicine; Cardiopulmonary Unit, San Francisco General Hospital, San Francisco, California. Page* 306

MANUEL RIVEROS, *Department of Clinical Surgery, School of Medicine of Asuncion; Clinical Surgery Service, Primera Catedra Universitaria of the School of Medicine of Asuncion, Paraguay. Page* 638

JOSEPH E. ROBERTS, *Cardiology Section, Veterans Administration Hospital; Department of Medicine, University of Buffalo School of Medicine, Buffalo, New York. Page* 280

SIMMON RODBARD, *University of Buffalo Chronic Disease Research Institute; Department of Experimental Medicine, University of Buffalo School of Medicine, Buffalo, New York. Page* 31

LEO A. SAPIRSTEIN, *Department of Physiology, Ohio State University, Columbus, Ohio. Pages* 66, 135, 207, 496, 521

ALEXANDER SCHIRGER, *Section of Medicine, Mayo Clinic; Department of Medicine, Mayo Foundation, Graduate School, University of Minnesota, Rochester, Minnesota. Page* 719

* ROBERT L. SCHULTZ, *Department of Anatomy, College of Medical Evangelists, Loma Linda, California. Page* 233

* Collaborator.

EDWARD E. SELKURT, *Department of Physiology, Indiana University School of Medicine, Indiana University Medical Center, Indianapolis, Indiana. Pages* 330, 369, 382

* JOHN W. SEVERINGHAUS, *Department of Anesthesiology, University of California School of Medicine, Cardiovascular Research Institute, University of California Medical Center, San Francisco, California. Page* 306

* HYMAN E. SHISTER, *Department of Pharmacology, McGill University; Reddy Memorial Hospital, Montreal, Canada. Pages* 77, 209, 340, 380, 411, 523, 670

RALPH R. SONNENSCHEIN, *Department of Physiology, School of Medicine, University of California Medical Center, Los Angeles, California. Pages* 71, 241

JEREMIAH STAMLER, *Department of Medicine, Northwestern University Medical School; Heart Disease Control Program, Chicago Board of Health, Chicago, Illinois. Page* 557

NORMAN C. STAUB, *Department of Physiology, Research Institute, University of California Medical Center, San Francisco, California. Page* 299

LEON H. STRONG, *Department of Gross Anatomy, Chicago Medical School, Chicago, Illinois. Page* 221

C. BRUCE TAYLOR, *Department of Pathology, Northwestern University Medical School, Chicago, Illinois; Evanston Hospital Association, Evanston, Illinois. Page* 572

LOUIS TOBIAN, *Department of Medicine, The Medical School, University of Minnesota, Minneapolis, Minnesota. Page* 595

W. CURTIS WORTHINGTON, JR., *Department of Anatomy, Medical College of South Carolina, Charleston, South Carolina. Page* 428

* LUCIANO ZAMBONI, *Department of Anatomy, McGill University, Montreal, Canada. Page* 537

* Collaborator.

Preface

The plan of this volume has been to assemble, consolidate, and interrelate current data on the embryology, anatomy, physiology, pharmacology, biochemistry, and pathology of blood vessels and lymphatics. Information has thus been compiled which has previously been available only through exhaustive and time-consuming perusal of a large number of original papers and review articles, disseminated in a great many journals. It is hoped that the final product will provide a critical, provocative, and authoritative summary of our present knowledge of vascular and lymphatic responses, which will be of value to the research worker and the advanced student in the field and to the specialist desiring background and sources on subjects other than his immediate concern. At the same time, it should help focus attention on the existence of gaps in our knowledge and perhaps act as a springboard for projection and exploration into new avenues and spheres of research.

Since the broad scope of the subject matter precludes its treatment by a single or even several authors, the book has been formulated by a large number of investigators, each of whom has contributed one or several sections, limited to the areas of his endeavor and research interests. In most instances individual chapters have been written by a number of different authors, many of whom have sections in other portions of the book. In order to maintain a certain degree of similarity in format, rigid criteria have been followed with regard to the types of headings and subheadings and to the style of presentation of the material. The task of the Editor has been to integrate and mold the various accounts of the contributors into a systematic and coordinated volume.

At no time have the viewpoints expressed by a contributor been altered. However, the presence of these views in the content does not necessarily indicate that they are those held by the other authors or by the Editor. It has been felt that any hypothesis which is provocative and is supported by substantial experimental and clinical data should be included, even though the available evidence is not sufficient for its unquestioned acceptance. Alternative, and perhaps unorthodox, concepts have also been presented where differences of opinion exist.

It is believed that this approach to the organization of a multiauthored treatise will minimize the objection frequently directed at such a project that the result is not an integrated account but actually a compilation of a large number of papers, designated as chapters, lacking continuity and containing repetition of material and differences of opinion in various portions. Under the present plan, the intimate contact each contributor has with his field permits a comprehensive survey of available material, with a minimum of speculation and of sweeping generalities. A possible objection is that the subject matter may conceivably be slanted more than is warranted in the direction of the interests of the various authors. However, this would appear preferable to the frequently uncritical evaluation of the literature by an author who has been assigned a subject regarding which he has only a theoretical knowledge.

Because of the large number of subjects considered and the relatively small size of the work, it has been necessary to provide an extensive bibliography of current research, as well as of reviews permitting access to previous papers, in order to produce adequate coverage of the literature. Attention has also been paid to papers beyond national boundaries.

The subject matter has been divided into four parts. In the first part is presented, in broad terms, a discussion of the various types of blood vessels: the large and small arteries and arterioles, the microcirculation, and the venous system. In the second part, the vascular tree in the different organs is described, so as to make readily available to the reader information regarding a specific local circulation. The third part deals with different pathologic vascular states, while the final one is devoted to the lymphatic system. Descriptions of the clinical manifestations of the various disorders affecting blood vessels and lymphatics have not been included, for if this had been done, it would have resulted in an unwieldy volume, at the same time detracting from the intended emphasis on the basic approach to the problem.

I wish to acknowledge my indebtedness to Doctor Geoffrey H. Bourne, who originally pointed out that a book of this type was not available and that such an undertaking would fill a void in our knowledge of the field of vascular responses. I am grateful to my secretary, Miss Naomi McCutcheon, who was involved in the voluminous correspondence required first, in obtaining a large group of contributors willing to participate in the project and second, in supporting them in their determination to complete their assignments within the deadline set up for publication. Finally, I wish to express my appreciation to the staff of Academic Press, for their patience, cooperation, and advice during the preparation of this volume.

DAVID I. ABRAMSON

February, 1962

Contents

PART ONE

GENERAL CONSIDERATIONS OF THE CIRCULATION OF THE BLOOD

CHAPTER I

ARTERIAL AND ARTERIOLAR SYSTEMS: Embryology and Gross, Microscopic and Submicroscopic Anatomy

CHAPTER II

ARTERIAL AND ARTERIOLAR SYSTEMS: Biophysical Principles and Physiology

CHAPTER III

ARTERIAL AND ARTERIOLAR SYSTEMS: Pharmacology and Biochemistry

CHAPTER IV

SYMPATHETIC INNERVATION OF ARTERIAL TREE

CHAPTER V

MICROCIRCULATION

CHAPTER VI

MICROCIRCULATION

CHAPTER VII

VENOUS SYSTEM

PART TWO

SPECIAL VASCULAR BEDS

CHAPTER VIII

CIRCULATION TO THE BRAIN AND SPINAL CORD

CHAPTER IX

CORONARY CIRCULATION

CHAPTER X

PULMONARY CIRCULATION

CHAPTER XI

GASTROINTESTINAL CIRCULATION

CHAPTER XII

HEPATOPORTAL-LIENAL CIRCULATION

CHAPTER XIII

RENAL CIRCULATION

CHAPTER XIV

THE BLOOD VESSELS OF THE PITUITARY AND THE THYROID

by W. C. Worthington, Jr.

CHAPTER XV

PLACENTAL CIRCULATION
by E. M. Ramsey

CHAPTER XVI

CUTANEOUS CIRCULATION

CHAPTER XVII

CIRCULATION IN SKELETAL MUSCLE

CHAPTER XVIII

CIRCULATION OF BONE

PART THREE

DISORDERS AFFECTING THE ARTERIAL OR VENOUS CIRCULATION

CHAPTER XIX

ARTERIOSCLEROSIS:
Atherosclerosis; Mönckeberg's Sclerosis

CHAPTER XX

CONDITIONS AFFECTING BLOOD PRESSURE

CHAPTER XXI

PERIPHERAL VASCULAR DISORDERS

PART FOUR

LYMPHATIC SYSTEM

CHAPTER XXIII

LYMPHATIC VESSELS AND LYMPH

CHAPTER XXIV

LYMPHATIC DISORDERS

by A. Schirger and E. G. Harrison, Jr.

PART ONE

General Considerations of the Circulation of the Blood

Chapter I. ARTERIAL AND ARTERIOLAR SYSTEMS: Embryology and Gross, Microscopic and Submicroscopic Anatomy

A. EMBRYOLOGY

By R. Licata

1. Primitive Plan

a. Vascular arcs of prenatal circulation: Prenatally, three primary vascular arcs represent the basic plan upon which the adult circulation is constructed (Keibel and Mall, 1912). Two of these, designated as vitelline (visceral) and umbilical (placental), are regarded extraembryonic, due to their peripheral distribution. The remaining arc gives rise to the systemic circulation and is considered intraembryonic. The earliest indication of vascular development may be seen in the wall of the rudimentary yolk sac, where blood islands composed of hemangioblastic elements appear *in situ.* Initially, a wide unconnected vascular mesh is developed precociously from these isolated vasculogenic foci and subsequently becomes consolidated into a plexus. The vitelline plexus, thereby developed, moves centrally via the yolk stalk, consequently encroaching upon the embryo. The resulting network foreshadows the establishment of the alimentary or omphalomesenteric arc of circulation. Simultaneously, a similar process is generated in the body stalk, from where a primitive vascular network extends peripherally into the outlying chorionic membranes. Centripetally the plexus gains entrance into the embryonic body at the root of the body stalk. In this manner an umbilical or placental arc is established, carrying the embryo into functional relations with the maternal tissues.

b. Intraembryonic systemic circulation: Prior to formation of a systemic circulation, vascular channels are developed by *in situ* differentiation and extensions of pre-existing vessels, generally along lines paralleling the body axis. From this system of undeclared endothelial tubes are etched the main somatic postnatal circuits. Foremost is a paired dorsal system of paraxially arranged plexiform vessels, running coextensively with the developing neural tube. The longitudinal channels represent the primordia of paired dorsal aortae. In the cephalic region these vessels curve forward within the substance of the mandibular arches to become continuous, by ventral extension, with the endocardial anlagen of the developing heart. The duplicated ventral segments of this circuit are frequently referred to as the ventral aortae, which become confluent at the arterial end of the embryonic tubular heart. With the first heart beats, the circulation of blood is activated, and formerly undeclared channels can be distinguished as arteries or veins, depending on the given direction of flow of the contained vascular elements.

As in the case of the mandibular arch, paired serially arranged aortic arches appear within each set of pharyngeal arches. The sequence of development of the homologous segments proceeds in a cephalocaudal direction. Ideally, six such sets of aortic arches are developed, but at no one time do the paired vessels appear simultaneously, since some are periodically abandoned for preferential channels. The first arch, for example, begins to degenerate at a time when the second is completing its development. The gradual transformation of the heart from its tubular configuration to a septated four-chambered organ necessitates a rerouting of the intracardiac streams. The partitioning of the truncus arteriosus plays an active major role in determining the fate of the aortic arch derivatives.

2. Development of Aorta

a. Partitioning of truncus arteriosus: The truncus arteriosus represents an originally common trunk, through which the outlet end of the heart communicates with the aortic arches via the ventral aortae. During the second month of development the truncus is partitioned by the action of a spiral aortic septum and is consequently divided into an aortic and a pulmonary trunk (Patten, 1953). Dorsally, axially oriented channels presage the development of paired primary dorsal aortae. The single aorta recognized postnatally is developed from fusion of the two dorsal vessels. This process is initiated in the thoracic region and involves all segments except the first eight presumptive cervical levels; it extends caudally, terminating at lower lumbar landmarks. In the postnatal condition the point of termination represents the bifurcation of the main line vessel and origin of the common iliac arteries.

b. Derivation of postnatal aorta: From the truncus arteriosus and proximal segment of the left ventral aorta (Fig. 1) is derived the ascending aorta which leads into the corresponding left fourth aortic arch. By continuity with the normally unfused segment of the dorsal aortic root of

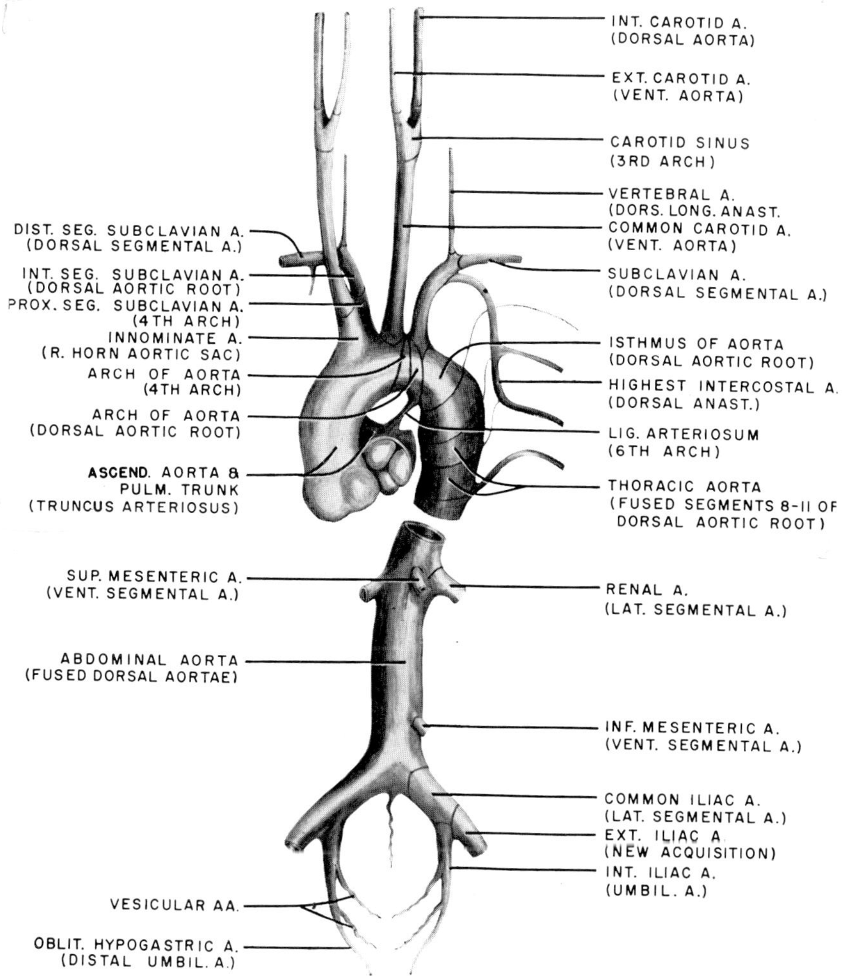

FIG. 1. Postnatal plan of aorta and major divisions. The derivation of the different outlined segments of the aorta is indicated. Embryonic rudiment in parentheses. (Adapted from several sources, principally A. Barry, 1951.)

its respective side, the enlarging left fourth arch completes the groundwork for differentiation of much of the definitive arch of the aorta. That part of the descending aorta which curves over the left bronchus in the adult develops from the remainder of the unfused segment. The homologue of the dorsal aortic root, namely, the distal part of the right fourth arch, becomes progressively attenuated and finally lost, although a vascular rudiment may sometimes persist.

c. Intersegmental branches: Dorsally oriented rootlets, arranged in symmetrical pairs, develop from the main aortic trunk and are fixed at intersegmental somite levels. Invariably, these arterial offshoots bear a close relationship to the costal element at the corresponding level. At all levels the pre- or postcostal position of the main arterial stem is dependent upon the mode of final differentiation of the costal element. With the establishment of a vertebral column, a shift in relations occurs, so that the intersegmentally placed vessels lose their original position and finally overlie the vertebral centra. The dorsal segmental artery essentially represents a somatic feeder which is divided into dorsal and lateroventral primary rami. The former division supplies the dorsal body wall and vertebral column, the latter, the lateral and ventral body wall areas via extensions forming the intercostolumbar system of arteries.

3. Main Aortic Branches (Medium-Sized Vessels)

The idealized schematization illustrated in Figs. 1 and 2 (Barry, 1951), represents the architectural plan of the ancestral lineage of the major segments and branches of the aorta. In the following account each branch shall be considered individually.

a. Subclavian artery: Bilaterally, the distal segment of this vessel is derived from the ventral ramus of the seventh dorsal intersegmental artery. An apparent ascent, countenanced by caudal migration of the developing heart, carries the vessel to a position overlying the first rib in the adult. Concomitant with reduction of the left fourth arch and dorsal aortic root, the base of the left subclavian climbs the arch of the aorta, finally to implant itself at a point close to the origin of the common carotid. During this migration, segments 8 and 9 of the dorsal aortic root of the same side contribute considerably to the mass of the aortic trunk. In its proximal extent the right subclavian artery is derived from a greatly reduced fourth arch, forming a composite segment with an equally foreshortened dorsal aortic root. The remainder of the dorsal aortic root below the level of the subclavian artery completely regresses, thereby losing all connection with the dorsal aorta. A vestige of this vessel may sometimes be found within the mediastinum, as the arteria aberrans.

b. Innominate artery: The right horn of the aortic sac elongates to form a stout trunk, the innominate artery, which serves as the main pathway of supply to the upper ipsolateral part of the body. In configuration this artery appears as a conical sacculation from the aortic wall, serving as a base-stem support for the carotid and subclavian

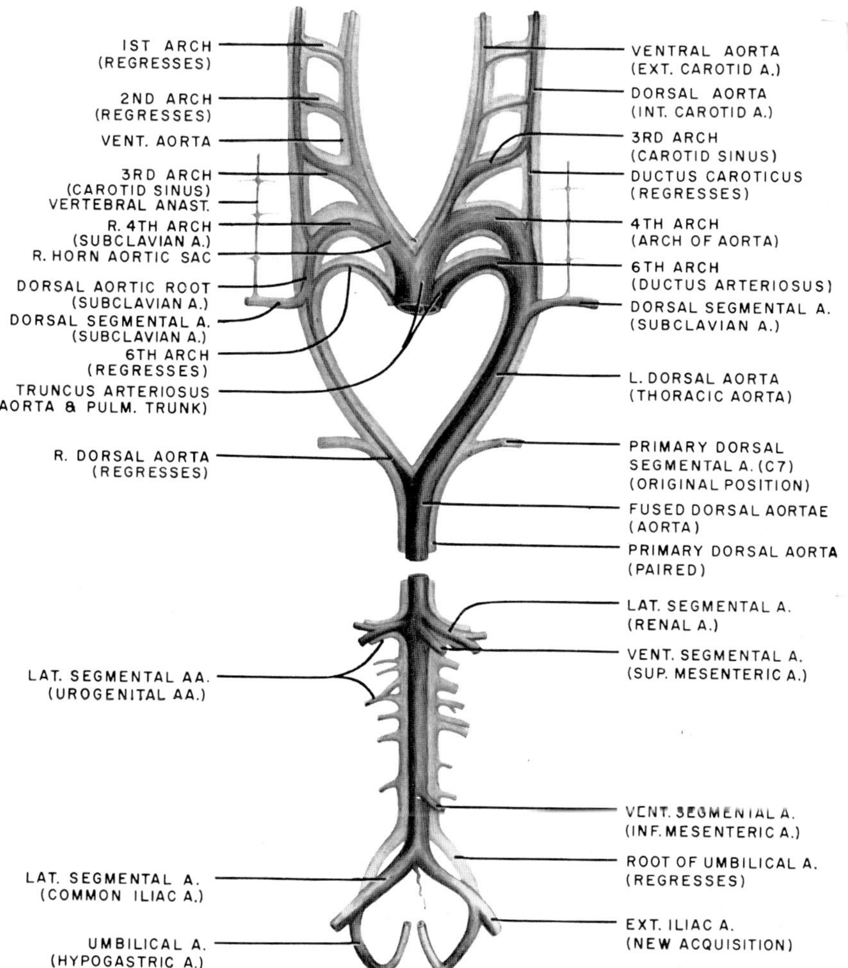

FIG. 2. Silhouette of primary arch system in background (light). Rudiments of derived definitive vessel shown in relief over background diagram and indicated in parentheses.

arteries. The homologous left horn area is absorbed into the wall of the ascending aorta, adjacent to the insertion of the left common carotid artery. That segment of the aortic arch, indicated in Fig. 1 as a wedge-shaped projection, comprises the foreshortened fourth arch and dorsal aortic root which are crowded into this aortic territory.

c. Common carotid artery: Because of rapidly changing relations of left and right fourth arch patterns, the common carotid arteries issue differently out of the arch of the aorta. In each case, the common stem is derived from the corresponding ventral aortic root, but the two vessels finally come to hold different positional relations with respect to each other. The right common carotid arises from the innominate artery, whereas the left takes its origin as a centrally implanted vessel off the main arch. The bifurcation at carotid sinus levels represents the markedly reduced but dilated remains of the third aortic arch. With regression of arches one and two, the distal segments of the ventral and dorsal aortae are freed to follow separate pathways of development. On each side the internal carotid is carried forward into the carotid canal as a prolongation of the original dorsal aorta. The external carotid proximally comprises the free segment of the ventral aortic root, and, by continued outgrowth, divides into the multiple branchings supplying the superficial head area. In consequence of re-routing the blood along preferential lines, the dorsal aortic segment (ductus caroticus), which intervenes between the third and fourth arches, is completely abandoned and eventually lost through disuse.

d. Vertebral arteries: The dorsal rami of the serially arranged segmental arteries become longitudinally cross-linked into a common vessel. The vertebral artery is formed from this interanastomosed chain. In each instance the anastomotic union is accomplished dorsally to the costal element. As a result, with completion of the costal arch, the vessel comes to occupy an osseous canal comprised of the costotransverse foramina of the upper six cervical vertebrae. The vertebral foramen of cervical seven is spared because of the ventrally inclined path the artery must take in gaining the dorsally-placed arch. With formation of the vessel, the six associated intersegmental feeders become progressively attenuated and finally severed from the main channel.

e. Highest intercostal arteries: In line with the anastomoses giving rise to the vertebral artery, a thoracic counterpart extends below the level of the subclavian artery. By contrast, the descending limb is formed by precostal anastomoses which unite with the ventrolateral rami of dorsal segmentals 8 and 9. These rami differentiate into the intercostal arteries of the adult thorax. The main connecting stem develops into the highest intercostal branch of the costocervical trunk, and in its dependent position postnatally receives the intercostal arteries coursing within the first two or three interspaces.

B. GROSS ANATOMY

By N. A. Michels

1. Elastic Arteries

The aorta and its main branches are typical of an elastic artery. The aorta arises from the vestibule of the left ventricle and, after forming an arch behind the manubrium, descends on the left side of the vertebral column to the aortic hiatus of the diaphragm, through which it passes. It then assumes a midline position, to divide into the two common iliacs at the level of the fourth lumbar vertebra. At this point, a middle sacral artery is also given off. The root of the aorta is firmly attached to the fibrous ring of the cardiac skeleton. At its dilated commencement (aortic bulb), the aorta has three secondary dilatations (sinuses of Valsalva), situated behind the three semilunar cusps which, as an aortic valve, guard the entrance to the arterial tree. The sinuses are the sites of origin of the coronary arteries: the right coronary, from the right anterior sinus and the left coronary, from the left anterior sinus. The diameters of the orifices vary from 0.7 to 1.5 mm but may reach 3 mm (Roberts, 1959).

The segments of the aorta comprise: (1) ascending aorta in the middle mediastinum; (2) aortic arch in the superior mediastinum; and (3) descending aorta. The thoracic part of the latter courses in the posterior mediastinum, while its abdominal portion enters the abdomen through the aortic hiatus, a triangular space at the level of the twelfth thoracic vertebra, formed by the middle arcuate ligament that arches across the midline, connecting the two crura of the diaphragm. Actually, therefore, the aorta does not pass through the diaphragm.

The ascending portion of the aorta is enclosed in a common sheath with the pulmonary trunk, made by the visceral pericardium. Having an average length of 5 cm and a diameter of 3 cm, it courses upward anteriorly and to the right, to the level of the upper border of the second costal cartilage, where it is closest to the anterior thoracic wall. The two coronary arteries arise near the origin of this vessel. (For anatomy of the coronary arteries, see Sections B. and C., Chapter IX.)

The aortic arch is situated behind the lower half of the manubrium, from whose right border it courses upward and backward to about 2.5 cm below the upper border of the sternum. The aorta then descends posteriorly to the left border of the disc between T_4 and T_5, to become the descending thoracic aorta. In its course, it makes two curvatures, the

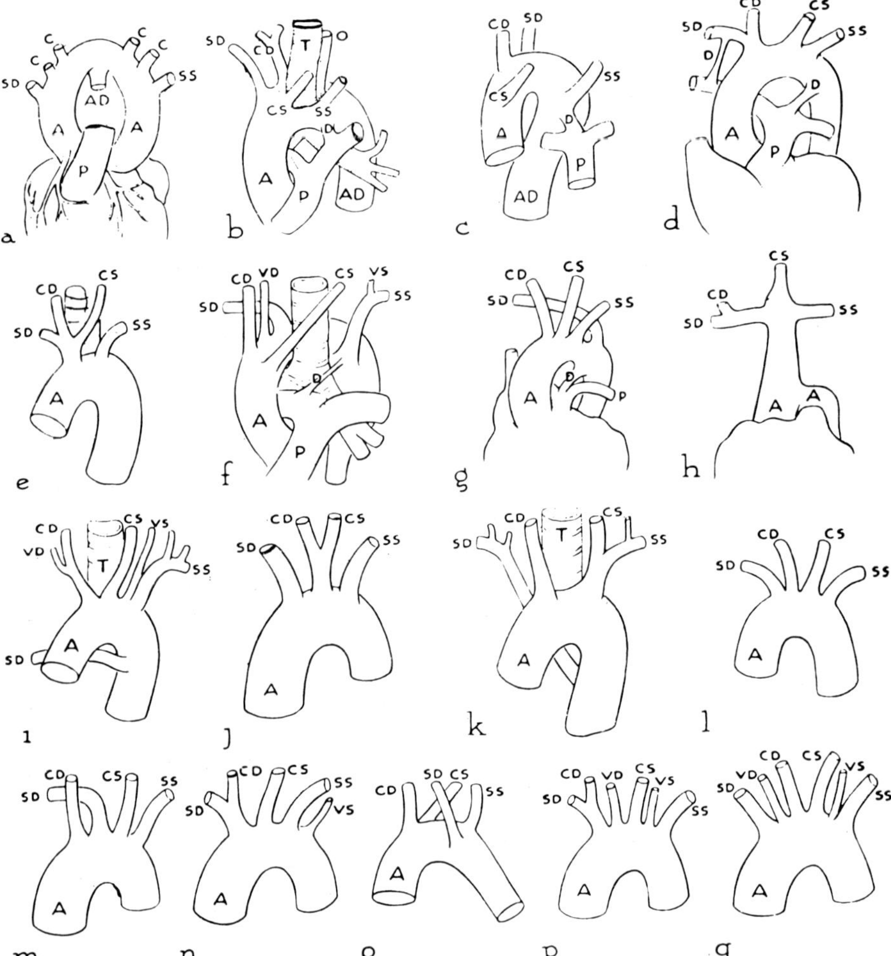

FIG. 3. Examples of irregular branching of the aortic arch. *a*, double aortic arch; *b*, division of aortic arch; *c*, right aortic arch; *d*, aortic arch with persistent right and left arterial ducts; *e*, left common carotid from innominate trunk; *f*, right aortic arch, common trunk for right subclavian and left common carotid; *g*, right subclavian as last branch from arch; *h*, single branch of arch from which all other branches arise; *i*, right subclavian crossing posterior to ascending aorta; *j*, single stem for right and left common carotid; *k*, low origin for right subclavian; *l*, independent origin of main vessels; *m*, right subclavian as second branch of arch; *n*, left vertebral as last branch; *o*, right subclavian as last branch of arch; *p*, arch with five branches; *q*, arch with six branches. *A*, arch; *C*, common carotid; *CD* and *CS*, right and left common carotids; *SD* and *SS*, right and left subclavian; *VD* and *VS*, right and left vertebral; *P*, pulmonary. (Redrawn from Poynter's Monographic Collection; with shortened subscript from Anson and Maddock, 1958; reproduced with the permission of W. B. Saunders Co., Philadelphia.)

convexity of the proximal one being upward and that of the distal one being to the left, over the root of the left lung.

Typically, the aortic arch, from right to left, gives off three branches: (1) brachiocephalic (innominate) which, behind the right sternoclavicular joint, divides into the right common carotid and right subclavian; (2) left common carotid; and (3) left subclavian. Variations in the number and site of origin of branches from the aortic arch are manifold and have been investigated repeatedly (Poynter, 1922; McDonald and Anson, 1940; Edwards, 1948; Liechty *et al.*, 1957; Anson, 1959).

The number of primary branches may be reduced to one or two or increased to four to six. With one arch stem, all branches spring from one innominate; with two arch stems, there may be two innominates (right and left) or a pattern in which the left common carotid arises with the innominate. With four arch stem arteries, the left vertebral is usually the additive vessel, arising from the arch: (a) between the left common carotid and left subclavian; (b) between the innominate and left subclavian; (c) distal to the left subclavian; (d) from the innominate; and (e) from the left subclavian. Occasionally, the additive derivative from the arch is the thyroidea ima instead of the vertebral. It supplies the lower part of the thyroid gland and, in instances, arises from the right common carotid, subclavian or internal mammary.

A clinically important pattern (1.3%) of four arch stem arteries is the one in which the right subclavian artery arises as the last branch of the arch and courses behind the esophagus (retroesophageal right subclavian). Clinically, it may cause compression and thus give rise to a condition known as dysphasia lusoria. Surgically considered, this anomalous artery often occurs in tetralogy of Fallot. When it is present, the right recurrent laryngeal nerve is not recurrent but passes directly to its area of innervation, a fact to be remembered in thyroidectomy and in tracheostomy.

The number of aortic arch branches may be increased to five or six, due to a separate origin of the internal and external carotid arteries from the arch (the common carotid being absent on one or both sides), or to a separate origin of the right and left vertebral artery from the arch. Many more patterns of anomalous branches of the aortic arch could be cited, a total of 24 types having been presented by Poynter (1922) (Fig. 3). (For descriptions of various subdivisions of the aorta, see sections on gross anatomy of specific vascular beds, Chapters VIII, IX, XI, XIII.)

C. MICROSCOPIC AND SUBMICROSCOPIC ANATOMY

By D. C. PEASE

1. Fine Structure of Elastic Arteries

a. Constituents of "elastic" arteries: The large "elastic" conducting arteries, such as the aorta and its main subdivisions, differ histologically from the muscular distributing arteries in having very much more connective tissue in their walls. Yet, as the excellent review of Benninghoff (1930) indicates, there is no sharp line of demarcation. As their name implies, there are substantial quantities of elastin, organized in general into a repeating system of concentric lamellae, in the walls of such vessels. As any suitable polychrome stain easily demonstrates, there is also a great deal of collagen permeating the whole thickness of the wall. Bjoling (1911) and Schultz (1922) have further shown by suitable staining procedures that much "mucoid" material likewise is present. Along with these extracellular components of the tunica media and intima, there is, of course, a population of cells. There has been doubt as to whether or not fibroblasts and histiocytes are included along with the smooth muscle. (For histochemical and biochemical studies of mucopolysaccharides in arterial tissues, see Section G., Chapter III.)

b. Ultramicroscopic structure: Elastic arteries have proved particularly difficult to preserve adequately and section for electron microscopy. Aside from reports which deal with isolated extracellular components, it has principally been Buck (1958, 1959) who has achieved a measure of success with sections of the intimal region, and Pease (1955) who first commented on the media. Only Pease and Paule (1960) and Keech (1960) have attempted comprehensive reports which are, however, limited to the thoracic aorta of the rat. Pease (1960), has also worked to a limited extent with the monkey aorta, and it is upon this experience that part of the following presentation is based.

In a small animal like the rat there is a regular alternating pattern of elastic lamellae and cellular layers throughout the substance of the media. Each elastic lamella, although occasionally fenestrated, is nearly a continuous sheet of elastin. Between adjacent lamellae a single layer of smooth muscle cells is found, which connects from one lamella to the next by spanning the distance diagonally (Figs. 4 and 5). In any given cellular layer the muscle cells are all arranged in parallel array, so that they would pull in the same direction. Successive layers are oriented in different directions, however, so that forces are complexly balanced.

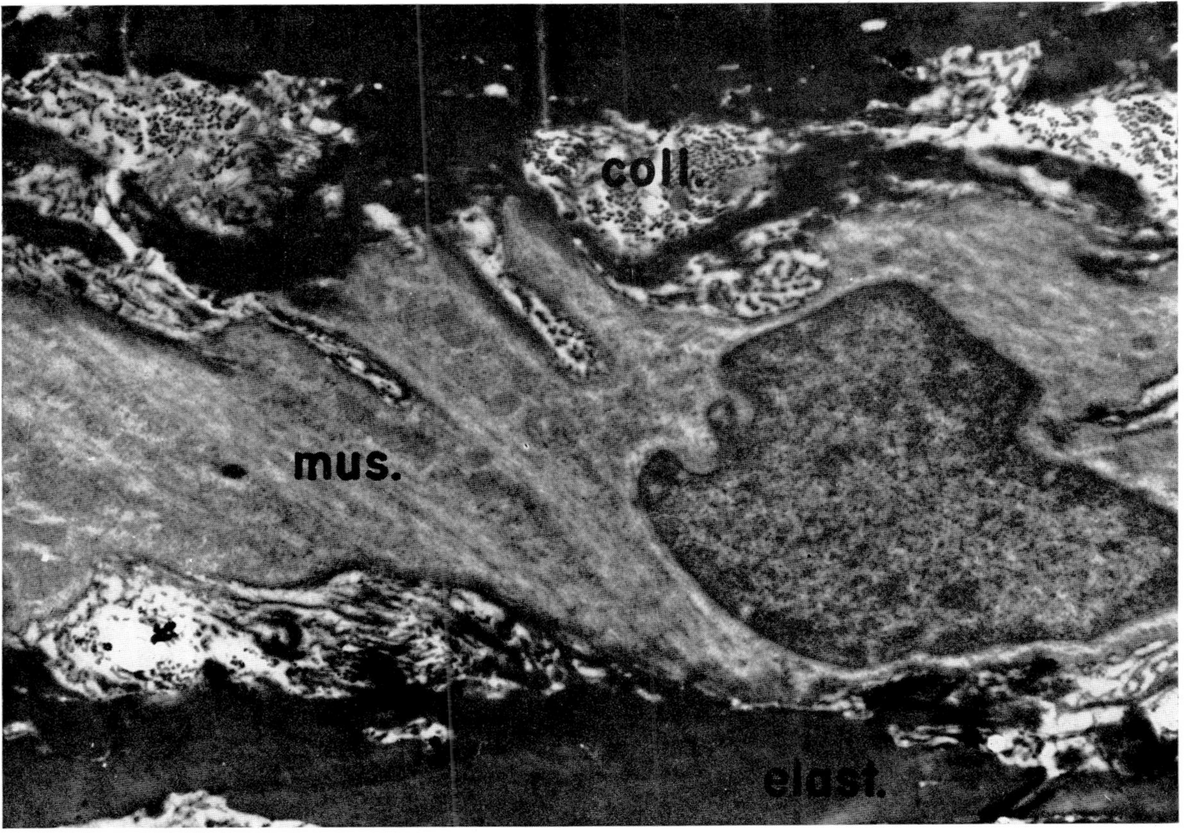

FIG. 4. The thoracic aorta of the rat has closely spaced and nearly continuous elastic lamellae alternating with layers of muscle cells. Here muscle cells (mus.) span diagonally from one elastic lamella (elast.) to the next. Substantial quantities of collagen (coll.) are present, particularly immediately adjacent to the elastic lamellae. (Phosphotungstic acid staining.)

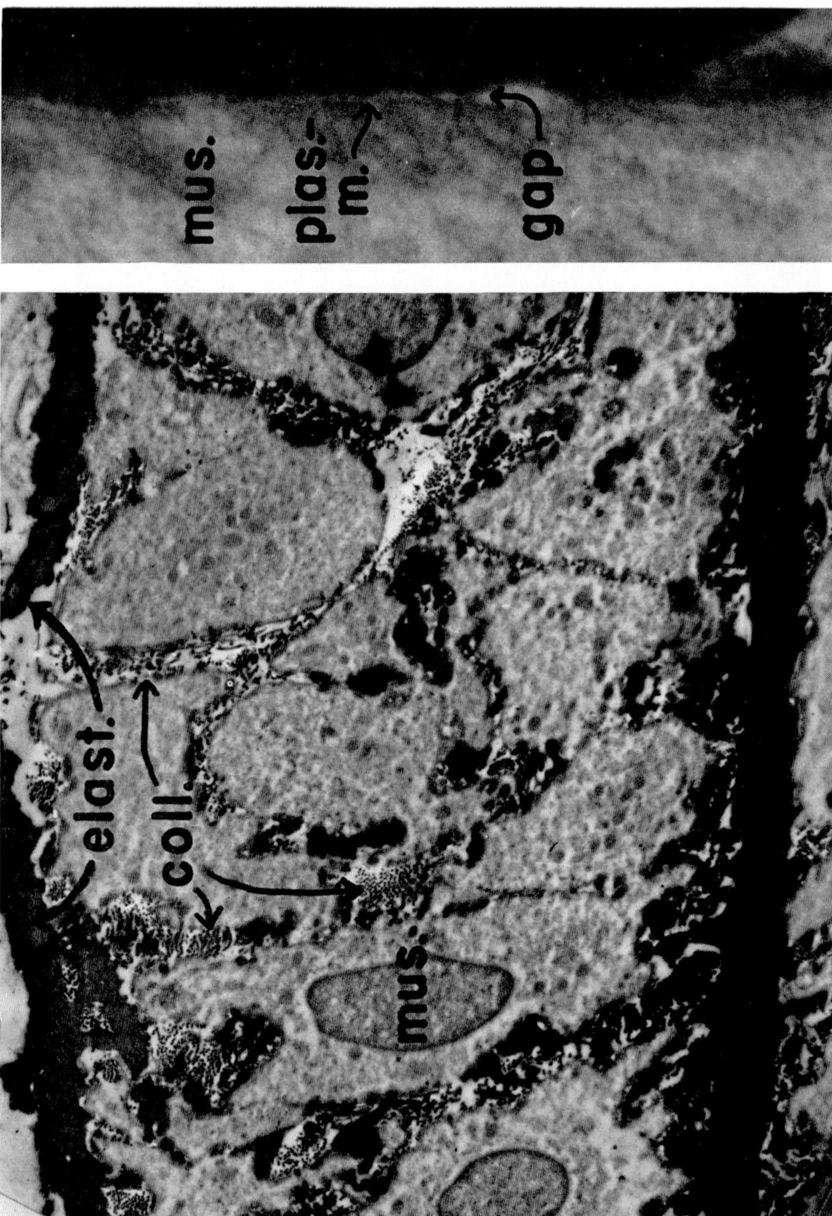

Fig. 5. *Left.* Similar to Fig. 4 except sectioned in a plane transverse to the long axis of the muscle cells. Two nearly continuous elastic lamellae (elast.) are visible, and also something of the elastic network which penetrates between muscle cells

Strands of elastin are also present in the interstitial spaces between muscle cells. In transverse sections they are seen as irregularly shaped profiles (Fig. 5), but they probably connect with the principal elastic lamellae and in all likelihood form a continuous three-dimensional elastic network in which the muscle cells are embedded.

Much collagen is found in the connective tissue spaces, particularly immediately adjacent to the principal elastic lamellae. Collagen is also present in substantial quantities surrounding and between muscle cells (Fig. 5).

The muscle cells of elastic arteries are extensively attached to the elastic lamellae and strands, but not to the collagen. In muscular arteries, it can easily be demonstrated that elastin actually can be incorporated into the sarcolemmal basement membrane of smooth muscle (see Section 2, below). In the aorta the sarcolemmal system is not well developed, but it is quite apparent that the elastin substitutes for it at many places of attachment. Figure 6 shows such a site, where a gap of relatively electron translucent material is seen, interposed between the plasma membrane of the muscle cells and the adjacent piece of elastin. This gap is characteristic of a typical basement membrane and has been referred to as a layer of "cement substance."

c. Size related to structure: It is evident from conventional microscopy that the human aorta does not have as simple and regular a pattern as the rat aorta, as described above. However, the thoracic aorta of the monkey is more nearly comparable with the human, although obviously derived from the simpler pattern. In this aorta the concentric layers of principal lamellae can still be recognized, but they are separated by much greater distances. It seems most unlikely that individual muscle cells span from one to the other. Interspersed between the major lamellae are less regularly layered masses and networks of elastin along with much collagen. Such a zone is shown in Fig. 7. These rather irregularly arranged belts almost have the status of secondary lamellae and serve to anchor many of the muscle cells. No doubt this is an adaptation related to the larger size of the vessel wall and the greater internal pressure.

It is, perhaps, at the junction of the media with the intima that in-

---

from one lamella to the next. Substantial quantities of collagen (*coll.*) are evident between muscle cells (mus.). (Phosphotungstic acid staining.)

FIG. 6. *Right.* A high magnification of the attachment of a muscle cell (mus.) to a piece of elastin (elast.). A gap of quite constant dimension separates the plasma membrane of the muscle cell (plas. m.) from the elastin. This corresponds to the gap of "cement substance" which typically underlies all basement membranes, and in essence it can be said that the elastin substitutes for a basement membrane. (Phosphotungstic acid staining.)

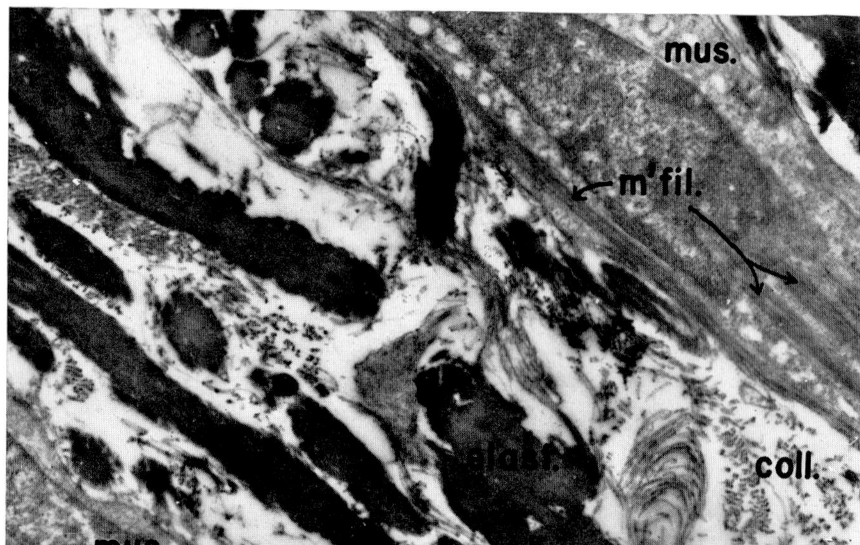
mus.
m'fil.
elast.
coll.
mus.

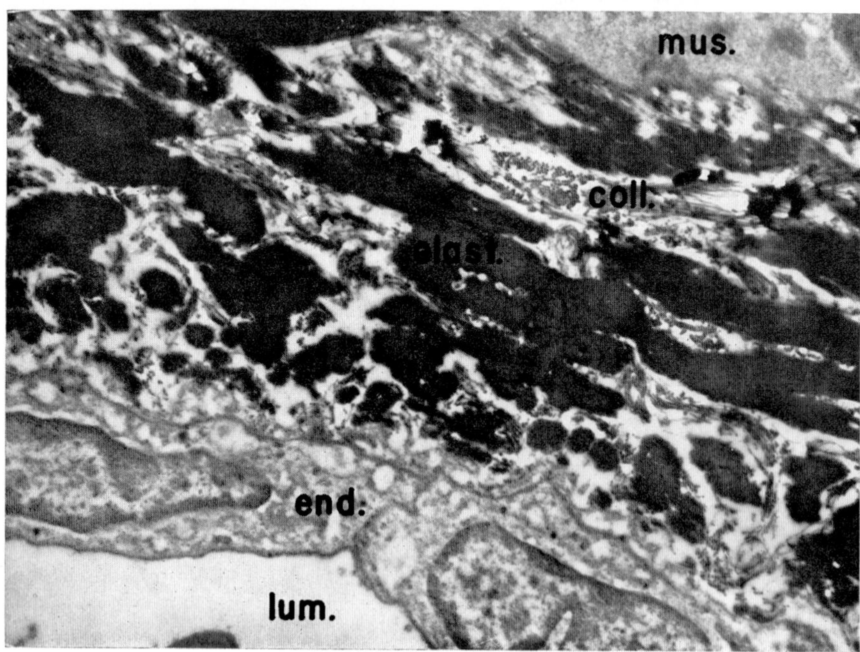
mus.
coll.
elast.
end.
lum.

creasing complexity with greater size is most dramatically observed. In the thoracic aorta of the rat the innermost elastic lamella is somewhat more massive than any of the others, but it is continuous except for occasional fenestrations. Often the endothelium is directly attached to the inner margin of this layer. Sometimes, however, there are minor connective tissue spaces, which particularly contain mucopolysaccharide and, under some circumstances, a little collagen. Buck (1958) reports a similar relationship in the rabbit aorta. However, in the thoracic aorta of the monkey the elastic lamella is broken up into a network of elastin, to form a very thick layer (Fig. 8). Interspersed between the elastic trabeculae is much collagen, and the endothelium rests upon a bed of loose connective tissue.

d. Cell types: The tunica media of normal rat and monkey aortae consists of one cell type only, the somewhat atypical smooth muscle cells. It has already been mentioned that these lack well defined sarcolemmal envelopes. Likewise they do not have the well defined dense masses of cytoplasm which attach the myofilaments to the cell surface and which are so characteristic of smooth muscle in muscular arteries (Pease and Molinari, 1960). Furthermore, their shape is not that of a simple spindle. Rather, it is evident that they tend to have multiple extensions or arms which attach to the elastic framework in different places. In several of these features, then, they have characteristics remindful of fibroblasts. However, well defined myofilaments are present (Fig. 7).

e. Development of connective tissue elements: Except for the endothelial layer, the tunica intima is acellular in the aortae of young, normal animals. Thus, it appears that the massive development of connective tissue elements throughout the media and intima must be accomplished

---

FIG. 7. *Upper.* Transverse section through the tunica media of the thoracic aorta of a monkey. Widely spaced major lamellae are present, in between which are poorly arranged layers of smooth muscle (mus.) and major aggregates of connective tissue. The connective tissue is in the form of extensively branching networks of elastin (elast.) interspersed with much collagen (coll.). Myofilaments (m'fil.) are visible within the smooth muscle. Sarcolemmal envelopes are not evident in most places. (Phosphotungstic acid staining.)

FIG. 8. *Lower.* Intimal zone of the thoracic aorta of a young adult monkey. The vessel lumen (lum.) is in the lower left. The endothelium (end.) rests on a bed of loose connective tissue containing nearly amorphous, faintly fibrillar material thought to be mucopolysaccharide. The internal elastic lamella consists of a network of elastin (elast.). Much collagen (coll.) courses through its trabeculae. Part of the innermost layer of muscle (mus.) is visible in the upper right corner. (Phosphotungstic acid staining.)

by the activities and regulation of the muscle cells, perhaps aided in the intima by the endothelium. The neonatal development of these elements is spectacular, and there can be no doubt about their formation *in situ*.

f. Question of innervation: Nerve bundles, of course, are present in the adventitial layer of the aorta, but they seem totally absent in the media and intima. Thus the nerve fibers presumably are simply *en passage*, and the musculature of the aortic wall is without specific innervation of its own.

g. Vasa vasorum: The larger arteries, of course, are known to have a vasa vasorum, consisting of small vessels penetrating the more superficial part of the media from the adventitial side. Woerner (1959) has studied the extent and circumstances of this vascularization and occasionally finds that vessels communicate from the aortic lumen. Genuine intimal vascularization, however, always seems to be associated with atherosclerotic change. (For further discussion on this point, see 1. Atherosclerosis, Section F.-1, Chapter XIX.)

2. Fine Structure of Muscular Arteries and arterioles[1]

a. Microscopic structure: Traditional histologists have not paid much attention to the connective tissue skeleton of muscular arteries beyond observing the specialized patterns of elastin which are evident after suitable staining procedures. An occasional, thoughtful investigator, however, has realized that this framework differs from usual connective tissue in being deficient in collagen. White fibers are scanty indeed in the media and intima. Even the ubiquitous "reticular fibers" have not been well demonstrated in the arterial wall (cf. the excellent review of Benninghoff, 1930). Bjoling (1911) anticipated currently developing concepts in emphasizing that the arterial framework was predominantly "mucoid," a conclusion reached on the basis of comparative staining reactions. Schultz (1922) refined the histochemical approach and emphasized that the binding tissue showed much metachromasia; he even suggested that this material might be a precursor of elastin, a thought subsequently borne out by the electron microscopic study of Pease and Molinari (1960).

[1] For discussion of submicroscopic structure of specific vascular beds, see Section C.-3, Chapter IX (coronary vessels); Section C.-1b, Chapter X (pulmonary vessels); Section C. Chapter XIII (renal vessels); (1) The Pituitary, Section E.-2, Chapter XIV (pituitary vessels); (2) The Thyroid, Section C.-3, Chapter XIV (thyroid vessels); Section B.-2, Chapter XVI (cutaneous vessels); Section B.-4c, Chapter XVIII (vessels in bone).

b. Ultramicroscopic structure: Current knowledge of arterial fine structure is very limited, perhaps mainly because of unusual difficulties in the adequate preservation and subsequent ultrathin sectioning of large vessels. There are, to be sure, occasional published micrographs of small arterioles which usually have been included in relation to other aspects. Kisch in a long series of papers (summarized, 1957) is the one worker who has persistently studied blood vessels, but his research is mainly confined to capillaries and the smallest arterioles found in the heart. Policard *et al.* (1955) investigated the subject of small arterioles, but before currently acceptable techniques were available to them. Moore and Ruska (1957) limited their work and remarks to vessels of small caliber. Although Fawcett (1956) has been studying arterial vessels, there has been no full report as yet. More recently Buck (1958) has discussed the endothelial lining of the aorta, and Pease and Paule (1960) and Keech (1960) have considered comprehensively the thoracic aorta of the rat.

The first paper which dealt in a reasonably complete way with the wall structure of muscular arteries is the report of Parker (1958), who studied the coronary arteries of rabbits with considerable success. Subsequently Pease and Molinari (1960) made a detailed study of the largest arteries to be found in the pia of the cat and the monkey. The tunica media of these vessels often has as many as five to six muscle layers. Thus far larger arteries have not been successfully studied.

c. Smallest arterioles: Aside from an adventitial layer, the smallest arterioles have a skeleton made up only of a continuous system of basement membranes. Thus, in Fig. 9 it can be seen that there is a basement membrane shared by both the endothelial lining of the vessel and the overlying smooth muscle (cf. the discussion of Altschul, 1954, concerned with classical ideas of the relationship of basement membranes to endothelium). A continuation of this basement membrane reflects over each smooth muscle cell to form a sarcolemmal envelope. Basement membranes may also be shared by adjacent cells. It is this system alone which completes the connective tissue framework of the tunica media in the smallest and simplest arterioles, no collagen being found within it. Small patches of elastin may be evident, however, between endothelium and muscle.

d. Larger arterioles: When the arteriole is of more than minimum caliber, the shared basement membrane between endothelium and muscularis may be substantially thickened, and cavities will appear within it (Fig. 9, "cav."). Small patches of elastin are apt to appear here first, giving it the character of a presumptive internal elastic membrane.

sarcolem.
mus.
b.m.
end.
cav.
cap.
advent.
mus.
attach.
coll.
coll.
elast.
b.m.
b.m.
elast.

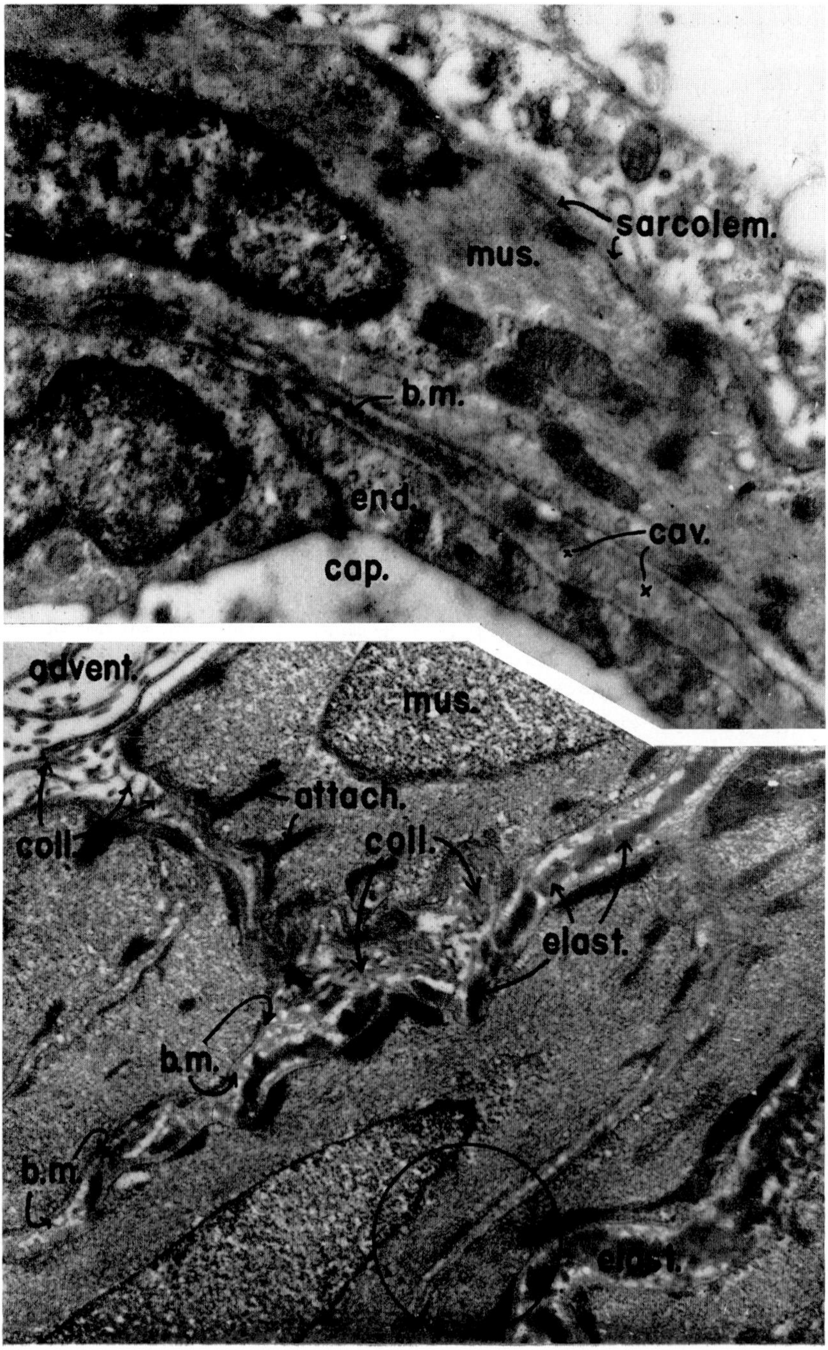

Larger connective tissue spaces are readily formed by the bifurcation of the shared basement membrane. In arteries with more than a single layer of smooth muscle, extensive connective tissue spaces are developed in this way (Figs. 10 and 11). Large masses, even continuous sheets of elastin, then develop in these spaces. Finally, as one considers still larger arteries, small bundles of collagen appear in the connective tissue spaces (Fig. 10).

e. Connective tissue skeleton: The heaviest layer of elastin develops, of course, as the internal elastic membrane between the endothelium and the muscularis. Curiously, here the elastin invades and breaks up the endothelial basement membrane, but ordinarily remains isolated from the basement membrane of the overlying muscle (Fig. 12). Thus it is polarized, which is further evident by the finding that the inner surface is deeply trabeculated and very rough, while the outer side remains relatively smooth. This sheet of elastin may be almost continuous with only occasional fenestrations.

Within the media, the most massive components of the arterial skeleton are arranged in horizontal planes parallel to the surface of the vessel (Fig. 11). Radial parts of the skeleton, between adjacent smooth muscle cells, remain relatively simple. Often they consist of nothing more than shared sarcolemmal membranes, even in arteries of substantial size, but here and there are penetrating cords of elastin and sometimes bundles of collagen (Fig. 10). Careful studies of tangential sections indicate that the elastin forms a nearly (or perhaps completely) continuous system. This is in the form of a three dimensional net, with the muscle cells filling the interstices.

It has been demonstrated that elastin can be incorporated into the sarcolemmal basement membranes. In this way smooth muscle cells are literally attached to the elastic framework. In muscular arteries of the

---

FIG. 9. *Upper.* Transverse section of a small pial arteriole of a rat. A shared basement membrane (b. m.) is evident between the endothelium (end.) and the muscle (mus.). This shows evidence of an incipient cavitation (cav.) toward the right where it is thickest. Basement membranes (b. m.) cover individual muscle cells as sarcolemmal envelopes (sarcolem.). (Phosphotungstic acid staining.)

FIG. 10. *Lower.* Outermost muscular layers of a cat pial artery. Fenestrated elastic membranes (elast.) are visible in places. Bundles of collagen (coll.) are to be found within the media of the larger vessels and are conspicuous in the adventitia (advent.). Basement membranes (b. m.) surrounding the muscle cells (mus.) are evident in places. The encircled area includes a place where a single basement membrane, shared by two adjacent muscle cells, starts to bifurcate and so produce a patent connective tissue space. Conspicuous dense masses within the cytoplasm of smooth muscle cells apparently serve to anchor groups of myofilaments to the cell surface (attach.). (Phosphotungstic acid staining.)

advent.
mus.
elast. m.
end.
b.m.
mus.
end.
lum.

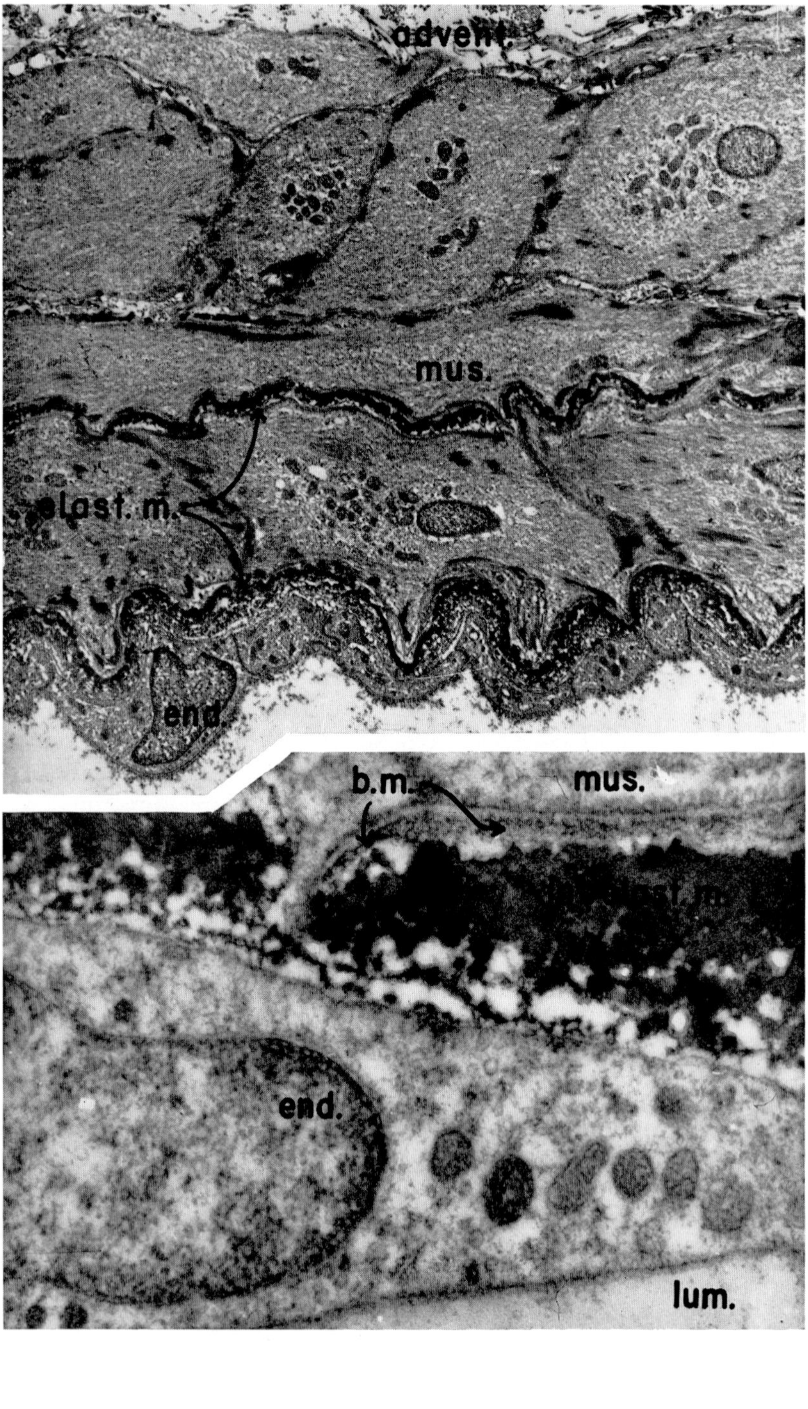

size so far studied these attachments are sporadic. They clearly suggest, however, a most intimate functional relationship between the elastic skeleton and the contractile cells. The system of collagenous fibers is entirely distinct. Collagen has never been seen incorporated into a basement membrane, nor does it have more than a fortuitous relationship to elastin.

Pease and Molinari (1960) have emphasized that basement membranes, when suitably stained, have a reticulated appearance, as does the diffusely distributed mucopolysaccharide. Elastin apparently is formed in such a mucopolysaccharide matrix by the addition of a second material. Following phosphotungstic staining, globular units of elastin, approximately 200 Å in diameter, can be discerned. In elastin which is no longer actively growing, these may have fused into a continuum, but their less dense centers remain visible, producing a stippled effect, with a spacing of about 200 Å from center to center (barely visible in Fig. 12). (For further discussion of normal structure of connective tissue of blood vessels, see Section D.-6, Chapter II.)

f. Cytology of muscle: The muscle cells of arteries have particularly dense and conspicuous masses of cytoplasm which apparently serve to anchor groups of myofilaments to the cell surface (Fig. 10). Muscle cells in other body organs, such as the gall bladder, the urinary bladder, and the uterus, usually show only traces of similar specialized attachment plaques (Caesar *et al.*, 1957). Also the muscle cells of arteries have unusually large numbers of pinocytotic vesicles of unknown function (for further discussion of this point, see Section E.-1, Chapter VI).

Syncytial anastomoses between smooth muscle cells have been observed particularly well by electron microscopy in the gastrointestinal tract by Thaemert (1959). However, Parker (1958) doubted that they existed in the arterial wall, and Pease and Molinari (1960) did not find any.

It is interesting to note that the smooth muscle cells of pial vessels, at least, are without specific innervation. Although unmyelinated nerves

Fig. 11. *Upper.* Longitudinal section through the wall of a pial artery from a cat. The predominantly collagenous adventitia (advent.) is apparent at the top of the figure. Three layers of smooth muscle (mus.) and endothelium (end.) are evident. Principal elastic membranes (elast. m.) are indicated. (Phosphotungstic acid staining.)

Fig. 12. *Lower.* Intimal region of a pial blood vessel from a monkey. The internal elastic membrane (int. elast. m.) is conspicuous, being distinct from the basement membrane (b. m.) of the overlying muscle (mus.). It has, however, invaded and largely destroyed the basement membrane of the endothelium (end.). (Phosphotungstic acid staining.)

have been seen in the adventitia, none has ever been observed to penetrate the tunica media.

It should be emphasized, too, that the tunica media consists of a pure population of smooth muscle cells. Fibroblasts are limited to the adventitial layer of normal vessels. Thus it seems evident that the connective tissue skeleton is formed and regulated by the activities of the smooth muscle cells themselves.

g. Adventitia: The adventitia, of course, consists principally of bundles of collagenous fibers (Figs. 10 and 11). The complex 3-dimensional patterns of their distribution (as well as those assumed by muscle cells) have been studied by Schultze-Jena (1939). Interspersed with the collagen is a scattered population of fibroblasts. Nerve bundles may be found also, with their associated Schwann cells.

References

Altschul, R. (1954). "Endothelium," p. 18. Macmillan, New York.

Anson, B. J. (1959). The aortic arch and its branches. *In* "Cardiology" (A. Luisada, ed.), Vol. 1, p. 119. McGraw-Hill (Blakiston), New York.

Anson, B. J., and Maddock, W. G. (1958). "Callander's Surgical Anatomy." Saunders, Philadelphia, Pa.

Barry, A. (1951). Aortic arch derivatives in the human adult. *Anat. Record* 3, 221.

Benninghoff, A. (1930). Blutgefässe und Herz. *In* "Handbuch der mikroskopischen Anatomie des Menschen" (W. von Möllendorff, ed.), Vol. VI, Part 1. Springer, Berlin.

Bjoling, E. (1911). Über mukoides Bindegewebe. *Virchows Arch. pathol. Anat. u. Physiol.* **205**, 71.

Buck, R. C. (1958). The fine structure of endothelium of large arteries. *J. Biophys. Biochem. Cytol.* **4**, 187.

Buck, R. C. (1959). The fine structure of the aortic endothelial lesions in experimental cholesterol atherosclerosis of rabbits. *Am. J. Pathol.* **34**, 897.

Caesar, R., Edwards, G. A., and Ruska, H. (1957). Architecture and nerve supply of mammalian smooth muscle. *J. Biophys. Biochem. Cytol.* **3**, 867.

Edwards, J. E. (1948). Anomalies of the derivatives of the aortic arch system. *Med. Clin. N. Am.*, Mayo Clinic p. 925.

Fawcett, D. W. (1956). Observations on the submicroscopic structure of small arteries, arterioles and capillaries (Abstract). *Anat. Record* **124**, 401.

Keech, M. K. (1960). Electron microscopic study of the normal rat aorta. *J. Biophys. Biochem. Cytol.* **7**, 533.

Keibel, F., and Mall, F. P. (1912). "Manual of Human Embryology," Vol. II. Lippincott, Philadelphia, Pa.

Kisch, S. (1957). "Der ultramikroskopische Bau von Herz und Kapillaren." Steinkopff, Darmstadt.

Liechty, J. D., Shields, T. W., and Anson, B. J. (1957). Variations pertaining to the aortic arches and their branches. *Quart. Bull. Northwestern Univ. Med. School* **31**, 136.

McDonald, J. J., and Anson, B. J. (1940). Variations in the origin of arteries derived from the aortic arch, in American whites and Negroes. *Am. J. Phys. Anthropol.* **27**, 91.

Moore, D. H., and Ruska, H. (1957). The fine structure of capillaries and small arteries. *J. Biophys. Biochem. Cytol.* **3**, 457.

Parker, F. (1958). An electron microscope study of coronary arteries. *Am. J. Anat.* **103**, 247.

Patten, B. M. (1953). "Human Embryology," 2nd ed. McGraw-Hill, New York.

Pease, D. C. (1955). Electron microscopy of the aorta (Abstract). *Anat. Record* **121**, 350.

Pease, D. C. (1960). Electron microscopy of the monkey aorta. In preparation.

Pease, D. C., and Molinari, S. (1960). Electron microscopy of muscular arteries; pial vessels of the cat and monkey. *J. Ultrastruct. Research* **3**, No. 4, 447.

Pease, D. C., and Paule, W. J. (1960). Electron microscopy of elastic arteries; the thoracic aorta of the rat. *J. Ultrastruct. Research* **3**, No. 4, 469.

Policard, A., Collet, A., and Giltaire-Ralyte (1955). Observations au microscope electronique sur la structure inframicroscopique des arterioles des mammifers. *Bull. microscop. appl.* [2] **5**, 3.

Poynter, C. W. M. (1922). Congenital anomalies of the arteries and veins of the human body, with bibliography. *Nebraska Univ. Studies* **22**, Nos. 1–2.

Roberts, J. T. (1959). Arteries, veins and lymphatic vessels of the heart. *In* "Cardiology" (A. Luisada, ed.), Vol. 1, p. 85. McGraw-Hill (Blakiston), New York.

Schultz, A. (1922). Über die Chromotropie des Gefässbindegewebes in ihrer physiologischen und pathologischen Bedeutung, insbesondere ihre Beziehungen zur Arteriosklerose. *Virchows Arch. pathol. Anat. u. Physiol.* **239**, 415.

Schultze-Jena, B. S. (1939). Über die schraubenformige Struktur der Arterienwand. *Gegenbaurs Morphol. Jahrb.* **83**, 230.

Thaemert, J. C. (1959). Intercellular bridges as protoplasmic anastomoses between smooth muscle cells. *J. Biophys. Biochem. Cytol.* **6**, 67.

Woerner, C. A. (1959). Vasa vasorum of arteries, their demonstration and distribution. *In* "The Arterial Wall" (A. I. Lansing, ed.), p. 1. Williams & Wilkins, Baltimore, Md.

Chapter II. ARTERIAL AND ARTERIOLAR SYSTEMS: Biophysical Principles and Physiology

D. BIOPHYSICAL PRINCIPLES

By R. H. Haynes[1]; S. Rodbard[2]

1. Rheological (Viscous) Properties of Blood

a. Anomalous viscosity and the circulatory system: The presence of the formed elements in blood, in particular the red cells, gives rise to its "anomalous," or non-Newtonian, viscous properties; that is, its viscosity depends on the flow rate (or rate of shear) and tube size as well as the hematocrit. Thus, the rheological behavior of blood resembles that of certain colloidal suspensions rather than ordinary liquids, such as water or glycerine (see Bayliss, 1952, for a recent review). However, it would appear that under normal conditions, in most branches of the circulatory system, these non-Newtonian properties are not of great importance, and blood can often be considered as a Newtonian fluid (Haynes and Burton, 1959a,b). For example, in the large arteries, Taylor (1959) has found, rather surprisingly, that the oscillatory flow is largely unaffected by the shear dependent viscosity of blood: an error of only 2 per cent is introduced in the calculation of the flow from the pressure gradient by assuming a constant coefficient of viscosity. An important exception to this generalization is in the capillaries where "plug flow" occurs, since the cell and vessel diameters are nearly equal. However, only the non-Newtonian properties of blood in steady flow in tubes of diameter greater than about 50μ will be considered here. (A thorough discussion of the biophysical aspects of oscillatory blood flow is to be found in the recent monograph of McDonald, 1960.)

Most of the *in vitro* viscosity measurements have been made using defibrinated blood (Kümin, 1949), blood to which an anticoagulant has been added, or simply red cells suspended in acid-citrate-dextrose solution (Haynes and Burton, 1959a). Although the rheological properties of such modified suspensions do not differ greatly from those of whole blood (Bingham and Roepke, 1944), nevertheless, one must be cautious in extrapolating the results of the *in vitro* studies to the intact circula-

[1] Author of Subsection 1.

[2] Author of Subsections 2–8.

tion.[3] A further complication is that the apparent viscosity is affected by surface films (e.g., fibrin) that may be present on the tube or vessel walls (Copley and Scott Blair, 1958).

b. Pressure-flow curves of erythrocyte suspensions: The rheology of blood may be studied by making absolute pressure-flow measurements in

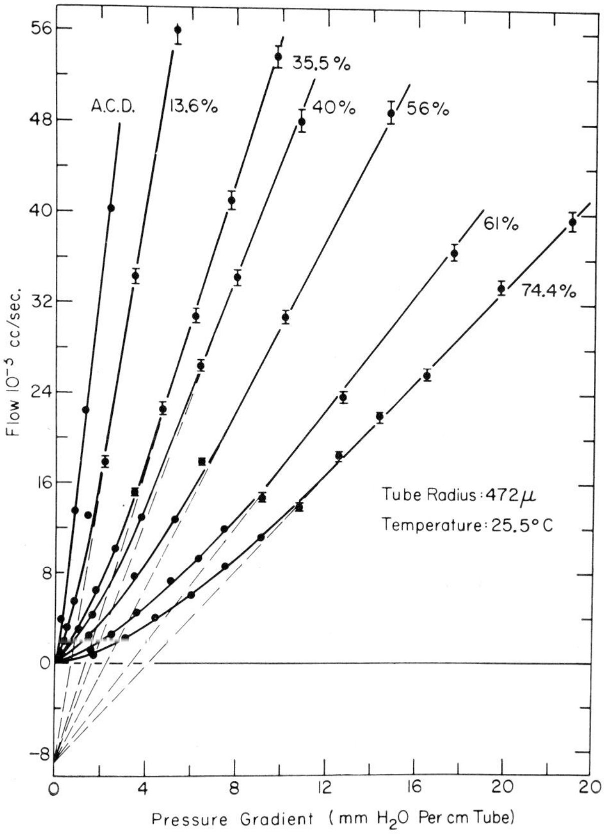

FIG. 13. Pressure-flow curves measured in a glass tube for human erythrocytes suspended in acid-citrate-dextrose solution for various hematocrits, as indicated in volume per cent. (From Haynes and Burton, 1959a), reproduced with the permission of the American Journal of Physiology.)

glass tubes. A typical set of pressure-flow curves for various suspensions of human red cells in acid-citrate-dextrose is shown in Fig. 13. Such

[3] Wells and Merrill (1961) have recently found that fresh plasma, to which no anticoagulant has been added, does have a shear dependent viscosity, as measured in a Couette viscosimeter, but that this shear dependence effectively vanishes upon the addition of oxalate or heparin.

curves have the same basic shape for all hematocrits and tube diameters at ordinary temperatures: that is, they are effectively linear at moderate pressures and flows; at low flow rates they become nonlinear and appear to converge toward the origin. The linear segments can be extrapolated back to a common "nodal point" on the negative flow axis. The slope at any point on these curves is inversely proportional to the "differential" viscosity of the suspension. Thus, the direction of curvature shows that

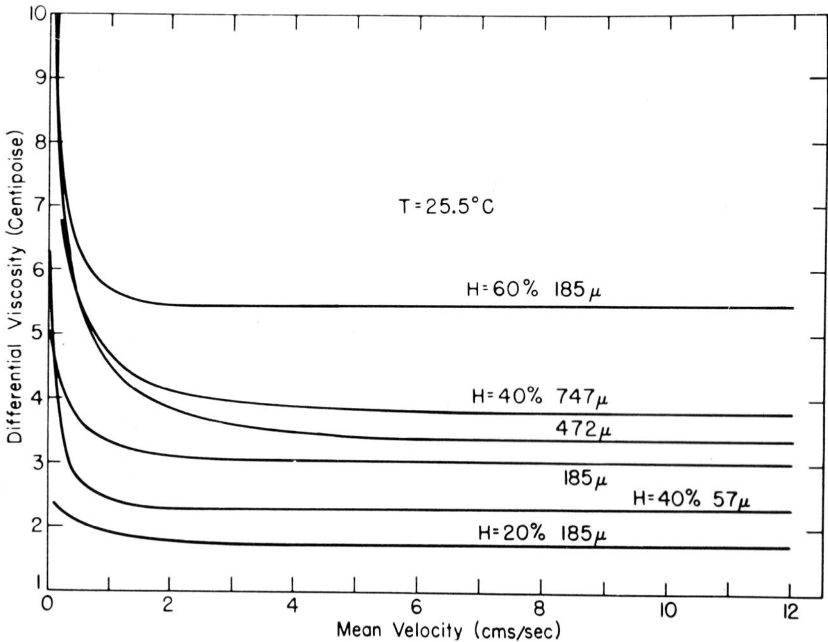

Fig. 14. Shear dependence of human erythrocyte suspensions of three hematocrits in glass tubes of various radii.

the viscosity decreases rapidly at first, but then more slowly, as the flow increases. This is shown explicitly in Fig. 14, where the differential viscosity is plotted as a function of the mean velocity of flow for a variety of tube diameters and hematocrits.[4]

The curves of Figs. 13 and 14 also show that the viscosity increases very rapidly with hematocrit; the viscosity-hematocrit relation is, in fact, almost exponential (Whittaker and Winton, 1933; Haynes, 1960b).

[4] In an Ostwald viscosimeter one would measure the "apparent" viscosity, which is inversely proportional to the slope of the line joining a given point on the pressure-flow curve to the origin. At infinite and zero flow rates, the two coefficients are equal.

It is sometimes stated that blood is a plastic fluid, which implies the existence of a finite yield pressure below which no flow can occur. Furthermore, this alleged yield pressure has been invoked to account for the phenomenon of critical closing pressure in a vascular bed. The curves of Fig. 13 show no evidence of any yield pressure, although it is clear from their shape that, if no measurements were made at very low flow rates, one might well be tempted to extrapolate the linear segments back to a "yield value" on the positive pressure axis (cf. Burton, 1953; Haynes and Burton, 1959a; Haynes, 1960a).

c. Axial accumulation of red cells: The physical mechanism probably responsible for the shear dependent viscosity is the axial accumulation of the red cells: as the flow rate increases, the cells near the wall tend to drift toward the tube axis. Thus, the hematocrit is reduced in the region of maximum shear where the viscous drag is most potent in limiting the flow rate through the tube. The flow rate is therefore greater, and the viscosity is lower than it would be with the cells evenly distributed across the tube. However, the axial drift cannot proceed indefinitely, since it is opposed by mutual collisions among the cells; thus, the effect ultimately saturates out at moderate flow rates. In this qualitative way, one can account for the shapes of the curves of Figs. 13 and 14. Although the nature of the transverse force that gives rise to axial accumulation has been the subject of some controversy, it might be a simple Magnus effect arising from "Bernoulli" forces (Tollert, 1954; Starkey, 1956).

d. Fahraeus-Lindqvist effect: The other anomaly of major interest is the dependence of blood viscosity on tube diameter (Fahraeus and Lindqvist, 1931). This effect is exhibited in Fig. 14, where the curves for 40 per cent hematocrit do not coincide for all tube radii; it is also shown in Fig. 14, where the viscosity at high shear (the "asymptotic" viscosity) is plotted against the radius; the viscosity was reduced by 10 per cent from the plateau value at a tube radius of 100μ. In another series of experiments the tube radius corresponding to a similar 10 per cent fall in viscosity was found to vary from 130μ at 10 per cent hematocrit to 633μ at 80 per cent hematocrit (Haynes, 1960b).

Two theories have been proposed to account for this effect: the first considers it to be a "sigma phenomenon" arising from the finite size of the suspended particles (Dix and Scott Blair, 1940; Scott Blair, 1958), while the second attributes it to the presence of an essentially cell-free marginal zone, adjacent to the tube wall, whose thickness is independent of tube radius. The equations of each theory predict the shape of curves such as that in Fig. 15 equally well; however, it should be noted that the two theories are not mutually exclusive and, in fact, both mechanisms may be operating simultaneously.

The characteristic parameter in the former theory is the thickness of the unsheared laminae that are produced by the particles in suspension. The thickness of these laminae would have to vary from 3.5μ at 10 per cent hematocrit to 34μ at 80 per cent, i.e., by a factor of 10 over this hematocrit range (Haynes, 1960b). On the other hand, only a very thin marginal zone is necessary to account for the effect: 6μ at 10 per cent hematocrit, 3μ at 40 per cent, and 1.5μ at 80 per cent.

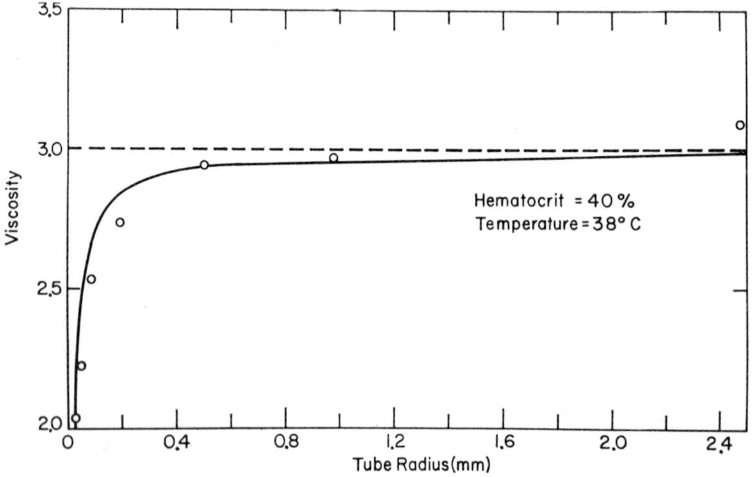

FIG. 15. Viscosity (in centipoise) of erythrocyte suspensions at high shear (asymptotic viscosity), as a function of tube radius. (Points calculated from data of Kümin, 1949; Haynes, 1960b.)

cent. This marginal zone could be produced by axial accumulation, and indeed, just such a zone has been observed optically by Taylor (1955). It is difficult to determine the relative importance of the two mechanisms in producing the Fahraeus-Lindqvist effect, although the fact that the required marginal zone is so thin and varies only by a factor of 4 over the hematocrit range of 10–80 per cent is undoubtedly a strong point in its favor.

e. Summary: All of the rheological phenomena that have been discussed here can be summarized by considering the curves in Fig. 14. An ordinary Newtonian liquid has a unique coefficient of viscosity that would correspond to a single horizontal line in such a plot. In the case of red cell suspensions, there is no unique coefficient of viscosity, and whether the differential viscosity or the apparent viscosity is plotted, its value will depend on the hematocrit, the flow or shear rate, and the diameter of the tube in which the measurements are made. However, for a given hematocrit and tube radius, the curves of Fig. 15 become effectively horizontal at moderate flow rates and to this extent, at least, the behavior of such suspensions approximates that of a Newtonian liquid.

2. Physics of Blood Flow[1]

The complex dynamics of the cardiovascular system are described in part by the idealized formulae for steady laminar flow in rigid pipes (Rouse, 1950; Goldstein, 1938; Prandtl, 1952). Such rigorous relationships can seldom be established for the cardiovascular system since the pulsatile flow of the nonhomogeneous blood takes place in curved elastic tubes of variable cross sections with numerous irregularly-spaced side branches. Nevertheless, a review of the physical principles involved can aid greatly in the study of some of the dynamic relationships between pressure and flow.

a. Sedimentation: Particles with specific densities greater than that of the plasma tend to settle to the lowest part of the system. The rate of fall is proportional to the difference in density between the particle and the fluid; denser particles therefore tend to settle more rapidly. Settling is markedly slowed by the viscosity of the fluid, especially if the particle is as small as a red blood cell. When red cells form rouleau or clump together, the settling velocity increases with the square of the diameter of the particle (Stoke's Law). The tendency to sediment is diminished in rapidly moving streams, as in large arteries. When blood moves relatively slowly, settling of the cells is more rapid. The erythrocyte sedimentation rate (ESR) is a clinical measure of this tendency in a static column.

b. Pressure: Blood and other liquids are incompressible, i.e., volume does not change significantly on application of physiologic pressures. The pressure in a vessel is the force which results from the continuous bombardment by the molecules of the fluid against the surface; this varies with the number of molecules striking a unit of surface at any instant.

Pressure is equal at all points on the same horizontal level in a static liquid system (Pascal's Law). Pressure increases with depth; thus, the pressure on the wall of blood vessels (lateral pressure) is greater with the dependency of a part of the body. In a standing man, the pressure

[1] By S. Rodbard.

on the wall of the dorsalis pedis is higher than that on the carotid artery, in proportion to the difference in the elevation of the two sites. Under conditions of no flow, as when an artery is occluded, all the energy of the stream is manifested as lateral pressure.

Pressure acts perpendicularly to the vessel lining, compressing and condensing the elements of this structure, distending the wall and driving the ultrafiltrate into the vascular tissues. The pressure is greatest at the vessel lining, falling off to a "zero" or ambient level at the outer

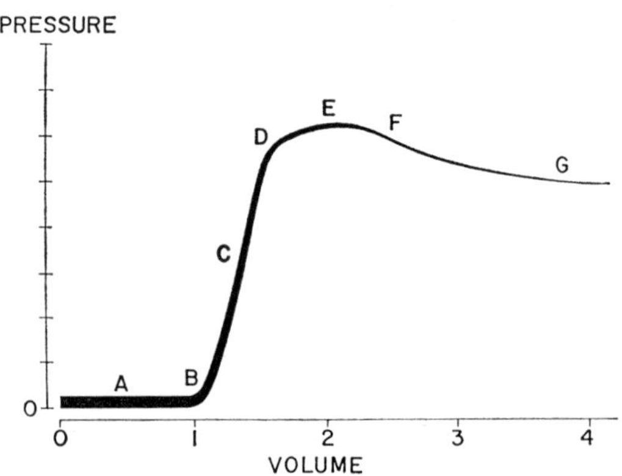

FIG. 16. Pressure/volume relations of an elastic tube. Vertical axis-pressure; horizontal axis-volume in tube. Thickness of line suggests thickness of wall as fluid is injected into tube. Increments of volume produce no pressure (A) until an unstressed volume has been injected at B. Further increments result in a pressure rise (C) which is steep at first, leveling off at D, and reaching a maximum at approximately twice the unstressed volume (E). Further increments result in a progressive fall in pressure at F and G. Curves of this general type are obtained with elastic tubes of the body, including arteries, veins, and the heart. Fall in pressure occurs as a result of (1) thinning of the vessel as it is stretched, and (2) the increasing surface area, the latter effect being due to distribution of the force on the wall (equal to the pressure × surface area). Since surface area increases with square of the radius and force increases only with the radius, a fall in pressure must occur.

adventitia. The zero level varies with the site, being approximately atmospheric in the thorax, at the level of tissue pressure in the parenchymatous organs, and at still higher levels in specialized sites, such as in the cerebrospinal and intraocular spaces. The tendency toward passage of an ultrafiltrate through the vessel wall is proportional to the difference between the intravascular and the ambient pressure. Oscillation of the pressure between systolic and diastolic levels probably assists in the movement of filtrates into the collecting vasa of the media.

c. Pressure/volume relationship: When fluid enters an empty vessel, no pressure is generated until the vessel fills and rounds out with an unstressed volume (indicated at A in Fig. 16). This may be compared with blowing into a paper bag in which no pressure is present and no tension is developed in the walls until it is filled. Injection of more fluid into the vessel, while distending it but little, results in a rise in pressure (Fig. 16-B), the height of which is a reflection of the elasticity of the wall for each unit of volume injected (Fig. 16-CDE). The injection of even small volumes of fluid into a vessel with a thick wall or one composed of relatively nonextensible materials results in a sharp rise in pressure. A peak pressure is achieved when the total volume is approximately twice the unstressed volume.

d. Tension: Tension is a pulling force which strains (stretches) the molecules or fibers of a tissue, as occurs when the intravascular pressure

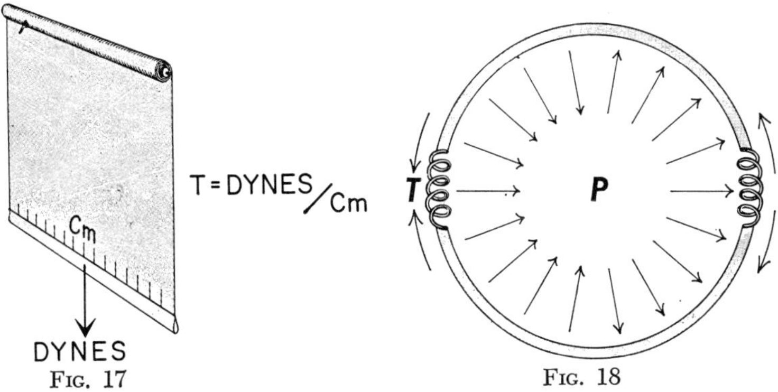

FIG. 17 FIG. 18

FIG. 17. Tension represented as a pull, as on the cord of a window shade. Tension is distributed over the length of the roller and is expressed as dynes/cm.

FIG. 18. Forces in a container produce tensile forces in the wall. Shortening of the wall, as by the shortening of the spring at left indicated by arrows, compresses the contents (arrows pointing to the center) and produces a pressure, P. P is transmitted uniformly in all directions, tending to drive the fluid in the lumen into the wall (filtration) and compressing the layers of the wall. P also distends the vessel and tends to rend the wall, as indicated by the arrows at right, which show a stretching force on the spring. This is proportional to the product of the pressure and the radius of the container.

distends the vessel walls. The strength of the materials of the wall is that property by which it resists excessive deformation or rupture. Tension may be visualized as the stretching of the elements of a window shade as a pull is applied on its cord (Fig. 17). Its magnitude is expressed as pull per unit length of the roller. In the blood vessel wall, tension (T) varies with the product of blood pressure (P) and the radius (r) of the vessel: T = a Pr, where "a" is a constant. Thus, tension varies directly with blood pressure or with vessel radius, even though the other factor remains unchanged. In vessels with small radii, such as the capillary, the tensile forces are minimal; vessels with larger radii may be under greater tension even at low pressures (Fig. 18).

As the vessel is stretched, the fibers of the wall are elongated. This stretch is balanced by the tensile force in the fibers. Injection of fluid into a distended vessel increases the tension through three related effects: (1) the rise in pressure elevates the tension directly, (2) distention of the wall enlarges the radius and, (3) the wall is thinned and weakened as radius increases.

When the pressure rises relatively slowly in a living vessel, inherent mechanisms adapt the strength of its connective tissues to the force acting on it, and hypertrophy results. However, if the rise in tension is rapid, hypertrophy cannot keep pace, or if the wall is weakened by pathologic processes, stretching of the wall may become progressive and an aneurysm or even a rupture may result. The stress is great at the junction of the normal wall with the dilated segment; as a result, the normal vessel tends to be incorporated in the aneurysm. Aneurysmal dilatation therefore stretches the wall and elongates the vessel. When the ends of an ectatic vessel are at fixed distances from each other, lengthening becomes apparent as an increasing tortuosity of the vessel.

e. Compression: The outward push of the blood pressure compresses the endothelial layer and condenses the elements of the media. Compression is greatest at the inner lining, diminishing progressively with the pressure gradient as the outer walls are approached and becoming negligible at the adventitia. The rate at which this compressive force is applied in each beat varies not only with the pressure level but also with the distensibility of the wall.

A shearing force is a combination of compression and tension which makes one part of an object tend to slide over an immediately adjacent part. This type of force occurs as the pulse wave passes and the sleeves of blood vessels are caused to move with respect to each other.

Parallel, but oppositely directed forces of equal magnitude, with different lines of action, constitute a couple. When a couple is applied perpendicularly to the long axis of an object, such as a calcified plaque in a blood vessel, it produces torsion or twisting, causing one part of a tissue to be rotated with respect to the remainder.

f. Stress and strain: Stress is the effect of an outside force on the molecular arrangement within a tissue. The magnitude of the stress is in force per unit area. Strain is the ratio of the change in length to the resting length of a molecule or fiber which results from the action of compressive or stretching forces (stress). Although stress and strain are terms with entirely distinct meanings, they are sometimes erroneously used as synonyms.

g. Flow: A difference in hydraulic energy in two parts of a system produces a flow which eliminates this difference. The rate of flow is proportional to the difference in energy between the two points. With flow, some of the energy of the total pressure head appears as

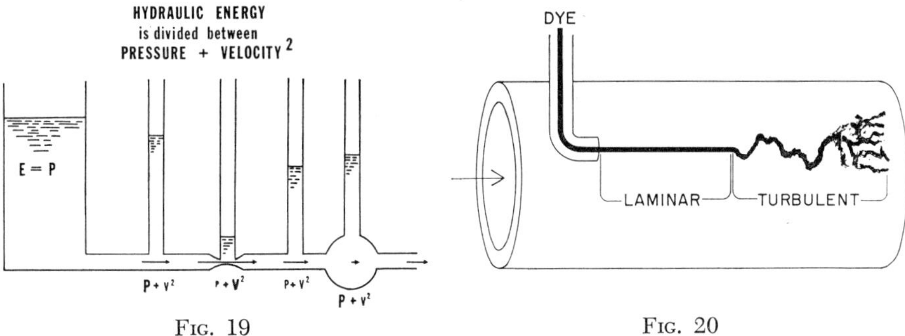

FIG. 19 FIG. 20

FIG. 19. Bernoulli's principle represents the law of conservation of energy as applied to hydraulic systems. Hydraulic energy in a system remains unchanged except for frictional losses. In a tank (at left) in which flow is negligible, the hydraulic energy (E) is equal to the pressure (P). Flow through a pipe results in conversion of some of this energy to velocity (indicated by length of arrow), with a fall in pressure in the manometer. When velocity (V) is increased greatly (long arrow), as at the narrowing, the pressure energy may be very small or even less than ambient. In wider pipes the velocity is reduced and the pressure is higher.

FIG. 20. Patterns of flow.—If dye is injected into a flowing stream indicated by the arrow, it will form a uniform thread as it is carried in a stable streamline of laminar flow. If the velocity is excessive, the stable pattern will give way to turbulent flow, indicated by the wavy line, and the dye will tend to be distributed widely in the stream.

kinetic energy, and the distending pressure falls accordingly (Bernoulli's principle) (Fig. 19). At high velocities this effect may so reduce the distending pressure that a negative pressure gradient (suction) may result. Such effects have been demonstrated in catheterization of stenotic orifices, when the velocity becomes maximal during systole and the pressure falls to negative values.

Fluid flow is of two general kinds: laminar and turbulent (Fig. 20). Laminar flow is a motion in which each filament of liquid retains its identity and flows smoothly alongside its neighbor. Normally, flow in the arteries and veins is laminar. This type of streamline motion curves smoothly around irregularities in its path, rather than setting up whorls and eddies as it moves past. By contrast, in turbulent flow, each fluid element moves irregularly with reference to its neighbor, with a considerable energy loss.

The quantity of laminar flow (Q) through a uniform tube of circular cross section is directly proportional to the difference in pressure (ΔP) between the inlet and outlet of the tube. Resistance to flow increases with the length (L) of the vessel and with the viscosity of the fluid.

For a given pressure gradient, the volume, Q, of laminar flow through a vessel increases with the fourth power of its radius, r, so that $Q \propto r^4$. This relationship depends on the combined effects of a geometric factor and a dynamic factor. The cross section area (A) of a pipe increases as the square of its radius (r^2), thereby accounting for the geometric factor affecting laminar flow through pipes ($A \propto r^2$). Laminar flow accounts for the dynamic factor. Thus, fluid in the boundary layer adjacent to the vessel wall moves very slowly because of the friction at the wall, whereas fluid in the next adjacent layer slides slightly more freely; the more central laminae slide even more rapidly on those adjacent to them. As a result, blood in the central axis of the tube moves many times faster than that in layers nearer the wall. The velocity of the central laminae is greater when the radius is large. The velocity of a particular lamina is proportional to the square of its distance from the wall, i.e., the velocity distribution is parabolic.

h. Vascular resistance: The relationship between pressure gradient and rate of flow, Q, is expressed as resistance, R. When flow is laminar, the resistances through various vascular beds may be compared by means of Poiseuille's Law which states, in part, that:

$$\text{Resistance is proportional to } \frac{\text{pressure gradient}}{\text{rate of flow}}$$

Resistance may also be expressed as:

$$\frac{\Delta \text{ Pressure}}{\text{Rate of Flow}} = \frac{\text{dynes/cm}^2}{\text{cm}^3/\text{sec}} = \frac{\text{dynes sec}}{\text{cm}^5} = \text{dynes sec cm}^{-5}$$

This value is useful for the comparison of the resistance of vascular beds and in evaluating changes in resistance from one test period to another. However, Poiseuille's Law assumes a steady laminar flow of a fluid of unchanging viscosity through a vessel of constant cross section. If these conditions are not fulfilled, the calculated results will be approximate rather than exact. Further, some of the characteristics of flow through blood vessels and through valves suggest that the flow patterns are similar to those observed to take place through collapsible tubes.

i. Pressure at the outlet: One of the striking differences between the behavior of rigid pipes and soft-walled collapsible vessels is the effect of an increased pressure at the outlet of a collapsed tube (Rodbard, 1955). In a rigid pipe, flow is proportional to the difference in height between inlet and outlet; an increased pressure at the outlet will reduce flow in accord with the reduction in pressure gradient from inlet to outlet. In soft-walled collapsible vessels an increase in outlet pressure may result in an augmented flow. When extramural pressure on a collapsible vessel exceeds the intravascular pressure, the vessel tends to collapse. This occurs, for example, when the pressure in a sphygmomanometer cuff exceeds that in the brachial artery, with collapse of the vessel and obstruction to flow. An elevation in pressure in the outlet tends to distend the collapsed segment of the brachial artery, increases its radius, and thereby reduces resistance to flow through it. A rise in outlet pressure may therefore result in an increase in flow even though the pressure gradient is reduced. This effect occurs in the systemic arteries and veins as well as in the pulmonary system.

j. Flow through an orifice: The quantity Q of flow through an orifice is proportional to its cross sectional area A, ($Q \propto A$). As fluid passes through an orifice, the stream tends to contract somewhat so that the effective orifice is slightly less than its actual area. This narrowed segment is the *vena contracta.* The coefficient of contraction (c) with a value less than 1 is therefore introduced into the proportionality: ($Q \propto cA$). If the lips of the opening are rough or of irregular shape, as in valvular stenosis, flow is reduced further. These variants require the addition of other modifying coefficients (c), each with a value of less than 1, and the formula becomes $Q \propto c_1c_2A$. The greater the number of such coefficients, the less will be the flow through a given orifice for a given pressure head. Usually the several coefficients are collected into a single constant, C, and the statement is given simply as $Q \propto CA$.

The combination of the geometrical factors related to area (CA) and the dynamic factor due to pressure ($\sqrt{P}$) gives the general rule for flow through an orifice: $Q = CA\sqrt{P}$. This hydraulic relationship has been used to estimate the area of orifices of stenotic mitral or aortic valves. It is evident from the formula that small pressures are adequate for considerable flow through an orifice. However, as pressure increases, flow through the opening does not rise proportionately, but only according to the square root of the pressure head. For example, if a pressure of 1 mm Hg produces a given flow through an orifice, elevation of the pressure to 10 mm Hg increases the flow by only approximately three times ($\sqrt{10} = 3.2$), a rise in pressure gradient to 100 mm Hg produces only ten times as much flow. It can be seen that flow through an orifice does not keep pace with the rise in pressure gradient. This is an important consideration in flow through a stenotic segment, since an elevation in driving pressure may not result in an appropriately increased flow.

k. Kinetic energy: During flow, many molecules are moving forward and fewer beat against the wall, i.e., much of the energy of the system is manifested as kinetic energy and the distending pressure is reduced. The kinetic energy (KE) in the forward movement of the stream is proportional to the square of its average velocity ($KE \propto V^2$). When flow is obstructed, all the energy appears as pressure. This conversion of flow energy to pressure energy may be demonstrated with the clinical oscillometer, an instrument which responds to the pulsatile volume changes in the vascular segments compressed by the cuff. When blood flow into the extremity is obstructed by a tourniquet or an embolus, the pulsations central to the obstruction increase in amplitude, since the energy of flow now augments the distending pressure.

l. Skimming: Particles and blood corpuscles tend to move into the axis of the stream. This effect also depends on the interrelation between pressure and velocity. In laminar flow, the velocity of the stream on the side of a cell which faces the wall is less than on the axial side. Hence, the higher pressure on the mural side of the cell drives it into the axial stream. Since the cells tend to be pushed to the central, fastest layers of the stream, the cell mass may move more rapidly than would be indicated by measurement of *mean* blood flow or plasma flow. Furthermore, branches which come off the main vessel at a sharp angle, as in the case of the renal arteries, may tend to skim off a predominance of plasma, while others which leave tangentially, such as at the carotids, may have a higher hematocrit and oxygen content than in blood from other sites.

m. Turbulent flow: Viscosity tends to hold the laminae in their relative positions. When the lining of a vessel is rough or the velocity is increased beyond a critical value, laminarity of the stream may become disturbed (Fig. 21).

At critical velocities the normal balance of forces acting on each lamina become disturbed and the normal streamline pattern is disrupted (Coulter and Pappenheimer, 1949). A cell may then be jogged from one side of the stream to the other as it is thrown from one tiny vortex to another. An important feature of turbulent motion is its high rate of mixing, with a resultant improved transfer of heat, momentum, energy, and suspended solids across the line of flow. A significant portion of the energy of the stream may be dissipated in this manner and the mean forward velocity and delivery for a given pressure head are diminished. The energy losses which result are of sufficient magnitude that engineers take care to prevent turbulence when they wish to maintain the highest possible efficiency of their pumping systems. Energy losses due to turbulence can also have physiologic importance when the load of the heart is increased.

For a given velocity, the tendency to turbulence is greater in large vessels, such as the aorta, than in small channels. The likelihood of turbulence is also greater in fluids of low viscosity.

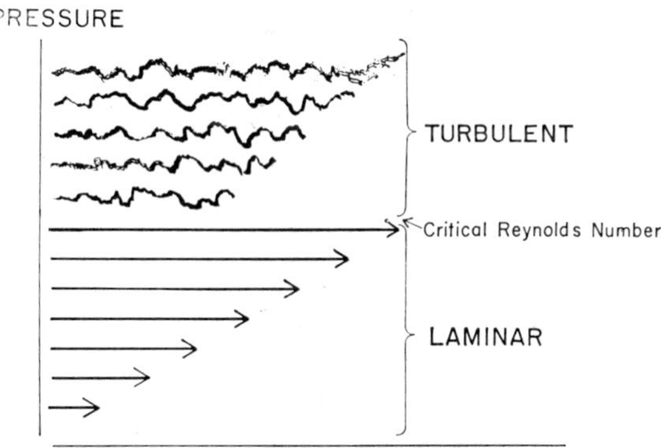

FIG. 21. Pressure and flow in laminar and turbulent flow. Vertical scale-pressure; horizontal scale-velocity. As pressure rises, flow increases proportionately, as shown by the length of the arrows. At a critical velocity, the Reynolds number of the system is exceeded and the flow becomes turbulent; this is associated with a reduced flow, as indicated by the length of the irregular lines. At still higher pressures, flow increases and the turbulence may become more marked.

In low velocity laminar flow a minimal disturbance which produces a minor undulation in a streamline tends to be dissipated. Turbulence occurs when a critical Reynolds number is exceeded:

$$\text{Reynolds number} = \frac{\text{radius of the vessel} \times \text{average velocity of the blood} \times \text{density}}{\text{viscosity of the blood}}$$

Once laminar flow has become unstable, eddies generated in a zone of instability spread rapidly through the fluid and the entire laminar flow pattern is disrupted.

With turbulence, the numerous vortices produce vibrations which may become evident as high-pitched sounds with frequencies of about

1500 cycles per second. Such blowing sounds are heard occasionally in aortic or mitral insufficiency; they differ markedly from the coarse rumbling of stenotic murmurs which are generated by flutter of thc valve leaflets.

As the velocity of the stream increases, more energy is utilized in movement and less acts as distending pressure. At critical velocities the lateral pressure may be less than extramural pressure, which tends to collapse the vessel. When velocity becomes excessive, as at narrowings of valves or vessels, the fall in distending pressure may result in an apparent lift or suction. The potential magnitude of this force may be appreciated in the recognition that the same lifting force produced by the high velocity of airflow across the upper surface of an airplane wing makes it airborne.

n. Chatter: The lifting effect noted above can displace a valve leaflet or the vessel wall in the direction of the axial stream and thereby lead to transitory closure of the lumen and obstruction of blood flow (Rodbard and Saiki, 1953). As flow stops, the velocity falls to zero and all the energy of the stream becomes manifest as lateral pressure, forcing the leaflets apart and flow begins again. The walls of the channel then snap shut recurrently as the cycle is repeated. Somewhat similar phenomena are evident in the flapping of a flag in the breeze and in the vibration of the reeds of musical instruments and of the vocal cords. In the vascular system, coarse, low-pitched murmurs result from such patterns of flow through stenosed valves or through the partially-compressed brachial artery during indirect blood pressure measurement.

o. Jet: A high velocity jet is produced as high pressure extrudes fluid through a constriction. This may strike a small area of a vessel wall with considerable force, proportional to the square of its velocity. Thus, a doubling of the jet velocity increases the energy beating on the wall fourfold. If the velocity were increased ten times, the energy of the jet would be amplified a hundredfold. The destructive potential of jets may thus be appreciated.

p. Water hammer: Vascular pressure waves and sounds are sometimes attributed to water hammer effects. This phenomenon is commonly heard when the closing of a faucet produces a sudden obstruction to the flowing stream. The momentum of the moving column of fluid generates a pressure surge at the closed end of the pipe which becomes abnormally distended. The elastic rebound of a rigid pipe quickly restores its original form, ejects the excess fluid and generates a knocking noise and a pressure wave which are transmitted away from the faucet at the speed of sound. The pressure wave may be reflected from a bend in the pipe, causing a second pressure build-up at the closed faucet and a second knock. Sudden closure of a valve may thus generate a series of knocks (water hammer) before its force is dissipated.

Even slight damping of a tubular system reduces the energy of water hammer shocks. This is commonly accomplished in household systems by inclusion of a small air pocket near the faucets.

Water hammer phenomena may occur in patients with severe generalized arteriosclerosis and high systolic pressures. The excessive pressure surges may contribute to the development of progressive stiffness and fragility of the wall and may thus thereby open the way to ultimate rupture.

In arterial trees with moderate sclerosis, the residual effect of the change in direction of the stream on closure of the aortic valve can produce a "standing wave" which may superimpose undulations on the diastolic portion of the arterial pulse. However, the rubberlike distensibility of the normal arterial tree tends to dampen the pulse waves, reduce their force, and slow their rate of transmission.

3. Effect of Mechanical Forces on Structure of Vascular System[1]

Like the other connective tissues which arise from the embryonic mesenchyme, the vascular system adapts in specific ways to the mechanical forces which impinge on it (Table I) (Rodbard, 1956; Bremer,

TABLE I
Physical Forces and Connective Tissue Structure

Force	Response
Pressure	Filtration, tension, compression
Transmural pressure	Wall strength
Tension	Reticulin, collagen, muscular hypertrophy
Rate of change of tension	Hyaline, elastica
Compression	Osteoid, bone
Rate of change of compression	Cartilage
Increased flow	Enlargement of lumen
Decreased flow	Subendothelial proliferation

1932). For example, stretching of fibroblasts in tissue culture results in an increased rate of multiplication, with the cells becoming oriented in lines parallel to the direction of the strain (P. Weiss, 1939; Fell, 1951).

[1] By S. Rodbard.

As a consequence, the collagen which is formed is aligned with the tensile forces, and the cells are relieved of the stretching force. This constitutes a feed-back mechanism whereby strain on the fibroblasts leads to the production of a material which takes up the strain and removes this force from the cell. Similar cybernetic mechanisms appear to be operative in response to other forces.

a. Origin of capillary: An introduction to the contribution of mechanical forces to vascular structure may be exemplified in the embryonic origin of the capillary. Throughout the tissues of the embryo, mesodermal cells, which adhere to each other because of their proteinaceous

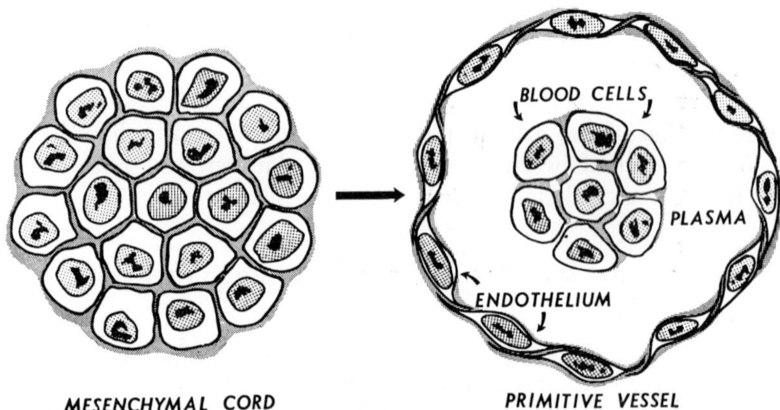

FIG. 22. A representation of the origin of the capillary. The cord of mesenchymal cells is shown in cross section at the left. The intercellular proteins (shaded region) secreted by the cells draw fluid osmotically from the surrounding tissues. A pressure then develops in the cord which causes separation of the outer ring of the cells (figure at the right), differentiating these into the endothelial lining of the vessel. The enclosed cells differentiate into blood corpuscles.

secretions, develop into solid hemangioblastic cords (Patten, 1953). Intercellular proteins draw water osmotically into the cord and the rising pressure which results acts with greatest force against the outer ring of cells of the cord, i.e., the layer of cells with the greatest radius. A coherent cellular cylinder is thus caused to separate and to form a capillary tube which encloses the cells of the core. The latter differentiate into blood corpuscles (Fig. 22).

b. Origin of blood pressure: Since the proteinaceous materials retained within the semipermeable endothelial barrier continue to exert an osmotic pull, a blood pressure becomes evident which distends the capillary. Thus, capillaries, blood cells, plasma, and blood pressure appear suddenly in the embryo, full-blown from the head of the osmotic force generated by the plasma proteins. The discrete capillaries join to form the vascular system. The primitive blood pressure mechanism continues to be of prime importance in the definitive vascular system: fluid transuded at the arterial ends of the capillaries is returned to the circulation at their venous ends by the osmotic pull of the plasma proteins (Starling, 1909). In its turn, the pressure generated within the vascular space determines the characteristic of the endothelium.

4. Intima[1]

The intima, consisting of endothelium and subendothelium, is limited by the internal elastic membrane.

a. Endothelium: This layer has been estimated by Altschul (1954) to be one of the major constituents of the body, with a mass twice that of the liver. The endothelial cells may vary in form from flattened plates to columnar palisades. Where the pressure is low, as in the veins, the endothelial cells may be cuboidal, elliptical, or rounded. With the higher pressure in the arterial system, the greater compression of the endothelium is associated with its appearance as a very thin smooth lining of characteristically thin flat hexagons, whose long axes are aligned with the direction of blood flow. The form of the endothelial cells thus appears to be associated with the distending pressure and the direction of the flow of blood (Fig. 23). (For further discussion of normal structure of endothelium, see Section E.-1; of pathologic changes in endothelium, see Section G.-2b, Chapter VI.)

Only bone, cartilage, and the inner layers of the blood vessels can survive the continuous application of high pressures. In all other tissues, persistent compression collapses the nutrient capillaries and leads to local necrosis. Thus, continued application of pressure will quickly destroy glandular epithelium and nerve and also produce localized ulcerations of the skin (Kosiak *et al.*, 1958).

b. Intimal perfusion: The special privilege of the inner lining of the vascular system results from the continuous perfusion of the vessel wall. The blood pressure drives an ultrafiltrate of the plasma through the semipermeable membrane of the endothelium, supplying the intima and inner media with oxygen and substrates. The endothelial cells are tightly interlocked to form a smooth membrane which probably controls the exchange of materials between blood and tissues. The ground substance, which surrounds the endothelium, modifies vascular permeability and

[1] By S. Rodbard.

the rate of ultrafiltration. Transudates normally find their way into the media where they satisfy the local metabolic requirements, with the modified perfusate ultimately being returned to the circulation via the vasa vasorum (Winternitz, 1938). (For movement of materials through the capillary wall, see Section E., Chapter VI.)

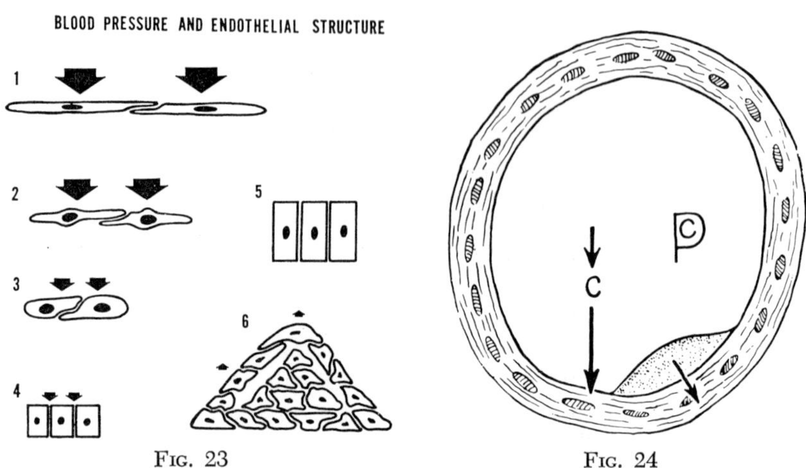

FIG. 23 FIG. 24

FIG. 23. Relation of endothelial structure to distending pressure. The heights of the pressures are represented by the length and width of the arrow acting on the cells. 1, cells lining arteries are flattened by high pressures. 2 and 3, cells are more rounded, in accord with the lower pressures in veins and capillaries. 4, with lower pressures in the lymphatic vessels, the cells may appear cuboid or 5, even cylindrical. At regions of high velocity of the stream, the distending pressure may fall below the ambient value and a suction effect may result, illustrated by the arrows pointing up into the lumen. At such sites, the reduced pressure may permit proliferation of the cells and result in the formation of a cushion.

FIG. 24. A scheme to show certain aspects of the mechanisms of atherosclerosis. Cholesterol (C) bound to protein macromolecules circulating in the plasma filters into the vessel wall as a result of the driving force of the blood pressure (arrows). The vessel lining is caused to metamorphose into plaques from which cholesterol may disappear; the lesion may then regress. The administration of estrogen causes cholesterol to be bound in macromolecules which contain much phospholipid (P); these molecules do not infiltrate the coronary and cerebral arteries, although the aorta may be involved more severely.

c. Atherogenesis: When lipids and lipoproteins of the proper shape, size, and charge are present in the plasma in high concentrations, the lateral blood pressure which drives them into the vascular lining may produce an excess in specific segments of the intima, to initiate regions of lipid infiltration (Fig. 24) (Wilens and McCluskey, 1952; Rodbard,

1959; Blumenthal, 1956). Lipid deposition in a localized region of the vessel wall can be expected to change the solubility characteristics of the intima in such a manner that further lipids will tend to be deposited in this region. Localized accumulations of fatty materials may then stimulate the metamorphosis of mesenchymal cells into lipid-laden macrophages which ingest the excess materials. An aggregation of such macrophages is an atheroma. Atheromatous plaques distort the local patterns of tensile forces and stimulate the local formation of collagen.

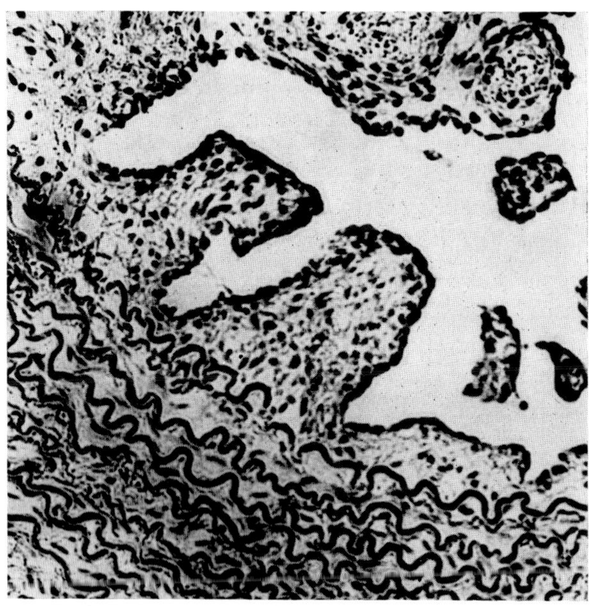

FIG. 25. Fingerlike projections of hyperplastic intima consisting of oval to spindle shaped fibroblasts are seen in a vessel which has been partially constricted. The surface is lined by endothelial cells.

When the lipids are digested and the fat-soluble character of the atheroma is washed away, a fibrous lesion may persist. If the mechanical conditions return to normal, the fibromatous lesion will gradually regress and the vessel will return essentially to its pristine state. Under other circumstances, calcium salts may be deposited to form calcific and even bony plaques, which can further disturb the mechanics of the vessel wall. (For a detailed discussion of atherosclerosis, see (1) Atherosclerosis, Chapter XIX.)

d. Subendothelium: This layer of connective tissue, varying greatly in thickness, separates the endothelial cells from the internal elastic membrane. In the normal vessel it may be present as an almost undetectable film of cytoplasm. In other conditions, the subendothelium proliferates until it virtually obliterates the lumen.

An unobstructed blood flow is associated with a thin subendothelial layer, the endothelium being virtually in contact with the internal elastic membrane. When the blood flow through a vessel is decreased, as by chronic arteriolar spasm or partial constriction, the subendothelial connective tissue may proliferate, pushing the endothelium into folds and reducing the vascular lumen (Fig. 25) (Mehrotra, 1953). Processes of this type, sometimes termed endarteritis obliterans, may thus be a response to decreased flow through a vessel rather than its cause. By this means, the subendothelium may adapt the lumen of the vessel to the volume of blood flowing through it.

5. Ground Substances[1,2]

In the course of embryonic development and in growth and aging, the vascular pressure rises and the radius of vessels enlarges. Both factors increase the tensile forces on the wall, this type of response being associated with a proliferation of structural supports for the vessel (Wolff, 1892; Keith, 1918). The endothelium is first invested with a supporting sleeve of basement membrane which merges imperceptibly with the surrounding connective tissue. Adjacent fibroblasts become aligned and stretched by the various tensile forces and materials are secreted which become organized as gels and fibers. The most deformable of these materials, the ground substance, plays a minor role in resisting tension, but it has other mechanical properties which contribute to the strength and function of the vessel wall.

Ground substance, composed of translucent jellies, fills the spaces between the other elements of the connective tissues, playing an important role as a variable filter, hygroscopic agent, adsorption column, ion exchange medium, lubricant, cement, and prosthesis. It varies greatly in form, composition, and physiologic role, depending on the region in which it is found. Despite a gelatinous appearance and the ease with which it may be deformed by pressure, ground substance has a definitive structure, as shown by electron microscopy and by return to its original shape when pressure on it is relieved. It takes up water avidly but compression drains it readily. Its high viscosity provides variable resistances to the movement of fluid across the vessel wall. Metabolites and ions passing through this barrier may be adsorbed, bound, or exchanged,

[1] By S. Rodbard.

[2] For a more detailed discussion of this subject, see Section F.-1, Chapetr VI.

as in an ion-exchange column. By binding ions, ground substance modifies osmotic balance and contributes to the facility of movement of solutes across the vessel wall. These mechanisms contribute to the control of fluid exchange, to the regulation of the circulating and extracellular fluid volumes, and ultimately to the control of the blood pressure. (For role of ground substance in exchange of materials between blood vessels and cells, see Section F., Chapter VI.)

The consistency and properties of ground substance are capable of remarkable changes (Zimmerman, 1957). For example, the relatively firm *gel* state of ground substance responds to allergic reactions as in rheumatic fever, by imbibition of water, with its conversion into a more fluid or *sol* state. Such *sol*ation occurs in association with the remarkable water uptake of the premenstrual phase; the excess water is lost with the endocrine changes of menstruation (Duran-Reynals, 1958), as ground substance reverts to the relatively dehydrated gel form. While ground substance contributes a limited structural support for the vessel wall, its more important role is in supplying the substratum in which the major extracellular structural supports, the fibers, can develop (Smith, 1957).

6. Fibers[1,2]

In embryonic development, repair, and aging, the ground substances of the blood vessels become progressively infiltrated with extracellular fibers of three general types: reticulin, collagen, and elastica. Each has its specific mechanical properties, functions, and geometrical relations.

a. Reticulin: Early in differentiation, sharply drawn nets of reticular fibers appear in the walls of the capillaries. These delicate branching and anastomosing fibers course irregularly in all directions in the young ground substance, surrounding and supporting the capillary endothelial cells and participating in the formation of a coherent vascular wall. The reticulin is not readily visible in the fresh state, but impregnation with silver nitrate solution brings it into view as a fine black network (Robb-Smith, 1957).

The role of the reticulin in the mechanical function of the vessels remains to be elucidated in detail. The rich development of reticular meshes in the choroidal vascular plexuses provides a morphologic

[1] By S. Rodbard.

[2] For a discussion of the biochemistry of the normal constituents of arterial wall, see Section G., Chapter III; for ultramicroscopic structure, see Section C.-1 and 2, Chapter I.

basis for the suggestion that, like ground substance, these meshes play a special role in the great selectivity of the blood-brain barrier.

The reticular fibers probably represent a transitional supporting material with properties intermediate between ground substance and collagen. This belief is supported by the gradual reduction in reticular fibers in the embryo as collagen becomes more evident. Since microdissection has shown the reticular fibers to be relatively inelastic, these probably serve to prevent excessive distention of the otherwise unsupported fragile capillaries. In the larger vessels, mechanical strength is contributed by other fibers, especially collagen.

b. Collagen: This threadlike protein, common to the connective tissues of all animals, appears wherever a tissue tends to be placed under mechanical stress. Fibroblasts under such tensile stress secrete the components of the collagen macromolecules into the surrounding ground substance where these molecules aggregate into fibrils oriented along the lines of the tensile forces. The prevailing patterns of mechanical tension determine the orientation of the fibrillar aggregates into networks, sheets, or columns which contribute to the strength of the fibrous fabrics of the blood vessel walls.

Fibrils of collagen are composed of three helically coiled polypeptide chains which are strongly linked by hydrogen bonds. They are characterized by a high proportion of hydroxyproline, an amino acid found nowhere else in the body. Fibrils of precisely ordered axial periods of 640 Å can be reconstituted from extracts of connective tissues without the intervention of cells. The elongated, polar molecules aggregate progressively in one of several combinations to form filaments, fibrils, and finally the definitive collagen fibers. Coarser collagenous fibers are composed of many fibers bound together by cement substance.

In the vessel wall, collagen supplies the means for the establishment of an outer limit of distensibility (Burton, 1954; Roach and Burton, 1957), in a manner similar to an automobile tire. The amount of collagen in a vessel varies with the peak mechanical tension on the vessel wall. Since the vessel walls are under longitudinal as well as radial tension, a feltwork tends to develop. Enhanced amounts of collagen and a more uniform array of fibrils are evident at sites of increased stress, as at branchings of vessels and the roots of the valves. The aorta is reported to consist of 20 per cent collagen. When the pressure remains at persistently high levels, as in hypertension, the greater stress is balanced by an increased strength of the collagen sleeves of the vessel wall.

The quantity of collagen in vessels increases with age. However, some of this accumulation may be ascribed to adaptation to the greater tension which results from a rising blood pressure or an increase in the radius of the aorta rather than to aging per se.

While collagen limits the degree of distention of a vessel, it does little to modify or dampen the pulsatile forces of the arterial system.

c. *Elastica:* The elastica provides the reversible extensibility characteristic of normal blood vessels, offering long range movement, rebound, and shock-absorption, i.e., a damping or decrease in the magnitude of vibration. Like collagen, it is organized from the secretions of the fibroblasts. Elastic fibers are straight, freely branching fibers with a high refractile index; they consist of insoluble, inert, heat-stable, durable materials which betray no apparent organization under ordinary conditions. The elastin of which it is composed is believed to be long protein molecules with covalent cross-linkages similar to those of vulcanized rubber. Recent work suggests that the fibrils contain a core of such protein surrounded by a sheath of mucopolysaccharide (Moore and Schoenberg, 1959).

Elastica may appear as isolated fibrils, as a wickerwork or as well formed fenestrated sleeves which surround the arterial wall. The interlarded ground substance facilitates movement of the elastic sleeves on each other.

Elastica appears to be associated with the pumping action of the heart. It is absent in mollusks and related species with low blood pressures and poorly developed vascular systems. A hyaline limiting membrane with the optical properties of elastica, but which does not take the characteristic stains, may be seen in the vessels of some of the more advanced invertebrates (Hass, 1942).

In the developing human embryo, a pre-elastic hyaline membrane appears only after the development of the pulsations imparted by the heart, at which time long lines of fibroblasts containing Verhoeff-staining cytoplasm are seen in juxtaposition to the new elastic fibers. A comparable metamorphosis of hyaline membrane into elastica has been observed in tissue culture of cardiac muscle (Lansing, 1959).

Examination of vascular dynamics suggests that the rate of change of tension with respect to time is the adequate stimulus for the elaboration of the elastica (Table I). The compliant properties of elastica dampen the thrust of the pulse upstroke and reduce the height of the pressure surge during systole.

The development of elastica varies considerably in the larger arteries of the body. It is most marked in vessels subjected to the greatest variation in circumference during each pulsation. For example, nearly half of the dry weight of the thoracic aorta, which pulsates more than any other structure, consists of elastica. Through its distensibility, elastica provides a resilient reservoir which stores much of the potential energy of the heart beat in the walls of the aorta and large arteries. During ventricular diastole, the stretched elastic elements of the large arteries shorten as blood escapes through the arterioles. The strong intermittent cardiac pulsations are thereby converted by the elastica into a modulated pressure pulse and relatively continuous and smooth delivery of blood to the tissues. Intracranial arteries, which are prevented from distending by the rigidity of the cranium and the relatively high cerebrospinal fluid pressure, have limited development of elastica.

The arterioles, subjected to lesser pulsations, show a thin internal elastic membrane composed of a network of delicate fibrils, together with a variable development of medial elastica, depending on its site. As might be expected, elastica is not present at the precapillary arterioles where the pulse pressure is almost entirely attenuated.

The arterial elastica, called into being by the pulsations generated by the heart, thus acts as a mechanical auto-regulatory device which controls the rate at which the vascular wall and the surrounding tissues are stretched.

d. Cartilage: Cartilage consists of spheroid chondrocytes lying in special cavities (lacunae) surrounded by a matrix of chondroitin sulfate and related compounds. This material, with the resilient properties of hard rubber, is usually found strategically placed, where it reduces the rate of change of compression of adjacent tissues, as in the joints. It is seen in the walls of the blood vessels as a glassy or hyaline material in normal, as well as pathologic, conditions. While mammals show a fibrous aortic ring, a ring of hyaline cartilage is seen in this site in the heart of certain birds, including the domestic chicken; in large mammals, such as the cow, bone is present in the aortic ring. Hyaline cartilage has been produced experimentally in the arterial wall by reducing the distensibility of a segment of the wall, as by placement of a steel wire through the lumen (Rodbard, 1958b). The metamorphosis of the connective tissue of the wall into cartilage appears to be a response to a high rate of change of compression which results when the pulse wave hammers against the tissues of the wall held in position by a wire or a firm plaque (Fig. 26). Further evidence for this point of view is the finding that removal of the hearts of amphibian embryos results in the cessation of cartilage formation even though the animal may continue to grow for two weeks or more (Kemp and Quinn, 1954).

e. Bone: This tissue prevails in sites where a continuous high compressive force is in effect. When a vessel segment is subjected to continuous compression, osteoid and/or calcific changes may occur. Thus, the bony aortic ring in the heart of the cow and of other large mammals may be accounted for on the basis that the continuous compression by the aortic pressure leads to the osteoid transformation and subsequent

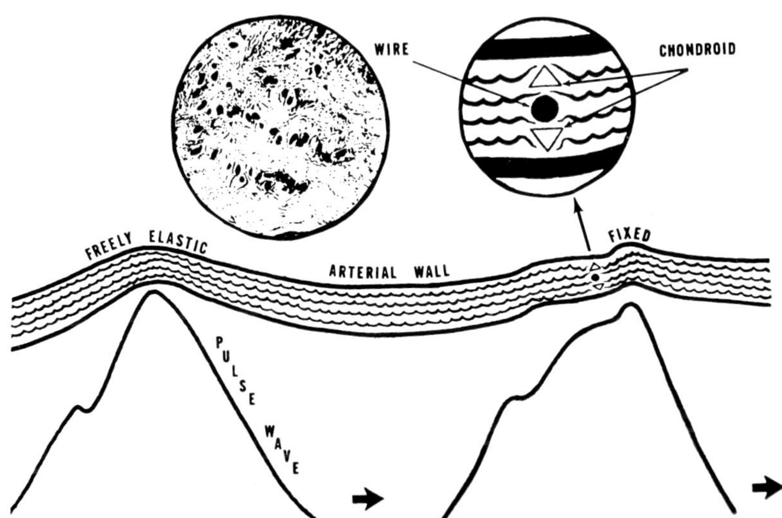

FIG. 26. Induction of chondroid by pulse wave. The pulse wave (below) moves along the vessel to right, distending the normally distensible wall. Wherever the wall is fixed, as at right (the black dot represents a cross section of a steel wire through the vessel), the pulse wave tends to hammer the point of fixation with increased force. Elastica is disrupted and small islands of chondroid appear, upper left.

calcification. Bone formation near atherosclerotic plaques may also result from similar mechanisms.

7. CELLS[1]

a. Muscle: Early in vascular development, mobile mesenchymatous cells join together to form continuous spirals of smooth muscle around the endothelium. Muscles occur only in the media, being absent from the intima and adventitia. The function of a muscle is to generate tension and thereby to draw the origin closer to its point of insertion. By this means, tension on the other tissues encompassed by the cell is relieved. In a blood vessel, contraction of the muscle reduces the lumen; distention of the vessel by the blood pressure re-elongates the muscle. Since the

[1] By S. Rodbard.

pull of each contracting vascular muscle cell is transmitted to the elastica, the muscles may participate in the damping waves in the arterial system. Sympathetic and parasympathetic nerve fibers course along the blood vessels and terminate in close proximity to the smooth muscle cells.

An important protective function of vascular muscle lies in its potential for spasm. Local trauma may initiate spasm which reduces the radius of the vessel; the danger of blood loss or of rupture of the injured vessel is diminished thereby, since a contracted vessel wall can easily withstand pressures which would cause rupture if it were in its normal state of distention. Injury to the wall also releases tissue thromboplastin to form a local clot. Through these means, spasm increases the effectiveness of the clot in preventing exsanguination, since a diminution in the vascular radius reduces the likelihood that the blood pressure will dislodge the clot. The marked fall in blood pressure which sometimes accompanies localized vascular injuries also tends to limit the loss of blood through the injured site.

Vessels in sites where elongation and shortening occur usually have a tortuous appearance, and the well developed smooth muscle is longitudinally or spirally oriented. Longitudinal and circular muscles in the main coronary arteries facilitate adaptation of the length of the vessel to the intermittent contraction and relaxation of the heart muscle. Longitudinal muscle fibers are also dominant in the veins and in the main pulmonary artery. In the lung, they may contribute to the ease of elongation of the pulmonary vessels during inspiration and shortening during expiration.

b. Fibroblasts: The fibroblasts, or the closely related mesenchymal cells, are believed by many workers to have the potential of conversion into other cells of the connective tissue series and of participation in the elaboration and possibly the digestion of extracellular fibers. When abnormally high concentrations of colloidal aggregates enter the blood stream, endothelium in some sites becomes phagocytic. Perhaps similar mechanisms are called forth when lipids infiltrate the vessel wall in excess and macrophages take up this material to form an atheroma.

c. Other cells: Almost any type of connective tissue cell is seen in blood vessels, including eosinophils, mast and plasma cells. The specific function of these cells is not known. Lymphoid cells appear in lytic processes. True fat cells are never observed in the inner layers of blood vessels, perhaps because the high pressures in these regions compress the cells and prevent the accumulation of the large lipid deposits.

8. Vascular Organization[1]

a. Intimal growth: The cells lining a blood vessel always have a large available space into which they can grow: the vascular lumen. The capacity for growth of the lining cells in arteries becomes clearly evident after local injury when the endothelial cells slowly extend cytoplasmic film over the denuded area and undergo mitotic multiplication (Poole *et al.*, 1958). More striking evidence of such a proliferative potential is seen when a segment of the common carotid artery is stripped of its blood and is then tied off; the lumen of such a vessel becomes filled within a few days by a rich ingrowth. However, where such a response does occur, the ensuing obstruction of blood flow destroys the primary function of the vessel.

What prevents these intimal cells, with such a frightening capacity for rapid growth, from proliferating and filling the lumen of all the vessels of the body? Such an unhappy prospect may be prevented because the cells lining the vessels are under continuous inhibition by the compression exerted by the blood pressure, i.e., by a form of "pressure atrophy." Similar flattening and inhibition of growth of connective tissue cells occur in other regions of high pressure. Thus, fibroblasts tend to be relatively compressed and in a nonproliferative stage in tendinous capsules, in cysts, and at pressure points in the subcutaneous tissue. By contrast, large succulent fibroblasts in a stage of rapid proliferation may be seen at sites of reduced pressure, such as occur in the subcutaneous tissues where the skin has been cut.

These compressive forces may determine whether growth of the intima of blood vessels will take place. Where the pressure is high, the cells are flattened and in a nonreproductive phase. However, where the pressure is low, the cells, freed from pressure atrophy, may proliferate rapidly.

b. Pressure and flow: The pressure on the wall of a blood vessel may differ strikingly at immediately adjacent sites. Thus, the compressive force at one site in a vessel may be very high while a much lower pressure may be in evidence nearby (Rodbard and Saiki, 1953). This discontinuity in pressure, a consequence of the law of conservation of energy, may account for vascular changes such an intimal cushion and valve formation, as well as for the progression and limitation of stenotic processes.

c. Cushion formation: High blood flow velocities, which reduce the local distending pressure in arteries and other soft walled vessels, can

[1] By S. Rodbard.

cause the apposing walls of these vessels to be drawn together (Rodbard, 1956). Flow at such critical velocities through channels lined with materials which can be deformed can cause the lining to "grow" at sites of reduced pressure (Rodbard and Harasawa, 1959), producing cushions which narrow the lumen (Fig. 27). These effects tend to occur especially at convergences, bends, or at sites of rapid pressure drop. Increases

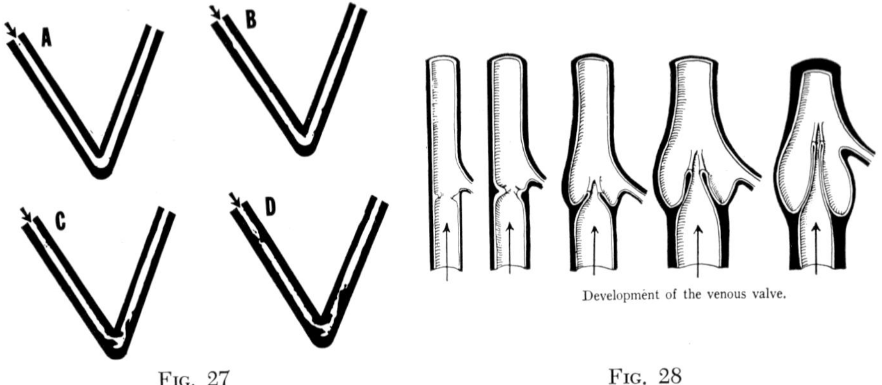

FIG. 27 FIG. 28

FIG. 27. Deformation of the lining of a channel by flow. A, the walls of the channel are shown in black. Flow passes from the upper left between the layers of deformable silicone to the bend below and thence out at the upper right. The lining has been made as smooth as possible. B, after flow for about five minutes, small cushions protrude into the lumen at various sites. C, effect of about three more minutes of flow, with the development of well-developed cushions, especially at the bend in the tube. The lumen is clearly reduced. Beyond the bend a valvelike structure is seen. D, further progression of these changes, with a stenosis near the inlet and modification of the cushions and the valvelike structure.

FIG. 28. Development of the venous valve. Flow indicated by the arrow is associated with occurrence of a cushion which increases in size until it becomes formed into leaflets. The increased pressure immediately above the valve is associated with formation of sinuses. (From Kampmeier and Birch, 1927; reproduced with the permission of the American Journal of Anatomy.)

in velocity of the blood stream can produce similar local reductions in the distending pressure of blood vessels, especially at a convergence of veins, or at an outfall, as when the blood "falls" into the relaxed ventricle during the phase of rapid filling.

When the pressure is reduced to a low value on a segment of the wall of a vessel, the intimal cells, released from pressure atrophy, may proliferate and form cushions. The development of even a small cushion tends to increase the likelihood of further growth, since the cushion deviates and accelerates the streamlines of blood flow; the locally increased velocity is associated with a further reduction in the local pressure. Proliferation of the intima may thus tend to become progressive. Such intimal cushions appear to be important transitional phases in the genesis of valves, atheromata, and stenoses.

d. Valvulogenesis: If the sites of placement of the multitudinous valves of the body, as well as their specific structure, were determined by purely genetic forces, a great burden would be placed on the biologic memory mechanisms in the DNA of the cellular nucleus, a structure with many other urgent tasks to perform during embryonic development. This burden would be obviated if the site of placement, as well as the definitive structure of the valves, were determined by the action of hydromechanical forces on the connective tissues of the vessels. Such a mechanism would also provide an elegant means of vascular regulation since valves would then be situated at sites of hydraulic stress.

Analysis of the mode of valve formation in veins and in the heart suggests that such hydrodynamic forces actually contribute to the selection of the sites of valvulogenesis. Patten's motion pictures and reports on the developing heart (Patten *et al.*, 1948) clearly demonstrate that bilateral intimal cushions serve as the first cardiac valves of the embryo. Similarly, as Kampmeier and Birch (1927) have shown, valves in the veins and the lymphatics arise as intimal cushions, which are then molded progressively into the characteristic valve form (Fig. 28).

Experiments on channels lined with deformable silicone have shown that mechanical forces can modify the walls of a tube in a similar fashion, with the production of valvelike structures, oriented in such a way as to inhibit regurgitant flow. Thus, after flow has brought the development of cushions, the downstream ends of the cushions may be observed to be suddenly hollowed out, producing characteristic valvular leaflets (Fig. 29). This striking change in contour may also be attributed to the high velocity of the stream. As the flowlines are deviated by the cushions, the filaments of the stream gain such great momentum that they cannot follow the downstream curve of the cushion. The streamlines in the lee of the cushion then separate from the channel wall, leaving a pocket of relatively stationary fluid. A local rise in pressure results at this site, compressing the cells and blunting the lee of the cushion. The region of separation thus becomes more marked and the relatively high local pressure brings about regression of this aspect of the cushion, hollowing out a pocket. A valvelike structure with delicately shaped leaflets thus emerges. These newly formed valvelets tend to vibrate in the forward-flowing stream, being constructed in a fashion that causes them to approximate when flow stops or is reversed. Structures with a valvular semblance have been produced in the arteries of experimental animals by modifying the vessel so that the blood streamlines are deviated and accelerated. The action of the stream on the tissues of the vessel wall may therefore be considered to participate in valvulogenesis in the heart, veins, and lymphatics.

e. The definite valve: After valvulogenesis has been completed in the embryo, new stresses are placed on the vessel wall at the root of the valves, and tough fibrous tissues develop there and in the leaflets, and the supple yet firm valves of the veins and of the heart appear.

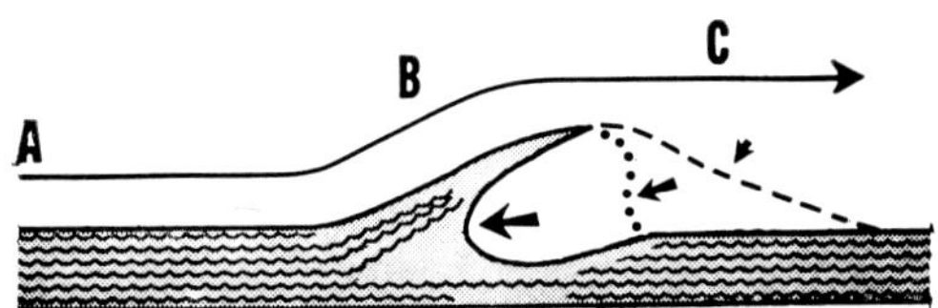

FIG. 29. Flow patterns around a cushion. Streamline filament, A, is deviated and accelerated at B by a cushion in the vessel wall which extends to the dashed line. The resulting increase in momentum prevents the filament from following the curvature of the cushion and continues it in a straight line to C, separating from the wall in the region of the dashed line. Fluid in this region of separation moves slowly and therefore has greater pressure, with a resultant compression (small arrow) at the lee of the cushion. As a result, the cushion recedes in the direction of the dotted line. The separation becomes more complete, the local pressure rises higher (middle arrow), and recession continues toward the solid line. A valvelike structure is thus formed.

Valves probably continue to grow throughout life, for their margins would otherwise be worn away by the incessant clapping together of the lines of coaptation. This adaptive replacement may also depend on the action of the stream. For example, if the aortic leaflets are not properly apposed during diastole, a high velocity stream generated by the arterial pressure will flow between the insufficient valves, lowering the pressure locally. The lining cells at the valvular margins, released from pressure atrophy, may then grow until the edges of the valves again come into apposition to accomplish the normal valve function. This interpretation is supported by the finding of elongated aortic valves in patients with dilation of the aortic ring and regurgitation (Gouley and Sickel, 1943). When growth of the margins has achieved an apposition of the leaflets, regurgitant flow ceases and intimal growth is inhibited. Such a mechanism would maintain valvular competence. The tissues at the lines of coaptation are compressed recurrently by the forces that slap the leaflets into closure. These compressive forces apparently contribute to the condensation of tissues which results in the formation of the corpora aurantii.

f. Stenosis: The mechanical processes associated with valvulogenesis and with valvular growth also appear to participate in the genesis of stenotic lesions and in their progression. Analysis of the stenotic process discloses relationships which may be considered paradoxical. Curiously, the continuously high arterial pressure on the cardiac side of a coarctation of the aorta is often sufficient to induce sclerosis and even rupture of the vessel wall; yet somehow, it never acts to dilate the coarctated arterial ring and to relieve the stenosis. A similar situation is present in the stenotic lesions of the cardiac valves. For example, the hypertrophied left ventricle in aortic valvular stenosis generates high pressures which beat recurrently against the contracted outlet; yet the stenosis is inexorably progressive, leading to failure of the pump. Similarly, the contracted ring in pulmonic stenosis is not known to become less stenotic, in spite of the persistence of high ventricular pressures.

These apparent paradoxes have been highlighted by repeated demonstrations that gentle digital dilatation of stenotic mitral valves is sometimes sufficient to free the agglutinated valve leaflets and thereby to lead to clinical relief from the symptoms of obstructed blood flow. It is remarkable that the continuous action of the relatively high atrial pressure never achieves the same dilating effect as the momentary insertion of the surgeon's finger. Instead, the lesion is progressive and the stenosis always becomes more marked.

The interaction of hydrodynamic and biologic factors implicated in the process of cushion formation and valvulogenesis given in the discussion above may be involved in the genesis of stenotic lesions. Thus, disturbances of ground substances of the valves during the active phases of rheumatic fever appear to revert the normal firm structure of valves and of some other connective tissues into a relatively gelatinous state resembling that seen in embryonic vessels and valves. Under these circumstances, the leaflets may become easily deformed by the blood stream, forming cushions and nodules which interfere with normal coaptation. When the valves cannot approximate properly, regurgitation results and the high velocity of the regurgitant stream may further deform the valve leaflets. Thus, the high velocities of the stream at the regurgitation site tend to lower the lateral pressure locally and proliferation of the cells at the valvular surface may be expected. The margins of the leaflets may then become agglutinated by the new growth, with a progressive narrowing of the orifice. A stenotic lesion may thereby provide a site of high velocity flow which increases the tendency to further narrowing.

Experimental evidence of the progressive tendency for stenosis has been demonstrated in animals (Rodbard, 1958a). Thus, the deviation of the blood streamlines by partial narrowing of a vessel, such as the aorta or a large artery, results in the development of a progressive stenotic process as ingrowth of intima takes place. The medial supporting structures also become modified and ultimately a pattern remarkably characteristic of a coarctation-like stenosis develops (Fig. 30). Similar mechanisms may be operative in the stenosis of coronary arteries and other vessels.

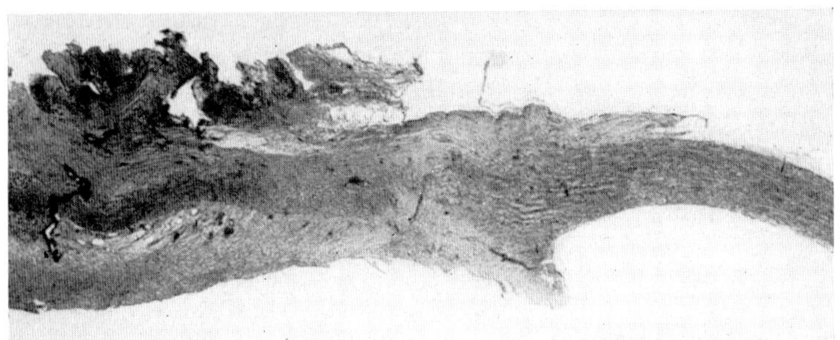

FIG. 30. A longitudinal section of the aorta (of a chicken) which was partially narrowed by a suture. The cardiad portion, where pressures are high, shows intimal and medial thickening. The site of the suture, which has been removed prior to sectioning, is seen as clefts in the intimal-medial border and in the adventitia in the left third of the specimen. Tissue which bears a superficial resemblance to a valvelike structure was part of a coarctationlike ingrowth in the fresh specimen. At the site of the protuberance into the lumen, a reorganization of direction of the medial elastic fibers can be discerned. A thinner vessel wall at the site of reduced pressure is evident beyond the site of narrowing. (From Rodbard, 1958a; reproduced with the permission of Grune & Stratton, Inc.)

g. *Orifice:* Probe-patency, the usual endpoint of the process of progressive stenosis, is sometimes seen in a coarctation, foramen ovale, and in ductus arteriosus. Similar tiny orifices are observed in coronary and other arteries subject to marked intimal ingrowth. The persistence of the tiny orifice can be shown to be determined by the mechanical action of the stream. Thus, as the lumen becomes progressively narrowed, the flow through this part of the channel falls sharply as the fourth power of the diminishing radius (Poiseuille). The local velocity is thereby reduced. At critical degrees of narrowing, the flow and velocity are minimal; the distending pressure is therefore relatively high and local ingrowth tends to be inhibited. An equilibrium is thus achieved between lumen diameter and velocity. The tiny remaining orifice may have no hydraulic importance, yet may call attention to itself by its acoustic assertiveness.

h. Adventitia: The adventitia is the site of entry of nerves and vessels which contribute to the nourishment and organization of the vessel wall. The loose connective tissue of which it is comprised contains a relatively small number of elastic and collagenous fibers as compared to the media. While the adventitia can contribute in only a minor way to the mechanical support of the wall, its removal results in fusiform dilatation of the vessel or even rupture (Schlichter, 1946). However, the main function of the adventitia appears to be its role in the defense of the territory of the vessel. As a vessel enlarges, any tissues adjacent to its adventitia become eroded, opening a pathway adequate for its growth. Adjacent adventitial boundaries also maintain the separation of vessels, as in the sharing of a sheath by an artery and a vein. When the adventitia is the seat of damage and inflammation, the isolationist property is lost and adjacent vessels may fuse to form an arteriovenous fistula.

That the organization of the blood vessel wall is determined to a great extent by the physical forces which act on it, must be concluded from the marked changes which occur when transplanted veins take on the characteristics of arteries and, in appropriate circumstances, transplanted arteries take on the appearance of veins. Similarly, at arteriovenous fistulae, the artery leading into the anastomosis, which has a relatively low pressure, appears "venous," while the adjacent vein, which is subjected to a higher-than-normal pressure, develops a thicker wall and becomes "arterialized."

i. Wall thickness: The force which distends the vessel wall and causes its elements to be under tension appears to determine the responses which strengthen it sufficiently to prevent excessive distention or rupture (D. L. Weiss and Linde, 1960). The bursting tendency is proportional to the product of the blood pressure and the radius of the lumen; the thickness of the arterial wall at any given point is related to this product. Since the pressure in the arteries is relatively constant, the wall thickness is proportional to the radius of the vessel. The intima and adventitia have little strength and they contribute minimally to the strength of the wall. Most of the stress on the vessel wall is met by the media. The larger the radius of the vessel, the greater the tension on its wall and the thicker it will become. For example, large veins have thicker walls than small veins (Hughes, 1936). Likewise, the more dependent arteries, such as those in the feet, have thicker walls, in proportion to the higher pressures in these sites, than do vessels such as those of the thorax. This is dramatized in the thick leg arteries of the giraffe which has a very high arterial pressure at this dependent level (Goetz *et al.,* 1960). Wherever the vessel wall is supported, as when a vein abuts on an artery in a fibrous sheath, it may be only half as thick as at other adjacent unsupported sites (Hughes, 1936). Venous valves and primary thrombi are also oriented with respect to the adjacent artery.

The normal pulmonary artery, which has a low pressure, also has a very thin wall which resembles that of the large veins. In pulmonary hypertension, the arterial wall may be as thick as the aorta and the pulmonary arterioles are thicker and contain significant quantities of muscle. The amount of elastica and collagen in the vessel wall varies with the characteristics of the force acting against it. Much of the tension is taken up by the extracellular fibrils. For example, much of the thrust of the pulse is taken up by the stretched elastic sleeves of the thoracic aorta. Collagenous segments, as at aneurysms, tend to be considerably thinner since the collagen can withstand considerable tension (Holman, 1950). The smooth muscle also contributes to the elastic properties of the vessel wall, although this force undoubtedly varies with the state of contraction of its myofibrils. Vessels which contain much muscle are usually thicker than those with little muscle.

j. The lumen: The cross section area of the lumen of a vessel is related to the size and activity of the vascular bed it serves; the lumen area thus reflects the quantity of blood flowing through it (Thoma, 1893). Arteries become progressively smaller as branches divert blood from them. The artery supplying a limb dwindles in size as it courses to the digits, while the vein increases in size as more and more tributaries join it. The mechanisms of continuous adaptation of these vessels during growth, hypertrophy, and atrophy of the parts they serve have received little attention. It is self-evident that the blood pressure level plays no role in determining the lumen of a vessel, since pressure is relatively constant throughout the arterial system, even though great variations in size from aorta to arterioles are seen.

If flow were the adequate adaptive stimulus, the enlargement and diminution of vessels during physiologic rhythms, as in the uterus, might be explained. To respond in such a manner, the vessel must sense the volume flowing through it. However, the blood stream contacts the vessel only at the endothelial layer where the drag or friction on the lining cells is proportional to the velocity of flow. It has been suggested that an increased drag may initiate mechanisms which lead to an increase in the lumen, while a decreased drag tends to be followed by a reduction. The change in flow must continue over relatively long periods of time to inaugurate the adaptive process. When flow is laminar, the velocity near the boundary layers is inversely related to the fourth power of the radius. Thus, the drag is greater for a given volume flow when the vessel is small. Enlargement of the vessel would diminish the drag on the endothelium. When streamlines are disturbed and flow is turbulent, as occurs when blood passes beyond a narrow orifice such as a stenotic valve or a coarctation, the drag on the wall of the affected segment of the vessel is increased. This produces an enlargement of the lumen out of proportion to the size which might be expected under normal circumstances.

E. PHYSIOLOGY

By F. J. HADDY[1]; L. A. SAPIRSTEIN[2]; R. R. SONNENSCHEIN[3]

1. HEMODYNAMICS

a. Variable resistance to flow: Arteries provide a highly variable resistance to flow of blood, thus permitting alteration in the rate of blood flow or blood pressure. If resistance to flow through a bed changes independently of that through other beds, the rate of blood flow in that bed will be affected, without any significant variation occurring in systemic arterial blood pressure. An example is muscular activity in a limb or injection of a vasodilator into the femoral artery. If, on the other hand, resistance to flow through many beds vary concomitantly and in the same direction, systemic arterial pressure may change without significant alterations in the rate of blood flow. A possible example is essential hypertension.

In the physiologic state, the arteriole provides the greatest and the most variable resistance to blood flow, because it is long and narrow and contains an abundance of muscle in its wall. A somewhat similar situation exists in the case of small and large arteries, hence they too, provide a variable resistance to flow. In fact, under unusual conditions,

[1] Author of Subsections 1 and 2.
[2] Author of Subsections 3 and 4.
[3] Author of Subsection 5.

most of the resistance to flow through a vascular bed may reside in large arteries (Haddy, 1958; Davis and Hamilton, 1959) (for further discussion, see Subsection 3, below). The resistance to flow through arteries and arterioles may change because of variations in the geometric and the viscous components of resistance.

The geometric component is a function of vessel length and radius. The internal radius of an artery is the most important dimension, and it may be altered through both passive and active mechanisms. (For further discussion of the influence of radius on rate of flow, see Section D.-2g of this chapter.)

Passive mechanisms include transmural pressure, organic change in the vessel wall and intraluminal plugging. Transmural pressure is the gradient of pressure across the vessel wall, that is, the difference between intraluminal and extraluminal pressures. A rise of transmural pressure, which tends to distend arteries, may result from increased intraluminal pressure subsequent to either, (1) rise of flow rate, (2) active arteriolar constriction, or (3) dependent position in relation to the heart. It may also follow reduction of extraluminal pressure, as from application of a suction cup to the skin. A fall of transmural pressure, which tends to collapse arteries, may result from a decrease of intraluminal pressure due either to, (1) fall of flow rate, (2) active arteriolar dilatation, or (3) elevation above heart level. It may likewise arise from an increase of extraluminal pressure following application of a tourniquet, contraction of skeletal or cardiac muscle, tissue edema, or growth of an extravascular tumor. Radii may also vary passively due to organic change in the vessel wall, such as narrowing due to wall edema, subintimal hemorrhage, fibrous intimal thickening, atherosclerosis, inflammation, and coarctation. Dilatation following wall dehydration, arteriosclerosis, and pyogenic or syphilitic infection may likewise be thought of as passive, inasmuch as it is not the result of change of muscle tone. The radius may narrow passively by means other than change in the vessel wall, such as luminal obstruction by clotted blood (thrombosis, embolism) or a foreign material.

The radius is influenced actively by alteration in the state of contraction of the smooth muscle in the medial layer of the vessel wall. The factors which influence muscle contraction are manifold and can be divided into 3 groups, namely; nervous, chemical and physical, although such categories are not necessarily discrete.

The nervous factors are the vasoconstrictor and vasodilator nerves. The vasoconstrictor nerves are part of the sympathetic division of the autonomic nervous system and have a much more important role in the control of resistance than the vasodilator nerves. They maintain the smooth muscle in arteries and especially arterioles in a state of partial contraction, by means of a continuous bombardment of action potentials ("tonic activity"). There is good evidence indicating that these action potentials are effective by virtue of noradrenaline production at the myoneural junction. A decrease or increase in their number results in vasodilatation or vasoconstriction, through a decrease or an increase, respectively, in the production of noradrenaline. The vasoconstrictor fibers appear to be especially important in regulating flow through the splanchnic and cutaneous vascular beds. Vasodilator nerves are more difficult to demonstrate and are more complex anatomically. They are probably cholinergic and without tonic activity. (For discussion of sympathetic control of blood vessels, see Chapter IV.)

The chemical substances which can actively alter the radius of a vessel range in structural complexity from ions to very large molecules. A partial list includes sodium, potassium, calcium, magnesium, hydrogen, oxygen, carbon dioxide, adrenaline, noradrenaline, acetylcholine, serotonin, angiotensin, histamine, pitocin, pitressin, adenyl compounds, steroids, parathormone and thyroxin. The rapidity of action varies widely. For example, the constrictor effect of the catacholamines is almost instantaneous, whereas that of the steroids is very slow. More and more evidence is accumulating to indicate that the state of contraction is a function of the concentrations of ions within and around the smooth muscle cell and that chemical agents change the state of contraction by altering these concentrations (Friedman *et al.*, 1959a,b; Haddy, 1960b). Hence, the rapidity of action of an agent may be related to the speed with which it accomplishes such a change. The direct action of an agent may differ from its remote effect. For example, the direct action of histamine is arteriolar dilatation, whereas at the same time arteriolar constriction is produced via the adrenal (Haddy, 1960a). Moreover, the interplay of direct and remote actions may result in different responses along the length of arteries. For example, a generalized elevation of the hydrogen ion concentration may dilate arterioles through a direct action but constrict larger arteries through a sympathoadrenal discharge (Fleishman *et al.*, 1957). (For a discussion of humoral control of blood vessels, see Subsections 3 and 4, below.)

The physical factors which influence radius, and consequently, resistance by altering the state of contraction of smooth muscle, are temperature and transmural pressure. The direct effect of lowering temperature is relaxation of smooth muscle and thereby a decrease of resistance, a response which is buffered by a sympathoadrenal discharge. The reverse may result by raising temperature. Since smooth muscle tends to resist extension by active contraction (Folkow and Lofving, 1956), an increase of transmural pressure through elevation of arterial pressure distends arteries less than it would if they were simple passive elastic tubes. Indeed, in some instances, it may even result in constriction of smaller arteries [Johnson *et al.*, 1960]. There is also some evidence to indicate that elevation of arteriolar transmural pressure through a rise of venous pressure results in arteriolar constriction in some beds (Yamada and Burton, 1954; Burton and Rosenberg, 1956; Rosenberg, 1956; Haddy and Gilbert, 1956; Selkurt and Johnson, 1958).

The resistance to flow through arteries and arterioles may also change because of variation in blood viscosity. Viscosity and therefore, vascular resistance, increase with polycythemia and decrease with anemia. Viscosity also varies with velocity of flow in small vessels. Increase of velocity results in axial streaming of red cells and reduction of viscosity, whereas decrease of velocity does the reverse. Such changes probably have little effect upon resistance, however, except at extremely low rates of flow (Haynes and Burton, 1959a). (For further discussion of the role of viscosity, see Section D.-1, this chapter.)

b. Variable storage of blood: Blood vessels function as variable reservoirs for blood. Changes in storage may thereby influence the cardiac filling and output. Other things remaining equal, the change in output is reflected in an appropriate alteration of arterial blood pressure.

The veins provide the main variable reservoir. Venous dilatation or constriction, whether active or passive, results in a large change in the return of blood to the heart. In contrast, arteries serve as a minor variable reservoir. Arteriolar constriction results in an increase in the volume of blood in small and large arteries because they are passively distended by the rise of pressure (Shadle *et al.*, 1958). If constriction is limited to one vascular bed, the rise in arterial volume is accompanied by fall in capillary volume. The latter change occurs because of a decrease of capillary pressure subsequent to a fall in rate of flow; this results in passive collapse of capillaries. On the other hand, active arteriolar dilatation passively decreases the volume in larger arteries and increases the volume in capillaries, provided the response is limited to a single bed. Active constriction or dilatation along the entire length of the arterial system results in a decrease or an increase of arterial volume, respectively.

2. Material Distribution

a. Variable vessel patency: The purpose of the circulation is to distribute materials rapidly as well as to transport heat. Oxygen, carbon dioxide and other substances leave and enter the circulation by diffusion at the capillary level. Arteries exert control over the rates of movements of these materials by changing the degree of patency of the bed and thus varying diffusion distances. For example, many arterioles are normally closed, resulting in no flow through connecting capillaries, and a consequent increase in the distance that materials must diffuse to or from the nearest open capillary. When materials are needed rapidly, as in exercise, such arterioles will open, causing a flow through connecting capillaries and, hence, a reduction of diffusion distances.

b. Variable filtration pressures: Arteries, in part, control the storage of water by varying capillary hydrostatic pressure. Since water leaves and enters the capillary by filtration, a rise of effective filtration pressure increases water storage in the tissue, whereas a fall does the reverse. The most variable component of effective filtration pressure is capillary hydrostatic pressure, which, in turn, is affected by aortic pressure, arterial caliber, venous caliber and right atrial pressure. Considering in each case a change in only one variable at a time, capillary hydrostatic pressure will increase as a result of a rise in aortic pressure, dilatation of arteries, constriction of veins or an elevation of right atrial pressure. The reverse changes result in a fall of capillary pressure. With aortic and right atrial pressures constant, a sustained alteration in capillary hydrostatic pressure can be accomplished only by a change of arterial or venous caliber. The highest capillary pressures, and hence, filtration rates, result from simultaneous local arterial dilatation and venous constriction (Haddy, 1960a). Capillary pressure may change even if arteries and veins simultaneously constrict or simultaneously dilate. For example, it will rise if venous constriction is proportionately greater than arterial constriction (Kelly and Visscher, 1956; Haddy *et al.*, 1959a).

The control of the storage of water exerted by arterioles can be exemplified by the action of histamine. Injection or release of this substance beneath the skin in low concentrations causes maximal arteriolar dilatation and a large local increase in the rate of blood flow, as long as aortic pressure remains unchanged. The resulting rise in capillary pressure to high levels, in turn, produces a rapid rate of filtration out of the capillary (Haddy, 1960a).

In the kidney, arterioles exert control over water excretion and retention by varying glomerular capillary pressure. Afferent arteriolar constriction and efferent arteriolar dilatation promote water retention, whereas afferent arteriolar dilatation and efferent arteriolar constriction produce excretion. (For further discussion, see Section D, Chapter XIII.)

Arterioles also influence water storage by indirectly changing the nature of the filtering membrane. For example, arteriolar dilatation, by raising capillary pressure, increases capillary membrane surface area (by opening closed capillaries and distending those already open) and permeability (by increasing pore size), thus promoting filtration.

3. Humoral Control of Arteries[1]

It has usually been held that the drop in blood pressure occurring in the arterial system is very small in comparison with that which takes place in the arterioles. If this view is correct, it would be difficult to understand how humoral control of arteries could be of physiologic significance so far as the flow of blood is concerned.

Nevertheless, there is considerable evidence to indicate that naturally occurring chemical substances can influence the smooth muscle of the walls of both large and small arteries, with the latter appearing to represent a much more significant fraction of the peripheral resistance than has heretofore been believed. This fact and the responsiveness of the arteries to changes in the chemical environment suggest that pre-arteriolar areas of the circulation may be of considerable significance in determining the pattern of blood flow distribution.

In line with this view one finds that, although reduction in pH of the circulating blood results in decreased arteriolar resistance in the dog's foreleg, this change may be compensated for, in part, by an increase in small artery resistance (Fleishman *et al.*, 1957). Similarly, elevation of plasma potassium produces dilation of the arterioles, but constriction of the small arteries (Emanuel *et al.*, 1959). Furthermore, it has been found that vagal stimulation reduces the pulse wave velocity more than would be expected from the fall in diastolic pressure, a change which may result from a decrease in muscle tone in the walls of the larger arteries, due to nervous or endocrine factors (Dow and Hamilton, 1939). When allowance is made for the level of diastolic pressure, epinephrine is also found to reduce the pulse wave velocity, thus suggesting that this hormone probably causes relaxation of arterial musculature (Kraner *et al.*, 1959). It produces dilation of the smaller arteries in the dog's leg, while norepinephrine constricts them (Haddy, 1957).

Some of the inconstant effects of serotonin (5-hydroxytryptamine) on certain territorial resistances may be explained by the finding that it dilates the arterioles while constricting the small arteries; either effect may predominate (Haddy *et al.*, 1959b).

It is possible that a number of the changes in arterial tone described above are not directly attributable to the chemical agent which produces

[1] By L. A. Sapirstein.

them, but instead may be mediated indirectly, either through autonomic activity, through unknown intermediary hormones, or perhaps through passive luminal changes which occur as the arterial pressure alters. Nevertheless, the findings quoted indicate the need for a thorough reassessment of the concept that the major portion of the peripheral resistance and the one which is most significant physiologically is in the arterioles.

4. Humoral Control of Arterioles[1]

The arterioles, representing the site of greatest vascular resistance, are the portion of the vascular bed most subject to nervous and humoral control. As a consequence, those nervous and humoral factors which influence the status of the arterioles will play a primary role in determining the apportionment of the cardiac output to the tissues.

In considering the effects of humoral agents on the arterioles, it will be convenient first to take up those nonspecific agents produced by tissue activity and injury, and second, the specific hormones which affect the arterioles at distant sites.

a. Products of tissue activity and injury: It is well known that blood flow is augmented in tissues which are active or have been rendered temporarily ischemic. Some of the increase in circulation is undoubtedly mediated by nervous vasodilation, while a part is probably due to the altered chemical environment in the active tissue. Among the possible chemical changes which may be implicated are: raised hydrogen ion and carbon dioxide concentrations, a fall in oxygen tension, an increase in lactic acid and other products of intermediary metabolism, acetylcholine, bradykinin, and many other substances.

Tissue injury likewise is followed by local arteriolar dilation, succeeded by increased local perfusion. Among the substances implicated in this response are the products of tissue metabolism listed above, histamine and histamine like substance, and, possibly, serotonin.

In the muscle, a minor increase in hydrogen ion concentration, such as a shift in pH of as little as .05 units, is followed by vasodilation (Fleisch and Sibul, 1933); a decrease, however, has the same effect (Kester *et al.*, 1952). In the case of the skin, a low pH has a dilator effect, while a high pH is vasoconstrictor (Deal and Green, 1954). Cerebral vascular resistance is decreased in acidosis (Kety *et al.*, 1948), while hyperventilation, producing alkalosis, is vasoconstrictor (Kety and Schmidt, 1946).

[1] By L. A. Sapirstein.

In the intact animal, on breathing higher than normal concentration of carbon dioxide, the cerebral vascular resistance is decreased (Kety and Schmidt, 1948), as is mesenteric vascular resistance (Brickner *et al.*, 1956). However, it is necessary to point out that the powerful central actions of carbon dioxide confuse the interpretation of experiments in which this gas is inhaled (McDowall, 1938). In experiments in which the head is perfused with blood from a normal animal during carbon dioxide inhalation, the blood pressure falls (McDowall, 1938), suggesting overall vasodilation. It is not clear whether the effects of carbon dioxide are greater than those attributable to the changed pH, although most evidence suggests that the effect is nonspecific.

In the heart-lung preparation, slight falls in oxygen saturation of the arterial blood result in prompt coronary vasodilation (Anrep, 1936). Oxygen lack also reduces cerebral resistance and increases cerebral blood flow (Kety and Schmidt, 1948). Though many studies on the circulatory effects of hypoxia are available, there is surprisingly little information on the local effects of oxygen lack in circumstances where the central effects are not prepotent.

The hyperemia of exercising muscles has, from time to time, been attributed to lactate, adenylic acid derivatives, peptones, and amino acids, all products of intermediary metabolism. However, no convincing studies implicating any of these with certainty are available.

Since vasodilation after nerve stimulation in many areas is partially prevented by atropine (Hilton and Lewis, 1957), the possibility that cholinergic nerves simultaneously accomplish an increase in activity of the structure stimulated and a decrease in arteriolar resistance, resulting in increased blood flow to active tissue, is an attractive one. The cholinergic sympathetic vasodilators to muscle may also synergize with the motor outflow, to increase the blood flow to active muscle. In general, acetylcholine appears to be vasodilator to arterioles. A possible exception is the coronary circulation; the vasoconstrictor effect here, however, may be a secondary one, due to the decreased work load on the heart, with the development of metabolic circumstances favoring vasoconstriction.

Vasodilation in the submandibular gland following stimulation of the chorda tympani persists after the administration of atropine and is due, at least in part, to the presence of a stable polypeptide substance (Hilton and Lewis, 1957), which may be formed during activity, through the release of a tissue proteolytic enzyme, kallikrein. The latter acts on a plasma globulin (callidinogen) to produce a polypeptide with vasodilator properties (bradykinin, callidin). The significance of such a system in arteriolar regulations is not yet clear, but the possibility that it may be of general importance must be considered because of the wide distribution of kallikrein.

The similarity between the vascular effects of intradermal histamine and skin injury suggests the possibility that histamine or a histamine-like substance produced by injury is responsible for local vasodilation (Lewis, 1927). Histamine dilates arterioles generally, except in the pulmonary circulation where it is constrictor. The physiologic significance of this drug in general arteriolar regulation is, at present, obscure.

The circulatory effects of serotonin (5–HT) have been found to be vasodilator for arterioles in most areas (see review by Page, 1958). It is, however, vasoconstrictor in the pulmonary circulation and in the kidney. When vasodilation is produced, it is most marked in areas where there has been previous vasoconstriction. It has therefore been suggested that 5–HT may act as a humoral buffer against vasoconstriction induced by the autonomic nervous system (Page, 1958). The significance of 5–HT in circulatory regulation awaits further elucidation. As Page (1958) points out, "the cardiovascular actions of HT are so varied it would be rash indeed on the basis of current evidence to predict what its predominant role, if any, will turn out to be."

b. Humoral substances active at distant sites: Unlike the substances considered above, these materials are produced at sites different from their points of vascular action. Whereas substances in the first group generally have essentially similar effects on most arteriolar resistances, those in the second group produce a pattern of selective vasoconstriction and vasodilation. Through their action, the cardiac output may be redistributed in response to the requirements of the whole animal rather than to the needs of local tissues.

It is convenient to consider epinephrine and norepinephrine in this section, despite the fact that they are also produced at sympathetic nerve endings, as well as in the adrenal medulla. An enormous amount of literature concerning their vascular effects is available, much of which has been summarized in a recent excellent review (Green and Kepchar, 1959). (For a discussion of the pharmacologic action of these two substances, see Section F.-2, Chapter III.)

Vasopressin (pitressin, antidiuretic hormone), secreted by the posterior lobe of the pituitary gland, is usually considered to be a potent vasoconstrictor in all vascular beds. Convincing evidence for this effect is available in the case of the coronary circulation (Green *et al.*, 1942). Although the changes in the splanchnic bed are less well established, there is some evidence (Sapirstein, 1957) to suggest that the hepatic arterial bed is actually dilated, while the vessels of the portal circulation are constricted. The well known blanching and cooling of the skin after local administration of vasopressin implies that the cutaneous resistance is radically increased. The effect of vasopressin on renal blood flow is equivocal (Corcoran and Page, 1939).

It is far from clear, however, whether these effects are of physiologic rather than of pharmacologic significance. The quantities of vasopressin required to produce vascular effects are at least two orders of magnitude greater than those which elicit manifestation of the antidiuretic effects. The vagohypophyseal reflex of Chang *et al.* (1937), in which vasoconstriction follows hypotension in preparations with intact pituitary glands but not after hypophysectomy, indicates that a significant role may be played by vasopressin in circulatory homeostasis.

The renal pressor system, originally elucidated in an attempt to clarify the mechanism of experimental renal hypertension, appears to have little or no relationship to this condition.[2] It seems probable, however, that it does play a role in the homeostasis of blood pressure, particularly after the induction of hypotension (Sapirstein *et al.*, 1941; Huidobro and Braun-Menendez, 1942). As presently understood, the important components of the renal pressor system are (1) renin, a proteolytic enzyme released by the kidney; (2) renin substrate, an alpha–2 globulin secreted by the liver, circulating in the plasma and hydrolyzed by renin; and (3) angiotensin I and II, the former a decapeptide, the latter an octapeptide formed from angiotensin I by a converting enzyme. The decapeptide is inactive, while the octapeptide is a powerful vasoconstrictor (Hoobler, 1958). Angiotensin is subsequently broken down by angiotonase (hypertensinase), a polypeptidase widely distributed in tissues, but physiologically active only in the plasma (Sapirstein *et al.*, 1946). Although angiotensin is an exceedingly powerful pressor agent and can be shown to increase the total peripheral resistance very strikingly, the nature of its effects on arterioles in specific vascular beds has been only incompletely investigated. Renal vascular resistance is known to be increased (Corcoran and Page, 1940). The blood flow through the isolated perfused rabbit's ear is reduced (Abell and Page, 1942), as is that through the mesentery (Abell and Page, 1946). Since no specific information is available regarding the effects of angiotensin on the cerebral, cutaneous, or myocardial arterioles, it seems premature, at this time, to speculate on the role of the renal pressor system in the regulation of the circulation from the standpoint of systemic blood flow distribution.

[2] For a somewhat different viewpoint, see (1) Hypertension, Section B.-3b, Chapter XX.

5. Alterations of Physiologic Responses of Arteries with Aging[1]

The major change in large arteries occurring with age, in the absence of arteriosclerosis, is an increase in the elasticity, defined roughly as the resistance to stretch (Butcher and Newton, 1958; Roach and Burton, 1959). Thus, at pressures around 100 mm Hg, the per cent change in volume of a segment of aorta per unit change in pressure is 3–4 times as great for age 25 as for age 70. The difference is less at higher pressure levels, but the trend remains the same as a function of age. In the aorta and the iliac arteries, resistance to deformation, both circumferentially and tangentially, increases with age. Moreover, a greater external pressure is required to cause collapse of the older vessel when empty. The greater elasticity manifested in these ways has been ascribed to increase and alteration in collagenous material in the arterial wall (Roach and Burton, 1959). An unfortunate error in usage has caused confusion between "distensibility" and "elasticity," which have, properly, essentially opposite meanings. Thus, while arterial elasticity increases with age, distensibility decreases. (For a discussion of the physics of blood flow, see Section D.-2 of this chapter.)

A result of the increased resistance to stretch of the aorta and large arteries is that their accommodation to the cardiac stroke volume by distension is reduced with advancing age. To the extent that compensatory mechanisms do not develop, such a change leads to alteration in characteristics of the arterial pressure wave. The velocity of the pulse wave is greater in older individuals (Landowne, 1958). (For a discussion of the biochemical changes observed with aging, see Section G.-3, Chapter III.)

References

Abell, R. G., and Page, I. H. (1942). The reactions of peripheral blood vessels to angiotonin, renin, and other pressor agents. *J. Exptl. Med.* **75**, 305.

Abell, R. G., and Page, I. H. (1946). The reaction of the vessels of the mesentery and intestine to angiotonin and renin. *Am. J. Med. Sci.* **212**, 166.

Altschul, R. (1954). "Endothelium." Macmillan, New York.

Anrep, G. V. (1936). Lane Medical Lectures: Studies in Cardiovascular Regulation. Stanford Univ. Press, Stanford, Calif.

Bayliss, L. E. (1952). *In* "Deformation and Flow in Biological Systems" (A. Frey-Wyssling, ed.), Chapter 6. North-Holland, Amsterdam.

Bingham, E. C., and Roepke, R. R. (1944). The rheology of blood. III. *J. Gen. Physiol.* **28**, 79.

Blumenthal, H. T. (1956). Response potentials of vascular tissues and the genesis of arteriosclerosis; hemodynamic factors. *Geriatrics* **11**, 554.

[1] By R. R. Sonnenschein.

Bremer, J. L. (1932). The presence and influence of two spiral streams on the heart of the chick embryo. *Am. J. Anat.* **49,** 409.

Brickner, E. W., Dowds, E. G., Willitts, B., and Selkurt, E. E. (1956). Mesenteric blood flow as influenced by progressive hypercapnia. *Am. J. Physiol.* **184,** 275.

Burton, A. C. (1953). *In* "Visceral Circulation" (G. E. W. Wolstenholme, ed.), p.70. Churchill, London.

Burton, A. C. (1954). Relation of structure to function of the tissues of the wall of blood vessels. *Physiol. Revs.* **34,** 619.

Burton, A. C., and Rosenberg, E. (1956). Effect of raised venous pressure in the circulation of the isolated perfused rabbit ear. *Am. J. Physiol.* **185,** 465.

Butcher, H. R., Jr., and Newton, W. T. (1958). The influence of age, arteriosclerosis and homotransplantation upon the elastic properties of major human arteries. *Ann. Surg.* **148,** 1.

Chang, H. C., Chia, K. F., Hsu, C. H., and Lim, R. K. S. (1937). Reflex secretion of the posterior pituitary elicited through the vagus. *J. Physiol.* (*London*) **90,** 87P.

Copley, A. L., and Scott Blair, G. W. (1958). Comparative observations on adherence and consistency of various blood systems in living and artificial capillaries. *Rheol. Acta* **1,** 170.

Corcoran, A. C., and Page, I. H. (1939). The effects of renin, pitressin, and pitressin and atropine on renal blood flow and clearance. *Am. J. Physiol.* **126,** 354.

Corcoran, A. C., and Page, I. H. (1940). The effects of renin on renal blood flow and glomerular filtration. *Am. J. Physiol.* **129,** 698.

Coulter, N. A., Jr., and Pappenheimer, J. R. (1949). Development of turbulence in flowing blood. *Am. J. Physiol.* **159,** 401.

Davis, D. L., and Hamilton, W. F. (1959). Small vessel responses of the dog paw. *Federation Proc.* **18,** 34.

Deal, C. P., Jr., and Green, H. D. (1954). Effects of pH on blood flow and peripheral resistance in muscular and cutaneous vascular beds in the hind limb of the pentobarbitalized dog. *Circulation Research* **2,** 148.

Dix, F. J., and Scott Blair, G. W. (1940). On the flow of suspensions through narrow tubes. *J. Appl. Phys.* **11,** 574.

Dow, P., and Hamilton, W. F. (1939). An experimental study of the velocity of the pulse wave propagated through the aorta. *Am. J. Physiol.* **125,** 60.

Duran-Reynals, S. (1958). The dermal ground substance of the mesenchyme as an element of natural resistance against infection and cancer. *In* "Frontiers in Cytology" (S. L. Palay, ed.). Yale Univ. Press, New Haven, Conn.

Emanuel, D. A., Scott, J. B., and Haddy, F. J. (1959). Effect of potassium on small and large blood vessels of the dog's forelimb. *Am. J. Physiol.* **197,** 637.

Fahraeus, R., and Lindqvist, T. (1931). The viscosity of the blood in narrow capillary tubes. *Am. J. Physiol.* **96,** 562.

Fell, H. B. (1951). Histogenesis in tissue culture. *In* "Cytology and Cell Physiology" (G. H. Bourne, ed.). Oxford Univ. Press (Clarendon), London and New York.

Fleisch, A., and Sibul, I. (1933). Über nutritive Kreislaufregulierung. II. Die Wirkung von pH, intermediären Stoffwechselprodukten und anderen biochemischen Verbindungen. *Pflügers Arch. ges. Physiol.* **231,** 787.

Fleishman, M., Scott, J., and Haddy, F. J. (1957). Effect of pH change upon systemic large and small vessel resistance. *Circulation Research* **5,** 602.

Folkow, B., and Lofving, B. (1956). The distensibility of the systemic resistance blood vessels. *Acta Physiol. Scand.* **38,** 37.

Friedman, S. M., Jamieson, J. D., Hinke, J. A. M., and Friedman, C. L. (1959a).

Drug induced changes in blood pressure and in blood sodium as measured by glass electrode. *Am. J. Physiol.* **196,** 1049.

Friedman, S. M., Jamieson, J. D., and Friedman, C. L. (1959b). Sodium gradient, smooth muscle tone and blood pressure regulation. *Circulation Research* **7,** 44.

Goetz, R. H., Warren, J. V., Gauer, O. H., Patterson, J. L., Jr., and Doyle, J. T. (1960). Circulation of the giraffe. *Circulation Research* **8,** 1049.

Goldstein, S. (1938). "Modern Developments in Fluid Dynamics." Oxford Univ. Press (Clarendon), London and New York.

Gouley, B. A., and Sickel, E. M. (1943). Aortic regurgitation caused by dilatation of the aortic orifice and associated with a characteristic valvular lesion. *Am. Heart J.* **26,** 24.

Green, H. D., and Kepchar, J. H. (1959). Control of peripheral resistance in major systemic vascular beds. *Physiol. Revs.* **39,** 617.

Green, H. D., Wegria, R., and Boyer, N. H. (1942). Effects of epinephrine and pitressin on the coronary artery inflow in anesthetized dogs. *J. Pharmacol. Exptl. Therap.* **76,** 378.

Haddy, F. J. (1957). Effects of epinephrine, norepinephrine and serotonin upon systemic small and large vessel resistances. *Circulation Research* **5,** 247.

Haddy, F. J. (1958). Vasomotion in systemic arteries, small vessels and veins determined by direct resistance measurements. *Minn. Med.* **41,** 162.

Haddy, F. J. (1960a). Effect of histamine on small and large vessel pressures in the dog foreleg. *Am. J. Physiol.* **198,** 161.

Haddy, F. J. (1960b). Local effects of sodium, calcium and magnesium upon small and large blood vessels of the dog forelimb. *Circulation Research* **8,** 57.

Haddy, F. J., and Gilbert, R. P. (1956). The relation of a venous arteriolar reflex to transmural pressure and resistance in small and large systemic vessels. *Circulation Research* **4,** 25.

Haddy, F. J., Fstensen, R. D., and Gilbert, R. P. (1959a). Studies on the mechanism of edema produced by prolonged infusion of norepinephrine. *J. Lab. Clin. Med.* **54,** 821.

Haddy, F. J., Gordon, P., and Emanuel, D. A. (1959b). The influence of tone upon responses of small and large vessels to serotonin. *Circulation Research* **7,** 123.

Hass, G. M. (1942). Elastic tissue. *Arch. Pathol.* **34,** 807 and 971.

Haynes, R. H. (1960a). Blood flow through narrow tubes. *Nature* **185,** 679.

Haynes, R. H. (1960b). Physical basis of the dependence of blood viscosity on tube radius. *Am. J. Physiol.* **198,** 1193.

Haynes, R. H., and Burton, A. C. (1959a). Role of the non-Newtonian behavior of blood in hemodynamics. *Am. J. Physiol.* **197,** 943.

Haynes, R. H., and Burton, A. C. (1959b). *In* "Proceedings of the First National Biophysics Conference" (H. Quastler and H. J. Morowitz, eds.), p. 452. Yale Univ. Press, New Haven.

Hilton, S. M., and Lewis, G. P. (1957). Functional vasodilatation in the submandibular salivary gland. *Brit. Med. Bull.* **13,** 189.

Holman, E. H. (1950). Thoughts on the dynamics of blood flow. *Angiology* **1,** 530.

Hoobler, W. W., ed. (1958). Proceedings of the Conference on Basic Mechanisms of Arterial Hypertension. *Circulation* **17,** 641.

Hughes, A. F. W. (1936). Studies on the area vasculosa of the chick. *J. Anat.* **70,** 76.

Huidobro, F., and Braun-Menendez, E. (1942). The secretion of renin by the intact kidney. *Am. J. Physiol.* **137,** 47.

Johnson, P. C., Knockel, W. L., and Sherman, T. L. (1960). Pressure volume relationships of small arterial segments. *Federation Proc.* **19**, 91.

Kampmeier, O. F., and Birch, C. L. F. (1927). The origin and development of the venous valves. *Am. J. Anat.* **38**, 451.

Keith, A. (1918). Hunterian lecture on Wolff's law. *Lancet* **i**, 250.

Kelly, W. D., and Visscher, M. B. (1956). Effect of sympathetic nerve stimulation on cutaneous small vein and small artery pressures, blood flow and hindpaw volume in the dog. *Am. J. Physiol.* **185**, 453.

Kemp, N. E., and Quinn, B. L. (1954). Morphogenesis and metabolism of amphibian larvae after excision of heart. *Anat. Record* **118**, 773.

Kester, N. C., Richardson, A. W., and Green, H. D. (1952). Effect of controlled hydrogen-ion concentration on peripheral vascular tone and blood flow in innervated hind leg of the dog. *Am. J. Physiol.* **169**, 678.

Kety, S. S., and Schmidt, C. F. (1946). The effects of active and passive hyperventilation on cerebral blood flow, cerebral oxygen consumption, cardiac output, and blood pressure of normal young men. *J. Clin. Invest.* **25**, 107.

Kety, S. S., and Schmidt, C. F. (1948). The effects of altered arterial tensions of carbon dioxide and oxygen on cerebral blood flow and cerebral oxygen consumption of normal young men. *J. Clin. Invest.* **27**, 484.

Kety, S. S., Polis, B. D., Nadler, C. S., and Schmidt, C. F. (1948). The blood flow and oxygen consumption of the human brain in diabetic acidosis and coma. *J. Clin. Invest.* **27**, 500.

Kosiak, M., Kubicek, W. G., Olson, M., Danz, J. N., and Kottke, F. J. (1958). Evaluation of pressure as a factor in the production of ischial ulcers. *Arch. Phys. Med. Rehabil.* **39**, 623.

Kraner, J. C., Ogden, E., and McPherson, R. C. (1959). Immediate variations in pulse wave velocity caused by adrenaline in short, uniform parts of the arterial tree. *Am. J. Physiol.* **197**, 432.

Kümin, K. (1949). Bestimmung des Zähigkeitskoeffizienten μ' für Rinderblut bei Newton'schen Strömungen in verschieden weiten Röhren und Kapillaren bei physiologischer Temperatur. Inaugural Dissertation, University of Bern. Paulusdruckerei, Freiburg in der Schweiz.

Landowne, M. (1958). The relation between intra-arterial pressure and impact pulse wave velocity with regard to age and arteriosclerosis. *J. Gerontol.* **13**, 153.

Lansing, A. I., ed. (1959). "The Arterial Wall." Williams & Wilkins, Baltimore, Md.

Lewis, T. (1927). "The Blood Vessels of the Human Skin and Their Responses." Shaw & Sons, London (2nd ed., 1956).

McDonald, D. A. (1960). "Blood Flow in Arteries." Williams and Wilkins, Baltimore.

McDowall, R. J. S. (1938). "The Control of the Circulation of the Blood." p. 341. Dawson & Sons, London.

Mehrotra, R. M. L. (1953). An experimental study of the changes which occur in ligated arteries and veins. *J. Pathol. Bacteriol.* **65**, 307.

Moore, R. D., and Schoenberg, M. D. (1959). The relation of mucopolysaccharides of vessel walls to elastic fibers and endothelial cells. *J. Pathol. Bacteriol.* **77**, 163.

Page, I. H. (1958). Cardiovascular Actions of Serotonin (5-hydroxytryptamine). *In* "5-Hydroxytryptamine" (G. P. Lewis, ed.). Pergamon, London.

Patten, B. M. (1953). "Human Embryology." McGraw-Hill (Blakiston), New York.

Patten, B. M., Kramer, T. C., and Barry, A. (1948). Valvular action in the embryonic chick heart by localized apposition of endocardial cushions. *Am. J. Anat.* **102**, 299.

Poole, J. C. F., Sanders, A. G., and Florey, H. W. (1958). The regeneration of aortic endothelium. *J. Pathol. Bacteriol.* **75**, 133.

Prandtl, L. (1952). "Fluid Dynamics." Hafner, New York.

Roach, M. R., and Burton, A. C. (1957). The reason for the shape of the distensibility curves of arteries. *Can. J. Biochem. Physiol.* **33**, 681.

Roach, M. R., and Burton, A. C. (1959). The effect of age on the elasticity of human iliac arteries. *Can. J. Biochem. and Physiol.* **37**, 557.

Robb-Smith, A. H. T. (1957). The reticulin riddle. *J. Mt. Sinai Hosp.* **24**, 1155.

Rodbard, S. (1955). Flow through collapsible tubes: Augmented flow produced by resistance at the outlet. *Circulation* **11**, 280.

Rodbard, S. (1956). Vascular modifications induced by flow. *Am. Heart J.* **51**, 926.

Rodbard, S. (1958a). Physical factors in the progression of stenotic vascular lesions. *Circulation* **17**, 410.

Rodbard, S. (1958b). A method for the induction of intravascular structure. *Circulation* **18**, 771.

Rodbard, S. (1959). Atherosclerosis and nutrition. *J. N. Y. Med. Coll., Flower Fifth Ave. Hosp.* **1**, 65.

Rodbard, S., and Harasawa, M. (1959). Stenosis in a deformable tube inhibited by outlet pressure. *Am. Heart J.* **57**, 544.

Rodbard, S., and Saiki, H. (1953). Flow through collapsible tubes. *Am. Heart J.* **46**, 715.

Rosenberg, E. (1956). Local character of the veni-vasomotor reflex. *Am. J. Physiol.* **185**, 471.

Rouse, H. (1950). "Engineering Hydraulics." Wiley, New York.

Sapirstein, L. A. (1957). Effect of hemorrhage on distribution of cardiac output. *Federation Proc.* **16**, 111.

Sapirstein, L. A., Ogden, E., and Southard, F. D., Jr. (1941). Renin-like substances in blood after hemorrhage. *Proc. Soc. Exptl. Biol. Med.* **48**, 505.

Sapirstein, L. A., Reed, R. K., and Page, E. W. (1946). The site of angiotonin destruction. *J. Exptl. Med.* **83**, 425.

Schlichter, J. G. (1946). Experimental medionecrosis of the aorta. *A.M.A. Arch. Pathol.* **42**, 182.

Scott Blair, G. W. (1958). The importance of the sigma phenomenon in the study of the flow of blood. *Rheol. Acta* **1**, 123.

Selkurt, E. E., and Johnson, P. C. (1958). Effect of acute elevation of portal venous pressure on mesenteric blood volume, interstitial fluid volume and hemodynamics. *Circulation Research* **5**, 592.

Shadle, O. W., Zukof, M., and Diana, J. (1958). Translocation of blood from isolated dogs hindlimb during levarterenol infusion and sciatic nerve stimulation. *Circulation Research* **6**, 326.

Smith, R. H. (1957). Biosynthesis of some connective tissue components. *Progr. in Biophys.* **8**, 217.

Starkey, T. V. (1956). The laminar flow of streams of suspended particles. *Brit. J. Appl. Phys.* **7**, 52.

Starling, E. H. (1909). "The Fluids of the Body." Constable Press, London.

Taylor, M. G. (1955). The flow of blood in narrow tubes. II. The axial stream and its formation as determined by changes in optical density. *Australian J. Exptl. Biol. Med. Sci.* **33**, 1.

Taylor, M. G. (1959). The influence of the anomalous viscosity of blood upon its oscillatory flow. *Phys. Med. Biol.* **3**, 273.

Thoma, R. (1893). "Untersuchungen über die Histogenese und Histomechanik des Gefässsystems." Enke, Stuttgart.

Tollert, H. (1954). Die Wirkung der Magnus-Kraft in laminaren Strömungen. *Chem.-Ingr.-Tech.* **26**, 141.

Weiss, D. L., and Linde, R. (1960). Stretch capacity of the component layers of the aortic wall. *A.M.A. Arch. Pathol.* **70**, 640.

Weiss, P. (1939). "Principles of Development." Henry Holt, New York.

Wells, R. E., and Merrill, E. W. (1961). Shear rate dependence of the viscosity of whole blood and plasma. *Science* **133**, 763.

Whittaker, S. F. R., and Winton, F. R. (1933). The apparent viscosity of blood flowing in the isolated hindlimb of the dog, and its variation with corpuscular concentration. *J. Physiol. (London)* **78**, 339.

Wilens, S. L., and McCluskey, R. T. (1952). The comparative filtration properties of excised arteries and veins. *Am. J. Med. Sci.* **224**, 540.

Winternitz, M. C. (1938). "Biology of Arteriosclerosis." Charles C Thomas, Springfield, Ill.

Wolff, J. (1892). "Das Gesetz der Transformation der Knochen," A. Hirschwald. Berlin.

Zimmerman, L. F. (1957). Demonstration of hyaluronidase sensitive acid mucopolysaccharide. *Am. J. Ophthalmol.* **44**, 1.

Yamada, S., and Burton, A. C. (1954). Effect of reduced tissue pressure on blood flow of the finger; The veni-vasomotor reflex. *J. Appl. Physiol.* **6**, 501.

Chapter III. ARTERIAL AND ARTERIOLAR SYSTEMS: Pharmacology and Biochemistry

F. PHARMACOLOGY

By K. I. Melville and H. E. Shister

1. Introduction

a. Factors affecting action of drugs: In considering the effects of drugs on the arterial vascular tree, one cannot speak of a uniform or selective action of a given agent on all types of blood vessels. Fundamental determinants are the size of the arteries and their relative distribution of elastic and muscular tissues (for discussion of this point, see Section C, Chapter I). The vast portion of studies in this field deals with medium-sized and small arteries and arterioles. On the other hand, the aorta and its immediate large ramifications (with their predominance of elastic tissue) constitute a different unit which is not so well-defined from the pharmacologic point of view.

Many other factors which can also modify the net effects of drugs on vasomotion include: (a) cardiac influences, e.g., changes in cardiac output; (b) shifting of blood from one area to another, e.g., from muscle to skin; (c) rapid alterations in external or internal environment, e.g., temperature and humidity changes; and (d) miscellaneous changes associated with endocrine functions, body metabolism and localized or systemic disease.

The exact role of each of these factors in respect to the overall alterations in vasomotion may be difficult to determine, and attempts are constantly being made to establish some integrated concept of the regulation of the cardiovascular system (see stimulating symposium by Peterson *et al.*, 1960a).

More recent basic researches on certain aspects of these problems, which are also of special interest, include the following: effects of water and electrolytes on vascular reactivity (Furchgott and Bhadrakon, 1953; Braun-Menendez, 1954; Gillman and Gilbert, 1956; Zsoter and Szabo, 1958); structural and functional changes of blood vessels in hypertension (Folkow, 1960; Redleaf and Tobian, 1958; Conway, 1958; Bohr *et al.*, 1958); mechanical properties of arteries and the influence of epinephrine, norepinephrine, acetylcholine, and the autonomic nervous system (Peterson *et al.*, 1960b).

b. Local hormones: The term "local hormone" (Gaddum, 1955) refers to substances which occur in tissues, usually in a physiologic inert state, but which can be activated under certain circumstances to induce vasodilatation or vasoconstriction. It can be stated that in some instances they may also be involved in vascular reactivity to drugs, although much remains to be known regarding their physiologic importance. (For further discussion of such substances, see Section E.-4, Chapter II.)

2. Adrenergic Drugs

a. Effects on blood vessels: In general terms, epinephrine and levarterenol (norepinephrine) are the most important agents in this group. They produce constriction of the arterioles in the splanchnic area, skin and mucous membranes, while the vessels of skeletal muscles are dilated; the coronaries usually respond in a similar manner.

The changes elicited by the above catecholamines vary with the type of vascular bed. Thus, in the cutaneous area, epinephrine is 2–10 times more potent than norepinephrine; in mucous membranes, both amines are of about equal potency; in skeletal muscle, epinephrine acts as a dilator, while norepinephrine is a constrictor. In the cerebral vessels epinephrine causes increased blood flow, while the effect of norepinephrine is variable.

Several other factors, such as dosage, route of administration, etc., influence the circulatory changes produced by these two amines. For example, Barcroft and Swan (1953) have shown that intravenous injections in man (10–20 μg/min) produce variable responses in liver, kidneys, skeletal muscle and brain, with the result that, as far as total peripheral resistance is concerned, epinephrine is a net vasodilator, whereas levarterenol is a net vasoconstrictor. With either amine after a therapeutic dose is given intravenously instead of subcutaneously, the blood pressure may rise to excessive levels (300/200 mm/Hg), and ventricular arrhythmia may be precipitated. Similarly, the state of the circulation, viz. the existing level of the blood pressure at the time of administration of the catecholamines, can influence the changes in blood flow through certain organs, e.g., kidney, brain (Moyer and Mills, 1953; Moyer *et al.*, 1954). Isoproterenol (Isuprel) is, on the whole, similar in action to epinephrine.

b. Mechanism of action: Ahlquist (1948) postulated two main types of adrenergic receptors, α and β, which belong functionally to the innervated effector cell. The α receptor mediates excitatory responses, while the β receptor controls inhibitory ones. Briefly, norepinephrine is the strongest α effector, Isuprel, the most potent β activator, while epinephrine occupies a somewhat intermediary position. These interactions may involve some enzymatic processes, and in this regard the role of monoamine oxidase in the action of adrenergic drugs has been suggested although not established. There is also some evidence that these catecholamines may have some sympathetic ganglionic blocking effect, since a fall in blood pressure, with flushing of skin, can occur after abrupt cessation of prolonged infusions in man.

c. Other agents: Many other compounds which exhibit adrenergic action have been synthesized. The most typical of these is ephedrine, which was introduced into clinical use by Chen and Schmidt (1930) and which can be administered orally. It is both an adrenergic agent and a cortico-medullary stimulant. While the general vascular effects of ephedrine are similar to those of epinephrine, this drug is distinctly less potent ($1/250$), but its effect lasts considerably longer, leading to tachyphylaxis or tolerance. In man, ephedrine also increases arterial pressure, both by peripheral vasoconstriction and cardiac stimulation. In general, amphetamine and phenylephrine exert similar vascular actions to those of ephedrine.

d. Pheochromocytoma: Pheochrome tumors or pheochromocytoma may arise in the adrenal glands or in the para-aortic rests of medullary tissue. They contain both epinephrine and norepinephrine, the content of the latter being as high as 95 per cent of the total, while that of epinephrine ranges from 5 to 15 per cent. The vascular changes are, therefore, similar to those obtained by infusions of epinephrine and norepinephrine in man (see above). (For reference to the clinical manifestations and pharmacologic methods of diagnosis of pheochromocytoma, see review of Roth *et al.*, 1960.)

3. Cholinergic Drugs

a. Acetylcholine (Ach.) and related choline esters: These drugs in general lead to hypotension and dilatation of the smooth muscle of the vasculature (arterioles). This effect (muscarinic) is independent of innervation and can be counteracted by atropine. Because of rapid destruction of Ach. (by cholinesterase), its effects are fleeting. Therefore, for therapeutic use, stable choline esters, such as methocholine and carbachol, are used. Similar to acetylcholine in action are muscarine, pilocarpine and neostigmine, the latter exerting its effects by an anticholinesterase action. Acetylcholine also has a "nicotinic" action (see below, Subsection 4c).

b. Nicotine: The peripheral responses of nicotine are complex. Thus, it resembles Ach. by stimulating ganglia from which cholinergic fibers arise. It causes adrenergic effects by stimulating sympathetic ganglia and by liberating epinephrine from the adrenal gland. Recent work has produced some interesting concepts on the relationships of nicotine to epinephrine (Burn and Rand, 1958). Large doses of nicotine also have a ganglionic-blocking action.

4. Blocking Agents

a. Ganglion-blocking drugs: There are a large number of drugs which, by competition with Ach., produce a blocking action in ganglia. Among such agents are tetraethylammonium, hexamethonium, pentolinium (Ansolysen), mecamylamine (Inversine), chlorisonamide (Ecolid), and trimethaphan (Arfonad). These vary in potency and side reactions, but their general mode of action is the same.

The effects of these drugs on the circulation are due principally to interference with sympathetic vasoconstrictor pathways. The result is a dilatation of both arteries and veins and also the abolition of reflexes controlling the blood pressure, thus producing orthostatic hypotension. Factors which influence this action are: existing level of blood pressure, posture and previous medication. Various vascular beds differ in their responses; thus, the blood flow is increased in the extremities (especially the feet) and skeletal muscles but is diminished in the splanchnic and renal areas. In the cerebral and coronary circulation, the vascular response is variable. The ganglion-blocking agents are also capable of blocking parasympathetic ganglia.

b. Adrenergic-blocking drugs: These comprise either the natural ergot derivatives (ergotamine, ergocornine, ergonovine, and Hydergine) or the synthetic agents (Dibenamine, phenoxybenzamine (Dibenzyline), benodaine (Piperoxan), phentolamine (Regitine), tolazoline (Priscoline), and others). Some of these substances probably act as competitive blocking agents for the epinephrine receptors. In others the exact mode of action is not clearly understood. These agents inhibit or paralyze the α effects of epinephrine and, to a lesser extent, β adrenergic effects. Hence, after their injection, epinephrine causes a fall rather than a rise in blood pressure, since the vasoconstrictor receptors in the arterioles are paralyzed, while the vasodilator receptors are not.

c. Cholinergic-blocking agents: The principal antagonist of Ach. and allied drugs is atropine. It does not prevent the liberation of Ach. but neutralizes its action once liberated. There is evidence to show that after atropine administration, large doses of Ach. can cause elevation of blood pressure (nicotinic action), due to the liberation of epinephrine from stimulation of the adrenal gland and adrenergic nerves throughout the body. Atropine also appears to cause vasodilatation by a direct independent peripheral action. In man, small doses of this drug have no effect on blood pressure; with large doses there is an increased heart rate and systolic blood pressure and the well-known flushing and reddening of the skin. In general, homatropine, methantheline (Banthine), propantheline (Probanthine), and a large number of related synthetic agents have similar actions to atropine.

5. Pharmacologic Responses Mediated through Other Routes

a. Nitrites and nitrates: These drugs are vasodilators, especially of the coronary vessels. Their exact mode of action is not known, but there is evidence that they may interfere with the metabolic activities of smooth muscle tissue. The basic process is believed to be an enzymatic reduction of organic nitrates to nitrites. The mode of action may also be due to a specific molecular configuration. A large variety of preparations are available, which differ simply in intensity and rapidity of action. The degree of vasodilatation also varies in different vascular beds. Nitroglycerin is still the major agent in this group.

b. Nicotinic acid (Niacin): This substance leads to transient vasodilatation, as manifested by intense flushing of skin within 15–60 seconds after an oral dose, with subsidence in about 30 minutes. Nicotinamide does not show such a reaction.

c. Caffeine and related xanthine derivatives: The action of these drugs on peripheral blood vessels is twofold: (a) vasoconstriction, through stimulation of the vasomotor center and (b) vasodilatation, through a more lasting relaxant effect on the blood vessels. Cardiac stimulation, with increased cardiac output, may also modify the response.

d. Papaverine: This drug has the property of inhibiting smooth muscle and is a powerful vasodilator agent, although of short action.

e. Ethyl alcohol: This is a well known vasodilator agent of the skin vessel; splanchnic vessels are constricted with small doses. Peripheral vasodilatation also results from central vasomotor depression.

G. BIOCHEMISTRY

By J. E. KIRK

Although the number of systematic studies dealing with the composition and metabolism of the arterial wall is somewhat limited, sufficient data are available from various publications to provide fairly adequate information about this subject. Since the composition and the metabolic activities of the normal arterial wall vary to some extent with age (see Subsection 3, below), the data listed in this section refer to observations made on normal arterial samples from young adult human subjects (20–39 years) and animals.

1. NORMAL CONSTITUENTS OF ARTERIAL WALL

a. Organic constituents: The main constituents of the human aorta and the aorta of horse and cattle are listed in Table II. It will be seen

TABLE II
COMPOSITION OF NORMAL ARTERIAL TISSUE IN YOUNG ADULT INDIVIDUALS AND ANIMALS[a]

Components	Human aorta	Human femoral artery	Human brachial artery	Cattle aorta	Horse aorta
Dry matter	28.0	25.3	26.3	26.3	22.5
Nitrogen	4.1	3.5	—	4.2	3.8
Total lipids	1.680	—	—	—	—
Cholesterol	0.290	0.135	0.185	0.102	0.128
Total ash	0.730	0.675	0.670	—	—
Calcium	0.070	0.147	0.144	0.012	0.011
Total PO_4	0.375	—	—	—	—

[a] Values expressed in percentage of wet tissue weight.

from the recorded values that the major part of the dry matter in the arterial wall consists of protein (nitrogen $\times$ 6.25). According to Lowry *et al.* (1941), the elastin content of the human aortic tissue is 7.7 per cent and the collagen content, 7.1 per cent; a somewhat lower collagen content for the aorta (4.0%) has been reported by Myers and Lang (1946). The proportion of free cholesterol to ester cholesterol in aortic tissue varies considerably between individual samples, and the number of samples analyzed is too small to permit definite conclusions with regard to this point. On the basis of data published by Meeker and Jobling (1934), the average ratio: free sterol/ester sterol in lipid material extracted from the intima is calculated to be 0.96, whereas values reported by Weinhouse and Hirsch (1940) show a mean ratio of 2.87 for lipids in the media. An average ratio of 3.88 was found by Buck and Rossiter (1951) for cholesterol extracted from intima-media samples. A concentration of total phosphatides in aortic tissue of 0.33 per cent has been reported by Bürger (1954), although a somewhat higher level (0.56%) was observed by Buck and Rossiter (1951). The latter authors found the following average concentrations of individual phospholipids: cephalin, 0.24 per cent; lecithin, 0.15 per cent; and sphingomyelin, 0.20 per cent.

b. Inorganic constituents: Investigations on the inorganic components of the aortic wall have shown a total ash content for the human aorta of about 0.73 per cent (Weinhouse and Hirsch, 1940; Kvorning, 1950). The calcium content observed for the human aorta (Bürger, 1939) is several times greater than the calcium concentrations found in cattle and horse aortas (Gerritzen, 1932; Keuenhof and Kohl, 1936), and even higher calcium values are reported for the human femoral and brachial arteries (Hevelke, 1954a,b). Further investigations on the human aorta have revealed the following average percentile tissue concentrations: total PO_4, 0.375 (Bürger, 1939); acid-soluble PO_4, 0.255 (Bürger, 1939); potassium 0.039 (Hevelke, 1958); magnesium, 0.010 (Rechenberger and Hevelke, 1955); iron, 0.010 (Page and Menschick, 1931); and silicate (SiO_2), 0.015 (Kvorning, 1950). Somewhat lower values for total PO_4 (0.285%) and acid-soluble PO_4 (0.132%) were observed by Südhof (1950) for the carotid artery of cattle.

c. Other constituents: The presence in the arterial tissue of other constituents deserves attention. For example, the normal human aorta contains 0.31 per cent of hexosamine and 0.085 per cent of acid-hydrolyzable sulfate (SO_4) (Kirk and Dyrbye, 1956b), both of which compounds represent mucopolysaccharides. Extensive studies on acid mucopolysaccharide material isolated from the human aorta have shown that chondroitin sulfate constitutes the major part of the sulfated mucopolysaccharides (Dyrbye and Kirk, 1957; Kirk and Dyrbye, 1957; Dyrbye *et al.*, 1958; Kirk *et al.*, 1958b). The isolated acid mucopolysaccharide material has been found to possess a definite, but low anticoagulant activity (Kirk, 1959d). A thorough review of the arterial mucopolysaccharides in human and animal arterial tissue has recently been published (Kirk, 1959e). Determinations have further been made of the creatine (Myers and Lang, 1946) and total glutathione (Wang and Kirk, 1960) contents of the human aortic tissue, and these have shown average values of 0.090 and 0.019 per cent, respectively.

2. Metabolism of Arterial Tissue

The human aorta is characterized by having a low respiratory rate, the Q_{O_2} of intima-media tissue sections being about 0.30 (Kirk *et al.*, 1954). The rate of glycolysis is somewhat higher, with the $Q_g^{O_2}$ value close to 1.00. The glycolysis rate of human aortic tissue is relatively independent of the oxygen tension; in comparative studies the average anaerobic glycolysis rate was found to be only about 30 per cent higher than the aerobic rate of glycolysis. It is estimated that the glycolysis accounts for approximately 50 per cent of the total energy production by the human aortic tissue. The respiratory rate and rate of glycolysis of various animal arterial tissues are listed in Table III.

TABLE III
Respiration and Glycolysis by Animal Arterial Tissue

Animals	Age	Q_{O_2}	$Q_g^{O_2}$	$Q_g^{n_2}$	References
Rat	Young	1.06	—	—	Briggs *et al.* (1949)
Rat	Adult	1.09	—	—	
Rabbit	Adult	1.14	1.70	—	Michelazzi (1938)
Rabbit	Newborn	2.00	—	—	Costa *et al.* (1950)
Rabbit	Adult	0.76	—	—	
Dog	Adult	0.24	—	—	Hiertonn (1952)
Dog	Adult	0.29	0.62	0.76	Kirk *et al.* (1954)
Cattle	Adult	—	0.32	0.71	Südhof (1950)

Studies carried out by Kirk *et al.* (see references, Table IV) on freshly prepared homogenates of human arterial samples have revealed the presence of enzymes of the glycolytic pathway, direct oxidative shunt, tricarboxylic acid cycle, and malate shunt in the tissue. The average activity values observed for intima-media samples of normal arterial tissue from young adults are recorded in Table IV, as well as the activities of various phosphatases and other enzymes. The high activity of adenylpyrophosphatase in the arterial tissue (Baló *et al.*, 1948; Banga and Nowotny, 1951; Kirk, 1959a) deserves attention because the activity of this enzyme represents an energy-producing mechanism. In view of the low oxidative metabolism of the arterial tissue, the high tissue concentration of adenylpyrophosphatase is somewhat remarkable; in this connection it should be mentioned that recent studies also have revealed

TABLE IV

MEAN ENZYME ACTIVITIES OF NORMAL HUMAN ARTERIAL TISSUE

Enzyme	Aorta	Pulmonary artery	Coronary artery	References
Glycolytic pathway	mM substrate metab./gm wet tissue/hour			
Hexokinase	0.013	0.011		Brandstrup *et al.* (1957)
Phosphoglucoisomerase	1.85	1.62	2.05	Brandstrup *et al.* (1957); Kirk *et al.* (1958a)
Aldolase	0.056	0.073	0.099	Kirk and Sørensen (1956)
Enolase	0.24	0.27	0.24	Wang and Kirk (1959)
Lactic dehydrogenase	1.06	0.86	0.68	Matzke *et al.* (1957); Kirk *et al.* (1958a)
Direct oxidative shunt				
Glucose-6-phosphate dehydrogenase	0.115	0.092	0.102	Kirk *et al.* (1959)
6-Phosphogluconate dehydrogenase	0.010	0.013	0.019	Kirk *et al.* (1959)
Ribose-5-phosphate isomerase	0.0034	0.0045	0.0045	Kirk (1959c)
Tricarboxylic acid cycle				
Aconitase	0.021			Laursen and Kirk (1955)
Isocitric dehydrogenase	0.063	0.071	0.082	Kirk and Kirk (1959)
Fumarase	0.178	0.328		Sørensen and Kirk (1956)
Malic dehydrogenase	0.57	0.71	0.98	Matzke *et al.* (1957); Kirk *et al.* (1958a)
Malate shunt				
TPN-malic enzyme	0.011	0.010	0.012	Kirk and Kirk (1959)
Phosphatases	mM PO_4 liberated/gm wet tissue/hour			
Adenylpyrophosphatase	0.24	0.26	0.44	Kirk (1959a)
Inorganic pyrophosphatase	0.20	0.24	0.38	Kirk (1959a)
Phosphomonoesterase	0.050	0.034	0.034	Kirk (1959a)
5-Nucleotidase	0.16	0.07	0.10	Kirk (1959b)
Other enzymes	mM substrate metab./gm wet tissue/hour			
Beta-glucuronidase	0.00017	0.00015	0.00015	Dyrbye and Kirk (1956)
Phenolsulfatase	0.000023	0.000025	0.000020	Kirk and Dyrbye (1956a)
Glyoxalase I	1.23	1.70	1.00	Kirk (1960b)
	E.U./mg wet tissue			
Carbonic anhydrase	0.01			Kirk and Hansen (1953)
	μg naphthylamine formed/gm wet tissue/hour			
Leucine aminopeptidase	1.37	1.49	1.60	Kirk (1960a)

the presence of an appreciable activity of phosphocreatine kinase in the tissue (Kirk, unpublished). It should further be pointed out that the activity of 5–nucleotidase in the arterial tissue is higher than that found in ossifying cartilage. Since this enzyme exhibits maximal activity near the physiologic pH, it has been suggested by Reis (1950, 1951) that the arterial 5–nucleotidase may be of importance for the process of tissue calcification.

The ability of the arterial tissue to synthesize cholesterol from acetate has been demonstrated through isotope experiments (Siperstein *et al.*, 1951), but activity measurements of the various enzymes involved in the lipid synthesis have not as yet been reported.

Investigations on the coenzyme concentrations of human aortic tissue have shown comparatively high values for both riboflavin (Schaus *et al.*, 1955) and nicotinic acid (Chang *et al.*, 1955), the determination of the latter compound serving as a measurement of the tissue level of phosphopyridine nucleotides. The total riboflavin concentration recorded in Table V is about 10 per cent, and the nicotinic acid value, 20 per cent of the corresponding values reported for human liver tissue. The presence of cytochrome c in human aortic tissue has also been demonstrated (Kirk, 1959f), but because of the low concentration of this compound in the tissue, exact determination of the cytochrome c level in the arterial tissue has not as yet been possible.

TABLE V

Coenzyme Concentrations of Human Aortic Tissue

Coenzyme	Micrograms/gm wet tissue
Free riboflavin + flavin mononucleotide	0.40
Flavin adenine dinucleotide	0.98
Total riboflavin	1.38
Nicotinic acid	20.0

Several studies have been reported on the rate of incorporation of radioactive-labeled sulfate in the arterial tissue of animals (Layton, 1950, 1951; Odeblad and Boström, 1952, 1953; Boström and Odeblad, 1953); these investigations have shown a surprisingly high rate of sulfate uptake in the mucopolysaccharides of the tissue. A high incorporation of sulfate was likewise observed by Dyrbye (1959) in *in vitro* experiments on human aortic tissue. Although the results of experiments with labeled sulfate do not necessarily correlate with the turnover of other parts of the mucopolysaccharide molecule, the reported findings are of considerable interest since they indicate that the sulfated mucopolysaccharides cannot be considered as inert constituents of the vessel wall.

3. Changes Observed with Aging

a. Changes in the composition of arterial tissue with age: The composition of normal human arterial tissue is altered with age in many respects. Although the changes occurring in different types of arteries in general are of a similar character, the degree of the alterations varies to some extent. The observations made on age changes in the composition of normal human arterial tissue are listed in Table VI. It will be seen from the recorded data that the available studies cover several, but not all, of the major components of the arteries.

Both the dry matter and the nitrogen content of the tissue show a tendency to decrease with age in large and medium-sized arteries (Table VI). The change in the nitrogen values reflects mainly a reduction in the protein content of the arterial wall. It has been reported by Myers and Lang (1946) that the elastin content of the intima-media layers of the aorta decreases with age, but studies conducted by Lansing *et al.* (1950a) on elastin isolated from the media of the human aorta have shown that the elastin content of the media layer remains essentially constant throughout life (after the first decade), with a mean value of about 42 per cent of the defatted dry tissue. A change with age in the amino acid composition of the aortic elastin was, however, observed by Lansing *et al.* (1951). The elastin from old aortas was found to contain more aspartic acid and glutamic acid and less glycine, proline, and valine than the elastin from young arteries. In contrast to the aorta, the elastin from the human pulmonary artery showed no significant amino acid changes with age (Lansing *et al.*, 1950b).

Whereas in the normal aorta the tissue content of dry matter and nitrogen decreases with age, the total lipid, cholesterol, calcium, total PO_4, and acid-soluble PO_4 concentrations increase significantly (Table VI). No certain variation in the arterial content of phosphatides was observed by Bürger (1954). In contrast to this, Buck and Rossiter (1951) reported an increase in the phospholipid content of the aorta from 0.56 per cent at the age of 30 years to 0.75 per cent at the age of 60 years. According to these authors, the increase in the phospholipid level is caused by an accumulation of sphingomyelin in the tissue, whereas the cephalin and lecithin concentrations have been found to remain remarkably constant over the age range studied. They also noted no significant change in the aortic content of neutral lipids with age. According to observations by various authors (Schönheimer, 1926; Meeker and Jobling, 1934; Weinhouse and Hirsch, 1940; Buck and Rossiter, 1951), the ratio: free sterol/ester shows a tendency to decrease with age. With regard to the content of calcium in the aorta, it has been shown by

TABLE VI

VARIATION WITH AGE IN COMPOSITION OF NORMAL ARTERIAL TISSUE[a,b]

1. Human arterial tissue

Aorta

Age group (years)	Dry matter	Nitrogen	Total lipids	Cholesterol	Total ash	Calcium	Total PO_4	Acid-soluble PO_4	Potassium
0– 9	30.0	4.60	—	0.10	—	0.02	—	—	—
10–19	29.5	4.38	1.23	0.15	—	0.03	0.25	0.16	0.055
20–29	28.7	4.27	1.54	0.20	0.65	0.05	0.32	0.24	0.038
30–39	28.5	4.03	1.75	0.28	0.81	0.09	0.40	0.29	0.040
40–49	28.3	3.93	1.87	0.34	1.27	0.16	0.47	0.35	0.044
50–59	28.0	3.67	1.90	0.48	1.55	0.21	0.54	0.41	0.039
60–69	28.0	3.78	1.80	0.61	2.00	0.26	0.61	0.49	0.044
70–79	28.0	3.38	—	0.71	2.80	0.39	—	—	0.033

	Pulmonary artery					Femoral artery				
Age group (years)	Dry matter	Nitrogen	Cholesterol	Calcium	Potassium	Dry matter	Nitrogen	Cholesterol	Total ash	Calcium
0– 9	27.9	3.71	0.10	—	—	27.8	4.01	0.09	0.51	0.13
10–19	26.5	3.91	0.12	0.025	0.033	26.9	3.84	0.11	0.59	0.14
20–29	26.8	4.02	0.15	0.032	0.032	26.2	3.55	0.12	0.65	0.18
30–39	25.7	3.71	0.17	0.028	0.031	24.4	3.30	0.15	0.71	0.18
40–49	26.9	3.67	0.21	0.029	0.031	22.8	3.01	0.19	0.86	0.27
50–59	24.9	3.45	0.22	0.027	0.026	22.8	2.83	0.23	1.15	0.40
60–69	24.5	3.43	0.26	0.048	0.025	23.8	2.77	0.33	2.21	0.55
70–79	23.0	3.25	—	0.060	0.025	25.3	2.90	0.55	3.17	1.07

2. Animal arterial tissue

	Cattle aorta				Horse aorta			
Age group (years)	Dry matter	Nitrogen	Cholesterol	Calcium	Dry matter	Nitrogen	Cholesterol	Calcium
0– 0.2	25.8	3.85	0.117	0.010	—	—	—	—
1– 5	26.3	4.20	0.102	0.012	22.5	3.82	0.128	0.011
6–10	26.0	3.90	0.102	0.012	22.4	3.96	0.169	0.013
11–15	26.1	3.97	0.110	0.017	24.3	4.09	0.194	0.019
16–20					26.2	4.14	0.213	0.022
21–25					28.0	4.08	0.237	0.025
					25.2	3.90	0.245	0.029

[a] Values expressed in percentage of wet tissue weight.

[b] The values recorded in the table for human arterial tissue were calculated on the basis of data reported by Bürger (1939, 1954), Weinhouse and Hirsch (1940), and Hevelke (1954a, 1958), or represent observations made by Kirk. The values for cattle and horse aortas were calculated on the basis of data published by Gerritzen (1932) and Keuenhof and Kohl (1936).

Lansing *et al.* (1950a) that the average percentage of this mineral in dry elastin prepared from the media increases from 0.4 at age 11–20 years to 6.6 in samples obtained from 51–60 year old individuals. The changes observed in the femoral artery (Hevelke, 1954a) resemble those occurring in the aorta, except that the accumulation of calcium takes place at a faster rate in the femoral artery (Table VI). In connection with the change in calcium concentration, a six-fold increase in the total ash content of the femoral artery tissue has been recorded.

The available information concerning the variation with age in the magnesium content of normal aortic tissue is controversial. A significant increase in the magnesium concentration from about 0.010 per cent at the age of 30 years to 0.041 per cent at the age of 60 years has been reported by Buck (1951), whereas Rechenberger and Hevelke (1955), over the same age range, have observed a decrease in the tissue magnesium level from 0.011 per cent to 0.003 per cent. Because magnesium is an important cofactor for various enzymes of the arterial tissue, further investigations of the arterial content of this mineral compound would be desirable.

Extensive studies on the variation with age in the composition of normal cattle and horse aortas have been carried out by Gerritzen (1932) and Keuenhof and Kohl (1936). Their results (Table VI) are of particular interest because the aging changes differ markedly from those observed in the human arterial tissue. Thus the composition of the cattle aorta showed practically no changes with age, while only a moderate increase in the concentrations of dry matter, cholesterol, and calcium was found in analyses of the horse aorta.

b. Changes in metabolic activities of arterial tissue with age: Although the reported studies on the respiratory and glycolysis rates of human aortic tissue have not revealed any significant correlation between these metabolic activities and age, the number of experiments conducted is not sufficient to permit conclusions with regard to this point.

The relation of enzyme activities and coenzyme concentrations of aortic tissue to the age of the subjects from whom the samples were obtained has been studied in detail by Kirk and his associates (see references listed in Table IV). If the values for children are excluded, an analysis of the data shows that the activities of several enzymes, namely hexokinase, phosphoglucoisomerase, enolase, glucose–6–phosphate dehydrogenase, 6–phosphogluconate dehydrogenase, ribose–5–phosphate isomerase, isocitric dehydrogenase, malic dehydrogenase, TPN–malic enzyme, adenylpyrophosphatase, inorganic pyrophosphatase, phosphomonoesterase and leucine aminopeptidase, remain essentially unchanged with age, whereas the activities of some other enzymes, such as fumarase, glyoxalase I, and phenolsulfatase, show a significant tendency to decrease with age. A definite reduction in activity with age was also noted by Maier and Haimovici (1957) for succinic dehydrogenase and cytochrome c oxidase of intact human aortic tissue. In the case of four other enzymes; aldolase, lactic dehydrogenase, 5–nucleotidase, and beta-glucuronidase, a significant increase in activity was observed until the age of 60–65 years.

With regard to the coenzymes studied, the concentrations of both riboflavin (Schaus *et al.*, 1955) and nicotinic acid (Chang *et al.*, 1955)

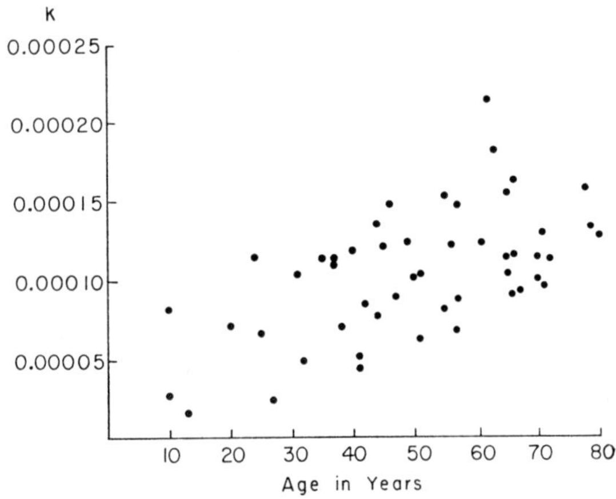

FIG. 31. Variation with age in the diffusion coefficient of glucose for membrane preparations of human aortic intima (with attached subintimal tissue). k = Diffusion coefficient, defined, according to Hill (1928), as the number of units of a substance diffusing through 1 cm^2 of a membrane in 1 min at a concentration gradient of 1 unit per ml per cm. (From Kirk and Laursen, 1955; reproduced with the permission of the Journal of Gerontology.)

in aortic tissue were found to decrease with age. The coefficient of correlation between age and the total riboflavin concentration of the tissue was —0.39 (t = 4.20, N = 100), and between age and the nicotinic acid concentration —0.37 (t = 3.17, N = 61).

In the *in vitro* studies by Dyrbye (1959) on human aortic samples, a definite decrease in the rate of sulfate incorporation by the tissue mucopolysaccharides was found.

c. Variation with age in the permeability of normal human aortic tissue: Extensive observations on the diffusion coefficients of various solutes for membranes prepared from human aortic tissue have been provided by Kirk and Laursen (1955), using the procedure for diffusion coefficient determination developed by Johnsen and Kirk (1955). Fifty-one samples of the aorta, obtained from individuals ranging in age from 10 to 80 years were employed for the investigations, and membrane preparations were made separately of the intima and the media from these samples. For each membrane preparation, determinations were made of the diffusion coefficients of nitrogen, oxygen, carbon dioxide, lactate, iodide, and glucose. The results showed a definite tendency for the diffusion coefficient values to increase with the age of the subjects, especially for the compounds of higher molecular weight.[1] For example, the correlation between age and the diffusion coefficient values of glucose for the aortic intima is illustrated in Fig. 31.

References

Ahlquist, R. P. (1948). A study of adrenotropic receptors. *Am. J. Physiol.* **153**, 586.

Baló, J., Banga, I., and Josepowits, G. (1948). Enzymic activity of the aorta. Adenylpyrophosphatase of the aorta. *Z. Vitamin-, Hormon- u. Fermentforsch.* **2**, 1.

Banga, I., and Nowotny, A. (1951). Comparative studies about adenosinetriphosphatase activity of human muscles, aorta, and arteria femoralis. *Acta Physiol. Acad. Sci. Hung.* **2**, 317.

Barcroft, H., and Swan, H. J. C. (1953). "Sympathetic Control of Human Blood Vessels." Edward Arnold, London.

Bohr, D., Brodie, D. C., and Cheu, D. H. (1958). Effect of electrolytes on arterial muscle contraction. *Circulation* **17**, 746.

Boström, H., and Odeblad, E. (1953). Autoradiographic observations on the incorporation of S^{35}-labeled sodium sulfate in the rabbit fetus. *Anat. Record* **115**, 505.

Brandstrup, N., Kirk, J. E., and Bruni, C. (1957). The hexokinase and phosphoglucoisomerase activities of aortic and pulmonary artery tissue in individuals of various ages. *J. Gerontol.* **12**, 166.

Braun-Menendez, E. (1954). Water and electrolytes in experimental hypertension. *In* "Hypertension: Humoral and Neurogenic Factors," Ciba Foundation Symposium (G. E. W. Wolstenholme, ed.), p. 238. Little, Brown, Boston, Mass.

Briggs, F. N., Chernick, S., and Chaikoff, I. L. (1949). The metabolism of arterial tissue. I. Respiration of rat thoracic aorta. *J. Biol. Chem.* **179**, 103.

Buck, R. C. (1951). Minerals of normal and atherosclerotic aortas. *A.M.A. Arch. Pathol.* **51**, 319.

Buck, R. C., and Rossiter, R. J. (1951). Lipids of normal and atherosclerotic aortas. *A.M.A. Arch. Pathol.* **51**, 724.

Bürger, M. (1939). Die chemischen Altersveränderungen an Gefässen. *Z. ges. Neurol. Psychiat.* **167**, 273.

Bürger, M. (1954). "Altern und Krankheit," 2nd ed. Thieme, Leipzig.

[1] These findings are of interest in view of the possibility that the permeability of the arterial wall may be a factor of importance in the pathogenesis of atherosclerosis (see Sections D. and G. of (1) Atherosclerosis, Chapter XIX).

Burn, J. H., and Rand, M. J. (1958). Action of nicotine on the heart. *Brit. Med. J.* **i,** 137.

Chang, Y. O., Laursen, T. J. S., and Kirk, J. E. (1955). The total nicotinic acid and pyridine nucleotide content of human aortic tissue. *J. Gerontol.* **10,** 165.

Chen, K. K., and Schmidt, C. F. (1930). Ephedrine and related substances. *J. med. Lyon* **9,** 1.

Conway, J. (1958). Vascular reactivity in experimental hypertension measured after hexamethonium. *Circulation* **17,** 807.

Costa, A., Weber, G., and Antonini, F. (1950). Lineamenti di biologia delle arteriopatie sperimentali. *Arch. "De Vecchi" anat. patol. e med. clin.* **14,** 29.

Dyrbye, M. O. (1959). Studies on the metabolism of the mucopolysaccharides of human arterial tissue by means of S^{35}, with special reference to changes related to age. *J. Gerontol.* **14,** 32.

Dyrbye, M., and Kirk, J. E. (1956). The beta-glucuronidase activity of aortic and pulmonary artery tissue in individuals of various ages. *J. Gerontol.* **11,** 33.

Dyrbye, M., and Kirk, J. E. (1957). Mucopolysaccharides of human arterial tissue. I. Isolation of mucopolysaccharide material. *J. Gerontol.* **12,** 20.

Dyrbye, M., Kirk, J. E., and Wang, I. (1958). Mucopolysaccharides of human arterial tissue. III. Separation of fractions by paper electrophoresis. *J. Gerontol.* **13,** 149.

Folkow, B. (1960). Role of the nervous system in the control of vascular tone. *Circulation* **21,** 760.

Furchgott, R. F., and Bhadrakon, S. (1953). Reaction of strips of rabbit, aorta to epinephrine, isopropylarterenol, sodium nitrite and other drugs. *J. Pharmacol. Exptl. Therap.* **108,** 129.

Gaddum, J. H. (1955). Polypeptides which stimulate plain muscle. "Pharmacology," 4th ed. p. 247. Oxford Univ. Press, London and New York.

Gerritzen, P. (1932). Beiträge zur physiologischen Chemie des Alterns der Gewebe. V. Untersuchungen an Rindeaorten. *Z. ges. exptl. Med.* **85,** 700.

Gillman, J., and Gilbert, C. (1956). Calcium, phosphorus and vitamin D as factors regulating the integrity of the cardio-vascular system. *Exptl. Med. Surg.* **14,** 136.

Hevelke, G. (1954a). Die chemischen Alterswandlungen der Arteria femoralis. *Z. Altersforsch.* **8,** 130.

Hevelke, G. (1954b). Beiträge zur Funktion und Struktur der Gefässe. I. Vergleichende angiochemische Untersuchungen der Arteria brachialis und Arteria femoralis. *Z. Altersforsch.* **8,** 219.

Hevelke, G. (1958). Die Angiochemie der Gefässe und ihre physiologischen Alterswandlungen. *Verhandl. deut. Ges. Kreislaufforsch.* **24,** 131.

Hiertonn, T. (1952). Arterial homografts. An experimental study in dogs. *Acta Orthopaed. Scand.* Suppl. No. 10.

Hill, A. V. (1928). The diffusion of oxygen and lactic acid through tissues. *Proc. Roy. Soc. (London)* **B104,** 39.

Johnsen, S. G., and Kirk, J. E. (1955). Procedure for determination of diffusion coefficients of gases and nongaseous solutes for membranes. *Anal. Chem.* **27,** 838.

Keuenhof, W., and Kohl, H. (1936). Beiträge zur Physiologie des Alterns. IX. Chemische und histologische Untersuchungen an Pferdeaorten. *Z. ges. exptl. Med.* **99,** 645.

Kirk, J. E. (1959a). The adenylpyrophosphatase, inorganic pryophosphatase, and phosphomonoesterase activities of human arterial tissue in individuals of various ages. *J. Gerontol.* **14**, 181.

Kirk, J. E. (1959b). The 5-nucleotidase activity of human arterial tissue in individuals of various ages. *J. Gerontol.* **14**, 288.

Kirk, J. E. (1959c). The ribose-5-phosphate isomerase activity of arterial tissue in individuals of various ages. *J. Gerontol.* **14**, 447.

Kirk, J. E. (1959d). Anticoagulant activity of human arterial mucopolysaccharides. *Nature* **184**, 369.

Kirk, J. E. (1959e). Mucopolysaccharides of arterial tissue. *In* "The Arterial Wall" (A. I. Lansing, ed.), Chapter 6. Williams & Wilkins, Baltimore, Md.

Kirk, J. E. (1959f). Enzyme activities of human arterial tissue. *Ann. N.Y. Acad. Sci.* **72**, 1006.

Kirk, J. E. (1960a). The leucine aminopeptidase activity of arterial tissue in individuals of various ages. *J. Gerontol.* **15**, 136.

Kirk, J. E. (1960b). The glyoxalase I activity of arterial tissue in individuals of various ages. *J. Gerontol.* **15**, 139.

Kirk, J. E., and Dyrbye, M. (1956a). The phenolsulfatase activity of aortic and pulmonary artery tissue in individuals of various ages. *J. Gerontol.* **11**, 129.

Kirk, J. E., and Dyrbye, M. (1956b). Hexosamine and acid-hydrolyzable sulfate concentrations of the aorta and pulmonary artery in individuals of various ages. *J. Gerontol.* **11**, 273.

Kirk, J. E., and Dyrbye, M. (1957). Mucopolysaccharides of human arterial tissue. II. Analysis of total isolated mucopolysaccharide material. *J. Gerontol.* **12**, 23.

Kirk, J. E., and Hansen, P. F. (1953). The presence of carbonic anhydrase in the media of the human aorta. *J. Gerontol.* **8**, 150.

Kirk, J. E., and Kirk, T. E. (1959). The isocitric and TPN-malic dehydrogenase activities of human arterial tissue. *Federation Proc.* **18**, 261.

Kirk, J. E., and Laursen, T. J. S. (1955). Diffusion coefficients of various solutes for human aortic tissue, with special reference to variation in tissue permeability with age. *J. Gerontol.* **10**, 288.

Kirk, J. E., and Sørensen, L. B. (1956). The aldolase activity of aortic and pulmonary artery tissue in individuals of various ages. *J. Gerontol.* **11**, 373.

Kirk, J. E., Effersøe, P. G., and Chiang, S. P. (1954). The rate of respiration and glycolysis by human and dog aortic tissue. *J. Gerontol.* **9**, 10.

Kirk, J. E., Matzke, J. R., Brandstrup, N., and Wang, I. (1958a). The lactic dehydrogenase, malic dehydrogenase, and phosphoglucoisomerase activities of coronary artery tissue in individuals of various ages. *J. Gerontol.* **13**, 24.

Kirk, J. E., Wang, I., and Dyrbye, M. (1958b). Mucopolysaccharides of human arterial tissue. IV. Analysis of electrophoretically separated fractions. *J. Gerontol.* **13**, 362.

Kirk, J. E., Wang, I., and Brandstrup, N. (1959). The glucose-6-phosphate and 6-phosphogluconate dehydrogenase activities of arterial tissue in individuals of various ages. *J. Gerontol.* **14**, 25.

Kvorning, S. A. (1950). The silica content of the aortic wall in various age groups. *J. Gerontol.* **5**, 23.

Lansing, A. I., Alex, M., and Rosenthal, T. B. (1950a). Calcium and elastin in human arteriosclerosis. *J. Gerontol.* **5**, 23.

Lansing, A. I., Rosenthal, T. B., and Alex, M. (1950b). Significance of medial age changes in the human pulmonary artery. *J. Gerontol.* **5**, 211.

Lansing, A. I., Roberts, E., Ramasarma, G. B., Rosenthal, T. B., and Alex, M.

(1951). Changes with age in amino acid composition of arterial elastin. *Proc. Soc. Exptl. Biol. Med.* **76**, 714.

Laursen, T. J. S., and Kirk, J. E. (1955). The presence of aconitase and fumarase in human aortic tissue. *J. Gerontol.* **10**, 26.

Layton, L. L. (1950). Quantitative differential fixation of sulfate by tissue maintained *in vitro. Cancer* **3**, 725.

Layton, L. L. (1951). The anabolic metabolism of radioactive sulfate by animal tissues *in vitro* and *in vivo. Cancer* **4**, 198.

Lowry, O. H., Gilligan, D. R., and Katersky, E. M. (1941). The determination of collagen and elastin in tissues, with results obtained in various normal tissues from different species. *J. Biol. Chem.* **139**, 795.

Maier, N., and Haimovici, H. (1957). Metabolism of arterial tissue. Oxidative capacity of intact arterial tissue. *Proc. Soc. Exptl. Biol. Med.* **95**, 425.

Matzke, J. R., Kirk, J. E., and Wang, I. (1957). The lactic and malic dehydrogenase activities of aortic and pulmonary artery tissue in individuals of various ages. *J. Gerontol.* **12**, 279.

Meeker, D. R., and Jobling, J. W. (1934). A chemical study of arteriosclerotic lesions in the human aorta. *A.M.A. Arch. Pathol.* **18**, 252.

Michelazzi, L. (1938). Alcuni dati sopra il metabolism delle pareti vasali. *Arch. ital. med. sper.* **3**, 43.

Moyer, J. H., and Mills, L. C. (1953). Hexamethonium—Its effect on glomerular filtration rate, maximum tubular function and renal excretion of electrolytes. *J. Clin. Invest.* **32**, 172.

Moyer, J. H., Morris, G., and Snyder, H. (1954). A comparison of the cerebral hemodynamic response to aramine and norepinephrine in the normotensive and hypertensive subject. *Circulation* **10**, 265.

Myers, V. C., and Lang, W. W. (1946). Some chemical changes in the human thoracic aorta accompanying the aging process. *J. Gerontol.* **1**, 441.

Odeblad, E., and Boström, H. (1952). An autoradiographic study of the incorporation of S^{35} labeled sodium sulfate in different organs of adult rats and rabbits. *Acta Pathol. Microbiol. Scand.* **31**, 339.

Odeblad, E., and Boström, H. (1953). A quantitative autoradiographic study on the uptake of labelled sulphate in the aorta of the rabbit. *Acta Chem. Scand.* **7**, 233.

Page, I. H., and Menschick, W. (1931). Der Eisengehalt normaler und verkalkter Aorten. *Virchows Arch. pathol. Anat. u. Physiol.* **283**, 626.

Peterson, L. H., Jensen, R. E., and Parnell, J. (1960a). Mechanical properties of arteries *in vivo. Circulation Research* **8**, 622.

Peterson, L. H., Rushmer, R. F., and Folkow, B. (1960b). Symposium on regulation of the cardiovascular system in health in disease. *Circulation* **21**, 739–768.

Rechenberger, J., and Hevelke, G. (1955). Der Magnesiumgehalt der Aorta beim Diabetiker und seine Abhängigkeit vom Lebensalter. *Z. Altersforsch.* **9**, 309.

Redleaf, P. D., and Tobian, L. (1958). The question of vascular hyper-responsiveness in hypertension. *Circulation Research* **6**, 185.

Reis, J. L. (1950). Studies on 5-nucleotidase and its distribution in human tissues. *Biochem. J.* **46**, 21.

Reis, J. L. (1951). The specificity of phosphomonoesterase in human tissues. *Biochem. J.* **48**, 549.

Roth, G., Flock, E., Kvale, W. F., Waugh, J. M., and Ogg, J. (1960). Pharmacological and chemical tests as an aid in the diagnosis of pheochromocytoma. *Circulation* **21**, 769.

Schaus, R., Kirk, J. E., and Laursen, T. J. S. (1955). The riboflavin content of human aortic tissue. *J. Gerontol.* **10**, 170.

Schönheimer, R. (1926). Zur Chemie der gesunden und der atherosklerotischen Aorta. I. Über die quantitativen Verhältnisse des Cholesterins und der Cholesterinester. *Hoppe-Seylers Z. physiol. Chem.* **160**, 61.

Siperstein, M. D., Chaikoff, I. L., and Chernick, S. S. (1951). Significance of endogenous cholesterol in arteriosclerosis; synthesis in arterial tissue. *Science* **113**, 747.

Sørensen, L. B., and Kirk, J. E. (1956). Variation with age in the fumarase activity of human aortic and pulmonary artery tissue. *J. Gerontol.* **11**, 28.

Südhof, H. (1950). Über den Kohlenhydratstoffwechsel der Arterienwand. *Pflügers Arch. ges. Physiol.* **252**, 551.

Wang, I., and Kirk, J. E. (1959). The enolase activity of arterial tissue in individuals of various ages. *J. Gerontol.* **14**, 444.

Wang, I., and Kirk, J. E. (1960). The total glutathione content of arterial tissue in individuals of various ages. *J. Gerontol.* **15**, 35.

Weinhouse, S., and Hirsch, E. F. (1940). Chemistry of atherosclerosis. I. Lipid and calcium content of the intima and of the media of the aorta with and without atherosclerosis. *A.M.A. Arch. Pathol.* **29**, 31.

Zsoter, T., and Szabo, M. (1958). Effect of Sodium and Calcium on Vascular Reactivity. *Circulation Research* **6**, 476.

Chapter IV. SYMPATHETIC INNERVATION OF ARTERIAL TREE[1]

A. CENTRAL NERVOUS CONTROL OF VASOMOTOR MECHANISMS

By C. N. PEISS

1. ANATOMIC LEVELS OF REPRESENTATION

A survey of the literature reveals a striking discrepancy between the relatively simple functional concepts of central vasomotor control and the extensive distribution of vasomotor components throughout most of the central nervous system. The following anatomic outline[2] is a condensed survey of this distribution and is based on a modified table of embryologic derivatives of the neural tube (Arey, 1942):

Spinal Cord
- Existence of vasomotor tone in spinal animals (Goltz, 1864).
- Vasomotor responses to cord stimulation (Bernard, 1852; Budge, 1853; Waller, 1853).
- Localization of spinal centers in lateral gray column (Hoeben, 1896; Biedl, 1897).
- Intact vasomotor reflexes in spinal animals (Sherrington, 1906; Langley, 1924; Brooks, 1933; Heymans *et al.*, 1936).
- Intact vasomotor reflexes in spinal man (Gilliatt *et al.*, 1948; Pollock *et al.*, 1951).

Rhombencephalon
- Cerebellum
  - Depressor effects presumably mediated via inhibition of bulbopontine and hypothalamic discharge to sympathetic vasoconstrictor fibers (Moruzzi, 1950).
- Medulla Oblongata (including Caudal Pons)
  - Transection experiments establishing medulla as area necessary for initiation of tonic discharge of sympathetic vasoconstrictor fibers and thus, maintenance of blood pressure (Dittmar, 1870, 1873; Owsjannikow, 1871).
  - Concept of reciprocally-acting vasoconstrictor and vasodilator centers (Bayliss, 1901, 1908, 1923).
  - Detailed mapping of lower brain stem for pressor and depressor areas, and establishment of concept of medullary vasomotor area comprising a vasoconstrictor center and a vasodepressor area, the latter operating through

[1] For discussion of sympathetic control of venous tree, see Section B.-3, Chapter VII.

[2] Details concerning most of the points enumerated in the outline can be found in recent reviews of Folkow (1956), Uvnäs (1960b) and Ingram (1960).

inhibition of sympathetic vasoconstrictor outflow (Ranson and Billingsley, 1916; Scott and Roberts, 1924; Chen *et al.*, 1936, 1937a,b; Wang and Ranson, 1939a,b; Monnier, 1939; Alexander, 1946; Bach, 1952; McQueen *et al.*, 1954).

Sympathetic vasodilator pathways to skeletal muscle (Lindgren and Uvnäs, 1953a,b).

Mesencephalon

Mapping for pressor and depressor activity (Allen, 1931; Kabat *et al.*, 1935; Hess, 1948; Thompson and Bach, 1950; Lindgren, 1955).

Mesencephalic location of descending pathways from hypothalamus involved in cardiovascular regulation (Beattie *et al.*, 1930; Magoun *et al.*, 1938; Magoun, 1940).

Mesencephalic representation of cholinergic sympathetic vasodilator system to skeletal muscle (Lindgren, 1955).

Prosencephalon

Diencephalon

Cardiovascular and other autonomic responses to stimulation of hypothalamus (Karplus and Kreidl, 1909, 1910, 1911, 1927; Hess, 1938, 1948; Ranson and Magoun, 1939).

Description of types of cardiovascular response to hypothalamic stimulation (Rushmer *et al.*, 1959; Manning and Peiss, 1960).

Description of discrete depressor area in hypothalamus just ventral and caudal to anterior commisure (Folkow *et al.*, 1959).

Telencephalon

Vasomotor responses from motor and premotor cortex (Hoff and Green, 1936; Green and Hoff, 1937; Ström, 1950; Kaada, 1951; Wall and Davis, 1951).

Vasomotor responses from temporal lobe (Kaada *et al.*, 1949; Chapman *et al.*, 1950; Kaada, 1951; Wall and Davis, 1951; Anand and Dua, 1956).

Vasomotor responses from frontal lobe (Chapman *et al.*, 1949; Bailey and Sweet, 1940; Livingston *et al.*, 1948; Sachs *et al.*, 1949; Kaada, 1951; Wall and Davis, 1951).

Vasomotor responses from rhinencephalon (Smith, 1945; Ward, 1948; Pool and Ransohoff, 1949; Kaada, 1951; Anand and Dua, 1956).

Vasomotor responses from insula (Kaada *et al.*, 1949; Kaada, 1951; Hoffman and Rasmussen, 1953).

On the basis of this outline, then, one can consider the correlation between specific anatomic structure of the central nervous system and its functional representation. Physiologic studies have provided a multitude of central pathways and areas which, in some manner, are related to peripheral vasomotor activity. In general, this information is of three types: (1) The description of certain areas (e.g., medullary reticular formation, posterior hypothalamus) from which very marked vasomotor activity can be elicited by electrical stimulation; (2) The description of relatively gross pathways in the central nervous system through which vasomotor responses are mediated (for the most part derived from stimulation experiments and from studies utilizing lesions and partial transections); and (3) The description of levels of the central nervous system at which various reflex vasomotor activities are accomplished (based chiefly on transection experiments).

However, it is necessary to call attention to the limitations imposed by all such techniques (Brodal, 1956), as well as to the additional complications imposed by the use of anesthesia in most studies. Furthermore, electrophysiologic methods cannot be considered infallible guides to functional activity (Peiss, 1958). Finally, in a large number of central vasomotor studies, the question of afferent vs. efferent activity was not treated adequately, since many previously described responses to stimulation of the medullary reticular formation involved significant activation of pathways to higher levels of the central nervous system (Peiss, 1960).

a. Neuroanatomic substrate: With very few exceptions, the specific neuroanatomic substrate involved in vasomotor reactions has not been identified. Brodal (1956) has analyzed the degree of correlation between the structure of the reticular formation and the motor, inspiratory and cardiovascular effects from these areas. Depressor and inspiratory areas, as well as the inhibitory region described by Magoun and Rhines (1946), appear to be confined chiefly to the nucleus reticularis gigantocellularis and the region of the nucleus reticularis ventralis and lateralis, from which originate the reticulospinal fibers. The correlation is not as good for pressor, expiratory and facilitatory responses.

Olszewski (1958) has proposed that the anatomic term "reticular formation" be discarded, and that the physiologic term "central internuncial system" be substituted. He states, "Today the central internuncial system emerges as an incredibly complicated but precisely organized system of neurons. Its input consists not only of sensory collaterals, but also of axons descending from higher centers, all orderly and precisely arranged. This output is directed both downward and upward, in the latter case ending predominantly in the diencephalon." Such a concept has great merit if for no other reason than to supply a descriptive basis for the evaluation of central nervous integration of vasomotor activity. It provides a reasonable substitute for the term "center," with its unfortunate connotations. In the following section, then, the term, vasomotor components of the central internuncial system, will be used.

2. Concepts of Functional Integration

In 1908, Bayliss proposed an integrative concept of central vasomotor regulation, comprising reciprocally-innervated constrictor and dilator centers in the bulb. This view was subsequently elaborated upon in his monograph (Bayliss, 1923). However, Uvnäs (1960a) has presented a convincing summary of the evidence against the validity of a medullary dilator center mediated through dorsal root fibers. In the half century since this concept was introduced, it has been established more and more firmly that vasodilatation mediated through bulbar mechanisms is accomplished by inhibition of sympathetic vasoconstrictor outflow. It is not known whether this inhibition operates directly on the tonically active cell bodies of the vasoconstrictor "center" or whether it is accomplished at the preganglionic cell bodies in the cord.

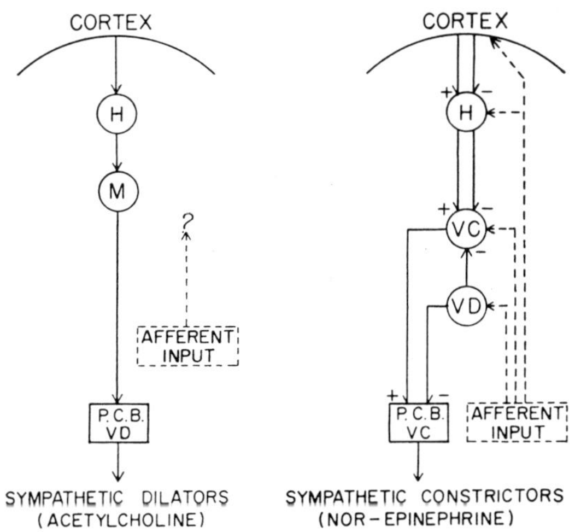

FIG. 32. Current concepts of central vasomotor integration. H = hypothalamus; M = mesencephalon; VC = vasoconstrictor center; VD = vasodepressor area; P.C.B., VC = preganglionic cell bodies, sympathetic constrictor fibers; P.C.B., VD = preganglionic cell bodies, sympathetic dilator fibers. See text for details.

The influence of higher levels of the central nervous system is generally considered to operate by modification of the activity of the prepotent medullary vasomotor center. Widespread acceptance of this "modifier" mechanism is somewhat surprising, in the face of the large number of reports dealing with vasomotor responses from higher levels whose descending pathways appear to bypass the integrative area in the medulla.

a. Current concept: Figure 32 is a simplified schematic diagram which attempts to summarize the chief features of the integration of vasomotor control on the basis of current concepts. The right side of the diagram indicates the overall plan of a central mechanism regulating discharge in sympathetic vasoconstrictor fibers. It provides a medullary vasomotor area comprising a vasoconstrictor center (VC) and a vasodepressor area (VD). Provision is included for the depressor area to operate by inhibition either of cells in the vasoconstrictor center or of preganglionic cell bodies (P.C.B., VC) in the spinal cord. Modifying influences from the hypothalamus (H) are exerted by inhibitory (—) and facilitatory (+) pathways synapsing in the vasomotor area of the medulla. The hypothalamus, in turn, is subject to modifying influences from cortical areas. An afferent input is provided, with little specific information concerning its intracerebral distribution. Although much is known about sensory pathways to the central nervous system, their interrelation with efferent vasomotor outflow is vague. It is well accepted, of course, that pressoreceptor and chemoreceptor reflexes continue to operate in animals with transections above the medullary vasomotor area. However, even in these well-established reflexes, there are components in which the afferent input operates at higher levels of the central nervous system.[3]

The left side of Fig. 32 is a simplified representation of the sympathetic vasodilator system (Lindgren, 1955; Uvnäs, 1960b). In essence it consists of a pathway from the motor cortex to a hypothalamic synapse. The neuron arising in the hypothalamus synapses in the tectal region of the mesencephalon (M), from which another neuron arises and passes uninterrupted to the spinal level (P.C.B., VD). The final terminations of this outflow are postganglionic fibers to the blood vessels of skeletal muscle, whose activation results in vasodilatation mediated through a cholinergic transmitter, probably acetylcholine. It thus appears that two mechanisms for blood vessel dilatation must be considered; one operating through inhibition of discharge in sympathetic constrictor fibers, and the other, through increased activity in sympathetic dilator fibers.

Available evidence from cats and dogs indicates that the sympathetic dilator outflow is limited to skeletal muscle. The hypothesis was advanced (Uvnäs, 1960a) that this mechanism might be involved in the integrative control of muscle blood flow in such situations as exercise, emergency reactions, etc. The hypothesis seems reasonable and consistent with many known facts. However, recent observations by Uvnäs (1960b) indicate that the functional significance of the sympathetic vasodilator outflow remains to be elucidated. He has shown that, despite the large increase in total blood flow through skeletal muscle when the central dilator system is stimulated, there is no significant change in nutritive (capillary) flow. The increased total circulation thus appears to be the result of augmented flow in shunt vessels. (For further discussion of this subject, see Sections E.-1d, F.-2a, Chapter XVII.)

[3] Akers and Peiss, unpublished data.

The diagram of Fig. 32, therefore, provides a reasonable but incomplete concept of vasomotor integration. It is probably adequate for explaining such extreme reflex activations as hemorrhagic shock, etc.; in fact, the bulbar mechanism appears primarily to serve the function of blood pressure homeostasis. The sympathetic dilator system, on the other hand, presents a beautifully worked out analysis of an efferent vasomotor mechanism. However, its place in an overall integrative concept cannot be determined until further information is available on the nature of its afferent input and it physiologic significance.

b. Proposed concept: The greatest defect in the concept of integration of vasomotor outflow lies in its neglect of the large amount of evidence for vasomotor components at higher levels of the central nervous system, which appear to bypass the medullary vasomotor area. Of interest in this regard are the following:

The data reviewed by Magoun (1940) in support of the anatomic location of descending vasomotor pathways from the hypothalamus. Recent studies by Peiss and Manning (1959) on the effect of curare on medullary and hypothalamic vasomotor excitability, although indirect, appear to support the anatomic findings.

The separate and distinct vasomotor pathway of the sympathetic dilator system, and the conclusion that vasoconstrictor fibers from the tectal mesencephalon are functionally discrete from the vasomotor area of the medulla (Lindgren, 1955).

Evidence for pyramidal, as well as extrapyramidal, vasomotor fibers descending from the cortex (Wall and Davis, 1951).

The recent description of a discrete hypothalamic depressor area (Folkow *et al.*, 1959). Although the descending pathway from this site is not known, the possibility exists that it may pass directly to spinal levels, directly to the medullary vasomotor area, or to both.

The existence of possible anatomic substrates for such bypass systems in the brain stem. Scheibel and Scheibel (1958) have traced single ascending reticular neurons from the spinal cord to the diencephalon and have indicated the wealth of arborizations from such a single fiber. They have also pointed out that most reticular neurons appear to be long and project to distant points up and down the brain stem. Similar axons in the descending reticular formation could easily provide substrates for bypass systems, as well as the interconnections between various levels of the central nervous system involved in vasomotor activity.

It seems, therefore, that some broader conceptualization of central vasomotor integration is necessary, if for no other reason than to provide a working hypothesis and a schema for the evaluation of presently available evidence and for planning future experiments.

The diagram shown in Fig. 33 is essentially such a working hypothesis and should be considered as nothing more than an attempt to synthesize an integrative concept from data in the literature and from a broad consideration of the physiologic requirements of the organism's

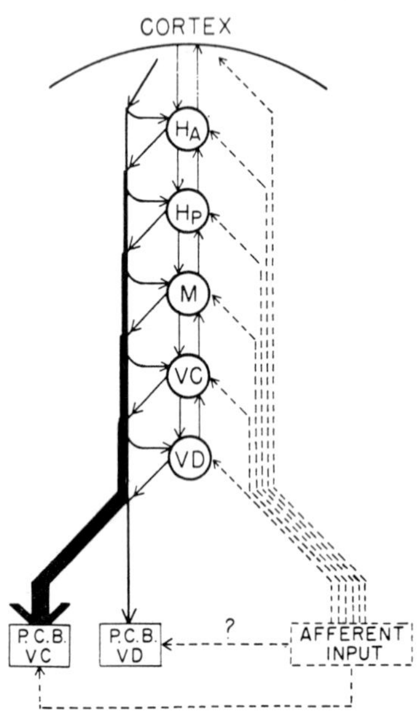

FIG. 33. Working hypothesis for overall integration of central vasomotor mechanisms. H_A = anterior hypothalamus; H_P = posterior hypothalamus; M = mesencephalon; VC = medullary vasoconstrictor area; VD = medullary vasodepressor area; P.C.B., VC = preganglionic cell bodies, sympathetic constrictor fibers; P.C.B., VD = preganglionic cell bodies, sympathetic dilator fibers. See text for details.

vasomotor apparatus. The figure may thus be considered as a representation of the vasomotor components of a central internuncial system. Detailed examination of the diagram can be divided into three main components, viz., the ascending pathways, the descending pathways and the interconnecting pathways.

As indicated in Fig. 33, the pathway for afferent input to areas regulating cardiovascular function is essentially similar to the broad concept of sensory input in general. In this regard, provision is made for ascending tracts which send off collaterals at all levels of the central nervous system. The integrative role of the bulbar vasomotor area in blood pressure homeostasis is consistent with such an input. Specific control of efferent discharge is a function of the number of cell bodies in each area, their specific excitability, and the number of afferent fiber arborizations terminating on them. This concept of ascending pathways also provides possibilities for more subtle control of efferent vasomotor discharge from higher levels. Specific afferent inputs (e.g., thermoreceptors) may have a direct ascending pathway to the hypothalamus and result primarily in vasomotor reactions in cutaneous blood vessels subserving the function of temperature regulation. This general mechanism could then serve other specific vasomotor requirements, such as the redistribution of blood locally to a working muscle, in which the afferent input may originate as collaterals from the classical receptors in skeletal muscle.

Also depicted in Fig. 33 is an equally broad descending vasomotor outflow, in which all levels of the brain can send fibers directly to preganglionic cell bodies in the spinal cord. In addition, collaterals or separate fiber tracts from any site may synapse at a lower level, providing other possibilities for integration. These intracerebral connections can then, in turn, influence the direct spinal outflow from each level of the brain. It is not implied, however, that the outflow from all levels of the central nervous system is widely or equally distributed in the spinal cord. More consistent with most of the experimental data available is the concept of a broad distribution at the spinal level of medullary vasomotor outflow. It is proposed that distribution from higher levels may be more discrete and restricted to less comprehensive vasomotor functions, and possibly to the control of regional vasomotor activity.

The sympathetic vasodilator system is included in the general vasomotor outflow, although its specific physiologic activation and integration are not known. For the present, one must consider that the preganglionic cells involved in this system are subject to the same synaptic properties as other spinal cell, i.e., facilitation, inhibition, etc., and that there may be interplay between these cells and those subserving vasoconstrictor discharge. The major differentiation between the two outflows at present resides in the difference of transmitter substance at their respective postganglionic endings. Currently, at least, it seems wise to consider the sympathetic dilator system as part of a unified central integrative mechanism.

The interconnecting pathways provide an additional mechanism for overall integration of vasomotor discharge from the central nervous system. As shown in Fig. 33, they are merely intended to represent functional possibilities by which the excitability of various levels of the brain may be facilitated or inhibited by discharge from higher and/or lower levels. Together with the collaterals from descending pathways, these interconnecting neurons complete feedback loops between all levels at which integration may occur, and conceptually one should consider their operation as including both negative and positive feedback.

In essence, then, this schema for central control of vasomotor activity consists of two major areas of integration:

A central internuncial system (excitatory and inhibitory), which is activated by the afferent inflow to all levels of the brain. The efferent vasomotor outflow from each area within the central internuncial system will be the resultant of the interaction of afferent input, specific excitability, cell population, and the excitatory and inhibitory interconnections between each level.

Preganglionic cell bodies in the spinal cord, which are the final common pathway for vasomotor activity. The net efferent outflow from the central internuncial system is finally integrated here, as well as the distribution of vasomotor activity to the peripheral end-organs.

In summary, it can be stated that the scope of the task which remains in this area is becoming more and more evident. The bulk of experimental evidence in the past has provided information primarily on responses to relatively large stimuli, requiring massive activation of vasomotor mechanisms. Such changes appear to be integrated chiefly at the bulbar level. The physiologic role of higher levels of the central nervous system is not clear, but it seems reasonable to expect that these levels must in some manner exert a finer regulation over vasomotor activity and provide controls for specific vascular requirements evolving at more recent times in the phylogenetic scale. Within this latter group of requirements, one must consider the needs imposed on the vasomotor apparatus by exercise, temperature regulation, and, in general, the problem of control of distribution of blood flow in the organism.

B. PERIPHERAL VASOMOTOR MECHANISMS

By W. C. RANDALL[4]; K. CHRISTENSEN[5]

1. SPINAL VASOMOTOR REGULATION

The cell bodies of the first neurones in the sympathetic pathway, situated in the intermediolateral cell column between the first thoracic and the third or fourth lumbar levels, are affected by a variety of influences. For example, it has been known for many years that afferent impulses may elicit reflex vasomotor responses by way of the sensory input to the spinal cord. The excitatory state of these sympathetic cells is affected by impulses descending from higher levels in the brain stem. In the absence of such influence (as immediately following transection of the cord), the activity of these cells is greatly depressed and vascular tone is decreased. In time, however, excitability is partially restored and vasomotor reflexes may be elicited in areas innervated solely by the decentralized spinal cord. Sherrington (1906) described such spinal vasomotor reflexes (dog, 300 days post-transection at C8), in which sensory nerve stimulation elicited a mean blood pressure elevation of 118 mm Hg. There is little evidence that preganglionic neurones initiate an impulse flow in the preganglionic fibers spontaneously (Alexander, 1945), although the possibility remains that the chemical environment of these cells might furnish an excitatory drive.

a. Mediation of vasomotor reflexes by the isolated spinal cord: Distension of the abdominal viscera, particularly the urinary bladder, in patients sustaining complete transection of the spinal cord elicits a reflex vasoconstriction in the toes of the lower extremities. If the lesion is above T5 or T6, this type of change is accompanied by a similar response in the fingers and a very large rise in both systolic and diastolic pressure. With lesions at T1 or C7, vasoconstriction in the fingers is so powerful that flow almost stops, while circulation through forearm muscles is greatly increased and the face and neck become intensely flushed (Guttmann, 1954). The mediation of thermoregulatory reflex sweating by the transected spinal cord was recently reported (Seckendorf and Randall, 1958). It is evident, therefore, that the isolated spinal cord is capable of mediating reflex vasomotor and sudomotor activity. Whether there is organization of cells in the intermediolateral cell columns into

[4] Author of Subsections 1–5.
[5] Author of Subsection 6.

specific "centers" is not clear. Guttmann (1954) is of the opinion that two major sudomotor centers exist in the spinal cord, one at the cervico-thoracic junction and the other at the thoraco-lumbar junction.

2. Preganglionic Pathways

Thinly myelinated axons of the cells in the intermediolateral columns, the preganglionic fibers of Langley, pass by way of the anterior root from the spinal cord. The majority of preganglionic fibers leave the spinal nerve, not the ventral root as generally depicted, in the communicating (white) ramus, to join the paravertebral ganglion trunk, or in some instances, by longer pathways to the prevertebral ganglia or to the adrenal medulla. Within the paravertebral ganglia, the preganglionic fibers may synapse with many postganglionic cell bodies, and indeed, may pass either up or down the sympathetic trunk to make synaptic junction with cells in other regions of the trunk. This represents a divergent system and one which might elicit a very diffuse spread of impulses from a relatively few preganglionic fibers. It has been stated that as many as eight ganglia may be excited by the stimulation of a single anterior root (Patton, 1948). However, simultaneous excitation of all the preganglionics within a single anterior root is not common, and probably never occurs with normal reflex excitation of the sympathetic trunk. Connections of a single preganglionic fiber are certainly less diffuse, and the response to direct electrical stimulation of the sympathetic trunk, both in animals and in man, indicates the existence of relatively discrete pathways to the periphery.

a. Lumbar accessory sympathetic pathways: Anatomic studies in man reveal accessory ganglia lying within the communicating (white) rami, the sympathetic roots (gray rami), the sites of origin of these two structures in the spinal nerve, and the ventral primary ramus of the spinal nerve (Gruss, 1932; Wrete, 1941, 1943; Skoog, 1947; Alexander *et al.*, 1949). Fibers may be traced distally from the ganglion cells beyond the origin of the communications between spinal nerves and the paravertebral ganglia.

In a series of studies in the trained, unanesthetized dog, vascular, sweating, and temperature responses were elicited in the lower extremity following verified complete removal of the lumbar sympathetic trunk (Randall *et al.*, 1950). Precise anatomic and histologic analysis of the spinal nerve roots was correlated with the functional observations. Immediately following operation, relatively high blood flow (recorded plethysmographically) and surface temperatures were observed in the pad of the hind foot. Within 2–7 days, however, progressively declining flows indicated a remarkable recovery of vascular tone. The return of positive sweating responses was frequently observed as well. The existence of accessory vasomotor pathways after complete removal of the lumbar sympathetic trunk was demonstrated in three ways: (1) the presence of vasoconstrictor reflexes after complete sympathectomy, (2) the production of intense vasoconstriction in the footpad by electrical stimulation of the tibial and peroneal nerves, and (3) prompt vascular relaxation following procain block of these nerves. From such evidence it was assumed that a major part of the recovery in tone following operation was related to the accessory fibers. Serial sections of the spinal nerves and their intercommunicating rami in the lumbar segments, taken from the operated animals at autopsy, showed ganglion cell aggregates (with variation from animal to animal) in, or closely associated with, the primary divisions of the second, third, fourth and sixth lumbar nerves. The postganglionic fibers of these cells passed directly along the ventral primary ramus of the spinal nerve without entering the paravertebral ganglion chain and were not interrupted by lumbar ganglionectomy.

Kuntz and Alexander (1950) have found accessory sympathetic ganglia, varying in size from a few to many thousands of cells, in one or more segments from T12 to L3, inclusive, in all cadavers examined. In some bodies, such ganglia were present in all these segments; in others, in only two or three. The ganglion cells were comparable cytologically to those in the sympathetic trunk ganglia. Boyd and Monro (1949) have reported accessory ganglia in the fourth and fifth lumbar segments, particularly in relation to the sympathetic roots of the spinal nerves.

Cone (cited by Ray and Console, 1948) resected the ventral roots of the twelfth thoracic and the first and second lumbar nerves, in addition to extirpation of the sympathetic trunk from T8 to T12. This operation, plus thoracolumbar sympathectomy, in the hands of Ray and Console (1948), resulted in no recognizable motor defects in the lower extremities, although the cremasteric reflex was abolished in male patients.

Such anatomic studies of accessory sympathetic ganglia related to all the spinal nerves from T12 to L5, inclusive, corroborate the functional demonstration of residual sympathetic activity following surgical extirpation of the sympathetic trunk.

b. Thoracic accessory pathways: The frequency of occurrence of accessory ganglia in the thoracic segments from T1 to T10 has been reported by Ehrlich and Alexander (1951). These structures are present more frequently in the four rostral thoracic segments than in the fifth to tenth segment. Thus, sympathetic conduction pathways which do not traverse the sympathetic trunk are conveyed into the upper extremity, being sufficiently abundant in many instances to account for extensive sympathetic activity following extirpation of the cervico-thoracic trunk.

Contrary to the opinions of many, it can be shown that fibers from T1, as well as from T2 and T3, contribute significantly to the innervation of sweat glands and blood vessels of the upper extremity in a majority of patients (Coldwater *et al.*, 1957). Furthermore, the C8 nerve may transmit vasoconstrictor fibers in some individuals. Coldwater *et al.*, reported the finding that extirpation of T1, T2 and T3 ganglia bilaterally resulted in remission of Raynaud's phenomenon, which, however, recurred. At re-operation a fragment of the inferior cervical ganglion with a communicating ramus to spinal nerve C8 was found, electrical stimulation of the ramus producing marked vasospasm in the digital finger pads. Extirpation of these remnants again caused relief of vasospastic activity, which persisted to date of publication. It is of interest that Horner's syndrome was not present following the initial operation but did appear on the left side after re-operation.

c. Direct electrical stimulation of the sympathetic trunk: The responses to direct electrical stimulation may be employed to study the level of emergence of preganglionic fibers from the spinal cord and their entry into the sympathetic trunk. Conventional concepts hold that preganglionic fibers leaving the spinal nerve pass directly to the corresponding segmental vertebral ganglion. Hence, it is believed that the preganglionic inflow to the sympathetic trunk is regular and predictably distributed from the vertebral levels of T1 or T2 to L3 or possibly L4. Geohegan *et al.* (1942) and Ray *et al.* (1943) stimulated the ventral roots in animals and in man and found considerable variation in the spinal emergence levels of preganglionic fibers in the maintenance of sympathetic activity. For example, one intact ventral root maintained marked sudomotor activity in the hand, which was eliminated by peripheral nerve block. Langley's classic reports (1891, 1894, 1923) described color changes in the foot pad due to vascular changes induced by electrical stimulation of the ventral roots and the sympathetic trunk. Sheehan and Marrazzi (1941) and Patton (1948) examined the preganglionic distribution to the lumbar sympathetic trunk using stimulation techniques, but recorded action potentials which could not differentiate pilomotor, sudomotor or vasomotor influences. Except for the very early studies of Langley and those of Bayliss and Bradford (1894), who measured volume changes in the dog's foot and leg, adequate descriptions of specific vasomotor pathways to the lower extremity are not available. Of interest in this regard are the studies carried out in the dog (Cox *et al.*, 1951, Randall *et al.*, 1953), as a preliminary to similar observations in man. These are presented below.

d. Preganglionic entry into the sympathetic trunk: Bipolar electrodes were placed successively above and below each ganglion along the lumbar trunk, and photoelectric plethysmograms were recorded from the footpad during electrical stimulation. The vascular responses to stimulation at each level were later correlated with histologic reconstruction of the trunk, in order to compare sites at which increments in functional response were elicited with demonstrable entry of type B fibers into the trunk. An increase in the intensity of vasoconstriction elicited by trunk stimulation was interpreted to indicate the presence of additional preganglionic fibers under the electrodes.

Figure 34 illustrates this procedure, as the electrodes were moved from a position on the trunk at the L2 vertebral level through L7. The first four stimulations (A, B, C, and D) failed to elicit any change in baseline or in amplitude of the volume pulses recorded from the footpad, but successively greater intensity of constriction in records E, F, and G indicated the entry of preganglionic vasomotor fibers innervating the footpad at vertebral levels L5, L6, and L7, respectively. Serial sections demonstrated histologically the entry of small, thinly myelinated fibers at levels appropriate to account for each successive increment in vasoconstrictor response. Immediately following dissection of the large ganglion located at the brim of the pelvis, the amplitude of the volume pulse increased dramatically (H), indicating a marked increase in blood flow to the footpad. Prolonged stimulation of the intact sacral portion of the trunk resulted in only slight constriction, and suggested only a minimal outflow of postganglionic fibers from the sacral chain in the dog. This minimal response was in contrast to profound constrictions which were frequently induced by stimulation of the intact sacral trunk. These studies, therefore, clearly revealed a significant variability in anatomic levels of entry of preganglionic vasomotor fibers and progressively increasing intensity of vasoconstriction as the lumbar trunk was stimulated at successively lower levels. However, it is necessary to point out that they were limited to observations on a single vascular bed in the extremity.

In order to obtain knowledge of the vasomotor supply to multiple beds, highly sensitive photoelectric plethysmographs were employed to examine the simultaneous cutaneous responses from the thigh, calf, and footpad skin, together with gastrocnemius muscle, during electrical stimulation of the lumbar trunk (Rawson and Randall, 1960). Figure 35 illustrates the results in the dog and also demonstrates the kind of observations currently being made in man. Five separate stimulations of the lumbar trunk in the dog elicited the specific local vasoconstrictor responses which indicated a common pattern of distribution of vasomotor fibers to the lower extremity. No detectable change in pulse amplitude occurred in any area when the electrodes were placed on the trunk at the superior border of the L3 vertebra. Movement of the electrodes to a position immediately below the ganglion (L3), however, elicited a progressive constriction in the cutaneous vessels of the thigh and footpad, with no alteration in those of the calf skin or gastrocnemius muscle. Stimulation at L4, on the other hand, induced intense responses both in thigh and footpad, with distinct constriction appearing in calf

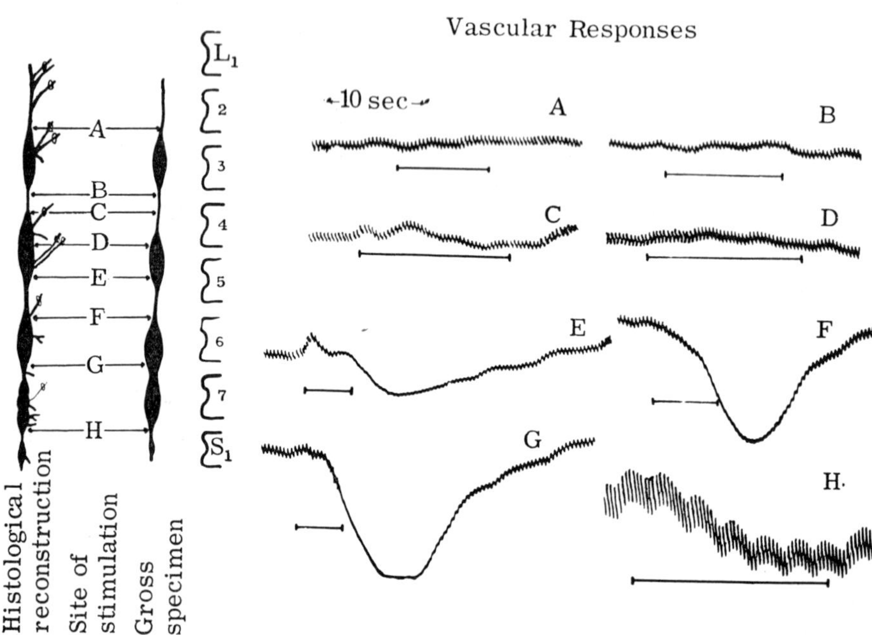

FIG. 34. The correlation of vasoconstrictor responses in the central footpad of the dog to electrical stimulation of the lumbo-sacral sympathetic trunk. The precise vertebral level of stimulation is shown on the gross specimen of the trunk sketched at time of operation, this, in turn, being related to its histologic reconstruction. Levels of entry of predominantly myelinated fibers were differentiated from levels of exit of predominantly nonmyelinated sympathetic roots. The letters A-H designate the levels of application of stimuli and also the corresponding records of vasoconstrictor response. Horizontal bars indicate the period of stimulation. (From Randall *et al.*, 1953; reproduced with the permission of Circulation Research.)

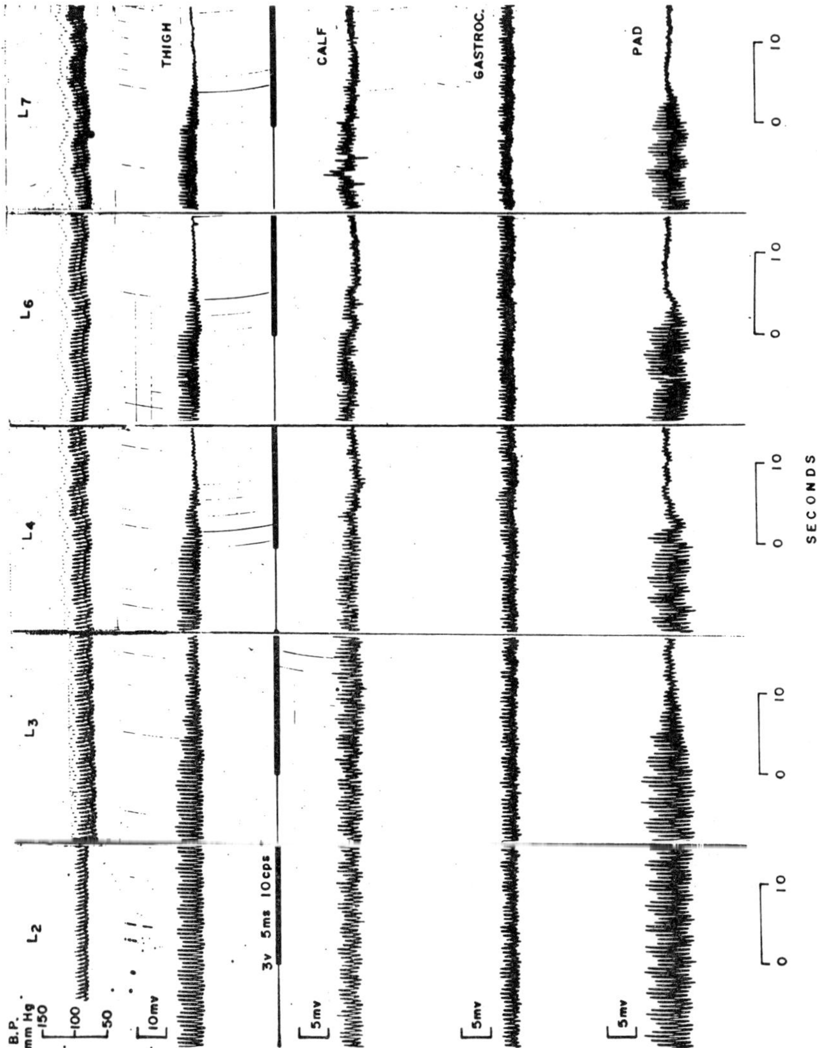

FIG. 35. Simultaneous vascular responses in three cutaneous and one muscle bed of the lower extremity of the dog during electrical stimulation of the lumbar sympathetic trunk. The vertebral level of stimulation is indicated at the top, with the systemic blood pressure changes shown immediately beneath. The period of stimulation is indicated by the thickened signal line between the second and third recording graphs. Recovery periods are not included. (From Rawson and Randall, 1960; reproduced by permission of the American Journal of Physiology.)

skin. Stimulation at L6 produced almost complete obliteration of the pulses in all three cutaneous areas, but still no response in muscle. Finally, stimulation at L7 again sharply constricted all cutaneous beds and elicited a modest (28%) reduction in gastrocnemius pulses. The latter change appeared late in the excitation period and was not prominent (Fig. 35).

From these experiments it becomes evident that a large variation exists in levels of preganglionic entry into the sympathetic trunk between L2 and L7. Fibers which ultimately innervate cutaneous vascular beds in the hind extremity may enter at any level between these extremes, and there is no way of predicting in any given animal at what level such fibers may be found in the trunk. Although there is uniformity in agreement that the spinal outflow is from T-11 through L-3 or L-4 in the dog, the above data establish the fact that preganglionic vasomotor fibers may actually enter the lumbar trunk as low as L-7, and that many segmental levels in the spinal cord may contribute to the innervation of the skin vessels of the hind extremity of the dog (Rawson and Randall, 1960). In some animals there is no significant supply from rostral portions of the lumbar trunk, the outflow from the cord originating at levels lower than L-3 or, as a possible alternative, taking a long and devious course through the psoas muscle for several segments before entering the sympathetic trunk. These observations are significant because they indicate that in such animals one could not be certain of preganglionic denervation (exclusive of direct pathways through the spinal nerves) of the blood vessels unless practically all of the lumbar sympathetic trunk were excised.

e. Stimulation of the sympathetic trunk in man: In the absence of adequate techniques for recording blood flow on different skin surfaces in man, sweating responses were obtained from multiple areas during electrical stimulation of the lumbar sympathetic trunk (Randall *et al.*, 1955). All observations were made under spinal block anesthesia to prevent reflex activation of the sweat glands. Figures 36 and 37 illustrate the mapping of sudomotor pathways through the trunk. When the electrodes were applied to the trunk and individual rami at L1 and L2 (Fig. 36), little sweating was observed on any of the cutaneous surfaces. Successive stimulations at more caudal positions on the trunk induced distinct alterations in sweating patterns, indicating the entry of sudomotor preganglionic fibers between the points of stimulation. Marked sweating appeared on all the test areas when the electrodes were applied to the trunk at the L3–L4 interganglionic segment. All skin areas were completely free from sweating in the absence of nerve stimulation.

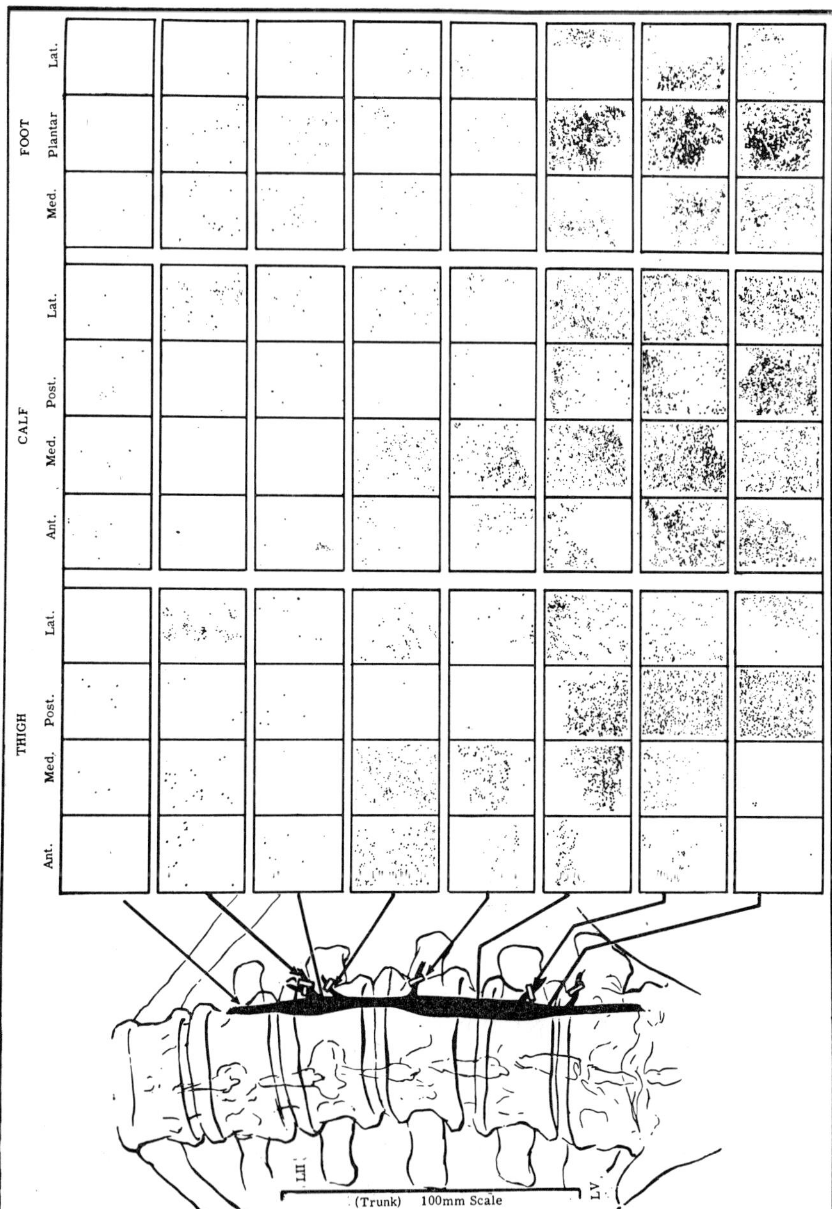

FIG. 36. Composite representation of stimulation of the lumbar trunk in man. Sweat records are shown from several areas of the thigh, calf and foot and related by interconnecting line to the site of stimulation on the trunk. Silver clips, placed during operation, aided in accurate identification of stimulation sites. Note the sparce sweating when rostral portions of the trunk were stimulated, as contrasted with the results illustrated in Fig. 37. (From Randall *et al.,* 1955, reproduced with the permission of the Journal of Applied Physiology.)

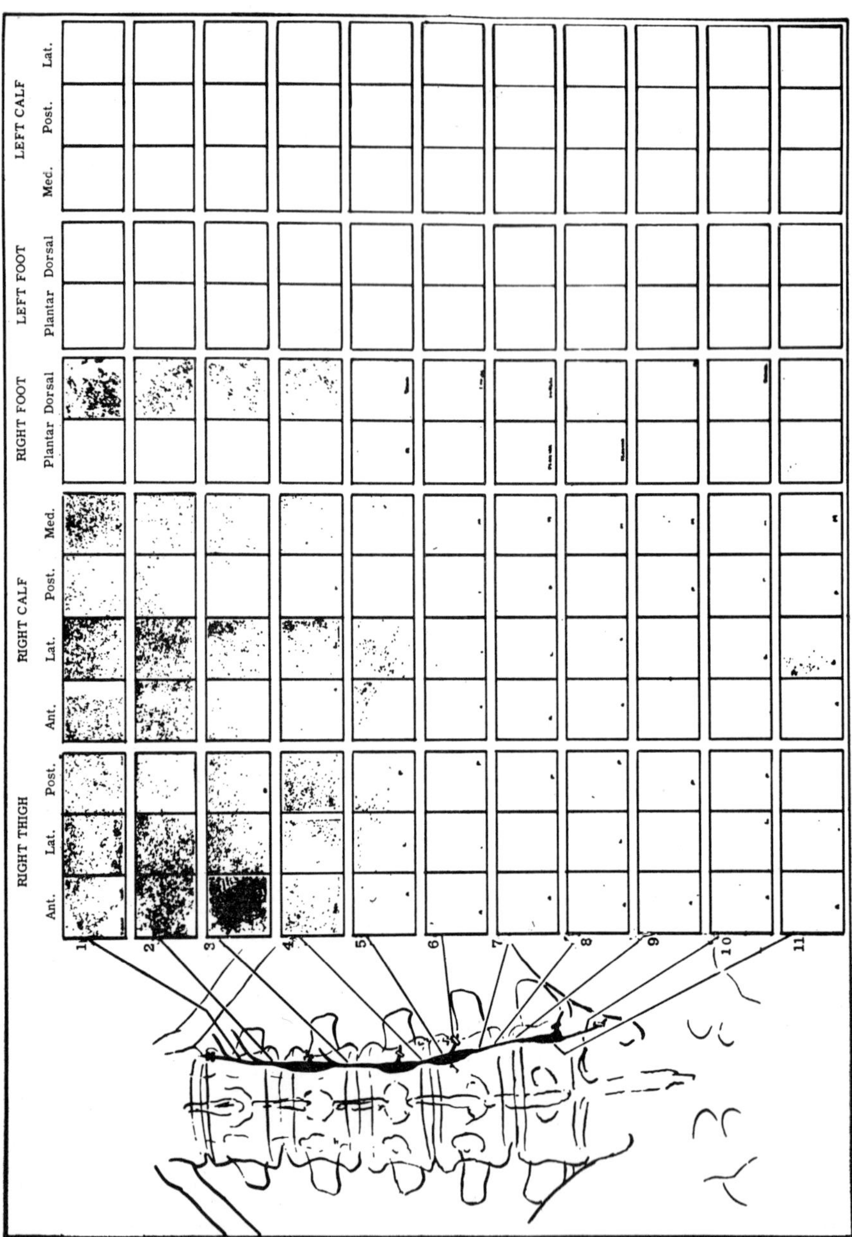

FIG. 37. Stimulation of the rostral end of the lumbar sympathetic trunk in this patient elicited profuse sweating on all the test areas, while stimulation of its caudal portion did not. (Compare with Fig. 36.) Note also the complete absence of response on the contralateral extremity. This evidence argues against the functional significance of cross over from one sympathetic trunk to the opposite side. (From Randall *et al.,* 1955; reproduced with the permission of the Journal of Applied Physiology.)

In contrast to the pattern of preganglionic sudomotor entry into the lumbar trunk, described above, are the results depicted in Fig. 37. Stimulation at L1 now elicited profuse sweating on all surfaces of the lower extremity, with variable changes in sweating patterns as the electrodes were moved caudally. Virtually no change was noted during stimulations below L3. It is evident from such observations that marked variations exist in the levels of entry of sudomotor pathways in different patients.

Both functional and anatomic evidence for the entry of preganglionic fibers into the trunk as low as L5 in man has occasionally been obtained. In a patient with severe causalgic pain, a communicating ramus consisting primarily of small, thinly myelinated fibers was found entering a large ganglion at L5 (Randall *et al.,* 1955). Direct electrical stimulation of this ramus elicited sparce but distinct sweating on the lower extremity. Its removal was accompanied by the permanent disappearance of the severe pain which had arisen from the great toe. Hence, it was concluded that a major fraction of the myelinated fibers in this ramus was pain afferents, but a few fibers were definitely sudomotor efferents.

3. The Sympathetic Trunk and Its Postganglionic Outflows

Anatomic studies have demonstrated marked variations in distribution of sympathetic ganglia of the lumbar sympathetic trunk in man and dog, and rarely is there a systematic association of one ganglion with each vertebral body as conventionally shown in text figures. In fact, there may be a condensation of nearly all cell bodies into a single large ganglion. More frequently, however, some modification between these two extremes is noted, but, here again, it is impossible to predict before actual dissection what the ganglion distribution may be in a given patient or animal. The presence of cross communications between the two lumbar trunks has been emphasized by many authors, but careful dissection and histologic examination have generally revealed the absence of actual nervous connections (Webber, 1958). In several hundred direct stimulation experiments, both in man and the dog, neither contralateral vasoconstriction nor sudomotor response was observed. The absence of contralateral sweating is illustrated in man (Fig. 37).

The sympathetic ganglion is the site of large, postganglionic cell bodies which give rise to nonmyelinated (or very thinly myelinated) class C fibers. These may extend up or down the sympathetic trunk but eventually take exit by way of the sympathetic root (grey ramus), which re-enters the spinal nerve at a point close to the origin of the communicating (white) ramus. Those postganglionic fibers arising from cell bodies within the spinal nerve, itself, presumably extend distally within the same structure. Interconnecting rami occasionally extend from the second to the third lumbar nerves and from the third lumbar through the second sacral nerve in the usual pattern of the lumbosacral plexus. Postganglionic sympathetic fibers may traverse these rami, to reach lower levels and consequently contribute to the widespread distribution of sympathetic vasomotor outflows.

In view of the great variation in level of preganglionic inflow to the sympathetic trunk, it is not surprising that considerable difference also exists in levels of postganglionic outflows. As expected, the postganglionic fiber generally takes exit from the trunk one or two segments caudal to the point of entry of the preganglionic fiber with which it synapses. In some instances, however, postganglionic fibers course many segments before they exit, and in others, they apparently take extremely devious routes to the periphery after leaving the sympathetic trunk. The distribution of postganglionic fibers to peripheral cutaneous areas is not strictly in accordance with the sensory dermatomes. In some patients the most caudal parts of the lumbar sympathetic trunk send fibers to the distal-most portions of the lower extremity, while in others the caudal segment of the trunk supplies only the proximal regions of the limb.

4. Problems of Surgical Sympathectomy

A most significant conclusion resulting from these observations relates to the requirements for complete sympathetic denervation of the lower extremity. Extirpation of the L2–L3 segment of the sympathetic trunk is certainly inadequate to accomplish removal of all sudomotor and vasomotor pathways through the sympathetic trunk in most patients. Rami have been observed entering the lumbar trunk at all levels from L1 through L5 in man and through L7 in the dog. It appears probable that much confusion has been introduced into current literature because of the fundamental fact, that with the conventional lumbar sympathectomy many sympathetic pathways still remain intact and functional. In addition, there are direct accessory fibers which synapse within, or are closely associated with, the spinal nerves. Clearly, completely verified extirpation of the lumbar and upper sacral paravertebral ganglion chain does not always eliminate vasomotor control over blood vessels in the lower extremity in either man or dog.

It is well known that therapeutic failure frequently characterizes lumbar or thoracic sympathectomy, even when only minimal organic changes are present in the blood vessels of the extremities. Several possible explanations have been advanced for such results: (1) incomplete section or the removal of inadequate amounts of sympathetic trunk (Kuntz and Alexander, 1950; Coldwater *et al.*, 1957; Randall *et al.*, 1955, 1958b), (2) the presence of accessory sympathetic pathways outside the thoracic and abdominal distribution of the sympathetic trunks (Randall *et al.*, 1950), (3) the syncytial nature of the vascular neuroeffector units which would permit a few fibers extensive neural control (Hillarp, 1946), and (4) denervation hypersensitivity to a circulating constrictor substance (Simeone and Felder, 1951). Any or all of these factors may combine to render a given surgical procedure ineffective in a particular patient. However, the fact that careful surgical sympathectomy is successful in a significant number of patients requires that each of these factors be examined further in an effort to establish a sound understanding of the vascular innervation and its control.

5. The Question of Denervation Hypersensitivity

The determination of how completely a small residuum of fibers can assume the function of the original vasomotor innervation is a fascinating problem, but it remains essentially unsolved. Until the intimate nature of the peripheral autonomic neuroeffector terminations has been thoroughly and accurately described, it is doubtful that this question can be answered. Whether the postganglionic fibers lose their individuality and pass into a common terminal nervous network, or retain their identity and end in separate terminal structures in the form of distinct "motor units" (Randall *et al.*, 1958a), has a great bearing on the manner in which a small residual group of fibers may reinnervate a given vascular bed. In similar fashion, the relationship of a small residual innervation to the possibility of sensitization remains open. One should remain skeptical of conventional concepts involving "denervation hypersensitivity," since it appears that most sympathectomies, as generally performed, are incomplete.

Following unilateral lumbar sympathectomy, the reactivity of the vascular beds of the two hind feet of the dog to a standard physiologic dose of intravenously administered 1–epinephrine was evaluated by photoelectric plethysmography (Cooper *et al.*, 1959). The operation consisted of complete removal of the chain from a level above the renal vessels to a point below the sacral prominence. Postoperative observations were made from 10 days to 20 weeks after operation. A standard dose of 1×10^{-8} gm/kg body weight consistently elicited a small but measurable constriction in the footpad but did not alter central arterial pressure pulses. In such a preparation, no significant differences could be detected in responses of the sympathectomized and intact vessels to this physiologic dose of l–epinephrine. However, a larger dose (1.3×10^{-6} gm/kg) induced a stronger reaction on the operated side. Ederstrom *et al.* (1958) also failed to show increased sensitivity to either intra-arterially or intravenously infused epinephrine and norepinephrine in chronically sympathectomized dogs. Nor was sensitivity to epinephrine found in isolated vessels or in perfused isolated limbs after sympathectomy. Ederstrom *et al.* (1956) also observed that temperature of chronically sympathectomized feet was generally indistinguishable from that of nonoperated control feet in the same animal when observed several days postoperatively. However, under deep anesthesia the vessels of the control intact extremity dilated significantly, while those in the operated limb failed to do so. Thus, greater "tone" marks the sympathectomized vessels, but it is unrelated to increased sensitivity to epinephrine (Ederstrom *et al.*, 1950). Felder *et al.* (1951) also found that following lumbar sympathectomy in man, the operated foot became colder than the nonoperated foot when the patient was anesthetized for subsequent surgery.

In contrast to the results of lumbar sympathectomy, pad vessels of limbs which were totally denervated by femoral and sciatic neurectomy show augmented reactivity to epinephrine in physiologic doses (Cooper *et al.*, 1959). It is conceivable, therefore, that completeness of denervation, including cholinergic fibers (Armin *et al.*, 1953), is an all-important factor in the ultimate determination of the development and intensity of "denervation hypersensitivity."

6. Sympathetic Vasodilator Fibers[6]

a. Identification of vasodilator nerve fibers: This group of autonomic nerve fibers has been identified and defined mainly from physiologic and pharmacologic observations in laboratory mammals. Recent investigations of the central control of sympathetic vasodilator pathways have provided almost as significant a role in establishing the nature of these nerves as previous peripheral vasomotor studies. Anatomic studies, however, have hardly kept pace with the functional ones and have had little to do with the morphologic identification of these fibers peripherally. Much credit is due the Swedish physiologists who have made important contributions to the understanding of the function of these

[6] By K. Christensen.

nerves. Since adequate review articles have traced the steps in the development of the concept (Burn, 1938; Uvnäs, 1954; McDowall, 1938; Folkow, 1956; Uvnäs, 1960b), only the most recent conclusions will be reviewed here.

b. Physiologic evidence for the function of vasodilator nerves: In order to demonstrate peripheral vasodilatation, it is necessary to stimulate the sympathetic vasodilator nerve fibers in the peripheral nerves which transport them to the blood vessels. However, since peripheral nerves are a mixture of nerve fibers and may include both sympathetic vasoconstrictor and sympathetic vasodilator fibers, it is necessary first to block the former with a sympatholytic drug. Under such circumstances, if vasodilator fibers are present, stimulation of the mixed nerves produces vascular dilatation, resulting in an increase in volume of the tissue, as recorded with a plethysmograph. Recently Uvnäs, (1960a,b) has selected the direct blood flow recording technique as an accurate and more regularly reproducible method for determining vasodilatation.

c. Areas involved: In the head, sympathetic vasodilator fibers from the cervical sympathetic trunk have been reported to innervate blood vessels of the mucous membrane of the mouth, the nostrils, certain areas of cheek skin, and muscles of the face. Sympathetic vasodilator fibers have also been localized in the nervous rami from the stellate ganglion to the coronary arteries (dog). Abdominal viscera, especially the intestines, receive these fibers through the splanchnic nerves, while in the case of skeletal muscles of the extremities, the fibers run in the nerves of the brachial and lumbar plexuses. At the present time, Uvnäs (1960b) states that convincing evidence for vasodilator fibers is found only in the nerves to the blood vessels supplying skeletal muscles of the extremities. (For further discussion of vasodilator fibers in muscle, see Sections E.-1d and F.-2a, Chapter XVII.)

d. Anatomic pathway for vasodilator fibers: The peripheral sympathetic vasodilator pathways presumably have the same characteristics as other nerves of the sympathetic nervous system. All of the central neurones (general visceral efferent) are located in the intermediolateral gray matter of the spinal cord of thoracic and upper lumbar segments of the spinal cord. Their fibers synapse in the sympathetic ganglia with cells whose axones are distributed to the blood vessels. Many of these postganglionic fibers are distributed to the blood vessels by way of the peripheral nerves. Any anomalous location of sympathetic ganglionic neurones would affect vasodilator pathways as well as vasoconstrictor ones. Thus far no morphologic differences have been recorded regarding vasodilator fibers, although some atypical myelinated sympathetic fibers to the lower extremity have been considered to be vasodilator.

e. Neurohumoral mechanism controlling vasodilator fibers: In recent studies Uvnäs (1960b) used reserpine to block vasoconstrictor fibers. Reserpine rids the tissues of noradrenaline, and therefore peripheral nervous stimulation in a reserpinized animal provides no vasoconstrictor influence to blood vessels of skeletal muscle; instead vasodilatation occurs. It was found that atropine blocks the resulting vasodilatation completely. From studies of this type, it is reasoned that the sympathetic vasodilator fibers are entirely of the cholinergic type of sympathetic fibers.

f. Clinical implications: Clinically, vasodilation is of interest for its physiologic implications. Uvnäs has suggested that sympathetic vasodilators become active in emergency states which require sudden muscular activity. His study and that of Hilton (1959) suggest how effector tissues may react to stress phenomena.

C. MECHANISMS INVOLVED IN THE ESTABLISHMENT OF VASCULAR TONE

By B. Folkow

1. General Considerations

While "vascular tone" is a widely used expression, the mechanisms responsible for it have been exposed to surprisingly little quantitative analysis. There is, for instance, much ignorance concerning the factors which create vascular tone (still more regarding their exact interrelationship), how vascular tone should be measured, and even exactly what is implied in the expression.

Obviously "vascular tone" should refer to the average activity level of the vascular smooth muscle cells. However, the vascular bed is a highly complex structure, with a number of very specialized "parallel-coupled" circuits supplying the different tissues. Further, each of these circuits consists of a number of "series-coupled," functionally differentiated sections, which can be called: *Windkessel vessels, pre- and postcapillary resistance vessels, sphincter vessels, shunts, and capacitance vessels,* all controlled by smooth muscle. In their specialized functions, however, these various sections subserve the main purpose of the cardiovascular system, namely, the establishment of a properly balanced contact between blood and tissues across the walls of the *exchange vessels,* the true capillaries.

There are at present, good reasons to assume that smooth muscle activity neither runs closely parallel in all these specialized vascular sections and circuits, nor has it exactly the same quantitative background. It is, therefore, rather meaningless to talk about "vascular tone" with relation to the vascular tree *as a whole.* Generally, the term is used to denote only the average contraction level of the resistance vessels, mainly because these have so far been exposed to the greatest study and have attracted the most interest. However, this does not mean that they actually constitute the most important vascular section. In fact, the control of tone of the capacitance vessels is of the utmost hemodynamic importance, although rather little is known about this function. Because of the relative lack of knowledge about the other portions of the vascular tree, the present discussion will deal mainly with the tone of the resistance vessels, its measurement, and when possible, the regulation of tone in some of the other vascular sections.

2. Means of Measuring "Vascular Tone"

Measurement of the actual flow resistance alone is of limited value in determining the extent of smooth muscle activity in the resistance vessels; nor does it give any information regarding other portions of the vascular tree. Actually all it implies is the degree of restriction of the *lumina* of the resistance vessels; unfortunately, in most instances in which knowledge about smooth muscle tone is essential, the reduction in lumen is partly or wholly due to structural changes. Therefore, if smooth muscle tone is to be judged from the actual flow resistance, the latter must have some sort of reference point, such as the resistance remaining after complete smooth muscle inactivation, i.e., at maximal dilatation. The *ratio* of the actual resistance to this minimal resistance may then give some information about smooth muscle tone of the resistance vessels, just as the activity state of an isolated muscle strip can be judged from the relationship between its active and resting lengths. Although only relative figures are obtained in this manner, derived in a roundabout way via the average luminal shifts, they are, nevertheless, quite useful for comparisons of vascular tone between various "parallel-coupled" circuits of normal organisms. However, the procedure is invalid for comparisons between vascular beds in which structural shifts in wall-lumen ratio have taken place. This is because the thicker the wall is in relation to the lumen at maximal dilatation (regardless of whether this is due to medial hypertrophy, waterlogging or intimal proliferation), the more pronounced the luminal reduction becomes in response to a given smooth muscle shortening (Folkow, 1957; Conway, 1958; Redleaf and Tobian, 1958; Folkow *et al.*, 1958). For this reason, the raised flow resistance in well-established hypertension, for example, may at least partly be a consequence of the adaptive increase of vascular wall mass, whenever this is present, and in rough proportion to this change.

Estimations of the flow resistance with complete smooth muscle inactivation reveal striking differences among various tissues. This is understandable, since the maximal blood flow capacity can be expected to be roughly at a level sufficient to meet maximal nutritional demands and other needs for blood flow in the various tissues, e.g., as raw material for urine production, sweat secretion, etc., with such demands varying considerably. However, the maximal flow capacity is only exceptionally utilized, and usually the flow through the vascular bed is more or less restricted by the smooth muscle tone to meet current needs, thus creating an easily mobilized blood flow reserve for each tissue. Hence, estimation of "vascular tone" implies a measurement of the magnitude of this "blood flow reserve," which in many vascular diseases may be just as important as estimations of the cardiac reserve in heart disease.

3. Mechanisms Responsible for "Blood Flow Reserve"

Among the mechanisms which create this "blood flow reserve" are excitatory nervous influences, blood-borne or locally released vasoconstrictor substances, myogenic automaticity and, as proposed by Bayliss in 1902, the steady stimulation of muscle cells by the distention offered by the blood pressure.

a. Sympathetic nervous impulses: Most vessels are supplied by vasoconstrictor fibers, which by their discharge variations, can centrally direct the tone within a wide range. In "resting equilibrium" these fibers contribute to vascular tone in most areas by way of their low-frequency excitatory drive (Folkow, 1955, 1956; Uvnäs, 1960a). Their range of influence is, however, not the same in different tissues, partly because they are unequally distributed and partly for other reasons (Folkow, 1960). To mention some extremes, they appear to exert little influence on such vital tissues as the brain and the myocardium, while they completely dominate such vessels as cutaneous arteriovenous shunts, specialized to serve centrally controlled heat balance. At various levels between these extremes of nervous control, the other vascular circuits are found; in some the neurogenic contribution to vascular tone in the resting equilibrium is negligible, in others it is moderate, while in a few it is dominant. This becomes clear if the constrictor fibers to the different tissues are regionally blocked in the resting state and the effect of such a procedure on the respective flow resistance is measured. Vascular tone which remains can be called the *basal* tone. Its extent can be quantitatively estimated by studying how much the flow resistance can be further decreased by arterial injections of supramaximal amounts of a dilator drug. It is then evident that basal vascular tone tends to be more pronounced in a region in which the constrictor fiber influence is less dominating. To cite some examples, resting muscle blood flow increases approximately 2–3 times (to about 8–10 ml/min/100 ml of tissue) after regional block of tonic vasoconstrictor fiber discharge, but it can be increased another 4–5 times by intense muscle work. Obviously, these vessels have a pronounced basal vascular tone which is also true of the coronary vessels. Upon acute sympathetic denervation of the cutaneous arteriovenous shunts, on the other hand, a nearly maximal dilatation is obtained in the resting organism if the hormone discharge from the adrenal medulla is normally small. Obviously then, these specialized, centrally dominated shunt vessels possess very little basal tone.

b. Circulating vasoconstrictor agents: The question arises as to what extent circulating vasoconstrictor agents, particularly the hormones from the adrenal medulla, normally do contribute to basal vascular tone. If this is significant, then it follows that the cutaneous shunts are *less* sensitive to such stimuli than the muscle vessels, since the former have almost no basal tone while the muscle vessels have considerable. However, exactly the opposite is the case. The cutaneous shunts are more sensitive to biogenic vasoconstrictor agents and will, in fact, reveal even minute concentrations of catechol amines by their increased tone. From such observations the conclusions may be drawn that (1) basal tone is an essentially local affair, and (2) in the resting, normal organism there are only insignificant plasma concentrations of vasoconstrictor agents, for, otherwise a constrictor response of the cutaneous shunts would reveal their presence (Löfving and Mellander, 1956). Normally such agents are released episodically in hemodynamically significant amounts with intense increases of sympathetic activity. Even then the vasoconstrictor influence of the hormonal sympathetic link is generally quite small when related to the direct vasoconstrictor fiber innervation (Celander, 1954). Another question arises in the case of pathophysiologic conditions in which substances like angiotensin can contribute considerably to basal vascular tone.

c. Myogenic automaticity: Another interesting subject is the phenomenon of myogenic automaticity and the possibility that excitatory agents may be locally released. Rhythmic, unsynchronized contractions (vasomotion), independent of extrinsic nerves or of blood-borne substances, are known to occur in the small vessels of many tissues (see Nicoll and Webb, 1946). They are also found in larger vessels, even when isolated, although not to the same degree. It has been suggested that vasomotion is not of myogenic origin but is due rather to activity of ganglion cells situated in the vascular walls. However, the existence of such ganglion cells, specifically connected to the vascular smooth muscles, is seriously doubted (Hillarp, 1960). On the other hand, myogenic automaticity is a well-established characteristic of most, if not all, smooth muscles (Bozler, 1948; Furchgott, 1955; Bülbring, 1957–1958).

It is also possible that myogenic automaticity might be facilitated by excitatory agents, released from non-nervous cells situated within the smooth muscle layer. In this regard it has been found that smooth muscle stimulating substances can be extracted from nerve-free intestinal smooth muscles. However, the few studies that have dealt with vascular smooth muscles indicate that their "spontaneous" activity remains relatively unaffected even after the muscle cells are rendered insensitive to practically all known vasoexcitatory biogenic agents by blocking drugs (Furchgott, 1955). The catechol amines, extractable from the vascular walls, appear to be almost entirely derived from the extrinsic sympathetic innervation (von Euler, 1956), although a minor fraction might come from the chromaffine cells that recently have been found along the walls of many small vessels (Adams-Ray *et al.,* 1958). Little is known at present about the functional significance of these cells and what they normally release. It is possible that they are sympathetically innervated; if so, they form a sort of a scattered adrenal medullary gland. There are, therefore, few reasons to doubt that most vascular smooth muscles exhibit myogenic activity, although its extent evidently varies widely from one region to another. This myogenic activity forms the very foundation of basal vascular tone.

d. Stimulation of smooth muscle by distention: Most smooth muscles are promptly stimulated by distention, and the study of Bülbring (1955) on intestinal strips makes it clear that such a stimulus causes facilitation of the *rate* of myogenically generated action potentials and hence produces an increase of active tension. In 1902, Bayliss proposed that continuous distention, offered by the blood pressure, might form an effective vascular smooth muscle stimulus and contribute to the maintenance of tone. This interesting hypothesis has recently been supported by Fog (1937), Folkow (1949, 1952), Hilton (1953), and Wood *et al.* (1955). Even so, Bayliss' hypothesis has not been generally accepted, simply because it appeared to lead to a paradox. It was argued that if the intraluminar blood pressure should cause an increase of vascular tone, pronounced enough to narrow the vascular lumen, the distention stimulus would thus eliminate itself (Gaskell and Burton, 1953). However, with Bülbring's data as background, together with observations of the unsynchronized contractions of the smallest vessels which often lead to rhythmic luminal closures, the paradox seems to vanish. As is the case with intestinal smooth muscles, it is highly probable that continuous distention of the small vessels leads to an increase in the *rate* of their myogenically generated contractions. In fact, such a response has been directly observed in the small veins of the bat's wings (Wiedeman, 1957). This probably means that the contraction phase will occupy an increased fraction of the contraction-relaxation cycle of each muscle cell. When applied to the thousands of muscle cells of all the resistance vessels, the net effect will be increased tone wherever myogenic activity is present, induced in a way which is in harmony with findings in other myogenically active smooth muscles. The theory also explains why the phenomenon is virtually absent in vessels with little or no basal tone and hence, insignificant myogenic activity (Folkow, 1952). However, other mechanisms can contribute markedly to the regional increase of tone, often obtained when blood pressure is raised. The latter response is also generally associated with an increased flow, which, in turn, raises local oxygen tension and decreases the local concentrations of so-called vasodilator metabolites. These chemical changes constitute a type of shift in vascular smooth muscle environment which tends to enhance the tone of blood vessels to the extent that flow may increase proportionately *less* than pressure. Exactly the opposite reactions will occur with a drop in pressure, when smooth muscle relaxation is revealed by the resulting reactive hyperemia.

4. Control of Tone in Other Series-coupled Vascular Sections

With regard to control of tone in other vascular sections, the specialized cutaneous shunt vessels have already been dealt with; virtually nothing is known about tone regulation in other types of shunt vessels.

The precapillary sphincters, which form a specialized section of the resistance vessels, have also been mentioned. The muscle cells of these sphincters appear to be regulated locally through myogenic activity, which seems to form the basis for rhythmically occurring obstructions of flow through the corresponding capillaries. With regard to the capacitance vessels, corresponding mainly to the veins, little is known about the exact background of their tone, although it appears that the muscles of the smallest veins also exhibit a pronounced myogenic activity. It is probable that the extrinsic nervous control of the capacitance section as a whole has a relatively more dominant influence than exists in the case of the resistance vessels. At least this is true in the case of tissues like the skeletal muscles, where the basal tone is fairly pronounced and where a more moderate nervous influence is superimposed.

5. Summary

It is reasonable to assume that essentially the same excitatory elements can be traced in all vascular regions, with their influence varying quantitatively among the different parallel-coupled vascular circuits and among the series-coupled sections. The expression "vascular tone," if used in its widest sense, therefore covers many levels of vascular smooth muscle activity, with considerable differences existing in different parts of the vascular bed.

References

Adams-Ray, J., Nordenstam, H., and Rhodin, J. (1958). Über die chromaffinen Zellen der menschlichen Haut. *Acta Neuroveget. (Vienna)* **18,** 304.

Alexander, R. S. (1945). The effects of blood flow and anoxia on spinal cardiovascular centers. *Am. J. Physiol.* **143,** 698.

Alexander, R. S. (1946). Tonic and reflex functions of medullary cardiovascular centers. *J. Neurophysiol.* **9,** 205.

Alexander, W. F., Kuntz, A., Henderson, W. P., and Ehrlich, E. (1949). Sympathetic conduction pathways independent of sympathetic trunks—their surgical implications. *J. Intern. Coll. Surgeons* **12,** 111.

Allen, W. F. (1931). An experimentally produced premature systolic arrhythmia (pulsus bigeminus) in rabbits. IV. Effective areas in the brain. *Am. J. Physiol.* **98,** 344.

Anand, B. K., and Dua, S. (1956). Circulatory and respiratory changes induced by electrical stimulation of the limbic system (visceral brain). *J. Neurophysiol.* **19,** 393.

Arey, L. B. (1942). "Developmental Anatomy." Saunders, Philadelphia, Pa.

Armin, J., Grant, R. T., Thompson, R. H. S., and Tickner, A. (1953). An explanation for heightened vascular reactivity of the denervated rabbit's ear. *J. Physiol.* **121,** 603.

Bach, L. M. N. (1952). Relations between bulbar respiratory, vasomotor and somatic facilitatory and inhibitory areas. *Am. J. Physiol.* **171,** 417.

Bailey, P., and Sweet, W. H. (1940). Effects on respiration, blood pressure and gastric motility of stimulation of orbital surface of frontal lobe. *J. Neurophysiol.* **3,** 276.

Bayliss, W. M. (1901). On the origin from the spinal cord of the vaso-dilator fibres of the hind-limb and on the nature of these fibres. *J. Physiol. (London)* **26,** 173.

Bayliss, W. M. (1902). On the local reactions of the arterial volume to changes in internal pressure. *J. Physiol. (London)* **28,** 220.

Bayliss, W. M. (1908). On reciprocal innervation in vaso-motor reflexes and on the action of strychnine and chloroform thereon. *Proc. Roy. Soc.* **B80,** 339.

Bayliss, W. M. (1923). "The Vaso-Motor System." Longmans, Green, New York.

Bayliss, W. M., and Bradford, J. R. (1894). The innervation of the vessels of the limbs. *J. Physiol. (London)* **16,** 10.

Beattie, J., Brow, G. R., and Long, C. N. H. (1930). Physiological and anatomical evidence for the existence of nerve tracts connecting the hypothalamus with spinal sympathetic centers. *Proc. Roy. Soc. (London)* **B106,** 253.

Bernard, C. (1852). De l'influence du système nerveux grand sympathique sur la chaleur animale. *Compt. rend. acad. sci.* **34,** 472.

Biedl, A. (1897). Beiträge zur Physiologie der Nebenniere. Erste Mitteilung: Die Innervation der Nebenniere. *Pflügers Arch. ges. Physiol.* **67,** 443.

Boyd, J. D., and Monro, P. A. G. (1949). Partial retention of autonomic function after paravertebral sympathectomy. *Lancet* **ii,** 892.

Bozler, E. (1948). Conduction, automaticity, and tonus of visceral muscles. *Experientia* **4,** 213.

Brodal, A. (1956). "The Reticular Formation of the Brainstem: Anatomical Aspects and Functional Correlations." Charles C Thomas, Springfield, Ill.

Brooks, C. M. (1933). Reflex activation of the sympathetic system in the spinal cord. *Am. J. Physiol.* **106,** 251.

Budge, J. (1853). De l'influence de la moëlle épinière sur la chaleur de la tête. *Compt. rend. acad. sci.* **36,** 377.

Bülbring, E. (1955). Correlation between membrane potential, spike discharge and tension in smooth muscle. *J. Physiol. (London)* **128,** 200.

Bülbring, E. (1957–1958). Physiology and pharmacology of intestinal smooth muscle. *In* "Lectures on the Scientific Basis of Medicine" (Brit. Postgrad. Med. Federation), Vol. 7, p. 374. Oxford Univ. Press (Athlone), London and New York.

Burn, J. H. (1938). Sympathetic vasodilator fibers. *Physiol. Revs.* **18,** 137.

Celander, O. (1954). The range of control exercised by the sympathico-adrenal system. *Acta Physiol. Scand.* **32,** Suppl. 116.

Chapman, W. P., Livingston, R. B., and Livingston, K. E. (1949). Frontal lobotomy and electrical stimulation of the orbital surfaces of the frontal lobes; effect on respiration and on blood pressure in man. *A.M.A. Arch. Neurol. Psychiat.* **62,** 701.

Chapman, W. P., Livingston, K. E., and Poppen, J. L. (1950). Effects upon blood pressure of electrical stimulation of tips of temporal lobes in man. *J. Neurophysiol.* **13,** 65.

Chen, M. P., Lim, R. K. S., Wang, S. C., and Yi, C. L. (1936). On the question of a myelencephalic sympathetic centre. I. The effect of stimulation of the pressor area on visceral function. *Chinese J. Physiol.* **10,** 445.

Chen, M. P., Lim, R. K. S., Wang, S. C., and Yi, C. L. (1937a). On the question of a myelencephalic sympathetic centre. II. Experimental evidence for a reflex sympathetic centre in the medulla. *Chinese J. Physiol.* **11,** 355.

Chen, M. P., Lim, R. K. S., Wang, S. C., and Yi, C. L. (1937b). On the question of a myelencephalic sympathetic center. III. Experimental localization of the centre. *Chinese J. Physiol.* **11,** 367.

Coldwater, K. B., Alexander, W. F., Cox, J. W., and Randall, W. C. (1957). The functional significance of the first thoracic ganglion in sympathectomy of the upper extremity in man. *Ann. Surg.* **145,** 530.

Cone, W. V. (1948). Cited by Ray, B. S., and Console, A. D., in *J. Neurosurg.* **5,** 23.

Conway, J. (1958). Vascular reactivity in experimental hypotension measured after Hexamethonium. *Circulation* **17,** 807.

Cooper, T., Willman, V. L., and Hertzman, A. B. (1959). Responses of dog's pad vessels to epinephrine following sympathectomy or denervation. *Circulation Research* **7,** 574.

Cox, J. W., Randall, W. C., Alexander, W. F., Franke, F. E., and Hertzman, A. B. (1951). A method for the study of cutaneous vascular responses: Its application in mapping pre- and postganglionic vasomotor supply to the foot. *Bull. St. Louis Univ. Hosp.* **3,** 72.

Dittmar, C. (1870). Ein neuer Beweis für die Reizbarkeit der centripetalen Fasern des Rückenmarks. *Ber. Verhandl. K. sächs. Ges. Wiss.* **22,** 4.

Dittmar, C. (1873). Über die Lage des sogennanten Gefässcentrums der medulla oblongata. *Ber. Verhandl. K. sächs. Ges. Wiss.* **25,** 103.

Ederstrom, H. E., and Porter, D. A. (1950). Blood flow changes in extremities of dogs following sympathectomy and exercise. *Federation Proc.* **19,** 94.

Ederstrom, H. E., Vergeer, T., Rohde, R. A., and Ahlness, P. (1956). Quantitative changes in foot blood flow in the dog following sympathectomy and motor denervation. *Am. J. Physiol.* **187,** 461.

Ederstrom, H. E., Vergeer, T., and Oppelt, W. W. (1958). Effects of l-epinephrine and l-norepinephrine on foot vasculature after sympathectomy in dogs. *Am. J. Physiol.* **193,** 157.

Ehrlich, E., and Alexander, W. F. (1951). Surgical implications of upper thoracic independent sympathetic pathways. *A.M.A. Arch. Surg.* **62,** 609.

Felder, D. A., Linton, R. R., Todd, D. P., and Banks, C. (1951). Changes in the sympathectomized extremity with anesthesia. *Surgery* **29,** 803.

Fog, M. (1937). Cerebral circulation. The reaction of the pial arteries to a fall in blood pressure. *A.M.A. Arch. Neurol. Psychiat.* **37,** 351.

Folkow, B. (1949). Intravascular pressure as a factor regulating the tone of the small vessels. *Acta Physiol. Scand.* **17,** 289.

Folkow, B. (1952). A study of the factors influencing the tone of denervated blood vessels perfused at various pressures. *Acta Physiol. Scand.* **27,** 99.

Folkow, B. (1955). Nervous control of the blood vessels. *Physiol. Revs.* **35,** 629.

Folkow, B. (1956). Nervous control of the blood vessels. *In* "The Control of the Circulation of the Blood," Suppl. Vol. (R. J. S. McDowall, ed.), pp. 1–85. Dawson & Sons, London.

Folkow, B. (1957). Structural, myogenic, humoral and nervous factors controlling peripheral resistance. *In* "Hypotensive Drugs," Symposium, London, 1956 (M. Harrington, ed.), p. 163. Pergamon, New York.

Folkow, B. (1960). Role of the nervous system in the control of vascular tone. *Circulation* **21,** 760.

Folkow, B., Grimby, G., and Thulesius, O. (1958). Adaptive structural changes of the vascular walls in hypertension and their relation to the control of the peripheral resistance. *Acta Physiol. Scand.* **44,** 255.

Folkow, B., Johansson, B., and Öberg, B. (1959). A hypothalamic structure with a marked inhibitory effect on tonic sympathetic activity. *Acta Physiol. Scand.* **47,** 262.

Furchgott, R. F. (1955). The pharmacology of vascular smooth muscle. *Pharmacol. Revs.* **7,** 183.

Gaskell, P., and Burton, A. C. (1953). Local postural vasomotor reflexes arising from the limb veins. *Circulation Research* **1,** 27.

Geohegan, W. A., Wolf, G. A., Adair, O. J., Hare, K., and Hinsey, J. C. (1942). The spinal origin of the preganglionic fibers to the limbs in the cat and monkey. *Am. J. Physiol.* **135,** 324.

Gilliatt, R. W., Guttmann, L., and Whitteridge, D. (1948). Inspiratory vasoconstriction in patients after spinal injuries. *J. Physiol.* (*London*) **107,** 67.

Goltz, F. (1864). Über den Tonus der Gefässe und seine Bedeutung für die Blutbewegung. *Virchows Arch. pathol. Anat. u. Physiol.* **29,** 394.

Green, H. D., and Hoff, E. C. (1937). Effects of faradic stimulation of the cerebral cortex on limb and renal volumes in the cat and monkey. *Am. J. Physiol.* **118,** 641.

Gruss, W. (1932). Über Ganglien im ramus cummunicans. *Z. Anat. Entwicklugsgeschichte* **97,** 464.

Guttmann, L. (1954). *In* "Peripheral Circulation in Man," Ciba Foundation Symposium (G. E. W. Wolstenholme, ed.), p. 192. Little, Brown, Boston.

Hess, W. R. (1938). "Das Zwischenhirn und die Regulation von Kreislauf und Atmung." Thieme, Leipzig.

Hess, W. R. (1948). "Die Funktionelle Organization des Vegetativen Nervensystems." Benno Schwabe, Basel, Switzerland.

Heymans, C., Bouckaert, J. J., Farber, S., and Hsu, F. Y. (1936). Spinal vasomotor reflexes associated with variations in blood pressure. *Am. J. Physiol.* **117,** 619.

Hillarp, N. A. (1946). Structure of the synapse and the peripheral innervation apparatus of the autonomic nervous system. *Acta Anat.* **2,** Suppl. 4, 1.

Hillarp, N. A. (1960). Peripheral autonomic mechanisms. *In* "Handbook of Physiology," Section 1: Neurophysiology (J. Field, H. W. Magoun, and V. E. Hall, eds.), Vol. II, p. 979. Am. Physiol. Soc., Washington, D. C.

Hilton, S. M. (1953). Experiments on the postcontraction hyperaemia of skeletal muscle. *J. Physiol.* (*London*) **120,** 230.

Hilton, S. M. (1959). A peripheral arterial conducting mechanism underlying dilatation of the femoral artery and concerned in functional vasodilatation in skeletal muscle. *J. Physiol.* **194,** 93.

Hoeben, G. W. M. (1896). Cited in Mitchell, G. A. G. (1956). "Cardiovascular Innervation." Livingstone, Edinburgh and London.

Hoff, E. C., and Green, H. D. (1936). Cardiovascular reactions induced by electrical stimulation of the cerebral cortex. *Am. J. Physiol.* **117,** 411.

Hoffman, B. L., and Rasmussen, T. (1953). Stimulation studies of insular cortex of *Macaca mulatta*. *J. Neurophysiol.* **16,** 343.

Ingram, W. R. (1960). Central antonomic mechanisms. *In* "Handbook of Physiology," Section 1: Neurophysiology (J. Field, H. W. Magoun, and V. E. Hall, eds.), Vol. II, Chapter 37. Am. Physiol. Soc., Washington, D. C.

Kaada, B. R. (1951). Somato-motor, autonomic and electrocorticographic responses to electrical stimulation of "rhinencephalic" and other structures in primates, cat and dog. *Acta Physiol. Scand.* **20,** Suppl., 83.

Kaada, B. R., Pribram, K. H., and Epstein, J. A. (1949). Respiratory and vascular responses in monkeys from temporal pole, insula, orbital surface and cingulate gyrus. *J. Neurophysiol.* **12,** 347.

Kabat, H., Magoun, H. W., and Ranson, S. W. (1935). Electrical stimulation of points in the forebrain: the resultant alterations in blood pressure. *A.M.A. Arch. Neurol. Psychiat.* **34,** 931.

Karplus, J. P., and Kreidl, A. (1909). Gehirn und Sympathicus. I. Mitteilung: Zwischenhirnbasis und Halssympathicus. *Pflügers Arch. ges. Physiol.* **129,** 138.

Karplus, J. P., and Kreidl, A. (1910). Gehirn und Sympathicus. II. Mitteilung: Ein Sympathicuszentrum im Zwischenhirn. *Pflügers Arch. ges. Physiol.* **135,** 401.

Karplus, J. P., and Kreidl, A. (1911). Gehirn und Sympathicus. III. Mitteilung: Sympathicusleitung im Gehirn und Halsmark. *Pflügers Arch. ges. Physiol.* **143,** 109.

Karplus, J. P., and Kreidl, A. (1927). Gehirn und Sympathicus. VII. Mitteilung: Über Beziehungen der Hypothalamuszentren zu Blutdruck und innerer Sekretion. *Pflügers Arch. ges. Physiol.* **215,** 667.

Kuntz, A., and Alexander, W. F. (1950). Surgical implications of lower thoracic and lumbar independent sympathetic pathways. *A.M.A. Arch. Surg.* **61,** 1007.

Langley, J. N. (1891). On the course and connections of the secretory fibers supplying the sweat glands of the feet of the cat. *J. Physiol. (London)* **12,** 347.

Langley, J. N. (1894). Further observations on the secretory and vasomotor fibers of the foot of the cat, with notes on other sympathetic nerve fibers. *J. Physiol. (London)* **17,** 296.

Langley, J. N. (1923). The vascular dilatation caused by the sympathetic and the course of vasomotor nerves. *J. Physiol. (London)* **58,** 70.

Langley, J. N. (1924). Vaso-motor centres. Part III. Spinal vascular (and other autonomic) reflexes and the effect of strychnine on them. *J. Physiol. (London)* **59,** 231.

Lindgren, P. (1955). The mesencephalon and the vasomotor system. *Acta Physiol. Scand.* **35,** Suppl., 121.

Lindgren, P., and Uvnäs, B. (1953a). Vasodilator responses in the skeletal muscles of the dog to electrical stimulation in the oblongate medulla. *Acta Physiol. Scand.* **29,** 137.

Lindgren, P., and Uvnäs, B. (1953b). Activation of sympathetic vasodilator and vasoconstrictor neurons by electrical stimulation in the medulla of the dog and cat. *Circulation Research* **1,** 479.

Livingston, R. B., Chapman, W. P., Livingston, K. E., and Kraintz, L. (1948). Stimulation of orbital surface of man prior to frontal lobotomy. *Research Publs., Assoc. Research Nervous Mental Disease* **27,** 421.

Löfving, B., and Mellander, S. (1956). Some aspects of the basal tone of blood vessels. *Acta Physiol. Scand.* **37,** 134.

McDowall, R. J. S. (1938). Vasodilator mechanisms. *In* "The Control of the Circulation of the Blood" (R. J. S. McDowall, ed.), pp. 58–65. Dawson & Sons, London (2nd ed., 1956).

McQueen, J. D., Browne, K. M., and Walker, A. E. (1954). Role of the brain stem in blood pressure regulation in the dog. *Neurology* **4**, 1.

Magoun, H. W. (1940). Descending connections from the hypothalamus. *Research Publs., Assoc. Research Nervous Mental Disease* **20**, 270.

Magoun, H. W., and Rhines, R. (1946). An inhibitory mechanism in the bulbar reticular formation. *J. Neurophysiol.* **9**, 165.

Magoun, H. W., Ranson, S. W., and Hetherington, A. (1938). Descending connections from the hypothalamus. *A.M.A. Arch. Neurol. Psychiat.* **39**, 1127.

Manning, J. W., and Peiss, C. N. (1960). Cardiovascular responses to electrical stimulation in the diencephalon. *Am. J. Physiol.* **198**, 366.

Monnier, M. (1939). Les centres végétatifs bulbaires. *Arch. intern. physiol.* **49**, 455.

Moruzzi, G. (1950). "Problems in Cerebellar Physiology." Charles C Thomas, Springfield, Ill.

Nicoll, P. A., and Webb, R. L. (1946). Blood circulation in the subcutaneous tissue of the living bat's wing. *Ann. N. Y. Acad. Sci.* **46**, 697.

Olszewski, J. (1958). quoted by Scheibel, M. E., and Scheibel, A. B. Structural substrates for integrative patterns in the brain stem reticular core. *In* "Reticular Formation of the Brain" (H. H. Jasper *et al.*, eds.), Chapter 2, p. 58. Little, Brown, Boston.

Owsjannikow, P. (1871). Die tonischen und reflektorischen Centren der Gefässnerven. *Ber. Verhandl. K. sächs Ges. Wiss.* **25**, 103.

Patton, H. D. (1948). Secretory innervation of the cat's footpad. *J. Neurophysiol.* **11**, 217.

Peiss, C. N. (1958). Cardiovascular responses to electrical stimulation of the brain stem. *J. Physiol.* (*London*) **141**, 500.

Peiss, C. N. (1960). Central control of sympathetic cardioacceleration in the cat. *J. Physiol.* (*London*) **151**, 225.

Peiss, C. N., and Manning, J. W. (1959). Excitability changes in vasomotor areas of the brain stem following *d*-tubocurarine. *Am. J. Physiol.* **197**, 149.

Pollock, L. J., Boshes, B., Chor, H., Finkelman, I., Arieff, A. J., and Brown, M. (1951). Defects in regulatory mechanisms of autonomic function in injuries to spinal cord. *J. Neurophysiol.* **14**, 85.

Pool, J. L., and Ransohoff, J. (1949). Autonomic effects on stimulating rostral portion of cingulate gyri in man. *J. Neurophysiol.* **12**, 385.

Randall, W. C., Alexander, W. F., Hertzman, A. B., Cox, J. W., and Henderson, W. P. (1950). Functional significance of residual sympathetic pathways following verified lumbar sympathectomy. *Am. J. Physiol.* **160**, 441.

Randall, W. C., Alexander, W. F., Cox, J. W., and Hertzman, A. B. (1953). Functional analysis of the vasomotor innervation of the dog's hind footpad. *Circulation Research* **1**, 16.

Randall, W. C., Cox, J. W., Alexander, W. F., Coldwater, K. B., and Hertzman, A. B. (1955). Direct examination of the sympathetic outflows in man. *J. Appl. Physiol.* **7**, 688.

Randall, W. C., McDonald, R., and Stalzer, R. (1958a). A functional study of sympathetic neuroeffector terminations. *Anat. Record* **130**, 39.

Randall, W. C., Pickett, W. J., Folk, F. A., and McNally, H. J. (1958b). The effective level of lumbar sympathectomy as determined by direct electrical stimulation at operation. *Ann. Surg.* **148**, 51.

Ranson, S. W., and Billingsley, P. R. (1916). Vasomotor reactions from stimulation of the floor of the fourth ventricle. Studies in vasomotor reflex arcs. III. *Am. J. Physiol.* **41,** 85.

Ranson, S. W., and Magoun, H. W. (1939). The hypothalamus. *Ergeb. Physiol. u. exptl. Pharmakol.* **41,** 56.

Rawson, R. O., and Randall, W. C. (1960). Cutaneous and muscle vascular responses to electrical stimulation of lumbar sympathetic trunk in the dog. *Am. J. Physiol.* **199,** 112.

Ray, B. S., and Console, A. D. (1948). Residual sympathetic pathways after paravertebral sympathectomy. *J. Neurosurg.* **5,** 23.

Ray, B. S., Hinsey, J. C., and Geohegan, W. A. (1943). Observations on the distribution of the sympathetic nerves to the pupil and upper extremity as determined by stimulation of the anterior roots in man. *Ann. Surg.* **118,** 647.

Redleaf, P. D., and Tobian, L. (1958). The question of vascular hyper-responsiveness in hypertension. *Circulation Research* **6,** 185.

Rushmer, R. F., Smith, O., and Franklin, D. (1959). Mechanisms of cardiac control in exercise. *Circulation Research* **7,** 602.

Sachs, E., Jr., Brendler, S. J., and Fulton, J. F. (1949). The orbital gyri. *Brain* **72,** 227.

Scheibel, M. E., and Scheibel, A. B. (1958). Structural substrates for integrative patterns in the brain stem reticular core. *In* "Reticular Formation of the Brain" (H. H. Jasper *et al.*, eds.), Chapter 2. Little, Brown, Boston.

Scott, J. M. D., and Roberts, F. (1924). Localization of the vaso-motor centre. *J. Physiol.* (*London*) **58,** 169.

Seckendorf, R., and Randall, W. C. (1958). Thermal reflex sweating in paraplegic man. *Federation Proc.* **17,** 144.

Sheehan, D., and Marrazzi, A. S. (1941). Sympathetic preganglionic outflow to limbs of monkeys. *J. Neurophysiol.* **4,** 68.

Sherrington, C. S. (1906). "The Integrative Action of the Nervous System." Yale Univ. Press, New Haven, Conn.

Simeone, F. A., and Felder, D. A. (1951). The supersensitivity of denervated digital blood vessels in man. *Surgery* **30,** 218.

Skoog, T. (1947). Ganglia in the communicating rami of the cervical sympathetic trunk. *Lancet* **ii,** 457.

Smith, W. K. (1945). The functional significance of the rostral cingular cortex as revealed by its responses to electrical excitation. *J. Neurophysiol.* **8,** 241.

Ström, G. (1950). Vasomotor responses to thermal and electrical stimulation of frontal lobe and hypothalamus. *Acta Physiol. Scand.* **20,** Suppl. 70, p. 83.

Thompson, W. C., and Bach, L. M. N. (1950). Some functional connections between hypothalamus and medulla. *J. Neurophysiol.* **13,** 455.

Uvnäs, B. (1954). Sympathetic vasodilator outflow. *Physiol. Revs.* **34,** 608.

Uvnäs, B. (1960a). Central cardiovascular control. *In* "Handbook of Physiology," Section 1: Neurophysiology (J. Field, H. W. Magoun, and V. E. Hall, eds.), Vol. II, Chapter 44. Am. Physiol. Soc., Washington, D. C.

Uvnäs, B. (1960b). Sympathetic vasodilator system and blood flow. *Physiol. Revs.* **40,** Suppl. 4, 69.

von Euler, U. S. (1956). "Noradrenaline." Charles C Thomas, Springfield, Ill.

Wall, P. D., and Davis, G. D. (1951). Three cerebral cortical systems affecting autonomic function. *J. Neurophysiol.* **14,** 507.

Waller, A. V. (1853). Neuvième mémoire sur le système nerveux: l'influence exercée par le nerf sympathique sur la circulation du sang. *Compt. rend. acad. sci.* **36**, 378.

Wang, S. C., and Ranson, S. W. (1939a). Autonomic responses to electrical stimulation of the lower brain stem. *J. Comp. Neurol.* **71**, 437.

Wang, S. C., and Ranson, S. W. (1939b). Descending pathways from the hypothalamus to the medulla and spinal cord. Observations on blood pressure and bladder responses. *J. Comp. Neurol.* **71**, 457.

Ward, A. A., Jr. (1948). The cingular gyrus: area 24. *J. Neurophysiol.* **11**, 13.

Webber, R. H. (1958). A contribution on the sympathetic nerves in the lumbar region, *Anat. Record* **130**, 581.

Wiedeman, M. P. (1957). Effect of venous flow on frequency of venous vasomotion in the bat wing. *Circulation Research* **5**, 641.

Wood, J. E., Litter, J., and Wilkins, R. W. (1955). The mechanisms of limb segment reactive hyperemia in man. *Circulation Research* **3**, 581.

Wrete, M. (1941). Die Entwicklung und Topographie der intermediären vegetativen Ganglien bei gewissen Versuchstieren. *Z. mikroskop.-anat. Forsch.* **49**, 503.

Wrete, M. (1943). Die intermediaren vegetativen Ganglien der Lumbalregion beim Menschen. *Z. mikroskop.-anat. Forsch.* **53**, 122.

Chapter V. MICROCIRCULATION

A. INTRODUCTION

By G. P. FULTON

The primary function of the heart, distributing arteries, and collecting veins is to conduct the blood to and from the regions of the microcirculation within the tissues and organs. The microcirculatory patterns provide the morphologic and functional basis for exchange of oxygen, nutrients, metabolites and humoral products between the extravascular and intravascular spaces. The small precapillary arterioles and postcapillary venules are permeable at times and therefore should be included, along with the capillaries, as components of the microcirculation.

The precise pattern or microarchitecture of the minute vascular bed is different in diverse organs and is related to the specific function of the structure. It is unlikely that the preferential shunt (Zweifach, 1937) or any special pattern may be regarded as typical, although the arcuate arrangement has been frequently identified (Webb and Nicoll, 1954). The amount of endothelial surface area may be important in relation to the function of the organ. Also the relative amount of endothelial surface in capillaries, as compared with venules, may be a significant measure in view of the greater susceptibility of the latter to intravascular thrombosis and petechial extravasation (Fulton, 1957). Although the rate and output of the heart may have definite effects on the microcirculation, intrinsic factors also have significant regulatory importance, effective mechanisms existing for local control of blood flow.

B. METHODOLOGY

By L. A. SAPIRSTEIN

Several areas of the microcirculation lend themselves to direct observation in living preparations. These include the microcirculation of the mesentery, the bat wing, the hamster cheek pouch, the brain and meninges, the rabbit ear, and the tongue and foot web of the frog. In man, the vessels of the nail bed, the bulbar conjunctiva, and the eye grounds may be observed directly. The techniques used in such studies have been described in the *First Conference on Microcirculatory Physiology and Pathology* (1954). It is necessary to point out, however, that in most sites the microcirculation is still not susceptible to such study. Furthermore, overenthusiastic extrapolation of findings in the observable beds to the microcirculation as a whole is hazardous. This is the more evident when it is recalled that the beds studied show major differences from each other, and that even the same bed may demonstrate variation in architecture from species to species.

1. TECHNIQUES

a. Electron microscopy: Major new additions to information concerning the microcirculation have resulted from electron microscopy. Bennet *et al.*, (1959) have presented a morphologic classification of capillaries based on differences revealed by this procedure. (For further results obtained with the electron microscope, see Section E.-1 of Chapter VI.)

b. Use of disappearance curves: Another powerful tool for the study of the microcirculation is the analysis of "disappearance curves" for substances injected into the blood stream. These may be a diffusible ion, a protein, a radioactively-labeled red cell, India ink, or glass or plastic microspheres. By appropriate choice of substance, the disappearance of the indicator from the blood and its subsequent localization can give information regarding transcapillary exchanges for small or large molecules, the existence of functioning arteriovenous anastomoses in areas not readily available for *in vivo* examination, patency of capillaries, and capillary permeability.

When small molecules are used as the indicator substance, the results may be misinterpreted through failure to recognize the existence of flow limitations on transcapillary transfer rates. It is quite obvious that no material can leave the circulation more rapidly than it is presented to the part of the circulation through which it will leave. Slow disappearance of a small molecule from the circulation may indicate impermeability of the capillary blood vessels to the molecule; but it can equally be a consequence of a diminished circulation which fails to put the material before these vessels (Sapirstein *et al.*, 1955). In the case of K^{42}, transcapillary exchange is so rapid in all areas other than the brain that the label is initially distributed in proportion to blood flow in every area (Sapirstein, 1958). Clearly, it would be erroneous to consider that unusually high accumulations of this label in a well-perfused organ represented excessive permeability of the capillaries in this structure. The same applies to other small molecules and ions, such as D_2O, thiocyanate, labeled sodium, etc.

The disappearance of larger molecules from the blood may, on the other hand, give useful information on capillary permeability. By combining data on blood disappearance and lymph appearance of dextran molecules of graded sizes (Wasserman *et al.*, 1955), it has been possible to estimate the size and distribution of capillary "pores." A study of the early disappearance of I^{131}-labeled-albumin and Cr^{51}-labeled-red cells in the dog (Vidt and Sapirstein, 1957) suggested the possibility that the capillary endothelium might not restrain the movement of the plasma albumins, and that the movement of all materials from the blood to tissue cells might be a function of the properties of the ground substance rather than the anatomically visible portions of the capillary bed.

The injection of India ink permits estimates of the number of capillaries patent at any one time (Krogh, 1929). The number of such capillaries in a muscle has been shown by this technique to vary with the activity of the muscle.

The recovery of microspheres in tissues after intra-arterial injection can supply valuable information regarding the size of vessels in a circulatory bed *in vivo*. The fraction of the injected spheres which remains in the bed corresponds to the fraction of the small vessels which are smaller than the spheres. There is a possible error due to deformation of the vessels by the spheres. Microspheres have also been used in a study of the distribution of arteriovenous anastomoses in the small intestine (Grim and Lindseth, 1958).

c. Perfusion experiments: Important information regarding the properties of vascular endothelium may also be obtained from perfusion experiments in which the hydrostatic and colloid osmotic pressures of the blood are subject to control. From such experiments (Pappenheimer, 1953), it has been possible to make deductions concerning the ultramicrostructure of the capillary wall. The size of the pores has been estimated, as has also the total area occupied by them. These studies have demonstrated the existence of pores in the capillaries of the tissues of the perfused leg which have not yet been revealed by electron microscopy, though their calculated size is well within the resolving power of the instrument (Bennet *et al.*, 1959).

C. STRUCTURAL BASIS OF THE MICROCIRCULATION

By B. R. LUTZ AND G. P. FULTON

1. VASCULAR PATTERNS

a. The pattern in the mesentery: A double set of small vessels is found in the mesentery of the frog (*Rana pipiens*), in the region of the gut and at the apical region where the mesenteric arteries branch and the veins merge. Away from the gut the blood vessels merge into a single layer within the connective tissue of the so-called "mesentery," commonly used for physiologic experiments. This region receives an arterial supply both from the radiating mesenteric arteries outward and from the arteries on the mesenteric side of the gut inward. While this arterial supply consists of mere twigs and is frequently difficult to demonstrate, the origin of the blood and the direction of flow indicate the source. The rate of flow is greater in the small arterial twigs than in other small vessels nearby, and only with difficulty can one determine the proportion of oxygenated blood delivered to this region, an area definitely served by a venous net (Fig. 38) as ascertained by the origin of the vessels. The rather small arterial twigs assure "mixed blood" in the mesentery.

b. Precapillary sphincters: At the point of origin of the capillary there is frequently a group of muscle cells which, according to Zweifach (1937), "acts as a proximal muscular sphincter and regulates the blood flow through the capillary branches of the vessel." This precapillary sphincter has been confirmed in the retrolingual membrane by Fulton and Lutz (1940), and similar structures have often been observed by others in the urinary bladder of the frog. They have also been found by Clark and Clark (1943) in the rabbit, and by Zweifach *et al.* (1944)

in the rat. In the cheek pouch of the hamster (*Mesocricetus auratus*) these structures have been cinephotographed and reported by Lutz and Fulton (1958). Zweifach (1937) has further stated, "where capillaries drain into muscular vessels occasional muscle cells occur at the point of junction." Prevenular sphincters have been described definitely by Knisely (1939) for the liver of the frog. Because sphincters rarely act in unison, and because there are capillary anastomoses with other arteriolar systems, one hardly expects sphincters to be very effective in

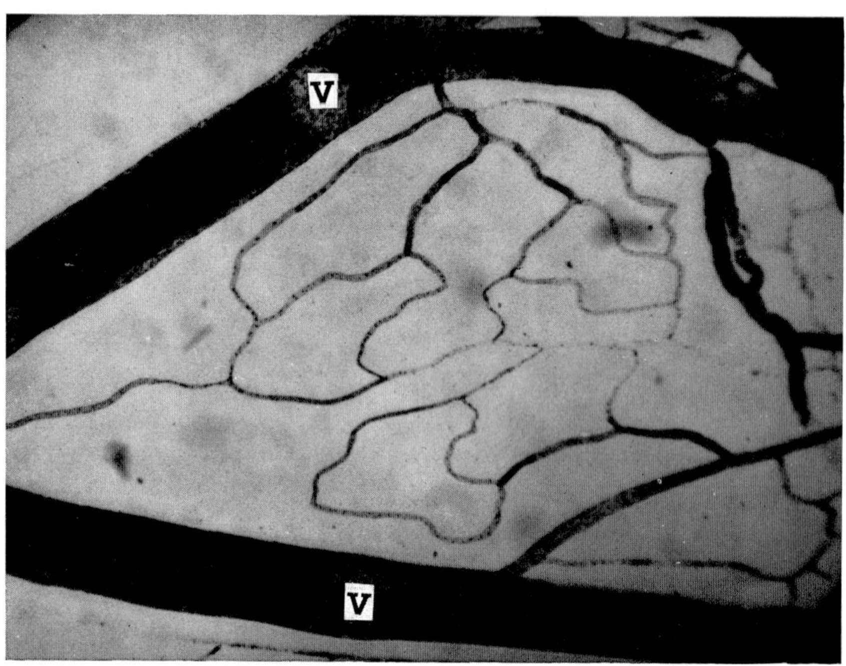

FIG. 38. Venous net in the mesenteric area between the large veins (V).

regulating blood flow. Furthermore, it is doubtful that they are a part of any basic vascular pattern. Krogh (1929) proposed the concept of independent capillary contractility accomplished by means of contractile pericapillary Rouget cells, thus providing a mechanism for local regulation of blood flow. However, Zweifach (1937) has found no muscle cells on capillaries, and since Krogh offered no objective evidence for capillary contraction, he probably confused an arteriolar net with a capillary net. An arteriolar network was reported by Nicoll and Webb (1946) for the bat's wing and by Lutz and Ward (1960) for the stomach of the frog and the jejunum of the hamster. In all three sites this structure is a part of the "arcuate system."

c. "Thoroughfare channel": Zweifach (1937) has conceived the peripheral vascular bed as a composite of structural and functional units, each having a capillary-like channel, the arteriovenous, preferential or "thoroughfare channel," from which the capillaries are side branches. The concept of the "thoroughfare channel" has been extended by Zweifach (1939, 1940) and Zweifach and Kossman (1947) to the mesentery of the mouse, the mesoappendix of the rat, and the omentum of the dog. Chambers and Zweifach (1944), after extensive work on the frog's mesentery, the rat's mesoappendix, and the dog's omentum, have confirmed the earlier work of Zweifach (1937). Others (Lee and Holze, 1950), have considered that the human conjunctiva has a similar pattern. Zweifach (1937) has written, "In certain organs the preferential channel lost its anatomical identity just proximal to the venous side of the bed." He also stated (1937) that, "In other structures, such as the urinary bladder and the skin, the thoroughfare type of channel was not a major feature." Although Zweifach has implied that the "thoroughfare channel" as described for mesenteric microcirculation is typical, this idea has not been accepted for the bat's wing (*Myotis lucifugus*) by Nicoll and Webb (1946) and by Webb and Nicoll (1954), nor for the human conjunctiva by Grafflin and Bagley (1953) and Grafflin and Corddry (1953).

d. The pattern in the nictitating membrane: Zweifach (1937), ascribed to the nictitating membrane of the frog the same structural units which he found in the mesentery. However, functionally, this structure has little in common with the mesentery. Its arterial supply is mainly from the outer upper borders, and the venous outflow is frequently toward the ventral side by a single large vein. Sometimes a small artery may enter ventrally on each side of the large vein. Nothing resembling a "thoroughfare channel" can be seen. Branching appears haphazard, and the pattern varies from eye to eye. Occasionally, as in many "low blood pressure" animals, a long vessel will pass to the venous side without branching. At the pigmented upper border the capillaries are short and wide like those of the lungs, while in the center large meshes lead to the large vein.

e. Retrolingual membrane: In the retrolingual membrane, blood vessels, skeletal muscle fibers and connective tissue form the thin wall of a lymph sac. Two to four arterioles enter at the sides from under skeletal muscle and divide into anastomosing capillaries which converge laterally into venules. Occasionally an arteriovenous anastomosis is seen but no "thoroughfares," as reported by Zweifach (1937). Precapillary sphincters are commonly followed by a considerable length of endothelial tube, with frequent perivascular cells, most of which are inert (Lutz *et al.*, 1950a).

f. Tissues with an "arcuate system": An "arcuate system" at the micro-level, designated as such in the bat's wing by Webb and Nicoll (1954), with anastomosing arterioles offers collateral pathways and insures a uniform pressure and flow within the capillaries supplied by terminal arterioles. A similar arrangement, on two levels, has been found in the stomach of the frog and the jejunum of the hamster (Lutz and Ward, 1960). In the latter, one or many arterioles or capillaries arise from the arteriolar net to each papillus, then form a capillary net, which converges into a single large vein passing down to an underlying venular net. The skin of the frog (Fig. 39) offers an excellent example of a two-level arcuate system and illustrates the common condition that the venular net is more profuse and obviously of greater capacity than the arteriolar net.

Because Lee and Holze (1951) had accepted the idea of "thoroughfare channels" without reservation and explained their findings in the bulbar conjunctive on the basis of this concept, Grafflin and Corddry (1953) repeated the work. However, they failed to find any recognizable structural unit among an endless variety of vascular branching. Numerous arteriovenous anastomoses were characteristic and there was an arterial, as well as a venous, net. No sphincters were reported, but intermittence was common without complete cessation of flow.

g. Cheek pouch of hamster: The cheek pouch of the hamster (Fulton *et al.*, 1946), offers a thin transparent membrane covered with stratified squamous epithelium but without glands or hairs, which is more vascular than mesenteric preparations or transparent chambers.[1] The vascular pattern shows no "thoroughfare channels" or arteriovenous anastomoses. It is an arcuate system on one level. There are many arterio-arterial anastomoses and a very profuse venous network. The former are able to supply the entire pouch when any given artery, normally serving as a source of blood flow, is tied off (Poor and Lutz, 1958). Branching of arterioles into capillaries is terminal, and convergence into venules results. Venular junctions are very susceptible to hemorrhaging and thrombosis.

h. Labile patterns in growing tissues: The "round chamber" technique devised by Sandison (1924) was modified by Clark *et al.* (1930) so that preformed blood vessels could be examined immediately. Earlier, Clark and Clark (1925a) had reported that the endothelium of the blood vessel wall of the amphibian larva was contractile, without the need for implicating the "Rouget" cells for this purpose. The latter structures had been demonstrated by Vimtrup (1922) in the nictitating membrane of the adult frog. Clark and Clark (1943) found that the endothelium

[1] See Fig. 45, Chapter VI, for photograph of the cheek pouch of the hamster.

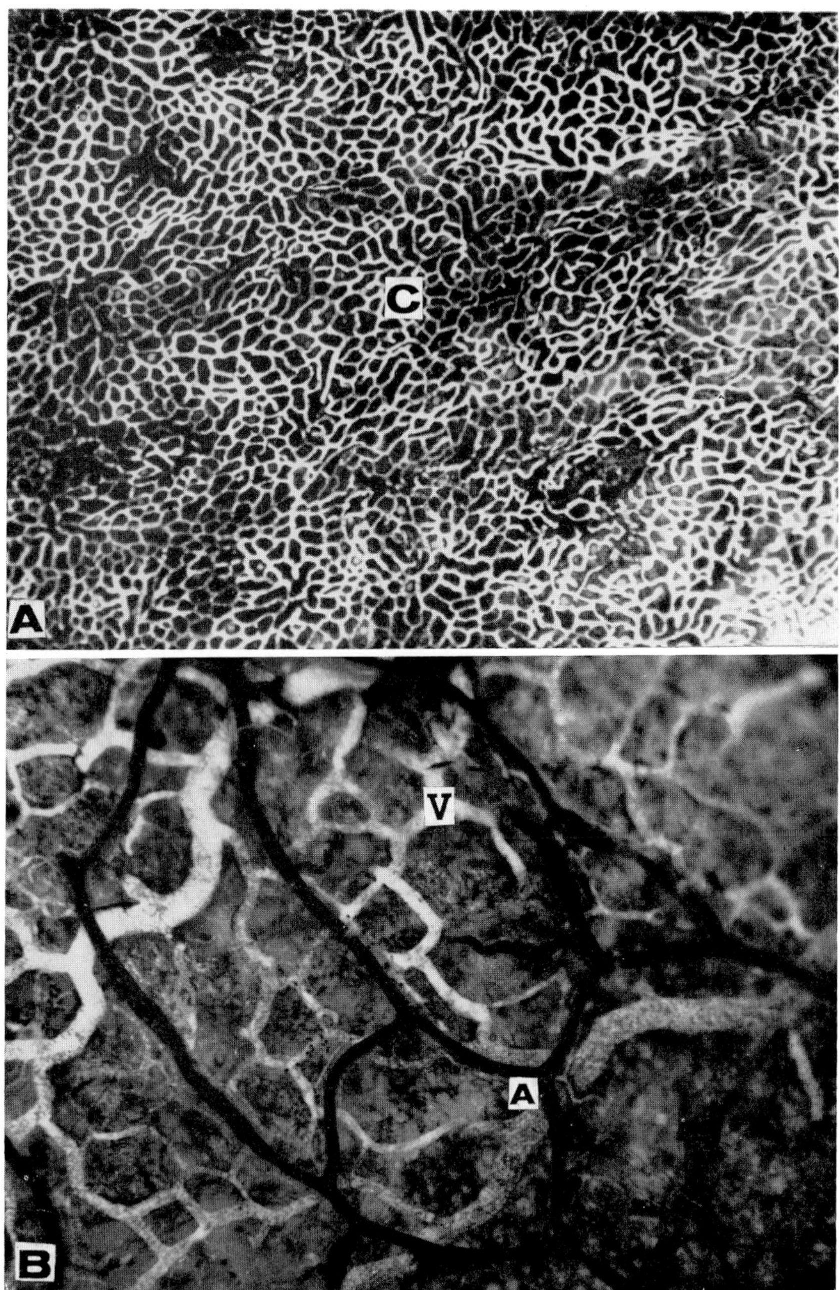

FIG. 39. An arcuate system of the two-level type as observed in the skin of the frog. A. Surface capillary net (C). B. Under side of the skin. Dark vessels (A) make up an arteriolar net sending branches to the surface net. Light vessels (V) form a venular net receiving branches from the surface.

of the mammalian capillary did not contract, but instead all constriction of small vessels occurred in regions provided with smooth muscle. The perivascular cells on capillaries were inert to mechanical and electrical stimulation, but if these small vessels were subjected to increased pressure for at least 48 hours, they became arterioles and developed many arteriovenous anastomoses. This "extreme lability" occurred on slight alteration of the environment. Although Clark and Clark (1932) demonstrated "lability" under artificial conditions and in regenerating blood vessels, the ability to grow and change appears to be a normal anatomic characteristic of all blood vessels. For example, a tumor transplant is revascularized in the cheek pouch within four days (Lutz *et al.*, 1951).

2. Morphology of Vascular Smooth Muscle

The vascular smooth muscle cells are arranged in circular or spiral alignment and comprise a single layer at the level of the microcirculation. The distribution is diffuse on the precapillary arterioles where the typical spindle-like shape may be lost. No contractile cells are found on the true capillaries in mammals. However, a few smooth muscle cells are frequently grouped at points of origin of small precapillary arterioles from larger vessels where they may function as a vascular sphincter.

The possibility of a vascular smooth muscle syncytium is suggested by certain physiologic responses and by the general occurrence of protoplasmic bridges in tissue cultures and in embryogenesis. Although the smooth muscle of the uterus has been described as a protoplasmic entity without nerves, in the case of the mouse, evidence from electron microscopy is against a syncytium in the smooth muscle of this organ, as well as the urinary bladder and gall bladder (Caesar *et al.*, 1957). However, some investigators still consider smooth muscle to be syncytial. For example, unipolar stimulation of the vascular smooth muscle in the frog's retrolingual membrane produces a segmental conducted response which persists after blocking the nerve fibers with cocaine, thus suggesting conduction by smooth muscle cells (Fulton and Lutz, 1941b). Nevertheless, conduction by protoplasmic contact is an alternate explanation to that of syncytial arrangement.

3. Connective Tissue Components

Volterra (1925) has stressed the importance of a pericapillary connective tissue sheath, and Fulton *et al.* (1953) have observed the temporary containment of petechial hemorrhaging and even frank bleeding within a sheath-like structure paralleling the vascular wall. The existence of reticular or argyrophil fibers, as demonstrated by silver impregnation methods, has been confirmed by electron microscopy (Fawcett, 1959).

Tissue mast cells are particularly numerous on the vessel walls in the microcirculatory bed. Their significance remains to be determined, but the possibilities are intriguing in view of the evidence for humoral and hormonal properties of such structures (Fulton *et al.*, 1957).

4. Innervation of Small Blood Vessels

a. Origin of innervation: In the retrolingual membrane of the frog and the cheek pouch of the hamster, the vasomotor nerves are derived from entering nerve bundles and form a loosely-meshed primary plexus of nonmyelinated fibers made visible by differential staining with methylene blue (Lutz *et al.*, 1950a). As the fibers approach the arterioles, they form a long-meshed secondary plexus paralleling the vessels. The secondary plexus is continuous with a fine beaded perivascular network of delicate strands which pass beyond the range of vision with white light. Woollard (1926) described nerve endings on the surface of vascular smooth muscle cells, but Hill (1927) reported intracellular terminations. With electron microscopy, Caesar *et al.* (1957) described endings in minute pockets within the sarcolemma. However, irrespective of the cytologic detail of the myoneural junction, the perivascular plexus is sufficiently abundant to innervate all the vascular smooth muscle and the vascular sphincters.

b. Tertiary perivascular plexus: This plexus seems to resemble the "ground plexus" of Boeke (1940) and the "plasmodial strands" of Stöhr (1938), as demonstrated in reduced silver impregnation methods. It does not appear to be comparable with Boeke's "periterminal network" which is intracellular, or Stöhr's "terminal reticulum" which extends throughout the tissues. Since methylene blue does not usually stain these structures, the "terminal reticulum" and the "periterminal network" may be non-nervous. Supporting evidence for this was published by Nonidez (1936), who was unsuccessful in demonstrating the "terminal reticulum" with the silver nitrate method of Cajal. However, he succeeded with the Rio-Hortega silver carbonate method, which is regarded as specific for connective tissue. The Bielschowsky procedure, used by Boeke and Stöhr, combines the features of the Cajal and Rio-Hortega techniques, and consequently may favor impregnation of nerve fibers and connective tissue strands as well.

In methylene blue preparations, the perivascular plexus becomes sparse on the capillaries, which are without smooth muscle cells and lack contractile properties (Lutz *et al.*, 1950a). Zweifach (1937) reported "Myofibrils" within endothelial cells of capillaries in the frog and also nerve "fibers" which were embedded in the endothelial cells. Fawcett (1959) described oriented filaments in electron micrographs of endothelium, suggesting inherent contractile power. Although it is generally conceded that capillaries are not contractile in mammals, Sanders *et al.* (1940) reported that, on stimulation of the cervical sympathetic nerve, an occlusion of the capillary lumen took place by swelling of endothelial cell nuclei, as determined within the rabbit ear chamber.

c. Functional discontinuity of perivascular plexus: Although the perivascular plexuses have been described as anatomically continuous (Woollard, 1926), discrete segmental vasomotor responses have been obtained from stimulation of various portions of the nerve network (Lutz *et al.*, 1950a). Presumably, the plasmodial strands serve as cables to conduct the axons of vasomotor neurons which innervate the vascular smooth muscle. In this respect, Leontowitsch (1930) distinguished two nerve networks in the frog's palate and speculated that one might be vasodilator and the other vasoconstrictor.

d. Denervation: The effect of denervation on the perivascular nerve plexus is not entirely clear. Boeke (1940) reported persistence of the plexus in the tongue after section of the lingual nerves. He directed attention to the large "interstitial" sheath cells which might serve a nutritional purpose, although it is doubtful that these are comparable with ganglion cells. Joftes (1951) has attempted to eliminate the perivascular nerve plexus in the frog's retrolingual membrane by denervation of the tongue, but the results have been contradictory.

e. Role of perivascular nerve plexuses: The rich perivascular nerve plexuses suggest an important role of the innervation in control of the circulation through small blood vessels. Nevertheless, how significant a factor it may be in comparison with humoral factors is not known. In specific instances, the microcirculation has been found not to participate in compensatory vascular reflexes (Yudkofsky, 1960). Furthermore, it is not known how much, if any, of the perivascular plexus is sensory. It is entirely possible that sensory endings in blood vessels may be more extensive than those described in the carotid sinus and homologous vascular structures.

5. Summary

The foregoing brief presentation implies that patterns in the microcirculation vary among animals and in various tissues in the same animal.

Arteriolar components frequently form a net before losing their musculature to become the endothelial tubes making up the capillary net. Similarly, veins and venules commonly form a plexus, often more profuse than the arterial net (Berman, 1953). Arteriovenous anastomoses and sphincters are found in certain tissues, there generally being an obvious close correlation with function. An arcuate system or some modification of it is common to many tissues. Thus far, however, no typical vascular pattern at the microlevel has been reported. Little is definitely known of the afferent innervation and less about the part played by the central nervous system in peripheral vascular reflexes. (For an excellent review of the motor aspects at the level of the nerve supply to smooth muscle, see Green and Kepchar, 1959.)

D. PHYSIOLOGY[2]

By G. P. FULTON

1. CHARACTERISTICS OF NORMAL FLOW

By the use of *in vivo* microscopy, micromanipulation, and documentation by cinephotomicrography, the characteristics of blood flow and the properties of the formed elements in the microcirculation have been described in transillumined membranes of laboratory animals and in opaque tissues of animals and man examined with reflected light. The fundamental features of normal flow, as summarized by Knisely *et al.*, (1947) are as follows:

The red cells do not circulate in aggregates and fail to adhere to each other under normal conditions.

The endothelial lining of normal small blood vessels is generally free of adhering red cells, platelets and white cells. The formed elements are confined intravascularly and petechial hemorrhages are absent.

In the small arteries and veins, the blood cells are circulated in an axial stratum, with laminar or streamlined flow, and within a peripheral layer of plasma which may contain platelets and a few leukocytes. However, laminar flow characteristics are less evident in the microcirculation,

[2] For detailed discussion of the microcirculation of the pituitary, see (1) The Pituitary, Section E., Chapter XIV; of the thyroid, see (2) The Thyroid, Section C., Chapter XIV.

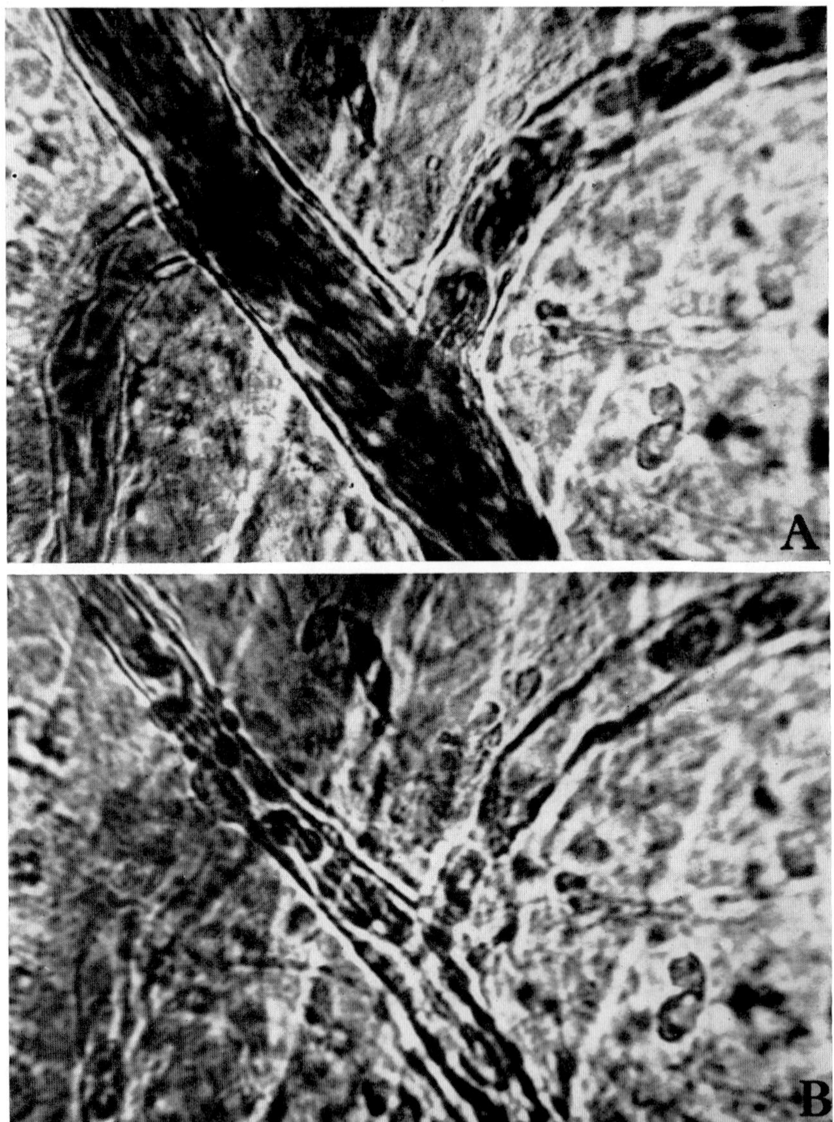

FIG. 40. Enlargements from cinephotomicrographic record of spontaneous vasomotor activity in the retrolingual membrane of the frog. A. Vascular sphincter relaxed. B. Contraction of vascular sphincter. Original microscopic magnification, 220×.

and, even in small arteries, high speed cinephotomicrography has revealed turbulence, random orientation of rotating cells, and absence of lamination (Bloch, 1960). (For further discussion of types of flow, see Section D.-1 and 2, Chapter II.)

The formed elements normally circulate freely and rapidly in "nonleaky" vessels, as contrasted with the apparent hemoconcentration and slow flow within vessels in areas of tissue injury. Stasis is lacking or minimal.

Using the method of Hugues (1953), the range of rate of flow in normal arterioles of the hamster cheek pouch has been found to be from 0.5 to 12.8 mm/sec, with an average of 4.6 mm/sec; in venules it is from 0.4 to 6.5 mm/sec, with an average of 2.6 mm/sec (Slechta, 1960).

The progressive narrowing, which is characteristic of the bifurcating "arterial tips" in the microcirculation, has been regarded as a conical bottleneck with important physiologic implications (Jeffords and Knisely, 1956). In small arterioles possessing normal tone, the lumen is narrow and individual formed elements may be distorted during passage through the vessel.

Vasomotion may occur in arteriolar segments of the microcirculation, spontaneous intermittency is common in capillaries, and vascular sphincters may operate to regulate the flow of blood in limited areas (Lutz and Fulton, 1958) (Fig. 40).

2. Variations in Flow and Differential Distribution

a. Spontaneous intermittency of flow: This type of response is a characteristic phenomenon within small arterioles and capillaries of the microcirculation in most of the preparations which have been examined for such activity. The cessation of flow may last for only a fraction of a second or for several minutes. Sometimes the intermittency is merely a variation in rate of flow without actual stoppage, but at times the direction of flow in the capillaries may reverse suddenly. Intermittency of flow may be due to the activity of vascular sphincters, changes in blood pressure differential in various areas of the microcirculatory bed, temporary lodging of intravascular aggregations of formed elements at vascular junctions, slow rhythmic changes in caliber of the walls of small arterioles (vasomotion), or perhaps local changes in permeability which may induce stasis.

b. Vasomotion: Spontaneous rhythmic contractility is characteristic of segments of the arterioles. The term "vasomotion" has been used to describe this activity, especially the slow responses requiring time-lapse photography for detection, as in the mesoappendix of the rat (Chambers and Zweibach, 1946). The rhythmic response may also be greater, as shown by the beating action of the vascular sphincters in the retrolingual membrane of the frog (Fulton and Lutz, 1940). This phenomenon is not necessarily abolished by section of the supplying nerve trunks, but stimulation of the components of the perivascular nerve plexus with a microelectrode may produce vasomotor responses which alter the rhythm of the spontaneous activity. The precise cause of rhythmic contractility is unknown at present.

c. Variations in number of active capillaries: In skeletal muscle the number of capillaries containing circulating blood during activity may be 200 times greater than that during rest (Krogh, 1929). In addition, an increase may occur in capillary vascularity (capillary counts in cross section of muscle after intravascular perfusion with India ink) as an adaptation to functional demand, demonstrated in the muscles of guinea pigs native to the Peruvian mountains, in comparison with sea level controls (Valdivia, 1958). Further evidence for this concept is the dynamic lability of vascular endothelium in response to injury or to injection of foreign protein subcutaneously, as shown by the extensive vascularization of tumor transplants (Lutz *et al.*, 1950b). Since the distribution of blood within the microcirculatory bed may be changed appreciably, as in muscle contraction or in hemorrhagic shock (Zweifach and Metz, 1955), the mechanisms which regulate the flow at the local level are most important.

3. Effector Mechanisms for Adjusting the Microcirculation

The following four mechanisms have been proposed as the devices which operate to adjust the blood flow in the individual capillaries:

a. Swelling of the endothelial cells with protrusion of the nuclei into the lumen to block blood flow: Although endothelial swelling was first described many years ago by Stricker (1865), this phenomenon is not usually mentioned as a mechanism involved in the regulation of blood flow. However, Sanders *et al.* (1940) have been impressed by the frequency and rapidity of such a response, which may be obtained in the capillaries in the rabbit's ear following stimulation of the cervical sympathetic nerve. The exact nature of the mechanism remains unexplained, and its occurrence appears to be limited to the microcirculation in the rabbit's ear.

b. Contraction of capillary endothelium: Although capillary endothelial cells possess irritability and may contract under special conditions (Zweifach, 1934) or in certain species (Federighi, 1928; Clark and Clark, 1925b), endothelial contractility is considered to be negligible and of little significance in the control of blood flow in the microcirculation of the mammal. This phenomenon may be more important in lower vertebrates, and the property of endothelial contractility may be an evolutionary attribute which has been lost or replaced by smooth muscle contractility in mammals (Clark and Clark, 1943).

c. Question of contraction of pericapillary Rouget cells: Krogh (1929) is largely responsible for popularizing the concept that pericapillary cells, such as those described by Rouget (1873), contract to regulate the flow of blood in individual capillaries. Fulton and Lutz (1940) and Chambers and Zweifach (1942) reported that active contraction is confined to vessels with muscular elements on their walls, and Clark and Clark (1943) stated, "that the control of the peripheral circulation in the mammal resides in the vessels which have muscular walls and intact nerve supplies."

d. Contraction of vascular sphincters: The sphincter-like contractility of small blood vessels at their point of origin from a supplying arteriole has been reported by Golubew, Tarchanoff, Tannenberg, and others (for reference to these articles and for a review of literature, see Lutz *et al.*, 1950a). Cinephotomicrographic records have been made showing the spontaneous beating of the smooth muscle cells which comprise the vascular sphincters in the retrolingual membrane of the frog and the cheek pouch of the golden hamster (Lutz and Fulton, 1958). Vascular sphincters may occur at the point of origin of capillaries from precapillary arterioles, as in the frog's retrolingual membrane, or they may be confined to the points of branching of precapillary arterioles, as in the hamster cheek pouch, thus controlling blood flow in several capillaries or in an entire capillary network. The vascular sphincter, as influenced by neural and/or humoral factors, must be regarded as a device of considerable significance for the regulation of the microcirculation.

4. Factors Controlling the Microcirculation

a. Vasomotor innervation: The possible importance of vasomotor innervation as a factor in the control of the microcirculation is frequently overlooked in favor of the concept of local humoral factors, which are considered to act directly on the vascular smooth muscle (Akers and

Zweifach, 1955). However, an abundant perivascular nerve plexus has been demonstrated in various tissues, and it seems reasonable to implicate the innervation in the mediation of vascular responses at the microcirculatory level. Nevertheless, most of the published findings on the function of vasomotor and vasosensory nerves pertain to regulation of components of the macrocirculation rather than the microcirculation *per se*. Even so, much of the evidence concerning the existence of vasomotor innervation and the nature of such is indirect and somewhat inferential, based upon the interpretation of plethysmography, the results of perfusion studies, and pharmacologic experimentation. The occurrence of exclusively vasoconstrictor or vasodilator fibers, or a mixture of both, may depend upon the region under study or perhaps the species being utilized.[3]

Three types of receptor are hypothesized to explain the autonomic mediation in vascular smooth muscle: first, alpha constrictor receptors, which are sympathetically innervated and respond to noradrenalin and related pressor substances; second, beta dilator receptors, which are not innervated but respond to adrenergic depressants; and third, gamma dilator receptors, which are parasympathetically innervated and respond to acetylcholine and related cholinergic materials. The degree and nature of vasomotor response induced by injection of autonomic drugs and by autonomic nerve stimulation are thus explained on the basis of the distribution and relative proportion of the three kinds of smooth muscle receptors (Green and Kepchar, 1959).

Using microelectrodes, Fulton and Lutz (1940) stimulated the components of the perivascular nerve plexus in the retrolingual membrane of the frog and recorded the vasomotor responses of the microcirculation by means of cinephotomicrography. A diphasic response, dilation followed by constriction, was usually produced, involving a limited portion of the microcirculation. Although the perivascular nerve plexus appeared to be continuous anatomically, it was discontinuous physiologically. The region which constricted was frequently only a part of that which dilated originally. Furthermore, some fibers gave dilation only and a few, constriction only. For such reasons, a dual innervation was inferred. Physiologic evidence has been presented for the arrangement of vascular smooth muscle in the form of syncytial motor-units (Fulton and Lutz, 1941a).

[3] The physiology of the vasomotor nerves of the macrocirculation has been ably reviewed as follows: first, sympathetic vasodilators by Uvnäs (1954); second, nervous control of blood vessels by Folkow (1955); third, regulation of cutaneous circulation by Hertzman (1959); and fourth, control of peripheral resistance by Green and Kepchar (1959); also see Chapter IV.

Although the assumption has sometimes been made that the small blood vessels respond comparably with the larger ones during vasomotor reactions, conclusive observational evidence is lacking for the most part. In fact, Yudkofsky (1960) could not demonstrate compensatory changes in vessel diameters in the hamster cheek pouch during the expression of typical reflex cardiac and blood pressure responses to centrifugation, tilting, or following carotid occlusion. Furthermore, the microcirculation did not appear to participate in the usual cardiovascular compensatory reflexes involving the carotid sinus and aortic arch baroreceptors. Reflex changes seem to be lacking in the cheek pouch microcirculation in response to cold (Lynch and Adolph, 1957), although other microcirculatory beds might be more responsive.

b. Direct humoral control: Humoral factors are usually considered dominant in the adjustment of the microcirculation, although the possibility of modification by nervous impulses is generally recognized (Burn, 1950). In this respect, the concentration of circulating vasoactive substances, such as epinephrine in stress, may be highly significant. Metabolic products, such as adenylic acid, adenosine triphosphate, lactic acid, carbon dioxide and others, probably exert a vasodilator effect, at least in skeletal muscle. Another important point is the strategic proximity of large numbers of mast cells to small blood vessels, in view of the known vascular effects of substances found in these cells, such as heparin, histamine, hyaluronic acid and serotonin. Although a mechanism for regulation of humoral production or release by mast cells has not been demonstrated, the discharge of granules from these structures appears to be a characteristic response to stress. Histamine or histamine-like substances have long been implicated as vasoactive products of tissue injury, with particular importance in inflammation. According to the literature, histamine "dilates the capillaries," but this statement might be amended to read, with greater accuracy, "dilates the small arterioles and thereby increases blood flow in the capillaries" (Fulton *et al.*, 1959). In addition, statements that histamine "opens" capillaries which were previously "closed" might be better worded as "initiates circulation" in capillaries which had "no circulation" prior to treatment with this drug. The implication in the terms "open" and "closed" that capillaries either constrict actively or collapse passively is inconsistent with the behavior of the microcirculation as observed and photographed through the microscope at high magnifications.

Evidence for the production of local hormones, especially acetylcholine, in the rabbit auricles and possibly in the walls of blood vessels has been discussed by Burn (1950). The intramural synthesis of acetylcholine is said to be responsible for the maintenance of arterial tone. The vasodilator or "inhibitory" effects observed in the vessels of the rabbit's ear during perfusion with acetylcholine is explained on the basis of a possible depression of synthesis by an excess of acetylcholine. The vasoconstrictor "motor" or "stimulating" effect of acetylcholine, described after prolonged perfusion, is more difficult to understand, but may be "due to the synthesis being at that stage suboptimal." The interrelation of acetylcholine, noradrenalin, adrenalin in some tissues, and histamine is stressed as providing a delicate balance between stimulation and inhibition.

According to this concept, as proposed by Burn, only a single receptor is necessary to explain vasomotor responses. Using micropipettes, Lutz *et al.* (1950a) have observed the effects of minute amounts of acetylcholine and also epinephrine on individual smooth muscle cells comprising vascular sphincters in the retrolingual membrane of the frog. The identical cells dilated in response to acetylcholine and constricted with epinephrine. Furthermore, dilation could be obtained with dilute solutions of epinephrine and contraction of the same smooth muscle cells, with strong solution. These direct microscopic observations appear to be more in agreement with the unireceptor concept of Burn than the multireceptor concept of Green and Kepchar.

c. Indirect humoral control: Although humoral substances undoubtedly exert an important direct effect on vascular smooth muscle, the probable significance of indirect action through the rich perivascular plexus should be considered. For example, esters of nicotinic acid, essential oils and histamine solutions have been applied directly on the walls of small blood vessels in the hamster cheek pouch by means of micropipettes, and the responses have been predominantly vasodilation extending beyond the point of contact (Fulton *et al.*, 1959). In each instance the response involved an appreciably greater portion of the arteriolar supply than that directly exposed to the rubefacient. In addition, all portions of the reacting vascular segment appeared to dilate simultaneously. Treatment of the preparation with xylocaine abolished the conducted response in at least 50 per cent of the experiments. Therefore, it was concluded that the rubefacient action is mediated by a conducting mechanism with nerve-like properties. Sharply localized vasoconstriction was produced occasionally at the point of contact, especially with mustard oil, although dilation was usually also noted in an extensive segment beyond the site of constriction. Consequently, the direct action of the rubefacients on vascular smooth muscle appeared to elicit contraction.

d. Summary: The significance of direct humoral action on vascular smooth muscle, as compared with the role of the perivascular innervation, remains to be elucidated. Although conclusive evidence is lacking for significant control of the microcirculation by means of the central nervous system, the abundant perivascular plexus is presumably available for nerve regulation at the local level. Further work is necessary to ascertain the proportional degree of sensory, vasodilator and vasoconstrictor functions which may be mediated in the various microcirculatory patterns of specific tissues and organs of the same and different species.

References

Akers, R. P., and Zweifach, B. W. (1955). Effect of vasoactive materials on the peripheral blood vessels in the hamster. *Am. J. Physiol.* **183,** 529.

Bennet, H. S., Luft, J. H., and Hampton, J. C. (1959). Morphological classification of vertebrate capillaries. *Am. J. Physiol.* **196,** 318.

Berman, H. J. (1953). Observations on the blood pressure of the golden hamster (Mesocricetus auratus). Ph.D. Thesis, Boston University, Boston, Mass.

Bloch, E. H. (1960). High speed cinephotomicrography of hemodynamics. *Federation Proc.* **19,** 89.

Boeke, J. (1940). "Problems of Nervous Anatomy." Oxford Univ. Press, London and New York.

Burn, J. H. (1950). Relation of motor and inhibitor effects of local hormones. *Physiol. Revs.* **30,** 177.

Caesar, R., Edwards, G. A., and Ruska, H. (1957). Architecture and nerve supply of mammalian smooth muscle tissue. *J. Biophys. Biochem. Cytol.* **3,** 867.

Chambers, R., and Zweifach, B. W. (1942). Caliber changes of the vessels of the capillary bed. *Federation Proc.* **1,** 14.

Chambers, R., and Zweifach, B. W. (1944). Topography and function of the mesenteric capillary circulation. *Am. J. Anat.* **75,** 173.

Chambers, R., and Zweifach, B. W. (1946). Functional activity of the blood capillary bed, with special reference to visceral tissue. *Ann. N. Y. Acad. Sci.* **46,** 679.

Clark, E. R., and Clark, E. L. (1925a). The development of adventitial (Rouget) cells on the blood-capillaries of amphibian larvae. *Am. J. Anat.* **35,** 239.

Clark, E. R., and Clark, E. L. (1925b). B. The relation of "Rouget" cells to capillary contractility. *Am. J. Anat.* **35,** 265.

Clark, E. R., and Clark, E. L. (1932). Observations on living preformed blood vessels as seen in a transparent chamber inserted in the rabbit's ear. *Am. J. Anat.* **49,** 441.

Clark, E. R., and Clark, E. L. (1943). Caliber changes in minute blood vessels observed in the living mammal. *Am. J. Anat.* **73,** 215.

Clark, E. R., Kirby-Smith, H. T., Rex, R. O., and Williams, R. G. (1930). Recent modifications in the method of studying living cells and tissues in transparent chambers inserted in the rabbit's ear. *Anat. Record* **47,** 187.

Fawcett, D. W. (1959). The fine structure of capillaries, arterioles and small arteries. *In* "Microcirculation: Factors Influencing Exchanges of Substances across Capillary Wall" (S. R. M. Reynolds and B. W. Zweifach, eds.), p. 1. Univ. of Illinois Press, Urbana, Ill.

Federighi, H. (1928). The blood vessels of annelids. *J. Exptl. Zool.* **50,** 257.

First Conference on Microcirculatory Physiology and Pathology (1954). *Anat. Record* **120,** 239.

Folkow, B. (1955). Nervous control of the blood vessels. *Physiol. Revs.* **35,** 629.

Fulton, G. P. (1957). Microcirculatory terminology (editorial). *Angiology* **8,** 102.

Fulton, G. P., and Lutz, B. R. (1940). The neuromotor mechanism of small blood vessels of the frog. *Science* **92,** 223.

Fulton, G. P., and Lutz, B. R. (1941a). The control of small blood vessels. *Am. J. Physiol.* **133,** 284.

Fulton, G. P., and Lutz, B. R. (1941b). Smooth muscle motor-units in small blood vessels. *Am. J. Physiol.* **135,** 531.

Fulton, G. P., Jackson, R. G., and Lutz, B. R. (1946). Cinephotomicroscopy of normal blood circulation in the cheek pouch of the hamster, *Cricetus auratus. Anat. Record* **96,** 537.

Fulton, G. P., Akers, R. P., and Lutz, B. R. (1953). White thromboembolism and vascular fragility in the hamster cheek pouch after anticoagulants. *Blood* **8,** 140.

Fulton, G. P., Maynard, F. L., Riley, J. F., and West, G. B. (1957). Humoral aspects of tissue mast cells. *Physiol. Revs.* **37,** 221.

Fulton, G. P., Farber, E. M., and Moreci, A. P. (1959). The mechanism of action of rubefacients. *J. Invest. Dermatol.* **33,** 317.

Grafflin, A. L., and Bagley, E. H. (1953). Studies of peripheral blood vascular beds. *Bull. Johns Hopkins Hosp.* **92,** 47.

Grafflin, A. L., and Corddry, E. G. (1953). Studies of peripheral vascular beds in the bulbar conjunctiva of man. *Bull. Johns Hopkins Hosp.* **93,** 275.

Green, H. D., and Kepchar, J. H. (1959). Control of peripheral resistance in major systemic vascular beds. *Physiol. Revs.* **39,** 617.

Grim, E., and Lindseth, E. O. (1958). Measurement of regional flow in dog intestine. *Circulation* **18,** 728.

Hertzman, A. B. (1959). Vasomotor regulation of cutaneous circulation. *Physiol. Revs.* **39,** 280.

Hill, C. J. (1927). A contribution to our knowledge of the enteric plexuses. *Phil. Trans. Roy. Soc.* **B215,** 355.

Hugues, J. (1953). Contribution a l'etude des facteurs vasculaires et sanguins dans l'hemostase spontanee. *Arch. intern. physiol.* **61,** 565.

Jeffords, J. V., and Knisely, M. H. (1956). Concerning the geometric shapes of arteries and arterioles. *Angiology* **7,** 105.

Joftes, D. L. (1951). Some effects of the degeneration of the perivascular nerve plexus. Ph.D. Thesis, Boston University, Boston, Mass.

Knisely, M. H. (1939). Microscopic observations of the circulatory conditions in living frog liver lobules. *Anat. Record* **73,** Suppl., 69.

Knisely, M. H., Bloch, E. H., Eliot, T. S., and Warner, L. (1947). Sludged blood. *Science* **106,** 431.

Krogh, A. (1929). "The Anatomy and Physiology of Capillaries." Yale Univ. Press, New Haven, Conn.

Lee, R. E., and Holze, E. A. (1950). The peripheral vascular system in the bulbar conjunctiva of young normotensive adults at rest. *J. Clin. Invest.* **29,** 146.

Lee, R. E., and Holze, E. A. (1951). Peripheral vascular hemodynamics in the bulbar conjunctiva of subjects with hypertensive vascular disease. *J. Clin. Invest.* **30,** 539.

Leontowitsch, A. W. (1930). Über die Ganglionzellen der Blutgefässe. *Z. Zellforsch. mikroskop. Anat.* **11**, 23.

Lutz, B. R., and Fulton, G. P. (1958). Smooth muscle and blood flow in small blood vessels. *Proc. 3rd Microcircul. Conf., Am. Physiol. Soc., Washington, D.C.* p. 13.

Lutz, B. R., and Ward, B. A. (1960). Vascular patterns in the frog (Rana pipiens). *Anat. Record* **136**, 341.

Lutz, B. R., Fulton, G. P., and Akers, R. P. (1950a). The neuromotor mechanism of the small blood vessels in membranes of the frog (Rana pipiens) and the hamster (Mesocricetus auratus) with reference to the normal and pathological conditions of blood flow. *Exptl. Med. Surg.* **8**, 258.

Lutz, B. R., Fulton, G. P., Patt, D. I., and Handler, A. H. (1950b). The growth rate of tumor transplants in the cheek pouch of the hamster (Mesocricetus auratus). *Cancer Research* **10**, 231.

Lutz, B. R., Fulton, G. P., Patt, D. I., and Stevens, D. F. (1951). The cheek pouch of the hamster as a site for transplantation of a methylcholanthrene-induced sarcoma. *Cancer Research* **11**, 64.

Lynch, H. F., and Adolph, E. F. (1957). Blood flow in small blood vessels during deep hypothermia. *J. Appl. Physiol.* **11**, 192.

Nicoll, P. A., and Webb, R. L. (1946). Blood circulation in the subcutaneous tissue of the living bat's wing. *Ann. N. Y. Acad. Sci.* **46**, 697.

Nonidez, J. F. (1936). The nervous "terminal reticulum." A critique. I. Observations on the innervation of the blood vessels. *Anat. Anz.* **82**, 348.

Pappenheimer, J. R. (1953). Passage of molecules through capillary walls. *Physiol. Revs.* **33**, 387.

Poor, E., and Lutz, B. R. (1958). Functional anastomotic vessels of the cheek pouch of the hamster (*Mesocricetus auratus*). *Anat. Record* **132**, 121.

Rouget, C. (1873). Mémoire sur le développment, la structure et les proprietés physiologiques des capillaires sanguins et lymphatiques. *Arch. physiol. norm. et pathol.* **5**, 603.

Sanders, A. G., Ebert, R. H., and Florey, H. W. (1940). The mechanism of capillary contraction. *Quart. J. Exptl. Physiol.* **30**, 281.

Sandison, J. C. (1924). A new method for study of tissues in mammals. *Anat. Record* **28**, 281.

Sapirstein, L. A. (1958). Regional blood flow by fractional distribution of indicators. *Am. J. Physiol.* **193**, 161.

Sapirstein, L. A., Buckley, N. M., and Ogden, E. (1955). The rate of extravasation of intravenously injected thiocyanate in the dog. *Am. J. Physiol.* **183**, 178.

Slechta, R. F. (1960). Unpublished data.

Stöhr, P. (1938). Die mikroskopische Innervation der Blutgefässe. *Ergab. Anat. u. Entwicklungsgeschichte* **32**, 1.

Stricker, S. (1865). Untersuchungen über die capillaren Blutgefässe in der Nickhaut des Frosches. *Sitzber. Wien. Akad. Wiss.* **52**, Abt. 2, 379.

Uvnäs, B. (1954). Sympathetic vasodilator outflow. *Physiol. Revs.* **34**, 608.

Valdivia, E. (1958). Total capillary bed in striated muscle of guinea pigs native to the Peruvian mountains. *Am. J. Physiol.* **194**, 585.

Vidt, D. G., and Sapirstein, L. A. (1957). Volumes of distribution of Cr^{51} labelled red cells and T-1824 immediately after injection in the dog. *Circulation Research* **129**, 1957.

Vimtrup, B. (1922). Beiträge zur Anatomie der Capillaren. I. Über contractile elemente der Gefässwand der Blutcapillaren, Z. *Anat. Entwicklungsgeschichte* **65,** 150.

Volterra, M. (1925). Einige neue Befunde über die Struktur der Kapillaren und ihre Beziehungen zur "sogenannten" Kontraktilität derselben. *Zentr. inn. Med.* **46,** 876.

Wasserman, K., Loeb, L., and Mayerson, H. S. (1955). Capillary permeability to macromolecules. *Circulation Research* **3,** 594.

Webb, R. L., and Nicoll, P. A. (1954). The living bat's wing as a subject for studies in homeostasis of capillary beds. *Anat. Record* **120,** 253.

Woollard, H. H. (1926). The innervation of blood vessels. *Heart* **13,** 319.

Yudkofsky, P. (1960). The gross and microcirculatory effects of tilting and acceleration on the golden hamster. Ph.D. Thesis, Boston University, Boston, Mass.

Zweifach, B. W. (1934). A micromanipulative study of blood capillaries. *Anat. Record* **59,** 83.

Zweifach, B. W. (1937). The structure and reactions of small blood vessels in amphibia. *Am. J. Anat.* **60,** 473.

Zweifach, B. W. (1939). The character and distribution of blood capillaries. *Anat. Record* **73,** 475.

Zweifach, B. W. (1940). The structural basis of permeability and other functions of blood capillaries. *Cold Spring Harbor Symposia Quant. Biol.* **8,** 216.

Zweifach, B. W., and Kossman, C. E. (1947). Micromanipulation of small blood vessels in the mouse. *Am. J. Physiol.* **120,** 23.

Zweifach, B. W., and Metz, D. B. (1955). Selective distribution of blood through the terminal vascular bed of mesenteric structures and skeletal muscle. *Angiology* **6,** 282.

Zweifach, B. W., Lowenstein, B. E., and Chambers, R. (1944). Responses of blood capillaries to acute hemorrhage in the rat. *Am. J. Physiol.* **142,** 80.

Chapter VI. MICROCIRCULATION

E. EXCHANGE OF MATERIALS ACROSS CAPILLARY WALL

By H. S. MAYERSON

Capillaries serve as the functional unit of the circulation, since the essential function of the movement of blood is to insure a rapid turnover of tissue fluid in amounts adapted to the activity of various parts of the body. The capillaries are a closed system of vessels, except in the liver, where the blood comes in actual contact with the cells of the tissues (due to a deficiency of endothelial lining of the vascular capillaries) and perhaps also in the spleen. The cells everywhere else are bathed by tissue fluid (interstitial fluid) which acts as an intermediary, supplying nutritive materials and receiving the products of metabolic activity.

1. STRUCTURE OF ENDOTHELIUM AS RELATED TO PERMEABILITY

As seen with the light microscope (Fig. 41), the capillary wall is composed of several layers (from lumen outward): (a) the endocapillary layer, which is believed to be an adsorbed layer of some protein constituent of blood plasma; (b) the endothelium, composed largely of endothelial cells held together by the intercellular cement, possibly a calcium proteinate; and (c) a pericapillary sheath consisting of a closely woven investing layer of silver-staining connective tissue. The traditional concept of capillary permeability, based on these morphologic characteristics, is that passage of water and water-soluble molecules takes place through channels or pores penetrating the capillary wall (Pappenheimer, 1953). Since the total cross-sectional area of the pores seems to comprise less than 0.2 per cent of the histologic surface of the capillaries, these structures are thought to be limited to the area between endothelial cells and to range in diameter from 30–50 Å. On the other hand, the entire endothelial surface is available for passage of lipid-soluble molecules, oxygen, carbon dioxide, and other substances that are soluble in lipid as well as in water.

Recent work with the electron microscope (Fawcett, 1959) has, however, questioned the validity of the above concepts and suggests that they be revised (Fig. 42). For example, this method has failed to show an endocapillary layer, which may be due to the fact that either one does not exist or it is not preserved by current methods of specimen preparation. A few slender villi are occasionally noted projecting from the endothelial surface into the blood stream, but they do not seem common enough to be functionally significant. A layer is seen on the outer surface of the capillary which is 500–600 Å thick, composed principally of mucopolysaccharide and which corresponds to the basement membrane in other epithelia. It resembles ground substance of common

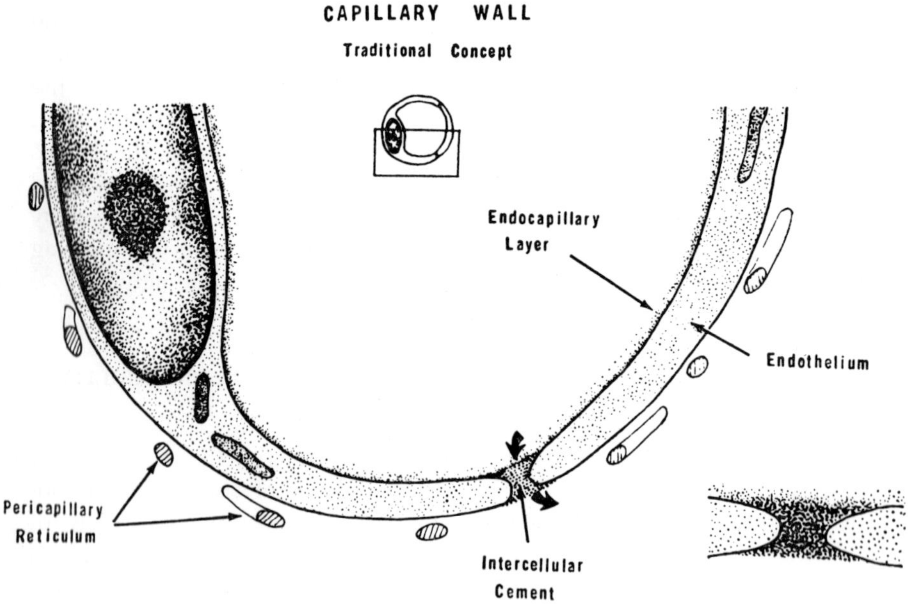

FIG. 41. Diagram illustrating traditional concept of structure of capillary wall. See text for details. (From Fawcett, 1959, reproduced with the permission of the University of Illinois Press, Urbana.)

connective tissue but differs from it in solubility characteristics. It is probably formed by the endothelial cells, themselves. The space between cells has been found to be much narrower than previously believed, and no pores have been observed in the intercellular substance. Injected colloid particles that pass through the capillary wall have not been seen between cells even though particulate matter has been found to accumulate intercellularly in other epithelia engaged in transport.

One of the most interesting findings in micrographs of capillaries is that of numerous inpocketings of the endothelial plasma membranes. Some of these communicate by a narrow neck with the lumen of the vessel, while others form small closed vesicles immediately beneath the cell membrane. Minute vesicles of this kind are often seen along the basal, as well as the luminal, surface of the endothelial cells. It has been suggested that these structures are important in the active transport of materials by pinocytosis. This process consists of fluids being taken into vesicles on the luminal surface, followed by migration of these structures across the cell to the basal surface, where the contents are released into the pericapillary interstitial fluid. Moore and Ruska (1957) have

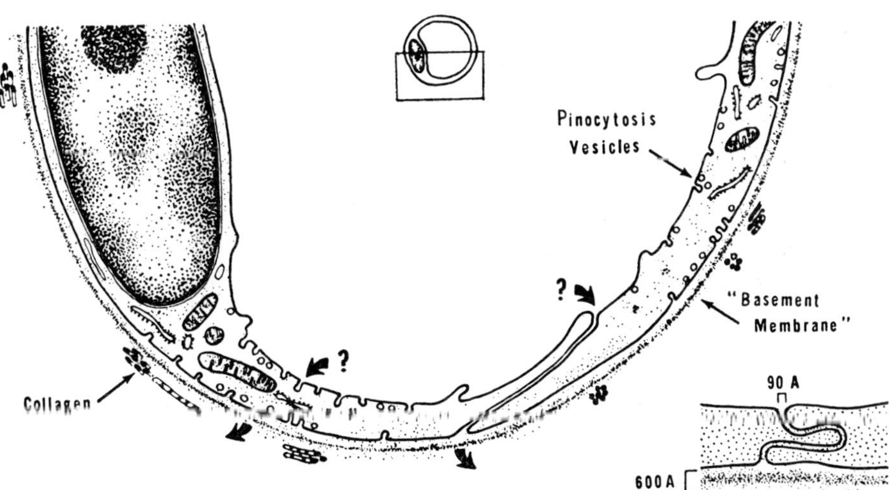

FIG. 42. Diagram illustrating current concept of structure of capillary wall. See text for details. (From Fawcett, 1959; reproduced with the permission of the University of Illinois Press, Urbana.)

recently reported results which strengthen this concept. However, they have suggested that *pinocytosis* be used to describe the active incorporation of fluid by cells for their own use and that a new term, *cytopemphis*, be applied to the taking up of fluid into small vesicles for transport across the cell. (For discussion of pathologic changes in endothelium, see Section G.-2b of this chapter.)

2. Exchange Mechanisms

The composition and volume of the tissue fluid may be affected by exchanges of materials between it and the plasma through the capillary wall and by exchanges with intracellular materials through the cellular membrane. Although it is evident that metabolic activity of cells modifies the composition of the interstitial fluid, this aspect of the problem has received relatively little attention. Instead, more emphasis has been given to the exchange through capillary membranes, as in studies concerned with lymph composition. The exchange of materials between the blood and extravascular fluid is by diffusion or by bulk filtration.

a. Diffusion: This mechanism may operate even against the flow by filtration. Thus the infusion of a hypertonic solution of sodium chloride will result in movement of water from the extravascular spaces into the capillary, concomitant with an outward diffusion of the sodium chloride against the bulk movement of the fluid. Work with isotopes has shown that, under normal conditions, there is an extremely rapid diffusion of small molecules and water back and forth through the capillary wall and that this mechanism accounts for most of the transfer of materials to and from cells of the organism.

b. Bulk filtration: The exchange of fluid by net filtration is actually small as compared to that resulting from diffusion. The underlying factors concerned in this process were outlined by Starling in 1896. The force of capillary pressure, which tends to drive water outward, is visualized as being balanced by the osmotic pressure of plasma proteins, which draws water inward from the extracellular fluid outside the capillaries. The Starling hypothesis requires that at equilibrium there be a balance between the "effective" capillary pressure (capillary blood pressure minus tissue pressure) and the "effective" osmotic pressure of the plasma (plasma osmotic pressure minus the osmotic pressure of the extracellular fluid). A rise in "effective" capillary pressure would result in movement of water out of the capillary, whereas a rise in "effective" osmotic pressure would produce movement of water into the vessel.

Measurements of capillary pressures (Landis, 1934) have shown that pressures are higher at the arteriolar end of the capillary than at the venous end. Thus, in terms of the Starling concept, fluid tends to leave at the arteriolar end and return at the venous end. Hence, an increase in venous pressure raises mean capillary pressure and so increases outward passage of fluid, while at the same time it raises the pressure at the venous end of the capillary and so decreases reabsorption. Thus, there is a double reason for the accumulation of fluid in the tissue spaces (edema) when there is an obstruction of the venous return from some area of the body. Likewise, any factor which diminishes the level of plasma protein (deficiency in the diet or abnormal leakage of protein from the capillaries) reduces reabsorption at the venous end and results in edema formation.

3. Permeability of Capillary Wall to Macromolecules

Evidence for the traditional "pores" concept of capillary permeability has been obtained chiefly in experiments in which relatively small molecules were used in studies involving capillaries in muscle. It is only recently that the work has been extended to other tissues of the body, using plasma proteins, lipoproteins and other large molecules. The results of these experiments indicate that all of the plasma proteins leak from capillaries throughout the body and are found in lymph, the rate of leakage appearing to be related to the molecular sizes of these substances. Thus, albumin leaks out of the capillaries about 1.6 times faster than globulin (Wasserman and Mayerson, 1952a). Dextrans of molecular weights similar to those of the proteins are lost from capillaries at approximately the same rates, suggesting that the charge of the molecule is not an important factor in its leakage (Wasserman and Mayerson, 1952b). Likewise, all of the different lipoproteins present in plasma have been identified in lymph (Page *et al.*, 1953; Courtice and Morris, 1955). Studies in the cat, dog, and rabbit have shown that plasma lipoproteins (with alpha-lipoproteins predominating) are present in lymph from leg, cervical, hepatic, and thoracic ducts. When rabbits have been fed cholesterol, however, the beta-lipoprotein fraction may increase enormously in plasma, with a much smaller rise in the alpha-lipoproteins. Beta-lipoprotein appears in the lymph but the evidence suggests that it leaves the circulation much less rapidly than the alpha fraction (Morris and Courtice, 1955). Chylomicrons are also found in lymph draining from various areas other than the intestines, suggesting that they are carried in the blood stream and that capillaries may be permeable to these large molecules. The normal capillary wall may likewise permit the passage of blood cells and inanimate particles. Yoffey and Courtice (1956) calculated that there is a leakage of 0.2 cc blood or about 0.004 per cent of the circulating red cells per day, on the assumption that thoracic duct lymph contains an average of 500 red cells per cubic millimeter and there is a daily flow of about 2 liters.

Recent studies (Mayerson *et al.*, 1960), in which dextrans of different molecular weights were infused into dogs, show quite clearly that the permeability of capillaries to macromolecules varies in different parts of the body. Liver capillaries would seem to be much more permeable than capillaries in the leg and cervical and lung areas. Intestinal capillaries are more permeable than those in the latter three regions but less permeable than those in the liver. Grotte (1956) has suggested that the differences are due to the presence of varying numbers of small pores and larger "leaks." An alternative suggestion (Mayerson *et al.*, 1960) is that small molecules (crystalloids) probably cross the capillary membrane by diffusion through the traditional pores, while for larger colloid molecules, filtration is probably the primary method of transcapillary exchange because of the relatively low diffusion coefficients of these molecules and because net water flow is in the same direction. While water molecules exchange very rapidly across membranes by diffusion, net water flow is toward the outside of the capillary, as evidenced by the formation of lymph.

It is difficult to see how pores or leaks large enough to allow passage of large macromolecules and red blood cells would have escaped notice in electron microscope studies. It is for this reason that the suggestion has been made that these substances are transferred across the capillary membrane by some active process, possibly by cytopemphis, as mentioned above (Subsection 1).

4. Factors Influencing Capillary Permeability

a. Capillary pressure: If saline solution is infused into dogs in moderate or large amounts, protein leakage through the capillaries is increased (Wasserman and Mayerson, 1952c), with the magnitude of the change being directly proportional to the amount of saline infused. At the same time, there is a marked increase in lymph flow during and after the infusion period. This results in a greater return of albumin via the thoracic duct, thus replacing, in varying degrees, the amount of protein which has escaped from the capillaries. If this lymph return is eliminated, significant deficits in total circulating protein appear. Further experiments (Shirley *et al.*, 1957) suggest that under conditions of marked plasma volume expansion, capillary pores are stretched to permit increased leakage of substances of large molecular size.

It has previously been mentioned that an increase in venous pressure, producing an elevated capillary pressure, results in greater fluid and protein leakage. Landis *et al.* (1932) have shown that, in the human, at venous pressures of 80 mm Hg, protein is lost from the blood plasma in an amount indicating that the capillary filtrate contains an average of 1.5 per cent of protein. At 60 mm Hg, very little loss in protein can be detected, and only an average of 0.3 per cent protein is present in the capillary filtrate. Cachera and Darnis (1950) and Geyer *et al.* (1953) have confirmed these findings and have shown that globulins, as well as albumins, leak out of the capillaries when the venous pressures are high.

Recent experiments (LeBrie and Mayerson, 1960) have demonstrated that at venous pressures above 30 cm H_2O, more protein appears in renal capsular lymph than at normal venous pressures.

b. Damage to endothelium: All injuries result in a local accumulation of fluid in the extravascular spaces, and at the same time capillary permeability to proteins is increased. The swelling of damaged tissue in mechanical trauma is due, in part, to hemorrhage but, in part, also to the effusion of a protein-rich plasma-like fluid (Freedlander and Lenhart, 1932). Thermal injury or chemical burns are likewise followed by a similar exudation, due to increased capillary permeability (Menkin, 1938). After a thermal burn, globulin will leak from capillaries as rapidly as albumin (Cope and Moore, 1944). Exposure to X-rays results in a significant increase in capillary permeability, as indicated particularly by the presence of large numbers of erythrocytes in lymph (Bigelow *et al.*, 1951).

c. Oxygen deficiency: While there is general agreement that infusion of fluids increases filtration, mainly by raising capillary pressure, and that injury does so by altering the permeability of the capillary membrane, there is still disagreement regarding the effects of acute oxygen deficiency. Most of the experiments show that anoxemia results in an increased flow of lymph, but the results are probably due to changes in capillary pressure rather than increased permeability. It is doubtful that low oxygen tensions compatible with life produce any increase in capillary permeability (Schiller and Wool, 1952; Hendley and Schiller, 1954).

d. Various disease states: With the increased use of isotopes, evidence is beginning to accumulate relative to changes in permeability of capillaries in various disease states. Gordon (1959) and Schwartz (1959) recently reported studies of the disorder known as "hypercatabolic hypoproteinemia" in which I^{131}-labeled polyvinylpyrolidone was used to demonstrate an abnormal permeability of the gastrointestinal tract to macromolecules. This finding was associated with hypoproteinemia and unimpaired protein synthesis, suggesting the designation of "exudative enteropathy" for this disorder, since it is not a true disease entity.

A number of studies have been done on other conditions. The results on congestive heart disease have been equivocal (Ross and Walker, 1956; Remenchik and Kark, 1954; Bauman *et al.*, 1955) and more data are needed before any conclusions can be reached. This also applies to the state of capillaries in shock. Most of the evidence suggests that there is no significant increase in permeability in the early stages of shock produced by burns, trauma or hemorrhage. On the other hand, data from dogs subjected to severe hemorrhage indicate that capillaries become more permeable to proteins (Wasserman and Mayerson, 1952b). (For a discussion of the pathogenesis of shock, see (2) Shock, Chapter XX.) The results also are equivocal with regard to cirrhosis, capillary permeability being reported as decreased in some patients (Sterling, 1951) and not changed in others (Tyor *et al.*, 1952).

F. PRESENT CONCEPTS OF THE VASCULAR ENVIRONMENT

By H. R. CATCHPOLE

1. VASCULAR ENVIRONMENT AND VASCULAR EXCHANGE

The increased knowledge gained of the morphologic, biochemical and physicochemical aspects of connective tissue in the past twenty-five years appears to justify an attempt both to modify and to expand certain hitherto accepted concepts of the vascular environment and of exchange of materials between blood vessels and cells.

a. Anatomic relationships: Figure 43A depicts schematically the relationship of a capillary to the fibrillar and nonfibrillar elements of connective tissue. Collagen, elastic and reticular fibers are shown, coursing through and bathed by the ground substance (*stippled*). Figure 43B shows a modified, denser congener of the ground substance, the basement membrane, containing reticular fibrils; this structure closely invests the endothelial capillary tube. Material similar to, or identical with, the basement membrane may constitute the "intercellular cement" occurring between cells (*cem*). The ground substance has been considered to be the nonfibrillar, extra- (or inter-) cellular material, homogeneous in the light microscope (Gersh and Catchpole, 1949, 1960). It stains under certain conditions with the periodic acid leucofuchsin stain (PAS), metachromatically with toluidine blue, and intravitally with Evans blue. The basement membrane stains similarly but more densely.

In summary, contiguous with the vascular endothelium are the ground substance and its congener, the basement membrane. They represent the actual cellular environment and the route of exchange of metabolites, nutrients, water, ions, and gases between blood and cells.

How materials actually leave and enter the vessels does not strictly concern the present discussion. Substances which cross the endothelial cytoplasm by any process (diffusion, pinocytosis, cytopempsis) enter the ground substance at the basement membrane. Substances passing

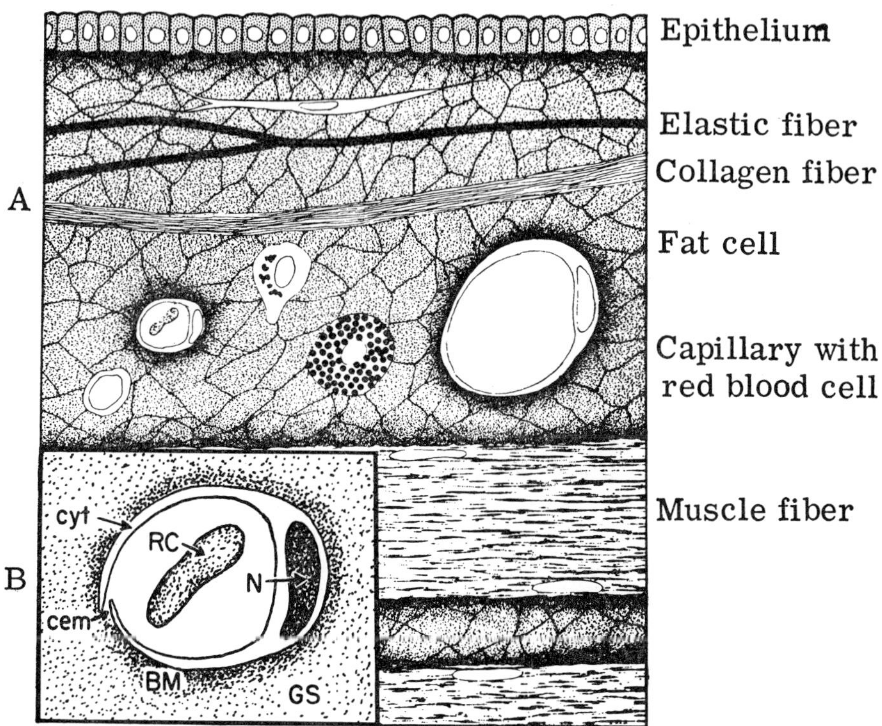

FIG. 43. A. Diagram of the relations of connective tissue ground substance (stippled) to cells and fibers. The more heavily stippled regions adjacent to certain cells are the homogeneous component of the basement membrane. In addition to labeled cells, additional cells are represented: fibroblast, mast cell, macrophage, small wandering cell. B. (Inset)—Capillary formed by an endothelial cell. RC, red cell; N, nucleus of endothelial cell; cyt, cytoplasm of endothelial cell; BM, basement membrane of capillary; GS, ground substance; cem, intercellular cement. (Gersh and Catchpole, 1960, reproduced with the permission of Perspectives in Biology and Medicine.)

between cells enter basement membrane or a closely related structure. In view of this, the extracapillary components must be given due weight in any concept of capillary permeability, for the nature and properties of these components may determine the kind and rate of passage of materials going from vessels to cells (Gersh and Catchpole, 1949).

b. Structure of the ground substance: Chemically, ground substance comprises a heterogeneous and probably largely unknown group of substances, some arising by local synthesis in fibroblasts, some derived from the blood. Table VII summarizes this information.

TABLE VII

SUMMARY OF MAJOR CONSTITUENTS OF GROUND SUBSTANCE OF CONNECTIVE TISSUE[a]

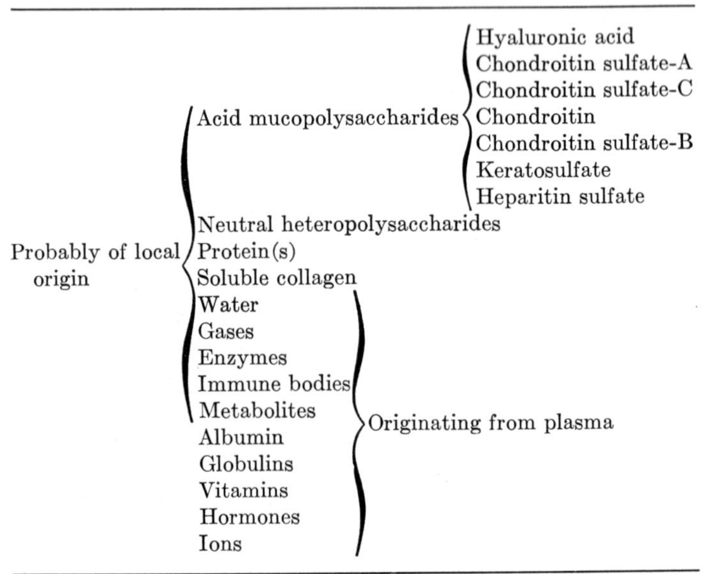

Origin	Constituent	Components
Probably of local origin	Acid mucopolysaccharides	Hyaluronic acid
		Chondroitin sulfate-A
		Chondroitin sulfate-C
		Chondroitin
		Chondroitin sulfate-B
		Keratosulfate
		Heparitin sulfate
Probably of local origin	Neutral heteropolysaccharides	
Probably of local origin	Protein(s)	
Probably of local origin	Soluble collagen	
Probably of local origin; Originating from plasma	Water	
Probably of local origin; Originating from plasma	Gases	
Probably of local origin; Originating from plasma	Enzymes	
Probably of local origin; Originating from plasma	Immune bodies	
Probably of local origin; Originating from plasma	Metabolites	
Originating from plasma	Albumin	
Originating from plasma	Globulins	
Originating from plasma	Vitamins	
Originating from plasma	Hormones	
Originating from plasma	Ions	

[a] From Gersh and Catchpole (1960).

Fundamental to the composition of ground substance is the group of mucoprotein and mucopolysaccharide compounds. Together with the more soluble constituents, they confer on the vascular and cellular environment a physical structure and consistency which varies from rigid to watery gels, depending on the location in the body.

In the electron microscope, the ground substance appears as a system of submicroscopic vacuoles of diameter 600–1200 Å, enclosed in a denser substance (Bondareff, 1957; Chase, 1959) (Fig. 44), findings which are in accord with a theoretical concept of this structure as a two phase system (see Subsection 1e below). The ground substance comprises a denser colloid-rich, water-poor phase, enclosing, and in equilibrium with, a looser colloid-poor, water-rich phase (vacuoles).

In summary, a complex chemical and physicochemical organization of the ground substance is described. This structure is conceived to be in a state of flux, in response to changing metabolic, hormonal, and physical conditions.

c. Colloidal charge of the ground substance: The colloidal charge density of the ground substance is an all important property. It represents "fixed" or relatively immobile charge. Under biologic conditions (i.e., at neutral pH) this charge is negative and may be attributed to the presence of acidic macromolecules of mucoproteins, mucopolysaccharides, and proteins. The charge density varies from tissue to tissue.

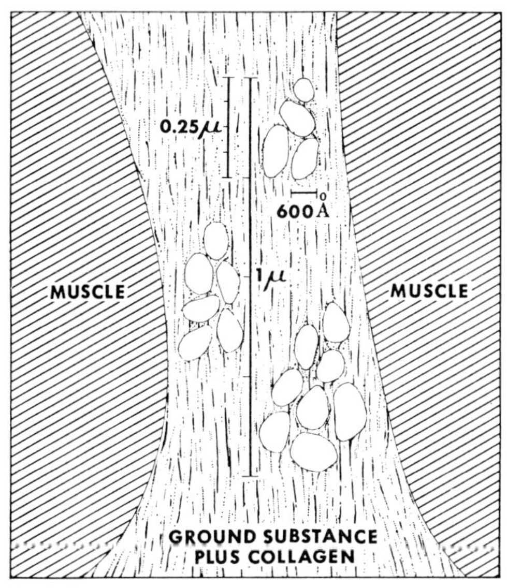

FIG. 44. Diagram of relations of submicroscopic droplets of ferrocyanide in the ground substance of the connective tissue of mouse diaphragm muscle. (Chase, 1959, reproduced with the permission of A.M.A. Archives of Pathology.)

Some characteristic values (Joseph *et al.*, 1954) are: cartilage, 160–170 mEq/liter water; subcutaneous connective tissue, 50–60 mEq/liter water; relaxed symphysis, 20 mEq/liter water. These immobile charge densities determine the electrolyte content of a tissue in accordance with the Donnan equilibrium. Thus, cartilage accumulates ions strongly, its Donnan ratio being of the order of 1.8. Other tissues accumulate ions more or less strongly in relation to their indiffusible charges. The connective tissue is thus an important cation reservoir. Further, it is selective in its ionic affinities.

Substances become associated with the ground substance in a variety of ways. Thus, divalent inorganic cations, such as calcium and magnesium, are selectively bound (Engel *et al.*, 1954; Catchpole *et al.*, 1956), and hence they lower the colloidal charge density. Ammonium ion and a wide variety of organic cations, including such biologically important substances as histamine, lysine, thiamin, and protamine, are also bound (Joseph *et al.*, 1959). The ionic mobility of potassium is reduced in ground substance (Joseph *et al.*, 1952) and it is, for this reason, preferentially accumulated. In addition, *anions* may combine with the ground substance and, when they simultaneously bind hydrogen ions, the charge density of the colloid may be modified in surprising ways (Joseph *et al.*, 1959).

Summarizing, the colloidal charge of the ground substance is regarded as one of its most fundamental properties. To ignore this primary reservoir of fixed negative charge would seem to be fatal to any concept of electrolyte distribution in tissues.

d. Passage of materials in ground substance: Substances passing from blood vessels to the ground substance do so at rates determined by molecular size and shape, among other factors. Strongly modifying the rate of passage are the interactions between the entering materials and the ground substance. For example, only under special circumstances, depending on the state of the ground substance, does the colloidal dye, Evans blue, leave the vessels and accumulate in the ground substance (Gersh and Catchpole, 1949; Catchpole, 1950). The behavior of the ferrocyanide ion is particularly interesting: it localizes promptly and specifically in the submicroscopic vacuoles, the water-rich phase of the ground substance. This site is thought to be related to the quadrivalent negative charge of the ferrocyanide ion, which tends to keep it off the more negatively charged vacuolar walls.

Cations entering the ground substance displace sodium and other ions from the ground substance in accordance with appropriate changes in free energy. The interactions between the physiologic cations, sodium, potassium, magnesium, and calcium and ground substance have been described nomographically. Certain foreign ions, such as lead, bind very strongly and tend to displace all the physiologic ions (Catchpole *et al.*, 1956).

As mentioned above, anions may enter into combinations with the basic groups of ground substance and may frequently bind additional hydrogen ions. The net effect is a displacement of mobile cations. Probably all familiar "inhibitors" of cellular respiration fall into this category. Simultaneously, by binding to tissue proteins generally, they affect electrolyte distribution by the above mechanisms. It is possible that neutral substances, like sugars and amino acids, or charged ampholytes, like polypeptides and proteins, may combine as electrical dipoles or in other types of binding with the ground substance and, in so doing, modify the properties of the latter.

e. Extravascular spaces and tissue fluid: The electron microscopic picture of the ground substance suggests a system of submicroscopic vacuoles enclosed in a denser continuum. This morphologic picture agrees with a purely physicochemical concept of a two phase system, a hypothesis derived to explain how blood may remain in equilibrium with connective tissues differing widely in their water content (Joseph *et al.*, 1952).

Most previous views have considered the tissue fluid to be a dialysate of blood, more or less protein-free, flowing in ill-defined channels or "spaces," or existing as surface films on fibers or cells. This fluid has been held to be distinct from the amorphous intercellular material (cf. Manery, 1954). The abandonment of such a concept, namely that of a separate tissue fluid, has been suggested (Gersh and Catchpole, 1960). The ground substance incorporates fluid and diffusible and nondiffusible components derived locally and from the plasma into the structure of a two phase system. Its two parts are the colloid-rich, water-poor phase, and the water-rich, colloid-poor phase, organized as a vacuolar structure. The relative amount of the two phases determines the density or looseness of the ground substance. Depending on the site, and conforming to adaptive processes, the system is relatively dense, as in cartilage, or relatively loose, as in skin. In either location, there exists a physiologic range through which the system may vary, yet continue to be homeostatic. This range covers normal hormonal and metabolic contingencies. The condition for homeostasis is that the two phases effectively remain in equilibrium with each other and with the blood.

f. Edema: The actual amount of fluid in the connective tissue at any given site will vary. It will be the resultant of such factors as intrinsic hydrophilic colloidal properties, tissue pressure, effective plasma filtration pressure, rate of lymphatic drainage, and metabolic activity of contiguous cells. All of these help to determine the relative amounts of the two phases in equilibrium.

Within limits, the water-rich phase may be expanded by a number of possible processes: (1) A partial disaggregation of tissue colloid will lead to an increase in molecular species and to fluid uptake. Such a reaction may be hormone mediated. (2) The cellular synthesis of hydrophilic colloid may locally increase fluid uptake. (3) Metabolite accumulation, possibly by affecting colloidal properties, may determine water (and electrolyte) holding capacity. These situations may expand the vacuolar structure but may not disturb effective equilibrium conditions.

In pathologic edemas, certain of the above processes are assumed to proceed to a nonequilibrium, irreversible stage. With increasing size and increasing disaggregation of colloid, the vacuoles would tend to become confluent. The enlarged pools of unstructured fluid are no longer in effective equilibrium with the dense phase or with blood. The system becomes, in fact, homogeneous. Colloids of the blood, normally removed by venous return or lymphatic drainage, tend to accumulate and to contribute their osmotic effect. Stasis also favors the accumulation of metabolites, which may further contribute to colloidal disaggregation.

In summary, edema represents a complex of events leading to a disruption of the normal equilibrated two-phase structure of the ground substance.

G. PATHOLOGY OF THE MICROCIRCULATION

By H. J. BERMAN

1. GENERAL CONSIDERATIONS

a. Introduction: The pathology of the microcirculation includes all intrinsic circulatory changes which interfere with or limit the functional capacity of this part of the circulatory system. The tolerable deviation of the internal environment from its optimal norm for a given tissue or organ depends not only on the inherent circulation (blood flow, functional vascularity, vascular pattern, etc.) but also on the susceptibility of the parenchymal cells to changes in the environment. The distinction between function and viability of cells is important. For example, excitability and contractility of cardiac muscle diminish rapidly with onset of anoxia. However, these properties can be partially re-established by perfusion even up to 20–30 hours after death (Kuliako, 1903; Ullrick *et al.*, 1957; Kardesch *et al.*, 1958; Bing and Michal, 1959).[1]

Pathologic changes which result from dysfunction of the micro-

[1] References at the end of a paragraph or subsection, especially if set off by a period, relate to the general topic or a particular aspect in that subunit.

circulation are most readily observable in the heart, brain, kidney, and retina because of the functional type of end arteries in these organs, the apparent susceptibility of their vessel walls to injury and degenerative changes, and the vital functions performed by them. Pathologic changes in the microcirculation are noted in arteriosclerosis, hypertension, glomerulonephritis, diabetes, shock and vascular collapse, "collagen" diseases of hypersensitive origin, inflammation, thromboembolism, and other abnormal states (Allen *et al.*, 1946; Adler, 1959; Danowski, 1957; Elwyn, 1946; Florey, 1958; Gilbert, 1960; Green and Kepchar, 1959; Jasmin and Robert, 1953; Lawrence, 1959; Menkin, 1956; Nordmann, 1955; Pickering, 1955; Ricketts, 1960; Stoner, 1960; Tobian, 1960; Wiggers, 1950).

2. Responses by Tissues

The magnitude and type of response to an injury depend on the kind, duration and intensity of the noxious agent, and on the local metabolism, environment, anatomy, and capacity for regeneration or repair. Although the components of the microcirculation are limited in their response to injury, usually several or all of the tissues become involved in varying degrees, to produce a relatively nonspecific spectrum of changes. Even in the relatively simple "flare" reaction, which may be produced by firmly stroking the skin, a triple response is observed: local dilation of blood vessels, increased vascular permeability, and swelling of connective tissue. The limited dilation of arteriolar vessels is mediated by a local axon reflex (Lewis, 1927).

a. Vascular smooth muscle and its innervation: Vascular smooth muscle and the nervous system are primarily involved in hypertension and shock and vascular collapse (Clark *et al.*, 1934; Duff, 1957; Folkow, 1960; Freis, 1960; Fulton *et al.*, 1960; Green and Kepchar, 1959; Grollman, 1958; Guyton and Bowlus, 1959; Haley and Anden, 1951; Hillarp, 1959; Levinson and Essex, 1943; McDowall, 1956; Oldham *et al.*, 1960; Pickering, 1955; Schmidt, 1960; Sommers *et al.*, 1958; Tobian, 1960)[2]—Barger, 1960; Deavers *et al.*, 1958; Fine *et al.*, 1959; Gilbert, 1960; Levy *et al.*, 1960; Negrete, 1956; Page and Abell, 1943; Wiggers, 1950; Zweifach, 1958).[3] (For a detailed discussion of hypertension and shock, see Chapter XX.) Since smooth muscle, by its varied states of contraction and responsiveness to stimuli, is a factor in the regulation of blood

[2] These references deal with the subject of hypertension.

[3] These references are concerned with shock and vascular collapse.

pressure and movement of fluids in the microcirculation (including tissue pressure, extravascular fluid flow and venous return), dysfunction of this tissue results in hemodynamic and tissue disturbances. It should be emphasized that capillary blood pressure is regulated, for the most part, independently of arterial blood pressure (Bazett, 1947; Gregersen, 1940; Lee and Visscher, 1957). In essential hypertension the capillary pressure is not elevated (Macleod, 1960). However, increased capillary pressure and edema occur at sites of inflammation (Florey, 1958) and in acute glomerulonephritis (Macleod, 1960). Furthermore, vascular beds in certain organs and tissues, such as the skin, skeletal muscle, mesentery, small intestine, and kidney, are more responsive to vasoconstrictor drugs than are those in other locations. This fact needs to be considered more carefully in the case of prolonged treatment of shock by such agents. They can, for a time, produce a satisfactory elevation of pressure in the large distributing arteries, but local blood pressure and blood flow through the more responsive beds in the body may actually still be at shock levels (Akers and Zweifach, 1955; Corday and Williams, 1960; Erlanger and Gasser, 1919; Freeman *et al.*, 1941; Nickerson, 1955; Pappenheimer, 1960; Richards, 1944; Robb *et al.*, 1958; Zweifach *et al.*, 1944).

b. Endothelium: This structure, together with either the basement membrane of the minute vessels or the internal elastic membrane of the larger ones, regulates, in large measure, vascular permeability and thus, the osmotic pressure of plasma and extravascular fluids (Bennett *et al.*, 1959; Buck, 1958; Mayerson *et al.*, 1960; Kisch, 1960; Moore and Ruska, 1957; Sinapius, 1958; Zweifach, 1959). The integrity of the endothelium also influences the reaction of blood to injury (O'Brien, 1959; Robertson *et al.*, 1959). In response to noxious stimuli, endothelial cells may round up, swell and either undergo degeneration or recover their normal properties. Their metabolism, secretions, phagocytic properties, permeability, and possible transcellular vesicle-transportation may change. The cells may also undergo mitosis and cover mural thrombi or denuded portions of the wall, or they may form new vessels (Abell, 1946; Altschul, 1954; Clark and Clark, 1935; Curran, 1957; McGovern, 1955; Margolis, 1959; Poole *et al.*, 1958; Robertson *et al.*, 1959; Spector, 1958; Wilhelm *et al.*, 1957; Zweifach, 1959). (For discussion of normal endothelium, see Section D.-4a, Chapter II and Section E.-1 of this chapter.)

The dense basement membrane of capillaries, which may also possibly be derived from connective tissue (Bohle and Herfarth, 1958; Gersh and Catchpole, 1949; Policard and Collet, 1959; Rothbard and Watson, 1959; Scott, 1957; Yamada, 1960), is especially prominent in the kidney glomerulus. There its thickening has been reported in systemic lupus erythematosus, malignant hypertension, glomerulonephritis, and especially, diabetes mellitus (see Subsection 3c, below) (Farquhar *et al.*, 1957; Farquhar, 1960; D. B. Jones, 1953; Pease, 1955; Rinehart *et al.*, 1953). (For normal structure of basement membrane, see Section C.-2c to e, Chapter I; Section C.-1 and 2, Chapter XIII.)

c. Blood: Blood may react to disease or to local or generalized injury by changes in its hemostatic and defense mechanisms, in the constancy of its contents, and in its suspension stability. Thrombi, emboli and endothelial coatings that form may be chiefly of platelet, leukocytic, fibrinous or mixed composition. Concentration and ease of activation of clotting, anticlotting and fibrinolytic factors may vary. The mobile phagocytic, antibody-producing, and concentrated protein sources of the body, both cellular (neutrophils, eosinophils, monocytes, lymphocytes, and plasma cells) and acellular (antibodies, opsinins, and properdin) may be concentrated at the site of injury, and aggregation or sludging of erythrocytes or gelation of plasma (cryoglobulinemia) may occur (Astrup, 1956; Biggs and Macfarlane, 1957; Billimoria *et al.*, 1959; Bloch, 1956 and 1959; Crowell and Read, 1955; Cruz and Oliveira, 1959; Florey, 1958; Foley, 1960; Fulton *et al.*, 1956; Henstell and Kligerman, 1959; Irwin *et al.*, 1955; Jasmin and Robert, 1953; Knisely *et al.*, 1945; Koppel and Olwin, 1959; Kugelmass, 1959; Lutz and Fulton, 1954; McDonald and Edgill, 1959; Menkin, 1956; Nour-Eldin, 1959; O'Brien, 1957; Pachter, 1959; Quick, 1960; Reich and Sternberger, 1956; Welch, 1899; Witte, 1958).

d. Connective tissue: Changes in structure and composition modify the supportive role of connective tissue in the vessel wall and in perivascular tissue by altering its permeability, tensile strength, flexibility, resiliency, and adsorptive and hemostatic properties (Angerine, 1959; Asboe-Hansen, 1954; Astrup, 1958; Buddecke, 1958; Dorfman, 1959; Florey, 1958; Gersh and Catchpole, 1949, 1960; Kohn, 1959; Lansing, 1959; McMaster and Parsons, 1950; Reynolds *et al.*, 1958). Such changes may also predispose to further pathologic alterations. (For normal structure of connective tissue of blood vessels, see Section C.-1e; 2e, Chapter I; Section D.-6, Chapter II.)

The reaction of vascular connective tissue may lead to the following observable alterations: changes in permeability, edema and swelling of ground substance, disruption of collagen and elastic fibers, purpura, rhexis and hemorrhage, local concentration of white cells, hyaline, amyloid or fibrinoid formation, aneurysm, thrombosis, arteriosclerosis, and narrowing of the vascular lumen (Amano, 1954; Baker and Iannone,

1959; Cruz *et al.*, 1959; Davis and Lawler, 1958; Elwyn, 1946; Florey, 1958; Fulton *et al.*, 1956; Lutz and Fulton, 1954; Mettier, 1959; More and Movat, 1959; Movat, 1960; Spaet, 1952). In addition, specific components of connective tissue are apparently intimately involved in reactions of a pathologic nature. For example, mast cells, which provide a discontinuous lining along the walls of many blood vessels and which may contain heparin, hyaluronic acid and histamine, have been implicated in antithrombic, edematous and hypersensitive reactions (Asboe-Hansen *et al.*, 1959; Beber *et al.*, 1960; Boréus and Chakravarty, 1960; McGovern, 1957; Stuart, 1952). Furthermore, accumulation and swelling of mucopolysaccharides around the internal elastic membrane, followed by disruption of its homogenous collagen-like core, may culminate in repair of the elastic membrane by fibrosis, thus constituting one of the important early changes in the genesis of certain vascular lesions (Gillman *et al.*, 1957; Gillman, 1959).

3. Responses in Selected Pathologic States

a. Inflammation: Inflammation is the response of living tissue to irritation or damage, in which the microcirculation participates by reacting ideally to neutralize and remove the noxious components, to isolate or limit the injury, and to provide a good environment for the processes of regeneration or repair. In and around the site of injury, vascular contraction may first occur. It is soon followed by local dilatation, increased blood flow and elevated capillary hydrostatic pressure, leukocytic sticking and emigration, increased vascular permeability and leakage of protein into the perivascular tissue. Locally the interstitial osmotic pressure is increased and concentration of antibodies follows. Other biochemical components of the defense, tissue clearance, and repair mechanisms collect in the inflamed site. In part, mobile cells bring with them the capacity for producing these substances. Edema and swelling of connective tissue occur. The lymphatics dilate and become more permeable, and flow of lymph increases.

In more severe injury, platelet thrombosis and occlusion of blood vessels may develop, as well as petechiae, frank hemorrhage, interstitial deposition of fibrin, and possibly local intravascular aggregation of erythrocytes. Damaged tissue may undergo necrosis and autolysis, after which the local vascularity increases and repair begins. There are many modifications of the process depending on the location, extent, severity and duration of the injury (Florey, 1958; Hunt, 1960; Jasmin and Robert, 1953; T. W. Jones, 1891; Karsner, 1947; Kelsall and Crabb, 1958, 1959; Menkin, 1956, 1957; Miles and Miles, 1958; Rich, 1936; Rinehart, 1951).

b. Hypersensitive states: An immunologic or hypersensitive reaction may damage cells or connective tissue and, as such, can produce a variety of "inflammatory" responses of an acute or protracted nature and of a focal or generalized pattern. The vascular manifestations are many and diverse and occur in anaphylaxis, the Arthus phenomenon and Shwartzman reaction, allergic dermatitis, asthma, hay fever, erythroblastosis fetalis, systemic lupus erythematosus, polyarteritis nodosa, rheumatoid arthritis, rheumatic fever, thrombocytopenic purpura, urticaria, and other conditions. These reactions include those of the blood, vessel wall, blood flow, and perivascular tissue (Gilbert, 1960; Goodman *et al.*, 1959; Ishizaka and Campbell, 1958; Lawrence, 1959; McMaster, 1959; Movat, 1960; Rich and Gregory, 1947; Seligman, 1958; Vazquez and Dixon, 1960). (For a discussion of clinical entities in which hypersensitivity may play a role, see (3) So-called Collagen or Systemic Connective Tissue Diseases, Chapter XXI; Section F.-4, Chapter XIII.)

In anaphylaxis the primary vascular response provoked by antibody antigen complexes seems to be a contraction of smooth muscle. In the guinea pig, bronchial contraction predominates, and in the dog, contraction of the hepatic veins, in each instance, because of the particular abundance of smooth muscle at these sites. In animals that survive the contraction phase, the following have been observed: hyperemia with slow blood flow and low pressure, increased vascular permeability, leukocytic sticking and emigration, thrombi and emboli, petechiae, and local stasis. In contrast, the Shwartzman reaction is characterized by hemorrhagic necrotic lesions, with thrombosis possibly acting to precipitate their formation, since anticoagulants can inhibit or prevent the development of the lesions (Shapiro and McKay, 1958). (Abell and Schenck, 1938; Irwin *et al.*, 1955; Koplik, 1937; McKay and Merriam, 1960; Rall and Kelly, 1956.)

Both vascular contraction and thrombosis, by causing local tissue anoxia and subsequent cellular damage, may precipitate further local nonspecific vascular reactions. In focal protracted reactions, such as in polyarteritis nodosa, smooth muscle may contract locally and accentuate the increase in vascular permeability. Increased edema in and around the arterial wall follows, and then swelling, fragmentation and necrosis of collagen and elastic fibers, and degeneration of smooth muscle. Fibroblasts are stimulated and collagen is produced to form the nodules of scar tissue that narrow or obliterate the lumen of the vessel. Degeneration at such sites of inflammation may also lead to thrombosis, hemorrhage, necrosis, and secondary aneurysms (Arkin, 1930; Davis *et al.*, 1960; Geschickter *et al.*, 1960; Goldberger, 1959; Goldgraber and Kirsner, 1959; Humphrey, 1955; Karsner, 1947; More and Movat, 1959; Piomelli *et al.*, 1959; Rinehart, 1951; Selye and Bajusz, 1959; Waksman and Bullington, 1956; Zeek, 1952).

c. Diabetes mellitus: In this disorder, vascular complications are now the leading cause of death. Subnutrition, suboxidation, and metabolic disorders, probably affecting mucopolysaccharides in the basement membrane of capillaries and in the ground substance of connective tissue, have been proposed to explain the various vascular effects (Ditzel, 1954). Atherosclerosis of medium-sized arteries and coronary thrombosis are common in diabetics (McCullagh, 1959), and the incidence of hypertension is reported to be increasing (Nonnenmacher, 1954). In the bulbar conjunctiva of patients with diabetes, Ditzel (1956) observed irregularities of arteriolar and venular walls, hyaline formation in perivascular tissue resulting from exudates, hemorrhages and degenerated connective tissue, and intravascular aggregation of red cells (see Subsection 3e, below), which in severe cases may plug arterioles and slow blood flow.

The disease process not only accelerates and exacerbates many commonly observed pathologic vascular changes, but it also produces distinguishing vascular lesions (Ashton, 1959; Ditzel, 1954, 1956; Elwyn, 1946) most noticeable in young diabetics. In the bulbar conjunctiva they consist of irreversible tortuous and angular elongations of small venules, reversible engorgement and distention of venules with blood, and reversible arteriolar narrowing (Ditzel, 1956). These alterations in a mild form, engorgement and distention of venules accompanied by perivascular edema and narrowing of arterioles, are the initial vascular changes observed in the conjunctiva of diabetics (Ditzel, 1956). In addition, in the fundus oculi characteristic microaneurysms appear early in the disease process, first, on the venous side and later, possibly on the arterial side, and actual closure of arterioles has been reported in the late stages (Ashton, 1959). In the renal glomeruli, the basement membrane is focally or diffusely, slightly or markedly, thickened and eventually is the site of intercapillary nodules (Kimmelstiel-Wilson lesion). (For further discussion of renal changes, see Section F.-2, Chapter XIII.)

In the lower extremities in diabetics, sweating, blood flow, the flare response of Lewis, and vascular reflexes to hot and cold may be reduced. Pickering (1960) concluded from these data that many people with diabetes mellitus have a lesion in the postganglionic fibers of the sympathetic nervous system.

d. Platelet thrombosis: Transilluminated living preparations provide an *in vivo* laboratory where the interactions of the component tissues of the microcirculation in the maintenance of hemostasis and in thrombosis and hemorrhage can be studied (Eberth and Schimmelbusch, 1886; Hugues, 1953; Roskam *et al.*, 1959; Welch, 1899). Such studies in the hamster cheek pouch (Fig. 45) have helped to place formation of platelet thrombi (Figs. 46 and 47) in proper perspective with fibrin clot formation and in *in vivo* thrombosis. White cells have been observed to adhere to or coat the endothelium in the pouch after infusion of plasma

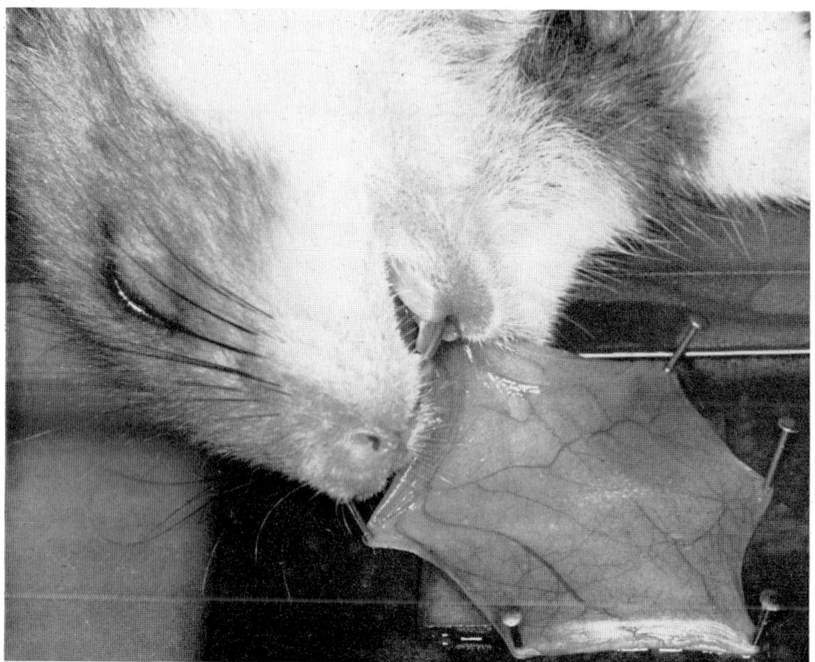

FIG. 45. Hamster cheek pouch preparation.

expanders, during infections, and following "gentle" physical manipulation of the vessels (Lutz and Fulton, 1954). Platelet thrombi (Figs. 46 and 47), on the other hand, have been seen forming at points of evident external injury and in response to local concentration of clotting factors.

Experiments (Berman *et al.*, 1954, 1955, 1956; Berman and Fulton, 1961) in which a number of micromanipulative techniques were applied to blood vessels in the hamster pouch have provided the following findings:

(1) Regardless of the type of noxious agent (mechanical, thermal, electrical or chemical), the first response observable with the

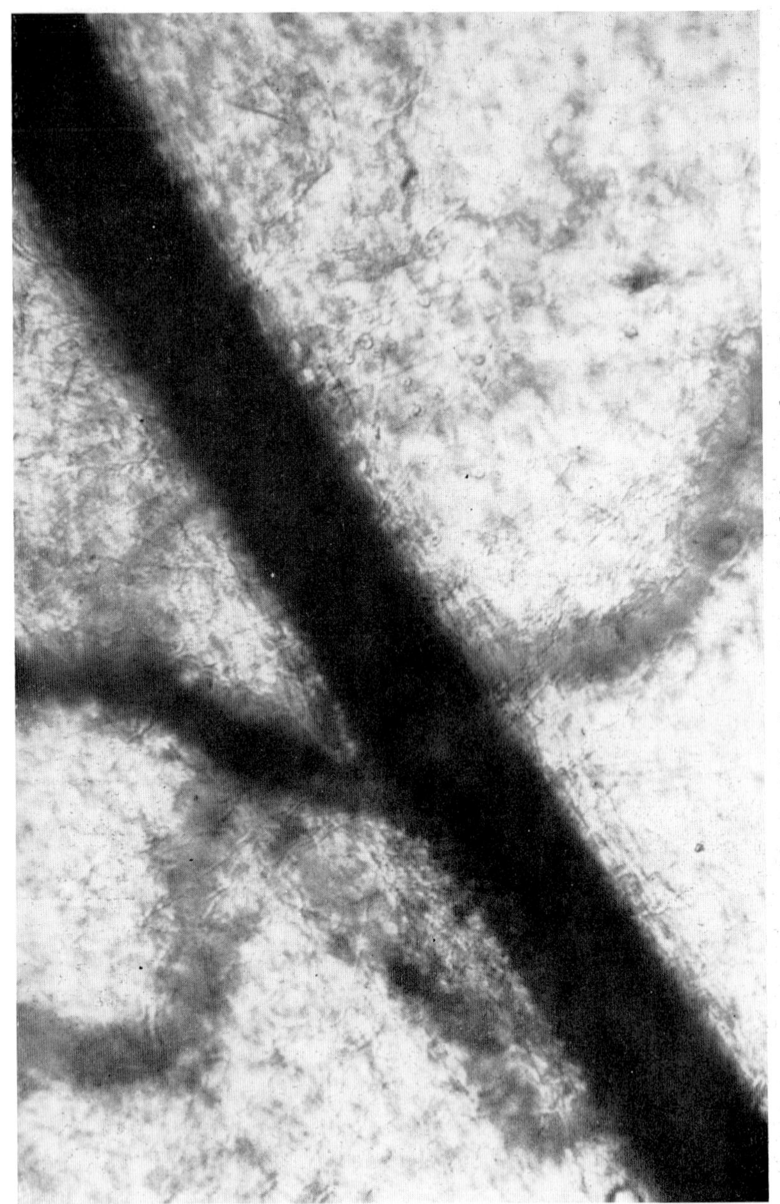

FIG. 46. Normal blood vessel (venule) before mechanical injury.

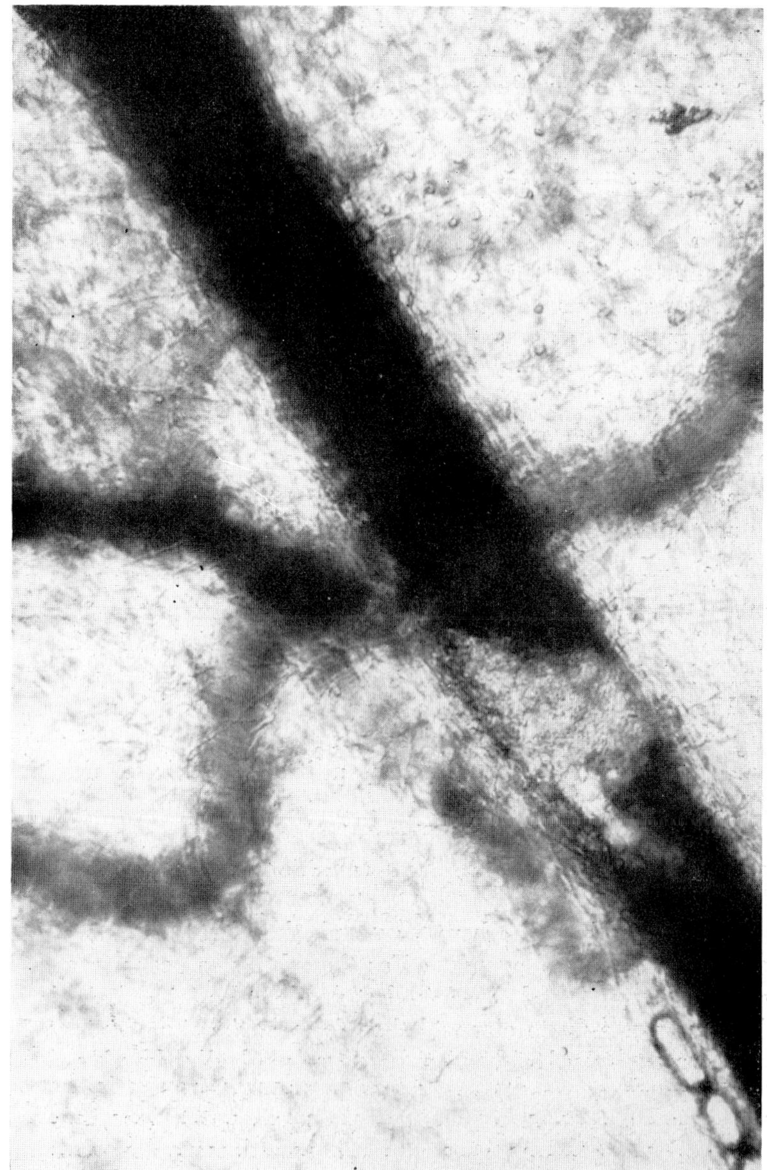

FIG. 47. Same vessel as in Fig. 46, with platelet thrombus after mechanical injury.

microscope is the same: formation of a platelet thrombus at the site of injury in vessels containing circulating blood.

(2) A fibrin clot may then form in the column of stationary blood that results from occlusion of a vessel by a platelet thrombus.

(3) Venous vessels are more susceptible than arterial vessels to platelet thrombosis.

(4) After a significant decrease in the rate of arterial flow, susceptibility of arterial vessels to readily observable platelet thrombosis approaches that of venous vessels.

(5) Topical applications of concentrates of clotting factors (thrombin or an incomplete thromboplastin, Russell's viper venom) first produce platelet thrombosis, and not the fibrin clot, in vessels containing circulating blood.

(6) The threshold for platelet thrombus formation *in vivo* appears to be much lower than that for noticeable fibrin clot formation.

(7) Extension of the area of thrombosis is by formation of additional platelet thrombi at points where stationary columns of blood contact circulating blood.

These findings supplement the observation made before the turn of the century, that as long as blood is in contact with an uninjured vessel, it remains fluid (Baumgarten, 1877; Brücke, 1857; Glénard, 1875: stated in Evans, 1936; Morawitz, 1905; Welch, 1899), and the more recent report of Wessler (1955) that the clot promoting potential in serum is rapidly neutralized when injected intravenously. For a fibrin clot to form in circulating blood into which serum has been infused, the blood flow must be stopped, or at least markedly slowed, immediately after the infusion (Wessler, 1955). A fibrin clot appears to form only in a stationary column of blood or under conditions where flow is exceptionally slow.

The *in vivo* work in the hamster pouch therefore supports the following concept of thrombogenesis: For thrombosis to occur in a vessel with rapidly flowing blood, the wall first needs to be injured. A platelet thrombus and not a fibrin clot will then form at the site of injury. The platelet thrombus may build up and stop or markedly slow blood flow and, in this manner, provide conditions favorable for fibrin clot formation. In arteries a marked reduction in local circulation, such as would occur during shock, would tend to unmask sites disposed to platelet thrombosis. (For a discussion of factors responsible for intravascular clotting, see Section B.-3b, Chapter XXII.)

e. Red cell aggregation (*sludged blood*—Knisely *et al.*, 1945): Circulating erythrocytes have been observed to form aggregates of varying size and firmness in a number of pathologic states. These aggregations may slow arteriolar flow by temporarily plugging terminal arterioles or capillaries or may slow venous return by forming large masses in venous vessels. However, they are more pronounced on the venous side of the circulation, since their formation is favored by slow flow. Their production has been correlated with increases in the concentration of globulin and fibrinogen in blood and with the presence of high molecular weight dextran. This tendency to clump probably reflects the composition of the coating on the surface of the erythrocytes. Aggregation can be partially inhibited or dispersed by injection of low-molecular weight samples of dextran or, more effectively, of heparin (Borgström *et al.*, 1957). The presence of these sludged aggregates, especially if they are of marked tenacity, may be an important complicating factor in abnormal states. They would tend to slow blood flow, accentuate conditions of tissue anoxia, and predispose to other forms of thromboembolism (Bloch, 1956, 1959; Fåhraeus, 1960; Gelin, 1956, 1957; Harders, 1957; Knisely *et al.*, 1945; Lutz, 1951; Snow, 1957; and Swindle, 1937).

References

Abell, R. G. (1946). The permeability of blood capillary sprouts and newly formed blood capillaries as compared to that of older blood capillaries. *Am. J. Physiol.* **147**, 237.

Abell, R. G., and Schenck, H. P. (1938). Microscopic observations on the behavior of living vessels of the rabbit during the reaction of anaphylaxis. *J. Immunol.* **34**, 195.

Adler, F. H. (1959). "Physiology of the Eye," 3rd ed., Chapter 9. Mosby, St. Louis, Mo.

Akers, R. P., and Zweifach, B. W. (1955). Effect of vasoactive materials on the peripheral blood vessels in the hamster. *Am. J. Physiol.* **183**, 529.

Allen, E. V., Barker, N. W., and Hines, E. A., Jr. (1946). "Peripheral Vascular Diseases." Saunders, Philadelphia, Pa.

Altschul, R. (1954). "Endothelium." Macmillan, New York.

Amano, S. (1954). Physio-path-morphology of so-called capillary bleeding. *Proc. 4th Intern. Congr. Intern. Soc. Hematol.* held in Mar del Plata, Argentina, 1952. p. 408

Angerine, D. M. (1959). Injury to connective tissue. *In* "Modern Trends in Pathology" (D. H. Collins, ed.), Chapter 3. Hoeber-Harper, New York.

Arkin, A. (1930). A clinical and pathological study of periarteritis nodosa. *Am. J. Pathol.* **6**, 401.

Asboe-Hansen, G., ed. (1954). "Connective Tissue in Health and Disease." Munksgaard, Copenhagen.

Asboe-Hansen, G., Dyrbye, M. O., Moltbe, E., and Wegelius, O. (1959). Tissue edema—a stimulus of connective tissue regeneration. *J. Invest. Dermatol.* **32**, 505.

Ashton, N. (1959). Diabetic retinopathy. A new approach. *Lancet* **ii**, 625.

Astrup, T. (1956). Fibrinolysis in the organism. *Blood* **11**, 781.

Astrup, T. (1958). The hemostatic balance. *Thromb. et Diath. Haemorrhag.* **2**, 347.

Baker, A. B., and Iannone, A. (1959). Cerebrovascular disease. III. The intracerebral arterioles. *Neurology* **9**, 441.

Barger, A. C. (1960). The kidney in congestive heart failure. *Circulation* **21**, 124.

Bauman, A., Rothschild, M. A., Yalow, R. S., and Berson, S. A. (1955). Distribution and metabolism of I^{131} labelled human serum albumin in congestive heart failure with and without proteinuria. *J. Clin. Invest.* **34**, 1359.

Baumgarten, P. (1877). "Die sagenannte Organization des Thrombus." Otto Wigand, Leipzig.

Bazett, H. C. (1947). Factors concerning the control of capillary pressure as indicated in a circulation schema. *Am. J. Physiol.* **149**, 389.

Beber, B. A., Landing, B. H., and Sutherland, J. M. (1960). Mast cell content of placental tissue. *A.M.A. Arch. Pathol.* **69**, 531.

Bennett, H. S., Luft, J. H., and Hampton, J. C. (1959). Morphological classifications of vertebrate blood capillaries. *Am. J. Physiol.* **196**, 381.

Berman, H. J., and Fulton, G. P. (1961). Platelets in the peripheral circulation. *Proc. Intern. Symposium on Blood Platelets, Henry Ford Hospital, Detroit, 1960* p. 7.

Berman, H. J., Lutz, B. R., and Fulton, G. P. (1954). White and red thrombi produced by thrombin applied to the hamster cheek pouch. *Federation Proc.* **13**, 12.

Berman, H. J., Fulton, G. P., Lutz, B. R., and Pierce, D. L. (1955). Susceptibility to thrombosis in normal young, aging, cortisone-treated, heparinized, and x-irradiated hamsters as tested by topical application of thrombin. *Blood* **10**, 831.

Berman, H. J., Fulton, G. P., and Lutz, B. R. (1956). Platelets in the peripheral circulation of the hamster (*Mesocricetus auratus*) (motion picture). *Federation Proc.* **15**, 17.

Bigelow, R. R., Furth, J., Woods, M. C., and Storey, W. H. (1951). Endothelial damage by X-rays disclosed by lymph fistula studies. *Proc. Soc. Exptl. Biol. Med.* **76**, 734.

Biggs, R., and Macfarlane, R. G. (1957). "Human Blood Coagulation and Its Disorders," 2nd ed. Blackwell Sci. Publs., Oxford.

Billimoria, J. D., Drysdale, J., James, D. C. O., and Maclagan, N. F. (1959). Determination of fibrinolytic activity of whole blood with special reference to the effects of exercise and fat feeding. *Lancet* **ii**, 471.

Bing, R. J., and Michal, G. (1959). Myocardial efficiency. *Ann. N. Y. Acad. Sci.* **72**, 555.

Bloch, E. H. (1956). Microscopic observations of the circulating blood in the bulbar conjunctiva in man in health and disease. *Ergeb. Anat. u. Entwicklungsgeschichte* **35**, 1.

Bloch, E. H. (1959). Visual changes in the living microvascular system in man and experimental animals as they are related to thrombosis and embolism. *Angiology* **10**, 421.

Bohle, A., and Herfarth, C. (1958). Zurfrage eines intercapillären bindegewebes im glomerulum der Niere des menschen. *Virchows Arch. pathol. Anat. u. Physiol.* **331**, 573.

Bondareff, W. (1957). Submicroscopic morphology of connective tissue ground substance with particular regard to fibrillogenesis and aging. *Gerontologia* **1**, 222.

Boréus, L. O., and Chakravarty, N. (1960). The histamine content of guinea pig mast cells. *Experientia* **16**, 192.

Borgström, S., Gelin, L-E., Wendeberg, B., and Zederfelt, B. (1957). Traumatic thrombosis: animal experiments. *Acta Chir. Scand.* **113**, 455.

Brücke, E. (1857). An essay on the causes of the coagulation of the blood. *Brit. and Foreign Med.-Chir. Rev.* **19**, 183.

Buck, R. C. (1958). The fine structure of endothelium of large arteries. *J. Biophys. Biochem. Cytol.* **4**, 187.

Buddecke, E. (1958). Untersuchungen zur Chemie der Arterienwand. II. Arteriosklerotische verander ungen am aortenbindegewebe des menschen. *Hoppe-Seylers Z. physiol. Chem.* **310**, 182.

Cachera, R., and Darnis, F. (1950). Capillary permeability in the normal subject. *Ann. méd. (Paris)* **51**, 509.

Catchpole, H. R. (1950). Solubility properties of some components of the ground substance in relation to intravital staining of connective tissue. *Ann. N. Y. Acad. Sci.* **52**, 989.

Catchpole, H. R., Joseph, N. R., and Engel, M. B. (1956). Homeostasis of connective tissues. III. Magnesium-sodium equilibrium and interactions with strontium and lead. *A.M.A. Arch. Pathol.* **61**, 503.

Chase, W. H. (1959). Extracellular distribution of ferrocyanide in muscle. *A.M.A. Arch. Pathol.* **67**, 525.

Clark, E. R., and Clark, E. L. (1935). Observations on changes in blood vascular endothelium in the living animal. *Am. J. Anat.* **57**, 385.

Clark, E. R., Clark, E. L., and Williams, R. G. (1934). Microscopic observations in the living rabbit of the new growth of nerves and the establishment of nerve-controlled contractions of newly formed arterioles. *Am. J. Anat.* **55**, 47.

Cope, O., and Moore, F. D. (1944). Study of capillary permeability in experimental burns and burn shock using radioactive dyes in blood and lymph. *J. Clin. Invest.* **23**, 241.

Corday, E., and Williams, J. H. (1960). Effect of shock and of vasopressor drugs on the regional circulation of the brain, heart, kidney and liver. *Am. J. Med.* **29**, 228.

Courtice, F. C., and Morris, B. (1955). The exchange of lipids between plasma and lymph of animals. *Quart. J. Exptl. Physiol.* **40**, 138.

Crowell, J. W., and Read, W. L. (1955). *In vivo* coagulation—a probable cause of irreversible shock. *Am. J. Physiol.* **183**, 565.

Cruz, W. O., and Oliveira, A. C. (1959). The effect on haemostasis of changes in the proportions of plasma and cells in the blood. *Brit. J. Haematol.* **5**, 141.

Cruz, W. O., Oliveira, A. C., and Magalhães, J. R. (1959). Crude cephalin as the active substance of organ extracts able to convert venous blood into hemostatic blood. *Arch. Biochem. Biophys.* **85**, 550.

Curran, R. C. (1957). The elaboration of mucopolysaccharides by vascular endothelium. *J. Pathol. Bacteriol.* **74**, 347.

Danowski, T. S. (1957). "Diabetes Mellitus," Chapter 23. Williams & Wilkins, Baltimore, Md.

Davis, M. J., and Lawler, J. C. (1958). The capillary circulation of the skin. *A.M.A. Arch. Dermatol.* **77**, 690.

Davis, R. B., Meeker, W. R., and McQuarrie, D. G. (1960). Immediate effects of intravenous endotoxin on serotonin concentrations and blood platelets. *Circulation Research* **8**, 234.

Deavers, S., Smith, E. L., and Huggins, R. A. (1958). Critical role of arterial pressure during hemorrhage in the dog on release of fluid into the circulation and trapping of red cells. *Am. J. Physiol.* **195**, 73.

Ditzel, J. (1954). Morphologic and hemodynamic changes in the smaller blood vessels in diabetes mellitus. 1. Considerations based on the literature. *New Engl. J. Med.* **250**, 540.

Ditzel, J. (1956). Angioscopic changes in the smaller blood vessels in diabetes mellitus and their relationship to aging. *Circulation* **14**, 386.

Dorfman, A. (1959). Role of connective tissue ground substance in degenerative disease. *Circulation* **19**, 801.

Duff, R. S. (1957). Adrenaline sensitivity of peripheral blood vessels in human hypertension. *Brit. Heart J.* **19**, 45.

Eberth, J. C., and Schimmelbusch, C. (1886). Experimentelle untersuchungen über thrombose. *Virchows Arch. pathol. Anat. u. Physiol.* **103**, 39.

Elwyn, H. (1946). "Diseases of the Retina." McGraw-Hill (Blakiston), New York (2nd ed., 1953).

Engel, M. B., Joseph, N. R., and Catchpole, H. R. (1954). Homeostasis of connective tissue. I. Calcium-sodium equilibrium. *A.M.A. Arch. Pathol.* **58**, 26.

Erlanger, J., and Gasser, H. S. (1919). Studies in secondary traumatic shock. III. Circulatory failure due to adrenalin. *Am. J. Physiol.* **49**, 345.

Evans, C. L., ed. (1936). *In* "Starting's Principles of Human Physiology," 7th ed., p. 1096. Lea & Febiger, Philadelphia, Pa.

Fåhraeus, R. (1960). Die Grundlagen der neueren Humoralpathologie. Die frühe Geschichte der Mikrozirkulation. *Virchows Arch. pathol. Anat. u. Physiol.* **333**, 176.

Farquhar, M. G. (1960). Electron microscopic studies in the renal glomerulus in the nephrotic syndrome. *In* "Edema: Mechanisms and Management" (J. H. Moyer and M. Fuchs, eds.). Saunders, Philadelphia, Pa.

Farquhar, M. G., Vernier, R. L., and Good, R. A. (1957). Studies on familial nephrosis. II. Glomerular changes observed with the electron microscope. *Am. J. Pathol.* **33**, 791.

Fawcett, D. W. (1959). *In* "Microcirculation: Factors Influencing Exchange of Substances across Capillary Wall" (S. R. M. Reynolds and B. W. Zweifach, eds.). Univ. of Illinois Press, Urbana, Ill.

Fine, J., Rutenburg, S., and Schweinburg, F. B. (1959). The role of the reticulo-endothelial system in hemorrhagic shock. *J. Exptl. Med.* **110**, 547.

Florey, H. W., ed. (1958). "General Pathology," 2nd ed. Saunders, Philadelphia, Pa.

Foley, W. T. (1960). Intravascular clotting. A biological error. *Am. J. Cardiol.* **6**, 456.

Folkow, B. (1960). Role of the nervous system in the control of vascular tone. *Circulation* **21**, 760.

Freedlander, S. O., and Lenhart, C. H. (1932). Traumatic shock. *A.M.A. Arch. Surg.* **25**, 693.

Freeman, N. E., Freedman, H., and Miller, C. C. (1941). The production of shock by the prolonged continuous injection of adrenalin in unanesthetized dogs. *Am. J. Physiol.* **131**, 545.

Freis, E. D. (1960). Hemodynamics of hypertension. *Physiol. Revs.* **40**, 27.

Fulton, G. P., Lutz, B. R., and Kagan, R. (1956). Effect of x-irradiation and beta emination on circulation in the hamster cheek pouch. *Circulation Research* **4**, 133.

Fulton, G. P., Lutz, B. R., and Callahan, A. B. (1960). Innervation as a factor in control of microcirculation. *Physiol. Revs.* **40**, Suppl. 4, 57.

Gelin, L-E. (1956). Studies in anemia of injury. *Acta Chir. Scand.* Suppl. 210.

Gelin, L-E. (1957). Intravascular aggregation and capillary flow. *Acta Chir. Scand.* **113**, 463.

Gersh, I., and Catchpole, H. R. (1949). The organization of ground substance and basement membrane and its significance in tissue injury, disease and growth. *Am. J. Anat.* **85**, 457.

Gersh, I., and Catchpole, H. R. (1960). The nature of ground substance of connective tissue. *Perspectives in Biol. and Med.* **3**, 282.

Geschickter, C. F., O'Malley, W. E., and Rubacky, E. P. (1960). A hypersensitivity phenomenon produced by stress: the "negative phase" reaction. *Am. J. Clin. Pathol.* **34**, 1.

Geyer, G., Keible, E., and Kolbl, H. (1953). Permeability of capillaries to proteins. *Z. ges. exptl. Med.* **122**, 1.

Gilbert, R. P. (1960). Mechanisms of the hemodynamic effects of endotoxin. *Physiol. Revs.* **40**, 245.

Gillman, T. (1959). Reduplication, remodeling, regeneration, repair, and degeneration of arterial elastic membranes. *A.M.A. Arch. Pathol.* **67**, 624.

Gillman, T., Hathorn, M., and Penn, J. (1957). Microanatomy and reactions to injury of vascular elastic membranes and associated polysacharides. *In* "Connective Tissue" (R. E. Tunbridge, ed.), p. 120. Blackwell Sci. Publs., Oxford.

Glénard, F. (1875). Contribution à l'étude des causes de la coagulation spontanée du sang à son issue de l'organisme; application à la transfusion. Thèse no. 50, Paris.

Goldberger, E. (1959). Etiology and pathogenesis of syndromes associated with periarteritis nodosa lesions. *Am. J. Cardiol.* **3**, 656.

Goldgraber, M. B., and Kirsner, J. B. (1959). The Arthus phenomenon in the colon of rabbits. *A.M.A. Arch. Pathol.* **67**, 556.

Goodman, H. C., Malmgren, R. A., Falen, J. L., and Brecher, G. (1959). Separation of factors in Lupus Erythematosus serum reacting with components of cell nuclei. *Lancet* **ii**, 382.

Gordon, R. S., Jr. (1959). Exudative enteropathy (abnormal permeability of the gastro intestinal tract demonstrable with labelled polyvinylpyrrolidone). *Lancet* **i**, 325.

Green, H. D., and Kepchar, J. H. (1959). Control of peripheral resistance in major systemic vascular beds. *Physiol. Revs.* **39**, 617.

Gregersen, M. I. (1940). Some effects of reflex changes in blood pressure on plasma volume. *Am. J. Physiol.* **129**, 369.

Grollman, A. (1958). Mechanisms of arterial blood pressure variations. *Texas Repts. Biol. and Med.* **16**, 277.

Grotte, G. (1956). Passage of dextran molecules across the blood-lymph barrier. *Acta Chir. Scand.* Suppl. 211.

Guyton, A. C., and Bowlus, W. E. (1959). Renal hypertension induced by a partial return of urine to the circulation. *Circulation* **20**, 707.

Haley, T. J., and Anden, M. (1951). The effect of topically applied adrenergic blocking agents upon the peripheral vascular system. *J. Pharmacol. Exptl. Therap.* **102**, 50.

Harders, H. (1957). Zur diatetisch induzierten und postoperativen thromboseneigung: das "blood-sludge phänomen" erfassung und thrombogenetische bedeutung. *Thromb. et Diath. Haemorrhag.* **1**, 482.

Hendley, E. D., and Schiller, A. A. (1954). Change in capillary permeability during hypoxemic perfusion of rat hindlegs. *Am. J. Physiol.* **179**, 216.

Henstell, H. H., and Kligerman, M. (1959). Hyperglobulinaemia thrombo-haemorrhagic diathesis. *Nature* **183**, 978.

Hillarp, N. A. (1959). The construction and functional organization of the autonomic innervation apparatus. *Acta Physial. Scand.* **46**, Suppl. 157.

Hugues, J. (1953). Contribution a l'étude des facteurs vasculaires et sanguins dans l'hémostase spontañee. *Arch. intern. physiol.* **61**, 565.

Humphrey, J. H. (1955). The mechanism of Arthus reaction. 1. The role of polymorphonuclear leucocytes and other factors in reversed passive Arthus reactions in rabbits. *Brit. J. Exptl. Pathol.* **36**, 268.

Hunt, A. H. (1960). Wound healing. *Proc. Roy. Soc. Med.* **53**, 41.

Irwin, J. W., Weile, F. L., and Burrage, W. S. (1955). Small blood vessels during allergic reactions. *Ann. Otol., Rhinol., Laryngol.* **64**, 1164.

Ishizaka, K., and Campbell, D. H. (1958). Biological activity of soluble antigen-antibody complexes. I. Skin reactive properties. *Proc. Soc. Exptl. Biol. Med.* **97**, 635.

Jasmin, G., and Robert, A., eds. (1953). "The Mechanism of Inflammation," Intern. Symposium. Acta, Inc., Montreal.

Jones, D. B. (1953). Glomerulonephritis. *Am. J. Pathol.* **29**, 33.

Jones, T. W. (1891). "Report on the State of the Blood and the Blood Vessels in Inflammation, and on Other Points Relating to the Circulation in the Extreme Vessels." Baillière, Tindall and Cox, London.

Joseph, N. R., Engel, M. B., and Catchpole, H. R. (1952). Interaction of ions and connective tissue. *Biochim. et Biophys. Acta* **8**, 575.

Joseph, N. R., Engel, M. B., and Catchpole, H. R. (1954). Homeostasis of connective tissues. II. Potassium-sodium equilibrium. *A.M.A. Arch. Pathol.* **58**, 40.

Joseph, N. R., Catchpole, H. R., Laskin, D. M., and Engel, M. B. (1959). Titration curves of colloidal surfaces. II. Connective tissues. *Arch. Biochem. Biophys.* **84**, 224.

Kardesch, M., Hogancamp, E. C., and Bing, R. J. (1958). The survival of excitability, energy production and energy utilization of the heart. *Circulation* **18**, 935.

Karsner, H. T. (1947). "Acute Inflammations of Arteries." Charles C Thomas, Springfield, Ill.

Kelsall, M. A., and Crabb, E. D. (1958). Lymphocytes and plasmacytes in nucleoprotein metabolism. *Ann. N. Y. Acad. Sci.* **72**, 293.

Kelsall, M. A., and Crabb, E. D. (1959). "Lymphocytes and Mast Cells." Williams & Wilkins, Baltimore, Md.

Kisch, B. (1960). "Electron Microscopy of the Cardiovascular System" (translated from the German by Arnold I. Kisch). Charles C Thomas, Springfield, Ill.

Knisely, M. H., Eliot, T. S., and Bloch, E. H. (1945). Sludged blood in traumatic shock. I. Microscopic observations of the precipitation and agglutination of blood flowing through vessels in crushed tissues. *A.M.A. Arch. Surg.* **51**, 220.

Kohn, R. R. (1959). Age and swelling in acid of perivascular connective tissue in human lung. *J. Gerontol.* **14**, 16.

Koplik, L. H. (1937). Experimental production of hemorrhage and vascular lesions in lymph nodes: an extension of the Schwartzman phenomenon. *J. Exptl. Med.* **65**, 287.

Koppel, J. L., and Olwin, J. H. (1959). Physiologic aspects of intravascular clotting. *Am. J. Cardiol.* **4**, 585.

Kugelmass, I. N. (1959). "Biochemistry of Blood in Health and Disease." Charles C Thomas, Springfield, Ill.

Kuliako, A. (1903). Sur la reviviscence du coeur. Rappel des battements du coeur humain trente heures après la mort. *Compt. rend. acad. sci.* **136**, 63.

Landis, E. M. (1934). Capillary pressure and capillary permeability. *Physiol. Revs.* **14**, 404.

Landis, E. M., Jonas, L., Angevine, M., and Erb, W. (1932). The passage of fluid and protein through the human capillary wall during venous congestion. *J. Clin. Invest.* **11**, 717.

Lansing, A. I., ed. (1959). "The Arterial Wall." Williams & Wilkins, Baltimore, Md.

Lawrence, H. S., ed. (1959). "Cellular and Humoral Aspects of the Hypersensitive States." Hoeber-Harper, New York.

LeBrie, S. J., and Mayerson, H. S. (1960). Influence of elevated venous pressure on flow and composition of renal lymph. *Am. J. Physiol.* **198**, 1037.

Lee, J. S., and Visscher, M. B. (1957). Microscopic studies of skin blood vessels in relation to sympathetic nerve stimulation. *Am. J. Physiol.* **190**, 37.

Levinson, J. P., and Essex, H. E. (1943). Observations of the effect of certain drugs on the small blood vessels of the rabbit ear before and after denervation. *Am. J. Physiol.* **139**, 423.

Levy, M. N., Blattberg, B., and Barlow, J. L. (1960). Effect of hemorrhagic shock upon the properdin system of the dog. *J. Lab. Clin. Med.* **56**, 105.

Lewis, T. (1927). "The Blood Vessels of the Human Skin and Their Responses." Shaw & Sons, London.

Lutz, B. R. (1951). Intravascular agglutination of the formed elements of blood. *Physiol. Revs.* **31**, 107.

Lutz, B. R., and Fulton, G. P. (1954). The use of the hamster cheek pouch for the study of vascular changes at the microscopic level. *Anat. Record* **120**, 293.

McCullagh, E. P. (1959). Vascular complications of diabetes mellitus. *Cleveland Clin. Quart.* **26**, 97.

McDonald, L., and Edgill, M. (1959). Changes in coagulability of the blood during various phases of ischaemic heart disease. *Lancet* **1**, 1115.

McDowall, R. J. S. (1956). "The Control of the Circulation of the Blood," 2nd ed. Dawson & Sons, London.

McGovern, V. J. (1955). Reactions to injury of vascular endothelium with special reference to the problem of thrombosis. *J. Pathol. Bacteriol.* **69**, 283.

McGovern, V. J. (1957). The mechanism of inflammation. *J. Pathol. Bacteriol.* **73**, 99.

McKay, D. G., and Merriam, J. C. (1960). Vascular changes induced by bacterial endotoxin during generalized Shwartzman reaction. *A.M.A. Arch. Pathol.* **69**, 524.

Macleod, M. (1960). Systemic capillary pressure in acute glomerulonephritis estimated by direct micropuncture. *Clin. Sci.* **19**, 27.

McMaster, P. D. (1959). General and local vascular reactions in certain states of hypersensitivity. *In* "Cellular and Humoral Aspects of the Hypersensitive States" (H. S. Lawrence, ed.), p. 319. Hoeber-Harper, New York.

McMaster, P. D., and Parsons, R. J. (1950). The movement of substances and the state of the fluid in the intradermal tissue. *Ann. N. Y. Acad. Sci.* **52**, 993.

Manery, J. F. (1954). Water and electrolyte metabolism. *Physiol. Revs.* **34**, 334.

Margolis, J. (1959). Hageman factor and capillary permeability. *Australian J. Exptl. Biol. Med. Sci.* **37**, 239.

Mayerson, H. S., Wolfram, C. G., Shirley, H. H., Jr., and Wasserman, K. (1960). Regional differences in capillary permeability. *Am. J. Physiol.* **198**, 155.

Menkin, V. (1938). "Dynamics of Inflammation: An Inquiry into the Mechanism of Infectious Processes." Macmillan, London.

Menkin, V. (1956). "Biochemical mechanisms in inflammation," 2nd ed. Charles C Thomas, Springfield, Ill.

Menkin, V. (1957). Direct effect of corticotropin (ACTH) on capillary permeability in inflammation. *Am. J. Physiol.* **189**, 98.

Mettier, S. R., Jr. (1959). Retinal manifestations of systemic diseases. *Postgraduate Med.* **26**, 162.

Miles, A. A., and Miles, E. M. (1958). The state of lymphatic capillaries in acute inflammatory lesions. *J. Pathol. Bacteriol.* **76**, 21.

Moore, D. H., and Ruska, H. (1957). The fine structure of capillaries and small arteries. *J. Biophys. Biochem. Cytol.* **3**, 457.

Morawitz, P. (1905). "The Chemistry of Blood Coagulation" (English translation by R. C. Hartmann and P. F. Guenther). Charles C Thomas, Springfield, Ill., 1958.

More, R. H., and Movat, H. Z. (1959). Character and significance of the cellular response in the collagen diseases and experimental hypersensitivity. *Lab. Invest.* **8**, 873.

Morris, B., and Courtice, F. C. (1955). Lipid exchange between plasma and lymph in experimental lipemia. *Quart. J. Exptl. Physiol.* **40**, 149.

Movat, H. Z. (1960). Pathology and pathogenesis of the diffuse collagen diseases. *Can. Med. Assoc. J.* **83**, 683, 747, 797.

Negrete, J. M. (1956). Effects of ischemia on neuromuscular transmission. *Communs. 20 Intern. Physiol. Congr.* held in Brussels, Belgium, 1956, St. Catherine Press, Brugge-Brussels, Belgium. p. 674.

Nickerson, M. (1955). Factors of vasoconstriction and vasodilation in shock. *J. Michigan State Med. Soc.* **54**, 45.

Nonnenmacher, H. (1954). Diabetes mellitus, diabetic retinopathy and hypertension in the postwar period. *Klin. Monatsbl. Augenheilk.* **124**, 579.

Nordmann, M. (1955). Die pathologische Anatomic der Kapillaren. *In* "Kapillaren und Interstitium" (H. Barthelheimer and H. Küchmeister, eds.), p. 41. Thieme, Stuttgart.

Nour-Eldin, F. (1959). Distribution of clotting factors in tissues. *Proc. 4th Intern. Congr. Biochem., Vienna, 1958* **10**, 138.

O'Brien, J. R. (1957). Fat ingestion, blood coagulation and atherosclerosis. *Am. J. Med. Sci.* **234**, 373.

O'Brien, J. R. (1959). Some properties of damaged endothelial cells. *Nature* **184**, 1580.

Oldham, P. D., Pickering, G., Roberts, F., and Sowry, G. S. C. (1960). The nature of essential hypertension. *Lancet* **i**, 1085.

Pachter, M. R. (1959). Macroglobulinemia of Waldenstrom. Clinical, laboratory and anatomic features. *Postgraduate Med.* **26**, 815.

Page, I. H., and Abell, R. G. (1943). The state of the vessels of the mesentery in shock produced by constricting the limbs and the behavior of the vessels following hemorrhage. *J. Exptl. Med.* **77**, 215.

Page, I. H., Lewis, L. A., and Plahl, G. (1953). The lipoprotein composition of dog lymph. *Circulation Research* **1**, 87.

Pappenheimer, J. R. (1953). Passage of molecules through capillary walls. *Physiol. Revs.* **33**, 387.

Pappenheimer, J. R. (1960). Central control of renal circulation. In proceedings on a symposium on central nervous system control of circulation. *Physiol. Revs.* **40**, Suppl. 4, 35.

Pease, D. C. (1955). Electron microscopy of the vascular bed of the kidney cortex. *Anat. Record* **121**, 701.

Pickering, G. W. (1955). "High Blood Pressure." Grune & Stratton, New York.

Pickering, G. (1960). The anatomical and functional aspects of the neurological lesions of diabetes. *Proc. Roy. Soc. Med.* **53**, 142.

Piomelli, S., Stefanini, M., and Mele, R. H. (1959). Antigenicity of human vascular endothelium: lack of relationship to the pathogenesis of vasculitis. *J. Lab. Clin. Med.* **54**, 241.

Policard, A., and Collet, A. (1959). Les membranes basales et leurs role dans la physiologie normale et pathologique des tissus. *Rev. franç. études clin. et biol.* **4**, 283.

Poole, J. C. F., Sanders, A. G., and Florey, H. W. (1958). The regeneration of aortic endothelium. *J. Pathol. Bacteriol.* **75**, 133.

Quick, A. J. (1960). Intravascular clotting. *Lancet* **i**, 774.

Rall, D. P., and Kelly, M. G. (1956). Mechanism of local Schwartzman reaction. *Communs. 20th Intern. Physiol. Congr.* held in Brussels, Belgium, 1956, St. Catherine Press, Brugge-Brussels, Belgium. p. 749.

Reich, T., and Sternberger, L. A. (1956). Postoperative thrombin production. *Surg., Gynecol., Obstet.* **102**, 463.

Remenchik, A. P., and Kark, R. M. (1954). Studies in endothelial capillary permeability: II. congestive failure. *J. Clin. Invest.* **33**, 959.

Reynolds, S. R. M., Kirsch, M., and Bing, R. J. (1958). Functional capillary beds in the beating, KCl-arrested and KCl-arrested-perfused myocardium of the dog. *Circulation Research* **6**, 600.

Rich, A. R. (1936). Inflammation in resistance to infection. *A.M.A. Arch. Pathol.* **22**, 228.

Rich, A. R., and Gregory, J. E. (1947). Experimental anaphylactic lesions of the coronary arteries of the "sclerotic" type, commonly associated with rheumatic fever and disseminated lupus erythematosus. *Bull. Johns Hopkins Hosp.* **81**, 312.

Richards, D. W., Jr. (1944). The circulation in traumatic shock in man. *Harvey Lectures, 1943–1944* **39**, 217.

Ricketts, H. T. (1960). *In* "Diabetes" (R. H. Williams, ed.), Chapter 38. Hoeber-Harper, New York.

Rinehart, J. F. (1951). The role of the connective tissue ground substance (mucopolysaccharide) in allergic injury. *Calif. Med.* **95**, 335.

Rinehart, J. F., Farquhar, M. G., Jung, H. C., and Abul-Haj, S. K. (1953). The normal glomerulus and its basic reactions in disease. *Am. J. Pathol.* **29**, 21.

Robb, H. J., Ingham, D. E., Nelson, H. M., and Johnston, C. G. (1958). Observations in vascular dynamics during hemorrhagic shock and its therapy. *Am. J. Surg.* **95**, 659.

Robertson, H. R., Moore, J. R., and Mersereau, M. A. (1959). Observations on thrombosis and endothelial repair following application of external pressure to a vein. *Can. J. Surg.* **3**, 5.

Roskam, J., Hugues, J., Bounameaux, Y., and Salmon, J. (1959). The part played by platelets in the formation of an efficient hemostatic plug. *Thromb. et Diath. Haemorrhag.* **3**, 510.

Ross, R. S., and Walker, W. G. (1956). Decreased permeability of capillaries to protein in chronic congestive heart failure. *J. Clin. Invest.* **35,** 732.

Rothbard, S., and Watson, R. F. (1959). Renal glomerular lesions induced by rabbit anti-rat collagen serum in rats prepared with adjuvant. *J. Exptl. Med.* **109,** 633.

Schiller, A. A., and Wool, G. N. (1952). Disappearance rate of fluorescein conjugated plasma proteins from the circulation of hypoxic rats. *Federation Proc.* **11,** 139.

Schmidt, C. F. (1960). Central nervous control of the circulation. *Circulation Research* **8,** 301.

Schwartz, M. (1959). Gastro-intestinal protein loss in idiopathic (hypercatabolic) hypoproteinemia. *Lancet* **i,** 327.

Scott, D. G. (1957). A study of the antigenicity of basement membrane and reticulin. *Brit. J. Exptl. Pathol.* **38,** 178.

Seligman, M. (1958). Études immunologique sur le lupus érythémateaux disséminé. *Rev. franç. études clin. et biol.* **3,** 558.

Selye, H., and Bajusz, E. (1959). Stress and cardiac infarcts. Effect of sodium deficiency upon necrotizing cardiopathies produced by various agents. *Angiology* **10,** 412.

Shapiro, S. S., and McKay, D. G. (1958). The prevention of the generalized Shwartzman reaction with sodium warfarin. *J. Exptl. Med.* **107,** 377.

Shirley, H. H., Jr., Wolfram, C. G., Wasserman, K., and Mayerson, H. S. (1957). Capillary permeability of macromolecules: stretched pore phenomenon. *Am. J. Physiol.* **190,** 189.

Sinapius, D. (1958). Über das Endothel der Venen. *Z. Zellforsch.* **47,** 560.

Snow, P. J. D. (1957). The sludging of blood in the retinal veins. *Lancet* **i,** 65.

Sommers, S. C., Relman, A. S., and Smithwick, R. H. (1958). Histologic studies of kidney biopsy specimens from patients with hypertension. *Am. J. Pathol.* **34,** 685.

Spaet, T. H. (1952). Vascular factors in the pathogenesis of haemorrhagic syndromes. *Blood* **7,** 641.

Spector, W. G. (1958). Substances which affect capillary permeability. *Pharmacol. Revs.* **10,** 475.

Starling, E. H. (1896). On the absorption of fluids from the connective tissue spaces. *J. Physiol.* (*London*) **19,** 312.

Sterling, K. (1951). Serum turnover in Laennec's cirrhosis as measured by I^{131} tagged albumin. *J. Clin. Invest.* **30,** 1238.

Stoner, H. B., ed. (1960). "The Biochemical Response to Injury." Blackwell Sci. Publs., Oxford.

Stuart, E. G. (1952). Mast cell responses to anaphylaxis. *Anat. Record* **112,** 394.

Swindle, P. F. (1937). Occlusion of blood vessels by agglutinated red cells, mainly as seen in tadpoles and very young kangaroos. *Am. J. Physiol.* **120,** 59.

Tobian, L. (1960). Interrelationship of electrolytes, juxtaglomerular cells and hypertension. *Physiol. Revs.* **40,** 280.

Tyor, M. P., Aikawa, J. K., and Cayer, D. (1952). Measurements of the disappearance of radioactive tagged albumin from serum and the excretion of I^{131} in the urine of patients with cirrhosis. *Gastroenterology* **21,** 79.

Ullrick, W. C., Lentini, E. A., and Sommers, S. C. (1957). Excitability and contractility of post-mortem human heart muscle. *Lab. Invest.* **6,** 528.

Vazquez, J. J., and Dixon, F. J. (1960). Immunopathology of hypersensitivity. *Ann. N. Y. Acad. Sci.* **86**, 1025.

Waksman, B. H., and Bullington, S. J. (1956). A quantitative study of the passive Arthur reaction in the rabbit eye. *J. Immunol.* **76**, 441.

Wasserman, K., and Mayerson, H. S. (1952a). Dynamics of lymph and protein exchange. *Cardiologia* **21**, 296.

Wasserman, K., and Mayerson, H. S. (1952b). Plasma, lymph and urine studies after dextran infusions. *Am. J. Physiol.* **171**, 218.

Wasserman, K., and Mayerson, H. S. (1952c). Mechanism of plasma protein changes following saline infusions. *Am. J. Physiol.* **170**, 1.

Welch, W. H. (1899). Thrombosis. *In* "Syst. Med." (Allbutt, T. C., ed.), Vol. 7, p. 155. Macmillan, London; also *in* "Papers and Addresses," Vol. 1, p. 110 (for importance of integrity of vascular wall, see p. 122). Johns Hopkins, Baltimore, Md., 1920; see also Brücke (1857), Glénard (1875), and Baumgarten (1877).

Wessler, S. (1955). Studies in intravascular coagulation. III. The pathogenesis of serum-induced venous thrombosis. *J. Clin. Invest.* **34**, 647.

Wiggers, C. J. (1950). "Physiology of Shock." Commonwealth Fund, New York.

Wilhelm, D. L., Mill, P. J., and Miles, A. A. (1957). Enzyme-like globulins from serum producing the vascular phenomena of inflammation. III. Further observations on the permeability factor and its inhibitor in guinea pig serum. *Brit. J. Exptl. Pathol.* **38**, 446.

Witte, S. (1958). Die steigerung der kapillarpermeabilität durch blutgerinnungsstörungen. *Thromb. et Diath. Haemorrhag.* **2**, 146.

Yamada, E. (1960). Collagen fibrils within the renal glomerulus. *J. Biophys. Biochem. Cytol.* **7**, 407.

Yoffey, J. M., and Courtice, F. C. (1956). "Lymphatics, Lymph and Lymphoid Tissue." Harvard Univ. Press, Cambridge, Mass.

Zeek, P. M. (1952). Periarteritis nodosa: a critical review. *Am. J. Clin. Pathol.* **22**, 777.

Zweifach, B. W. (1958). Microcirculatory derangements as a basis for the lethal manifestations of experimental shock. *Brit. J. Anaesthesia* **30**, 466.

Zweifach, B. W. (1959). Structure and behavior of vascular endothelium. *In* "The Arterial Wall" (I. A. Lansing, ed.), p. 15. Williams & Wilkins, Baltimore, Md.

Zweifach, B. W., Lowenstein, B. E., and Chambers, R. (1944). Responses of blood capillaries to acute hemorrhage in the rat. *Am. J. Physiol.* **142**, 80.

Chapter VII. VENOUS SYSTEM

A. EMBRYOLOGY

By K. Christensen

1. Early Vasculogenesis

The origin of intraembryonic veins is from vascular plexuses that have formed in the embryonic mesoderm. Enlargement of certain parts of the plexus leads to the establishment of channels that anticipate the primordial veins. The primordial vessels begin as endothelial-lined tubes, and during differentiation the surrounding mesoderm contributes the other tissues of the venous wall. The older literature on venous development has been reviewed by Evans (1912). Few studies have been made since the summary of Richardson (1937).

2. Venous Primordia

Present views of general venous development incorporate the investigations of a number of embryologists. Although Seib's (1934) purpose was to explain azygos, hemiazygos, and inferior vena caval development, he summarized the pattern of development in a diagrammatic way, not only with regard for the "venous lines" but also by careful reference to the intersegmental veins which have been described completely for thoracic and lumbar regions.

The "venous lines" are right and left members of pairs of primordial veins (usually without inclusion of the precardinal veins) which extend for variable distances throughout the embryonic body. These first venous primordia appear at about the 3 mm stage and all are formed before the end of the embryonic period (25 mm). The plan of development is as follows:

1. Right and left precardinal veins are the primitive systemic veins just rostral to the heart which are joined by the primary head veins. At the heart the precardinal and the postcardinal veins form the common cardinal veins which enter the sinus venosus.
2. Right and left postcardinal veins are primitive systemic veins caudal to the heart that extend from the pelvic region to the septum transversum. They appear posterolateral to the mesonephroi. Both pre- and postcardinal veins

receive the intersegmental veins (Reagan, 1926; Reagan and Robinson, 1926; McClure and Butler, 1925).

3. Right and left subcardinal veins are found in the thorax and abdomen, where they anastomose cranially and caudally with the postcardinal veins. They are located anteromedial to the mesonephroi, and they receive intersegmental veins and mesonephric veins. Intersubcardinal anastomoses appear near the middle of the embryonic period. At the 7–8 mm stage the right subcardinal joins the hepatic sinusoids, which is a step in the formation of the inferior vena cava (Lewis, 1903; Frazer, 1931).
4. Right and left thoracolumbar, supracardinal, or paraureteric veins are present in the thorax and abdomen posteromedial to the postcardinal veins and lateral to the aorta and the sympathetic trunk. There is free communication with the subcardinal, as well as with the intersegmental, veins (Gladstone, 1929; McClure and Butler, 1925; Reagan and Tribe, 1927).
5. Right and left lumbocostal veins, sometimes called the precostal veins, extend from the level of the 5th thoracic intersegmental vein to the upper lumbar region. They are situated posterior to the thoracolumbar veins and below the developing transverse processes of vertebrae (Gladstone, 1929).
6. Right and left median sympathetic veins are located in the anatomic position of the azygos and hemiazygos veins, medial to the sympathetic trunk and anterolateral to intercostal and lumbar arteries (Reagan, 1926; Reagan and Robinson, 1926; Reagan and Tribe, 1927).
7. Right and left subcentral veins are longitudinal venous trunks close to the midline, posterior to the aorta (dorsal aortic veins of Strong, 1929), and posteromedial to the intercostal and lumbar arteries.

A characteristic of all of these veins is the anastomoses which develop between members of a pair of veins, sometimes between branches of a pair, and between adjoining veins. The anastomoses of the "venous lines" (see sections of Gladstone, 1929) are reminiscent of the earlier venous plexuses.

Other venous primordia are the extensive vascular plexuses in the limb buds, in the head, and in somatic regions of the trunk. The primordia for the pulmonary veins include an outgrowth from the sinoatrial wall and the incorporation of the pulmonary part of the foregut venous plexus (Auer, 1948; Butler, 1952; Hamilton *et al.*, 1952).

3. Venous Differentiation

In venous differentiation, a variety of changes occurs: transformations of the primordial vessels and their anastomoses either in their entirety or in part; degeneration of parts of vessels not utilized; and sometimes complete new formation of vessels.

a. Superior vena cava and its tributaries: The superior vena cava is formed by the right common cardinal vein and the caudal part of the right precardinal vein. A transverse anastomosis between right and left precardinal veins gives most of the left innominate vein. The right precardinal vein above this transverse vessel becomes the right innominate vein. More distal parts of the right and left precardinal veins form the internal jugular veins; the external jugular vein on either side is a new vessel. The precardinal vein on the left side below the anastomosis contributes to the superior intercostal vein and to the oblique vein of Marshall; the left common cardinal vein becomes the coronary sinus.

b. Veins of upper extremities: These vessels form from the superficial vascular plexus in the limb buds which join the cardinal veins: first the postcardinals and then the precardinals. The persisting veins on the caudal border of the upper limb buds differentiate successively into the subclavian, axillary, and basilic veins. The cephalic vein on the cranial border of the limb bud is formed secondarily and connects with the axillary vein.

c. Azygos system: On the right side, the azygos vein differentiates from the right medial sympathetic vein. The right intersegmental veins which unite with it form the intercostal veins. The portion of the azygos opening into the superior vena cava develops from the most rostral part of the right postcardinal vein and its anastomosis with the right median sympathetic vein. On the left side the hemiazygos vein is the left median sympathetic vein, with the intersegmental veins contributing the intercostal veins that join with it. Cross anastomoses between median sympathetic veins form the various connections between hemiazygos and azygos veins which appear during differentiation.

The lumbocostal veins supplying the ascending lumbar veins which unite with the azygos veins form the so-called lateral roots. The median sympathetic veins may also provide an ascending vein arising from the inferior vena cava which anastomoses with the azygos vein, to form the so-called intermediate root. A median connection from the inferior vena cava to the azygos system is the sole remnant of the subcentral veins and contributes the medial root.

The superior intercostal vein which joins the left innominate vein is the left median sympathetic vein with the first 3 or 4 intersegmental (intercostal) veins, plus segments of the left post- and precardinal veins and an anastomosis between them.

d. Pulmonary system: In the differentiation of the pulmonary vein, there is not only fusion of the two primordia, but the cardiac primordium must be absorbed into the left wall if the pulmonary veins are to establish connections with the left atrium. Auer (1948) observed cranial and caudal parts of the cardiac primordium. If the cranial part differentiates, development is normal; differentiation of the caudal part leads to anomalous pulmonary venous connections. (For further discussion, see Section A.-1e, Chapter X.)

e. Inferior vena cava and its tributaries: The inferior vena cava with its immediate branches is a composite vessel derived from the postcardinal, subcardinal and thoracolumbar veins and certain of their anastomoses, chiefly on the right side. The latter are:

1. the right thoracolumbar vein below the kidney
2. the right thoracolumbar-subcardinal anastomosis
3. the right subcardinal vein
4. a right anastomosis between subcardinal vein and hepatic sinusoids
5. the intrahepatic portion of the inferior vena cava
6. the right primitive hepatic vein

There is a difference of opinion as to the contribution of the postcardinal veins and of the thoracolumbar veins in the formation of the right common iliac vein and the distal part of the left common iliac vein. The latter is derived mainly from an interthoracolumbar anastomosis.

f. Veins of lower extremities: Border veins in the vascular plexus of the lower limb buds which are continuous with the postcardinal veins account for the early venous development in the lower extremity. Their derivatives include the great saphenous vein, which is related to the origin of the femoral, and the posterior tibial. Below the knee, border veins on the fibular side contribute anterior tibial and small saphenous veins.

g. Renal veins: The right renal vein is derived from the right subcardinal vein. The left renal vein is largely an intersubcardinal anastomosis. Both right and left parts of the renal veins at the hilum of the kidney are thoracolumbar in origin. (For further discussion, see Section A.-1c, Chapter XIII.)

h. Cerebral veins: Streeter (1915, 1918) studied the venous development particularly with reference to the transformation of the primary head veins into dural sinuses and the formation of veins related to the brain. The most recent investigations are the detailed and excellently illustrated studies of Padget (1956, 1957). (For further discussion, see Section A.-2, Chapter VIII.)

4. Anomalies

The numbers of venous primordia and the paired condition of each make possible a wide variety of anomalies. Both superior and inferior vena cavae may be paired, in some cases with a nearly symmetrical development, in others, with a definite asymmetry of each vessel of the pair. Both superior and inferior vena cava may also show a reversal of the side of differentiation, with the main vessel being located on the left side. In the case of inferior vena cava primordia, the retention of various combinations of primordial vessels has been found, and the same characteristics hold true for other derivatives of the "venous lines." The azygos veins at times may be the main venous return to the heart. Both azygos and hemiazygos show great variability. Variations of the pulmonary veins involve the number and connections of their vessels with the lungs, with the heart, or with the large veins near the heart. (For reference to papers on venous anomalies, see Atwell and Zoltowski, 1938; Blasingame and Burge, 1937; Charles, 1889; Chouke, 1939; Compere and Forsyth, 1944; Davies and MacConaill, 1937; Dwight, 1901; Fraser and Brown, 1944; Givens, 1912; Halpert, 1930, 1942; Huseby and Boyden, 1941; Kaestner, 1900; Lewis, 1909; Neuberger, 1913; Smith, 1916; Thompson, 1929; Waring, 1894.)

Seib (1934) has used his diagram of the "venous lines" to analyze many of the venous anomalies in the thorax and abdomen. In a similar manner, Davis *et al.* (1958) have explained by the diagrammatic pattern the variations of lumbar, renal, and associated parietal and visceral veins.

5. Clinical Implications of Normal or Anomalous Development

A knowledge of venous development provides the clinician with a basis for understanding the more common arrangements of veins, any anomalous differentiation, and the clinically important venous anastomoses. To know about the early venous plexuses and to follow the primordial venous channels derived from them in differentiation are a means of anticipating departures from the normal developmental pattern. Normal, as well as anomalous, veins receive the attention of the surgeon, especially in certain regions, and equal consideration must be given to every type of venous development by clinicians who do venous catheterizations or who analyze the results of venography. Likewise, what is to be expected from a given venous anastomosis is an important problem in the prognosis of some cases.

B. PHYSIOLOGY

By A. D. M. GREENFIELD;[1] L. A. SAPIRSTEIN[2]

1. VEINS AS TRANSPORT VESSELS

a. Venous pressure: A horizontal reference plane may be imagined through the level to which blood rises in a tube connected to the right atrium of the heart (Fig. 48). If the veins were rigid valveless tubes, then with the circulation arrested, the blood in a manometer tube connected with any peripheral vein would also rise to the level of the reference plane (Fig. 48A). In such a system, the effect of circulation of the blood would be to increase the pressure in the distal veins, relative to that in the atrium, and hence the level of fluid in manometers connected to such vessels would rise above the reference plane (Fig. 48B and C). The amount of the excess pressure would depend on the rate of circulation and the resistance to flow between the measured distal veins and the atrium. All these statements are true, whatever the posture of the body and whatever the strength of the gravitational acceleration.

In fact, of course, the veins, far from being rigid valveless tubes, are collapsible vessels with reactive walls and frequently with valves. All of these characteristics lead to departures from the behavior expected of rigid tubes.

b. Conditions of flow in collapsed and distended veins: Because they are nonrigid, veins (with exceptions to be subsequently noted) are unable to support an internal subatmospheric pressure or to act as syphons. Those above the reference plane are, in general, in a semicollapsed condition. At all levels the pressure in the lumen is slightly in excess of the local tissue pressure, which is, itself, slightly in excess of atmospheric pressure (Ryder *et al.*, 1944). Hence there is a considerable hydrostatic pressure, amounting to 0.7 mm Hg per cm of vertical height, available to propel the blood heartwards through the semi-collapsed veins.

Within the skull, the dural sinuses are structurally unable to collapse. Further, the rigidity of the skull permits the intracranial pressure to fall below atmospheric when standing, and so veins in the skull, containing blood under such conditions, can remain fully distended provided the pressure in them is above intracranial pressure. These vessels, therefore, usually behave as rigid tubes and, by a syphon action, assist the cerebral

[1] Author of Subsections 1–3.
[2] Author of Subsection 4.

circulation. Thus they partly protect it against the effects of change of posture and of gravity (Green, 1950).

Veins below the reference plane are normally distended. It is probably correct to say that in a completely relaxed subject, the pressures are as they would be in rigid valveless tubes; relaxation is more

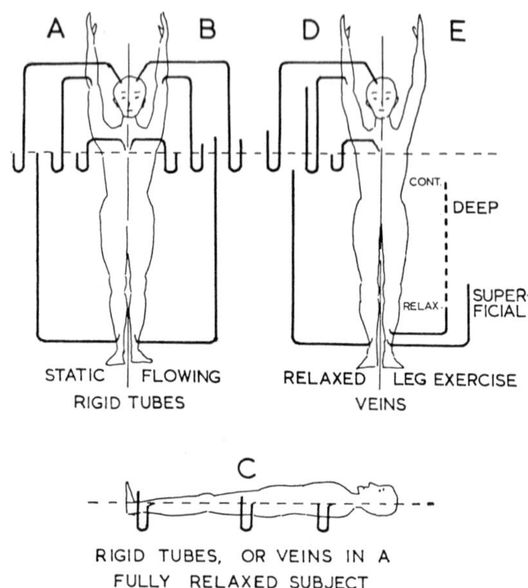

FIG. 48. The position of blood in blood filled manometer tubes connected to an intracranial vein, veins in the arm or leg, and the right atrium of the heart. The horizontal broken lines indicate the reference plane of right atrial pressure. A and B—a vertical model with rigid valveless tubes in place of veins with the blood static in A and flowing in B. C—both such a horizontal model with blood flowing and the condition in a fully relaxed horizontal subject. D and E—a vertical subject. In D he is relaxed and passively supported on a nearly vertical board. In E he is rhythmically contracting the leg muscles. The pressures in the deep veins during contraction and relaxation are indicated, although in practice a blood manometer would be incapable of registering accurately the large and rapid fluctuations of pressure.

easily attained in the horizontal (Fig. 48C) than in the vertical (Fig. 48D) position. Observations in the horizontal position show that the resistance to flow in the large veins is very small. At rest, the pressure drop from foot to groin is only 6 mm Hg (Ochsner *et al.*, 1951) and from the arm to the right atrium, 3 mm Hg (Richards *et al.*, 1942). In a person lying on his side, the pressure in the veins of the under arm hanging through a hole in the couch supports a column of saline extending 3.4 ± 0.9 cm above that supported by a cannula in the right atrium, the pressure difference being less than 3 mm Hg (Gauer and Sieker, 1956). Because resistance to flow in the large veins is so small, compared to that through the fine peripheral vessels, it has not been systematically determined. Strict measurement would require observations of pressure at two places and of the volume flow of blood in the intervening segment, a reading which is not easily obtained. However, total limb blood flow, to which the venous flow might be supposed to be proportional, is a fairly good alternative. Doupe *et al.* (1938) found no increase in the pressure in the basilic vein during reflex vasodilatation and little decrease during reflex vasoconstriction. This suggests, but in the absence of atrial pressure measurements does not prove, that downstream resistance between the veins and the heart and upstream resistance between the heart and the veins were undergoing proportional changes.

An important practical point is that the effect of the reactivity of the vein wall on venous resistance is usually small, compared with, for example, the response to nipping of the veins by fascial planes or by surface pressure. Peripheral venous pressure is far more likely to be raised by obstruction or a rise of atrial pressure (as in congestive heart failure or during the Valsalva maneuver) than it is by constriction of downstream veins.

c. Valves and the "muscle pump": Most observations on the relaxed vertical subject show that the pressure in the superficial veins of the foot is not quite sufficient to support a column of blood to the level of the reference plane (Fig. 48D). This departure from the "rigid tube" condition is probably due to the difficulty of securing full muscular relaxation. Rhythmic contractions of the leg muscles (Fig. 48E) reduce the venous pressure very considerably (Smirk, 1936; Pollack and Wood, 1949; Walker and Longland, 1950). This is due to the action of the "muscle pump." Contraction of the muscles in the leg increases the intramuscular pressure, especially in the deep muscles and those surrounded by a relatively nondistensible fascia (Wells *et al.*, 1938). Blood is squeezed from the veins in the same manner as blood is ejected from the ventricles by the contraction of the heart. Provided the valves are competent, the blood is forced to move centrally. The pressure in the veins, which is raised during muscular contraction, falls during relaxation (Fig. 48E). This permits blood to flow from superficial to deep veins, thus reducing the pressure in superficial vessels with competent valves. The magnitude of the reduction in pressure, which may amount to 60 mm Hg or more, depends on the vigor of the muscular contractions and the rate of inflow into the superficial veins. In a cold subject with a low limb blood flow, the pressure is more readily reduced than in a warm one with a large blood flow.

The limit to the drop in pressure in the superficial veins is set by the lowest level in the deep veins rather than by the mean pressure. During modest exercise, Höjensgärd and Stürup (1953) found the mean pressure to be more greatly reduced in the long saphenous (superficial) veins than in the posterior tibial (deep) vein, while in the popliteal vein the pressure was not altered at all. This state of affairs is analogous to that in the heart, where the mean pressure in the atria is below that in the ventricles.

In the upright subject, the "muscle pump" greatly assists the circulation to the leg muscles. These receive most of their blood inflow during periods of relaxation (Barcroft and Dornhorst, 1949), at a time when the pressure in the deep veins is most reduced and the arteriovenous pressure difference available for perfusion is the greatest.

d. Resistance to flow in veins: Haddy and associates (Haddy and Gilbert, 1956; Haddy *et al.*, 1957a,b; Haddy, 1960), using the dog's foreleg, have recently reported extensive simultaneous measurements of pressure in a large artery, in the peripheral collaterals of a small artery about 0.5 mm in diameter, in the peripheral collaterals of a vein about 0.5 mm in diameter, and in a large vein. In some experiments flow was held steady by pump perfusion. Haddy and his associates found the pressure drop, and hence the resistance to flow, between the venules and large veins to be an important fraction (normally about 10%) of the total in the limb. Not only did this resistance vary, but the change was independent of the alterations in resistance in other vessels. Thus, exposure to cold air, intra-arterial epinephrine and norepinephrine and venous congestion all raised the resistance between small arteries and small veins. However, cold air and epinephrine left the venous resistance almost unchanged, norepinephrine raised it, while venous congestion lowered it. Intra-arterial serotonin and histamine decreased small vessel resistance but often increased venous resistance. Further, when small vessel and venous resistance moved in the same direction, they did not necessarily do so proportionately or simultaneously. Wallace and Stead (1957) similarly demonstrated the importance of the resistance of small veins in the finger; the constriction of these vessels in cold fingers or in those in a shocked patient prevented the drainage of blood and color from a raised limb.

2. Veins as Capacity Vessels

a. Veins as blood reservoirs: From postmortem measurements, it has been estimated that of the blood in the dog's mesenteric vessels, 74 per cent is in veins and venules (Landis and Hortenstine, 1950). In man it is unfortunately not possible, during life, to measure this percentage or to divide it between veins and venules. However, the plethysmographic observations of McLennan *et al.* (1942) and of Litter and Wood (1954) indicate that in the human limbs most of the blood, and particularly the variable portion, lies in vessels having the low pressures characteristic of venules and veins. Thus, venous congestion at 30 mm Hg doubles the volume of blood in the forearm. It has also been estimated that while an extra 40 ml of blood raises the pressure in the systemic arteries of man by 40 mm Hg, the same volume added to the venous side of the circulation elevates the pressure by only 0.2 mm Hg (Gauer *et al.*, 1956). The veins, therefore, act as blood reservoirs, in the sense that they very often contain blood surplus to that required to fulfill efficiently their transport function as drainage channels. The reservoir function is especially developed in the pulmonary and hepatic veins (Sjöstrand, 1953). The blood in the venous reservoirs has the same composition as that actively circulating and is most probably wholly included in measurements of circulating blood volume. The splenic reservoir, important in some animals, is not so in man (Ebert and Stead, 1941).

The significance of the reservoir function of the veins is emphasized by the very considerable shifts of blood on changing posture. Asmussen (1943) found that of the total blood volume, the legs contain 13.5 per cent with the subject horizontal, 20.5 per cent while tilted 60° head up, 12 per cent while tilted 60° head down, 16 per cent while standing freely and 25.5 per cent while standing after working. Sjöstrand (1952), reported that 600 ml of blood may leave the legs when the subject, who has been standing, assumes the horizontal position.

When upright, the pooling of blood in the legs depends on how well the muscle pump deals with the arterial inflow. Of interest in this regard is the finding that the inflow may be raised for a long time (over 1 hr) after exercise (Halliday, 1959). Pooling is sufficient to cause arterial hypotension in one half, and syncope, in one quarter, of healthy men who, following severe exercise, are passively tilted to the 70° head up position (Eichna *et al.*, 1947).

b. Distensibility characteristics of veins: Of great significance to their reservoir function is the distensibility of veins. *In vitro*, a very modest elevation in pressure suffices to fill a collapsed vein, but once filled, further increase in volume depends on stretching the elastic tissue in the walls, a response which requires a considerable rise in pressure. Eventually, augmentation in volume is limited by nonelastic components (Ryder *et al.*, 1944). These three phases may occur in individual segments of veins *in vivo,* but they become blurred when the distensibility of all the low-pressure vessels in a limb is considered (Fig. 49). For example, on preventing venous return, the venous pressure beyond the obstruction promptly starts to rise steadily (Greenfield and Patterson, 1954), although little elevation would have been expected until all the collapsed veins had been filled.

c. Variation in the distensibility of veins: The control of the distensibility of the veins and other capacity vessels is similar in importance to the control of the volume of circulating blood. Some observations have been made on central venous pressure (Ross *et al.,* 1959), but most have been reported on the behavior of parts of the peripheral venous

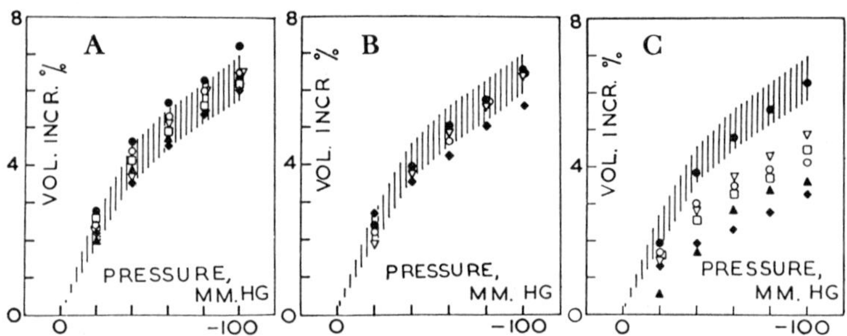

FIG. 49. Effect of vascular distension, by local exposure to subatmospheric pressure, on the volume of the calf of the leg. Hatched area—envelope of responses with the subject at a comfortable environmental temperature and the calf in water at 35°C. A—Temperature of water around the calf 45°C. B—Subject generally heated. C—Temperature of water around the calf 4°C. (From Coles *et al.,* 1957; reproduced with the permission of the Journal of Applied Physiology.)

system. The alterations in distensibility in response to what may be regarded as physiologic stimuli are modest and of the same order as the changes in volume in circulating blood which are compatible with life.

Nervous control appears to play a role in altering distensibility of veins. If the low pressure vessels as a whole are reflexly controlled, one might expect them to contract or at least to become less distensible when the return of venous blood to the heart is inadequate. The effective stimulus might be a change in pressure or a tissue distortion, probably in the thorax and related to cardiac filling. In support of such a speculation is the finding that inferior caval or mitral valve obstruction causes reflex constriction of the intestinal low pressure vessels in dogs (Alexander, 1956). Furthermore, breathing air at a pressure of 40 mm Hg above atmospheric produces a 25 per cent reduction in the distensibility of the low pressure vessels of the hand and a rise in pressure in an isolated venous segment in the forearm (Ernsting, 1957). Overbreathing, whether of air or carbon dioxide mixtures, causes a reduction in the distensibility of the low pressure vessels of the forearm (Eckstein *et al.*, 1958; Lyttle, 1961). In the case of all these conditions there is a mechanical disturbance in the thorax (Eckstein and Hamilton, 1958).

The limb veins react reflexly to a considerable variety of other stimuli, although the functional significance of the responses is not always clear (Doupe *et al.*, 1938). Duggan *et al.*, (1953) and subsequently others have measured the changes in pressure in a segment of a large subcutaneous forearm vein, temporarily isolated from the circulation and so containing a constant volume of blood. Rises in pressure, usually of a few mm Hg, but sometimes of 40 mm Hg or more, have been reported in response to apprehension, mental arithmetic, a single deep inspiration, inspiration against a resistance, the Valsalva maneuver, hyperventilation of air, breathing 5 per cent carbon dioxide, exercise of other parts of the body and tilting to the feet-down position (Page *et al.*, 1955; Burch and Murtadha, 1956; Martin *et al.*, 1959; Merritt and Weissler, 1959). A vasovagal attack is the sole reported cause of a fall in pressure (Burch and Murtadha). Pressure changes have, in the cases investigated, been reduced or abolished by blocking the efferent nervous pathway.

Although the reported reflex elevations in pressure in venous segments are modest compared with the hydrostatic increase in venous pressure due to alterations in posture, they appear to be of significance. Thus, Hickam and Pryor (1951) found that gravitational stress is not tolerated as well by men and by dogs when the venoconstrictor response is small.

Thermoregulatory reflexes have little effect on venous distensibility (Fig. 49). General body heating, which increases cutaneous blood flow in all parts of the limbs severalfold, produces practically no change in the distensibility of the low pressure vessels of the hand (Coles and Patterson, 1957), forearm (Greenfield and Patterson, 1956), or calf (Coles *et al.*, 1957). On the other hand, general body cooling decreases the distensibility of the low pressure vessels to about 70 per cent of the normal value (Coles and Patterson, 1957; Wood *et al.*, 1958).

Changes in distensibility can also be produced by chemical agents. For example, the distensibility of low pressure vessels in the hand and forearm is moderately reduced by infusion into the brachial artery of noradrenaline, adrenaline and 5–hydroxytryptamine. Histamine, however, is without effect (Glover *et al.*, 1958). Whether the various reactions are of physiologic significance remains to be evaluated.

The alteration in distensibility produced by physical agents will vary, depending upon the type of stimulus. Although light tapping causes local dilatation of veins (Franklin and McLachlin, 1936), on mechanical irritation the large subcutaneous vessels are capable of such powerful constriction that the lumen may be almost obliterated. Raising the temperature of the stirred water around the limb from about 34° to about 45°C increases the local rate of blood flow several times, but it has very little effect on the distensibility of the low pressure blood vessels, mainly veins and venules, of the forearm and calf (Fig. 49A) (Greenfield and Patterson, 1956; Coles *et al.*, 1957; Kidd and Lyons, 1958), and only a modest effect on those of the hand (Coles and Patterson, 1957).

Lowering the temperature to about 4–6°C, which causes recurrent vasodilatation and increased flow in skin (Lewis, 1930) and muscle (Clarke *et al.*, 1958), reduces the low pressure distensibility of the finger (Greenfield and Shepherd, 1950) and of the calf of the leg (Fig. 49C) (Coles *et al.*, 1957; Kidd and Lyons, 1958). Kidd and Lyons were unable to demonstrate increase in low pressure distensibility immediately after local exercise and at a time when flow was increased three- to fivefold. The reactions to chemical and physical agents, therefore, are of doubtful importance in relation to the economy of the circulation as a whole.

3. Sympathetic Innervation[3]

a. Criteria of active venoconstriction and venodilatation: Since veins are so readily made to fill or empty passively, in consequence of alterations in the circulation upstream or downstream, changes in venous diameter are, by themselves, no sure indication of the state of activity of the venous wall. To establish, for example, an active venoconstriction, it is necessary to observe: (a) narrowing of the veins while the internal (or, better, the transmural) pressure remains unchanged or rises; (b) an increase in the pressure gradient along the vein while the blood flow is unchanged or reduced and the transmural pressure is unchanged or elevated; (c) an alteration in the pressure/volume characteristics of a segment of vein temporarily isolated from the general circulation.

Unfortunately these studies are not very easy to make, and the literature of venous innervation is much less complete than that of the innervation of the total complex of peripheral vessels responsible for the resistance to flow between arteries and veins.

b. Evidence for sympathetic venoconstrictor nerves in animals: The earlier animal observations have been reviewed by Gollwitzer-Meier

[3] For sympathetic control of arterial tree, see Chapter IV.

(1932) and Franklin (1937) and a comprehensive review was made by Landis and Hortenstine (1950). Donegan (1921) found that stimulation of the sympathetic pathway caused narrowing, and section of it produced widening, of the mesenteric and superficial veins of the cat. He also reported changes in caliber of the mesenteric veins on stimulating afferent nerves. The distensibility characteristics of a temporarily isolated innervated region of the splanchnic circulation of the dog have been extensively investigated by Alexander (1954, 1955, 1956). It seems probable that the observed changes were mainly on the venous side, although responses of venules were not distinguishable from those of veins. There was reflex relaxation of the splanchnic veins on raising inferior vena caval pressure, and constriction, on eliciting pressor reflexes, and in response to hemorrhage.

Proof of vasoconstrictor innervation of the veins in the leg of the dog has been afforded by Kelly and Visscher (1956). Stimulation of the lumbar sympathetic trunk (Fig. 50) caused a considerable rise in pressure in the small veins (about 0.1–0.5 mm in diameter) in the foot pad, although the inferior vena caval pressure was unchanged and the femoral arterial blood flow was greatly reduced. On the reasonable assumption that the blood flow was also decreased in the venous system between the small veins and the inferior vena cava, these observations are evidence for an increase in the resistance to flow, and thus venoconstriction, at some or all places between the small veins and the vena cava. Salzman and Leverett (1956) found that on stimulation of the lumbar sympathetic chain, the pressure in a miniature balloon in the small saphenous vein of the dog rose more than did the pressure in the lumen of neighboring veins. This result seems to be clear evidence for an active constriction of the segment containing the balloon.

c. Evidence for sympathetic venoconstrictor nerves in man: There is now a good deal of evidence that the pressure in temporarily isolated segments of vein in the human forearm can be raised by such stimuli as apprehension, mental arithmetic, a deep inspiration, hyperventilation, the Valsalva maneuver, application of ice to the skin of the hand, and tilting into the foot down position (Duggan *et al.*, 1953; Page *et al.*, 1955; Burch and Murtadha, 1956; Merritt and Weissler, 1959; Martin *et al.*, 1959). Where it has been tested, perivenous infiltration of local anesthetic has abolished the rise in pressure. This is evidence for venoconstrictor innervation of the superficial veins of the human arm. Beaconsfield (1954) has reported that following sympathectomy, the superficial veins of the arm, as revealed by infrared photography, remain dilated at a time when the whole limb blood flow has returned to normal. This is sugges-

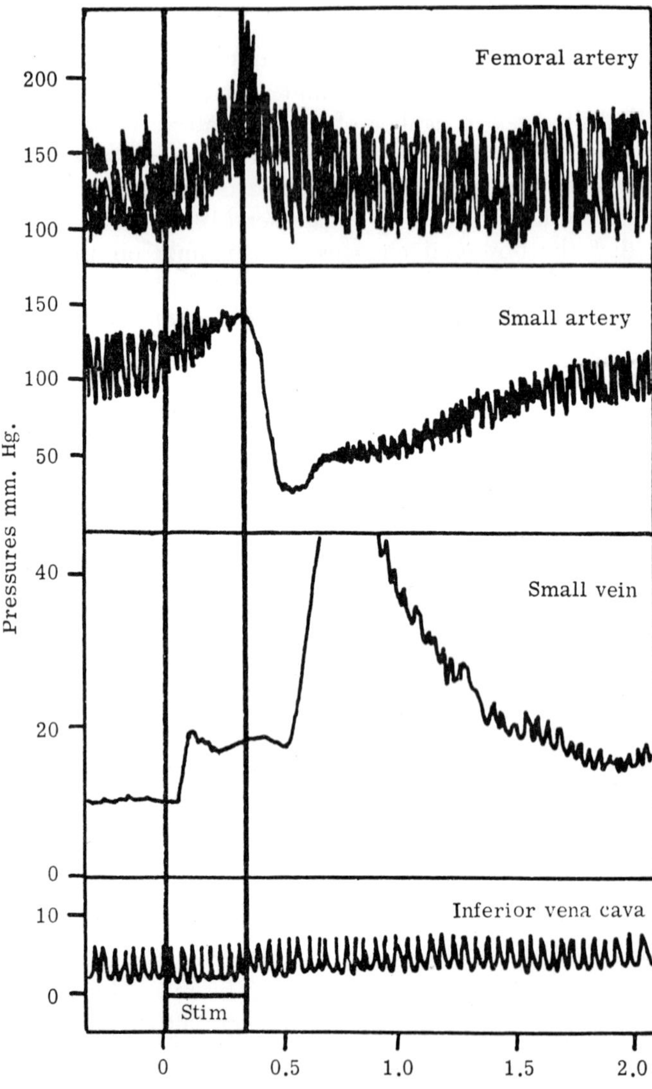

FIG. 50. Pressures in the veins of the hind leg and in the inferior vena cava of a dog. During electrical stimulation of the lumbar sympathetic chain (between the vertical lines) there is a rise of pressure in the small vein, with a secondary and larger rise after stimulation ceases. Inferior vena caval pressure is almost unchanged. (Redrawn from Kelly and Visscher, 1956.)

tive, but not conclusive, evidence that venomotor tone is abolished by sympathectomy for a longer period than is vasoconstrictor tone.

d. The possible existence of venodilator nerves: No evidence has so far come to light for the presence of sympathetic vasodilators to veins, similar to those well established for the resistance vessels of the skin (Roddie *et al.*, 1957) and muscle (Blair *et al.*, 1959) of the human forearm. Although Horiuchi (1924) observed that in animals there was widening of the superficial veins upon stimulation of the posterior nerve roots or the peripheral end of the somatic nerve five days after section, such an observation requires confirmation by modern techniques.

e. Innervation of venules: It is at present conjectured, but not proved, that venules share the sympathetic vasoconstrictor innervation now proved to exist in the large veins. Since the color of the skin is supposedly largely due to blood in the venules, the fact that changes in skin color occur readily suggests that these vessels are highly reactive. The "flare" of Lewis' triple response to injury may indicate that venules are innervated by axon-reflex vasodilator fibers.

4. Humoral Control of Veins[4]

Although, in the intact animal, a chemical agent may influence the condition of the veins through direct muscular action, indirect effects may be of equal or greater importance. For example, obvious dilatation of a vein may be a consequence of relaxation of the venous musculature and also of arteriolar vasodilatation, permitting the vein to be filled with blood at a higher than normal pressure. Conversely, narrowing of a vein may be due to the presence of a chemical agent which causes the muscle of its wall to contract or to a reduction in right atrial pressure which permits the vessel to empty its content of blood more readily into the heart. Also it is sometimes difficult to evaluate in an intact organism whether the effects of a chemical substance upon veins are due to local action on the vein wall or to a central effect on the venomotor centers. Nevertheless, despite these difficulties of interpretation, the responses of veins to some chemical agents are sufficiently well characterized to warrant discussion.

a. pH: A decrease in pH is followed by venous dilation, the effect being a local one (Voss and Gollwitzer-Meier, 1933). The vasomotor center, however, is simultaneously stimulated, and a general venoconstriction follows (Gollwitzer-Meier and Bohn, 1930). Thus, in the intact

[4] By L. A. Sapirstein.

organism, the veins draining an active tissue are presumably dilated, while others are constricted, as a result of the central stimulus (Franklin, 1937).

b. Epinephrine and norepinephrine: Summarizing the available literature, Franklin (1937) found few exceptions to the generalization that adrenaline is vasoconstrictor. However, the recent finding of Haddy (1957) that norepinephrine constricts the veins of the dog's forelimb, while epinephrine at the same dosage does not, suggests the need for reinvestigation of the effects of the catecholamines using purified compounds (for further discussion, see Section C.-2b, this chapter).

c. Acetylcholine: Fleisch (1930) noted that the characteristic response of veins to physiologic concentrations ($1{:}10^9$ or less) of acetylcholine is dilation, while higher concentrations are often constrictor.

d. Vasopressin: Large veins are quite unresponsive to this hormone (Portman and MacDonald, 1928). On the other hand, the intense pallor of the skin of subjects who have received parenteral vasopressin clearly demonstrates that the small veins of the subpapillary venous plexus are vigorously constricted.

e. Other agents: Applied to large veins, histamine in concentrations exceeding $1{:}10^9$ is predominantly constrictor (Fleisch, 1930). However, in the case of venules, characteristically it is vasodilator. Serotonin constricts the veins of the dog's forelimb (Haddy, 1957).

f. Significance of venous responses to chemical agents: The resistance of the venous system is very small in relation to the total resistance of any vascular bed. It therefore seems improbable that venous responses to chemical changes are of great importance in the regulation of local blood flow. It is more likely that the venous system responds as a whole, as a reservoir whose contraction and relaxation serve to make more or less blood available to the heart for its overall output. Whether, in this respect, the chemical control of the venous system makes a significant addition to the well known nervous regulation of the capacity of the venous system remains to be determined.

C. PHARMACOLOGY

By K. I. MELVILLE AND H. E. SHISTER

1. INTRODUCTION

The effects of drugs on the larger venous channels have not been systematically studied. There are few observations in the literature concerning the direct effect of autonomic agents on the venules and small veins, but in connection with the effect of cholinergic drugs there is a complete gap of information. The adrenergic drugs are generally regarded as constrictors of veins. The principal available studies are in connection with the problem of orthostatic or postural hypotension, associated with the actions of the ganglionic blocking agents. Some data are also present concerning various alterations induced by other types of agents, such as the nitrites and digitalis. It must, however, be borne in mind that venous vasomotion is also influenced by such conditions as anoxia (Eckstein and Horsley, 1959), mechanical changes (Wideman, 1959; Wood, 1959) electrolytes and physical or emotional stress, all of which must be considered in the evaluation of the net responses to drugs.

2. PRINCIPAL PHARMACOLOGIC RESPONSES OF VEINS[5]

a. Blocking agents: In general, ganglionic blocking agents produce a marked degree of orthostatic hypotension. The mechanism involved is somewhat complex, but it is generally assumed that the important factor is the interference with the normal venous adjustment of the continued return of the blood to the heart in the erect position. In this regard, it has been shown that hexamethonium abolishes vasoconstrictor reflexes (cold pressor test, Valsalva maneuver), from which it can be concluded that the compensatory vasomotor adjustment, which, in the erect position, normally elicits sympathetic vasoconstriction to prevent a drop in blood pressure, is abolished. This has also been shown to be so in patients with spontaneous orthostatic hypotension and in anesthesia. It is not certain whether veins, arterioles and capillaries are all involved in this process, although it appears that the veins will principally accommodate the pooling of blood following the administration of ganglionic blocking agents.

[5] For humoral control of veins, see Section B.-4, this chapter.

It is also interesting to note that in congestive heart failure (right or left), marked symptomatic improvement has been obtained with the use of hexamethonium. This is apparently brought about by: (a) decreasing the elevated pressure in the veins and right side of the heart (by abolishing normal sympathetic tone), with the drug acting as a venesection and thus restoring cardiac output to normal; and (b) reducing the raised peripheral resistance, usually found in cases with congestive failure, hence leading to a decreased demand for cardiac work.

b. Adrenergic agents: The beneficial effect of norepinephrine in hypotensive states is well established, particularly so, where there is deficient cardiac contraction, as in myocardial infarction or in pooling of blood. It has been shown that, in the early stages following hemorrhage in dogs, norepinephrine raises the blood pressure and cardiac output, the latter being due to increased venous tone, with consequent return of blood and improved stroke volume.

c. Digitalis and related glycosides: The effect of these drugs on the veins is not clearly understood. Early reports of spasm of hepatic veins in dogs following the use of digitalis have not been substantiated. The thesis of McMichael (1950) that digitalis leads to a primary reduction in venomotor tone is also not supported by later studies. It has been reported that this drug may have some constricting influence on veins, which may partly explain the diminished blood volume following digitalization (Walton, 1958).

d. Nitrites and nitrates: The principal effect of these agents is dilatation of veins. For example, it has been shown that sodium nitrite reduces venous (postarteriolar) tone. In man, small doses of this drug have no effect when the subject is in recumbency, whereas in the erect position it can produce a marked fall in venous pressure, as well as a general drop in blood pressure, with diminished blood flow in the hand. It is interesting to note that sodium nitrite or nitroglycerin does not produce cutaneous vasodilatation to the same degree as amyl nitrite. Furthermore, capillaries and venules of the skin are not normally dilated by these agents. (For detailed pharmacologic responses to nitrites, see Melville, 1958.)

D. BIOCHEMISTRY

By J. E. Kirk

1. Normal Constituents of Venous Wall

The anatomic structure of human veins differs in many respects from that of the arteries, and for this reason significant differences are also found in the chemical composition of the two types of blood vessels. Thus the veins contain less smooth muscle and elastic tissue than the arteries, but have a higher collagen content (80% of dry matter, or about 21% of wet tissue weight) (Hevelke, 1959b). The average composition of the normal human inferior vena cava and femoral vein is shown in Table VIII (Hevelke, 1958, 1959a,b).

TABLE VIII
Composition of Normal Human Venous Tissue in Young Adult Individuals[a]
Values expressed in percentage of wet tissue weight

	Inferior vena cava	Femoral vein
Dry matter	26.1	28.0
Nitrogen	—	4.08
Cholesterol	0.083	0.076
Total ash	—	0.590
Calcium	0.012	0.058
Potassium	0.065	—

[a] The values recorded in the table were calculated on the basis of data reported by Hevelke (1958, 1959a,b).

2. Changes Observed with Aging

Because the veins are rarely the site of significant pathologic lesions, the changes in the composition of these vessels with age as compared with those of the arterial tissue are of considerable interest. Valuable data based on analyses of 149 samples of the inferior vena cava and 96 samples of the femoral vein in human subjects of various ages have recently been reported by Hevelke (1958, 1959a,b). The average values listed in Table IX show that, similar to the findings for the arterial tissue, a decrease takes place with age in the venous tissue content of dry

TABLE IX

VARIATION WITH AGE IN COMPOSITION OF NORMAL HUMAN VENOUS TISSUE[a]

Values expressed in percentage of wet tissue weight

	Inferior vena cava				Femoral vein				
Age group	Dry matter	Cholesterol	Calcium	Potassium	Dry matter	Nitrogen	Cholesterol	Total ash	Calcium
0– 9	30.9	—	—	—	31.1	4.63	0.058	0.525	0.051
10–19	30.1	0.109	0.011	0.083	29.5	4.28	0.072	0.535	0.063
20–29	26.8	0.082	0.011	0.072	27.4	3.88	0.071	0.586	0.053
30–39	25.3	0.083	0.012	0.059	28.5	4.41	0.082	0.610	0.063
40–49	24.5	0.091	0.011	0.053	24.7	3.34	0.064	0.542	0.065
50–59	23.7	0.100	0.010	0.051	24.2	3.36	0.096	0.536	0.071
60–69	21.7	0.097	0.010	0.048	23.2	3.06	0.087	0.555	0.075
70–79	22.7	0.097	0.011	0.048	21.8	2.92	0.087	0.600	0.083

[a] The values recorded in the table were calculated on the basis of data reported by Hevelke (1958, 1959a,b).

matter, nitrogen, and potassium, whereas only minor changes occur in the cholesterol and calcium concentrations of the tissue.

REFERENCES

Alexander, R. S. (1954). The participation of the venomotor system in pressor reflexes. *Circulation Research* **2,** 405.

Alexander, R. S. (1955). Venomotor tone in haemorrhage and shock. *Circulation Research* **3,** 181.

Alexander, R. S. (1956). Reflex alterations in venomotor tone produced by venous congestion. *Circulation Research* **4,** 49.

Asmussen, E. (1943). The distribution of the blood between the lower extremities and the rest of the body. *Acta Physiol. Scand.* **5,** 31.

Atwell, W. J., and Zoltowski, P. (1938). A case of left superior vena cava without a corresponding vessel on the right side. *Anat. Record* **70,** 525.

Auer, J. (1948). Development of the human pulmonary vein and its major variations. *Anat. Record* **101,** 581.

Barcroft, H., and Dornhorst, A. C. (1949). The blood flow through the human calf during rhythmic exercise. *J. Physiol. (London)* **109,** 402.

Beaconsfield, P. (1954). Veins after sympathectomy. *Surgery* **36,** 771.

Blair, D. A., Glover, W. E., Greenfield, A. D. M., and Roddie, I. C. (1959). Excitation of cholinergic vasodilator nerves to human skeletal muscles during emotional stress. *J. Physiol. (London)* **148,** 633.

Blasingame, F. J. L., and Burge, C. H. (1937). A case of left postrenal inferior vena cava without transposition of viscera. *Anat. Record* **69,** 465.

Burch, G. E., and Murtadha, M. (1956). A study of the venomotor tone in a short intact venous segment of the forearm of man. *Am. Heart J.* **51,** 807.

Butler, H. (1952). Some derivatives of the foregut venous plexes of the albino rat, with reference to man. *J. Anat.* **86**, 95.

Charles, J. J. (1889). Notes of a case of persistent left superior vena cava, the right superior vena cava being in great part a fibrous cord. *J. Anat. and Physiol.* **23**, 649.

Chouke, K. S. (1939). A case of bilateral superior vena cava in an adult. *Anat. Record* **74**, 151.

Clarke, R. S. J., Hellon, R. F., and Lind, A. R. (1958). Vascular reactions of the human forearm to cold. *Clin. Sci.* **17**, 165.

Coles, D. R., and Patterson, G. C. (1957). The capacity and distensibility of the blood vessels of the human hand. *J. Physiol.* (*London*) **135**, 163.

Coles, D. R., Kidd, B. S. L., and Moffatt, W. (1957). Distensibility of blood vessels of the human calf determined by local application of subatmospheric pressures. *J. Appl. Physiol.* **10**, 461.

Compere, D. E., and Forsyth, H. F. (1944). Anomalous pulmonary veins. *J. Thoracic Surg.* **13**, 63.

Davies, F., and MacConaill, M. A. (1937). Cor biloculare, with a note on the development of the pulmonary veins. *J. Anat.* **71**, 437.

Davis, R. A., Milloy, F. J., and Anson, B. J. (1958). Lumbar, renal, and associated parietal and visceral veins based upon a study of 100 specimens. *Surg., Gynecol. Obstet.* **107**, 1.

Donegan, J. F. (1921). The physiology of the veins. *J. Physiol.* (*London*) **55**, 226.

Doupe, J., Krynauw, R. A., and Snodgrass, S. R. (1938). Some factors influencing venous pressure in man. *J. Physiol.* (*London*) **92**, 383.

Duggan, J., Love, V. L., and Lyons, R. (1953). A study of reflex venomotor reactions in man. *Circulation* **7**, 869.

Dwight, T. (1901). Absence of the inferior cava below the diaphragm. *J. Anat. and Physiol.* **35**, 7.

Ebert, R. V., and Stead, E. A. (1941). Demonstration that in normal man no reserves of blood are mobilized by exercise, epinephrin and haemorrhage. *Am. J. Med. Sci.* **201**, 655.

Eckstein, J. W., and Hamilton, W. K. (1958). Changes in transmural central venous pressure in man during hyperventilation. *J. Clin. Invest.* **37**, 1537.

Eckstein, J. W., and Horsley, A. W. (1959). Effect of hypoxia on peripheral venous tone in man. *Circulation* **20**, Part II, 688.

Eckstein, J. W., Hamilton, W. K., and McCammond, J. M. (1958). Pressure volume changes in the forearm veins of man during hyperventilation. *J. Clin. Invest.* **37**, 956.

Eichna, L. W., Horvath, S. M., and Bean, W. B. (1947). Post-exertional orthostatic hypotension. *Am. J. Med. Sci.* **213**, 641.

Ernsting, J. (1957). Effect of raised intrapulmonary pressure upon the distensibility of the capacity vessels of the upper limb. *J. Physiol.* (*London*) **137**, 52P.

Evans, H. M. (1912). The Development of the Vascular System. *In* "Manual of Human Embryology" (F. Keibel and F. P. Mall, eds.), Vol. 2. Lippincott, Philadelphia, Pa.

Fleisch, A. (1930). Die Wirkung von Histamin, Acetylcholin und Adrenalin auf die Venen. *Pflügers Arch. ges. Physiol.* **228**, 351.

Franklin, K. J. (1937). Veins and the nervous system. *In* "A Monograph on Veins" (K. J. Franklin, ed.), Chapter 10. Charles C Thomas, Springfield, Ill.

Franklin, K. J., and McLachlin, A. D. (1936). Dilatation of veins in response to tapping in man and in certain other mammals. *J. Physiol. (London)* **88,** 257.

Fraser, J. E., and Brown, A. K., (1944). A clinical syndrome associated with a rare anomaly of the venae portae system. *Surg., Gynecol. Obstet.* **78,** 520.

Frazer, J. E. (1931). "A Manual of Embryology." Baillière, Tindall, & Cox, London.

Gauer, O. H., and Sieker, H. O. (1956). The continuous recording of central venous pressure changes from an arm vein. *Circulation Research* **4,** 74.

Gauer, O. H., Henry, J. P., and Sieker, H. O. (1956). Changes in central venous pressure after moderate haemorrhage and transfusion in man. *Circulation Research* **4,** 79.

Givens, M. H. (1912). Duplication of the inferior vena cava in man. *Anat. Record* **6,** 475.

Gladstone, R. J. (1929). Development of the inferior vena cava in the light of recent research, with especial reference to certain abnormalities, and current descriptions of the ascending lumbar and azygos veins. *J. Anat.* **64,** 70.

Glover, W. E., Greenfield, A. D. M., Kidd, B. S. L., and Whelan, R. F. (1958). Reactions of the capacity blood vessels of the human hand and forearm to vaso-active substances infused intra-arterially. *J. Physiol. (London)* **140,** 113.

Gollwitzer-Meier, K. (1932). Venensystem und kreislaufregulierung. *Ergeb. Physiol. u. exptl. Pharmakol.* **34,** 1145.

Gollwitzer-Meier, K., and Bohn, H. (1930). Über die Venoconstrictorische Wirkung der Kohlensäure und ihre Bedeutung für den Kreislauf. *Klin. Wochschr.* **9,** 872.

Green, H. D. (1950). Circulatory system. *In* "Physical Principles in Medical Physics" (O. Glasser, ed.), p. 2. Year Book Publs., Chicago, Ill.

Greenfield, A. D. M., and Patterson, G. C. (1954). The effect of small degrees of venous distension on the apparent rate of blood inflow to the forearm. *J. Physiol. (London)* **125,** 525.

Greenfield, A. D. M., and Patterson, G. C. (1956). On the capacity and distensibility of the blood vessels of the human forearm. *J. Physiol. (London)* **131,** 290.

Greenfield, A. D. M., and Shepherd, J. T. (1950). A quantitative study of the response to cold of the circulation through the fingers of normal subjects. *Clin. Sci.* **9,** 323.

Haddy, F. J. (1957). Effects of epinephrine, norepinephrine, and serotonin upon systemic small and large vessel resistances. *Circulation Research* **5,** 247.

Haddy, F. J. (1960). Effect of histamine on small and large vessel pressures in the dog foreleg. *Am. J. Physiol.* **198,** 161.

Haddy, F. J., and Gilbert, R. P. (1956). Relation of a venous-arteriolar reflex to transmural pressure and resistance in small and large systemic vessels. *Circulation Research* **4,** 25.

Haddy, F. J., Fleishman, M., and Emanuel, D. A. (1957a). Effect of epinephrine, norepinephrine and serotonin upon systemic small and large vessel resistance. *Circulation Research* **5,** 247.

Haddy, F. J., Fleishman, M., and Scott, J. B., Jr. (1957b). Effect of change in air temperature upon systemic small and large vessel resistance. *Circulation Research* **5,** 58.

Halliday, J. (1959). Blood flow in the human calf after walking. *J. Physiol. (London)* **149,** 17P.

Halpert, B. (1930). Complete situs inversus of the vena cava superior. *Am. J. Pathol.* **6,** 191.

Halpert, B. (1942). Situs inversus of the superior vena cava. *A.M.A. Arch. Pathol.* **38,** 75.

Hamilton, W. J., Boyd, J. D., and Mossman, H. W. (1952). "Human Embryology," p. 143. Williams & Wilkins, Baltimore, Md.

Hevelke, G. (1958). Die Angiochemie der Gefässe und ihre physiologischen Alterswandlungen. *Verhandl. deut. Ges. Kreislaufforsch.* **24,** 131.

Hevelke (1959a). Zum Problem der Physiosklerose. Untersuchungen zum Verhalten des Trockenrückstandes venöser und arterieller Blutgefässe in den verschiedenen Altersstufen. *Z. Altersforsch.* **13,** 280.

Hevelke, G. (1959b). Zum Problem der Physiosklerose. Untersuchungen an menschlichen Venen. *Z. Altersforsch.* **13,** 337.

Hickam, J. B., and Pryor, W. W. (1951). Cardiac output in postural hypotension. *J. Clin. Invest.* **30,** 401.

Höjensgärd, I. C., and Stürup, H. (1953). Static and dynamic pressures in superficial and deep veins of the lower extremity in man. *Acta Physiol. Scand.* **27,** 49.

Horiuchi, K. (1924). Beiträge zur Frage der Venodilatatoren. *Pflügers Arch. ges. Physiol.* **206,** 473.

Huseby, R. A., and Boyden, E. A. (1941). Absence of the hepatic portion of the inferior vena cava with bilateral retention of the supracardinal system. *Anat. Record* **81,** 537.

Kaestner, S. (1900). Eintreten der hinteren Cardinalvenen fur die fehlende Vena cava inferior beim erwachsenen Menschen. *Arch. Anat. u. Physiol., Anat. Abt.* p. 271.

Kelly, W. D., and Visscher, M. B. (1956). Effect of sympathetic nerve stimulation on cutaneous small vein and small artery pressures, blood flow and hindpaw volume in the dog. *Am. J. Physiol.* **185,** 453.

Kidd, B. S. L., and Lyons, S. M. (1958). Distensibility of the blood vessels of the human calf determined by graded venous congestion. *J. Physiol. (London)* **140,** 122.

Landis, E. M., and Hortenstine, J. C. (1950). Functional significance of venous blood pressure. *Physiol. Revs.* **30,** 1.

Lewis, F. T. (1903). The intra-embryonic blood vessels of rabbits from 8½ to 13 days. *Am. J. Anat.* **3,** 12.

Lewis, F. T. (1909). On the cervical veins and lymphatics in four human embryos with an interpretation of anomalies of the subclavian and jugular veins in the adult. *Am. J. Pathol.* **9,** 33.

Lewis, T. (1930). Observations upon the reactions of the vessels of the human skin to cold. *Heart* **15,** 177.

Litter, J., and Wood, J. E. (1954). The volume and distribution of blood in the human leg measured in vivo. I. Effects of graded external pressure. *J. Clin. Invest.* **33,** 798.

Lyttle, D. (1961). Observations on the mechanism of the reduction in distensibility of the low pressure vessels of the human forearm during overventilation. *J. Physiol. (London)* **156,** 238.

Lyttle, D. (1960). The reduction in distensibility of the capacity vessels in the forearm during hyperventilation. *J. Physiol. (London)* in press.

McClure, C. F., and Butler, E. G. (1925). The development of the vena cava inferior in man. *Am. J. Anat.* **35,** 331.

McLennan, C. E., McLennan, M. T., and Landis, E. M. (1942). Effect of external pressure on the vascular volume of the forearm and its relation to capillary blood pressure and venous pressure. *J. Clin. Invest.* **21**, 319.

McMichael, J. (1950). "Pharmacology of the Failing Human Heart." Charles C Thomas, Springfield, Ill.

Martin, D. A., White, K. L., Vernon, C. R. (1959). Influence of emotional and physical stimuli on pressure in the isolated vein segment. *Circulation Research* **7**, 580.

Melville, K. I. (1958). *In* "Pharmacology in Medicine," 2nd ed. (V. A. Drill, ed.), Part 9, Section 32, p. 482. McGraw-Hill, New York.

Merritt, F. L., and Weissler, A. M. (1959). Reflex venomotor alterations during exercise and hyperventilation. *Am. Heart J.* **58**, 382.

Neuberger, H. (1913). Ein Fall von volkommener Persistenz der linken Vena cardinalis posterior bei fehlender Vena cava inferior. *Anat. Anz.* **43**, 65.

Ochsner, A., Colp, R., and Burch, G. E. (1951). Normal blood pressure in the superficial venous system of man at rest in the supine position. *Circulation* **3**, 674.

Padget, D. H. (1956). The cranial venous system in man in reference to development, adult configuration, and relation to arteries. *Am. J. Anat.* **98**, 307.

Padget, D. H. (1957). The development of the cranial venous system in man from the standpoint of comparative anatomy. *Carnegie Inst. Wash. Publ., Contribs. Embryol.* **36**, 79.

Page, E. B., Hickam, J. B., Sieker, H. O., McIntosh, H. D., and Pryor, W. W. (1955). Reflex venomotor activity in normal persons and in patients with postural hypotension. *Circulation* **11**, 262.

Pollack, A. A., and Wood, E. H. (1949). Venous pressure in the saphenous vein at the ankle in man during exercise and change in posture. *J. Appl. Physiol.* **1**, 649.

Portman, E. D., and MacDonald, A. D. (1928). The action of pituitary extracts upon isolated blood vessels. *J. Physiol. (London)* **65**, 13P.

Reagan, F. P. (1926). The earliest blood vessels of the mammalian embryo studied by means of the injection method. *Univ. Calif. Publs. (Berkeley)* **28**, 361.

Reagan, F. P., and Robinson, A. (1926). The later development of the inferior vena cava in man and in carnivora. *J. Anat.* **61**, 482.

Reagan, F. P., and Tribe, M. (1927). The early development of the postrenal vena cava in the rabbit. *J. Anat.* **61**, 480.

Richards, D. W., Cournand, A., Darling, R. C., Gillespie, W. H., and Baldwin, E. de F. (1942). Pressure of blood in the right auricle in animals and in man; under normal conditions, and in right heart failure. *Am. J. Physiol.* **136**, 115.

Richardson, K. (1937). The embryology of veins. *In* "A Monograph on Veins" (K. J. Franklin, ed.), Chapter 3, pp. 17–34. Charles C Thomas, Springfield, Ill.

Roddie, I. C., Shepherd, J. T., and Whelan, R. F. (1957). The vasomotor nerve supply to the skin and muscle of the human forearm. *Clin. Sci.* **16**, 67.

Ross, J. C., Hickam, J. B., Wilson, W. P., and Lowenbach, H. (1959). Reflex venoconstrictor response to strong autonomic stimulation. *Am. Heart J.* **57**, 418.

Ryder, H. W., Molle, W. E., and Ferris, E. B., Jr. (1944). The influence of the collapsibility of veins on venous pressure, including a new procedure for measuring tissue pressure. *J. Clin. Invest.* **23**, 333.

Salzman, E. W., and Leverett, S. D. (1956). Peripheral venoconstriction during acceleration and orthostasis. *Circulation Research* **4**, 540.

Seib, G. A. (1934). The azygos system of veins in American whites and American negroes, including observations on the inferior caval venous system. *Am. J. Phys. Anthropol.* **19,** 39.

Sjöstrand, T. (1952). The regulation of the blood distribution in man. *Acta Physiol. Scand.* **26,** 312.

Sjöstrand, T. (1953). Volume and distribution of blood and their significance in regulating the circulation. *Physiol. Revs.* **33,** 202.

Smirk, F. H. (1936). Observations on causes of oedema in congestive heart failure. *Clin. Sci.* **2,** 317.

Smith, W. C. (1916). A case of a left superior vena cava without a corresponding vessel on the right side. *Anat. Record* **11,** 191.

Streeter, G. L. (1915). The development of the venous sinuses of the dura mater in the human embryo. *Am. J. Anat.* **18,** 145.

Streeter, G. L. (1918). The developmental alterations in the vascular system of the brain of the human embryo. *Carnegie Inst. Wash. Publ., Contribs. Embryol.* **8,** 5.

Strong, L. H. (1929). A case of persistent subcentral vein in adult man. *Anat. Record* **42,** 64.

Thompson, I. M. (1929). Venae cavae superiores dextra et sinistra of equal size in an adult. *J. Anat.* **63,** 496.

Voss, E., and Gollwitzer-Meier, K. (1933). Einfluss der Wasserstoffionenkonzentration auf die Weite innervierter Venen. *Pflügers Arch. ges. Physiol.* **232,** 749.

Walker, A. J., and Longland, C. J. (1950). Venous pressure measurements in the foot in exercise as an aid to the investigation of venous disease in the leg. *Clin. Sci.* **9,** 101.

Wallace, J. M., and Stead, E. A. (1957). Spontaneous pressure elevations in small veins and effects of norepinephrine and cold. *Circulation Research* **5,** 650.

Walton, R. P. (1958). *In* "Pharmacology in Medicine," 2nd ed. (V. A. Drill, ed.), Part 9, Section 31, p. 451. McGraw-Hill, New York.

Waring, H. J. (1894). Left vena cava inferior. *J. Anat. and Physiol.* **28,** 46.

Wells, H. S., Youmans, J. B., and Miller, D. A. (1938). Tissue pressure (intracutaneous, subcutaneous and intramuscular) as related to venous pressure, capillary filtration and other factors. *J. Clin. Invest.* **17,** 489.

Wideman, M. (1959). Response of subcutaneous vessels to venous distention. *Circulation Research* **7,** 238.

Wood, J. E. (1959). Peripheral venous distensibility in essential hypertension. *Circulation* **20,** Part II, 787.

Wood, J. E., Bass, D. E., and Iampietro, P. F. (1958). Responses of peripheral veins of man to prolonged and continuous cold exposure. *J. Appl. Physiol.* **12,** 357.

PART TWO

Special Vascular Beds

Chapter VIII. CIRCULATION TO THE BRAIN AND SPINAL CORD

A. EMBRYOLOGY[1]

By L. H. STRONG

In the mesenchyme adjacent to the neural tube, cords of cells hollow out into vessels which form a plexus covering the brain except in the mid-dorsal and midventral lines. The primitive carotid artery and anterior cardinal vein connect with this plexus.

1. DEVELOPMENT OF ARTERIES[2]

a. Primitive internal carotid:[3] This vessel develops from the dorsal parts of the aortic arches and the interarch segments of the dorsal aorta, together with the third arch and a rostral[4] plexus which works into caudal and rostral divisions dorsal to the optic vesicle. Of the various branches, the anterior cerebral forms from the rostral division of internal carotid. It has an anterior communicating which arises from the inter-anastomosis of the primitive olfactory artery (terminal of the anterior cerebral). Another branch of the internal carotid, the ophthalmic, develops from the dorsal and ventral ophthalmic arteries (of the rostral division of the carotid), joined to the supraorbital division of the stapedial artery, which is derived from the hyoid artery arising from the second aortic arch. A third branch, the choroid, develops from the anterior cerebral. Its choroid plexus of the lateral ventricle originates from an arterial plexus in the hemispheric sulcus and also from the sulcate termination of the anterior choroid. A fourth, the medial striate, develops from the primary olfactory artery, and a fifth, the middle cerebral, from the rostral division of the primitive carotid. The latter has internal and external lenticulostriate branches, the origin of which is uncertain. A sixth branch, the posterior communicating, arises from the caudal division of the internal carotid. It gives off a posterior cerebral vessel

[1] For a detailed discussion of the embryology of the pituitary, see (1) The Pituitary, Section B., Chapter XIV.

[2] Terminology used in this section is from Padget's (1948, 1956) investigations.

[3] The words, artery and vein, will be omitted except for clarity.

[4] This term is substituted for "cranial."

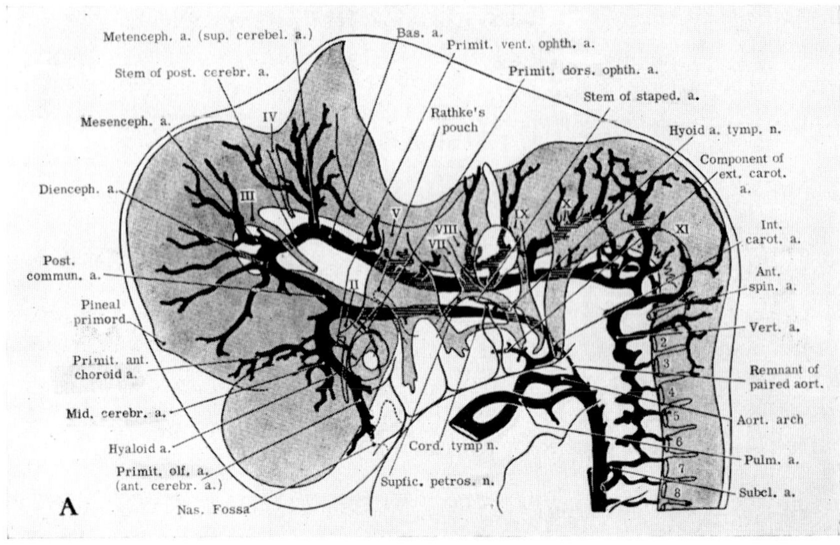

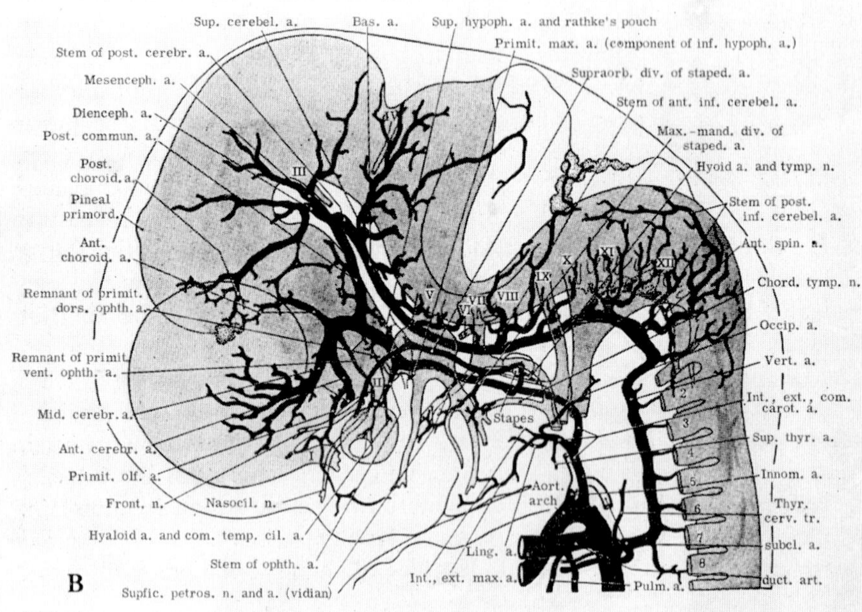

Fig. 51. Reconstructions of arterial channels of human development. A. Embryo of 12.5 mm; B. Embryo of 18 mm; C. Embryo of 24 mm; D. Embryo of 43 mm. (For description, see text.) (From Padget, 1948; reproduced with permission of the Carnegie Institute of Washington). Parts C and D are on facing page.

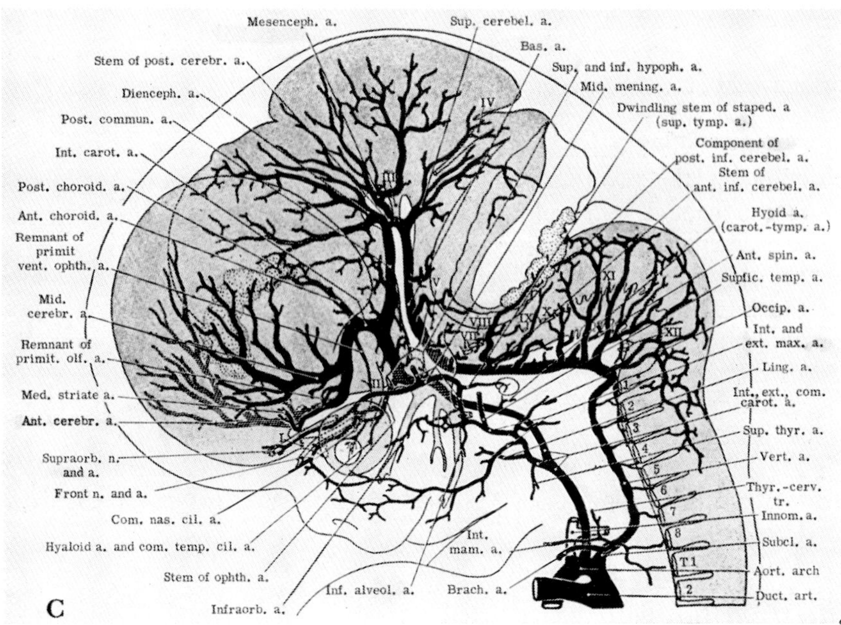
Mesenceph. a.
Sup. cerebel. a.
Bas. a.
Stem of post. cerebr. a.
Sup. and inf. hypoph. a.
Dienceph. a.
Mid. mening. a.
Post. commun. a.
Dwindling stem of staped. a (sup. tymp. a.)
Int. carot. a.
Component of post. inf. cerebel. a.
Stem of ant. inf. cerebel. a.
Post. choroid. a.
Ant. choroid. a.
Hyoid a. (carot.-tymp. a.)
Remnant of primit vent. ophth. a.
Ant. spin. a.
Supfic. temp. a.
Mid. cerebr. a.
Occip. a.
Int. and ext. max. a.
Remnant of primit. olf. a.
Ling. a.
Med. striate a.
Int., ext., com. carot. a.
Ant. cerebr. a.
Sup. thyr. a.
Supraorb. n. and a.
Vert. a.
Front n. and a.
Thyr.-cerv. tr.
Com. nas. cil. a.
Innom. a.
Hyaloid a. and com. temp. cil. a.
Int. mam. a.
Subcl. a.
Aort. arch
Stem of ophth. a.
Inf. alveol. a.
Brach. a.
Duct. art.
Infraorb. a.
C

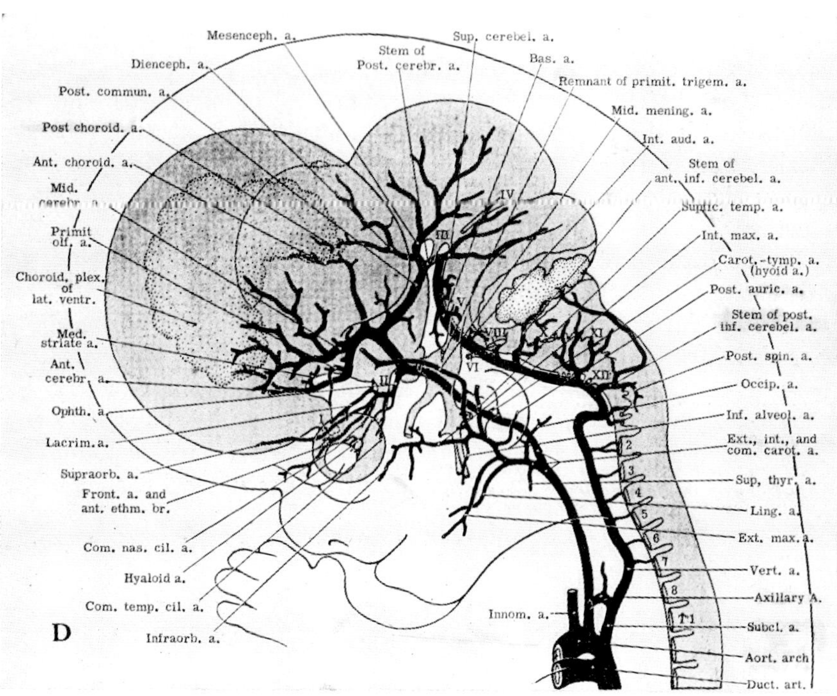
Mesenceph. a.
Sup. cerebel. a.
Stem of Post. cerebr. a.
Dienceph. a.
Bas. a.
Post. commun. a.
Remnant of primit. trigem. a.
Mid. mening. a.
Post choroid. a.
Int. aud. a.
Ant. choroid. a.
Stem of ant. inf. cerebel. a.
Mid. cerebr. a.
Supfic. temp. a.
Primit olf. a.
Int. max. a.
Carot.-tymp. a. (hyoid a.)
Choroid. plex. of lat. ventr.
Post. auric. a.
Stem of post. inf. cerebel. a.
Med. striate a.
Post. spin. a.
Ant. cerebr. a.
Occip. a.
Ophth. a.
Inf. alveol. a.
Lacrim. a.
Ext., int., and com. carot. a.
Supraorb. a.
Sup. thyr. a.
Front. a. and ant. ethm. br.
Ling. a.
Ext. max. a.
Com. nas. cil. a.
Vert. a.
Hyaloid a.
Axillary A.
Com. temp. cil. a.
Innom. a.
Subcl. a.
Aort. arch
Infraorb. a.
Duct. art.
D

which forms late or may even be completed postnatally. The basilar artery forms only by fusion of longitudinal neural arteries midventrally, and not by alternate segmental fusion, and gives off a number of arteries. The superior cerebellar arises from the metencephalic artery of the basilar; the anterior inferior cerebellar, from the otocyst artery of the basilar; the posterior inferior cerebellar and the posterior spinal, from the caudal trunc of the myelencephalic plexus of the vertebral artery; and the choroid plexus of the fourth ventricle, from the arterial plexus on the roof of the fourth ventricle. The anterior spinal branch of the internal carotid develops from the fusion of longitudinal neural arteries annexed by the vertebral artery.

b. Reconstruction of arterial channels: The changing patterns of development of the arterial channels in the human brain are demonstrated in Fig. 51. The various sections are presented in series from a typical early stage to that in which the adult configuration may be recognized. A description of the figure follows.

In the embryo of 12.5 mm (Fig. 51A) the internal carotid artery has been constituted from the aortic arches into a rostral and caudal (posterior communicating) division, which has fused in the midline into the basilar. The latter is connected with the vertebral artery, which is still showing the intersegmental plexus of origin. The ventral pharyngeal artery is migrating into the second and third aortic arch bars, as the external carotid. The hyoid artery is present. The primitive olfactory artery extends the anterior cerebral ventromedially. Later the olfactory arteries, between the hemispheres, will anastomose across the midline to form the anterior communicating artery. The middle cerebral artery is evident.

In the embryo of 18 mm (Fig. 51B), the primitive olfactory artery is dwindling, and at the same time the middle cerebral artery has grown larger than the anterior cerebral. The choroid plexus and its two arteries have developed, as well as the cerebellar arteries. The two spinal arteries have also appeared. The stem of the ophthalmic artery has migrated to its adult position, and the hyoid artery has passed through the stapes to become the stapedial, with its ventral (maxillomandibular) and rostral (supraorbital) divisions; the middle meningeal branch of the latter is shown cut off.

In the embryo of 24 mm (Fig. 51C), the external carotid system has now joined the maxillomandibular division of the stapedial artery which has broken from the hyoid. Thus, the stapedial no longer has any connection with the internal carotid, but is supplied wholly by the external carotid. The supraorbital division of the stapedial has joined the ophthalmic system.

In the fetus of 43 mm (Fig. 51D), most of the adult configuration is apparent. The growing hemisphere will, by birth, have forced the posterior communicating artery into its adult horizontal position, this vessel remaining large even postnatally. The cerebellar arteries will surround the cerebellum, and the anterior cerebral will pass over the corpus callosum. The skull condenses around the middle meningeal below the stapedial; the latter has anastomosed with the ophthalmic and has grown a lacrimal branch. The connection from lacrimal to stapedial may either dwindle to extinction (Fig. 51D) or, more often, may remain as the orbital (anastomotic) branch of the middle meningeal; the latter may take over, in part or wholly, the supply of the orbit.

2. Development of Veins

a. Principal cephalic veins: Of the vessels in this group, the anterior cardinal vein develops from the primitive head sinus and the anterior, middle and posterior dural plexuses. Another component of this group consists of the sinuses (vessels) of the dura mater involving: (1) the superior sagittal, which develops from the anterior dural and tentorial plexuses; (2) the transverse, which forms from the middle dural plexus, rostral to its stem, and the marginal sinus; (3) the sigmoid, which develops from anatomosis between the posterior and middle dural plexuses, dorsal to the otocyst, and from the stem of the posterior dural plexus; (4) the confluence of sinuses (torcular), which forms from the condensation of loops of the tentorial plexus (not sinus), at the medial end of the marginal sinus; (5) the occipital, which is formed from the tentorial plexus and the marginal sinus; (6) the straight, which develops by midline fusion of the anterior dural plexuses (tentorial plexus), between hemispheres, as they drain the lateral ventricular choroid plexus; and (7) the cavernous, which forms from an hypophyseal plexus, between the prootic sinuses, and the prootic sinus laterally, the ventral myelencephalic vein caudally and the superior ophthalmic vein rostrally. Outlets of the cavernous are: (1) the inferior petrosal, which arises from the ventral myelencephalic vein and its rostral plexus; and (2) the superior petrosal, which develops from the ventral mesencephalic vein and prootic sinus (the last of the adult sinuses to form); the connection with the cavernous sinus being wholly secondary. The superior ophthalmic vein, the main inlet vessel to the cavernous sinus, arises from the stem of the anterior dural plexus, the primitive supraorbital vein, the primitive maxillary vein, and the termination of the prootic sinus.

The cerebral veins can be divided into deep and superficial branches. In the former category is the great (Galen) cerebral vein which forms by the fusion of internal cerebral veins. The internal cerebral vein arises from the superior choroid vein and from the mid-dorsal interhemispheric capillaries of the anterior dural plexuses. The superior choroid forms from a marginal condensation of capillaries of the lateral ventricular choroid plexus.

The superficial cerebral veins arise from the superior, inferior, middle, and anterior cerebral veins. The superior vessels develop as four or five trunks from the medial and lateral superficial networks of the hemispheres to the superior sagittal sinus. Each has two tributaries. The inferior cerebral veins form as enlarged trunks from the basal and lateral superficial network entering into the formation of the middle cerebral vein, terminal vessel of the tentorial sinus. The middle cerebral is formed from the telencephalic vein, together with enlargements from the venous network on the lateral surface of its own hemisphere. Since it is the end of the tentorial sinus, it connects the cavernous with the transverse or sigmoid sinuses. The superior anastomotic (Trolard), an enlargement of the lateral cerebral network connecting the middle cerebral vein with the superior sagittal sinus, forms as late as birth or even later. The inferior anastomotic (L'Abbe), also an enlargement of the lateral cerebral network, connects the middle cerebral vein to the transverse sinus and likewise has a late formation. The anterior cerebral vein forms from an enlargement of the venous plexus in the medial surface of the hemisphere, drained by the middle cerebral and basal veins.

b. Pontomedullary veins: These vessels develop from a longitudinal plexus of dien-, mesen-, meten-, and myelencephalic veins and drain into transverse, sigmoid or inferior petrosal sinuses.

c. Cerebellar veins: The vessels consist of superior and inferior branches, the derivation of which is uncertain, and a great anterior cerebellar which develops from the ventral metencephalic vein and drains into the superior petrosal sinus.

d. Emissary veins: There are seven of these vessels. The hypoglossal emissary develops from the primary head sinus and runs into the internal jugular by way of the ventral myelencephalic vein. The condyloid originates from the stem of the posterior dural plexus. The mastoid arises from an anastomosis of the middle and posterior dural plexuses (dorsal segment of sigmoid sinus). The sinus (rete) of the foramen ovale is developed from the prootic sinus by anastomoses of the dorsal pharyngeal with the primitive maxillary (adult inferior ophthalmic) and with the ventral pharyngeal (adult maxillary), thus forming the pterygoid plexus. The occipital emissary is an extremely rare vessel which may arise from the tentorial plexus. The parietal emissary probably also has the same origin. The venous plexus of the internal carotid artery arises from the plexus which forms the inferior petrosal sinus.

e. Reconstruction of venous channels: The changing patterns of development of the venous channels are demonstrated in Fig. 52 and described below.

In the embryo of 14 mm (Fig. 52A) the anterior, middle, and posterior dural plexuses show their stems converging to form the head sinus, caudal to which the anterior cardinal has just become the internal jugular vein. The ventral pharyngeal vein, as the stem of the common facial system, is growing rostrally from the internal jugular. The dorsal pharyngeal vein is in the second pharyngeal bar. The primitive maxillary vein is draining the frontomandibular region. Around the hemisphere rostromedially is the marginal sinus, caudoventrally, the telencephalic vein (middle cerebral). A ventral myelencephalic vein drains into the jugular through the jugular foramen. The head plexuses are about to form an anastomosis dorsal to the otocyst.

In the embryo of 18 mm (Fig. 52B), the anastomosis dorsal to the otocyst is now complete, and the primitive head sinus, between the stems of the middle and posterior dural plexuses, degenerates. After this, the prootic sinus is constituted by the remains of the head sinus, together with its middle stem. Branches of the maxillary, dorsal pharyngeal and facial system are mutually approaching. The dorsal otocystic anastomosis, plus the stem of the posterior plexus, makes the sigmoid sinus. The channel of the middle plexus, continuous with the sigmoid sinus, is the lateral component of the transverse sinus. The tentorial sinus has enlarged and its middle cerebral tributary is turning first rostrally and later, dorsally.

The supraorbital vein approaches the maxillary, which will then capture it. The choroid plexus in the hemisphere is drained by the straight sinus whose plexal end is the future internal cerebral vein.

In the embryo of 40 mm (Fig. 52C), the anterior and middle dural plexuses have formed the tentorial plexus (not the sinus) across the dorsal midline. The latter vessel drains the internal cerebral vein. The marginal sinus (the torcular end of the transverse sinus) has moved in line with the lateral component of the primitive transverse sinus. The

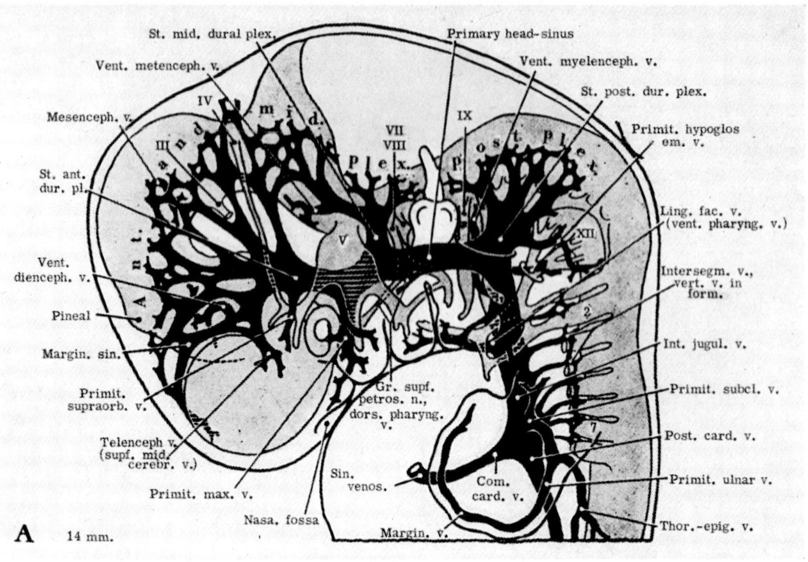

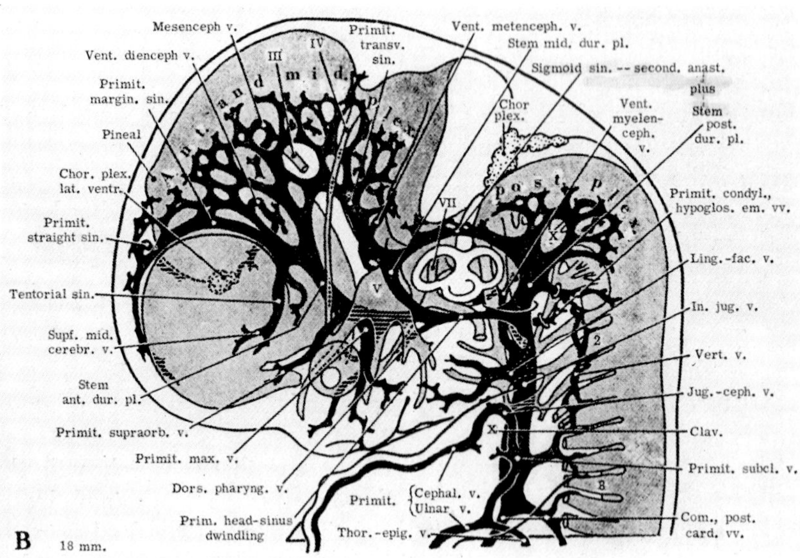

FIG. 52. Reconstruction of venous channels of human development. A. Embryo of 14 mm; B. Embryo of 18 mm; C. Embryo of 40 mm; D. Fetus of 60–80 mm (composite). (For description, see text.) (From Padget, 1956; reproduced with permission of the Carnegie Institute of Washington.) Parts C and D are on facing page.

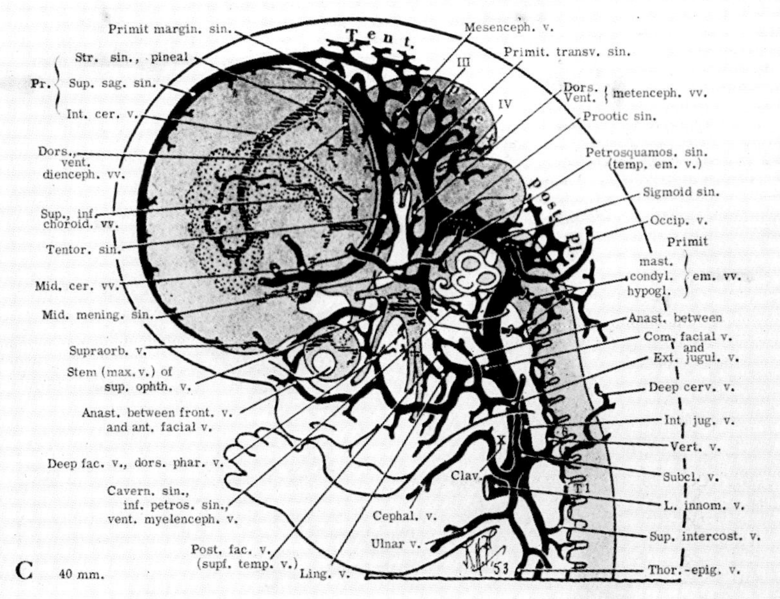
Primit margin. sin.
Mesenceph. v.
Tent.
Primit. transv. sin.
III
IV
Pr.
Str. sin., pineal
Sup. sag. sin.
Int. cer. v.
Dors. Vent. metenceph. vv.
Prootic sin.
Dors., vent. dienceph. vv.
Petrosquamos. sin. (temp. em. v.)
Post. pl.
Sigmoid sin.
Sup., inf. choroid. vv.
Occip. v.
Tentor. sin.
Primit mast. condyl. hypogl. em. vv.
Mid. cer. vv.
Mid. mening. sin.
Anast. between Com. facial v. and Ext. jugul. v.
Supraorb. v.
Stem (max. v.) of sup. ophth. v.
Deep cerv. v.
Anast. between front. v. and ant. facial v.
Int. jug. v.
Vert. v.
Deep fac. v., dors. phar. v.
X
Clav.
Subcl. v.
Cavern. sin., inf. petros. sin., vent. myelenceph. v.
T1
L. innom. v.
Cephal. v.
Sup. intercost. v.
Ulnar v.
Post. fac. v., (supf. temp. v.)
Ling. v.
Thor.-epig. v.
C
40 mm.

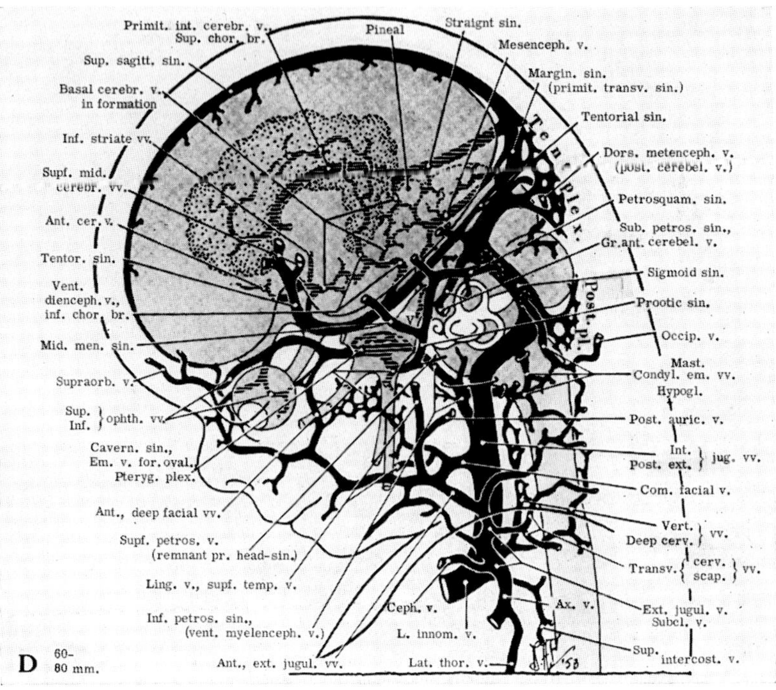
Primit. int. cerebr. v., Sup. chor. br.
Pineal
Straight sin.
Mesenceph. v.
Sup. sagitt. sin.
Margin. sin. (primit. transv. sin.)
Basal cerebr. v. in formation
Tentorial sin.
Tent. plex.
Inf. striate vv.
Dors. metenceph. v. (post. cerebel. v.)
Supf. mid. cerebr. vv.
Petrosquam. sin.
Ant. cer. v.
Sub. petros. sin., Gr. ant. cerebel. v.
Tentor. sin.
Sigmoid sin.
Vent. dienceph. v., inf. chor. br.
V
Prootic sin.
Post. pl.
Mid. men. sin.
Occip. v.
Mast. Condyl. em. vv. Hypogl.
Supraorb. v.
Sup. Inf. ophth. vv.
Post. auric. v.
Cavern. sin., Em. v. for. oval. Pteryg. plex.
Int. Post. ext. jug. vv.
Com. facial v.
Ant., deep facial vv.
Vert. Deep cerv. vv.
Supf. petros. v. (remnant pr. head-sin.)
Transv. cerv. scap. vv.
Ling. v., supf. temp. v.
Ax. v.
Ceph. v.
Ext. jugul. v. Subcl. v.
Inf. petros. sin., (vent. myelenceph. v.)
L. innom. v.
Sup. intercost. v.
Ant., ext. jugul. vv.
Lat. thor. v.
D
60-80 mm.

prootic sinus has captured the orbital drainage through its maxillary stem, rostral to which the stem of the anterior dural plexus has dwindled and has broken. The ventral component (that deep to the trigeminal ganglion) of the prootic sinus has anastomosed with an hypophyseal plexus, which drains into an inferior petrosal plexiform sinus.

In the fetus of 60–80 mm (Fig. 52D), the adult configuration is clearly discernible. The tentorial plexus will subsequently form the superior sagittal sinus and the confluence of sinuses, plus the occipital sinus. The emissary veins are present (also noted in Fig. 52A, B, and C). The prootic sinus, receiving the ophthalmic system, forms the lateral compartment of the cavernous sinus, which, through the prootic sinus, receives the pterygoid plexus.

The inferior petrosal sinus is complete. The prootic sinus just dorsal to the middle meningeal sinus (Fig. 52C) has degenerated completely (Fig. 52D), leaving a superior petrosal sinus which receives the great anterior cerebellar vein. Seen through the hemisphere, on the base of the brain, is the longitudinal anastomosis which becomes the basal vein, the tributaries of which still drain into the tentorial sinus. The latter will degenerate, except for its rostral end which remains as the spheno-parietal sinus, draining into the cavernous sinus and receiving the superficial middle cerebral vein of the adult. The two great anastomotic veins are late formations of the superficial venous network, extending from the middle cerebral vein to the superior sagittal sinus (Trolard) and the transverse sinus (L'Abbe).

B. GROSS ANATOMY

By H. C. H. Fang

1. Arterial Blood Supply to the Brain

a. Source of blood supply: The brain derives its arterial circulation from the paired internal carotid and vertebral arteries. Each communicates with the other at the arterial anastomatic ring at the base of the brain—the arterial Circle of Willis. Schema modified from that originally proposed by Foix and Hillemand (1926) generally also include the following: (1) paramedian vessels perforating the base of the brain to supply structures near the midline; (2) short circumferential vessels supplying the thalamus, lentiform nuclei and other basal nuclei; and (3) long circumferential branches (the anterior, middle, and posterior cerebral arteries) supplying the cortex and underlying white matter.

b. Collateral circulation: There are at least four sources of collateral blood supply to the brain: (1) meningeal arterial anastomoses among the anterior, middle, and posterior cerebral arteries in the pia-arachnoid space (Vander Eecken and Adams, 1953); (2) rich interconnecting anastomotic network at a capillary level within the brain parenchyma; (3) potential collateral channels between the major branches of the external and internal carotid arteries, as for example, those which exist between the branch of the middle meningeal of the external carotid and that of the nasociliary branch of the ophthalmic artery of the internal carotid system;[5] and (4) the arterial anastomotic ring of the Circle of Willis at the base of the brain, mentioned above (Subsection 1a).

2. Brain Stem and Cerebellum

a. Source of blood supply: The circulation to the brain stem and the cerebellum is derived from (1) the paramedian branches of the basilar artery, e.g., the perforating pontile divisions; (2) short circumferential branches; and (3) long circumferential arteries, e.g., the superior, anterior, and posterior inferior cerebellar arteries. Rich anastomoses exist between the surface branches of the cerebellar arteries, across the midplane, and between these and branches of the superior cerebellar and posterior cerebral arteries.

b. Collateral circulation: Collateral channels in the cerebellum are seldom sufficient to offset the irreversible effects of total curtailment of blood supply to this structure. Also in the case of the pons, the paramedian arteries, which supply the ventral median portion of the brain stem, show very poor cross anastomosis with each other. There is extreme variability in the exact territory of supply, particularly in the case of the tegmentum of the mesencephalon.

3. Spinal Cord

Of importance in supplying the spinal cord with blood is the anterior spinal artery, which is actually a single interconnecting anterior

[5] Such collateral channels may become highly developed and angiographically demonstrable under pathologic conditions, as in the case of stenosis and occlusion of the internal carotid artery, close to the origin of the ophthalmic branch.

anastomotic arterial channel representing the common anastomotic trunk linking the various segmental radicular arteries. With the passage of time, the earlier embryonic segmental pattern is lost, with some radicular arteries becoming rudimentary or disappearing. There are two small posterior communicating channels, the posterior spinal arteries.

The anterior spinal artery supplies the major portion of the white and gray of the spinal cord, with the exception of the posterior white column, which receives its circulation from the two posterior spinal arteries. At the cephalic end, the anterior spinal artery is formed by the spinal branches of the vertebral arteries and is then united by contribution from the radicular arteries throughout the remainder of the cord. At the caudal end, the posterior spinal arteries communicate directly with the anterior spinal artery and cephalad, with the spinal branches of the vertebral arteries.

The radicular arteries, which reach the spinal cord along the anterior and posterior spinal nerve roots, are branches of the lateral spinal arteries. The latter, in turn, are derived from parietal branches of the thoracic and abdominal aorta via the intercostals, lumbar, iliolumbar, and lateral spinal arteries (Tureen, 1937). Altogether there are eight anterior radicular arteries and eight to sixteen smaller posterior radicular arteries. Between the levels of thoracic ten and lumbar two of the cord, there is a large anterior radicular artery.

The cervical and lumbosacral portions of the cord are adequately supplied, but the intervening thoracic portion is relatively vulnerable to ischemic damage (Tureen, 1937; Suh and Alexander, 1939). For unknown reasons, the arterial system of the spinal cord in man is relatively sheltered from the deleterious effects of atherosclerosis and hypertension, as compared with arteries elsewhere in the brain.

4. Periarterial Innervation of Major Arteries

a. Source of nerve supply: The nervous innervation of the major arteries at the base of the brain is believed to be derived chiefly from the pericarotid plexus, formed by the branches of the upper portion of cervical sympathetic plexus, the upper cervical roots and the tympanic plexus. Nerve filaments from the vagus and upper two cervical roots also contribute to the supply of the vertebral and basilar arteries and their branches and possibly the third, fifth, seventh, ninth, eleventh and twelfth cranial nerves. However, Pool (1958) reported the presence of nerve strands penetrating vertebral arteries only from the eleventh and twelfth cranial nerves and not from the others.

b. Question of cerebral vasospasm: The concept of cerebral vasospasm has been a subject of much interest, although morphologic support for such a view has not been unequivocally demonstrated in the past, particularly with reference to the periarterial nerve supply of the major cerebral arteries. In the studies in which such structures had been described, they were visualized with the use of silver impregnation techniques, which, as is generally accepted, are sensitive and capricious and often not easily reproducible. Furthermore, their application is limited to microscopic sections.

Fang (1961) investigated the problem in man by employing the histochemical technique which specifically stains the plasmalogens in the nerves.[6] He found the distribution of periarterial nervous network to be confined mainly to the adventitia of cerebral arteries, which is in general agreement with the description of this structure in animal tissue, especially in the case of rats as determined by electron microscopy (Pease and Schultz, 1958). He also noted a marked variability in the size, extent and distribution of the periarterial network. This finding provides further support for the view that the cervical sympathetic plexus plays a relatively insignificant role in the regulation of cerebral blood flow.

C. SUBMICROSCOPIC ANATOMY

By D. C. Pease and R. L. Schultz

1. Question of Perivascular "Spaces"

a. Original work: It has been realized since the monumental work of Key and Retzius (1876), that leptomeningeal sleeves surround the larger arterioles and venules as these vessels penetrate into the tissue of the central nervous system. In the bigger channels it is obvious that an extension of the subarachnoid space continues along the penetrating

[6] The chemical basis for this staining reaction stems from the isolation of acetal phospholipide, a compound whose chemical properties differ from those of the native plasmalogens. According to Rapport *et al.* (1957), the sulfurous acid liberates the aldehyde from the nerves, the latter then staining purple with the Schiff's reagent.

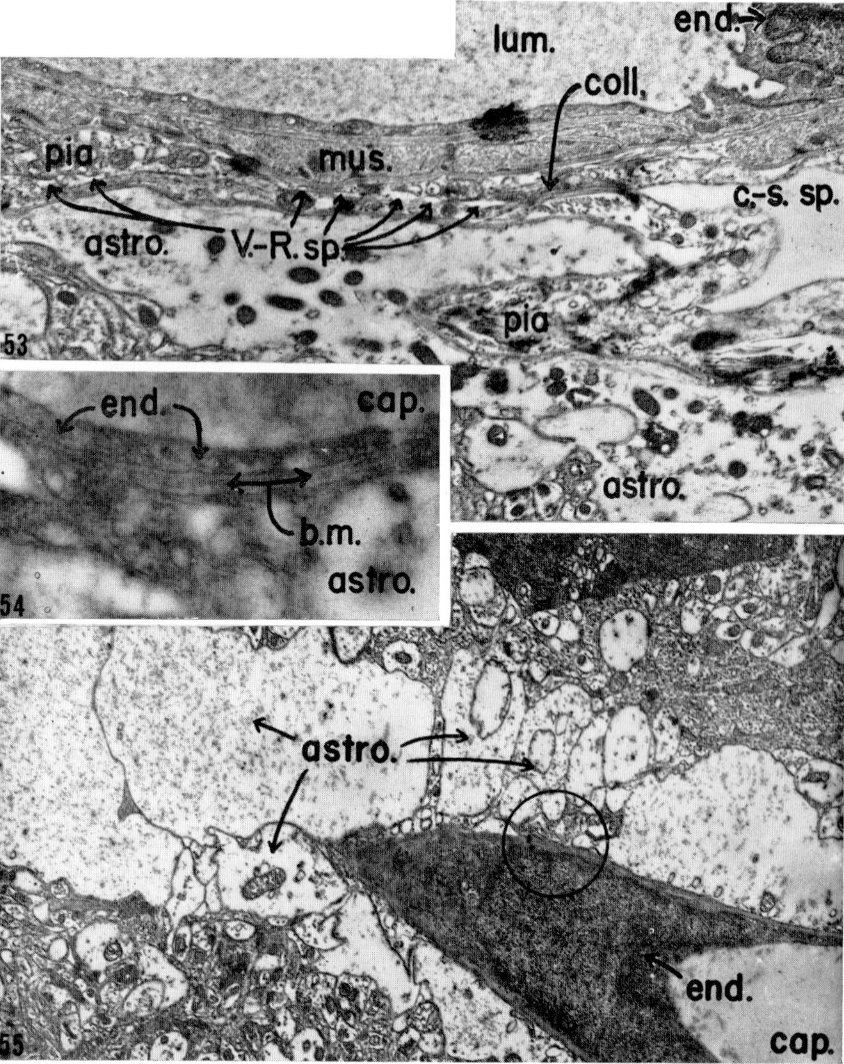

FIG. 53. Longitudinal section of wall of an arteriole penetrating the substance of the cortex. The surface of the brain, with its pial covering (pia) and subpial layer of astrocytic processes (astro.), is at the right. A portion of the cerebrospinal fluid-filled-subarachnoid space (c.-s. sp.) is evident there. A double layer of pia extends to the left, along the wall of the penetrating vessel. In places a patent Virchow-Robin space (V-R. sp.) can be seen. Collagen (coll.) is present with the pia. Astrocytic processes also reflect over the vessel.

FIG. 54. Transverse section of a cortical capillary at high magnification, showing the delicate basement membrane (b.m.) that is shared by endothelium (end.) and neuropil.

FIG. 55. Tangential section of a capillary in the cerebral cortex. The extent of

vessels for a considerable distance (Fig. 53) and that an adventitial sheath with collagen is likewise present. Originally most investigators thought that the subarachnoid space could be traced over all vessels, even those of capillary dimensions (the Virchow-Robin space), and even a perineuronal extension was postulated. The problem was further complicated by the description of still other "spaces" or potential ones, such as the spaces of His and Held, between nervous tissue and the leptomeningeal layer, and the space of Obersteiner, perinueuronal in location. These spaces have been regarded from time to time as analogous to a lymphatic drainage system. In more recent years, however, particularly as a result of the work of Patek (1944), there has been grave doubt about some of the spaces, and, most important, that the cerebrospinal fluid reached the capillary level. Woollam and Millen (1954) have published an excellent critical review of this subject, concluding that patent Virchow-Robin spaces are found normally only in relation to the large vessels and that "at the termination of the space the two opposing walls of the reticular perivascular sheath (leptomeningeal sheath) come into simple opposition and continue as the sheath of the vessel."

b. Electron microscopic studies: Electron microscopy has confirmed the fact that pial layers extend along penetrating arterioles and venules (Maynard *et al.*, 1957; Pease and Schultz, 1958) (Fig. 53). These coverings stop while arterioles still possess a complete muscle layer. In the case of venules, however, they extend over vessels whose walls are reduced to simple endothelium, the leptomeningeal layer being single, with no extension of the subarachnoid space (Fig. 54). There is a tendency for astrocytic feet to be applied to the surface of venules, but the arrangement is less evident than with capillaries. Under no circumstances do the spaces reach the capillary level.

2. Question of Existence of a "Blood-brain Barrier"

a. Original work: Interest in the detailed structure of the capillaries of the brain has received impetus in the past few years from the gradual realization by physiologists that an apparent barrier with unusual properties separates the blood plasma from the nervous tissue. In seeking morphologic peculiarities which might account for this "barrier," it was apparently Spatz (1933) who first postulated that the layer of peri-

some astrocytic feet (astro.) on the surface of the capillary is shown particularly well. It can be seen that in many places these make intimate contact with neighboring feet. In the encircled area, however, a nerve cell process makes brief contact with the capillary (cap.) surface.

vascular astrocytic feet might be involved. Tschirgi (1958) has developed this concept in detail. However, the extent to which brain capillaries are coated with astrocytic foot processes remained a moot question until observations became possible by electron microscopy.

b. Lack of connective tissue investment of capillaries: Dempsey and Wislocki (1955) first emphasized, as a result of electron microscopic observation, that capillaries in most tissue of the central nervous system lack a connective tissue investment, and that glial end-feet of unspecified variety are closely applied to their walls. Subsequently, Maynard *et al.*, (1957) published the only comprehensive study of this vascular bed. In the course of reporting other observations, Farquhar and Hartmann (1957) and Dempsey and Luce (1958),[7] included comments on blood vessels, as other authors have occasionally done.

c. Endothelium: There is general agreement that the capillary wall consists of an endothelial layer that is continuous. Where individual endothelial cells are in contact, there is often some overlapping, accompanied by special thickenings resembling "terminal bars." Thus, the endothelium appears as a much more substantial morphologic barrier than in the case of capillaries in many other parts of the body. Such an arrangement is at the opposite pole of specialization from the fenestrated endothelium of glomerular capillaries (see Section C.-1, Chapter XIII). (For an attempt to systematize capillary variations, see Bennett *et al.*, 1959.)

d. Basement membrane: A very delicate basement underlies the endothelial sheet (Fig. 54). At least when astrocytic foot processes make contact with the capillary surface, this membrane is shared with them. It is a typical basement membrane, in that electron translucent layers exist between the dense component and the plasma membranes of adjacent cells. The optically lucid zones can be regarded as "cement layers," and the dense lamina as the principal structural component.

e. Astrocytes: The cerebral cortex, of course, is well endowed with astrocytes, so that many feet make contact with the capillary surface. The astrocyte, at least in this location, is a very large, notably watery cell, and its cytoplasm is easily recognized. A tangential view of a capillary surface (Fig. 55) is particularly revealing in demonstrating the extent of individual astrocytic feet. Foot processes cover much of the

[7] Those who would read this paper of Dempsey and Luce must understand that these authors have reversed the identification of astrocyte and oligodendrocyte, as these cells have been characterized by almost every other investigator who has studied nervous tissue by electron microscopy.

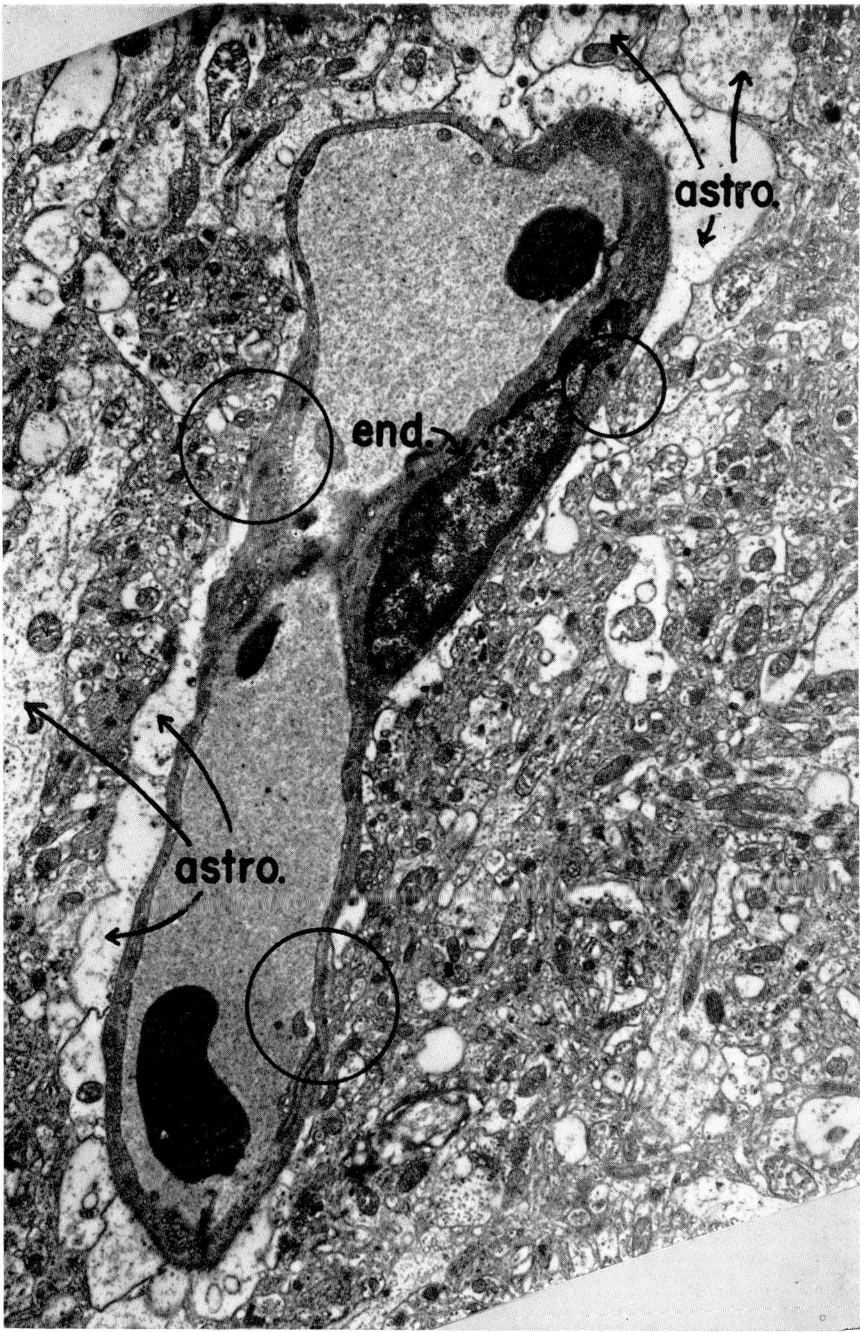

FIG. 56. Longitudinal view of a capillary in the cerebral cortex. An endothelial cell (end.) is obvious. The pale watery cytoplasm, characteristic of astrocytes, is indicated in several places (astro.). It is evident that much of the capillary surface is covered by astrocytic feet. However, in the encircled areas there are three places where there are gaps in the astrocytic layer, and other neuropil elements reach the surface of the capillary.

capillary surface, as noted in Fig. 56. It can be observed that most astrocytic feet are in intimate lateral contact with neighboring ones, so that there is a continuous "glial membrane" over much of the capillary surface. However, in the encircled regions of Figs. 55 and 56 the layer obviously is not complete. Maynard *et al.* (1957) made a visual estimate that the astrocytic feet covered approximately 85 per cent of the capillary surface in the cerebral cortex, although individual capillaries varied considerably. However, this figure is much too high for some other parts of the central nervous system, where astrocytes are not as abundant as in the cerebral cortex. For example, less than half of the capillary surface may be ensheathed with astrocytic feet in some parts of the cerebellar cortex, so that there is little likelihood of this layer serving in any simple mechanical way as a "blood-brain barrier." Where astrocytic feet do not exist on the capillary surface, any other component of the neuropil may be present, including even short contacts with nerve cells and their processes.

f. Evidence opposing existence of a blood-brain barrier: Maynard *et al.* (1957) came to the conclusion that the "blood-brain barrier" might be largely an illusion. Its existence is postulated on the assumption that there are considerable extracellular spaces which would be in equilibrium with blood plasma if a barrier did not exist. There is general agreement among all of the electron microscopists who have studied tissue of the central nervous system that living cell processes are tightly packed together, with negligible interstitial spaces between. Thus, it becomes necessary that the physiologic data on the distribution of ions and molecules be reinterpreted on the assumption that nearly all of these substances are within cytoplasmic compartments, if they are present at all. Under these circumstances, there is no expectation of a simple correlation with their distribution in blood plasma. Instead, the commonly accepted laws of cellular permeability, which apply elsewhere in the body, may also exist in the case of the brain, without important exception and without invoking a notably specialized barrier.

g. Metarterioles: These vessels are found interposed between typical arterioles and capillaries, having applied to their surface a cell of the type regarded by Maynard *et al.* (1957) as a poorly developed or primitive form of smooth muscle. The cell is enveloped in a basement membrane exactly comparable to the sarcolemma of smooth muscle, but its shape is fundamentally stellate, so that profiles of flattened arms are commonly seen in transverse sections of blood vessels. These ordinarily do not completely surround the capillary. In three dimensions they probably have a pattern resembling "pericytes," as described by Zimmermann (1923). They were first pointed out in electron micrographs by Farquhar and Hartmann (1956, 1957).

h. Arterioles: These vessels in the cerebral cortex seem to be without specific innervation. It is also true that the larger pial arteries do not have nerves which penetrate into the substance of the media, and thus, these, too, are without specific innervation. However, nerve bundles have been seen in the adventitial sheathes of pial arteries (Pease and Molinari, 1960).

i. Special vascular structures in specific sites: A few particular areas of the brain, such as the choroid plexus, neurohypophysis, area postrema, pineal body, supraoptic crest, etc., can be vitally stained, and hence, it has been felt, lack a typical "blood-brain" barrier (van Breemen and Clemente, 1955; Dempsey and Wislocki, 1955). It is evident that in these locations substantial quantities of connective tissue penetrate with the blood vessels. Under such circumstances, however, the connective tissue is walled off from neighboring nervous tissue by a basement membrane which everywhere underlies the nervous tissue. Silver proteinate reaches the connective tissue compartments but not the nervous tissue, suggesting that the basement membrane has some of the properties of a barrier.

D. AGING CHANGES

By H. C. H. FANG

1. MACROSCOPIC CHANGES

The aging process appears to be associated with definite gross changes in cerebral arteries. Large-sized vessels over the surface of the brain become wider and longer, more tortuous and stiff, and less easily collapsible. Furthermore, there is a greater degree of translucency of the cerebral vessel wall, with or without gross or microscopic evidence of atherosclerosis. In this regard it has been found by Fang (1958), using a photoelectric densitometer, that the milky opalescence of the vessel wall of extracerebral major arteries could be better correlated with the chronological age rather than with the degree of involvement by atherosclerosis.

2. Microscopic Changes

a. Internal elastic lamella: Hackel (1927), on the basis of a small series of patients ranging from 15 to 50 years, noted that with age there was an alteration in the internal elastic lamella of the intracranial arteries. Other evidence has been presented (Baker, 1937) to indicate that this structure is one of the first to show aging changes. As early as the latter part of the third decade, the lamella becomes fragmented, reduplicated and later frequently loses the uniformity of its tinctorial properties. In old age, portions of the internal elastic lamella show changes in their normal staining properties and appear to split into longitudinal and transverse fragments (Baker, 1937).

b. Media: In the smaller cerebral arteries the quantity of collagenous tissue in the media becomes greater with increasing age at the expense of elastic and muscular elements (Binswanger and Schaxel, 1917). During the fifth and sixth decades the heavy collagenous fibers may lose their outline and become hyalinized. The arterial fibrosis is considered to be a normal anatomic process, although hypertension and diabetes mellitus tend to accelerate it (Baker and Iannone, 1959). It may produce difficulty in differentiating the media from the adventitia because of the apparent merging of the two layers into a single structure (Baker, 1937).

In the smaller cerebral vessels, the deposition of calcium, either on the surface of, or within the elastic elements of, the media (elastocalcinosis) is relatively rare, only infrequently being observed before the fifth decade (Baker, 1937). However, it may increase in amount to the point that a solid ring of calcification is formed. The process may occur with or without hyalinization or fibrosis of the various elements of the media (Baker, 1937).

E. PHYSIOLOGY[8]

By R. R. SONNENSCHEIN

The volume flow of blood to the entire brain or to any part of it is primarily determined by the same basic physical factors that operate in other tissues: the perfusion pressure and the vascular resistance. Both factors, however, operate in ways somewhat peculiar to this vascular bed, as a result of specialized anatomic relations and physiologic behavior.

Strikingly constant under a wide range of physiologic conditions, the total cerebral blood flow (CBF) averages about 54 cm^3 per minute per 100 gm of brain tissue. For the average-sized adult brain, weighing 1300 gm, CBF is of the order of 700 cm^3 per minute. Thus, this organ, constituting about 2 per cent of the body weight, receives 17–18 per cent of the cardiac output under resting conditions.

1. FUNCTIONAL ANATOMIC PECULIARITIES

A consequence of the residence of the brain and its vasculature within a rigid, essentially closed skull is that the total volume of these contents must remain constant. Furthermore, pressure changes are readily transmitted across the highly distensible walls of the veins and the subarachnoid spaces and ventricles, which are in functional continuity, in this respect, with the brain tissue. Within some limit, an increase in volume of one of these compartments can occur, associated with compensatory reduction in one or both of the others. Beyond this point, the tendency toward increase in volume can only result, in such a constant volume system, in a rise in pressure, whether the process arises in brain tissue (e.g., tumor), ventricular system (obstruction to drainage) or venous system (venous occlusion of any source). The ensuing elevation in intracranial pressure may contribute to a significant increase in cerebral vascular resistance. The ultimate effect of the latter on cerebral flow will depend as well on secondary changes in arterial pressure (Lassen, 1959).

The Circle of Willis (see Section B. of this chapter) may in some individuals constitute a functional collateral channel in case of occlusion of one or another of its main supplying arteries. Pressure relations within the Circle are such that under ordinary circumstances blood

[8] For extensive reviews see Lassen (1959); Kety (1956); Schmidt (1950).

supplied by the carotid or vertebral artery of either side is distributed almost solely to the same side of the brain (List *et al.*, 1945).

2. Factors Affecting the Total Cerebral Blood Flow

a. Perfusion pressure: Over a wide range of mean arterial pressures, the steady-state total cerebral blood flow (CBF) is constant. This is true in a variety of clinical and experimental conditions, including essential hypertension and drug-induced hypotension (Fig. 57). The CBF is, however, reduced when the mean arterial pressure falls below 40–50 mm Hg. Experimental evidence (Schmidt, 1950) suggests that transient changes in arterial pressure may indeed be accompanied by

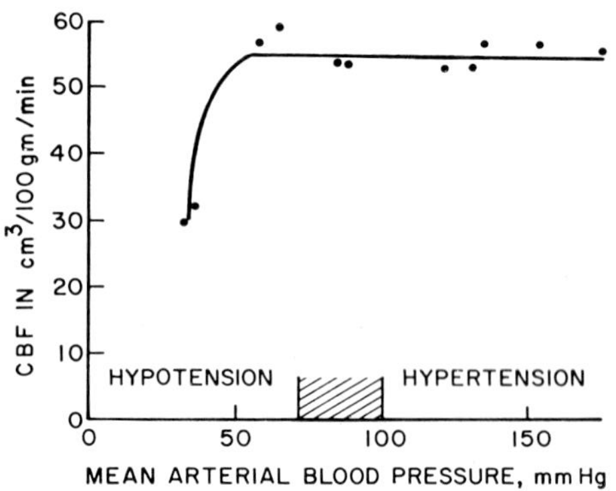

Fig. 57. Graph represents averages of 11 groups of subjects in a variety of clinical states. (Modified after Lassen, 1959; reproduced with permission of Physiological Reviews.)

parallel changes in CBF. Syncope apparently represents a clinical condition in which a decrease in CBF accompanies a fall in perfusion pressure. In general, however, the mechanisms discussed below which regulate the vascular resistance operate nicely to compensate within wide limits for changes in arterial pressure, thus allowing for relative constancy of the total CBF. Conversely, the elevation of cerebral vascular resistance associated with an increase in intracranial pressure is compensated for, to a considerable extent, by a parallel increment in arterial pressure.

b. Vascular resistance: Resistance to flow, defined as the ratio of pressure to flow, varies directly with the viscosity of the blood when the caliber of the resistance vessels (small arteries and arterioles) remains constant, and inversely with vessel caliber when viscosity is constant. In the case of cerebral circulation, changes in blood viscosity, whether due to anemia or polycythemia, are accompanied by the expected alterations in resistance (Heyman *et al.,* 1952; Schmidt, 1950). At least part of these changes in resistance may, however, be due to the effects of chemical factors acting directly on the vessels. By far the most important factor in the regulation of CBF is the activity of the cerebral small arteries and arterioles in altering the hindrance to flow. The mechanisms by which such activity is modulated may next be considered.

Smooth muscle, whether of blood vessels or other organs, may under some conditions show apparently spontaneous activity, often rhythmic in nature. By "spontaneous" is meant that no known external stimulus is involved and that the activity is generated intrinsically. A suggestion that this may occur in cerebral vessels is offered by Clark *et al.* (1958), who observed phasic changes in oxygen tension of the cat's brain at a rate of six to twelve per minute, unaccompanied by alterations in arterial pressure or, presumably, in cerebral oxygen utilization. These phasic vascular changes were asynchronous in areas 1 cm apart. As for total CBF in man, it appears that other extrinsic factors predominate.

The existence of both a sympathetic vasoconstrictor and a parasympathetic (facial nerve) vasodilator influence on cerebral vessels has been demonstrated in the experimental animal (Forbes and Wolff, 1928; Chorobski and Penfield, 1932). Likewise, a fairly extensive perivascular nervous network has been described in man, as well as other species. Despite the existence of such potential vasomotor mechanisms, there is no evidence to implicate them in normal regulation of the cerebral circulation. Clinical studies on the effect of block of the stellate ganglion, a procedure advocated in occlusive cerebrovascular accidents, have failed to reveal any increase in CBF (Lassen, 1959). This negative finding does not, of course, rule out the participation of nervous control mechanisms. It could mean that the sympathetic innervation is not tonically active in the case of cerebrovascular accidents but may be so under other conditions. One could further speculate that while no change in total CBF occurs during stellate block, regional changes could take place, either so localized as not to be detected by the methods used, or so balanced (some areas manifesting an increase and others a decrease in flow) as to result in no change in net flow. For the most part, however, the action of chemical factors, described below, has been shown to be adequate for the explanation of the regulation of total CBF. It is thus unlikely that nervous mechanisms play any important role.

Pronounced dilation of cerebral vessels and accompanying increase in CBF occurs in any condition in which arterial pCO_2 increases. The breathing of 5–10 per cent CO_2, for example, results in augmentation of

CBF by 50–100 per cent, while forced hyperventilation, which reduces arterial pCO_2, produces a marked reduction in CBF (Kety and Schmidt, 1948). The changes in CBF are considerably greater than can be accounted for by any alteration in perfusion pressure and hence represent primarily changes in vascular resistance. The action of CO_2 is apparently directly on the resistance vessels of the brain, or conceivably indirectly through the perivascular nerve plexuses, for total extrinsic denervation of the vessels has no significant effect on the response to this gas (Wolff, 1936). Of all known substances, CO_2 is the most potent cerebral vasodilator (for the physiologic significance of CO_2, see Subsection 4, below).

Diametrically opposed to the action of CO_2 is the effect of O_2 on cerebral vessels. Starting from a normal resting level of arterial pO_2, hypoxia, as produced by breathing 8–10 per cent O_2, brings about an increase in CBF of the order of 40 per cent (Kety and Schmidt). Small changes in partial pressure of inspired oxygen, insufficient to alter arterial pO_2, have no effect on CBF. Increase in arterial pO_2, as by breathing O_2 at partial pressure of 1 atmosphere or greater, causes moderate reduction in CBF (Kety and Schmidt). Again, these changes are primarily due to alterations in resistance, perhaps through direct effect of O_2 on the vessels, although the mechanism is not clear. While significant, the vasodilating effect of O_2 lack is not as potent as that of increased CO_2. In some clinical conditions such as asphyxia the two influences act in concert. In primary hypoxia, increase in CBF occurs even in the face of hypocapnia associated with hyperventilation, indicating that the dilating effect of low pO_2 can over-ride the constrictor action of low pCO_2.

Apparently, some degree of cerebral vasodilation may occur in acidosis, independent of arterial pCO_2 (Kety *et al.*, 1948a). Hypoglycemia may also cause vasodilation (Kety *et al.*, 1948b). In general, vasodilation and augmentation in CBF occur whenever cerebral metabolism is enhanced, as in seizures; the extent of increase is such as to maintain constancy of the cerebral venous pO_2 and pCO_2 (Schmidt, 1950). Conversely, CBF decreases when cerebral metabolism is reduced.

3. Regulation of Regional Flow

Persuasive data have been adduced from experimental animals that regional changes may occur in cerebral circulation, independent of blood flow to the brain as a whole (Schmidt and Hendrix, 1937; Sokoloff, 1957); some evidence for such a response in man also exists (Fulton, 1928). Apparently in those regions whose activity is selectively increased, circulation likewise rises. This localized vasodilation thus parallels the generalized dilation, described above, which accompanies greater activity of the brain as a whole or in large part. It is likely, although not established, that the important factors in the local response are increased tissue pCO_2 and decreased pO_2, reasonably presumed accompanants of increased metabolic activity. The possibility of local regulation of an axone reflex type, involving the perivascular plexuses, or of a localized reflex vasomotor response, implicating central autonomic mechanisms, has not been adequately investigated. Parenthetically, it may be noted that the reciprocal problem, that of the role of local circulation in regulating regional neuronal activity, has also received insufficient attention.

4. Some Considerations on Homeostatic Regulation of Cerebral Blood Flow

Whether the total cerebral circulation or that of a limited part of the brain is considered, the nicety of control is striking. No matter what the conditions, over a wide physiologic range, adjustments occur to the end that cerebral venous pCO_2 and pO_2, reasonably considered as reflections of tissue pCO_2 and pO_2, tend to remain constant. Both CO_2 and O_2 play critical roles in the activity of cerebral neurones. On the one hand, accumulation of CO_2 leads to marked alteration in function, as indicated by both electrical and behavioral changes. On the other, a continuously adequate O_2 supply to the cells must be maintained; the loss of consciousness following anoxia or circulatory arrest of but a few seconds' duration testifies to this. Davies and Bronk's (1957) observations on cerebral tissue pO_2 suggest that some zones at a distance from nutritive vessels exist normally just at the brink of anoxia. This situation may partially account for the extreme sensitivity of the brain to circulatory arrest.

Generally, it appears that total circulatory adjustment of the body is such as to maintain an adequate or optimal perfusion pressure to the cerebral vascular bed. The presence of the important barostatic receptors in the carotid sinus, in the main stream of the cerebral arterial supply, supports this contention. Moreover, cerebral vessels do not participate in the generalized vasoconstriction accompanying pressor reflexes, flow being maintained in cerebral vessels at the expense of other areas. The maintenance of a relatively constant arterial pressure allows for precise and rapid intrinsic control of cerebral circulation, according to the metabolic needs of the brain in whole or in part.

F. PHARMACOLOGY[9]

By R. R. SONNENSCHEIN

As would be expected, to a major extent the reaction of cerebral vessels to various pharmacologic agents can be predicted on the basis of physiologic characteristics of these vessels. In addition, however, alterations in cerebral circulation, as in the case of any organ, are also dependent upon general cardiovascular effects of the agents, since these ultimately affect the perfusion pressure acting on the cerebral vascular bed.

1. CARBON DIOXIDE AND OXYGEN

As pointed out in Section E. of this chapter, the tissue tensions of CO_2 and O_2 apparently represent the most important and the most potent regulators of cerebral vasomotor activity. In turn, alterations of these gases in the inspired air constitute the most effective pharmacologic means of changing cerebral blood flow (CBF). Numerous studies, recently reviewed by Sokoloff (1959), confirm the marked augmenting action of increased CO_2, at least above 2–5 per cent, and of reduced O_2. The effects of these two factors are somewhat additive, so that in the presence of hypoxemia, which has already produced some degree of cerebral vasodilation, hypercapnia of a given level does not cause increase in CBF proportionately as much as it otherwise would. While it appears that the reactivity of cerebral vessels to CO_2 may be somewhat diminished with age or arteriosclerosis (Fazekas *et al.*, 1953), a significant degree of dilation in response to CO_2 occurs in these states and in other known pathologic processes involving cerebral vessels. Likewise, some reduction in vasodilation and vasoconstriction, accompanying hyper–and hypoxemia, respectively, occurs in cerebrovascular disease (Heyman *et al.*, 1953).

Therapeutic use of CO_2 inhalation has been recommended for conditions in which hypoxemia exists without hypercapnia. In such situations, the augmented CBF tends to raise cerebral tissue pO_2. On the other hand, when CO_2 retention is associated with hypoxemia, little or no benefit is derived from CO_2 inhalation, since the cerebral vessels are already dilated by the increased cerebral pCO_2; in fact, such a procedure may lead to deleterious effects of high pCO_2 on cerebral activity.

[9] For an extensive review and complete references, see Sokoloff (1959).

2. Autonomic Drugs

Epinephrine may have a slight cerebral vasoconstrictor effect in some species. In man, intravenous administration of this agent generally results in an augmentation in CBF, proportionate to the increase in arterial pressure (King *et al.*, 1952). On the other hand, intravenous infusion of 1–norepinephrine into normotensive man produces a moderate reduction of CBF, in the face of an increased arterial pressure, thus indicating a definite vasoconstrictor effect; in hypotension, however, this drug elicits an increase in CBF (Moyer *et al.*, 1954). A variety of other sympathomimetic amines have effects on CBF similar to either epinephrine or norepinephrine (Sokoloff, 1959). Acetylcholine and methacholine produce a marked cerebral vasodilation (Schmidt and Hendrix, 1937). An initial transient fall in CBF may occur, associated with the decrease in arterial pressure produced by the choline esters; this is followed by an increase in CBF. The action of these agents on cerebral vessels can be blocked by atropine.

Whatever effects adrenergic or ganglionic blocking drugs may have on CBF, it would not be expected that they would be induced through functional blocking of cerebral vasomotor innervation, since the latter is apparently not tonically active (see Section E. of this chapter); rather, such responses would be the result of direct action on the blood vessels or of alterations in perfusion pressure brought about by effects on other vascular beds. The natural ergot alkaloids, such as ergotamine, have a variable and modest effect on cerebral vessels (Schmidt and Hendrix, 1937; Dumke and Schmidt, 1943). Administration of the dihydrogenated ergot alkaloids results in a fall in arterial pressure, with no significant change in CBF (Hafkenschiel *et al.*, 1950); the accompanying cerebral vasodilation appears to be a local homeostatic response secondary to the hypotension. Other adrenergic blocking agents, such as Dibenzyline, appear to affect CBF only to the extent that changes may occur in response to alterations induced in arterial pressure (Moyer *et al.*, 1954). Essentially the same situation appears to hold in the case of the ganglionic blocking agents (Sokoloff, 1959).

3. Other Vasoactive Drugs

Histamine appears to have a direct action upon cerebral vessels, causing a significant vasodilation. Since a similar response is produced in other peripheral beds, generally more pronounced than in the cerebral vessels, administration of this drug is accompanied by a marked fall in arterial pressure. Consequently, CBF tends to increase only slightly or may show a distinct decrease (Wolff, 1936). The unreliability and transient nature of the response to histamine make this agent unsuitable for clinical use in cases of insufficiency of CBF. The headache which commonly occurs after its administration is apparently related to distension of intracranial vessels.

The nitrites act similarly to histamine, generalized vasodilation and a fall in arterial pressure occurring. The latter commonly is sufficiently marked so as to produce a decrease in CBF (Norcross, 1938). Papaverine, in relatively large doses given orally or parenterally, may cause a moderate increase in CBF (Jayne *et al.*, 1952). Nicotinic acid has been recommended as a cerebral vasodilator on the basis of the intense facial flushing which it produces. This response in the skin and in other extracranial tissues is not, however, accompanied by any demonstrable dilation of cerebral vessels and increase in CBF (Scheinberg, 1950).

4. Drugs Acting on the Central Nervous System

a. Depressants: The most commonly used CNS depressants are the general anesthetics, and of these the barbiturates have been studied most thoroughly with respect to effects on CBF. The central depression produced by this group of drugs is accompanied by a reduction in cerebral oxygen uptake. This of itself would be expected to result in a reduction in CBF (see Section E. of this chapter). Nevertheless, while such a reduction does occur, it is proportionately less than the decrease in oxygen consumption (Sokoloff, 1959). The discrepancy is probably due to the usual hypercapnia resulting from hypoventilation. The effects of anesthetics other than the barbiturates are variable and have not been well defined. Diethyl ether seems to have a direct cerebral vasodilator effect which results in a moderate increase in CBF (Sokoloff, 1959). It is of interest that some recent studies indicate that changes in CBF during anesthesia may not be uniform throughout the brain (Sokoloff, 1957).

The central depressant action of the narcotics is accompanied by cerebral vasodilation and increase in CBF (Sokoloff, 1959). These changes can be accounted for by the respiratory depression and ensuing increase in arterial pCO_2, in the absence of a direct effect of the agents on the cerebral vessels. Moderate doses of ethyl alcohol, while producing cutaneous vasodilation, have no significant effect on CBF (Fazekas *et al.*, 1955). Doses high enough to produce coma, however, lead to cerebral vasodilation and increased CBF, partly, at least, as a result of respiratory depression (Battey *et al.*, 1953).

b. Excitants: Convulsant agents, such as Metrazol, picrotoxin and nikethamide, produce an increase in CBF which is directly related to the greater metabolic activity and oxygen consumption which they induce (Dumke and Schmidt, 1943). With doses below the convulsant threshold, no alteration in CBF occurs. CBF may not increase proportionately as much as oxygen consumption, so that cerebral tissue pO_2 may fall. In man, usual therapeutic doses of the xanthines, especially caffeine and theophylline ethylenediamine (Aminophylline), lower the CBF through an increase in cerebral vascular resistance. This effect may be related to the ability of these agents to relieve hypertensive headaches (Sokoloff, 1959).

5. General Comments

A wide variety of other pharmacologic agents has been studied, and the effects of these on CBF have been summarized and evaluated by Sokoloff (1959). In general, the cerebral vascular bed is more resistant to alteration than are other areas. Aside from the relatively potent effects of the respiratory gases, there is little in the pharmacologic armamentarium that affords significant effectiveness in the control of CBF.

G. PATHOLOGY

By H. C. H. Fang

1. Cerebral Atherosclerosis

a. Focal softening: In instances of widespread and extensive atherosclerosis, if the process is of sufficiently long duration, cerebral atrophy invariably results. Cerebral sulci become widened and ventricles dilate correspondingly. Small focal softening may occur, both grossly and microscopically. This is often found in the anterior part of the pons and less frequently, in the centrum semiovale and basal nuclei. Microscopically, the lesions usually contain compound granule cells located near the affected artery or arterioles. (For general discussion of atherosclerosis see (1) Atherosclerosis, Chapter XIX.)

b. État lacunaire and état criblé: The areas of focal softening are to be distinguished from *état lacunaire* and *état criblé,* which are similarly found in the senile arteriosclerotic brain (Betrand, 1923). *État lacunaire* (Pierre Marie) usually are multiple, clearly visible cavities with irregular shapes involving the basal nuclei. *État criblé* (Vogt) are similar cavities in the centrum semiovale. Within each of these cavities there is usually an artery or arteriole which appears quite normal; however, the perivascular spaces are often dilated as a result of deterioration of ground matrix of brain parenchyma. The frequent occurrence of *état lacunaire* and *état criblé* in hypertensive cerebral vascular disease is not fully understood. Nor is it clear as to whether the lesions are related to the functional state of the capillary network in their vicinity.

2. Hypertensive Encephalopathy

Cerebral lesions in hypertensive encephalopathy are predominantly arteriolar in distribution and probably related to the severe degree of hypertension (diastolic pressure in excess of 130–140 mm Hg and occasionally as high as 200 mm). Adams (1958) has suggested that the primary vascular abnormalities are probably functional in origin and that they take the form of intense arteriolar spasm arising from the Bayliss reflex (Byrom, 1954); only in severe cases do they lead to irreversible structural damage, such as arteriolar necrosis with protein exudation. As a consequence of these vascular abnormalities, widespread neuronal anoxia occurs, producing focal infarcts and diffuse cerebral edema, changes which are prominent in advanced cases.

3. Cerebral Thrombosis

Cerebral thrombosis with atherosclerosis accounts for more than half the cases of cerebral vascular disease. It has been shown that the lipid material tends to accumulate at the arterial branchings; hypertension appears to aggravate the process, leading to lipid deposition in smaller vessels, such as the penetrating arteries. The sites of predilection for atheromatous plaques are: (1) the internal carotid artery; (2) its continuation into the middle cerebral artery; (3) vertebral arteries at their intracranial end; and (4) the rostral third of the basilar artery. Luminal closure tends to be more frequent at the sites of bifurcation than elsewhere.

The mechanism by which a thrombus forms over an atheromatous plaque is not clearly understood. Factors which might predispose to such a process include ulceration of a plaque, stagnation of the blood, eddying, etc. Hypercoagulability of the blood has been suggested though not confirmed. (For discussion of the clinico-pathologic correlation of the lesions in the brain and types of clinical pattern caused by occlusion of a specific cerebral artery, see Stopford, 1916; Kubik and Adams, 1946; Tichy, 1949.)

4. Cerebral Embolism

The blood supply to the brain may be abruptly obstructed by an embolus, consisting either of a clot detached from vegetations over a diseased heart valve, of a mural thrombus from the left ventricle after myocardial infarction, of material from an atherosclerotic plaque in the aorta or carotid arteries or of thrombi arising in other sites of cerebral artery occlusion or in pulmonary veins. Circulating tumor cells, air or nitrogen may also cause infarction.

The embolus usually lodges at the bifurcation of an artery, producing acute ischemic infarction. The rapidity of onset of total arterial occlusion often precludes protection from collateral blood supply. Although emboli may obstruct any arterial channel, depending on the size of the embolus, the middle cerebral artery is the commonest site of occlusion. There are instances in which the embolic material may secondarily fragment further and come to lie in distal smaller vessels.

In the majority of instances, the pathologic diagnosis must rely upon indirect evidences; e.g., presence of a source of embolus, embolic phenomena affecting other viscera, atherosclerosis or other causes for arterial thrombosis in the cerebral arteries, and the presence of hemorrhagic infarctions. A clinical history of abrupt onset of maximal neurologic deficit is extremely helpful in most instances.

5. Cerebral Infarction

When the cerebral arterial supply is suddenly obstructed either by a rapidly forming thrombus adherent to an atheromatous plaque or by embolic material, infarction or necrosis of cerebral tissue results. Other conditions that are capable of producing cerebral infarction include cerebral venous thrombosis, systemic hypotension, arteritis, blood dyscrasias and complications of arteriography.

a. Gross changes: The early gross changes in cerebral infarction consist of swelling of both gray and white matter. In the case of hemorrhagic infarction, the gray matter appears congested, either diffusely or focally, and is often stippled with petechial hemorrhages. With ischemic infarction the cerebral tissue is pale.

b. Histologic changes: In the acute stage of infarction there is diffuse tissue necrosis. If this change is not maximal, varying stages of disintegration of affected nerve cells, myelin sheaths and glial elements can be seen, at times associated wtih perivascular extravasations. Small arterial channels are filled with neutrophils. In the first few days swelling may be observed, involving the oligodendroglia and astrocyte nuclei. Axons may appear swollen and myelin sheaths also stain poorly. There is an astrocytic reaction, and capillaries in the vicinity show endothelial hyperplasia.

In the ensuing days, the polymorphonuclear leukocytes become replaced by activated histiocytes. As the process progresses, the tissue begins to break down and phagocytosis removes tissue debris; subsequently cystic infarcts begin to appear. Gliosis is noted at the margins of the infarct.

6. Cerebral Aneurysm

Cerebral aneurysms can be divided into the following types: (1) congenital saccular, (2) mycotic, (3) atherosclerotic fusiform, and (4) more rarely, spontaneous and post-traumatic arteriovenous communications between the internal carotid artery and the cavernous sinus. Multiple saccular aneurysms of congenital type are sometimes formed on cerebral arteries in association with arteriovenous angiomas. In regard to the age incidence, although it has been reported at both ends of the life scale, arterial aneurysms occur commonly during the fifth decade of life (Carmichael, 1950). Of those that are clinically symptomatic, nearly half are diagnosed after the age of forty years (Meadows, 1951). Regarding the site of aneurysms of saccular type, almost all of them are right at, or close to, the points of bifurcation of arteries (Carmichael, 1950). The commonest site (30% of cases) is along the middle cerebral artery at its first or second point of branching in the Sylvian fissure. The next most frequent site (25% of cases) is at the junction between the anterior cerebral and anterior communicating artery. In approximately 20 per cent of cases, the lesion is at the juncture of the terminal portion of internal carotid with its branches. Involvement of the basilar artery and its branches accounts for about 12–15 per cent of cases, while in 12 per cent the lesion is found on the vertebral and posterior cerebral arteries. Concerning pathogenesis, it is generally accepted that the major etiologic factor in the development of congenital aneurysms is either hypoplasia or aplasia of the medial or muscular coat, at the points of branching and junction of arterial trunks.

7. Cerebral Hemorrhage

Intracranial hemorrhage accounts for nearly one-fourth of the cerebral vascular lesions seen at autopsy. Hypertension and ruptured saccular aneurysms are by far the most common causes of this condition.

Hemorrhage from diseased vessels may manifest itself in the following ways: (1) by bleeding into vessel walls, perivascular spaces or adjacent brain parenchyma; (2) in the form of slit hemorrhages; and (3) as massive hemorrhages in the cerebrum, pons, and cerebellum. Slit hemorrhages are often present in hypertension and are found in the superficial portions of the cerebral white matter.

The common sites for massive intracerebral hemorrhages are the lentiform nucleus and the deep white matter in the cerebellar hemispheres and throughout the pons and mid-brain. Courville (1950) gave the relative incidence in his 40,000 consecutive autopsies as 938 in the lentiform nucleus, 124 in the deep white matter of the cerebellum, and 73 in the pons and mid-brain. Hemorrhage into the lentiform nucleus area has been further subdivided by Courville and Friedman (1942) into the medial ganglionic type, which is usually fatal, and the lateral ganglionic type, which is not fatal. In the medial ganglionic variety, the bleeding arises medial to the putamen, usually destroying the hypothalamus and rupturing into the ventricular system. On the other hand, hemorrhages arising from the lateral ganglionic variety begin either in the putamen, external capsule or claustrum. Primary massive hypertensive hemorrhages in the pons, mid-brain or cerebellum are nearly always catastrophic and fatal in their clinical course. The underlying mechanism which results in the fatal rupture of an artery, particularly in massive hypertensive brain hemorrhage, remains obscure. (For the pathogenesis of cerebral hemorrhage, see Bouman, 1931; Stern, 1938; Globus and Epstein, 1952; and Byrom, 1954.) Whether the abnormally high systolic or diastolic pressure, the presence of cerebral vasospasm, structural defect in the vessel wall, or the presence of perivascular degenerative changes accounts for the final fatal blow-out in hypertensive cerebral hemorrhage still remains unresolved.

References

Adams, R. D. (1958). Pathology of cerebral vascular diseases. B. Cranial cerebral lesions. *In* "Cerebral Vascular Diseases," 2nd Conf., Princeton, 1957 (I. S. Wright and C. M. Millikan, eds.), p. 23. Grune & Stratton, New York.

Baker, A. B. (1937). Structure of the small cerebral arteries and their changes with age. *Am. J. Pathol.* **13**, 453.

Baker, A. B., and Iannone, A. (1959). Cerebrovascular disease: III. The intracerebral arterioles. *Neurology* **9**, 441.

Battey, L. L., Heyman, A., and Patterson, J. L., Jr. (1953). Effects of ethyl alcohol on cerebral blood flow and metabolism. *J. Am. Med. Assoc.* **152**, 6.

Bennett, H. S., Luft, J. H., and Hampton, J. C. (1959). Morphological classifications of vertebrate blood capillaries. *Am. J. Physiol.* **196**, 381.

Bertrand, I. (1923). "Les Processus de Désintegration Nerveuse." Masson, Paris.

Binswanger, O., and Schaxel, J. (1917). Beiträge zur normalen und pathologischen Anatomie der Arterien des Gehirns. *Arch. Psychiat. Nervenkrankh.* **58**, 141.

Bouman, L. (1931). Hemorrhage of the brain. *A.M.A. Arch. Neurol. Psychiat.* **25**, 255.

Byrom, F. B. (1954). The pathogenesis of hypertensive encephalopathy and its relation to the malignant phase of hypertension. Experimental evidence from the hypertensive rat. *Lancet* **ii**, 201.

Carmichael, R. (1950). The pathogenesis of non-inflammatory cerebral aneurysms. *J. Pathol. Bacteriol.* **62**, 1.

Chorobski, J., and Penfield, W. (1932). Cerebral vasodilator nerves and their pathway from the medulla oblongata. *A.M.A. Arch. Neurol. Psychiat.* **28**, 1257.

Clark, L. C., Misrahy, G., and Fox, R. P. (1958). Chronically implanted polarographic electrodes. *J. Appl. Physiol.* **13**, 85.

Courville, C. B. (1950). "Pathology of the Central Nervous System," 3rd ed. Pacific Press, Mountain View, Calif.

Courville, C. B., and Friedman, A. P. (1942). Hemorrhages into the lateral ganglionic region. Their relationship to recovery from cerebral apoplexy. *Bull. Los Angeles Neurol. Soc.* **7**, 137.

Davies, P. W., and Bronk, D. W. (1957). Oxygen tension in mammalian brain. *Federation Proc.* **16**, 689.

Dempsey, E. W., and Luce, S. (1958). Fine structure of the neuropil in relation to neuroglia cells. *In* "Biology of Neuroglia" (W. F. Windle, ed.), Chapter 7. Charles C Thomas, Springfield, Ill.

Dempsey, E. W., and Wislocki, G. B. (1955). An electron microscopic study of the blood-brain barrier in the rat, employing silver nitrate as a vital stain. *J. Biophys. Biochem. Cytol.* **1**, 245.

Dumke, P. R., and Schmidt, C. F. (1943). Quantitative measurements of cerebral blood flow in the macaque monkey. *Am. J. Physiol.* **138**, 421.

Fang. H. C. H. (1958). Pathology of cerebral vascular diseases. A comparison of blood vessels of the brain and peripheral vessels. *In* "Cerebral Vascular Diseases," 2nd Conf., Princeton, 1957 (I. S. Wright and C. H. Millikan, eds.), p. 17–22. Grune & Stratton, New York.

Fang, H. C. H. (1961). Cerebral arterial innervations in man. *Arch. Neurol.* **4**, 651.

Farquhar, M. G., and Hartmann, J. F. (1956). Electron microscopy of cerebral capillaries (Abstract). *Anat. Record* **124**, 288.

Farquhar, M. G., and Hartmann, J. F. (1957). Neuroglial structure and relationship as revealed by electron microscopy. *J. Neuropathol. Exptl. Neurol.* **16**, 18.

Fazekas, J. F., Bessman, A. N., Cotsonas, N. J., Jr., and Alman, R. W. (1953). Cerebral hemodynamics in cerebral arteriosclerosis. *J. Gerontol.* **8**, 137.

Fazekas, J. F., Albert, S. N., and Alman, R. W. (1955). Influence of chlorpromazine and alcohol on cerebral hemodynamics and metabolism. *Am. J. Med. Sci.* **230,** 128.

Foix, C., and Hillemand, P. (1926). Contribution à l'étude des ramollissements protubérantiels. *Rév. méd.* **43,** 287.

Forbes, H. S., and Wolff, H. G. (1928). Cerebral circulation. III. The vasomotor control of cerebral vessels. *Arch. Neurol. Psychiat.* **19,** 1057.

Fulton, J. F. (1928). Observations upon the vascularity of the human occipital lobe during visual activity. *Brain* **51,** 310.

Globus, J. H., and Epstein, J. A. (1952). Massive cerebral hemorrhage: spontaneous and experimentally induced. *Proc. 1st Intern. Congr. Neuropathol., Rome* **1,** 289.

Hackel, W. M. (1927). Über den Bau und die Altersveränderungen der Gehirnarterien. *Virchows Arch. pathol. Anat. u. Physiol.* **266,** 630.

Hafkenschiel, J. H., Crumpton, C. W., and Moyer, J. H. (1950). The effect of intramuscular dihydroergocornine on the cerebral circulation in normotensive patients. *J. Pharmacol. Exptl. Therap.* **98,** 144.

Heyman, A., Patterson, J. L., and Duke, T. W. (1952). The cerebral circulation and metabolism in sickle cell and chronic anemias, with observations on the effects of oxygen inhalation. *J. Clin. Invest.* **31,** 824.

Heyman, A., Patterson, J. L., Jr., Duke, T. W., and Battey, L. L. (1953). The cerebral circulation and metabolism in arteriosclerotic and hypertensive cerebrovascular disease. *New Engl. J. Med.* **249,** 223.

Jayne, H. W., Scheinberg, P., Rich, M., and Belle, M. S. (1952). *J. Clin. Invest.* **31,** 111.

Kety, S. S. (1956). *In* "The Control of the Circulation of the Blood," Suppl. to 1937 ed. (R. J. S. McDowall, ed.). Dawson & Sons, London.

Kety, S. S., and Schmidt, C. F. (1948). The effects of altered arterial tensions of carbon dioxide and oxygen on cerebral blood flow and cerebral oxygen consumption of normal young men. *J. Clin. Invest.* **27,** 484.

Kety, S. S., Polis, B. D., Nadler, C. S., and Schmidt, C. F. (1948a). The blood flow and oxygen consumption of the human brain in diabetic acidosis and coma. *J. Clin. Invest.* **27,** 500.

Kety, S. S., Woodford, R. B., Harmel, M. H., Freyhan, F. A., Appel, K. E., and Schmidt, C. F. (1948b). Cerebral blood flow and metabolism in schizophrenia. The effects of barbiturate semi-narcosis, insulin coma and electroshock. *Am. J. Psychiat.* **104,** 765.

Key, A., and Retzius, G. (1876). "Studien in der Anatomie des Nervensystems und des Bindesgewebe." P. A. Norstedt, Stockholm.

King, B. D., Sokoloff, L., and Wechsler, R. L. (1952). The effects of 1-epinephrine and 1-nor-epinephrine upon cerebral circulation and metabolism in man. *J. Clin. Invest.* **31,** 273.

Kubik, C. S., and Adams, R. D. (1946). Occlusion of basilar artery. A clinicopathological study. *Brain* **69,** 73.

Lassen, N. A. (1959). Cerebral blood flow and oxygen consumption in man. *Physiol. Revs.* **39,** 183.

List, C. F., Burge, C. H., and Hodges, F. J. (1945). Intracranial angiography. *Radiology* **45,** 1.

Maynard, E. A., Schultz, R. L., and Pease, D. C. (1957). Electron microscopy of the vascular bed of rat cerebral cortex. *Am. J. Anat.* **100,** 409.

Meadows, S. P. (1951). "Modern Trends in Neurology" (A. Feiling, ed.), Vol. 1. Butterworths, London.

Moyer, J. H., Morris, G., and Snyder, H. (1954). A comparison of the cerebral hemodynamic responses to aramine and norepinephrine in the normotensive and hypotensive subject. *Circulation* **10,** 265.

Norcross, N. C. (1938). Intracerebral blood flow: an experimental study. *A. M. A. Arch. Neurol. Psychiat.* **40,** 291.

Padget, D. H. (1948). The development of the cranial arteries in the human embryo. *Carnegie Inst. Wash. Publ.* **575,** 205.

Padget, D. H. (1956). The development of the cranial venous system in man, from the standpoint of comparative anatomy. *Carnegie Inst. Wash. Publ.* **611,** 79.

Patek, P. R. (1944). The perivascular spaces of the mammalian brain. *Anat. Record* **88,** 1.

Pease, D. C., and Molinari, S. (1960). Electron microscopy of muscular arteries: pial vessels of the cat and monkey. *J. Ultrastruct. Research* **3,** No. 4, 447.

Pease, D. C., and Schultz, R. L. (1958). Electron microscopy of rat cranial meninges. *Am. J. Anat.* **102,** 301.

Pool, J. L. (1958). Cerebral vasospasm. *New Engl. J. Med.* **259,** 1259.

Rapport, M., Lerner, B., Alonzo, N., and Franzl, R. E. (1957). The structure of plasmalogens. II. Crystalline lysophosphatidal ethanolamine (acetal phospholipide). *J. Biol. Chem.* **225,** 859.

Scheinberg, P. (1950). The effect of nicotinic acid on the cerebral circulation with observations on extracerebral contamination of cerebral venous blood in the nitrous oxide procedure for cerebral blood flow. *Circulation* **1,** 1148.

Schmidt, C. F. (1950). "The Cerebral Circulation in Health and Disease." Charles C Thomas, Springfield, Ill.

Schmidt, C. F., and Hendrix, J. P. (1937). The action of chemical substances on cerebral blood vessels. *Research Publs. Assoc. Research Nervous Mental Disease* **18,** 229.

Sokoloff, L. (1957). Local blood flow in neural tissue. *In* "New Research Techniques in Neuroanatomy" (W. F. Windle, ed.), p. 51. Charles C Thomas, Springfield, Ill.

Sokoloff, L. (1959). The action of drugs on the cerebral circulation. *Pharmacol. Revs.* **11,** 1.

Spatz, H. (1933). Die Bedeutung der vitalen Färbung für die Lehre von Stoffaustausch zwischen dem Zentralnervensystem und dem übrigen Körper. *Arch. Psychiat. Nervenkrankh.* **101,** 267.

Stern, K. (1938). Critical review. The pathology of apoplexy. *J. Neurol. Psychiat.* **1,** 26.

Stopford, J. S. B. (1916). The arteries of the pons and medulla oblongata. *J. Anat. and Physiol.* **50,** 131 and 255.

Suh, T. H., and Alexander, L. (1939). Vascular system of the human spinal cord. *A.M.A. Arch. Neurol. Psychiat.* **41,** 659.

Tichy, F. (1949). The syndromes of the cerebral arteries. *A.M.A. Arch. Pathol.* **48,** 475.

Tschirgi, R. D. (1958). The blood-brain barrier. *In* "Biology of Neuroglia" (W. F. Windle, ed.), Chapter 8. Charles C Thomas, Springfield, Ill.

Tureen, L. L. (1937). The circulation of the Brain and Spinal Cord. *Assoc. for Research Publs. Assoc. Research in Nervous Mental Disease* **18,** 394.

van Breemen, V. L., and Clemente, C. D. (1955). Silver deposition in the central nervous system and the hematoencephalic barrier studied with the electron microscope. *J. Biophys. Biochem. Cytol.* **1,** 161.

Vander Eecken, H. M., and Adams, R. D. (1953). The anatomy and functional significance of the arterial anastomoses of the human brain. *J. Neuropathol. Exptl. Neurol.* **12,** 132.

Wolff, H. G. (1936). The cerebral circulation. *Physiol. Revs.* **16,** 545.

Woollam, D. H. M., and Millen, J. W. (1954). Perivascular spaces of the mammalian central nervous system. *Biol. Revs. Cambridge Phil. Soc.* **29,** 251.

Zimmermann, K. W. (1923). Der feinere Bau der Blutcapillaren. Z. *Anat. u. Entwicklungsgeschichte* **68,** 29.

Chapter IX. CORONARY CIRCULATION

A. EMBRYOLOGY

By R. Licata

The wall of the early embryonic heart is characterized by being smoothly contoured externally and broken up into a mesh of irregular trabeculae internally. As a result, the interior of the developing heart consists of a spongy framework of muscular trabeculae, alternating with irregular sinusoidal spaces. Consequently, the most primitive means of nourishment for the heart wall is a labyrinthine system of sinusoids, the intertrabecular spaces, which intervene between the trabeculae. This intertrabecular circulation is followed in an orderly sequence of events by the establishment of a definitive circulation.

1. Superficial Circulation

a. Earliest phases: The earliest phase of the superficial circulation is represented by a primitive vascular plexus of endothelial channels and blood islands developed *in situ* within the epicardium. The main extensions of this epicardial plexus occur along the territory of the greatest mass of epicardial connective tissue lodged within the sulci. However, prior to development of a coronary circulation, the cardiac veins make their appearance as delicate endothelial diverticula from the wall of the presumptive coronary sinus, namely, the left horn of the *sinus venosus.* The first recognizable tributaries to be so formed are the middle cardiac vein and the posterior vein to the left ventricle. The greater and lesser cardiac veins follow somewhat later in development. The primordial venous channels produced in this manner rapidly link up with the epicardial plexus. Consequently, a cardiac drainage system is established in anticipation of an arterial supply, thus setting the stage preparatory to development of a coronary circulation.

b. Development of coronary vessels: The coronary arteries develop by means of endothelial budding from the lining adjacent to their respective aortic sinuses (Fig. 58). The left coronary artery makes its appearance somewhat in advance of the right, around the end of the second month. From the outset the left vessel is more strongly differ-

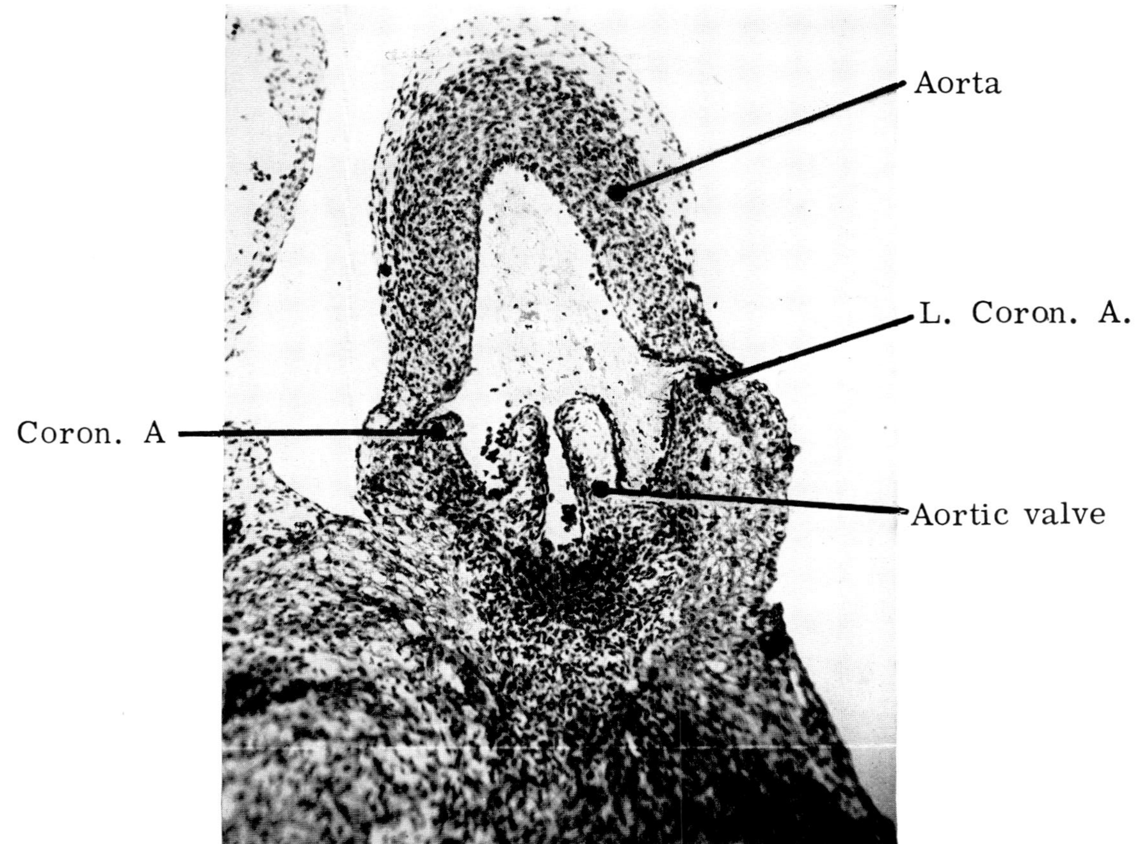

Fig. 58. Micrograph showing developing coronary arteries. Human embryo, end of second month.

entiated than its mate on the right although it follows closely in development. In both instances the arterial anlagen establish almost immediate connection with the primitive plexus which is thus capitalized in etching out the main trunks and their primary divisions. The first branch clearly derived in this manner is the anterior descending artery, followed in sequence by the posterior descending branch of the right coronary and the circumflex branch of the left vessel.

The anterior cardiac veins or *venae parvae cordis* are foreshadowed by endocardial diverticula arising directly from the right atrial wall close to the level of the sulcus. These endothelial channels at once hook up with the epicardial plexus to form the anterior system of cardiac veins.

2. Deep Circulation

The developmental basis of the deep or intramural circulation is represented by the previously described intertrabecular spaces. From these sinusoids are derived the principal intramural circuits of blood supply to the heart. In this process the intertrabecular spaces are reduced and remodelled to give rise principally to channels which open directly into the interior of the heart. Postnatally these vessels are categorized as *vasae minimae cordis* and, for the most part, include the Thebesian veins (*venae minimae cordis*) and the myocardial sinusoids. During late fetal stages the myocardial wall is altered from its sponge-like form into a compact outer zone, due to the rearrangement of the muscle constituents into fascicles. In addition, consolidation of the innermost trabeculae into muscle columns results in an inner trabeculated zone. Concomitantly an intramural capillary bed is laid down both *in situ* and by vascular sprouting from the preformed epicardial vessels. The capillary beds thus developed form connections between the previously mentioned deep channels and the penetrating branchings off the epicardial trunks. Special vascular primordia arch deeply through the wall, by-passing intermediary beds of capillaries to terminate by opening directly into the cardiac lumen. In the late fetus these short circuit vessels differentiate into arterioles and in the postnatal condition they are designated as arterioluminal vessels. Although well established during the last trimester, many of these vessels appear to degenerate prior to birth.

3. Circulation to the Conduction System

a. Sinuatrial node: The sequence of differentiation of the various divisions of the conduction system is chronicled by coincident development of the blood supply to these specialized local regions. The primordium of the definitive pace making center, or sinuatrial node, makes its appearance early in the second month, in advance of the other parts of the conduction apparatus. At this time, a specific blood vessel ascends the atrial wall from the right coronary artery to meet the node along its medial aspect. This vessel, designated as the right nodal artery or *ramus ostiae cavae superioris*, girdles the root of the superior vena cava and frequently terminates by passing into the *crista terminalis*. In some cases an homologous left nodal artery arises from the circumflex branch of the left coronary artery and travels dorsally over the atrial roof and septum also to penetrate the sinuatrial node.

b. Atrioventricular node and bundle of His: The primordium of the atrioventricular node and bundle of His can be clearly distinguished late in the second month as a sequestered bridge of atrioventricular muscle. At this time a distinct septal branch of the posterior descending artery serves to supply this part of the conduction mechanism. The vessel, commonly called the *ramus septi fibrosi* or first perforating artery, represents the highest septal branch developed off the posterior descending artery; throughout development it lags behind the sinal vessels in differentiation. During the third month of development a septal branch off the anterior descending artery penetrates the wall of the conus arteriosus and runs deeply in the crista supraventricularis. A terminal branch accompanies the right branch of the bundle of His, which it supplies as the ramus limbi dextri. Perforating septal branches from the anterior descending artery supply the remainder of the right bundle branch and the anterior division of the left branch of the atrioventricular bundle. Similar septal branches off the posterior descending artery supply the posterior division of the developing left bundle branch.

B. GROSS ANATOMY

By R. Licata

Morphologically, the blood supply to the heart can be conveniently divided into a superficial and deep circulation. The superficial circulation consists of the coronary arteries and their main epicardial branches, whereas the deep circulation is comprised of the intramural circuits.

1. Superficial Circulation

The paired coronary arteries can be regarded as modified vasa vasorum, in the broadest sense, specialized to meet the nutritional demands of the heart. The site of origin of the coronary orifice is situated in the wall of the corresponding aortic sinus, at a level in a plane somewhat above the occlusal margins of the aortic similunar valve. In this supravalvular position the ostia are rendered free of any mechanical impedance due to movements of the valve leaflets during systole or diastole. In addition, such an arrangement ensures the availability of a constant source of oxygenated blood to the coronary circulation. In general the primary branches of the main vessels can be subdivided into two main categories of secondary rami: the septal arteries or arteries to the interventricular septum and the parietal vessels, the arteries which nourish the remainder of the outlying heart wall.

a. Left coronary artery: This vessel consists of a relatively short trunk which exits from the wall of its respective aortic sinus, to pass through the interval formed by the root of the pulmonary trunk and the left atrial wall. After a brief course the main vessel breaks up into its primary divisions, namely, an anterior descending interventricular branch and a circumflex atrioventricular ramus. The former vessel descends apicalward within the anterior longitudinal sulcus, giving rise principally to anterior perforating septal arteries, which serve to supply approximately the ventral two thirds of the interventricular septum and, in addition, the apex. The circumflex atrioventricular ramus runs circuitously within the left atrioventricular groove, usually terminating along the diaphragmatic surface. Throughout its course it supplies the heart with atrial and preventricular arteries. At the *margo obtusus* strongly developed marginal branches may be found.

b. Right coronary artery: This vessel takes its origin from the corresponding aortic sinus and swings sharply rightward and, after a long excursion within the atrioventricular groove, terminates in the posterior longitudinal sulcus. In this location it is called the posterior descending artery, which extends to the apex. Throughout the proximal part of its course, atrial and preventricular branches enter the heart wall. Clearly defined marginal branchings are given off at the *margo acutus*. From its distal segment, namely the posterior descending branch, arise the posterior septal arteries which supply the remaining third of the interventricular septum. A stout conal artery is usually the first branch off the main trunk. In some instances this vessel exists from a separate sinal ostium, thus resulting in a supernumerary coronary artery.

2. Deep Circulation

Penetrating arteriolar type rami spring from the principal epicardial vessels, to enter the myocardium roughly at transverse angles, and within the heart wall, rapidly grade into the intramural capillary beds by extensive arborization. The basic unit of supply, the capillary bed, is formed by a system of endothelial channels generally oriented in the direction of the muscle bundles. Within the fascicles individual capillaries parallel single muscle fibers, periodically forming interanastomotic loops throughout their extent. A one-to-one ratio is preserved in the adult, although at birth there is about one capillary per five myocardial fibers. In spite of the fact that some correlation between vascular distribution and arrangement of muscle bundles exists, the blood supply to the heart wall on the whole is regional.

In addition to the aforementioned intramural capillary network, intrinsic collateral channels open directly into the cardiac chambers. Within the deeper strata of the myocardium fairly wide irregular sinusoidal spaces may be found. These sequestered vascular lakes communicate with the heart lumen by means of irregular endothelial channels. Arteriosinusoidal vessels augment the filling of these sinusoids, which, in addition, receive drainage from adjacent capillary beds. By far the most direct intramural circuit is represented by arteriolar short circuits which empty into the ventricular cavity. These channels, called the arterioluminal vessels, undergo a transition to endothelial tubes as they approach the lumen of the heart. Intracardiac collateral channels connecting the distal ends of cardiac veins and capillaries form venous arcs in open communication with the heart chambers. These vessels, commonly referred to as the Thebesian veins or *venae minimae cordis*, principally drain the deeper layers of the heart wall and are most frequently encountered in the interventricular septum and ventricular wall.

C. MICROSCOPIC AND SUBMICROSCOPIC ANATOMY

By R. LICATA

1. LAYERS OF BLOOD VESSEL WALL[1]

Typically, three basic tunics, namely, intima, media and adventitia (Fig. 59) are clearly delineated in the walls of the coronary arteries and their major subdivisions. From the outset, the intimal lining of the coronary vessels in the newborn undergoes a progressive histologic alteration. To a certain degree this process is initiated prenatally, particularly in the male. As a result, selection of a histologic condition representative of the normal becomes, at best, arbitrary. The coronary arteries are most sensitive in reflecting sequentially the periodic aging process, and to a remarkable degree the most significant changes involve the internal elastic membrane of the inner tunic (Gross *et al.*, 1934).

a. Intima: At birth this layer is simply constructed, consisting of an endothelium overlying a minimal amount of subendothelial connective tissue, supported by an intact internal elastic membrane. With advancing age an increasingly thickened subendothelial layer of connective tissue makes its appearance, accompanied by an exaggerated rearrangement of the internal elastic membrane. The originally single elastic membrane splits into a number of reduplicated layers, forming a complexity of lamellae. Consequently, from the intact membrane are developed secondary inner and outer limiting elastic membranes which circumscribe an intermediary musculoelastic layer. From the inner limiting, or secondary intimal, membrane is subsequently derived a complex but active lamina, the elastic hyperelastic layer which evolves to striking proportions in older age groups. A connective tissue layer is elaborated by fibrotic transformation from the elements comprising the inner aspect of this lamina.

b. Media: This layer consists of circularly disposed smooth muscle fibers intermingled with elastic tissue. Progressive collagenic consolidation of this tunic accompanies the age changes described for the intima. In the advanced age condition the intima finally exceeds the media in thickness due to preponderance of collagen. As a compensatory mechanism, patches of longitudinal bundles of smooth muscle simultaneously appear in the thickened media.

[1] For enzyme activities of normal coronary artery tissue, see Table IV, Chapter III.

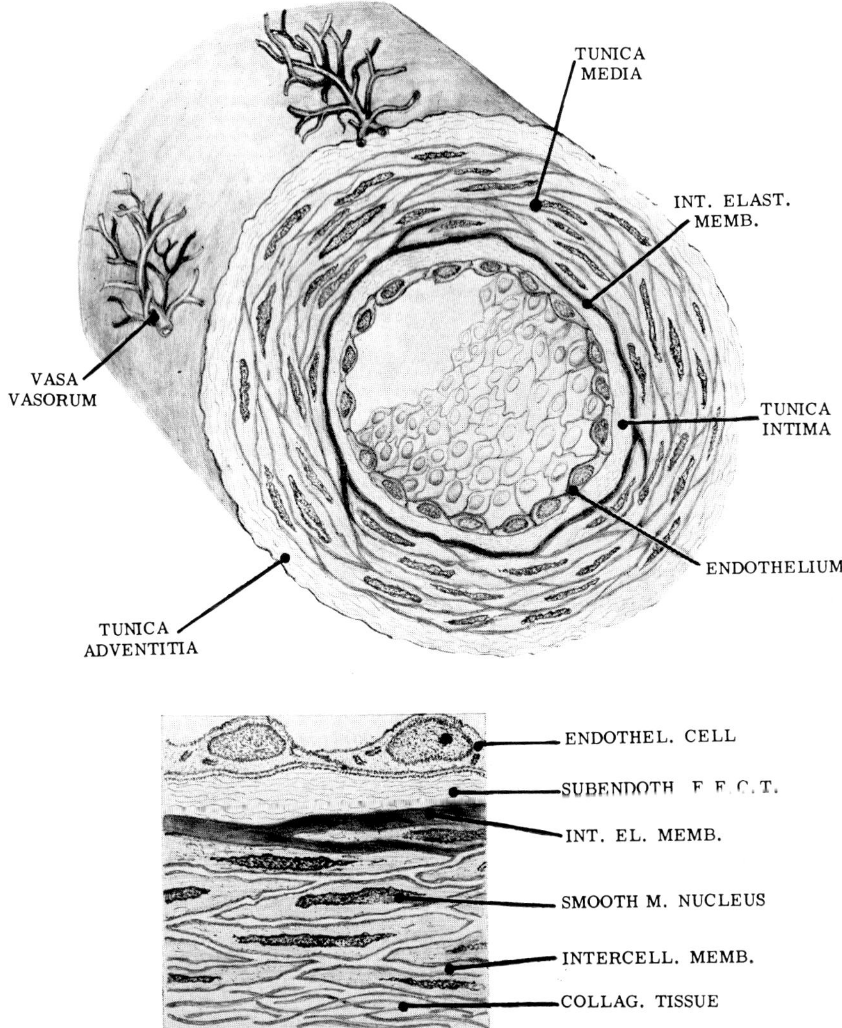

FIG. 59. Diagram of cross-sectional view of a normal coronary artery. (Adapted from F. Parker, 1958.)

c. Adventitia: Initially this layer is composed of relatively loose interwoven bundles of connective tissue. The component fibers are circularly and diagonally disposed, containing interspersed elastic fibers. A concentration of elastic elements accompanies compaction of the connective tissue fibers at the media-adventitial line, thus forming a poorly organized external elastic membrane.

2. Blood Supply of Vessel Wall

Nourishment of the vessel wall is accomplished by a rich network of minute adventitial vessels, namely, the vasa vasorum. The coronary arteries are characteristically richly supplied with vasa which may be found most frequently at the sites of side branchings from the main vessels. From the adventitial vascular plexus, penetrating offshoots ramify through the media to form an extensive network. A similar venous system of channels, including sinusoids, is distributed in the vessel wall and ultimately drains into adventitial veins. In addition, some vessels spring directly from stomae in the intima, to anastomose with the aforementioned plexus. By proximity, arteriolar *telae adiposae cordis* of the epicardial fat pads represent the most contiguous source of collateral supply through union with the vasa vasorum.

3. Ultrastructure[2]

a. Endothelium: Electron microscopic investigation (Parker, 1958) has yielded much concerning the ultrastructure of coronary vessels (Fig. 59). Although most of the studies conducted have been on lower mammals, it is nevertheless of advantage to examine the results obtained. In these studies profile views of individual endothelial cells have revealed a variety of intercellular junction types, namely, areas of simple mutual contact, areas of definite attachment, and areas of overlap. From the plasma membrane of the endothelial cell arise numerous small invaginations, the *caveolae intracellulares,* which appear related to small ovoid vesicles found internally. These structures have been assigned the function of nutritive transport. Under these conditions the internal elastic membrane appears as a fenestrated sheet composed of fibrillar elements imbedded in an homogeneous matrix. Through the openings of the membrane enter occasional filiform cytoplasmic processes which originate from the endothelial cells and extend toward the media.

b. Smooth muscle: On the basis of studies dealing with the fine structure of the media of the coronary arterial wall, a discrete layer of intercellular cement can be distinguished separating the sarcolemma of adjacent smooth muscle cells. From the sarcolemmal membrane also arise invaginations forming *caveolae intracellulares* associated with the internally situated vesicles. Delicate bundles of myofilaments traverse the individual smooth muscle cells. The presence of intercellular cytoplasmic

[2] For a general discussion of the ultrastructure of blood vessels, see Section C., Chapter I.

bridges has not yet been clearly demonstrated for the smooth muscle cells of the coronary arteries. The tunica adventitia is conventionally represented, consisting of loosely organized bundles of collagenous fibers containing vasa vasorum and neural elements.

D. AUTONOMIC INNERVATION[3]

By R. Licata

a. Embryologic considerations: Innervation of the heart is more fully understood when considered in the light of developmental events. The early embryonic heart develops as a simple pulsating tubular structure consisting of a venous inlet and an arterial outlet. The venous end develops into the sinuatrial region, whereas the arterial end becomes the truncus arteriosus. With the primary areas thus established, neural plexuses of vagal and sympathetic origin come into relationship along the lines of mesocardial fixation of the cardiac tube. Subsequently, a truncoconal plexus of nerves is elaborated around the arterial end and a sinuatrial plexus develops around the venous end. Intricate interconnections between the two plexuses may be recognized quite early in development.

b. Innervation in postnatal state: With the developmental background presented, the innervation of the coronary vessels in the postnatal condition can now be logically discussed. From the truncoconal plexus is derived principally the superficial plexus of adult morphology. The superficial plexus is related mainly to the arch of the aorta and becomes subdivided into two secondary arterial plexuses around the proximal segments of the coronary arteries. In this manner a left and right coronary periarterial plexus is established, extending at least up to the major divisions of the respective vessels. The main source of fibers to the plexuses involved descend through the interval formed by the trunks of the ascending aorta and pulmonary artery, whence both postganglionic and preganglionic components enter the plexus. Sometimes a discrete ganglionic mass, the ganglion of Wrisberg, can be distinguished in the primary plexus of origin. For the most part, scattered ganglion cells

[3] For a general discussion of the sympathetic nervous system, see Chapter IV.

lying within the pulmonoaortic space are the source of cell bodies for parasympathetic neurons. Terminal ganglion cells may also be encountered within the epicardial connective tissue adjacent to the routes of the main coronary branches. In all cases sympathetic and parasympathetic fibers are intimately co-mingled. Filamentous offshoots also reach the plexuses from the neural meshes around the roots of the pulmonary vessels and the atrial wall. Adventitial neuronal plexuses have been demonstrated in the outer coats of the coronary vessels. The precise mode of innervation of the inner tunics has not as yet been clearly defined.

c. Function: Functionally, the sympathetic fibers serve to dilate the coronary vessels (Mitchell, 1953). The vagal or parasympathetic fibers have a vasoconstrictor effect. The behavior of the terminal divisions of the vessels indicates that perhaps the innervation of the distal segments may be different from that of the proximal segments. That is to say, whereas both sympathetic and parasympathetic fibers innervate the main trunks and branches, the terminal branches are principally under vagal control. Afferent fibers leave the coronary plexuses and return to the central nervous system by way of the main splanchnic stems. Pain is transmitted via the lower cervical cardiac nerves and branches from the upper four to five thoracic segments of the sympathetic trunk through the corresponding white rami communicantes. Within the central nervous system, the parasympathetic preganglionic fibers originate in a group of cells adjacent to the nucleus ambiguus and cells intermingled with those of the dorsal efferent nucleus of the vagus. The sympathetic cardiac preganglionic fibers arise from cells located in the intermediolateral gray columns of the upper thoracic cord and enter the sympathetic trunk through the corresponding white rami.

d. Termination of fibers: The precise type of nerve ending in relation to individual muscle fibers has been difficult to demonstrate. In general the specialized endings appear to be in the form of netlike loops and boutons or cloverleaf terminations. A terminal reticulum or ground plexus has been described as being intimately associated with individual muscle fibers. Throughout, it is almost impossible histologically to distinguish parasympathetic from sympathetic fibers. The intrinsic or terminal ganglia consist of principally multipolar cell bodies.

E. PHYSIOLOGY

By D. E. GREGG AND J. D. COFFMAN

1. FUNCTIONAL ANATOMY

The coronary circulation is similar to that of most vascular beds, with its arteries, arterioles, capillaries, superficial veins, arterial and venous anastomoses and arteriovenous shunts, and with the veins emptying into the right atrium. It differs, in that arterioles, capillaries, and veins are also connected with both ventricular cavities (Wearn *et al.*, 1933). In dogs, however, almost all coronary artery inflow can be collected in the superficial veins of the left and right heart, leaving only a small volume to drain into the ventricular cavities (Gregg, 1950). Presumably, a similar utilization of anatomic pathways exists in man.

2. DETERMINANTS OF CORONARY FLOW AND OXYGEN UPTAKE

The objective of studies of the coronary circulation is knowledge of the determinants of coronary flow and oxygen uptake by the myocardium when it is contracting, as compared with when it is relaxed, and of the relation of flow and oxygen uptake to cardiac work in states of normalcy, of increased or decreased stress, and of disease in men. This is far from being realized. Coronary flow is regulated by the central coronary artery pressure, by the active state of the intravascular coronary smooth muscle throughout the cardiac cycle, and by the mechanical effect exerted during ventricular systole by the large muscle mass around the coronary vessels. Directional changes in coronary flow and oxygen usage are generally similar, although the mechanisms controlling the latter are not well understood. Experimentally and in man, left coronary arterial flow, together with the oxygen or other chemical differences between the arterial and coronary sinus blood, is used as an index of left ventricular metabolism (Rayford *et al.*, 1959). Similar studies for the right ventricle are generally not available. Ventricular contraction reduces coronary flow through the ventricular wall by about 50 per cent (Sabiston and Gregg, 1957). However, under conditions that augment coronary flow, the major portion of increase is through active vessel dilation and not through the decreased extravascular resistance (Lewis *et al.*, 1961). Oxygen consumption during systole averages about three times that during diastole for the same time period.

The metabolism of the beating heart is predominantly aerobic, only a small oxygen debt having been found for this organ (McKeever *et al.*, 1958).

3. Methodology

Coronary flow is determined (1) in the anesthetized dog, by the orifice meter and electromagnetic flow meter (phasic flow) and by the rotameter, bubble flow meter and nitrous oxide methods (mean flow); (2) in the unanesthetized and/or active dog, by the electromagnetic flow meter; and (3) in the resting human, by the nitrous oxide method. The oxygen saturation of the coronary arterial and venous bloods is obtained by Van Slyke, Beckman or oximeter techniques (Gregg, 1950; Khouri *et al.*, 1960; Shipley and Wilson, 1951).

4. Patterns of Responses to Stress States (Table X)

Intermediate metabolites increase coronary flow and decrease oxygen extraction, but whether normal concentrations of these chemicals within the coronary vessels effectively alter flow is not known, since intracoronary artery injection of coronary sinus blood does not affect coronary flow (Jellife *et al.*, 1957; Wolf and Berne, 1956). Reasonable changes in arterial carbon dioxide and pH have little effect on the coronary circulation. Coronary flow is markedly increased, and coronary resistance and arteriovenous oxygen difference are decreased, by short periods of ischemia, reduction in the oxygen content of the inspired air, coronary perfusion with somewhat unsaturated blood, and mild anemia; but these alterations have no effect on myocardial metabolism, cardiac work, heart rate, or systemic dynamics (Bing, 1951; Case *et al.*, 1955; Gregg, 1955, 1957, 1959a; Lombardo *et al.*, 1953). The augmentation in coronary flow noted in the exercising dog is largely on the basis of an elevated heart rate, since the stroke coronary flow may not increase (Khouri *et al.*, 1960). When the primary change in stress is caused by a rise in heart rate, cardiac output and work (whole blood transfusion), aortic coarctation (moderate), cardiac sympathetic nerve stimulation, and severe anemia or anoxia, then both cardiac work and myocardial oxygen usage increase. Concurrently, coronary flow rises while coronary vascular resistance falls (Alella *et al.*, 1956; Bing, 1951; Case *et al.*, 1955; Eckstein *et al.*, 1950; Laurent *et al.*, 1956; Lewis *et al.*, 1960). In situations of primary reduction in stress, such as hemorrhagic hypotension, hypothermia, and extreme heart failure, cardiac work and oxygen consumption decrease as does coronary inflow (Hackel and Goodale, 1955;

TABLE X

RESPONSE OF THE CORONARY CIRCULATION TO STATES OF STRESS[a]

State	Coronary							Systemic						
	L.C.F. cc/min/ 100 gm	Art. B.P. L.C.F.	Cor. sin. O_2 Vol. %	Cor. A-V O_2 Vol. %	O_2 cc/min 100 gm	Stroke cor. flow cc	Stroke cor. O_2 cc	Art B.P.	C.I.	C.W.I.	S.V.I.	S.W.I.	C.W. O_2	H.R.
Resting	76 (84)		4.0 (5.3)	13 (11 3)	9.8 (9.2)			106 (96)	3.4 (3.3)	54 (4.1)				119 (80)
Increased metabolites	+	−	+	−	+									
Inc. blood CO_2 and pH	0	0	0	0	0	0	0	0	0	0	0	0	0	0
Mild ischemia and hypoxia	+	−	+	−	0			0	0	0	0	0	0	0
Mild anemia	+ (+)	− (−)	− (−)	− (−	0 (−)	0	0	0	0	0	0	0	0	0
Exercise	+ (+)	− (−)	0 (0)	0 (0)	+ (+)	0 −	0 −	+ (+)	+ (+)	+ (+)	0 ($^{0}_{+}$)	+	(+)	+
Increased heart rate	+ (+)	− (−)	0 (0)	0 (0)	+ (+)	− (0)	− (0)	+ (0)	+ (0)	+ (0)	− (−)	− (−)	− (−)	+
Transfusion	+	−	0	0	+	+	+	+	+	+	+	+	+	−
Aortic coarct. (moderate)	+ (+)	−	0 (−)	0 (+)	+ (+)	+	+	+ (+)	0 (0) −	+ (+)	+	+	+	−
Cardiac sympathetic nerves	+	−	−	+	+	±	±	+	+	+	±	±	−	+ 0
Severe anemia and hypoxia	+ (+)	− (−)	− (−)	− (−)	+ (+)			+	+	+			0	+
Shock (hemorrhagic)	−	−	0	0	−	−	−	−	−	−	−	−	−	+
Hypothermia	−	0	0	0	−	−	−	−	−	−		−	0	−
Heart failure (extreme)	± (0)	−	−	+	±	±	±	−	− (−)	− (−)	−	−	− (−)	0
Aortic coarct. (complete)	+	−	0	0	+	+	+	+	−	−	−	−	−	−
Hypertensive cardiovascular disease	(0)	(+)	(0)	(0)	(0)	(0)	(0)	(+)	(0)	(+)	(0)	(+)		(0)
Aortic stenosis	+	−	−	+	+	+	+	−	−	−	−	−	−	0
Pulmonary stenosis[b]	+	−	−	+	+	+	+	−	−	−	−	−	−	0

[a] The data are subject to considerable error and, in some instances, directional changes may be wrong because of a general lack of a normal environment in dogs and difficulties of the N_2O method as used in humans and dogs. In various stress states using the latter method, each subject or dog does not generally serve as his own control and it has not been demonstrated that repeated measurements on the same day or some weeks apart are reasonably similar.

Key: + = increase. 0 = little or no change. − = decrease. Results in parentheses from man, others from dog.

[b] Directional coronary changes refer to right coronary circulation.

Hansen *et al.*, 1956; Starzl and Gaertner, 1955). Exceptions to this general picture of coronary compensation to changing stress are complete aortic coarctation, chronic hypertension, and aortic and pulmonary stenosis; in all these conditions coronary flow and oxygen usage may not be related to cardiac output and work (Bing, 1951; Gregg, 1950).

The basic mechanisms regulating coronary flow have been variously related to the oxygen tension of the arterial blood, oxygen tension within the myocardium, the action of local metabolites or vasodilating substances, and, at the cellular level, the rate of reduction of cytochrome oxidase and the needs of the hydrogen transport system. Similarly, the basic control of myocardial oxygen usage has been related to diastolic ventricular size, cardiac output, cardiac work, and ventricular tension. Present information does not permit a final decision (Berne *et al.*, 1957; Gregg, 1950; Olson and Piatnek, 1959; Sarnoff *et al.*, 1958).

5. The Collateral Circulation

The collateral circulation may be estimated from the injectable collateral bed, or better, by collecting the volume of blood flowing externally from a tube inserted into the peripheral end of a permanently occluded coronary artery. In the latter preparation, collateral blood flow averages about 3 cc/50 gm myocardium, with each cc of blood containing 0.6 cc oxygen. This level of retrograde flow does not naturally increase for a number of hours, although it can be raised by passive elevation of the arterial blood pressure and lowered by excessive myocardial stretch and reactive hyperemia in the other coronary artery branches. In most hearts collateral flow starts to increase within 12 hours, and within 3–4 weeks it may approximate 40–100 per cent of normal inflow into the occluded coronary artery (Gregg, 1950; Kattus and Gregg, 1959).

a. Factors affecting collateral circulation: The collateral circulation can be improved prophylactically without the stimulus of coronary insufficiency (Table XI). The causative factors naturally occurring in man are largely unknown but are illustrated by the increase in the incidence of the injectable collateral bed in the presence of hypertrophy, valvular heart disease, cor pulmonale, anemia, and possibly high altitude (Blumgart *et al.*, 1950; Zoll *et al.*, 1951). No good experimental evidence exists to indicate that physical exercise *per se* augments collateral flow. Experimentally, apparently many procedures increase the injectable and functional collaterals (Beck *et al.*, 1957; Day and Lillehei, 1959; Gregg, 1959b). Since, however, sham operations involving cardiac manipula-

TABLE XI
EFFECT OF VARIOUS SITUATIONS DESIGNED BY MAN AND NATURE ON THE CORONARY COLLATERAL CIRCULATION IN MAN AND DOG[a]

Situation	Tissue	Injectable collaterals	Retrograde flow ml/min	Angina	Work and exercise tolerance
Acute ligation of coronary artery ramus			3		
Hypertrophy, valvular disease, cor pulmonale		(+)			
Anemia		+ (+)	+		
Exercise			0		
Section cardiac sympathetic nerve fibers			0	(−)	(+)
Myocardial hypoxia	Coronary sinus constriction or ligation, *aorta-coronary sinus shunt*, pulmonary artery-left atrial fistula.	+	+	(−)	(+)
Mechanical and chemical irritants applied chronically to the myocardial surface	*Mechanical abrasion, phenol*, talc, powdered bone, *asbestos*, mica, gelatin, sponge, silver nitrate.	+	+	(−)	(+)
Extracardiac tissue applied to the heart	Spleen, omentum, fat, muscle, pedicle skin flap, intestine, lung, *carotid and internal artery implants.*	+	0	(−)	(+)
Internal mammary artery ligation		0	0	(−)	(+)
Sham operation	Handling heart, skin incision.	+	+	(−)	(+)
Coronary bypass and coronary endarterectomy	Internal mammary and carotid arteries.	0	0	(−)	(+)

[a] The above analysis must be used with caution since generally each dog does not serve as its own control. Although selection may be standardized by using as controls those dogs with maximal ECG changes following temporary coronary artery ligation, control coronary flows still show a rather wide range and, at times, may overlap with those obtained in the presence of an induced variable. Effects of all situations listed have been examined in both man and dog except for hypertrophy, valvular disease and cor pulmonale in the dog and exercise in man. Retrograde flow measurements refer primarily to procedures in italics.

Note: For key, see Table X.

tion may also augment collateral flow, the operative maneuvers may have no specific effect but may act by raising the ventricular fibrillation threshold, thus giving time for collaterals to develop.

b. Operative procedures in man: Most procedures, including sham operations, have been applied to the heart of man suffering from coronary artery disease (Adams, 1958; Beck *et al.*, 1957; Cannon *et al.*, 1959; Gregg, 1959a). These may increase the work and exercise tolerance and decrease cardiac pain. Such observations are not necessarily explained on the same basis as the improvement in collateral circulation of the dog which follows similar procedures, since in the latter, surgery is designed to promote collaterals in the presence of a normal coronary circulation, whereas in the human its purpose is to promote collateral circulation after coronary insufficiency has been naturally established. Such benefits, whose explanation is not clear, may arise from handling the heart, thus enhancing the ventricular fibrillation threshold, from obliteration of afferent pathways for pain by de-epicardialization, and from psychogenic effects (cf. sham operations with only a skin incision). Coronary endarterectomy or bypass should be effective, provided a gross coronary insufficiency of blood exists beyond the obstruction, thrombosis does not develop, and no sizeable atherosclerotic lesion is present distal to the region of the occluded coronary artery (Hall *et al.*, 1961). It is doubtful that these conditions prevail, which is probably why initial reports on endarterectomy are not too sanguine (Cannon *et al.*, 1959). In addition, it is possible that such an operation will cause the existing collateral flow to disappear, with the result that the patient will be in difficulties if another coronary occlusion should subsequently occur.

F. PHARMACOLOGY

By J. D. Coffman and D. E. Gregg

1. Methods of Study

The reactions of the coronary bed to various pharmacologic agents, electrolytes, metabolites and hormones have been investigated in the isolated heart, the heart-lung preparation, the fibrillating heart, and the *in situ* heart in anesthetized or unanesthetized dogs. Only in recent years, with the development of coronary sinus catheterization, have these studies been attempted in man. It is likely that information obtained on the *in situ* heart in unanesthetized subjects is the most useful concerning the over-all effects of the agent, though limited to the left coronary bed.

a. Mechanisms through which alterations occur: Coronary blood flow may be altered by a direct effect on the vasomotor state of the vessels, by an increase or decrease in central blood pressure, by myocardial stimulation or depression, by a change in the cardiac workload through extracardiac phenomena, and by electrolyte, pH or gaseous alterations of the blood perfusing the coronary bed. It is interesting to know whether a drug affects the extravascular and intravascular resistances of the coronary beds, but what is really desired is knowledge of the effect upon the supply of oxygen to the myocardium, the oxygen utilized by the myocardium, and the efficiency of the heart in the use of its oxygen for the work performed. In addition, a pharmacologic agent may be able to improve the oxygen utilization for external work of the heart without producing an increase in coronary blood flow or oxygen extraction.

b. Data necessary for evaluation of drug: In order to evaluate an agent properly, the following information is necessary: (1) coronary blood flow; (2) arteriovenous oxygen difference across the coronary bed; (3) blood pressure; (4) cardiac output; (5) myocardial contractility; and (6) heart rate. From these data, the myocardial oxygen availability and usage, cardiac work and efficiency can be calculated.

2. Effects of Drugs on Coronary Circulation

Tables XII and XIII list the agents for which the most information is available; it can be seen that few have been completely studied.

a. Nitrites: Experiments in man with nitroglycerin show a greater coronary blood flow, with an elevated myocardial oxygen uptake and decreased efficiency, while some studies in dogs demonstrate no effect on oxygen metabolism (Brachfeld *et al.*, 1959; Eckstein *et al.*, 1951). Furthermore, in patients with coronary artery disease, this drug does not increase coronary flow, and, with a steady oxygen extraction, it reduces cardiac work (Gorlin *et al.*, 1959). These data raise the old question of the applicability of knowledge obtained in normal animal or human experiments to the diseased states. If these studies should be confirmed in patients during anginal attacks, other theories for the action of nitroglycerin must be considered. One theory holds that nitroglycerin blocks the anoxic-inducing effect of the catecholamines on the heart (Raab and Lepeschkin, 1950), but an antiadrenergic action has not been

TABLE XII
THE ACTION OF VARIOUS AGENTS ON CORONARY BLOOD FLOW AND MYOCARDIAL OXYGEN METABOLISM[a]

Agents	Route	Coronary blood flow	Myocardial A-V O_2 difference	Myocardial O_2 uptake	Myocardial force of contraction	Cardiac output	Cardiac work	Cardiac efficiency	Total peripheral resistance	Blood pressure	Heart rate
	I.V.										
Nitroglycerin[b]	Sublingual	+(+)	−(0)	±(+)	0	(±)	(−)	(−)	(±)	−(−)	+(+)
Coronary artery disease	Sublingual	(0)	(0)	(0)		(−)	(−)	(−)	(±)	(−)	(+)
	I.C.	+	−	0						0	0
Xanthines[c]	I.V.	+	+	+	+	+(+)	+(+)	±	−(−)	−(−)	+(+)
	I.C.	+			+						
Papaverine[d]	I.V.	+	−	−	+		−				
Khellin[e]	I.V.	±	+	0		0	0	+	0	0	0
Nikethamide[f]	I.V.	+		+		+	+	−		−	+
	I.C.	+								0	+
Acetylcholine[g]	I.V.	+	−	0	−		0	0		−	+
	I.C.	+	−	0	−	0	0	0	0	0	0
Atropine[h]	I.V.	+(+)	−(0)	+(+)		0(0)	0(0)	−(−)	0(0)	0(0)	+(+)
Nucleic Acid Derivatives[i]	I.V.	+								±	0
	I.C.	+	−	+						0	0
Thyroid[j]	I.V.	+									
Hyperthyroid patients		(+)	(−)	(+)		(+)	(+)	(±)	(−)	(0)	(+)
Ethyl alcohol[k]	I.V.	+	−	−		+	+	+		±	0
Nicotine[l]	I.V.	±	±	±		±	±	±		+	+
	I.C.	+			+					±	0
Cigarette smoking		(+)	(−)	(0)		(0)	(−)	(−)	(±)	(±)	(+)
& coronary artery disease		(0)	(0)	(0)		(+)	(+)	(−)		(+)	(+)
Hydralazine[m]	I.V.	(+)	(−)	(0)		(+)	(±)	(±)	(−)	(−)	(+)
Ganglionic Blockers[n]	I.V.	−(−)	+	±		−(−)	−(−)	−	+(±)	−(−)	+(+)
	I.C.	+								0	
Digitalis—Normals[o]	I.V.	0(0)	0(0)	0(0)		±(−)	0(−)	0(−)	+(+)	±(0)	±(−)
Heart failure	I.V.	(0)	(0)	(0)		(+)	(+)	(+)	(±)	(0)	(−)
Pitressin[p]	I.V.	−	+			(0)	(0)		(0)	±(0)	−(0)
	I.C.	−				0	0			0	0

Key: + = increase. 0 = no change. − = decrease. ± = variable results. I.V. = intravenous. I.C. = intracoronary.

[a] Results are from dog experiments; results in parentheses are from man.
[b] Brachfeld *et al.*, 1959; Wégria, 1951.
[c] Foltz *et al.*, 1948, 1950.
[d] Foltz *et al.*, 1948; Wégria, 1951.
[e] Farrand and Horvath, 1959; Ratnoff and Plotz, 1946.
[f] Eckenhoff and Hafkenschiel, 1947; Foltz *et al.*, 1948.
[g] Schreiner *et al.*, 1957.
[h] Gorlin, 1958; Scott and Balourdas, 1959.
[i] Wolf and Berne, 1956.
[j] Rowe *et al.*, 1956; Wégria, 1951.
[k] Schmitthenner *et al.*, 1958.
[l] Kien and Sherrod, 1960; Wégria, 1951.
[m] Rowe *et al.*, 1955.
[n] Crumpton *et al.*, 1954; Grob *et al.*, 1953.
[o] Bing *et al.*, 1950.
[p] Green and Kepchar, 1959.

demonstrated for this drug (Eckstein *et al.*, 1951). It might be postulated that, in the presence of a stable oxygen consumption, a reduction in cardiac work (secondary to the decrease in blood pressure) may be helpful to the myocardium, despite a calculated decrease in myocardial efficiency. Present measurements only include evaluation of the external efficiency of the heart. If this organ were to use all the oxygen it could extract at a given workload, then a decrease in this work at the same oxygen consumption might be beneficial. That the inotropic and oxygen metabolic effects of an agent may be dissociated has been shown for acetylcholine (Schreiner *et al.*, 1957).

b. Other drugs: The xanthines, nikethamide, and the catecholamines increase coronary blood flow and the oxygen available to the myocardium, but at the expense of an increased oxygen usage and cardiac work (Foltz *et al.*, 1950; Eckenhoff and Hafkenschiel, 1947; Gerola *et al.*, 1959). It would probably be preferable to increase the oxygen availability without a stimulation of myocardial oxygen metabolism, such as has been shown for nitroglycerin and papaverine in animal studies (Eckstein *et al.*, 1951; Foltz *et al.*, 1948). Hydralazine, which may precipitate anginal attacks in hypertensive patients, possesses what are usually considered the desired properties for a coronary vasodilator: it increases coronary blood flow and oxygen availability and does not alter the oxygen uptake of the myocardium (Rowe *et al.*, 1955). It would appear, therefore, that our concepts of the "ideal" agent for coronary artery disease must be revised.

Certain agents which increase coronary blood flow are not included in the tables because of the scant available information. Isoproterenol, histamine, antihistamines, heparin, dicumarol, 5–hydroxytryptamine, amplivix, and certain electrolytes (potassium, calcium, chlorides) increase coronary flow in animal preparations (Ratnoff and Plotz, 1946; Wegria, 1951; Charlier, 1959). Reports are conflicting regarding metrazol, quinidine, quinine, and morphine (Ratnoff and Plotz, 1946; Wégria, 1951; Rowe *et al.*, 1957). Only two drugs, pitressin and angiotensin, decrease coronary blood flow without a reduction in central blood pressure (Green and Kepchar, 1959).

c. Clinical studies: For clinical use in angina pectoris, drugs are evaluated in patients by methods involving their ability to alter the electrocardiographic response to an exercise test or by drug-placebo (double blind) studies (Gregg *et al.*, 1956; Cole *et al.*, 1957). The selection of patients for these trials is very difficult, due to the variability of the disease and its response to many extraneous factors. Nitroglycerin remains the only universally preferred treatment. Other agents, including the long acting nitrates, have received some favorable but also many unfavorable reports. Alcohol, which benefits patients and possesses many desirable properties (Table XII), is considered to act only as an analgesic or sedative, for unlike nitroglycerin, it does not improve the exercise electrocardiogram (Schmitthenner *et al.*, 1958; Gregg *et al.*, 1956). The mode of action of the monoamine oxidases is unknown although the clinically beneficial, but toxic, iproniazid has been demonstrated to increase coronary flow and depress cardiac contraction in isolated hearts (Allmark *et al.*, 1953). Anticholesterol agents and thyroid are used in the hope of decreasing the atherosclerotic process, but long-term investigations are needed to assess their value. In this regard, study of hyperthyroid patients shows that their myocardial oxygen utilization is increased more than would be expected from the elevated heart rate (Rowe *et al.*, 1956). Also the coronary blood flow and myocardial oxygen extraction have been shown to be reduced during alimentary lipemia (Regan *et al.*, 1961).

d. Drugs in cardiogenic shock: Table XIII is a compilation of data concerning the action of the vasopressor agents most commonly used in cardiogenic shock. Since the great majority of the studies were performed in normal animals or man, the question again arises whether the information can be applied to the diseased state. Against such a transfer is the demonstration that drugs may act differently in a normal, as compared to a failing heart; for example, mephentermine may exert a myocardial oxygen-conserving effect in failing animal hearts but an oxygen-wasting action in normal ones (Welch *et al.*, 1958).

Since the etiology of cardiogenic shock is undetermined (decreased peripheral resistance and/or myocardial failure) (see (2) Shock, Sections B., C.-1 and D.-1, Chapter XX), a controversy exists regarding its treatment. The question concerns whether an agent should be used which only increases coronary flow and blood pressure (methoxamine), or one which elicits these responses but at the same time stimulates the myocardium and, therefore, raises myocardial oxygen consumption (epinephrine, levarterenol, metaraminol, and mephentermine) (Aviado, 1959). It appears that clinical results favor the latter concept, for levarterenol has met with the most success. Also, in animal studies, it has been found that the vasopressor agents which stimulate myocardial contractility and lower atrial pressure are beneficial to the "failing" heart (Sarnoff *et al.*, 1954). Ideally, the agents for use in cardiogenic shock should increase coronary flow, stimulate myocardial contractility, raise the blood pressure and cardiac output but not increase myocardial oxygen consumption

TABLE XIII
PRESSOR AGENTS USED IN CARDIOGENIC SHOCK[a]

Drug	Route	Blood pressure systolic/diastolic	Heart rate	Myocardial force of contraction	Coronary blood flow	Myocardial O_2 uptake	Coronary A-V O_2 difference	Cardiac output	Cardiac work	Cardiac efficiency	Total peripheral resistance	Right atrial pressure
Epinephrine[b]	I.V.	+/−	+(±)	(+)	(+)	(+)	(−)	+	+	(−)	−	
(Adrenalin)	I.C.			(+)	(+)	(+)	(+)					
Levarterenol[b]	I.V.	+/+	−(±)	(+)	(+)	(+)	(−)	−(±)	+(+)	(−)	+	(−)
(Levophed)	I.C.			(+)	(+)	(+)	(+)					
Ephedrine[c]	I.V.	+/0	+	(±)	(+)			+(±)	+		−	
	I.C.											
Metaraminol[d]	I.V.	+/+	−	(+)	(+)			0(±)	+		+	(−)
(Aramine)	I.C.			(+)	(+)							
Mephentermine[e]	I.V.	+/+	−	(+)	(+)	(+)	(−)	0(±)	+	(−)	+(0)	(−)
(Wyamine)	I.C.			(+)	(+)							
Phenylephrine[f]	I.V	+/+	−	(+)	(+)			±	+		+	+†
(Neosynephrine)	I.C.			(+)	(+)							
Methoxamine[f]	I.V.	+/+	−	(0)	(+)			(−)	(−)		(+)	(+)
(Vasoxyl)	I.C.			(0)	(0)							

Key: I.V. = intravenous. I.C. = intracoronary. + = increase. 0 = no change. − = decrease. ± = variable effects or conflicting data. † = venous pressure.

[a] Results in man. Figures in parentheses indicate if dogs react differently or if only dog results are available.

[b] Berne, 1958; Gerola *et al.*, 1959; Aviado, 1959; Wégria, 1951.

[c] Aviado, 1959.

[d] Livesay *et al.*, 1954; Sarnoff *et al.*, 1954.

[e] Brofman *et al.*, 1952; Welch *et al.*, 1958.

[f] Aviado, 1959.

in relation to its workload.[4] Such drugs may be available but more experimental proof is necessary.

G. PATHOLOGY

By J. T. Roberts

1. Introduction

Abnormalities in the structure of coronary vessels are very common findings, being noted soon after birth or even earlier. The changes may either be similar to those found in other vascular beds or they may be peculiar to the vessels of the heart. At present atherosclerosis is by far the most common and serious of all the conditions affecting the coronary circulation. At the same time it is the least understood because of the difficulty in determining whether each of the complicated phases in its pathogenesis is effect or cause (or coincidental association) in relation to the others (for further discussion of this point, see (1) Atherosclerosis, Section D.-Chapter XIX). Besides atherosclerosis, the following etiologic factors produce structural changes in coronary vessels: arteriosclerosis, medial sclerosis, adventitial changes, syphilitic effects, embolism, thrombosis, traumatic contusion and laceration, spasm, distention, elongation with hypertrophy, shortening with atrophy, tortuosity or straightening, congenital variations (in size, patterns, origin and anastomoses), infections (bacterial, parasitic, rickettsial or viral), and metabolic and "collagen" diseases. The involvement may be in the major or the smaller arteries, the anastomotic and penetrating branches, and probably also the coronary capillaries, veins and lymphatic vessels. Multiple factors are most likely present in today's aging population, with modifications of the underlying mechanisms occurring through treatment, dietary habits and stresses. Sensitization, aging, and diseases of unknown causes, as well as neoplasm and lymphomas, add other changes to the coronary vessels.

Not infrequently minimal clinical signs may be associated with very extensive lesions of coronary arteries, probably because of the re-

[4] In this regard, it has already been pointed out that myocardial contractility and oxygen consumption may be dissociated.

markable compensatory mechanisms of the vascular tree in the heart. Conversely, seemingly small structural changes sometimes cause sudden death either in apparently normal people or in ones suffering from some other disease. Since defective function of coronary vessels may produce heart failure of any type, the anoxic and other effects of such a state bring on either secondary complications or efforts at healing.

A detailed knowledge of the variable anatomy of coronary vessels is helpful in (1) understanding the causes of death and (2) in efforts to bring additional blood supply to ischemic heart muscle, as by arterializing the coronary sinus, anastomosis to coronary arteries, or endocardial puncturing (Roberts *et al.*, 1943; Roberts, 1959).

2. Pathogenesis

Each cellular and tissue type in coronary vessels can be injured by proliferations, degeneration, splitting, migration, compression, anoxia, impregnation with lipids and minerals (or other substances), clotting and hemorrhage, ingrowth of new nutrient vessels or loss of vasa vasorum, and, conceivably, neurovascular and harmonic vibratory phenomena affecting the subcellular or molecular constituents of the vessels. (For the many theories regarding clinical and experimental production of diseases of the coronary vessels, see thorough reviews of Morgan, 1956, and Boyd, 1959.)

a. Atherosclerotic coronary disease: This very common and widespread lesion is most severe in the arteries of the heart, as well as in the legs, a factor probably related to trauma and energy production in these sites. At present, confusion exists even on such a basic point as whether the two dominant lesions, plaque and thrombus, are either different stages of the same condition or different phenomena, with one predisposing to the other and either one preceding the other. (For the general discussion of atherosclerosis, see (1) Atherosclerosis, Chapter XIX.)

If accepted as a theory, thrombosis on the lining of coronary vessels, as a primary mechanism for coronary occlusion or narrowing, may lead to endothelialization of the clot, walling off by fibrosis, degeneration of enclosed red cells and transformation into a lipid or degenerated plaque, ingrowth of new vessels, infiltration with minerals, nearby degeneration or scarring or cellular reactions in the others coats, and formation of a site of predilection for several generations of the same processes (Wearn, 1923; Duguid, 1946; Morgan, 1956). Causes for thrombosing tendencies in these blood vessels remain a subject for further investigation. In this regard a rise in lipids or thrombosing substances or a fall in fibrinolytic or clot-blocking material may possibly be implicated.

Intimal and subintimal vascularization and hemorrhage often appear to be either the cause or the result of coronary artery occlusion, as either primary or complicating mechanisms (Paterson, 1936; Wartman, 1938; Winternitz *et al.*, 1937; Morgan, 1956). Capillaries in and near plaques may hemorrhage into these structures, making them bulge into the lumen of the artery. Bleeding may also occur between the plaque and its covering wall, thus causing projection of the latter into the blood stream. New or old capillaries may appear near a plaque, in an effort to "re-canalize" a thrombus or provide collateral channels past an occlusion in an artery. These small vessels may then, in turn, be injured by elastic lamellar formation and by other changes found in the parent vessel. Agreement is still lacking regarding the normal existence of intimal and subintimal capillaries, especially those which originate from the lumen of the parent artery rather than from either the vasa vasorum in the adventitia or from the walls of penetrating branches. If this question could be answered, it would be a major step in clarifying many problems.

b. Fibrinoid deposits: These changes are found in one or all three layers of coronary arteries in such "collagen" diseases as rheumatic fever, rheumatoid arthritis, systemic lupus erythematosus, polyarteritis nodosum, scleroderma, dermatomyositis, and thrombotic thrombocytopenic purpura. The alterations may be noted in the major coronary arteries and the arterioles in the myocardium, pericardium and endocardium. Healing phases of the fibrinoid deposits may merge into lesions typical of atherosclerosis. Treatment with steroids and antibiotics may greatly modify some of these changes. Relationships of so-called "stress" to the lesions suggest some solutions and emphasize many unresolved aspects of the problem (Selye, 1950; Roberts, 1958, 1961).

c. Adventitial inflammation, sclerosis, abscess or plaque formation: These are common primary changes in syphilis, bacteremia, or "collagen" diseases. They may also result from lesions in the intima or media.

d. Embolism: This condition is most often due to flaking of calcific aortic valvular or endocarditic lesions, and, less frequently, to aero-embolism (traumatic or that associated with compression or decompression phenomena) and to dislodgment of atrial, mitral or mural ventricular thrombi following myocardial infarction, "collagen" diseases, traumatic injuries of the heart or rheumatic heart disease. It probably happens more often than diagnosed, since it is a terminal event with major lesions of other types and thus may be overlooked. When it is survived, superimposed thrombosis, endothelialization, lipid degeneration, sclerosis and other changes may obscure the embolic origin of such a lesion. A dominant embolic role has recently been favored for many vascular lesions of the lower extremities, and a homologue may exist in the coronary system. Embolism, with reflux into the coronary sinus and veins from an engorged right atrium, seems a plausible consideration in the rare lesions of these vessels.

3. Gross Changes

Pathologic alterations in coronary arteries are often quite apparent even before the vessels are opened. The walls of the arteries may be distended, thickened, rigid, hardened, beaded, stony, straightened, tortuous,

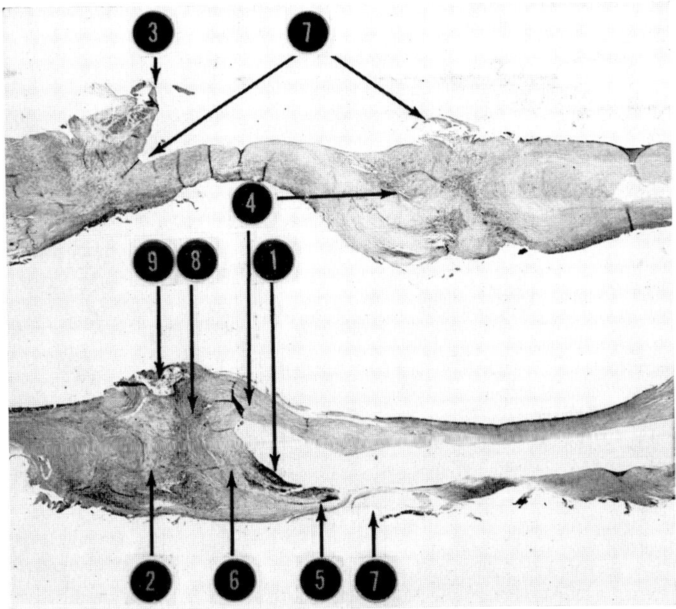

Fig. 60. Longitudinal sections through atherosclerotic coronary arteries, showing variabilities of lesions. (1) Complete thrombotic and (2) atheromatous obstruction of lumen; (3) irregular bulging of plaque; (4) irregular lamellar changes; (5) recent and (6) old thrombosis; (7) tears; (8) ingrowth of vessels; (9) patchy concentrations of foam cells, together with calcific changes. (H and E stain).

nodular, opaque, or even bony. In areas they may be narrowed and threadlike. Softened mushy bulgings of greenish, yellowish or reddish hue may occur. Dense adhesions, softened fatty excesses, extramural or intralumenal hemorrhages, and aneurysmal bulgings are rather common.

If the vessels are opened longitudinally (Fig. 60), all the known changes may be seen in sequence along a short strip of vessel, ranging from a new or old clot to a small mushy, greasy, fat-infiltrated plaque or a gritty and calcific ring. In many spots only yellow, fatty streaks are seen. If cut transversely, the lumen may be only variably narrowed or often the site of eccentric white plaques which may be again yellow, reddish or greenish. In some cross-sections the lumen of the vessel may be only a narrow irregular cleft or round hole, open or filled with clot. In others, only a clot may be seen. In still others the vessel may seem quite normal in one portion, although very severe abnormalities may exist only a minute distance above or below such a location. In slices of the myocardium, small white or red (sometimes yellow) pinhead-sized spots may be grossly visible in the case of active rheumatoid or rheumatic fever lesions or other myocarditic states. The Aschoff nodules of rheumatic fever may be difficult to distinguish grossly from some of the "collagen" diseases.

Anomalies of great importance are: (1) origin of a coronary artery from the pulmonary artery; (2) a single coronary artery; and (3) a fistula between a coronary artery and a cardiac chamber (Roberts and Loube, 1947). The latter anomaly, especially if it involves the right atrium or ventricle, may either simulate an arteriovenous fistula in its systemic effect or cause myocardial ischemia through abnormal coronary run-off.

4. Microscopic Changes

The microscopic alterations in coronary arteries assume a myriad of forms (Figs. 61 and 62), most obvious of which are the eccentric, cleft-like or circumferential changes in the walls around the narrowed lumen. Slight to massive thickening of the intima, fibrinous or fibrillar and fibrinoid changes in the subendothelial tissue, hemorrhagic and/or mucoid fatty infiltrations of the plaques are common. Thrombosis in various degrees of aging and secondary changes are notable and strongly suggest that the occlusion occurred in stages. Needle-shaped clefts or spicules are seen in the plaques or other zones and again often have a concentric or laminated distribution. Dilated small vessels are frequent and are grouped in several layers, often at different depths in the thrombus, plaque or nearby tissues. The internal elastic lamellae are fractured, split or replaced and may be hyperplastic at times. Calcific zones are common in any part of the plaques or adjacent media and adventitia. Where side branches are cut, mounding of the intima near the orifice in the parent artery is conspicuous. On closer search, numerous large foam cells of many types, as well as leucocytes, monocytes, lympho-

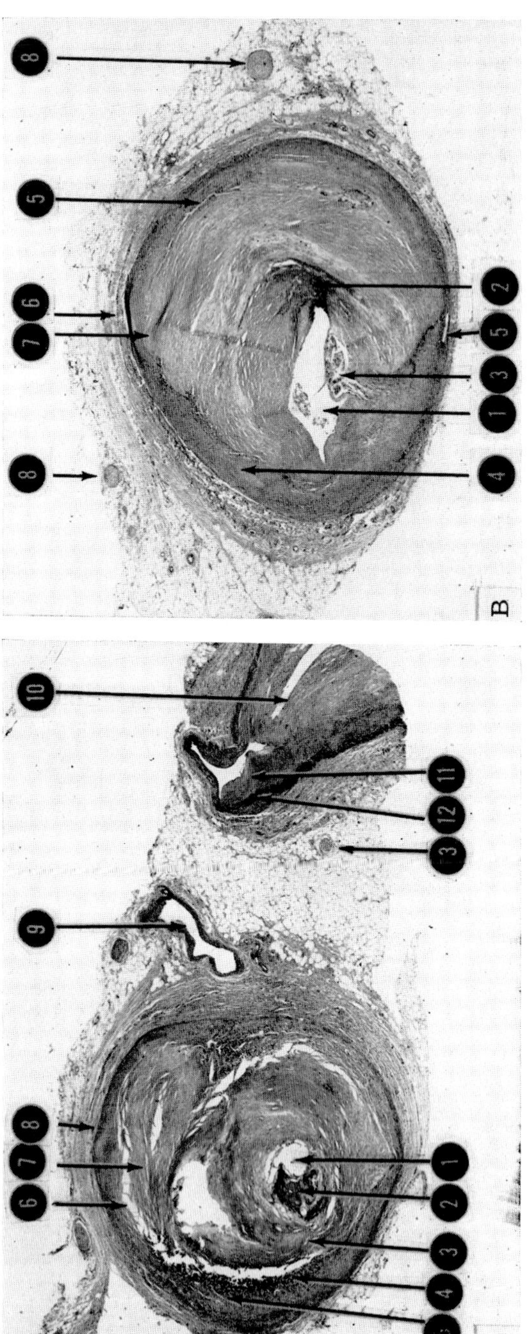

Fig. 61. Cross sections of atherosclerotic coronary artery, showing multiple changes commonly found in varying degrees.

A. Original lumen reduced to a small round hole (1) in lower center of vessel, with a partly vascularized recent thrombus (2) to the left of it. Numerous stratified differences of density (3) cellular content (4) and vascularization (5) are seen. Some clefts are artifacts (6), while many are spaces with needle-like accumulations of lipid products (7). (8) Media irregularity interrupted. Side branch (9) on the right clearly shows tendency for plaque-formation (10), with thickening of intima (11) near orifice. (12) Spotty calcification; (13) nerve in adventitia.

B. (1) Lumen as narrow irregular cleft, with recent highly eosinophilic thrombi (2), foam-cell concentrations (3), lamination (4), layers of vasa vasorum (5), pressure atrophy in media (6) and calcification (7). (8) Nearby coronary nerves in adventitia. (H and E stain).

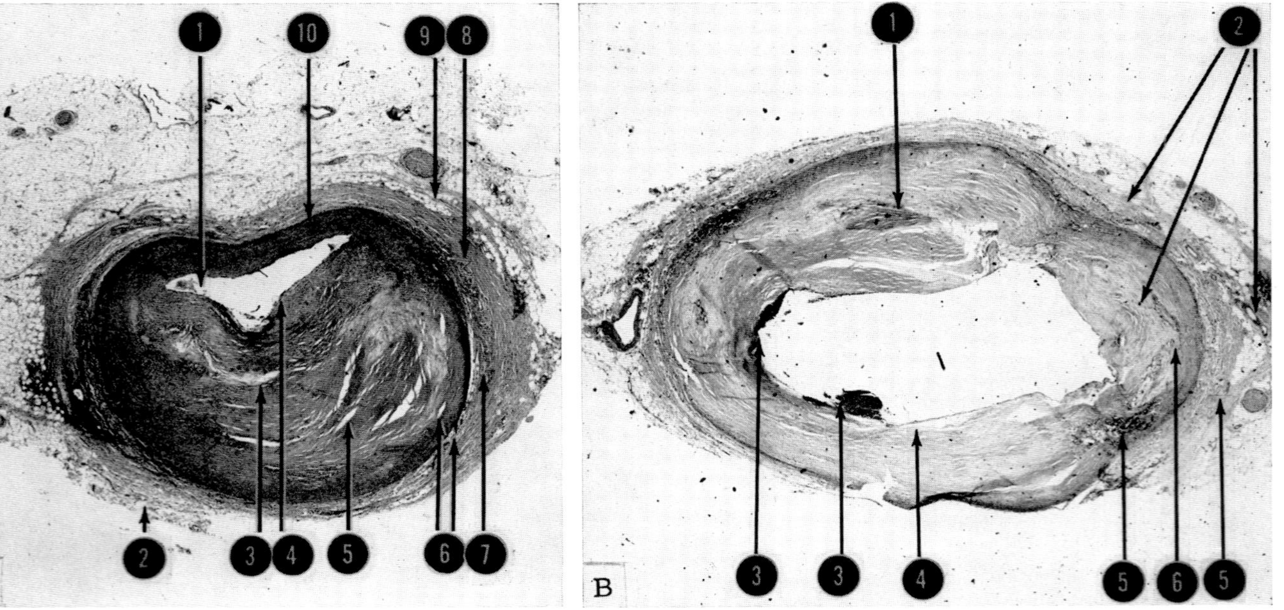

Fig. 62. Other cross sections of atherosclerotic coronary artery.

A. Original lumen reduced to a markedly eccentric opening (1) in area away from the myocardium (2). Layering in plaque (3), which could have been due to an old altered and endothelialized thrombus (4), is plainly seen. Many concentric clefts (5) are abundant in deeper center of plaque and in media near plaque. Some may be newly-formed vessels (6). (7) Media markedly distorted by encroachment from plaque. Changed staining reaction of muscle and elastic fibers (8), demonstrating spots of fatty infiltrations (9), is of an advanced degree. Signet-ring appearance, due to eccentric layering of encrusted thrombosis or plaque formation, is typical. (10) Thickening of intima.

B. (1) More concentric arrangement of thickened intimal layers is evident. (2) Newly-formed vessels or vasa vasorum near media or outer zone of intima. (3) Recent spots of thrombosis; (4) endothelium covering an old area of thrombosis. (5) Bulging of media, with filling in of cavity (6), another phase due to weakening of wall. (H and E stain).

cytes and other cells, are found either scattered or concentrated in the different zones of the lesion. Clefts with hemorrhage and apparently functional small vasa vasorum are seen in and near the plaques and thrombi. Occasionally small vessels in the plaque or thrombi contain elastic, collagen or reticular lamellae, or they may be only simple endothelialized tubules. New, darkly staining thrombi or polyps are common in the lumen of the vessels. Nerve fibers and bundles are abundant near the outer coats and sometimes are seen penetrating the walls of the arteries; these cardiac nerves have their own intrinsic blood supply (vasa nervorum) which, in turn, are seen to be affected by the vascular lesions of the generalized condition, a state which may well be related to the mechanisms of coronary pain (Roberts, 1948). At times, all structures of the walls of the coronary arteries may seem quite normal, although the lumen may be small in proportion to the thickness of the walls, thus suggesting a congenital or acquired generalized narrowing of the vessels.

In diseases other than atherosclerosis, the smaller coronary vessels are most affected. Fibrinoid changes (brightly eosinophilic), cellular infiltrates, ulcerative intimal changes, thrombosis, aneurysmal bulgings, or cord-like atrophy of sclerosed arterioles are common. The proportions of changes in the different layers vary with the type of "collagen" disease or necrotizing arteriolitis.

References

Adams, R. (1958). Internal mammary-artery ligation for coronary insufficiency. An evaluation. *New Engl. J. Med.* **258**, 113.

Alella, A., Williams, F. L., Bolene-Williams, C., and Katz, L. N. (1956). Role of oxygen and exogenous glucose and lactic acid in the performance of the heart. *Am. J. Physiol.* **185**, 487.

Allmark, M. G., Lu, F. C., Carmichael, E., and Lavallee, A. (1953). Some pharmacological observations on isoniazid and iproniazid. *Am. Rev. Tuberc.* **68**, 199.

Aviado, D. M., Jr. (1959). Cardiovascular effects of some commonly used pressor amines. *Anesthesiology* **20**, 71.

Beck, C. S., Brofman, B. L., and Mautz, F. R. (1957). Symposium on Coronary Artery Disease. *Diseases of Chest* **31**, 243.

Berne, R. M. (1958). Effect of epinephrine and norepinephrine on coronary circulation. *Circulation Research* **6**, 644.

Berne, R. M., Blackmon, R. J., and Gardner, T. H. (1957). Hypoxemia and coronary blood flow. *J. Clin. Invest.* **36**, 1101.

Bing, R. J. (1951). The coronary circulation in health and disease as studied by coronary sinus catheterization. *Bull. N. Y. Acad. Med.* **27**, 407.

Bing, R. J., Maraist, F. M., Dammann, J. F., Jr., Draper, A., Jr., Heimbecker, R., Daley, R., Gerard, R., and Calazel, P. (1950). The effect of strophanthus on coronary blood flow and cardiac oxygen consumption of normal and failing human hearts. *Circulation* **2**, 513.

Blumgart, H. L., Zoll, P. M., Paul, M. H., and Norman, L. R. (1950). The experimental production of intercoronary arterial anastomoses and their functional significance. *Circulation* **1**, 10.

Boyd, W. (1959). "Pathology for the Physician," 6th ed., reprinted. Lea & Febiger, Philadelphia, Pa.

Brachfeld, N., Bozer, J., and Gorlin, R. (1959). Action of nitroglycerin on the coronary circulation in normal and in mild cardiac subjects. *Circulation* **19**, 697.

Brofman, B. L., Hellerstein, H. K., and Caskey, W. H. (1952). Mephentermine—an effective pressor amine. *Am. Heart J.* 44, 396.

Cannon, J. A., Longmire, W. P., and Kattus, A. A. (1959). Consideration of the rationale and technic of coronary endarterectomy for angina pectoris. *Surgery* **46**, 197.

Case, R. B., Berglund, E., and Sarnoff, S. J. (1955). Changes in coronary resistance and ventricular function resulting from acutely induced anemia and the effect thereon of coronary stenosis. *Am. J. Med.* **18**, 397.

Charlier, R. (1959). Un nouveau dilatateux coronarien de synthese. Etude Pharmacologique. *Acta Cardiol.* **14**, Suppl. 7, 1.

Cole, S. L., Kaye, H., and Griffith, G. C. (1957). Assay of antianginal agents. I. A curve analysis with multiple control periods. *Circulation* **15**, 405.

Crumpton, C. W., Rowe, G. G., O'Brien, G., and Murphy, Q. R., Jr. (1954). The effect of hexamethonium bromide upon coronary flow, cardiac work and cardiac efficiency in normotensive and renal hypertensive dogs. *Circulation Research* **2**, 79.

Day, S. B., and Lillehei, C. W. (1959). Experimental basis for a new operation for coronary artery disease. A left atrial-pulmonary artery shunt to encourage development of interarterial intercoronary anastomoses. *Surgery* **45**, 487.

Duguid, J. B. (1946). Thrombosis as a factor in the pathogenesis of coronary atherosclerosis. *J. Pathol. Bacteriol.* **58**, 207.

Eckenhoff, J. E., and Hafkenschiel, J. H. (1947). The effect of nikethamide on coronary blood flow and cardiac oxygen metabolism. *J. Pharmacol. Exptl. Therap.* **91**, 362.

Eckstein, R. W., Stroud, M., III, Eckel, R., Dowling, C. V., and Pritchard, W. H. (1950). Effects of control of cardiac work upon coronary flow and oxygen consumption after sympathetic nerve stimulation. *Am. J. Physiol.* **163**, 539.

Eckstein, R. W., Newberry, W. B., McEachen, J. A., and Smith, G. (1951). Studies of the antiadrenergic effects of nitroglycerine on the dog heart. *Circulation* **4**, 534.

Farrand, R. L., and Horvath, S. M. (1959). Effects of khellin on coronary blood flow and related metabolic functions. *Am. J. Physiol.* **196**, 391.

Foltz, E. L., Wong, S. K., and Eckenhoff, J. E. (1948). Effects of certain "cardiac stimulant" drugs on coronary circulation and cardiac oxygen metabolism. *Federation Proc.* **7**, 219.

Foltz, E. L., Rubin, A., Steiger, W. A., and Gazes, P. C. (1950). The effects of intravenous aminophyllin upon the coronary blood-oxygen exchange. *Circulation* **2**, 215.

Gerola, A., Feinberg, H., and Katz, L. N. (1959). Role of catecholamines on energetics of the heart and its blood supply. *Am. J. Physiol.* **196**, 394.

Gorlin, R. (1958). Studies on the regulation of the coronary circulation in man. I. Atropine-induced changes in cardiac rate. *Am. J. Med.* **25**, 37.

Gorlin, R., Brachfeld, N., MacLeod, C., and Bopp, P. (1959). Effect of nitroglycerine on the coronary circulation in patients with coronary artery disease on increased left ventricular work. *Circulation* **19,** 705.

Green, H. D., and Kepchar, J. H. (1959). Control of peripheral resistance in major systemic vascular beds. *Physiol. Revs.* **39,** 617.

Gregg, D. E. (1950). "Coronary Circulation in Health and Disease." Lea & Febiger, Philadelphia, Pa.

Gregg, D. E. (1955). Some problems of the coronary circulation. *Verhandl. deut. Ges. Kreislaufforsch.* **21,** 22.

Gregg, D. E. (1957). Regulation of the collateral and coronary circulation of the heart. *In* "Circulation," Proc. Harvey Tercentenary Congr. (J. McMichael, ed.). Charles C Thomas, Springfield, Ill.

Gregg, D. E. (1959a). The coronary circulation. *In* "Encyclopedia of Cardiology" (A. A. Luisada, ed.), Vol. I, Chapter 23. McGraw-Hill (Blakiston), New York.

Gregg, D. E. (1959b). Physiopathology of different surgical approaches to human coronary atherosclerosis. *Acta Cardiol.* **14,** Suppl.8, 3.

Gregg, D. E., Batterman, R. C., Katz, L. N., Raab, W., and Russek, H. I. (1956). Experimental methods for the evaluation of drugs in various disease states. Part II: Angina Pectoris. *Ann. N. Y. Acad. Sci.* **64,** 494.

Grob, D., Scarborough, W. R., Kattus, A. A., Jr., and Langford, H. E. (1953). Further observations on the effects of autonomic blocking agents in patients with hypertension. *Circulation* **8,** 352.

Gross, L., Epstein, E. Z., and Kugel, M. A. (1934). Histology of the coronary arteries and their branches in the human heart. *Am. J. Pathol.* **9,** 143.

Hackel, D. B., and Goodale, W. T. (1955). Effects of hemorrhagic shock on the heart and circulation of intact dogs. *Circulation* **11,** 628.

Hall, R. J., Khouri, E. M., and Gregg, D. E. (1961). Coronary-internal mammary artery anastomosis in dogs. *Surgery* **50,** 560.

Hansen, A. T., Haxholdt, B. J., Husfeldt, E., Lassen, N. A., Munck, O., Sorenson, H. R., and Winkler, K. (1956). Measurement of coronary blood flow and cardiac efficiency in hypothermia by use of radioactive Krypton 85. (1956). *Scand. J. Clin. & Lab. Invest.* **8,** 182.

Jellife, R. W., Wolf, C. R., Berne, R. M., and Eckstein, R. W. (1957). Absence of vasoactive and cardiotropic substances in coronary sinus blood of dogs. *Circulation Research* **5,** 382.

Kattus, A. A., and Gregg, D. E. (1959). Some determinants of coronary collateral blood flow in the open chest dog. *Circulation Research* **7,** 628.

Khouri, E. M., Gregg, D. E., Hall, R. J., and Rayford, C. R. (1960). Regulation of coronary flow during treadmill exercise in the dog. *Physiologist* **3,** 93.

Kien, G. A., and Sherrod, T. R., Forte, I. E., *et al.,* and Regan, T. J., *et al.* (1960). Conference on Cardiovascular Effects of Nicotine and Smoking. *Ann. N. Y. Acad. Sci.* 90, 161–189.

Laurent, D., Bolene-Williams, C., Williams, F. L., and Katz, L. N. (1956). Effect of heart rate on coronary flow and cardiac oxygen consumption. *Am. J. Physiol.* **185,** 355.

Lewis, F. B., Coffman, J. D., and Gregg, D. E. (1961). The effect of heart rate and intracoronary isoproterenol, levarterenol, and epinephrine on coronary flow and resistance. *Circulation Research* **9,** 89.

Livesay, W. R., Moyer, J. H., and Chapman, D. W. (1954). The cardiovascular and renal hemodynamic effects of Aramine. *Am. Heart J.* **47,** 745.

Lombardo, T. A., Rose, L., Taeschler, M., Tully, S., and Bing, R. J. (1953). The effect of exercise on coronary blood flow, myocardial oxygen consumption and cardiac efficiency in man. *Circulation* **7**, 71.

McKeever, W. P., Gregg, D. E., and Canney, P. C. (1958). Oxygen uptake of the non-working left ventricle. *Circulation Research* **6**, 612.

Mitchell, G. A. G. (1953). "Anatomy of the Autonomic Nervous System." Livingstone, Edinburgh.

Morgan, A. D. (1956). "The Pathogenesis of Coronary Occlusion." Charles C Thomas, Springfield, Ill.

Olson, R. E., and Piatnek, D. A. (1959). Conservation of energy in cardiac muscle in metabolic factors in cardiac contractility. *Ann. N. Y. Acad. Sci.* **72**, 466.

Parker, F. (1958). An electron microscope study of coronary arteries. *Am. J. Anat.* **103**, 247.

Paterson, J. C. (1936). Vascularization and hemorrhage of the intima of arteriosclerotic coronary arteries. *A.M.A. Arch. Pathol.* **22**, 313.

Raab, W., and Lepeschkin, E. (1950). Antiadrenergic effects of nitroglycerine on the heart. *Circulation* **1**, 733.

Ratnoff, O. D., and Plotz, M. (1946). The coronary circulation. *Medicine* **25**, 285.

Rayford, C. R., Khouri, E. M., Lewis, F. B., and Gregg, D. E. (1959). Evaluation of use of left coronary artery inflow and O_2 content of coronary sinus blood as a measure of left ventricular metabolism. *J. Appl. Physiol.* **14**, 817.

Regan, T. J., Binak, K., Gordon, S., De Fazio, V., and Hellems, H. K. (1961). Myocardial blood flow and oxygen consumption during postprandial lipemia and heparin-induced lipolysis. *Circulation* **23**, 55.

Roberts, J. T. (1948). The effect of occlusive arterial diseases of the extremities on the blood supply of nerves. Experimental and clinical studies on the role of the vasa nervorum. *Am. Heart J.* **35**, 369.

Roberts, J. T. (1958). The heart in connective tissue diseases. *In* "Progress in Arthritis" (J. H. Talbott and L. M. Lockie, eds.), Chapter 35. Grune & Stratton, New York.

Roberts, J. T. (1959). Anatomy of the conduction system, and of the arteries, veins and lymphatics of the heart. In "Encyclopaedia of Cardiology" (A. A. Luisada, ed.), Vol. I, Chapters 6 and 7. McGraw-Hill, New York.

Roberts, J. T. (1961). Dynamics and circulation of heart muscle; cardiac pain; cardiac reserve; the cardiac cycle. *In* "Pathological Physiology and Mechanisms of Disease" (W. A. Sodeman, ed.), Chapter 15. Saunders, Philadelphia, Pa.

Roberts, J. T., and Loube, S. D. (1947). Congenital single coronary artery in man. *Am. Heart J.* **34**, 188.

Roberts, J. T., Browne, R. S., and Roberts, G. (1943). Nourishment of the myocardium by way of the coronary veins. *Federation Proc.* **2**, 90.

Crumpton, C. W. (1956). The hemodynamics of thyrotoxicosis in man with

Rowe, G. G., Huston, J. H., Maxwell, G. M., Weinstein, A. B., Tuchman, H., and Crumpton, C. W. (1955). The effects of 1-hydrazinophthalazine upon coronary hemodynamics and myocardial oxygen metabolism in essential hypertension. *J. Clin. Invest.* **34**, 696.

Rowe, G. G., Huston, J. H., Weinstein, A. B., Tuchman, H., Brown, J. F., and Crumpton, C. W. (1956). The hemodynamics of thyrotoxicosis in man with special reference to coronary blood flow and myocardial oxygen metabolism. *J. Clin. Invest.* **35**, 272.

Rowe, G. G., Emanuel, D. A., Maxwell, G. M., Brown, J. F., Castillo, C., Schuster, B., Murphy, Q. R., and Crumpton, C. W. (1957). Hemodynamic effects of quinidine: including studies of cardiac work and coronary blood flow. *J. Clin. Invest.* **36**, 844.

Sabiston D. C., and Gregg, D. E. (1957). Effect of cardiac contraction on coronary blood flow. *Circulation* **15**, 14.

Sarnoff, S. J., Case, R. B., Berglund, E., and Sarnoff, L. C. (1954). Ventricular function. V. The circulatory effects of Aramine; mechanism of action of "vasopressor" drugs in cardiogenic shock. *Circulation* **10**, 84.

Sarnoff, S. J., Braunwald, E., Welch, G. H., Jr., Case, R. B., Stainsby, W. N., and Marcruz, R. (1958). Hemodynamic determinants of the oxygen consumption of the heart with special reference to the tension-time index. *Am. J. Physiol.* **192**, 148.

Schmitthenner, J. E., Hafkenschiel, J. H., Forte, I., Williams, A. J., and Riegel, C. (1958). Does alcohol increase coronary blood flow and cardiac work? *Circulation* **18**, 778.

Schreiner, G. L., Berglund, E., Borst, H. G., and Monroe, R. G. (1957). Effects of vagus stimulation and of acetylcholine on myocardial contractility, oxygen consumption and coronary flow in dogs. *Circulation Research* **5**, 562.

Scott, J. C., and Balourdas, T. A. (1959). An analysis of coronary flow and related factors following vagotomy, atropine and sympathectomy. *Circulation Research* **7**, 162.

Selye, H. (1950). "The Physiology and Pathology of Exposure to Stress." Acta, Inc., Montreal, Canada.

Shipley, R. E., and Wilson, C. (1951). An improved recording rotameter. *Proc. Soc. Exptl. Biol. Med.* **78**, 724.

Starzl, T. E., and Gaertner, R. A. (1955). Chronic heart block in dogs. A method for producing experimental heart failure. *Circulation* **12**, 259.

Wartman, W. B. (1938). Occlusion of the coronary arteries by hemorrhage into their walls. *Am. Heart J.* **15**, 459.

Wearn, J. T. (1923). Thrombosis of the coronary arteries with infarction of the heart. *Am. J. Med. Sci.* **165**, 250.

Wearn, J. T., Mettier, S. R., Klumpp, T. G., and Zschiesche, L. J. (1933). The nature of the vascular communications between the coronary arteries and the chambers of the heart, *Am. Heart J.* **9**, 143.

Wégria, R. (1951). Pharmacology of the coronary circulation. *Pharmacol. Revs.* **3**, 197.

Welch, G. H., Jr., Braunwald, E., Case, R. B., and Sarnoff, S. J. (1958). The effect of mephentermine sulfate on myocardial oxygen consumption, myocardial efficiency and peripheral vascular resistance. *Am. J. Med.* **24**, 871.

Winternitz, M. C., Thomas, R. M., and LeCompte, P. M. (1937). Studies in the pathology of vascular disease. *Am. Heart J.* **14**, 399.

Wolf, M. M., and Berne, R. M. (1956). Coronary vasodilator properties of purine and pyrimidine derivatives. *Circulation Research* **4**, 343.

Zoll, P. M., Wessler, S., and Schlessinger, M. J. (1951). Interarterial coronary anastomoses in the human heart with particular reference to anemia and relative cardiac anoxia. *Circulation* **4**, 797.

Chapter X. PULMONARY CIRCULATION

A. EMBRYOLOGY

By R. Licata

1. Establishment of the Pulmonary Circulation

a. Early development: Coincident with development of the primary lung buds, two pulmonary arterial stems make their appearance distal to the truncus arteriosus (Keibel and Mall, 1912). At the outset endothelial sprouts grow out from the pulmonary stems to invade the thick stroma of the lung primordia, where they establish communication with the splanchnic endothelial plexus developing *in situ.* Throughout the period of early development, vascularization of the lung bed proceeds at a relatively leisurely rate. Prior to further advance of a pulmonary vascular bed, the developing sixth aortic arches, arising from each dorsal aorta, establish linkage with the primordia of the corresponding pulmonary vessels at the level of their proximal segments. During later stages of development, connection of the right pulmonary artery with the dorsal aorta, via its associated sixth arch, is completely lost. The corresponding distal segment of the sixth arch of the composite bridge of the left side is destined to divert the major portion of the right flow-tract stream into the systemic circulation. This segment is temporarily retained until term as a specialized channel, designated the ductus arteriosus. In the absence of a mechanism for activation of the pulmonary circulation, the increased resistance of the pulmonary field makes necessary the rerouting of the right stream into the ductus. Most important, however, is the availability of placental blood to the systemic circulation via this by-pass channel. The ductus reroutes the right ventricular blood away from the high resistance pulmonary field by shunting the blood almost entirely into the systemic aorta.

b. Later development: During later fetal stages the pulmonary circulation becomes increasingly activated, the degree of activation being apparently much greater than formerly believed, particularly during the last trimester of gestation. Concomitant with the first postnatal inspiratory efforts to ventilate the lungs, there is a marked decrease in pulmonary resistance, with an ensuing increase in circulation. At birth, spontaneous spasmodic muscular contraction attenuates the ductus, followed by physiologic closure. Complete morphologic closure occurs several months later, due to obliteration of the lumen by fibrotic intimal proliferation (Patten, 1953).

c. Histogenesis: Histogenetically the lung is prepared for the postnatal alteration in circulation because of a number of factors. First, prenatally the tunics of the major pulmonary branches contain a large quantity of elastic tissue. Secondly, the vessels, themselves, appear to be in a state of vasoconstriction, thus preserving the integrity of the potential lumina free of fluids in anticipation of the demands of immediate ventilation. Furthermore the smaller subdivisions are equipped with some smooth muscle formed in a spiral or helicoid fashion, an arrangement, which postnatally is conducive to progressively increasing expansion of the vessels.

d. Segmental supply: With the onset of the final phase of differentiation of lung during the last trimester, the intramural capillaries easily herniate through the lining of the air passages, which in many places are apparently denuded of epithelium. Such a histologic change introduces a significant prenatal phase, conducive to gaseous exchange, and initiates the final preparatory adjustments necessary for total functional activation of the pulmonary circulation. The development of the capillary bed of the pulmonary vasculature is along the lines of the basic segmental formation of the lung. The pulmonary artery and its subdivisions follow closely the pattern of distribution of the bronchial tree, preserving throughout the original segmental architecture.

e. Pulmonary veins: The pulmonary veins, in contrast, invade the lung along the intersegmental boundaries of connective tissue, thus retaining postnatally a peripheral relationship to the lung lobule as a whole. The pulmonary venous return reaches the heart by way of two pairs of pulmonary veins that drain directly into the left atrium. Embryonically an endocardial diverticulum from the left atrial wall establishes connection with the splanchnic vascular plexus of the lung territory. The single primitive vessel thus derived annexes adjacent venous radicles of the mediastinum that ramify in the pulmonary plexus. By confluence, extension of the atrial wall incorporates the vascular rudiments described, resulting finally in paired veins that empty into the atrial cavity.

f. Bronchial vessels: The bronchial arteries differentiate from reduced ventral visceral segmental vessels. Left and right sets usually arise from the aorta and intercostal artery, respectively. The bronchial veins develop as endothelial diverticuli from unpredictable sites. Generally they may issue from any one or more of the major veins in the vicinity of the bronchi.

B. GROSS ANATOMY

By W. H. Knisely

1. Arteries

a. Pulmonary arteries:[1] The comparative mammalian anatomy of the pulmonary artery has been studied (Huntington, 1919), as well as the gross anatomy of the vessel in the human newborn (Noback and Rehman, 1941).

In the adult the pulmonary artery is a short, relatively wide vessel, usually about 5 cm in length and 3 cm in diameter, arising from the conus arteriosus of the right ventricle, extending superiorly and posteriorly, and passing anterior and to the left of the ascending aorta. It divides inferior to the surface of the aortic arch, generally at the level of the interspace between the 5th and 6th vertebra, into right and left branches of approximately equal size. The full length of this vessel is contained within pericardium.

The right branch is the longer and larger of the two main divisions and runs horizontally to the right, posterior to the ascending aorta and superior vena cava and anterior to the right bronchus, to the root of the lung where it divides into two branches. The inferior of these two vessels is larger and goes to the middle and inferior lobes. The superior branch reaches the superior lobe.

The left branch of the pulmonary artery passes horizontally, anterior to the descending aorta and left bronchus, to the root of the lung, where it divides into two branches, one to each lobe. The left branch of the pulmonary artery is connected superiorly to the concavity of the aortic arch by the ligamentum arteriosum of the left vena cava.

In posterior-anterior relation, the bronchi, arteries and veins lie in sequence on both sides. However, in the superior-inferior relation on the right side, the superior lobe bronchus is superior while the artery is slightly inferior; next are the bronchi to the middle and inferior lobes,

[1] For mean enzyme activities of normal pulmonary artery, see Table IV, Chapter III; for variations in composition with age, see Table VI, Chapter III.

and then the pulmonary vein, which is most inferior. In the superior-inferior relation on the left side, the pulmonary artery is superior, the bronchus is in the middle and the pulmonary veins are inferior (Gray, 1959).

The branching pattern of pulmonary arteries is unique and characteristic; the divisions, both large and small, separate at more obtuse angles than do systemic vessels of similar diameters, and the angles of taper are greater. An early illustration of both the pulmonary branching pattern and the rapid taper of the pulmonary vessels can be seen in Fig. 14 of Miller's article (1893). The measured *in vivo* angles of taper of systemic vessels are approximately 1.0° (Jeffords and Knisely, 1956), while those of pulmonary vessels are several times this figure (Irwin *et al.*, 1954; Knisely, 1960).

The pulmonary arteries accompany the bronchi, running chiefly along their posterior surfaces, a functional interpretation of this gross relation being suggested by von Hayek (1960). He states that, "owing to the fact that the artery first crosses over the bronchus and then passes by the upper lobe bronchus on the left and the middle lobe bronchus on the right to arrive on the lateral side of the stem bronchus, this allows room for the pulsatile elongation of the arterial arch so that it does not collide with the bronchial tree. Furthermore, in their further course the branches of the pulmonary arteries appear to be so arranged that, upon bending, the concavity of the arterial arch is always turned toward the bronchus; a relationship which, as in the case of the trunks of the two pulmonary arteries, is in harmony with the pulsatile displacements. Finally, in the case of the smallest arteries which accompany the terminal and respiratory bronchioles, it turns out that where they reach the neighborhood of the lung surface, the artery and bronchiole course alongside one another at the same distance from the pleura. Thus, when there are unequal changes in length of the two structures a bending of the rod or support formed jointly by the bronchiole and artery does not occur toward the surface; there is only a shifting which is parallel to the pleura."

Main intersegmental branches of the pulmonary arteries are single for the most part and arise as common trunks to supply adjacent bronchopulmonary segments (Boyden, 1945). The artery for one segment usually supplies small branches to the neighboring segments. Distal to alveolar ducts, branches are distributed to each atrium from which arise smaller radicles terminating in the capillaries of alveoli.

Miller (1893) has reported the numbers of given sizes of pulmonary blood vessels and their measurements in the dog, along with similar data on the respiratory units. Such figures (Table XIV) are invaluable in the understanding of the relation between branching pattern and angles of taper, on the one hand, and the hemodynamics of pulmonary blood flow,

TABLE XIV
Comparison of Number and Size of Pulmonary Vessels in the Dog[a]

Pulmonary vessels	Number	Diameter	Area of section
Pulmonary artery	1	15.5 mm	181 sq mm
R. and l. branches	2	11.5 mm	208 sq mm
Lobar arteries	8	5.96 mm	223 sq mm
1st order arteries	24	3.96 mm	293 sq mm
2nd order arteries	164	2.26 mm	656 sq mm
3rd order arteries	1,021	1. mm	801 sq mm
Lobular arteries	16,000	.3 mm	1,120 sq mm
Atrial arteries	64,000	.165 mm	1,344 sq mm
Sac arteries	128,000	.165 mm	2,688 sq mm
Capillaries	600,000,000	.007 mm	23,000 sq mm
Sac veins	192,000	.23 mm	7,680 sq mm
Atrial veins	32,000	.45 mm	6,098 sq mm
Lobular veins	16,000	.4 mm	2,000 sq mm
3rd order veins	1,021	1.22 mm	1,194 sq mm
2nd order veins	164	2.44 mm	765 sq mm
1st order veins	24	4.18 mm	340 sq mm
Lobar veins	8	6.12 mm	299 sq mm
Venous trunks	4	13.75 mm	756 sq mm

[a] From Miller (1893).

on the other. Unfortunately, no comparable information exists in the case of the human lung.

b. Bronchial arteries: These vessels vary in number, size and origin. According to several authors (Miller, 1947; Verloop, 1948; Gray, 1959), as a rule, there is only one right bronchial artery, and its source, although somewhat variable, is usually from the first aortic intercostal or, less frequently, from the third intercostal, the right internal mammary or the right subclavian. The left bronchial arteries are generally two in number and arise directly from the thoracic aorta: the superior, opposite the fifth thoracic vertebra, and the inferior, just inferior to the level of the left bronchus. However, on the basis of other original data, it appears that both right and left systemic arterial supplies to the lungs are considerably more variable than the above (Nakamura, 1924; Hovelacque *et al.*, 1936; Cauldwell *et al.*, 1948). (For an extensive bibliography on the bronchial arteries of human and other species, see Ellis *et al.*, 1951.)

2. Veins

a. Pulmonary veins: There are usually two pulmonary veins from each lung, which have their major origin in the pulmonary capillaries from the alveoli. The smaller veins run through the substance of the lung independently of the pulmonary arteries and bronchi, and the larger ones are generally intersegmental in position and function, draining the blood from adjacent parts of two or more neighboring bronchopulmonary segments. As more and more veins join, they come into relation with the arteries and bronchi, accompany them to the hilum of the lung, and open into the left atrium.

b. Bronchial veins: The bronchial vein, formed at the root of the lung, receives superficial and deep veins from an area of the hilum. The larger part of the blood supplied by the bronchial arteries is returned by the pulmonary veins. On the right side, the bronchial veins end in the azygos vein, and on the left side, they usually terminate in the highest intercostal or in the accessory hemiazygos vein.

3. "Arteriovenous Anastomoses"

Several separable kinds of evidence have been presented regarding types of postulated vessels in the lungs, among which is the "arteriovenous anastomosis." What is usually assumed to be meant by this term is a direct connection, other than an alveolar capillary, between the pulmonary artery and pulmonary vein. It would appear, however, that many of the connections which have been found and labeled "pulmonary arteriovenous anastomoses" do not fit the assumed definition. To determine whether or not such anastomoses exist in normal human lungs is both a physiologic and morphologic problem. With regard to the morphologic aspects, it is necessary to point out that since the lungs contain two sets of blood vessels, pulmonary and bronchial, six different types of vascular anastomoses are theoretically possible between them, namely, arterioarterial, venovenal, and arteriovenous anastomoses from both sets of arteries to both sets of veins. Thus, restated, the morphologic problem is that of being certain that any connection which is found is called by its "right" name, by correctly identifying the parent vessels which are connected.

Von Hayek (1940a,b, 1942a,b, 1953, 1960) has been a major contributor to the idea that pulmonary arteriovenous anastomoses exist in normal human lungs and that other special types of vessels are also present. His evidence consisted mainly of the findings of thick-walled vessels with unique muscle cell arrangements in several areas, and of the tracing of such structures in serial sections to their abrupt entrance into thin-walled vessels which were identified as pulmonary veins. He also studied casts of vessels, some of which were so completely injected that he could not see the small vessels, while others were not filled enough to trace the anastomoses all the way. Nevertheless, he felt that the casts showed such connecting vessels. On the basis of the morphologic evidence, he named the thick-walled vessels "sperrarterien" ("control" or "sluice" arteries), and he stated that the arteriovenous anastomoses were branches from these vessels. From such evidence he developed several views with regard to the physiologic control of the pulmonary circulation (von Hayek, 1960). Support for the concept of pulmonary arteriovenous anastomoses in normal lung was found by Tobin and Zaraquiey (1950, 1953). These authors studied a number of human lungs and used several methods of investigation, including the dissection of plastic and latex vascular casts and the passage through the lungs of glass spheres up to 500 μ in diameter.

Verloop (1948) tried to duplicate von Hayek's work but came to different conclusions, namely, that the thick-walled vessels were altered bronchial arteries and that pulmonary arteriovenous anastomoses did not exist. Weibel (1958) found that thick-walled vessels appeared in normal individuals in the second decade and also that in animals, in which branches of the mesenteric artery were experimentally united with the diaphragm, typical thick walls developed. He concluded that these structures were a result of the stretching of the vessels and challanged the concept that thick-walled vessels, such as those present in the lung, could, in fact, shut, and thereby, act as "closing" or controlling arteries. In the human lungs, Weibel (1959) found twenty-two anastomoses between the pulmonary and bronchial arteries in the region of the small bronchi and two in the pleura. He noted, as have others, that the bronchial veins and the peribronchial veins are plexiform. He also found rare bronchial arteriovenous anastomoses between bronchial artery and vein. Similarly there were connections between bronchial and pulmonary veins, so that blood could flow from the peribronchial venous plexus into the pulmonary veins, thereby lowering the oxygen content of pulmonary venous blood. Weibel stated that the pulmonary arteries are endarteries and that the pulmonary veins are endveins, and, that there is no possibility for the formation of a collateral circulation in normal lung. He did not find precapillary pulmonary arteriovenous anastomoses.

4. Sympathetic Innervation

The sympathetic innervation of the pulmonary vessels is from the second, third, and fourth thoracic ganglia. Fibers from these ganglia contribute to the posterior pulmonary plexus and a few fibers go to the anterior plexus. Both plexuses give off fibers to the pulmonary artery.

There is a bilateral distribution of the fibers from each side (Gillilan, 1954). As to the effect of the sympathetic innervation on the pulmonary vessels, Cournand, in introducing a recent symposium on pulmonary circulation (Adams and Veith, 1959), considered the following quotation from François-Franck, in 1880, a timely remark, "In spite of many reports since Brown-Sequard became interested in the problem, there is still some doubt among physiologists as to the exact influence of the sympathetic system upon vasomotricity of the pulmonary vessels." (For additional references and discussion of the evidence regarding pulmonary innervation, see Gillilan, 1954.) (For innervation of pulmonary capillaries, see Section C.-2 of this chapter.)

C. MICROSCOPIC AND SUBMICROSCOPIC ANATOMY

By N. C. Staub

1. Individual Capillary Structure

a. Microscopic structure: Light microscope studies of the capillaries in the alveolar walls of routinely prepared sections show these vessels to be simple endothelial tubes (Bargmann, 1956), usually bulging into the alveolar lumen. The capillary diameter is given as 10–15 μ (Bargmann, 1956; Knisely, 1954) or 6–11 μ (Giese and Gieseking, 1957), approximately the same as capillaries in other organs (Zweifach, 1959). There is a very delicate pericapillary reticular and elastic supporting structure which presumably acts as a limiting net to prevent overdistention (Orsós, 1936). No other tissue support, beyond a few collagen fibers and the gossamer-thin alveolar epithelium, is noted.

b. Submicroscopic structure: The wonderful eyes of the electron microscope, when turned to the fine details of lung structure (Low, 1952; Karrer, 1956; Schulz, 1956), have demonstrated that the capillaries have a continuous, attenuated (average thickness 150mμ) endothelium without pores. The endothelial nuclei are conveniently located in the septal walls and so do not obstruct gas exchange. Surrounding the capillary tube is a continuous, homogeneous basement membrane (ground substance) in which are imbedded elastic and collagen fibrils (Schulz,

1956; Woodside and Dalton, 1958; Chase, 1959). There is some disagreement as to the extent of this perivascular supporting net, but most published electron micrographs do not show fibers as frequently as might be expected from the older light microscope studies of Orsós. Beyond the basement membrane, the alveoli are lined by a continuous thin epithelium (thickness 0.1–0.3 μ) (Low, 1952; Karrer, 1956). In high resolution studies the endothelium contains small vacuoles (Bargmann, 1956; Karrer, 1956) which are very suggestive of pinocytosis (see Section E.-1, Chapter VI), and, in acute experimental pulmonary edema, Kisch (1958) shows very large vacuoles in the capillary endothelial cells of rabbit lungs. (For an extensive atlas of normal and pathologic pulmonary capillaries by electron microscopy, see Schulz, 1959.) (For general discussion of ultramicroscopic structure of vessels, see Section C., Chapter I.)

2. Innervation of the Pulmonary Capillaries

After many years of study, Larsell and Dow (1933) stated that there are slender nerve filaments following the capillaries, which at intervals give off short terminal twigs. Nagaishi (1957) showed drawings with many bare nerve endings on the capillaries, while Honjin (1956) believed that there was a terminal syncytial nerve net extending to the capillaries.

Since the perivascular plexus in the lung is postganglionic sympathetic (Larsell and Dow, 1933; Elftman, 1943), any nerve terminations on or about the pulmonary capillaries must be the same. If Zweifach (1959) is correct that there is no motor activity of true capillaries, then any nerve terminations that do exist must be sensory—probably related to axon reflexes, since Honjin (1956) denies any true afferent fibers at the alveolar level.

If there are any nervous elements at the capillary level, the electron microscopists have not identified them. In fact, Schulz (1959) has stated that he has not found nerve fibers or synapses on the lung capillaries. (For innervation of large pulmonary vessels, see Section B.-4 of this chapter.)

3. The Alveolar Capillary Network

a. Methods for studying the capillary bed: In thick sections of lung, prepared by injection of India ink or warm gelatin solution into the pulmonary artery under pressure (Miller, 1947; Krahl, 1955), the tremendous extent of the capillary meshwork becomes apparent. Under

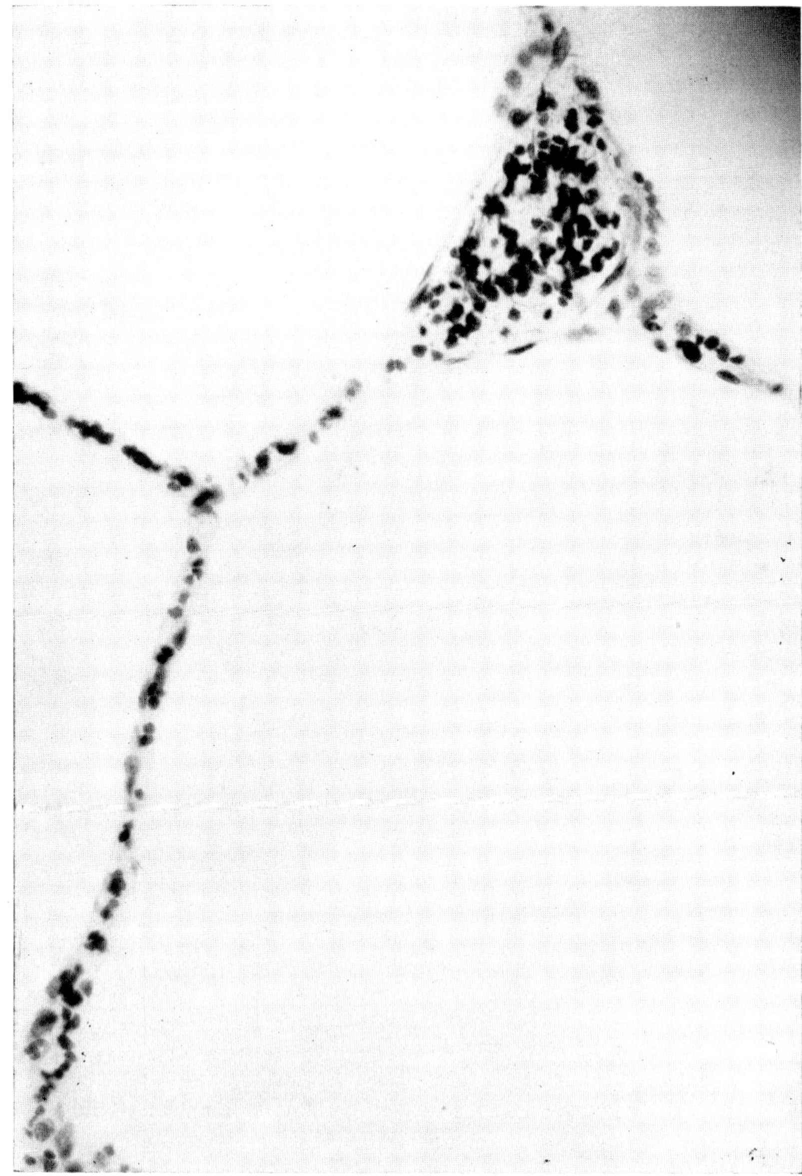

Fig. 63. Thin section (12 μ) of cat lung frozen in open chest at peak of inflation (+10 cm H_2O). The extent of the capillary net is difficult to visualize in this type of section. Red blood cells appear black or dark gray; tissue nuclei are pale gray. Note the smoothness of the alveolar walls and the larger vessel at the junction of three alveolar septa. 450×.

these conditions there is scarcely any space between the capillaries. When the lung in the intact animal is prepared by tracheal insufflation of liquid or gaseous fixatives, the capillary bed contains red cells, but the extent of filling is far less than suggested by injected preparations (see Subsection 4b, below).

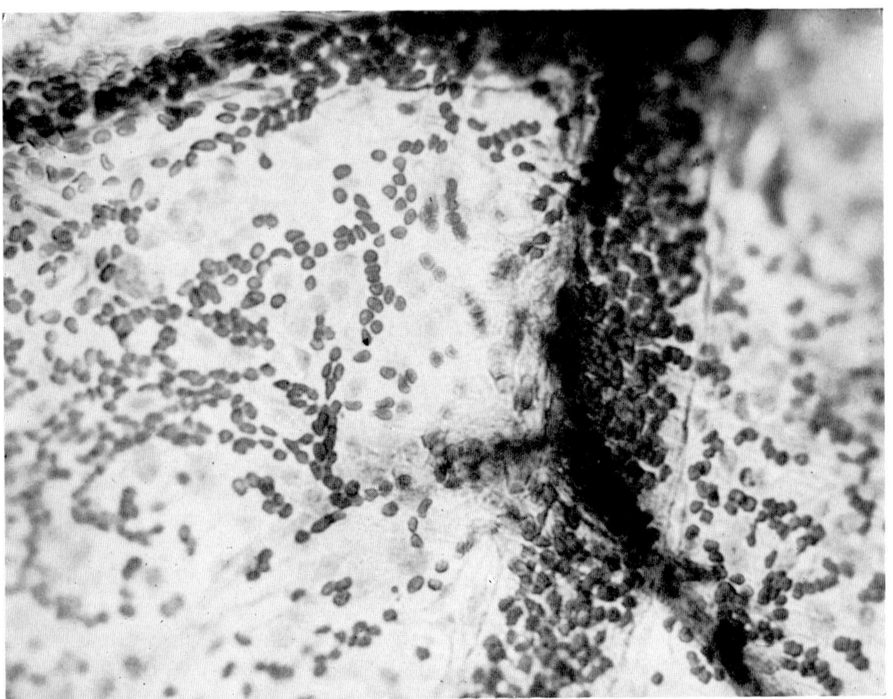

FIG. 64. Thick section (200 μ) of cat lung frozen in open chest at peak of inflation (+10 cm H_2O). The area between the two large vessels is a wall of an alveolus. Note capillaries joining larger vessels abruptly. This type of section gives a clear picture of the capillary net as it is in life. 280×.

Attempts to observe directly the fine circulation in the living lung (Hall, 1925; Wearn *et al.*, 1934; Vogel, 1947; Ramos, 1955) are by necessity limited to the pleural surfaces and lobar edges. The varied findings of investigators over the years are difficult to compare, owing to methodological differences and the tremendous subjective factors involved; for instance, Wearn *et al.* (1934) considered any vessel having one to three red cells flowing abreast as a capillary. Most authors agree that red cell flow in individual capillaries can be intermittent and reversible and that these vessels are not as well filled as in fixed sections.

The discrepancy is, in part, due to the differences between the pleural and deep capillary nets (Miller, 1947).

Since direct study of living lung is limited to the surface and since the usual fixation methods cannot prevent redistribution of blood within this organ because of compliance, osmotic and gravitational gradients (Sjöstrand and Sjöstrand, 1938), a new approach is necessary in order to obtain a true concept of capillary bed morphology in relation to function in life. The rapid freeze technique (Staub and Storey, 1960), in which living lung can be frozen solid to a depth of 0.1 cm in 500 msec and of 0.2 cm in 2000 msec, offers such a possibility (Figs. 63 and 64).

b. Relations of the capillary net to larger vessels: Knisely (1954), observing the living lung, has reported the abrupt formation of true capillaries from the sides of relatively large arterioles, and he has attributed the function of straining off intravascular debris to this arrangement. Study of thick sections prepared by the frozen lung technique (Fig. 64) has confirmed this observation.

On the other hand, von Hayek (1960) and Giese and Gieseking (1957) believe that the true capillary mesh arises from larger precapillaries that can act as short-circuits across or around an alveolus. Sjöstrand and Sjöstrand (1938) claim that, after crossing the alveoli, the true capillaries reunite in the alveolar septa, at the junction of three or more alveoli (Fig. 63), into vessels termed "capillary sinusoids"; to these they attribute a blood-depot function. Von Hayek (1960) calls these same vessels postcapillaries, while most other authors include them under venules.

c. Appearance of red cells in the capillaries: Generally, the red cells file singly through the capillaries and often appear to be sliding through with their biconcave surfaces facing the air spaces (Low, 1952; Ramos, 1955). This implies that the capillaries are slightly flattened. Figures 63 and 64 tend to support this, and, in addition, they show that the bulging of the capillaries into the air space, as stated in older literature, is artifactual (Macklin, 1950).

4. Physiologic Observations Relating to the Capillary Bed

a. Mean capillary pressure: Most workers accept +7 to +10 mm Hg as the mean capillary pressure in normal man lying supine for cardiac catheterization (Donnet and Ardisson, 1958). This is an indirect measure, based on mean pulmonary artery and left atrial pressure, and it is determined with respect to atmospheric pressure. Hydrostatic effects must play a considerable role in upright man, considering the distance between apical and basal capillaries of 20–30 cm (Miller, 1947). Differences in ventilation/perfusion between apex and base (Rahn *et al.*, 1956; West and Dollery, 1960) and changes in physiologic dead space with body position (Riley *et al.*, 1959) are attributed to the hydrostatic pressure.

b. Capillary blood volume: The volume of blood in the capillaries, V_C, of the sitting, resting subject, is 75–115 ml (McNeil *et al.*, 1958), as determined by carbon monoxide diffusion studies. This is less than the capacity of the bed, because on lying down the volume promptly increases approximately 25 per cent (Lewis *et al.*, 1958a); with exercise it also rises markedly (Forster, 1959). Piiper (1959) has used a stopped-flow gas-equilibration technique in isolated lungs to determine V_C and has undoubtedly obtained a maximal value with such a procedure. It is unfortunate that as of the moment, this method cannot be applied to intact animals, for it offers an alternate and more direct measure of V_C than do diffusion techniques. (For capacity of pulmonary vessels, see Section D.-2 of this chapter.)

c. Average time red cells remain in the capillaries: Using estimates of V_C and knowing cardiac output, the average time red cells remain in the lung capillaries can be obtained. Forster (1959) found times of 0.5–1.2 seconds at rest. However, Vogel (1947), who has made the only direct determination on living animals, obtained data which did not agree very well with these estimations. By high speed microcinematography of the surface of the living lung through a glass window in the thorax, he measured 0.1 second in cats, from which he calculated 0.1–0.3 second for man.

d. Capillary hematocrit: The pulmonary capillary hematocrit has been estimated (Rapaport *et al.*, 1956; Gibson *et al.*, 1946) to be considerably less than that of the pulmonary artery or peripheral artery blood. Other workers, using similar procedures, do not agree (Lilienfield *et al.*, 1956).

e. Compliance of the capillary bed: Engelberg and DuBois (1959) determined the compliance, $\Delta V/\Delta P$, of the pulmonary vascular bed and estimated that of the capillaries to be about 15 per cent of the total. Considering that only 10 per cent of the total pulmonary blood volume is ordinarily in the capillaries, their specific compliance must be greater than the rest of the bed. This agrees more with the known reserve capacity of the capillaries than do the conclusions of Lewis *et al.*

(1958b), based on failure to alter significantly diffusing capacity upon suddenly increasing pulmonary vascular volume and pressures by inflating a G-suit.

f. Nature of flow in the capillaries: Lee and DuBois (1955) have demonstrated pulsatile pulmonary capillary blood flow, measuring the instantaneous rate of nitrous oxide uptake in the lungs. They have found that there is marked systolic acceleration and that the end diastolic flow may approach zero. However, most workers directly viewing the lung surface have not noted pulsatile capillary flow normally. Since even the smallest arterioles definitely pulsate (Burrage and Irwin, 1953), a study of the actual dimensions in Fig. 63 suggests the possibility of soluble gas uptake by blood before it reaches the capillary level. Parenthetically, this also applies to CO_2 removal from the blood in the lungs.

g. Neurogenic and chemical responses of the capillaries: There is very little information about neurogenic or local chemical activity at the capillary level. Duke (1954) feels the receptor site for the pulmonary vascular response to anoxia must be in the region of the capillary bed. Wearn *et al.* (1934), on the other hand, could not see any effect of O_2, CO_2 or N_2 on transilluminated capillaries. Injections of adrenal medullary hormones and acetylcholine have shown widely varying effects on the capillaries, but these effects could not be interpreted apart from the drug actions elsewhere in the pulmonary circulation (MacGregor, 1934; Wearn *et al.*, 1934). Workers who directly view the living lung surface are about equally divided on the question of true capillary contractility.

h. Effect of breathing on the capillaries: A final perplexing problem concerns the effect of breathing on the capillaries. Are they squeezed during inspiration by stretching of alveolar walls (Olkon and Joannides, 1930; Riley, 1959), or are they distended by an increased transmural pressure gradient due to alveolar wall curvature and surface film properties (Macklin, 1950; Lloyd and Wright, 1960)? A solution to this problem will be most welcome.

D. PHYSIOLOGY

By E. Rapaport and J. Severinghaus

1. Introduction[2]

The main function of the pulmonary circulation is to conduct mixed venous blood through the alveolar capillaries where oxygen is added and excess carbon dioxide is eliminated. This system is composed of a pump (right ventricle), a conducting system (arteries and arterioles), a vast gas exchange area of about 50 M^2 (the capillary bed) and a collecting system (venules and veins). The pulmonary circulation differs from various systemic circulations in several respects: (1) the entire right ventricular output flows through the pulmonary bed; (2) the normal pulmonary capillary pressure cannot exceed colloidal osmotic pressure because the resulting fluid transudation would prevent intimate contact of alveolar gas with capillary blood and interfere seriously with proper gas exchange; and (3) a second circulation (bronchial) is responsible for the nutrition of the lung.

The special characteristics of the pulmonary circulation are that it is a low pressure, low resistance, readily distensible system without the intense neurogenic vasoconstrictor control that characterizes the systemic circulation, features which permit accomplishment of its function with minimal expenditure of energy. (For further discussion of some of the special characteristics of the pulmonary circulation, see reviews of Adams and Veith, 1959 and Marshall, 1959.)

2. Capacity and Distensibility of Pulmonary Vessels

The pulmonary arteries, capillaries and veins of an adult normally contain approximately 0.5 liters of blood. The pulmonary veins, along with the left atrium, serve as a reservoir for left ventricular filling. In animals, the pulmonary vessels, like other musculoelastic structures, exhibit stress relaxation: the pressure in the vessels increases when the vessels are suddenly distended and then falls during the next few seconds. The pulmonary blood volume varies considerably with posture and with changes in left atrial pressure, since the veins are easily collapsed and, up to a certain point, quite distensible. It may increase 30 per cent on lying down and 100 per cent when the vessels are fully distended, with only small accompanying rises in pressure. Increasing

[2] For discussion of lymph flow in the lungs, see Section D.-2e, Chapter XXIII.

the volume and pressure of gas in the alveoli at first increases and then decreases pulmonary vascular capacity. Of interest in this regard is the finding that both atelectatic and overdistended lungs contain little blood. (For capacity of pulmonary vessels, see Section C.-4b of this chapter.)

3. Pressure in Pulmonary Circulation

Three types of pulmonary pressure must be distinguished: (1) absolute pressure, usually referred to ambient air, which, in the normal, resting recumbent adult, averages 22/9 mm Hg, with a mean of 15 mm Hg in the pulmonary artery and 7 in the left atrium; (2) the driving pressure (the pulmonary arterial pressure minus the pulmonary venous pressure), which is the figure, normally around 8 mm Hg, used for computing pulmonary vascular resistance; and (3) the transmural pressure (the intravascular pressure minus the tissue pressure outside the vessel), which, since the pulmonary vascular bed is distensible, influences the vessel size and, hence, to some extent, resistance to flow.

With respiratory maneuvers, left atrial pressure approximately follows intrapleural pressure. The pressure in the airway has some effect on the pressure in the pulmonary capillaries, since the latter are suspended in alveolar walls. The absolute pressure in the vessels varies with height, being about 30 cm greater at the bases than at the apices. The vessels and capillaries at the bases are therefore distended and have a lower resistance, contain the most blood, and accept a greater share of the flow.

4. Resistance in Pulmonary Circulation

Pulmonary vascular resistance is ordinarily defined as the mean driving pressure divided by the average pulmonary blood flow. Normally, resistance is 1–1.5 mm Hg per liter of pulmonary blood flow per minute or about one-tenth that of the systemic circulation. Lloyd and Wright (1960) have suggested that a more conceptual definition is afforded by considering the rate of change in driving pressure with respect to flow at any moment, under the particular circumstances in question. This is necessary in a system which is distensible and therefore has resistance variable with pressure.

a. The effect of transmural pressure: Transmural pressure in the capillary is determined primarily by pulmonary venous pressure rather than by arterial pressure. Pulmonary arterial pressure changes in a pulsatile manner over a sufficiently wide range so that pulmonary blood flow apparently ceases in at least some capillaries with each diastolic minimal pressure. Therefore, the pulmonary vascular resistance is cyclically variable. Some investigators who perfused isolated lungs with steady pressure have found that at constant venous pressure the calculated resistance decreases as flow rate increases; others have noted a linear relationship between pressure and flow under these circumstances. All authors agree that elevating pulmonary venous pressure decreases pulmonary vascular resistance, owing to opening of new and/or distention of existing pulmonary capillaries. It seems probable that the pulmonary capillary can be considered as a tube which goes from the collapsed state to near maximum volume over a very small pressure range of perhaps 1 mm of Hg. Further increases in pressure result in only slight rises in capillary vascular volume. It may be assumed that this transition from closed to open state occurs when the absolute transmural pressure exceeds the critical closing pressure. The concept that critical closing pressure is significantly greater than zero in the pulmonary circuit has been questioned by Lloyd and Wright (1960). These authors found no cessation of arterial inflow at any positive arterial to venous pressure gradient when the effect of alveolar surface tension had been eliminated by using gas-free lungs and the effect of gravity had been removed by suspending the lungs in saline. They believed that the critical (transmural) closing pressure is effectively about zero. On the other hand, in the air-filled lung, Borst *et al.* (1956) demonstrated complete closure of the pulmonary vascular bed when left atrial pressures were below 7 mm Hg, since pulmonary arterial pressure failed to fall to left atrial pressure even with no flow.

b. The effect of airway pressure and lung inflation: During inspiration, the driving pressure increases considerably more than the flow and pulmonary vascular resistance is raised at increasing lung volumes. Resistance also becomes greater if the lung is allowed to collapse, in part, perhaps, because of kinking of vessels and, in part, because of the loss of the surface tension forces which previously tended to hold the capillaries open.

c. Exercise: With maximum exercise, pulmonary blood flow may be increased four or five times, and the calculated resistance falls appreciably. Either active dilatation of existing pulmonary capillaries occurs or a reserve of intermittently closed vessels opens up to accept the augmented flow; in either case, the mechanism is not known. Resistance appears to drop progressively during continued exercise, perhaps due to the additional factor of stress relaxation (see Section D.-2 of this chapter).

d. Vasomotor activity: It is known that there are sympathetic nerves to pulmonary arterioles which may cause vasoconstriction, independent of their effect on bronchomotor tone, and that there is smooth muscle in the walls of the smaller branches of the pulmonary arteries which can respond to appropriate stimulation. Extirpation of the sympathetic trunk reduces pulmonary resistance. There is, however, no evidence that these nerves have any role in the control of the pulmonary circulation under normal conditions. They may protect against pulmonary edema by raising pulmonary arteriolar resistance when pulmonary capillary pressure approaches plasma oncotic pressure, because of left ventricular failure or severe mitral stenosis.

Arterial anoxemia does cause a rise in pulmonary vascular resistance; presumably it is mediated, in part, by carotid and aortic body stimulation and activation of sympathetic vasoconstrictor nerves. In addition, low oxygen tension in the pulmonary alveoli apparently has an inconstant local effect, producing increase in resistance locally and hence shunting blood to better ventilated or less anoxic areas of the lung. This reflex has been difficult to demonstrate in man. Vasoconstriction in the isolated perfused lung occurs in response to the low oxygen tension in the respired gases rather than to the low oxygen tension in the perfusing fluid. Marked increases in pulmonary vascular resistance may be found in man and animals living at high altitudes.

Higher inspired concentrations of carbon dioxide elevate pulmonary vascular resistance. On the other hand, in the isolated cat lung an increased amount of carbon dioxide in the mixed venous blood may cause vasodilatation and a decrease in pulmonary vascular resistance (Nisell, 1951).

e. Pulmonary venoconstriction: The flow of blood from the capillaries through the pulmonary veins to the left atrium is accomplished without a detectable pressure drop, i.e., the pressure in the pulmonary capillaries and veins is for practical purposes identical to that in the left atrium. No clinical situations are recognized in which significant pulmonary venoconstriction occurs, resulting in a pressure gradient between the pulmonary capillaries and the left atrium, although such a response has been demonstrated in the experimental animal following production of pulmonary emboli and after the administration of E. coli endotoxin.

f. Miscellaneous factors: Changes in the viscosity of the blood may affect pulmonary vascular resistance. Profound hypothermia (15–18°C) in perfused dog lungs appears to increase pulmonary vascular resistance more than can be accounted for by viscosity changes alone. None of the usual anesthetic agents is known to have any direct constrictor effect on pulmonary circulation. Anesthesia, however, may affect pulmonary vascular resistance by changing left atrial pressure, since all anesthetics tend to depress the myocardium.

5. Effect of Pulmonary Artery Obstruction on Ventilation

The effects of airway obstruction (Subsection 4b, above) tend to shunt pulmonary blood flow to alveoli that are well ventilated. The converse response has been recently observed, diverting the ventilation away from alveoli that have no blood flow. The latter mechanism is dependent on the carbon dioxide concentration in the alveolus which falls nearly to zero when pulmonary blood flow stops. This low carbon dioxide tension produces bronchoconstriction, leading progressively to diminished ventilation and atelectasis. The mechanism becomes effective when the alveolar carbon dioxide falls to about 15 mm Hg (Severinghaus *et al.*, 1960).

6. The Lung as a Filter

An incidental function of the pulmonary vascular bed is to remove lymphocytes, blood clots and other material larger than perhaps 30 micra in diameter from the circulation. Mechanisms by which the lung deals with such material have not been adequately studied.

E. PHARMACOLOGY

By J. H. Comroe, Jr. and M. B. McIlroy

1. Physiologic Basis for Action of Drugs

Since drugs usually act through physiologic mechanisms, it is necessary to discuss the latter before presenting the pharmacologic responses of the pulmonary circulation. It is also important to point out that vascular mechanisms are different in this bed, as compared with the systemic circulation. For example, the latter requires regulatory control which attempts to secure a blood flow to all capillary beds of the body, adequate for the needs of each tissue at each moment. Local mechanisms dilate arterioles when local metabolism increases, while general redistributive mechanisms reduce blood flow to certain inactive or nonessential organs and increase or maintain flow in active or essential organs. In contrast, the pulmonary circulation has no need for local regulation in response to metabolic needs, since the bronchial circulation nourishes the lungs.

Until a decade ago, very little evidence was available to indicate that there was important vasomotor regulation in the pulmonary circulation or even much smooth muscle, capable of altering vascular tone. Hence, no mechanism appeared available for drug actions. It is now known, however, that there is smooth muscle in the pulmonary arterioles, that it can increase markedly in pathologic conditions, that it is influenced by nerve impulses and that it can respond to drugs. Pulmonary arteriolar constriction can occur generally (presumably to protect the capillary bed from high hydrostatic pressures which may result in pulmonary edema) or regionally (presumably to redistribute pulmonary arterial blood to the well-ventilated alveoli). Venular or venous constriction can also occur; however, there is still no conclusive evidence for independent capillary contractility. Drugs can therefore act on pulmonary arterioles, venules or veins. When arterioles constrict, pulmonary arterial pressure rises and the load on the right ventricle increases, although capillary pressure may remain normal; when venules or veins constrict, pulmonary capillary and arterial pressures and right ventricular pressure rise and pulmonary edema may result.

2. Effect of Drugs on Pulmonary Circulation

It is difficult to evaluate the effect of drugs on the pulmonary circulation in intact animals or man. Resistance cannot be measured directly but must be calculated from the relationship: resistance = pressure difference across vascular bed, divided by blood flow, or

$$\text{Resistance} = \frac{\Delta \text{ Pressure}}{\text{Flow}}$$

Flow must be measured simultaneously and, preferably, continuously. Further, because of the low pressures in the pulmonary circulation, small errors in measurement may affect calculated resistance considerably. If the true values for $R = \Delta P/F$ are $R = (15-6)/90 = 0.1$, small errors in measuring pressures may give

$$R = \frac{13-8}{90} = 0.055 \text{ mm Hg/ml/sec}$$

a change of 45 per cent. In addition, drugs which act on the pulmonary circulation may also affect respiration, bronchial smooth muscle, the heart and systemic blood vessels and so change intrathoracic pressure, systemic blood flow and systemic blood pressure. (For recent reviews on the actions of drugs on the pulmonary circulation, see Aviado, 1960; Fowler, 1960; Marshall, 1959.)

a. Sympathomimetic drugs: Generally the drugs in this series which are predominantly vasoconstrictor in action in the systemic circulation (norepinephrine, epinephrine) also produce pulmonary arteriolar constriction. This can be demonstrated in isolated, perfused lungs but less clearly in man, probably because increased pulmonary blood flow (caused by the inotropic effect on the heart) tends to dilate the small vessels of the lung. Sympathomimetic drugs which are predominantly vasodilator in action in the systemic circulation (isoproterenol) also seem to dilate pulmonary vessels. Isoproterenol, given by aerosol to patients with chronic pulmonary disease, has resulted occasionally in reduction of pulmonary arterial pressure; however, this may have been due to bronchodilatation and correction of anoxia rather than to a direct pharmacologic action on the pulmonary arterioles.

b. Sympathetic blocking drugs: These agents exert direct action on the pulmonary circulation only by decreasing sympathetic tone. Hexamethonium (typical of ganglionic blocking agents) and Dibenamine (typical of peripheral postganglionic or adrenergic blocking agents) have been used to decrease pulmonary vascular pressures in pulmonary edema, but the observed effects may well have been due to systemic peripheral vasodilatation and redistribution of blood volume from the pulmonary to the systemic vessels. Ganglionic blocking agents, in addition to producing both sympathetic and parasympathetic motor block, may act similarly on certain specialized sensory receptors; these include some in the pulmonary circulation which may initiate local or general vascular reflexes following local injury, such as embolization. The use of hexamethonium for this purpose is intriguing but not yet on a firm experimental basis.

c. Parasympathomimetic drugs: Acetylcholine is known to produce pulmonary vasodilation in isolated lungs. Recently it has been used in man, both to lower abnormally elevated pulmonary arterial pressure and as a diagnostic test to determine whether pulmonary hypertension is "fixed" or partly reversible. Approximately 0.5–1.5 mg are injected into the pulmonary artery via a catheter; the effects appear to be confined to the lung because no change in heart rate or systemic blood pressure occurs. Presumably this is due to the inactivation of acetylcholine by blood cholinesterase in a few seconds, before the drug reaches the coronary and other systemic arteries. The pulmonary vasodilator action of acetylcholine is blocked by atropine.

d. Other vasodilator drugs: Oxygen may cause vasodilation when anoxic constriction exists. Aminophylline, papaverine and nitrites are vasodilator in perfused lungs; however, there is insufficient evidence to justify their use in man as therapeutic agents. As noted previously, these drugs also affect bronchioles, respiration and the systemic circulation, which makes objective evaluation difficult.

e. Other constrictor drugs: Serotonin (5–OH–tryptamine) constricts pulmonary arterial and venous smooth muscle and bronchioles. Since it is a component of tissues and is liberated locally during blood clotting, it may be responsible, in part, for the bronchoconstriction, venular constriction and pulmonary capillary congestion sometimes observed following pulmonary embolism. Pituitrin is a weak vasoconstrictor. Certain bacterial endotoxins (Escherichia Coli), histamine and alloxan can cause both arterial and venous constriction; the experimental pulmonary edema induced by alloxan is presumably due to pulmonary venous constriction.

f. Inhaled gases and vapors: The action of O_2 and CO_2 has been discussed in Section D.-4d of this chapter. Little is known regarding the effect of smog, anesthetic gases or vapors, irritant fumes and toxic gases; these might be expected to affect the downstream vessels (capillaries, venules and veins) and might also diffuse to the precapillary vessels.

g. Pulmonary chemoreflexes: Injection of minute amounts (several micrograms) of veratridine, phenyl-diguanide, nicotine and serotonin into the pulmonary artery can cause reflex apnea, bradycardia, systemic vasodilation and hypotension in certain species. The receptors lie somewhere within the distribution of the pulmonary circulation, and the afferent impulses travel in the vagus nerves. The physiologic significance of this powerful reflex is not known; it appears to account for the reduction in systemic hypertension caused by the administration of the veratrum alkaloids (Dawes and Comroe, 1954; Aviado, 1960).

References

Adams, W. R., and Veith, I., eds. (1959). "Pulmonary Circulation." Grune & Stratton, New York.

Aviado, D. M. (1960). The pharmacology of the pulmonary circulation. *Pharmacol. Revs.* **12**, 159.

Bargmann, W. (1956). "Histologie und mikroskopische Anatomie des Menschen." 2nd ed. Thieme, Stuttgart.

Borst, H. G., McGregor, M., Whittenberger, J. L., and Berglund, E. (1956). Influence of pulmonary arterial and left atrial pressures on pulmonary vascular resistance. *Circulation Research* **4,** 393.

Boyden, E. A. (1945). The intrahilar and related segmental anatomy of the lung. *Surgery* **18,** 706.

Burrage, W. S., and Irwin, J. W. (1953). Microscopic observations of the pulmonary arterioles, capillaries and venules of living mammals before and during anaphylaxis. *J. Allergy* **24,** 289.

Cauldwell, E. W., Siekert, R. G., Lininger, R. E., and Anson, B. J. (1948). The bronchial arteries; an anatomic study of 150 human cadavers. *Surg., Gynecol. Obstet.* **86,** 395.

Chase, W. H. (1959). Distribution and fine structure of elastic fibers in mouse lung. *Exptl. Cell. Research* **17,** 121.

Dawes, G., and Comroe, J. H., Jr. (1954). Chemoreflexes from the heart and lungs. *Physiol. Revs.* **34,** 167.

Donnet, V., and Ardisson, J. L. (1958). Données récentes sur la circulation pulmonaire. *J. physiol. (Paris)* **50,** 587.

Duke, H. N. (1954). The site of action of anoxia on the pulmonary blood vessels of the cat. *J. Physiol. (London)* **125,** 373.

Elftman, A. G. (1943). Afferent and parasympathetic innervation of the lungs and trachea of dog. *Am. J. Anat.* **72,** 1.

Engelberg, J., and DuBois, A. B. (1959). Mechanics of pulmonary circulation in isolated rabbit lungs. *Am. J. Physiol.* **196,** 401.

Ellis, F. H., Grindlay, J. H., and Edwards, J. E. (1951). The bronchial arteries. *Surgery* **30,** 810.

Forster, R. E. (1959). The pulmonary capillary bed: Volume, area, and diffusing characteristics. *In* "Pulmonary Circulation" (W. R. Adams and I. Veith, eds.), p. 45. Grune & Stratton, New York.

Fowler, N. O. (1960). Effects of pharmacologic agents on the pulmonary circulation. *Am. J. Med.* **28,** 927.

François-Franck, C. A. (1880). Sur l'innervation des vaissseaux des poumons et sur les effets produits dans la circulation intracardiaque et aortique par le resserement de ces vaisseaux. *Séances et mém. soc. biol.* **7,** 231.

Gibson, J. C., Seligman, A. M., Peacock, W. C., Aub, J. C., Fine, J., and Evans, R. D. (1946). The distribution of red cells and plasma in large and minute vessels of the normal dog determined by radioactive isotopes of iron and iodine. *J. Clin. Invest.* **25,** 848.

Giese, W., and Gieseking, R. (1957). Die submikroskopische Structur des fibrilloren Grundgerüstes der Alveolarwand. *Beitr. pathol. Anat. u. allgem. Pathol.* **117,** 17.

Gillilan, L. A. (1954). "Clinical Aspects of the Autonomic Nervous System." Little, Brown, Boston, Mass.

Gray, H. (1959). *In* "Anatomy of the Human Body" (C. M. Goss, ed.). Lea & Febiger, Philadelphia, Pa.

Hall, H. L. (1925). A study of the pulmonary circulation by the transillumination method. *Am. J. Physiol.* **72,** 446.

Honjin, R. (1956). On the nerve supply of the lung of the mouse, with special reference to the structure of the peripheral vegetative nervous systems. *J. Comp. Neurol.* **104,** 587.

Hovelacque, A., Monod, O., and Evrard, H. (1936). Note au sujet des artères bronchiques. *Ann. anat. pathol. et anat. normale méd.-chir.* **13,** 129.

Huntington, G. S. (1919). The morphology of the pulmonary artery in the mammalia. *Anat. Record* **17**, 165.

Irwin, J. W., Burrage, W. S., Aimar, C. E., and Chesnut, R. W., Jr. (1954). Microscopical observations of the pulmonary arterioles, capillaries, and venules of living guinea pigs and rabbits. *Anat. Record* **119**, 391.

Jeffords, J., and Knisely, M. H. (1956). Concerning the geometric shapes of arteries and arterioles. *Angiology* **7**, 105.

Karrer, H. E. (1956). The ultrastructure of mouse lung. *J. Biophys. Biochem. Cytol.* **2**, 241.

Keibel, F., and Mall, F. P. (1912). "Manual of Human Embryology," Vol. II. Lippincott, Philadelphia, Pa.

Kisch, B. (1958). Electron microscopy of the lungs in acute pulmonary edema. *Exptl. Med. Surg.* **16**, 17.

Knisely, W. (1954). Microscopic Observation of the Architecture and Behavior of the Smaller Blood Vessels in Living Lungs of Mammals and Related Observations in other Vessels of Experimental Animals and Human Patients. Ph.D. Thesis. Medical College of S. Carolina.

Knisely, W. H. (1960). In vivo architecture of blood vessels supplying and draining alveoli. *Am. Rev. Respiratory Diseases* **81**, 735.

Krahl, V. W. (1955). The respiratory portions of the lung: An account of their finer structure. *Bull. School Med. Univ. Maryland* **40**, 101.

Larsell, O., and Dow, R. S. (1933). The innervation of the human lung. *Am. J. Anat.* **52**, 125.

Lee, G. deJ., and DuBois, A. B. (1955). Pulmonary capillary blood flow in man. *J. Clin. Invest.* **34**, 1380.

Lewis, B. M., Lin, T. H., Noe, F. E., and Komisaruk, R. (1958a). The measurement of pulmonary capillary blood volume and pulmonary membrane diffusing capacity in normal subjects; the effects of exercise and position. *J. Clin. Invest.* **37**, 1961.

Lewis, B. M., Forster, R. E., and Beckman, E. L. (1958b). Effect of inflation of a pressure suit on pulmonary diffusing capacity in man. *J. Appl. Physiol.* **12**, 57.

Lilienfield, L. S., Kovach, R. D., Marks, P. A., Hershenson, L., Rodman, G. P., Ebaugh, F. G., and Freis, E. D. (1956). The hematocrit of the lesser circulation in man. *J. Clin. Invest.* **35**, 1385.

Lloyd, T. C., Jr., and Wright, G. G. (1960). Pulmonary vascular resistance and vascular transmural gradient. *J. Appl. Physiol.* **15**, 241.

Low, F. N. (1952). Electron microscopy of the rat lung. *Anat. Record* **113**, 437.

MacGregor, R. C. (1934). Examination of the pulmonary circulation with the microscope. *J. Physiol. (London)* **80**, 65.

Macklin, C. C. (1950). The alveoli of the mammalian lung. An anatomical study with clinical correlations. *Proc. Inst. Med. Chicago* **18**, 78.

McNeil, R. S., Rankin, J., and Forster, R. E. (1958). The diffusing capacity of the pulmonary membrane and the pulmonary capillary blood volume in cardiopulmonary disease. *Clin. Sci.* **17**, **465**.

Marshall, R. (1959). The physiology and pharmacology of the pulmonary circulation. *Progr. in Cardiovasc. Diseases* **1**, 341.

Miller, W. S. (1893). The structure of the lung. *J. Morphol.* **8**, 165.

Miller, W. S. (1947). "The Lung." Charles C Thomas, Springfield, Ill.

Nagaishi, C. (1957). "The Structure of the Lung" (in Japanese). Igaku Shoin, Tokyo.

Nakamura, N. (1924). Zur Anatomie der Bronchialarterien. *Anat. Anz.* **58,** 508.

Nisell, O. (1951). The influence of blood gases on the pulmonary vessels of the cat. *Acta Physiol. Scand.* **23,** 85.

Noback, G. J., and Rehman, I. (1941). The ductus arteriosus in the human fetus and newborn infant. *Anat. Record* **81,** 505.

Olkon, A. M., and Joannides, M. (1930). Capillaroscopic appearance of the pulmonary alveoli in the living dog. *Anat. Record* **45,** 121.

Orsós, F. (1936). Die Gerüstysteme der Lunge und deren physiologische und pathologische Bedeutung. *Beitr. Klin. Tuberk.* **87,** 568.

Patten, B. M. (1953). "Human Embryology," 2nd ed. McGraw-Hill, New York.

Piiper, J. (1959). Grösse des Arterien-, des Capillar- und des Venen-Volumens in der isolierten Hundelunge. *Pflügers Arch. ges. Physiol.* **269,** 182.

Rahn, H., Sadoul, P., Farhi, L. E., and Shapiro, J. (1956). Distribution of ventilation and perfusion in the lobes of the dog's lung in the supine and erect position. *J. Appl. Physiol.* **8,** 417.

Ramos, J. G. (1955). On the dynamics of the lung's capillary circulation. *Am. Rev. Tuberc. Pulmonary Diseases* **71,** 822.

Rapaport, E., Kuida, H., Haynes, F. W., and Dexter, L. (1956). The pulmonary blood volume in mitral stenosis. *J. Clin. Invest.* **35,** 1393.

Riley, R. L. (1959). Effects of lung inflation on the pulmonary vascular bed. *In* "Pulmonary Circulation" (W. R. Adams and I. Veith, eds.), p. 147. Grune & Stratton, New York.

Riley, R. L., Permutt, S., Said, S., Godfrey, M., Cheng, T., Howell, J. B. L., and Shepard, R. H. (1959). Effect of posture on pulmonary dead space in man. *J. Appl. Physiol.* **14,** 339.

Schulz, H. (1956). Elektronenoptische Untersuchungen der normalen Lunge und der Lunge bei Mitralstenose. *Virchows Arch. pathol. Anat. u. Physiol.* **328,** 582.

Schulz, H. (1959). "The Submicroscopic Anatomy and Pathology of the Lung." Springer, Berlin.

Severinghaus, J. W., Swenson, E. W., Finley, T. N., and Lategola, M. T. (1960). Rapid shift of ventilation from unperfused regions of the lung. *Federation Proc.* **19,** 378.

Sjöstrand, E., and Sjöstrand, T. (1938). Über das Vorkommen von sinuoesen Blutgefässen in den Lungen der Maus. *Anat. Anz.* **87,** 193.

Staub, N. C., and Storey, W. F. (1960). On the possibility of freezing living lung rapidly. *Physiologist* **3,** 154.

Tobin, C. E., and Zaraquiey, M. O. (1950). Arteriovenous shunts in the human lung. *Proc. Soc. Exptl. Biol. Med.* **75,** 827.

Tobin, C. E., and Zaraquiey, M. O. (1953). Some observations on the blood supply of the human lung. *Med. Radiogr. and Photogr.* **29,** 9.

Verloop, M. C. (1948). The arteriae bronchiales and their anastomoses with the arteria pulmonalis in the human lung; a micro-anatomical study. *Acta Anat.* **5,** 171.

Vogel, H. (1947). Die Geschwindigkeit des Blutes in den Lungenkapillaren. *Helv. Physiol. et Pharmacol. Acta* **5,** 105.

von Hayek, H. (1940a). Ueber einen kurzchlusskreislauf (arteriovenose anastomosen) in der menschlichten lunge. *Z. Anat. Entwicklungsgeschichte* **110,** 412.

von Hayek, H. (1940b). Über verschlussfähige Arterien in der menschlichen Lunge. *Anat. Anz.* **89,** 216.

von Hayek, H. (1942a). Über Kurzschlüsse und Nebenschlüsse des Lungenkreislaufes. *Anat. Anz.* **93,** 155.

von Hayek, H. (1942b). Kurz- und Nebenschlüsse des menschlichen Lungenkreislaufes in der Pleura. *Z. Anat. Entwicklungsgeschichte* **112,** 221.

von Hayek, H. (1953). "Die menschliche Lunge." Springer, Berlin.

von Hayek, H. (1960). "The Human Lung," rev. ed. (English transl. by V. E. Krahl). Hafner, New York.

Wearn, J. T., Ernstene, A. C., Bromer, A. W., Ban, J. S., German, W. S., and Zschiesche, L. J. (1934). The normal behavior of the pulmonary blood vessels with observations on the intermittence of the flow of blood in the arterioles and capillaries. *Am. J. Physiol.* **109,** 236.

Weibel, E. (1958). Die Entstehung der Längsmuskulatur in den Ästen der arteria bronchialis. *Z. Zellforsch.* **47,** 440.

Weibel, E. (1959). Die Blutgefässanastomosen in der menschlichen Lunge. *Z. Zellforsch.* **50,** 653.

West, J. B., and Dollery, C. T. (1960). Distribution of blood flow and ventilation-perfusion ratio in the lung, measured with radioactive CO_2. *J. Appl. Physiol.* **15,** 405.

Woodside, G. L., and Dalton, A. J. (1958). Ultrastructure of lung tissue from newborn and embryo mice. *J. Ultrastruct. Research* **2,** 28.

Zweifach, B. W. (1959). The structural basis of the microcirculation. *In* "Cardiology" (A. A. Luisada, ed.), Vol. I, Part 1, Chapter 13. McGraw-Hill (Blakiston), New York.

Chapter XI. GASTROINTESTINAL CIRCULATION

A. EMBRYOLOGY

By R. Licata

1. Establishment of the Alimentary Circulation

a. Early development: Incident on transformation of an embryonic double aorta into one main trunk is coalescence of multiple ventral segmental (visceral) rootlets into single stems that traverse the primary dorsal mesentery (Keibel and Mall, 1912). Prior to this, arterial arcades arising from pallisade-like ventral branches pass around the primitive gut to form connections with the arteries of the yolk sac. In this fashion favored arterial routes are established. With differentiation of particular regions of the intestinal tract, special visceral branches develop to supply the various segments. At each level the unpaired intestinal trunks become displaced and elongated at their aortic attachments, due to migration of the associated viscera, as well as torsion of the primitive gut. The major vessels so formed are the superior mesenteric, celiac and inferior mesenteric arteries.

b. Superior mesenteric artery: At midgut levels a main line link is retained at the time of extroversion of the gut loop, which unites the yolk sac plexus with a central visceral branch. The arterial arc thus established is utilized in production of an omphalomesenteric or vitelline artery, which central arterial bridge, in turn, is capitalized in formation of a superior mesenteric trunk. A number of intestinal branches fan out from the main intestinal vessel and distally retain connection with the regressing vitelline arteries by means of arterial arcades passing around the midgut. With introversion of the primitive gut loop, all arterial connection is lost with the vitelline circulation of the yolk sac. Postnatally, persistent vitelline duct vestiges may sometimes be encountered in the form of a Meckel's diverticulum. The terminal branch normally accompanying such a structure represents the embryonic counterpart of the original omphalomesenteric circuit.

c. Celiac artery: Vessels are continually generated, which supply primary regions of the alimentary canal and their glandular derivatives. Abandonment of the rudimentary yolk sac follows the "pulling in" of the gut loop. Within the celom excessive rotation of the gut tract extends the original dorsal mesentery into a broad fold; at the same time there is loss of most of the ventral mesentery. The uppermost visceral branch derived from the ventral segmental vessels becomes the principal supply of the infradiaphragmatic portion of the foregut. This vessel, the celiac artery, develops as a foreshortened stalk. It is deeply imbedded within the retroperitoneum, due to the great torsion brought to bear on it as a result of exaggerated regional fixation of the gut. The sprawling, divided appearance of its branching reflects the manner in which it is prenatally caught in a vortex of rapid growth and migration of the glandular derivatives it supplies. Postnatally the region of anastomosis between the superior pancreaticoduodenal artery, a branch of the celiac artery, and the inferior pancreaticoduodenal artery constitutes the theoretical plane of vascular "divide" between foregut and midgut.

d. Inferior mesenteric artery: Caudally, the ventral segmental arteries normally distributed to the hindgut in the prenatal state fuse to form a short common trunk, called the inferior mesenteric artery. This vessel is directed retroperitoneally in development as a result of fixation of the hindgut. The remainder of the terminal hindgut receives its supply from arteries of dorsal intersegmental origin that differentiate into middle and inferior rectal branches. The region of anastomosis of left and middle colic arteries represents the developmental vascular "divide" between midgut and hindgut. At the time of fixation of the different regions of the primitive gut, fusion fasciae develop between the segment involved and the dorsal body wall. In these fascial planes, arterialization is suppressed prenatally, resulting in the "bloodless fields" found in such zones in the adult.

B. GROSS ANATOMY

By B. J. Anson;[1] N. A. Michels[2]

1. Blood Supply to the Stomach

a. Celiac artery: This vessel is the main source of blood supply to the stomach. It arises from the front of the abdominal aorta, immediately below the aortic orifice of the diaphragm and between the crura of the latter (Figs. 65a,b and 66). It is short and wide, and courses almost horizontally forward, below the caudate lobe of the liver, for a distance of about half an inch (12 mm). It ends, in typical instances, by dividing into three branches: the left gastric, the hepatic, and the splenic.

In 174 out of 287 specimens (60%), Cauldwell and Anson (1943) found the origin of the celiac artery to be opposite the upper third of the first lumbar vertebra (75 specimens), the middle third of the first lumbar (54 specimens), or the intervertebral fibrocartilage between the first lumbar and the twelfth thoracic vertebrae (45 specimens).

In the remaining specimens (40%) the sites of origin were spread out over vertebral areas cranial and caudal to the limits just described: cranialward to the level of the upper third of the eleventh thoracic vertebra, and caudalward to that of the upper third of the second lumbar, with no concentrated incidence in any of the vertebral zones.

The trunk lay behind the peritoneum of the omental bursa, below the caudate lobe of the liver and above the upper border of the pancreas, and was surrounded by the celiac plexus of autonomic nerves.

The left gastric artery is the smallest branch of the three regular divisions of the celiac artery. It runs obliquely upward and to the left and reaches the lesser curvature of the stomach, close to the junction of the latter with the esophagus (Fig. 65b). It then turns abruptly forward, downward, and to the right, and courses toward the pylorus of the stomach, where it anastomoses with the right gastric branch of the hepatic division of the celiac artery (Figs. 65a and 66). In the first part of its course, the left gastric artery lies on the left crus of the diaphragm behind the lesser sac (omental bursa).

The left gastric artery is continued to the stomach between the layers of the lesser omentum. It gives off esophageal branches that ascend on the esophagus to anastomose with esophageal branches of the thoracic aorta and with the phrenic artery. As it passes along the lesser

[1] Author of Subsection 1.
[2] Author of Subsections 2 and 3.

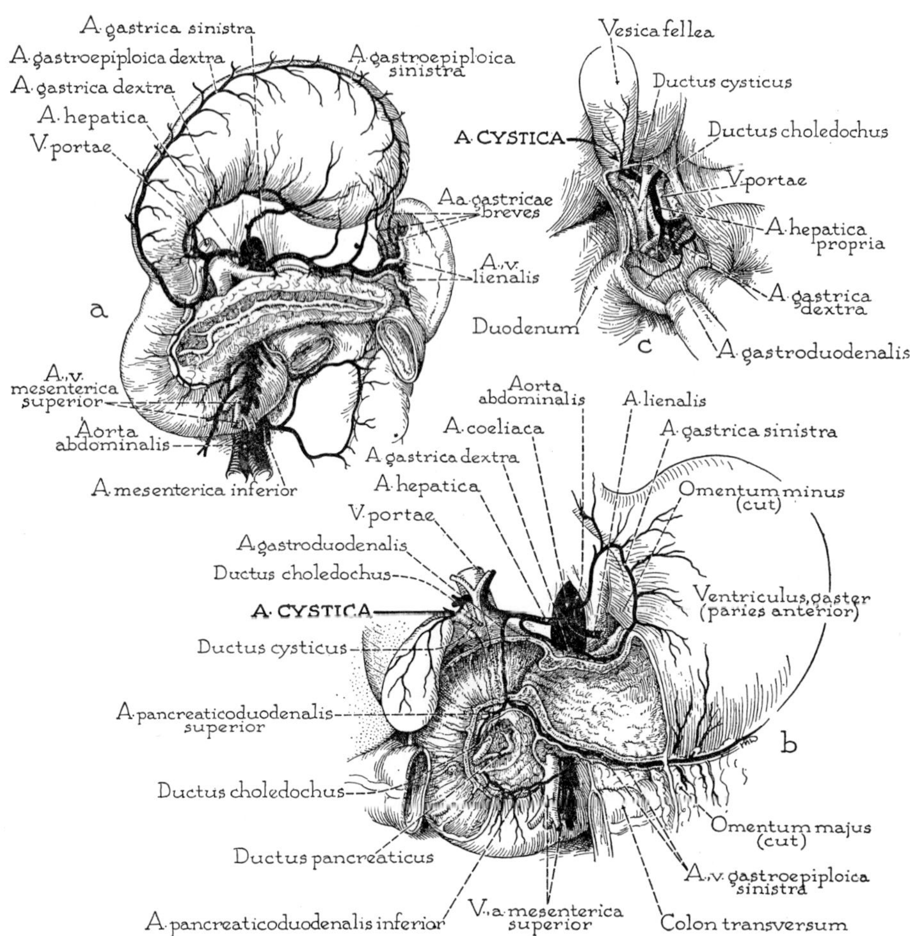

FIG. 65. Blood supply of the stomach and related organs. a. Branches of the celiac artery, exposed by lifting the stomach upward. The latter shown without its omental supports, in order to present more graphically the pattern of triradiate division and subsequent distribution of the celiac branches. b. Branches of the hepatic division of the celiac artery, shown with the stomach restored to natural position. c. Relation of the biliary ducts to the blood vessels, as seen upon dissection of the anterior layer of the hepatoduodenal ligament. (From Anson, 1956; reproduced with the permission of the Quarterly Bulletin of Northwestern University Medical School.)

curvature, it usually divides into two parallel vessels, from which branches are sent to both the anterosuperior and posteroinferior surfaces of the stomach. The latter vessels anastomose with the short gastric branches derived from the splenic artery and with branches of the gastroepiploic arterial arch that ascend from the greater curvature of the stomach (Figs. 65a and 66).

The splenic (lienal artery) is the largest branch of the celiac. Accompanied by the splenic vein, it follows a surprisingly tortuous course behind the peritoneum along the upper border of the pancreas (Fig.

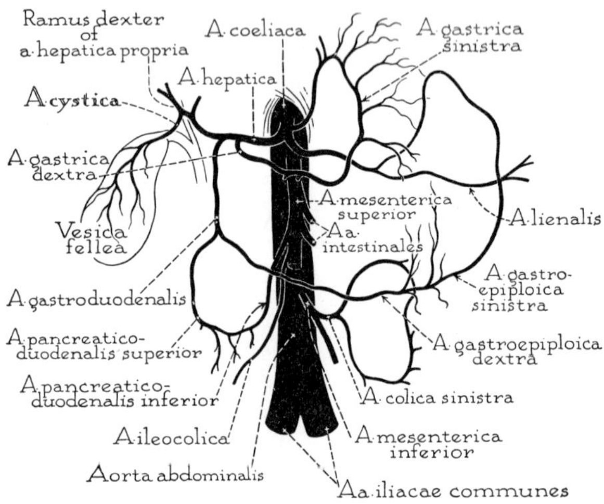

FIG. 66. Celiac artery, its aortic source and major branches. Semischematic. (From Anson, 1956; reproduced with the permission of the Quarterly Bulletin of Northwestern University Medical School.)

65a). It lies in front of the left suprarenal gland and the upper part of the left kidney and passes forward between the two layers of the phrenicolienal (lienorenal) ligament. Within the peritoneal reflection, it divides into five to eight terminal splenic branches which enter the hilum of the spleen to supply the splenic substance (for microscopic and submicroscopic anatomy of the splenic circulation, see Section C.-9, Chapter XII).

Numerous small pancreatic branches arise from the splenic artery. A larger branch is described as being occasionally present, which enters the upper border of the pancreas near the junction of its middle and left thirds. The pancreatic branches anastomose with one another and with branches of the pancreaticoduodenal arteries.

The short gastric arteries, four or five in number, arise either from the splenic artery or, more frequently, from some of the terminal branches of the latter (Fig. 65a). They pass between the layers of the gastrolienal (gastrosplenic) ligament to the left portion of the greater curvature of the stomach and anastomose with the esophageal, the left gastric, and the left gastroepiploic arteries.

The left gastroepiploic artery arises from the splenic (Fig. 66) or from one of its terminal branches and passes forward, between the layers of the gastrolienal ligament, to the left portion of the greater curvature of the stomach. Following this border, it is continued toward the right, between the layers of the greater omentum. It ends by anastomosing with the right gastroepiploic artery (Figs. 65a and 66). It gives off numerous branches to both surfaces of the stomach, which ascend, to anastomose with the short gastric and with branches of the left and right gastric arteries. Additionally, long, slender omental branches descend between the two ventral layers of the greater omentum (Fig. 65b).

The hepatic artery runs in retroperitoneal course along the upper border of the head of the pancreas to the first part of the duodenum, where, entering the lesser omentum, it ascends between its layers (Fig. 65). In the omentum it is accompanied by the portal vein and the common bile duct (*ductus choledochus*). Near the porta hepatis, it regularly divides into right and left terminal branches (for description of these vessels, see Sections B.-1 and C.-3, Chapter XII).

The gastroduodenal artery arises from the hepatic, where the source-vessel reaches the first part of the duodenum (Figs. 65b and 66), and descends behind the duodenum. In its course it lies between the neck of the pancreas and the first part of the duodenum and near the portal vein and the bile duct. The vessel ends at the junction of the duodenum and pancreas by dividing into the right gastroepiploic and superior pancreaticoduodenal arteries.

The right gastroepiploic artery is the larger of the two divisions (Figs. 65a and 66). It passes to the left, along the greater curvature of the stomach, between the ventral layers of the greater omentum, therein uniting with the left gastroepiploic branch of the splenic artery. From the arcade thus formed, branches pass upward on both surfaces of the stomach, to anastomose with branches of the right and left gastric arteries. Omental branches pass downward in the greater omentum, in series with those which are derived from the left gastroepiploic (Fig. 65b).

The right gastric artery is a small branch that arises as the hepatic enters the lesser omentum (Figs. 65a,b and 66). It runs between the layers of the lesser omentum to the pylorus and then along the lesser curvature of the stomach. It gives branches to both surfaces of the stomach and ends by anastomosing with the left gastric artery. It also gives a branch to the first part of the duodenum.

In a recent study of 100 consecutive specimens, it has been found that there are 34 types of branching in the vessels derived from the celiac trunk (Liechty and Anson, unpublished). The main artery itself may be absent, its branches arising separately from the aorta or from some other source. In those cases in which it gives off only two divisions, they are likely to be the left gastric and splenic. Rarely, it is the source of four branches, the additional one being either a second left gastric artery or a separate gastroduodenal artery. The left gastric artery is occasionally double; it may spring directly from the aorta, and it may give off the left hepatic or an accessory hepatic artery.

The splenic artery may arise from the middle colic, from the left hepatic, or from the superior or inferior mesenteric artery, while the hepatic artery may spring directly from the aorta or from the superior mesenteric artery. The left hepatic artery, or an accessory hepatic artery, arises occasionally from the left gastric artery and runs to the liver in the upper part of the lesser omentum. Accessory hepatic arteries are common; they originate from the superior mesenteric, the renal, or the inferior mesenteric artery. In some instances the accessory hepatic is as large as the regular vessel.

2. Blood Supply to the Small and Large Bowel[3]

Considerable extensive investigative work on the varied patterns of the arterial blood supply to the small and large bowel must yet be accomplished before a final, true concept of the variability of the arterial blood supply to these structures is attained. Of great surgical import is the conclusion reached by Sonneland *et al.* (1958) on the basis of 600 dissections, that the conventional textbook description of the blood supply to the colon from the superior mesenteric, through 3 branches (middle colic, right colic and ileocolic), occurs only in about one-fourth of the population. Furthermore, Robillard and Shapiro (1947) have stated that peritonitis due to dehiscence of enteroanastomoses is still a major cause of death in abdominal surgery and may often be accounted for by an inadvertently induced ischemic necrosis. The surgical hazards incident to the variational patterns of the arterial blood supply of the colon have recently been emphasized by Mayo (1955).

a. Superior mesenteric artery: This vessel supplies the third part of the duodenum, a portion of the head and, frequently, of the body of the

[3] By N. A. Michels.

pancreas, the small intestine (jejunum and ileum), the ascending colon and the transverse colon up to the splenic flexure. Typically it arises from the aorta at the vertebral level of L 1, but in many cases this may occur at the level of L 2. In rare instances, the vessel arises from the splenic or the celiac. Its caliber varies from 8 to 14 mm, the average being 10 mm (Kornblith, unpublished report). Its length to the junction point with the ileal branch of the ileocolic varies from 17 to 29 cm, with an average of 24 cm (Siddharth, unpublished report). It ends by uniting with the ileal branch of the ileocolic artery.

The constant branches of the superior mesenteric are: (1) inferior pancreaticoduodenal; (2) intestinal; (3) middle colic; (4) right colic; and (5) ileocolic. Inconstant branches are: dorsal pancreatic, transverse pancreatic, accessory middle colic, replaced or accessory right hepatic, and the entire hepatic trunk. In some instances, the splenic, gastroduodenal, right gastroepiploic and even the cystic may take origin from the superior mesenteric.

Predominantly (60%), the inferior pancreaticoduodenal artery is a single vessel arising from the right side of the superior mesenteric, and less frequently, from its left side or from the first or second jejunal branch. Coursing upwards behind the head of the pancreas, it divides into an anterior and posterior branch which, respectively, join the anterior and posterior pancreaticoduodenal arcades. In many instances (40%), each arcade joins the superior mesenteric separately via its own inferior pancreaticoduodenal. Occasionally, one or both arcades join an aberrant right hepatic stemming from the superior mesenteric and coursing behind the head of the pancreas.

Varying from 13 to 21 (Kornblith, unpublished report), the intestinal arteries arise from the left convex surface of the superior mesenteric, the jejunal group being much larger than the ileal. Radiating out into the mesentery, each vessel divides into two branches which anastomose with similar divisions from above and below. From the primary loops or arcades thus formed, secondary, tertiary, quaternary and even quinary loops are formed, their number being progressively increased at successive lower levels. Throughout the length of the small intestine, the most peripheral or terminal loop gives off a variable number of straight branches, the *vasa recta.* At the upper jejunal levels, the vasa recta are few in number and long, at lower ileal levels, numerous and short. As end arteries, the vasa recta, without intramesenteric communication, proceed to the gut wall, where some pass to the anterior or posterior side of the intestines, while others divide, sending branches to both sides. As mural trunks, the vasa recta ramify on the wall of the gut, either as single vessels with lateral branches or as rapidly forming groups of branches which subdivide in arborial fashion (Noer, 1943). The divisions of the mural trunks communicate with one another, thus forming a continuous network of delicate anastomosing arterial rings. Proximally, the network unites with the inferior pancreaticoduodenal, distally, with the ileocolic branch.

A possible etiologic relationship between the passage of blood vessels through the colonic wall and the development of colonic diverticula in diverticulosis has recently been discussed by Noer (1955). With latex injection studies, he showed that there was a marked concentration of blood vessels in the region of colonic diverticula, the size and distribution of the vessels in the wall of these structures being of such proportion as to make severe hemorrhage possible in the event of erosion or ulceration.

The middle colic artery arises from the concave side of the superior mesenteric, at the lower border of the pancreas, and passes into the right half of the transverse mesocolon. Here, at a variable distance from the gut wall (5–7 cm), it typically divides into two branches, one of which passes to the right, to anastomose with the ascending branch of the right colic artery, while the other turns to the left, to anastomose with the ascending branch of the left colic from the inferior mesenteric. Both main divisions of the middle colic undergo subsequent variant branching, forming primary and secondary arcades from which vasa recta are given off to the transverse colon. The vasa recta of the large bowel are fewer than those of the small intestine and far more precarious in surgical procedures, evulsion of more than two or three of them being sufficient to cause ischemic necrosis (Robillard and Shapiro). Instead of bifurcating, the middle colic may divide into three or more branches, a right one supplying the colic section when the right colic is absent (18%).

Of surgical import is the fact that in many instances variations may exist with regard to the middle colic artery. First, it may be absent, being replaced by a branch of the right or left colic (Steward and Rankin, 1933; Robillard and Shapiro, 1947; Sonneland *et al.*, 1958; Morgan and Griffiths, 1959; Quénu *et al.*, 1954). Secondly there may be an accessory middle colic (Steward and Rankin; Robillard and Shapiro; Sonneland *et al.*) arising, as a rule, from the aorta somewhat above the main middle colic, but also either from the trunk of the latter, the ascending branch of the left colic, or the dorsal pancreatic or transverse pancreatic. Finally, the middle colic may arise directly from the celiac or the aorta. In many instances, it gives rise to the dorsal pancreatic or transverse pancreatic before bifurcating, thus placing it in direct communication with the pancreatic blood supply (Michels, 1955).

As an independent branch of the superior mesenteric, the right colic artery is frequently absent (Steward and Rankin; Sonneland *et al.*). Under such circumstances, it is replaced by a branch from the middle colic or the ileocolic. The right colic, as it approaches the ascending colon, divides into two main branches: an ascending vessel that generally anastomoses very meagerly with a descending branch of the middle colic, and a descending one that anastomoses with the ascending or colic branch of the ileocolic artery.

The ileocolic artery usually arises separately from the superior mesenteric but may also do so via a common trunk with the right colic (33%, Steward and Rankin; 22.7%, Sonneland *et al.*). Its ascending division anastomoses with a descending branch of the right colic, while its descending one divides into the following vessels: colic, anterior cecal, posterior cecal, ileal, and appendicular.

The ascending colic branch constitutes the beginning of the marginal artery of Drummond (1913), which is continued with meager interruptions to the last sigmoid branch of the inferior mesentery artery, thereby affording a collateral circulation to various enteric segments, including the rectosigmoid junction. The anterior cecal branch courses through the superior cecal fold, while the posterior cecal passes behind the ileum. The ileal branch, running in the mesentery, anastomoses with the last intestinal branch of the superior mesenteric, forming a single or double arch from which vasa recta are supplied to the last six inches of the ileum.

Sites of origin of the appendicular artery are extremely varied. From an examination of 200 specimens, Anson (1950) noted the following different types of origin: (1) ileal branch of the ileocecal (35%); (2) the branching point of the ileocolic (28.5%); and (3) the anterior cecal (13.5%). Other less common sites were the posterior and anterior cecal, the right colic and the ileocolic. Double appendicular arteries occur frequently (30%), with one arising from the anterior, and the other, from the posterior cecal, or both may arise from either (Shah and Shah, 1946).

b. Inferior mesenteric artery: This vessel commonly arises from the aorta, 5–6 cm before its bifurcation point, at the level of the lower third of L 4. However, its vertebral level may vary from the middle of L 2 to the disk between L 3 and L 4, usually the middle of L 3 (Taniguchi, 1931). This vessel is much smaller than the superior mesenteric (3–5 mm, average, 4 mm—Kornblith, unpublished report).

The inferior mesenteric artery commonly divides into an ascending and descending branch in the left iliac fossa, at a variable distance from the gut wall (3–10 cm). The former vessel unites with the left descending branch of the middle colic, or in the rather frequent absence of the latter, it supplies the entire transverse colon as a very large branch, uniting ultimately with a right colic branch in those cases in which there is no middle colic present. The descending branch, after giving off two to four sigmoid arteries and a fairly constant rectosigmoid artery, becomes the superior rectal (hemorrhoidal) artery with average caliber of 3 mm (Kornblith). The latter anastomoses with the middle rectal artery from the hypogastric. An additional branch of the inferior mesenteric may effect an anastomosis with the middle colic, thus forming the arc of Riolan in the distal third of the transverse mesocolon. In rare instances, the inferior mesenteric gives rise to the middle colic (Robillard and Shapiro) or takes origin directly from the superior mesenteric (Balice, 1949; Ssoson-Jaroschewitsch, 1924). (For excellent papers on the varied blood supply of the distal colon and rectum, see Sunderland, 1942; Bacon and Smith, 1948; Goligher, 1949, 1954.)

3. Arterial Collateral Circulation of the Upper Abdominal Organs[4]

No other portion of the body presents more diversified routes of collateral blood supply to a specific region than the upper abdominal organs. For example, the pancreas lies in three interlocking arterial circles, formed by the hepatic, the splenic and the superior mesenteric arteries, none of which is the sole blood supply to the organ. The liver may have three separate sources of blood supply, namely, right, middle and left hepatic from the celiac, an accessory left hepatic from the left gastric, an accessory right hepatic from the superior mesenteric, a total of five supplying arteries. Invariably the stomach receives its main blood supply from six different arteries, and has a possible subsidiary supply from six other regional arteries. Due to the abundance of its blood vessels, the great omentum is exceptionally well adapted to function as a terrain of compensatory collateral circulation for both the liver and the spleen, when either the hepatic or the splenic artery is occluded.

Collateral arterial pathways to the upper abdominal organs, effected mainly through the splenic artery, comprise two types: (1) short ones, made by splenic branches communicating with themselves, such as union of the caudal pancreatic with the a. pancreatica magna; and (2) long ones, effected outside of the splenic system by regional neighboring arteries communicating with the splenic or its branches. This outside (extrasplenic) system of collateral pathways comprises the following twelve types:

[4] By N. A. Michels.

(1) The infragastric omental pathway is situated along the greater curvature of the stomach, in the anterior two layers of the great omentum, and is made by the union of the right and the left gastroepiploic.

(2) The supragastric omental pathway is situated along the lesser curvature and is made by the right and left gastric.

(3) The infracolic omental pathway, as the large epiploic arc of Barkow, is situated in the posterior layer of the great omentum below the transverse colon, and is formed by the left epiploic, derived from the left gastroepiploic, and the right epiploic, derived from the right gastroepiploic. The supply arteries in this collateral route consist of the hepatic, gastroduodenal, right gastroepiploic, right epiploic, left epiploic, left gastroepiploic and the inferior terminal of the splenic.

(4) The long transpancreatic pathway is routed through the entire length of the pancreas and is effected by the caudal pancreatic (an end branch of the splenic) uniting with the transverse pancreatic. The latter may communicate with the gastroduodenal, superior pancreaticoduodenal and right gastroepiploic and thus bring blood into the hepatic, or it may communicate with the superior mesenteric (in cases where it takes origin from this vessel), thereby effecting a collateral pathway which will function when either the hepatic or splenic is occluded.

(5) The pancreaticoduodenal route, a pathway outside of the channel of the transverse pancreatic, is effected by various branches of the gastroduodenal, the superior and the inferior pancreaticoduodenal uniting with branches of the splenic, especially those of the dorsal pancreatic, caudal pancreatic and large pancreatic arteries.

(6) The inferior coronary route is made by collaterals of the short gastrics from the splenic or its terminals uniting with branches from the infragastric omental route.

(7) The hepatogastric route consists of an arched anastomosis between the left gastric and the left hepatic. Cardioesophageal branches from the arc communicate with similar branches from the left gastric, short gastrics, left gastroepiploic and from the recurrent branch of the left inferior phrenic.

(8) The accessory gastrolienal pathway consists of short gastrics from the splenic terminals which unite with branches from an accessory left gastric coursing posteriorly along the fundic region of the stomach and arising either from the splenic, its terminal branches or the celiac.

(9) The accessory hepatolienal route consists of an accessory or a replaced right hepatic from the superior mesenteric, which is connected with the splenic, either through a branch of the dorsal pancreatic derived from the splenic, or through the transverse pancreatic and caudal pancreatic.

(10) The celiacomesenteric pathway consists of three routes. One is

a direct longitudinal communication between the splenic or the hepatic and the superior mesenteric, when the dorsal pancreatic (of splenic or hepatic origin) descends beyond the inferior border of the pancreas, to unite with the superior mesenteric, middle colic or an accessory middle colic. The second involves the anastomosis of the superior mesenteric with the celiac by a large solitary vessel coursing behind the pancreas. The final route consists of a communication between the superior mesenteric and the inferior pancreaticoduodenal, which, in turn, joins the anterior and posterior pancreaticoduodenal arcades. The latter connect with the gastroduodenal, which communicates with the splenic, via the right and left gastroepiploic, or with the celiac, via the common hepatic.

(11) The gastrolienophrenic collateral pathway may be effected by a direct communication between the cardioesophageal branches of the left inferior phrenic and similar branches stemming from the left gastric or its aberrant left hepatic branch; or by a connection between the short gastrics or an accessory left gastric from the splenic and the recurrent branch of the left inferior phrenic.

(12) The bifurcated circle pathway involves a hepatic arising separately from the aorta and giving rise to the gastroduodenal and a lienogastric trunk, both contributing equally to the blood supply of the upper abdominal organs. Collateral pathways between the two systems are effected by an anastomosis of the terminal branches of the right and left hepatics and by communications variously routed through the pancreatic, the gastric and the omental circles.

C. MICROSCOPIC ANATOMY

By E. E. SELKURT

1. MUCOSAL BLOOD SUPPLY OF THE STOMACH

a. Arterial circulation: The following description is based largely on the observations of Barlow *et al.* (1951), Barlow (1953), and Barclay and Bentley (1949). Branches from the arterial chains penetrate the muscle coat to the submucous layer where they form a main plexus of arteries of 200 μ diameter, the loops being connected by cross-anastomosing channels of 150 μ cross-section. From this submucosal plexus, a rich vascular network supplies the gastric mucosa. Thus, the latter literally lies in a vast vascular pool in which there exist channels capable of either bringing blood to this structure or transferring it rapidly to another point.

A series of separate branches from the main or subsidiary channels of the submucosal plexus takes an oblique course across the submucosa

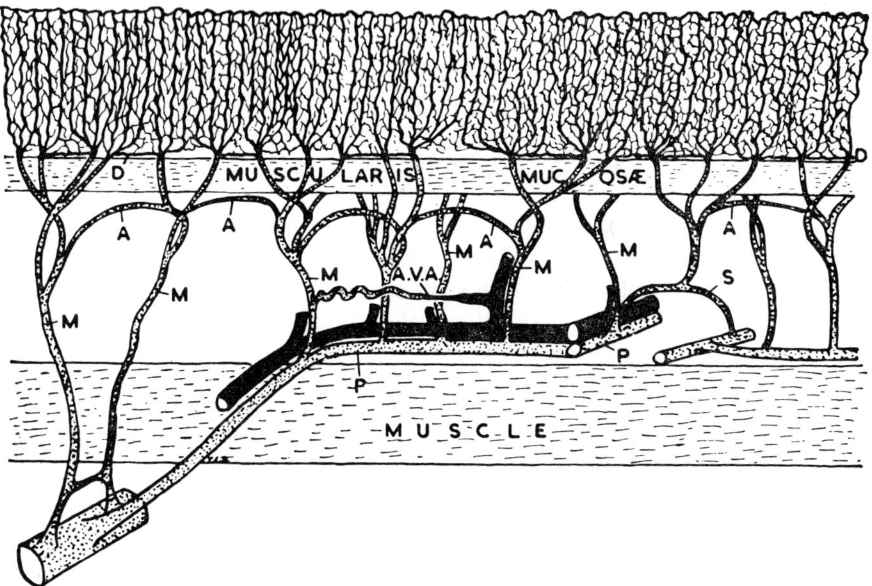

FIG. 67. Diagram of the vascular arrangement in the stomach wall. On the extreme left is shown the lesser curvature with mucosal arteries M arising from the left gastric artery outside the stomach wall. On the right, the pattern is the one seen in the anterior and posterior wall. In the center is shown an arteriovenous anastomosis (AVA) in relation to the other vessels. P: main supply channel of submucous plexus; S: anastomoses between arteries that pierce muscle; A: anastomoses between mucosal arteries on the under aspect of the muscularis mucosae; D: network of anastomotic channels on glandular aspect of muscularis mucosae from which capillaries of mucosa arise. (From Barlow *et al.*, 1951; reproduced with the permission of Surgery, Gynecology & Obstetrics.)

toward the muscularis mucosae (Fig. 67), the vessels being connected with slender anastomotic channels of 50 μ. Then, each divides into 3–4 branches which twist and coil upon each other before piercing the muscularis mucosae, about 90–120 perforating in every sq cm of muscularis mucosae.

Vessels of the mucosa form leashes of large capillaries about 20 μ in diameter. From these, smaller capillaries (8 μ) ramify among the glands like arborizations of a tree, joining to form loops (15 μ in diameter) around the opening of the glands.

In the region of the lesser curvature, mucosal arteries do not arise from a submucosal plexus in the stomach wall, as is the case with the rest of the stomach, but directly from the arterial chain within the lesser omentum outside the stomach. They therefore have to pierce the muscle coats of the stomach in their long passage from origin to distribution (Fig. 67). There is no submucous plexus in the lesser curvature region; the small communicating channels seen within the submucosa are the fine anastomotic branches which connect mucosal arteries in all parts of the stomach. Further distribution is similar to that above. Three to four small twisting branches pierce the muscle layer to form an anastomosing channel on its glandular aspect before splitting up into capillaries.

b. Venous circulation: The veins correspond to the arteries but are larger. In the mucosa, capillaries drain into large venules, 90 μ in diameter. Several of these unite to form mucosal veins, which run vertically through the mucosa to pierce the muscularis mucosae. The latter enter large veins which accompany mucosal arteries and drain to the wide venous channels of the submucous plexus.

c. Connective tissue plexus: A vast network of small arteries and veins (100 μ to capillary size) forms a complicated series of loops between the limbs and branches of the submucous plexus. The vessels do not penetrate the muscularis mucosae, are isolated from mucosal supply, and do not communicate with mucosal arteries. Their function is not known.

d. Arteriovenous anastomoses: These vessels usually arise from a mucosal artery of the direct channel type and appear to be controlled by muscle cells in the vessel wall. In this regard, at the junction with the branch from vein, a narrowed portion is seen with thickening of the wall (cells of musculoepithelial type), and the speculation is raised that this may be a control point. It has been calculated that the arteriovenous shunts may carry up to 1/20 of the blood supply to the stomach (Walder, 1953). An approximation of the caliber of the shunts was obtained by the injection of glass spheres into the arterial circulation; those of 40–140 μ were recovered on the venous side.

2. Blood Supply to the Small and Large Bowel

a. Arterial circulation: Arteriovenous anastomoses have been described in the villi of man and the rabbit but not in the dog (Jacobson and Noer,

1952). The mucosa is supplied from the submucosal vessels sweeping up the sides of the villi. The capillary network is immediately subepithelial, while arteries and veins are axial in position in the villus.

b. Venous circulation: Veins correspond to the arteries in distribution but generally are much larger. In the mucous membranes, blood from surface capillaries is collected by venules which enter mucosal veins. The latter then penetrate through the muscularis mucosa to large submucosal veins which drain into mesenteric veins supplying the portal vein.

D. PHYSIOLOGY

By E. E. Selkurt

1. Neurogenic Regulation of Flow to the Stomach[5]

Barlow *et al.* (1951) found that stimulation of the nerve bundle accompanying the left gastric artery along the lesser curvature caused a marked decrease in rate of perfusion and simultaneously, an increase in the outflow of spheres into the veins. A similar type of change was noted when adrenaline (10 μg/ml) was added to perfusing plasma. Thus, shunt flow was increased, while *total flow* was reduced. The opposite effect was observed with acetylcholine (10 μg/ml).

Walder (1952) also noted that adrenalin seemed to increase flow through arteriovenous anastomoses, while histamine had the opposite effect. However, it could not be decided if the response was one of variation in capillary resistance or of a particular controlling mechanism in the anastomoses. Arabehety *et al.* (1959) studied the sympathetic nerve influences on the circulation of the gastric mucosa of the rat by a technique of India ink deposition in the mucosal circulation. Splanchnic nerve blockade or section resulted in a more marked filling of the vessels, reflecting an increase in blood content. Thus, with removal of tonic sympathetic influences, there occurred a decrease in peripheral resistance, with smooth muscle relaxation of arterioles, metarterioles, and precapillary sphincters, and decreased vasomotion of the precapillaries.

[5] For a general discussion of sympathetic control of blood vessels, see Chapter IV.

Peters and Womack (1958) attempted to relate the control of the arteriovenous shunt mechanism to the production of gastric acid in dogs. When histamine was administered, the straight vessels to the villi coming off the submucosal plexus were generally filled and could be traced to the tip of the villi, where they looped around to return to the submucosa. This was correlated with increased gastric acid production. However, when adrenalin was given, the straight arteries either could not be observed at all, or when seen, they extended only to the base of the villi. Gastric acid production was depressed concomitant with constriction of these vessels. Simultaneously, the arteriovenous anastomoses could be observed, and glass beads up to 125 μ in size could be passed. With histamine, these shunts appeared to be closed, for glass beads larger than 25 μ could never be recovered.

Thompson and Vane (1953), in studying gastric blood flow and gastric secretion in cats, found that histamine significantly increased flow, and with it, gastric secretion, while superimposed adrenalin reduced both. Stimulation of the left celiac ganglion or splanchnic nerve also reduced the rate of histamine-induced gastric secretion and blood flow.

Supported by other physiologic studies, Peters and Womack concluded that when acid is secreted by the stomach, the capillary circulation of the mucosa is active; when it is not being formed, blood is shunted away through the arteriovenous anastomoses. Because acid production is an energy-requiring process, the resulting adjustment increases the necessary oxygen delivery to the acid producing cells. Alternatively, when the circulation is directed through the shunts, more arterialized blood enters the portal circuit to the liver, at a time when the need of this organ for oxygen increases because of the enhanced metabolism following the digestive and absorptive processes.

2. Neurogenic Regulation of Flow to the Small and Large Bowel[6,7]

a. Vasoconstrictor fibers: Many sympathetic fibers are distributed to the splanchnic area, exerting a tonic control over the vessels and providing possibilities of reflex regulation (Folkow, 1956). Stimulation of the splanchnic nerve decreases flow (Deal and Green, 1956), while ganglionic blocking agents (hexamethonium, chlorisondamine dimethochloride, and pentolinium tartrate) have the opposite effect (Trapold, 1956). Although both epinephrine and norepinephrine cause reduction in

[6] For a general discussion of sympathetic control of blood vessels, see Chapter IV.

[7] For a discussion of flow of lymph from intestines, see Section D.-2c, Chapter XXIII.

flow, it is more likely that the latter is the chemical mediator (Deal and Green, 1956).

b. Question of vasodilator fibers: It has been suggested that sympathetic vasodilator fibers are also distributed to the intestine, but at present the evidence is not conclusive. The small vasodilatation produced by splanchnic stimulation after the blockade of vasoconstrictor fibers may be a hemodynamic artifact, since sympatholytic drugs do not generally block the inhibitory action of splanchnic stimulation on the intestinal smooth muscles. Muscular relaxation may decrease intravascular tension and hence increase flow. Atropine is without effect on such apparent dilator responses, arguing for the absence of cholinergic fibers. The absence of specific vasodilator fibers does not eliminate the possibility of dilator reflexes, however. For example, a rise in inferior vena caval pressure has been shown to cause a reflex dilatation of the intestinal veins (Alexander, 1956). This response appears to be initiated by fibers ascending the vagus nerve, producing a reduction in sympathetic tone to the venous musculature.

3. Reflex and Humoral Regulation of Flow to Small and Large Bowel

a. Chemical agents: Chemoreceptor stimulation (carotid and aortic) by inhalation of 1.0 per cent oxygen in nitrogen causes a marked reduction in superior mesenteric flow in dogs (Bernthal and Schwind, 1945). When Hering's nerves and the vasosympathetic trunks are blocked by narcosis, this reflex is abolished.

An interesting relationship between skin temperature and colonic blood flow has been observed (Grayson, 1953). Using thermoelectric methods, with thermocouples placed into colostomies, caecostomies, and ileostomies, or into the rectum, cutaneous cooling of the lower extremities was observed to cause an increase in intestinal flow in excess of that anticipated on the basis of the elevation in blood pressure (cold pressor response); this response was interpreted as indicating active hyperemia. Body heating, causing vasodilation of the extremities, produced vasoconstriction of intestinal blood flow. Such evidence suggests a reciprocal interplay of the skin and intestinal blood reservoirs, which would aid in maintaining the homeostasis of the systemic blood pressure.

b. Effects of other agents: Epinephrine and norepinephrine, injected intra-arterially, produce vasconstriction (Deal and Green, 1956). Serotonin is a dilator in small dosages and a constrictor in large, while ATP, adenyllic acid, adenosine, and Substance P (a polypeptide isolated from the intestine) produce hyperemia (Selkurt *et al.*, 1958). Carbon dioxide is also an hyperemic agent (Mohamed and Bean, 1951; Brickner *et al.*, 1956). Low oxygen blood, perfusing an isolated (except for extrinsic nerves) dog intestinal loop, likewise results in a marked increase in flow, in absence of changes in tonus or motility, as does histotoxic anoxia

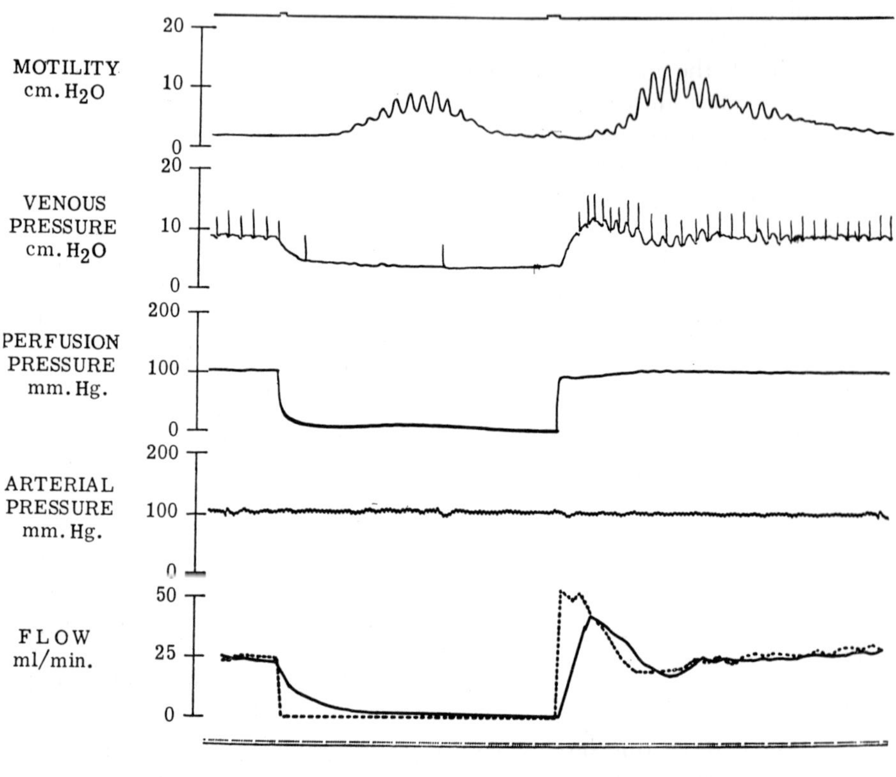

FIG. 68. Reactive hyperemia resulting from 2 min of arterial occlusion (isolated canine ileal loop). Flow curves have been reconstructed from original flow data: *solid:* venous outflow; *dashed:* arterial inflow. Note persistence of venous flow after occlusion of the artery, an emptying phenomenon. Although both arterial flow and venous flow are enhanced after release of occlusion, arterial inflow is initially greater as a result of re-filling of the vascular channels.

(NaCN) (Bean and Sidky, 1957). (For further discussion of responses of mesenteric vessels to pharmacologic agents, see Section E., this chapter.)

c. Effect of local anoxia: Reactive hyperemia in response to short bouts of complete ischemia can be readily demonstrated in the intestinal circulation, as illustrated in Fig. 68, based on an experiment done on an isolated segment of canine ileum. On the other hand, according to Turner *et al.* (1959), following relatively long periods of ischemia (artery or vein alone, or both clamped 30 min to 4 hr), reactive hyperemia (as measured in isolated loops by a gravimetric method) does not occur. Instead, reductions in flow follow release (particularly when the vein alone is occluded). Figure 69 shows an experiment in which intestinal ischemia of two hours' duration was performed. It will be noted

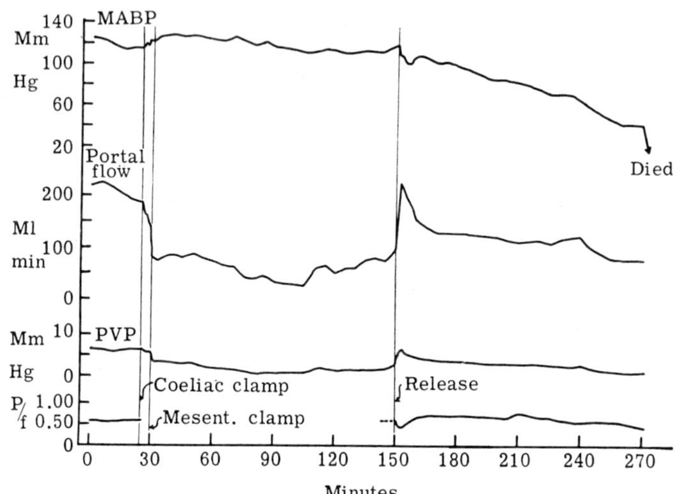

FIG. 69. Response to 2 hours of splanchnic ischemia in the dog. Occlusion of superior mesenteric artery at 30 min, and branches of the celiac artery supplying the gastrointestinal tract at 26 min. Hepatic artery supply left intact. Upper tracing: MABP, mean arterial blood pressure (carotid) in mm Hg: second: portal flow in ml/min (measured by rotameter in a portal vein-jugular vein shunt); third: PVP, portal venous pressure in mm Hg; fourth: P/F: vascular resistance in splanchnic bed as a ratio of A-V pressure over portal flow.

that although portal vein flow transiently increased after release, it soon fell because of decreasing arterial pressure and elevated vascular resistance. Dogs subjected to intestinal ischemia of two hours' duration regularly die in shock (Selkurt, 1959).

4. Hemodynamic Characteristics of Intestinal Blood Flow

Figure 70 demonstrates the relationship of intestinal blood flow in the dog to a wide range of perfusion pressures (Selkurt *et al.*, 1958), the experiments being performed in isolated, denervated loops of ileum. The pressure-flow relationship is evidently not linear, flow increasing progressively at a faster rate than the perfusion pressure. Two factors contribute to this: a passive enlargement in vessel caliber as intraluminal pressure is raised and a decrease in the apparent blood viscosity (so-called anomolous viscosity) at higher flow velocities. Flow ceases at about 15 mm Hg pressure, possibly the result of "critical closure" of the vessels. (For further discussion of the viscous properties of blood, see Section D.-1, Chapter II.)

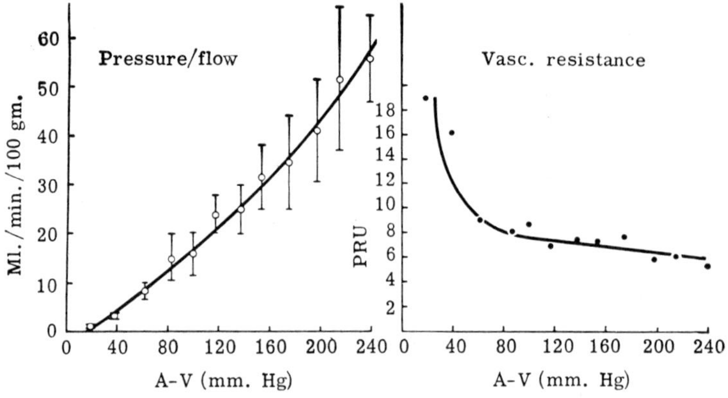

FIG. 70. Relationship of flow (left) and vascular resistance (right) in relation to perfusion pressure (A-V, mm Hg). Resistance is given in PRU (peripheral resistance units) (1 ml flow/1 mm Hg). Vertical lines (left) are the standard errors of the means of observations made in 20 mm Hg intervals of pressure (A-V). Results from 18 experiments. (From Selkurt *et al.*, 1958; reproduced with the permission of Circulation Research.)

To the right of Fig. 70, intestinal vascular resistance (given as P.R.U., peripheral resistance units) is related to the A-V pressure gradient. It is seen that this is also nonlinear, in nonconformity with Poiseuille's Law. This emphasizes the necessity of interpreting humoral and neurogenic influences on flow on the basis of a control pressure-flow curve, so that proper hemodynamic deductions can be made. Ideally, in order to quantitate vasomotor changes, experimental flow should be related to the control flow at the same perfusion pressure. To study neurogenic influences, preparations with intact innervation should, of course, be used.

Figure 70 also supplies information regarding the quantitative aspects of intestinal flow in the denervated preparation. However, denervation apparently has a small effect on flow (Lawson, 1941). Thus, at a normal arterial perfusion pressure, blood flow through intestinal loops of dogs averaged 36 ml/min/100 gms for the intact preparation, as compared to 42 ml/min for the denervated one.

5. Factors Modifying Intestinal Flow

a. Intestinal distention: Noer *et al.* (1951), studied the effect of acute distention in rabbits because of the anatomic similarity of the vascular supply to that of man. Blood flow was observed with the illuminated quartz rod technique during balloon distention. Slowing or stoppage of flow occurred first in venules (at 15 mm Hg pressure), then in small veins (at 20 mm) and venous capillaries (at 30 mm), followed by such a change in arterioles, small arteries and arterial capillaries, in that order. Blood flow stopped in most vessels at 50–70 mm Hg except for slight movement in mural veins and arteries. Distending pressures of 30–40 mm Hg created disturbances which were completely reversible, but ones of 50 mm Hg or above produced irreversible alterations in the mural circulation and intestinal wall, with permanent interruption of flow in many small vessels, interstitial hemorrhage, and pronounced intravascular agglutination of cells.

b. Blockage of veins: With such a change as the first event, increased capillary pressure causes leakage of plasma into the wall of the gut and into the lumen. If excessive, this will lead to shock, as often occurs with intestinal obstruction. Elevated venous pressure reduces intestinal blood flow (Selkurt and Johnson, 1958). The seriously impaired arterial circulation at the higher distending pressures promotes breakdown of the wall of the gut, furthered by proteolytic digestion of tissue and leading ultimately to gangrene and rupture. Distention of the dog intestine to 30 cm of water has been shown to reduce intestinal oxygen consumption, probably basic to these changes (Lawson and Ambrose, 1942).

6. Splanchnic Blood Volume

Delorme *et al.* (1951) reported that the visceral blood volume in the dog, based upon an exclusion technique utilizing P^{32}, was between 20 and 50 per cent of total blood volume (av 35%). The average volume was found to be 533 ml, but with a wide scatter (150–1400 ml). The volume of the mesenteric (intestinal) circuit averaged 56 per cent (range, 24–70%) of the total splanchnic volume. The remainder was roughly equally proportioned between the liver and the spleen, with somewhat more in the liver. Horvath *et al.* (1957), employing I^{131} albumin in a similar method, found the splanchnic blood volume of dogs to average only 21 per cent (11–42%) of total blood volume (av of 277 ml), with 6.2 per cent in the hepatic circuit, 6.4 per cent in the splenic circuit, and 8.7 per cent in the mesenteric circuit. Their estimates seem low in view of Friedman's (1954) report in mice of a splanchnic blood volume that was 30 per cent of the total volume.

E. PHARMACOLOGY

By K. I. Melville and H. E. Shister

1. Adrenergic Drugs

Epinephrine and norepinephrine show about equal potency in their constrictor effect on the mesenteric vascular bed of dogs. In man, intravenous epinephrine (0.1 μg/kg/min) decreases, while norepinephrine increases, the volume of the splanchnic vascular bed. Both amines produce constriction of the blood vessels of the colon. It has also been claimed that small amounts of epinephrine and norepinephrine dilate the intestinal vessels, but this is denied by other workers (see review by Green and Kepchar, 1959).

By eliminating the complicating factor of changes in the intestinal volume, direct measurements of blood flow can be obtained and vasoconstriction alone observed. Thus, intra-arterial injections of epinephrine in the mesenteric artery always causes vasoconstriction, and this is proportional to the dosage used (1–10 μg range). This effect is, however, often followed by moderate vasodilatation. Intravenous infusion of epinephrine can produce similar but less striking results. The vasoconstrictor phase appears to be associated with a reduced venous outflow, although there may be some initial increased flow. Norepinephrine has the same effect, but the secondary dilatation is not so marked. Isoproterenol (1–10 μg), given intra-arterially, causes marked vasodilatation.

2. Blocking Agents

Direct observations of mesenteric blood vessels in the mesoappendix of the rat have shown that the topical application of adrenergic blocking agents produces slowing and eventual complete cessation of blood flow and closure of the precapillary sphincters. These drugs also convert epinephrine constriction to dilatation. The constrictor response of norepinephrine is also abolished, but no vasodilator effect is revealed. This has likewise been shown to be true of ergotoxin, ergotamine, Dibenamine and azapetine (Ilidar) (see review of Green and Kepchar, 1959).

3. Other Agents

a. Cholinergic drugs: Acetylcholine (Ach) administered intra-arterially (0.1 μg) can increase venous outflow from intestines, but there is no evidence available of any cholinergically mediated nerve influences in the arteries (Celander and Folkow, 1951; Folkow, 1955). The cholinergic effect on the intestinal vessels requires further study.

b. Nitrites and nitrates: Experimentally these drugs produce dilatation of the splanchnic vessels with increased outflow from the splanchnic veins. However, they also depress intestinal smooth muscle.

4. Venous-Arterial Relationship

In evaluating the vascular effects of drugs, one must bear in mind the associated mechanical changes. In this regard, recently an increase in the vascular resistance of the intestines has been reported to occur with raised venous pressure. This venous-arterial response is apparently not due to either mechanical factors (encroachment of capacity vessels on resistance vessels, accumulation of interstitial fluid) or extrinsic nerve activity. Whether it is dependent upon some local reflex (Selkurt and Johnson, 1958) or upon the transmission of the increased venous pressure through the capillaries to the arterioles (Johnson, 1959) is not clear. (For a complete review of the complex problem of the control of resistance in the gastrointestinal vascular bed, see Green and Kepchar, 1959.)

References

Alexander, R. S. (1956). Reflex alterations in venomotor tone produced by venous congestion. *Circulation Research* **4**, 49.

Anson, B. J. (1950). "Atlas of Human Anatomy." Saunders, Philadelphia, Pa.

Anson, B. J. (1956). Anatomical considerations in surgery of the gallbladder. *Quart. Bull. Northwestern Univ. Med. School* **30**, 250.

Arabehety, J. T., Dolcini, H., and Gay, S. J. (1959). Sympathetic influences on the circulation of the gastric mucosa of the cat. *Am. J. Physiol.* **197**, 915.

Bacon, H. E., and Smith, C. H. (1948). The arterial supply of the distal colon pertinent to abdominoperineal proctosigmoidectomy with preservation of the sphincter mechanism. *Ann. Surg.* **127**, 28.

Balice, G. (1949). L'anatomia chirurgica dell'arteria mesenterica inferiore ed il punto del Sudeck. *Giorn. ital. chir.* **5**, 1.

Barclay, A. E., and Bentley, F. H. (1949). Vascularization of the human stomach: preliminary notes on shunting effect of trauma. *Brit. J. Radiol.* **22**, 62.

Barlow, T. E. (1953). Vascular patterns in the alimentary canal. *In* "Visceral Circulation," Ciba Foundation Symposium (G. E. W. Wolstenholme, ed.), Part I, p. 21. Little, Brown, Boston, Mass.

Barlow, T. E., Bentley, F. H., and Walder, D. N. (1951). Arteries, veins, and arterio-venous anastomoses in the human stomach. *Surg., Gynecol. Obstet.* **93**, 657.

Bean, J. W., and Sidky, M. M. (1957). Effects of low O_2 on intestinal blood flow, tonus and motility. *Am. J. Physiol.* **189**, 541.

Bernthal, T. G., and Schwind, F. J. (1945). A comparison in intestine and leg of the reflex vascular response to carotid-aortic chemoreceptor stimulation. *Am. J. Physiol.* **143**, 361.

Brickner, E. W., Dowds, E. G., Willits, B., and Selkurt, E. E. (1956). Mesenteric blood flow as influenced by progressive hypercapnia. *Am. J. Physiol.* **184**, 275.

Cauldwell, E. W., and Anson, B. J. (1943). The visceral branches of the abdominal aorta: topographical relationships. *Am. J. Anat.* **60**, 27.

Celander, O., and Folkow, B. (1951). Are parasympathetic vasodilator fibers involved in depressor reflexes elicited from the baroceptor regions? *Acta Physiol. Scand.* **23**, 1.

Deal, C. P., Jr., and Green, H. D. (1956). Comparison of changes in mesenteric resistance following splanchnic nerve stimulation with response to epinephrine and norepinephrine. *Circulation Research* **4**, 38.

Delorme, E. J., Macpherson, A. I. S., Mukherjee, S. R., and Rowlands, S. (1951). Measurement of visceral blood volume in dogs. *Quart. J. Exptl. Physiol.* **36**, 219.

Drummond, H. (1913). The arterial supply of the rectum and pelvic colon. *Brit. J. Surg.* **1**, 677.

Folkow, B. (1955). Nervous control of the blood vessels. *Physiol. Revs.* **35**, 629.

Folkow, B. (1956). Peripheral circulation. *Ann. Rev. Physiol.* **18**, 159.

Friedman, J. J. (1954). Organ plasma volume of normal unanesthetized mice. *Proc. Soc. Exptl. Biol. Med.* **88**, 323.

Goligher, J. C. (1949). The blood supply to the sigmoid colon and rectum. *Brit. J. Surg.* **37**, 157.

Goligher, J. C. (1954). The adequacy of the marginal blood supply to the left colon after high ligation of the inferior mesenteric artery during excision of the rectum. *Brit. J. Surg.* **41**, 351.

Grayson, J. (1953). Observation on blood flow in the human intestine. *In* "Visceral Circulation," Ciba Foundation Symposium (G. E. W. Wolstenholme, ed.), Part IV, p. 236. Little, Brown, Boston, Mass.

Green, H. D., and Kepchar, J. H. (1959). Control of peripheral resistance in major systemic vascular beds. *Physiol. Revs.* **39**, 617.

Horvath, S. M., Kelly, T., Folk, G. E., Jr., and Hutt, B. K. (1957). Measurement of blood volumes in the splanchnic bed of the dog. *Am. J. Physiol.* **189**, 573.

Jacobson, L. F., and Noer, R. J. (1952). The vascular pattern of the intestinal villi in various laboratory animals and man. *Anat. Record* **114**, 85.

Johnson, P. C. (1959). Myogenic nature of increase in intestinal vascular resistance with venous pressure elevation. *Circulation Research* **7**, 992.

Keibel, F., and Mall, F. P. (1912). "Manual of Human Embryology," Vol. II. Lippincott, Philadelphia, Pa.

Lawson, H. (1941). The mechanism of deflation hyperemia in the intestine. *Am. J. Physiol.* **134**, 147.

Lawson, H., and Ambrose, A. M. (1942). Utilization of blood O_2 by the distended intestine. *Am. J. Physiol.* **135**, 650.

Mayo, C. W. (1955). Blood supply of the colon. Surgical considerations. *Surg. Clin. North Am.* **35**, 1117.

Michels, N. A. (1955). "Blood Supply and Anatomy of the Upper Abdominal Organs, with a Descriptive Atlas." Lippincott, Philadelphia, Pa.

Mohamed, M. S., and Bean, J. W. (1951). Local and general alterations of blood CO_2 and influence of intestinal motility in regional regulation of intestinal blood flow. *Am. J. Physiol.* **167**, 413.

Morgan, C. W., and Griffiths, J. D. (1959). High ligations of the inferior mesenteric artery during operations for carcinoma of the distal colon and rectum. *Surg., Gynecol. Obstet.* **108**, 641.

Noer, R. J. (1943). The blood vessels of the jejunum and ileum. A comparative study of man and certain laboratory animals. *Am. J. Anat.* **73**, 293.

Noer, R. J. (1955). Hemorrhage as a complication of diverticulitis. *Ann. Surg.* **141**, 674.

Noer, R. J., Robb, H. J., and Jacobson, L. F. (1951). Circulatory disturbances produced by acute intestinal distention in the living animal. *A. M. A. Arch. Surg.* **63**, 520.

Peters, R. M., and Womack, N. A. (1958). Hemodynamics of gastric secretion. *Ann. Surg.* **148**, 537.

Quénu, L., Chabrol, J., and Herelemont, P. (1954). Le colon, ses variations, ses artères. *Compt. rend. assoc. anastomistes* **41**, 12.

Robillard, G. L., and Shapiro, A. L. (1947). Variational anatomy of middle colic artery; its significance in gastric and colonic surgery. *J. Intern. Coll. Surgeons* **10**, 157.

Selkurt, E. E. (1959). Intestinal ischemic shock and the protective role of the liver. *Am. J. Physiol.* **197**, 281.

Selkurt, E. E., and Johnson, P. C. (1958). Effect of acute elevation of portal venous pressure on mesenteric blood volume, interstitial fluid volume, and hemodynamics. *Circulation Research* **6**, 592.

Selkurt, E. E., Scibetta, M. P., and Cull, T. E. (1958). Hemodynamics of intestinal circulation. *Circulation Research* **6**, 92.

Shah, M. A., and Shah, M. (1946). The arterial supply of the vermiform appendix. *Anat. Record* **95**, 457.

Sonneland, J., Anson, B. J., and Beaton, L. E. (1958). Surgical anatomy of the arterial supply to the colon from the superior mesenteric artery based on a study of 600 specimens. *Surg., Gynecol. Obstet.* **106**, 385.

Ssoson-Jaroschewitsch, A. J. (1924). Zur chirurgischen Anatomie der A. mesenterica inferior. *Langenbecks Arch. klin. Chir.* **129**, 178.

Steward, J. A., and Rankin, F. W. (1933). Blood supply of the large intestine; its surgical considerations. *A. M. A. Arch. Surg.* **26**, 843.

Sunderland, S. (1942). Blood supply of the distal colon. *Australian New Zealand J. Surg.* **11**, 253.

Taniguchi, T. (1931). Beitrag zur Topographie der grossen Äste der Bauch aorta. *Folia Anat. Japon.* **9,** 201.

Thompson, J. E., and Vane, J. R. (1953). Gastric secretion induced by histamine and its relationship to the rate of blood flow. *J. Physiol.* (*London*) **121,** 433.

Trapold, J. H. (1956). Effect of ganglionic blocking agents upon blood flow and resistance in superior mesenteric artery of dog. *Circulation Research* **4,** 718.

Turner, M. D., Neely, W. A., and Barnett, W. O. (1959). The effects of temporary arterial, venous, and arteriovenous occlusion upon intestinal blood flow. *Surg., Gynecol. Obstet.* **108,** 347.

Walder, D. N. (1952). Arteriovenous anastomoses of the human stomach. *Clin. Sci.* **11,** 59.

Walder, D. N. (1953). Some observations on the blood flow of the human stomach. *In* "Visceral Circulation," Ciba Foundation Symposium (G. E. W. Wolstenholme, ed.), Part III, p. 210. Little, Brown, Boston, Mass.

Chapter XII. HEPATOPORTAL-LIENAL CIRCULATION

A. EMBRYOLOGY

By K. Christensen

Among the early investigators, His (1885) and Mall (1906) described the fundamental developmental characteristics of the hepatoportal venous system.[1] Streeter (1942) also studied early embryonic stages, while the embryonic differentiation which results in the adult pattern has been reinvestigated by Dickson (1957) in a study of the ductus venosus.

1. Embryonic Primordia

The earliest stages of development of hepatic vessels are found in human embryos at or before the 3 mm stage; by the 9 mm stage the differentiated pattern is completed. The primordial venous channels include: (1) the primitive vitelline venous plexus which carries blood from the yolk sac to the embryo; (2) the intrahepatic sinusoids which appear with the development of the liver; and (3) a venous plexus which forms along the developing gut (duodenum). Portions of the umbilical (allantoic) veins are intimately associated with the venous changes related to the liver.

2. Early Vasculogenesis

a. Vascular plexuses: Early vasculogenesis in the hepatic region is characterized by the appearance of a series of vascular plexuses from which the following primordial vascular channels are derived:

The vitelline plexus appears in the splanchnic mesoderm of the yolk sac and proceeds intraembryonically to join the heart at the sinus venosus. At the earliest sites of angiogenesis, blood islands develop in the splanchnic mesoderm; from these, vascular spaces are formed, and finally the vascular plexus is established. In 10– to 17–somite embryos (2.5 mm–4.3 mm) Streeter (1942) identified the intraembryonic vitelline venous plexus. In 20– to 24–somite embryos, by growth and adjustment

[1] For reviews of this subject see Evans (1912) and Lewis (1912).

within the vitelline plexus, enlarged channels form the vitelline veins which communicate with the sinus venosus of the heart.

The second vascular plexus appears during the development of the liver. In 25- to 30-somite embryos (3 mm–4 mm), downgrowths of epithelium from the duodenum form branching and anastomosing laminae of epithelial cells of the liver primordium. The hepatic laminae invade the ventral mesentery and the septum transversum, both of which provide the supporting tissues for the developing liver and the vascular tissues for the primordial sinusoids. Streeter (1942) is of the opinion that much of the primitive sinusoidal network develops *in situ,* beginning with the formation of blood islands. At the latter stage, he regards the liver as the most precocious organ of the digestive tube and suggests that angiogenesis in this region is stimulated by the developing hepatic laminae. During sinusoidal development, the channels of the vitelline veins on each side of the liver are tapped by the advancing hepatic tissues, and thereafter portions of the veins become included in the sinusoidal network. From this stage on, the vascular channels of the liver intervene between the remaining proximal and distal parts of the vitelline veins, and communication between them is only by way of the sinusoids.

The third venous plexus in the hepatic region develops about the emerging gut (duodenum), which differentiates as the yolk sac connection recedes. The enteric vascular plexus along the digestive tube is of importance in the establishment of paired embryonic veins on either side of the duodenum and for the cross anastomoses which develop between these vessels.

b. Symmetrical stage of development: Dickson (1957) refers to the 5 mm embryonic stage as the symmetrical stage of hepatic portal development because of the nearly complete bilaterality of all venous primordia. The original right and left vitelline veins connecting the liver with the heart have been named the hepaticocardiac channels by Streeter or the primitive right and left primary hepatic veins by Dickson. The caudal portions of the original right and left vitelline veins which enter the liver may properly be called the omphalomesenteric veins, because they now include the paired venous channels on either side of the developing gut. Between these vessels are four instead of three venous cross anastomoses (Dickson). The first, heretofore not described, is the subdiaphragmatic anastomosis between the right and left primary hepatic veins. The second anastomosis is the subhepatic, which joins the omphalomesenteric veins ventral to the duodenum, while the third is a dorsal anastomosis between these veins above the duodenum. The last and the most caudal is a similar but poorly differentiated anastomosis below the duodenum. At this symmetrical stage, the right umbilical vein still enters the sinus venosus near the right common hepatic vein, while the left umbilical vein has already lost its connections with the sinus venosus and opens into the liver near the entrance of the left omphalomesenteric vein. Functionally, this is important, because according to Dickson, the ductus venosus is already established as a midline shunt for transporting left umbilical venous blood between the subhepatic and the subdiaphragmatic anastomoses.

3. Differentiation of Hepatoportal System

In the 7.75 mm human embryo definitive changes of hepatoportal development have begun, and by the 9 mm stage they are essentially complete. Craniad, the right half of the subdiaphragmatic anastomosis and the right primary hepatic vein contribute to the definitive right common hepatic vein. Above the subdiaphragmatic anastomosis, the left primary hepatic vein, entering the sinus venosus from the liver, and the left omphalomesenteric vein, as far caudal as the dorsal anastomosis, have disappeared. As a replacement for the degenerated hepatic vein on the left side, a new left superior cranial hepatic vein is formed from the left half of the subdiaphragmatic anastomosis and several small new vessels which fuse with it.

The primordial veins which are retained in the formation of the hepatic portal vein are: (1) the persisting right omphalomesenteric vein that extends rostrally from the dorsal anastomosis to the liver; (2) the entire dorsal anastomosis; and (3) the left omphalomesenteric vein which is caudal to the dorsal anastomosis. The circuit for the hepatic portal vein thus established does not spiral around the duodenum, but is short and straight (Hamilton *et al.*, 1952); hemodynamically it satisfies the natural tendency for blood to seek the most direct route of flow. The superior mesenteric vein and the splenic veins are new vessels which arise *in situ* and join the left omphalomesenteric vein near the dorsal anastomosis. The remaining parts of the omphalomesenteric system degenerate. At the entrance to the liver the portal vein bifurcates into the right and left portal branches.

The channels for blood from the allantois to the heart now include (1) the left umbilical vein below the subhepatic anastomosis; (2) the left half of the subhepatic anastomosis; (3) the ductus venosus; (4) the right half of the subdiaphragmatic anastomosis; and (5) the common right hepatic vein. The right half of the subhepatic anastomosis connects the portal circulation in the embryo with the ductus venosus. This communication in the 9 mm embryo forms the sinus intermedia. At the 9 mm stage the differentiating inferior vena cava has reached the heart through the development of an intrahepatic segment and the utilization of the right hepaticocardiac channel. The right common hepatic vein from the liver makes its connection with the inferior vena cava at this stage.

After birth, the umbilical circulation degenerates. The sinus intermedia remains as a fibrous strand attached to the portal vein; the ductus venosus becomes the ligamentum venosum; and the left umbilical vein becomes the ligamentum teres.

4. Anomalies

Anomalies of the portal vein are rare. Cases reported include differences in relations between umbilical and portal circulations held over from fetal life and patency of some of the umbilical vessels. In another type of case, parts of the omphalomesenteric vessels, which usually degenerate, are retained in part and establish a duplication of vessels.

5. Clinical Implications

To the surgeon probably the most significant results of anomalous differentiation are the variations in major branches of the portal vein and the variability of their respective branches. (For representative papers describing anomalies involving the portal vein and the variations of its branches, see Douglass *et al.*, 1950; Fraser and Brown, 1944; Gilfillan, 1950; Hochstetter, 1886; Pernkopf, 1932.)

Of even greater interest developmentally are the numerous anastomoses between the portal vein and both the superior and the inferior vena caval systems. The interrelation of embryonic venous plexuses, making possible the anastomoses significant in disease or in surgical procedures involving the portal vein, is a remarkable feature of the primordial circulation.

B. GROSS ANATOMY

By N. A. Michels

1. Extrahepatic Distribution of Blood Vessels

a. Common hepatic artery: The blood supply to the liver comes from a common hepatic artery, which is derived from the celiac, along with the splenic and left gastric. In about 55 per cent of the population, the common hepatic artery, after giving off the gastroduodenal, divides into a right, a left and a middle hepatic, the latter supplying the quadrate lobe (Fig. 71A). In the remaining cases, these vessels are aberrant, being derived from a source other than the celiac. Under these circumstances, the right hepatic arises predominantly from the superior mesenteric; the left hepatic, from the left gastric. Aberrant hepatics comprise two types: replaced (substitutive) and accessory (additive). No aberrant hepatic artery, whether replaced or accessory, can be dispensed with, for, such a vessel supplies a definite liver region: the replaced variety, an entire lobe, the accessory, a section of a lobe (Healey and Schroy, 1953).

Replaced hepatics are formed when the celiac hepatic divides only into the gastroduodenal and left hepatic, the celiac right hepatic falling out and being replaced from the superior mesenteric. Conversely, when the celiac hepatic divides into the gastroduodenal and right hepatic, the left hepatic is replaced from the left gastric. Finally, when it divides into the gastroduodenal and middle hepatic, both right and left hepatics are replaced from other sources.

Accessory hepatics are those which supply specific regions of the liver but which do not receive a blood supply from the main hepatics. Such vessels may be derived from various sources, such as the superior mesenteric, left gastric, aorta and gastroduodenal. Rarely, the entire blood supply to the liver may come from the left gastric (Fig. 72D).

b. Right hepatic artery: Typically, this vessel arises from the common hepatic to the left of the hepatic duct, which, in most cases (85%) it crosses dorsally; purely ventral crossings occur in about 12 per cent. In the cystic triangle of Calot, formed by the angular union of the hepatic and cystic ducts, the right hepatic, in many instances, divides into two branches which supply the anterior and posterior segments of the right lobe of the liver, respectively. Division of the artery may occur early, in which case both branches cross the hepatic duct ventrally or dorsally, or they may have the hepatic duct between them. Of surgical importance is the fact that both inside and outside of the cystic triangle, the right hepatic makes characteristic caterpillar-like loops with convexity that points either upward, downward, to the left, or to the right. Such plexus formations are extremely vulnerable in cholecystectomy, for the cystic artery may arise from the proximal or distal end of the loop.

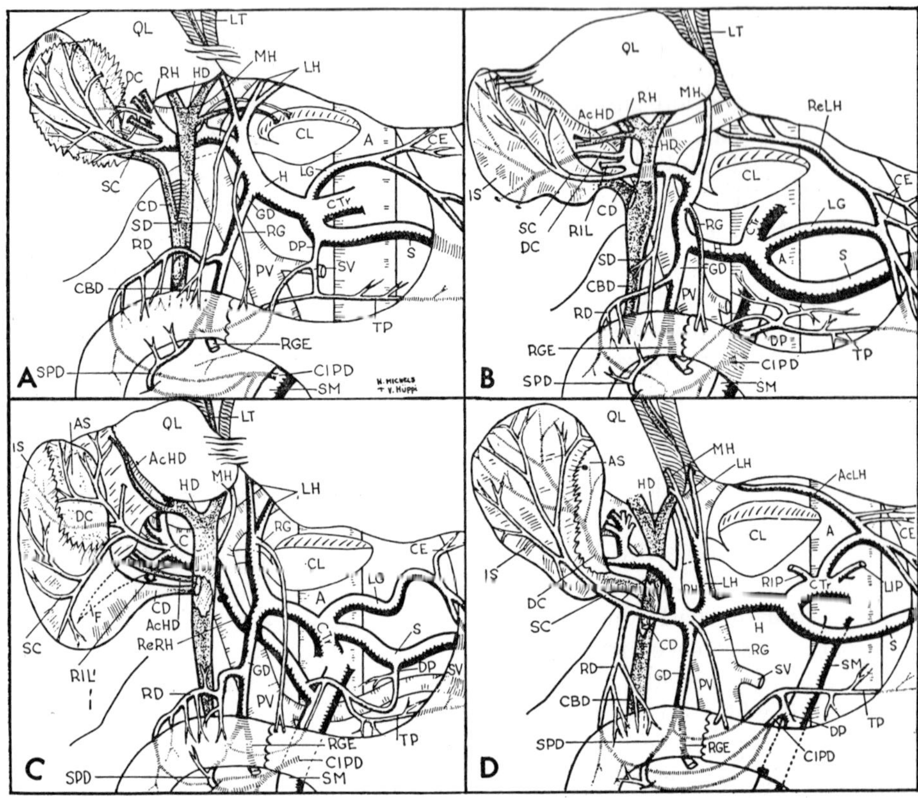

FIG. 71. Types of hepatic arterial blood supply as ascertained in 200 disections. A, right, left and middle hepatic (55%); B, replaced left hepatic (10%); C, replaced right hepatic (11%); D, accessory left hepatic (8%). (From Michels, 1955; reproduced with the permission of J. B. Lippincott Co.)

c. Aberrant right hepatic artery: Strikingly at variance with conventional descriptions is the fact that the superior mesenteric often gives rise to a right hepatic and, in some instances, to the entire hepatic, thereby constituting a hepatomesenteric trunk (Figs. 71C and 72C). Aberrant right hepatics from the superior mesenteric comprise two types: replaced right hepatic, with the only right hepatic present carrying the entire blood supply to the right lobe (11%), and accessory right hepatics, carrying, in most instances, the blood supply to the posterior segment of the right lobe (7%). Accordingly, no hepatic artery is accessory, for each is an end artery with selective distribution to a definite area of the liver. Therefore, it cannot be sacrificed without resultant necrosis of

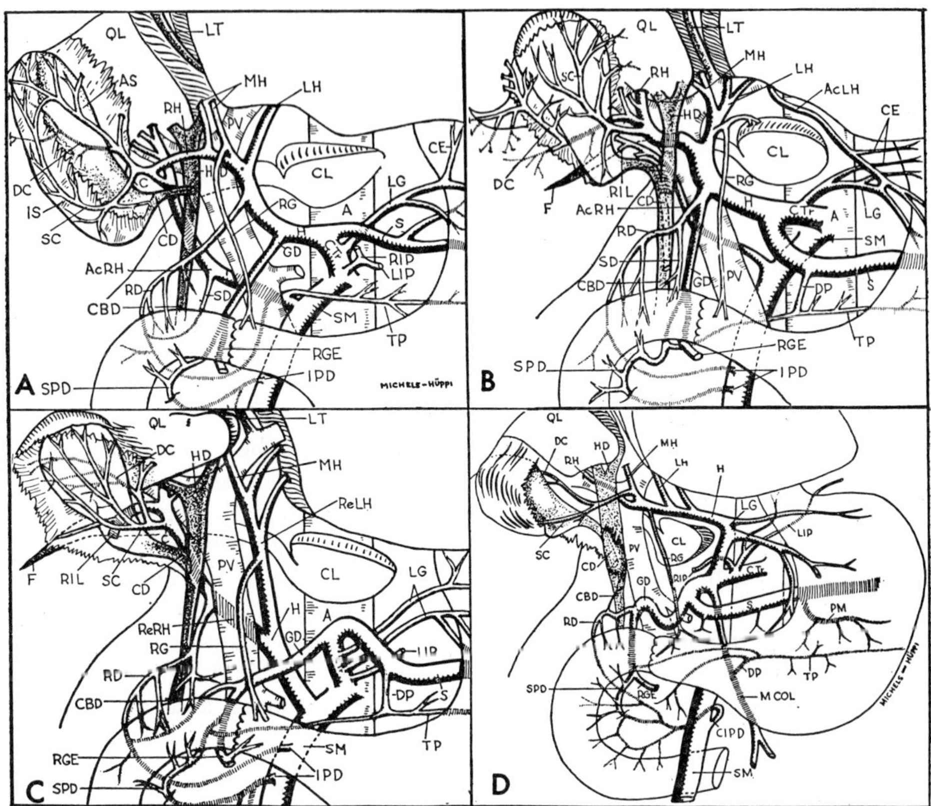

FIG. 72. Basic types of hepatic arterial blood supply. A, accessory right hepatic (7%); B, accessory right hepatic and accessory left hepatic (1%); C, common hepatic from superior mesenteric (5 cases); D, common hepatic from left gastric (1 case). (From Michels, 1955; reproduced with the permission of J. B. Lippincott Co.)

liver tissue (Gordon-Taylor, 1943). Aberrant right hepatics of superior mesenteric origin often give rise to one or two cystic arteries, a fact of surgical import in the search for the cystic.

d. Cystic artery: Typically, this vessel is single (75%), arising from the celiac right hepatic at any point of its varied course in the cystic triangle. Upon reaching the gallbladder, the cystic divides into a superficial and deep branch, the former being distributed to the peritoneal; the latter, to the attached surface of the gallbladder. In 20 per cent of cases, the cystic artery arises outside of the cystic triangle, taking origin from various arteries, such as the right (13%), common (1%), left and middle hepatic arteries (5%), or the gastroduodenal and its retroduodenal branch (4%). In 18 per cent of cases, the cystic takes origin in the triangle from an aberrant right hepatic derived from the superior mesenteric or aorta. In 25 per cent of subjects, the cystic artery is double, with its superficial and deep branches having separate origins (Figs. 71B, D and 72B). The dual cystics may arise from the same right hepatic at different levels, inside or outside of the triangle, from two different hepatic arteries (right, middle, left), or from a celiac right hepatic and some other regional vessel, such as the gastroduodenal, aberrant right hepatic from the superior mesenteric, or retroduodenal. A common pattern of dual cystics is the one in which the superficial cystic arises from the gastroduodenal and the deep, from a right hepatic in the triangle (Fig. 71D).

e. Left hepatic: Typically, this vessel is a branch of the hepatic artery proper and, distally, terminates into two branches for the respective superior and inferior areas of the lateral segment of the left lobe. Before dividing into its terminal branches, the left hepatic may give rise to the middle hepatic (45%), the right gastric, an accessory left gastric or one or two branches to the caudate lobe. Invariably, the left hepatic (via its superior area branch) gives off a falciform division which, upon leaving the umbilical fossa, courses in the falciform ligament to anastomose with the ensiform branch of the internal mammary. Thus a collateral circulation is established between the thorax and abdomen (Fig. 73).

f. Middle hepatic: This branch of the hepatic proper, so named by von Haller in 1756 and by Tiedemann in 1822, is a constant artery supplying the quadrate lobe or what is now known as the medial segment of the liver. In about equal proportion (45%), it arises from the right or left hepatic, and in 10 per cent from other sources, such as the celiac, gastroduodenal and right gastric. It may be the only celiac hepatic present, the right and left being replaced, i.e., derived from other arteries. As with the left hepatic, a branch of it is continued into the falciform ligament where it anastomoses with a branch of the internal mammary. In some instances the middle hepatic gives rise to a cystic artery, as does the left hepatic. Upon entering the liver, it subdivides

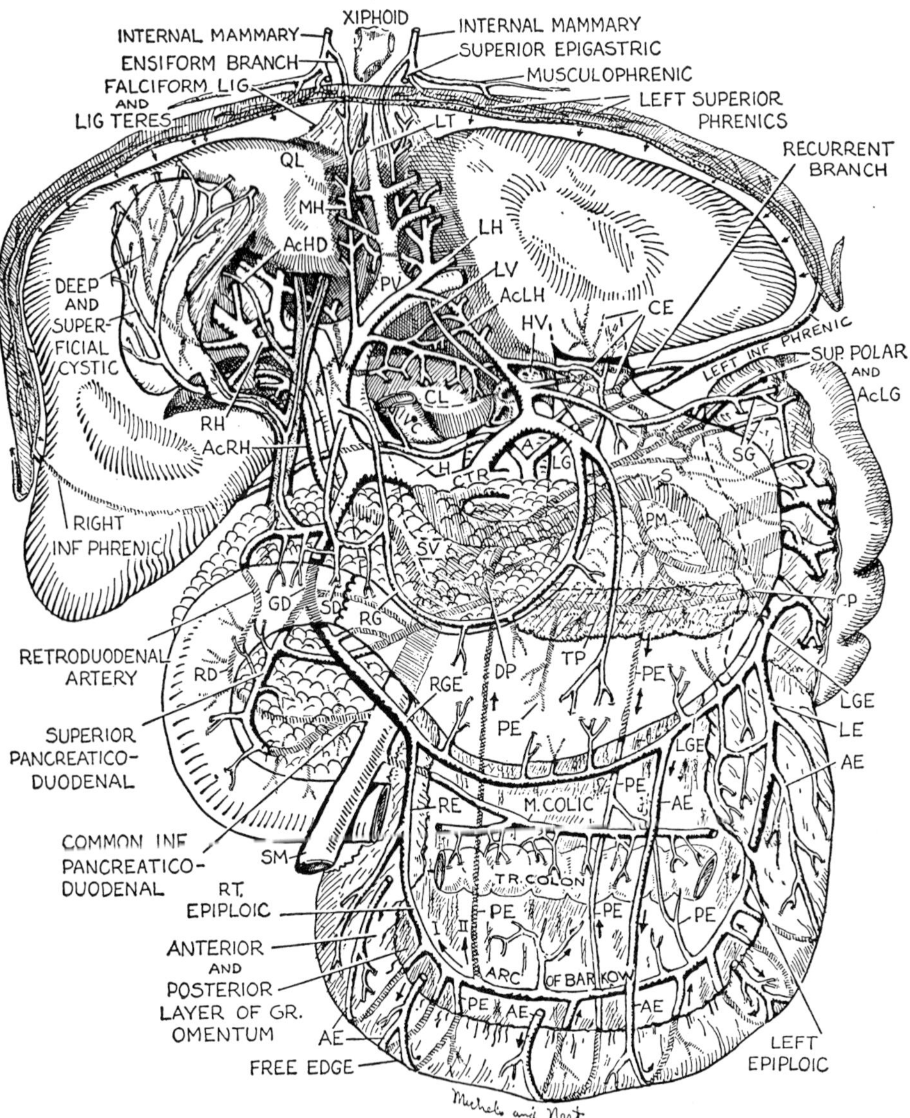

FIG. 73. Possible collateral arterial pathways (26) to the liver (composite of patterns found in dissections of 200 cadavers). (For detailed description of various pathways, see Michels, 1953.) (Reproduced with the permission of J. B. Lippincott Co.)

into two superior and two inferior area branches of the medial segment of the left lobe (Figs. 71 and 72).

g. Terminal hepatic branches: Before entering the liver, the three main branches of the hepatic give off penultimate and ultimate, short, extrahepatic branches which vary to such an extent that no two patterns are alike (Fig. 73). Their number varies from 20 to 65, the most common pattern comprising from 20 to 30 terminals. Of twig size and often tortuous, most of them run a subcapsular course. The terminal branches to the caudate lobe (3–10 in number) are markedly varied in origin and distribution, the right hepatic supplying, for the most part, the caudate process, the left, the papillary process, entirely or in part (for further details, see Michels, 1955).

2. Collateral Circulation to the Liver

There are at least 26 possible collateral pathways to the liver, as graphically presented in Fig. 73. Ten of these consist of basic routes via aberrant hepatic arteries (replaced or accessory), arising from the superior mesenteric, the left gastric or other sources. Six are nonhepatic routes, which are capable of connecting with severed hepatic arteries, while the remaining ten comprise arteries outside of the celiac blood supply.

It should be noted, however, that the blood supply of the liver is always unpredictable and that, because of existing anatomic variations, relatively few of the collateral pathways can definitely be relied upon to establish an adequate compensatory circulation when the main common hepatic trunk or celiac has been ligated. The fact that some patients have survived therapeutic ligation of the common hepatic trunk in operative procedure for portal hypertension is counterbalanced by cases in which the result has been fatal.

Attention must also be called to the fact that many of the collateral pathways are small and insignificant, and hence there is no basis for assuming that ligation of a major hepatic artery can be performed with impunity. In fact, determination of the actual contribution of the various collateral pathways to the economy of the liver cannot rest upon anatomic studies alone, but must be supported by animal experimentation, arteriograms and postmortem investigation. More and more, the latter method is becoming recognized by pathologists as a reliable means for detection and explanation of inadvertent operatively-induced ischemic necrosis, fatal gangrene and peritonitis.

3. Intrahepatic Distribution of Blood Vessels

a. Segmentation of the liver: A decade ago, liver surgery was seldom attempted because of the danger of fatal hemorrhage or the fear of producing biliary fistulas by cutting unknown intrahepatic arteries, veins and bile ducts. As a sequence to the development of modern surgical technique and the anatomic discovery that the liver, like the lung, is a segmental organ with definitely delimited vascular and biliary cleavage planes, partial removal of liver tissue and resection of one-half or an entire lobe of the liver is being successfully performed for the removal of benign and malignant tumors (Pack *et al.*, 1955; Brunschwig, 1953; Couinaud, 1957).

However, it is necessary to point out that the old concept that the liver is divided into a right and left lobe by the falciform ligament, the left sagittal fossa and the ligamentum venosum is erroneous. The correct morphologic division between the right and left lobe is made by an oblique main lobar fissure, about 1 cm wide. While there is no indication of this fissure on the external surface of the liver, it is clearly demonstrable in plastic corrosion casts of the organ; on the visceral surface it corresponds to a line extending from the fossa of the gallbladder to the fossa for the inferior vena cava. This type of bilaterality of the liver was first revealed by McIndoe and Counseller (1927) by means of injected specimens, and it became definitely established through the investigations of Hjortsjö (1948, 1951), who used plastic corrosion casts. The latter worker showed that the human liver is a segmental organ, the main lobar fissure (*Hauptgrenzspalte*) being the division line between the right and left lobe. Many different patterns of the segmental distribution of the bile ducts and arteries exist (Healey and Schroy, 1953) (Fig. 74). (For further discussion, see Elias and Petty, 1952; von Schmidt and Guttman, 1954; Couinaud, 1957; Bilbey and Rappaport, 1960.)

Each lobe of the liver is subdivided into segments by an additional fissure. A right segmental fissure, best seen when viewing the casts of the organ from the right side, divides the right lobe into an anterior and a posterior segment. A left segmental fissure divides the left lobe into a medial and lateral segment. In position this left fissure corresponds to the surface marking customarily used to divide the liver into a right and left lobe, i.e., on the parietal surface, the attachment of the falciform ligament, and on the visceral surface, a line running through the umbilical fossa below and through the fossa for the ligamentum venosum above.

The lateral segment of the left lobe corresponds to the left lobe of the liver, as described in textbooks. On the visceral surface, the medial segment corresponds to what has in the past been designated as the quadratc lobe, while on the parietal surface, it corresponds to all the area between the main lobar fissure and the left segmental fissure.

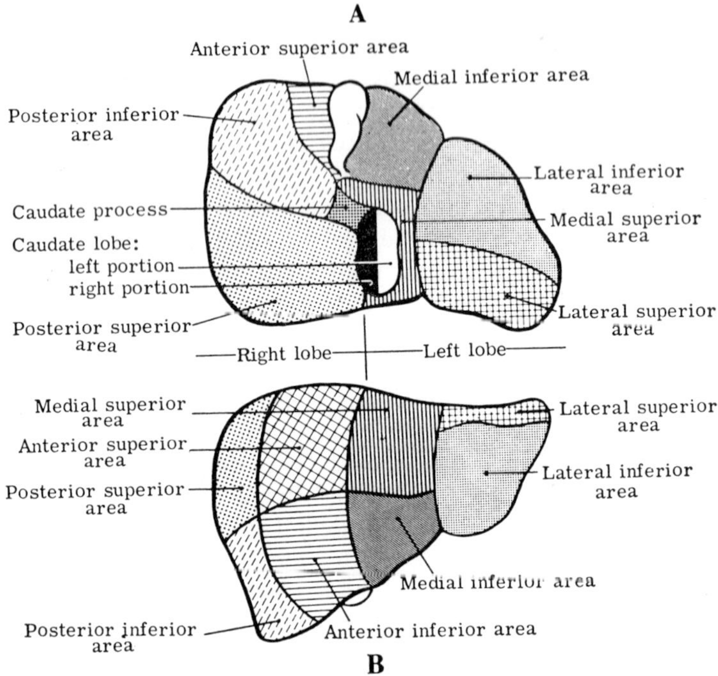

FIG. 74. Divisions of the liver, based on the biliary and arterial systems of the organ. A, visceral surface; B, parietal surface. Right lobe: 2 segments, anterior and posterior, each with a superior and inferior area. Lateral and medial segments of left lobe, with a superior and inferior area. (Redrawn with modifications in labeling and stippling from Healey and Schroy, 1953.) (From Michels, 1955; reproduced with the permission of J. B. Lippincott Co.)

The caudate lobe belongs to both lobes rather than to either one. It consists of three parts: (1) caudate process, (2) the right and (3) the left portion of the caudate lobe proper (Fig. 74).

b. Intrahepatic distribution of hepatic artery: Like the right hepatic duct, the right hepatic artery inside the liver divides into an anterior and a posterior segmental branch (anterior and posterior segmental artery), to supply, respectively, the anterior and posterior segments of the right lobe. Both of the latter portions have superior and inferior areas and, accordingly, are supplied by superior and inferior area branches of the segmental arteries.

The left hepatic artery, like the left hepatic duct, terminates by dividing into a medial and lateral segmental artery for the respective medial and lateral segments of the left lobe. The lateral segmental artery

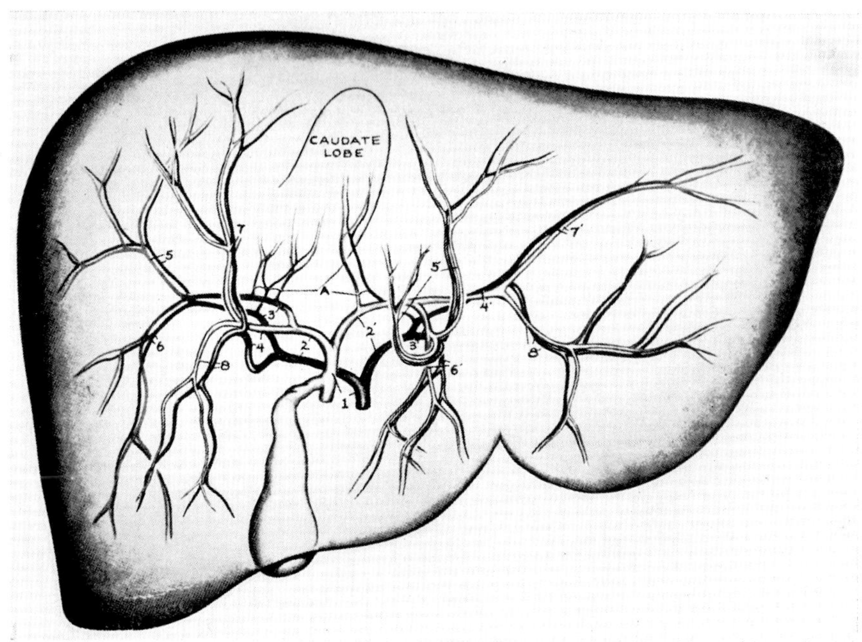

FIG. 75, Composite drawing of intrahepatic distribution of hepatic artery and bile ducts. 1, common hepatic artery and common bile duct; 2, right hepatic artery and right bile duct; 3, posterior segment artery and bile duct; 4, anterior segment artery and bile duct; 5, posterior-superior area artery and bile duct; 6, posterior-inferior area artery and bile duct; 7, anterior-superior area artery and bile duct; 8, anterior-inferior area artery and bile duct.

2^1, left hepatic artery and bile duct; 3^1, medial segment artery and bile duct; 4^1, lateral segment artery and bile duct; 5^1, medial-superior arteries and bile ducts; 6^1, medial-inferior arteries and bile ducts; 7^1, lateral-superior area artery and bile duct; 8^1, lateral-inferior area artery and bile duct; A, caudate lobe arteries. (From Healey *et al.*, 1953; reproduced by Michels, 1955, with the permission of J. B. Lippincott Co.)

subdivides into a superior and inferior area branch, while the medial segmental artery (for the quadrate lobe) gives off four area branches: two superior and two inferior. Three bile ducts drain and two arteries supply the caudate lobe, there being one bile duct for each of its parts, two arteries for the caudate process and the right half of the caudate lobe proper, and one artery for the latter's left portion. This most common pattern of the intrahepatic ramifications of the hepatic artery is illustrated in Fig. 75. However, there are many variant patterns as regards the origin of the segmental branches and their respective area branches.

c. Intrahepatic distribution of the portal vein: The segmental branching of the portal vein is, with minor modifications, the same as that of the bile ducts and arteries (Healey, 1954). The right portal vein is usually regarded as a continuation of the common portal trunk, situated

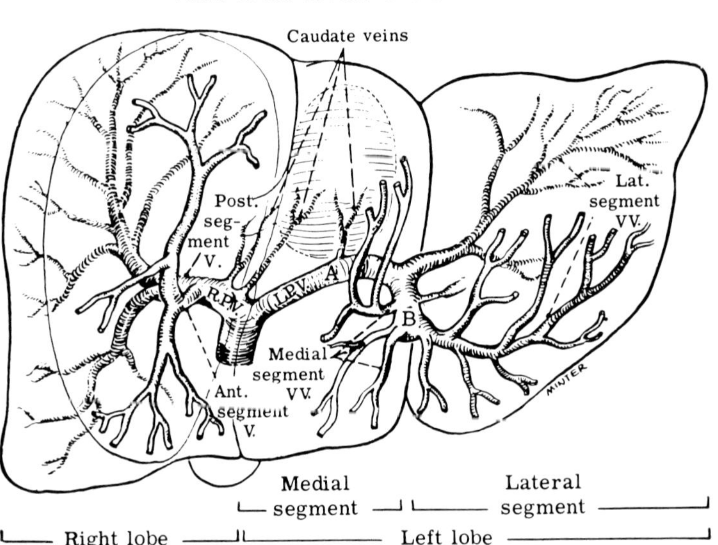

FIG. 76. Typical intrahepatic distribution of the portal vein in the human liver as seen in a vinyl acetate corrosion cast. Branches of portal vein inside liver arranged segmentally. (From Healey, 1954, reproduced by Michels, 1955, with the permission of J. B. Lippincott Co.)

toward the right end of the porta hepatis and having, at its beginning, a short transverse course. Like divisions of the right hepatic artery and right hepatic duct, it divides into an anterior and posterior segmental vein, each of which, as a rule, becomes subdivided into a superior and an inferior area branch (Fig. 76).

The left portal vein is considerably longer and narrower than the right trunk. After an oblique course (2–4 cm), it makes a 90° bend in a caudolateral direction. It usually exhibits two portions, a pars transversa (Fig. 76A), located in the porta hepatis, and a pars umbilicalis (Fig.

76B), which lies in the umbilical fossa, where it constitutes the bended section. The lateral segment of the left lobe has two main branches, one arising from the left side of the bend of the left portal trunk, the other, more distally, from the left side of the pars umbilicalis. The medial segment (quadrate lobe) of the liver receives four medial segmental veins from the right side of the umbilical section of the left portal vein, these being the two superior and the two inferior area branches which, predominantly, arise via a common stem. The caudate lobe has three veins: one for the caudate process from the right portal trunk and two for the papillary process from the left portal trunk (Fig. 76).

d. Intrahepatic distribution of hepatic veins: The distribution and subdivisions of the hepatic veins differ completely from those of the

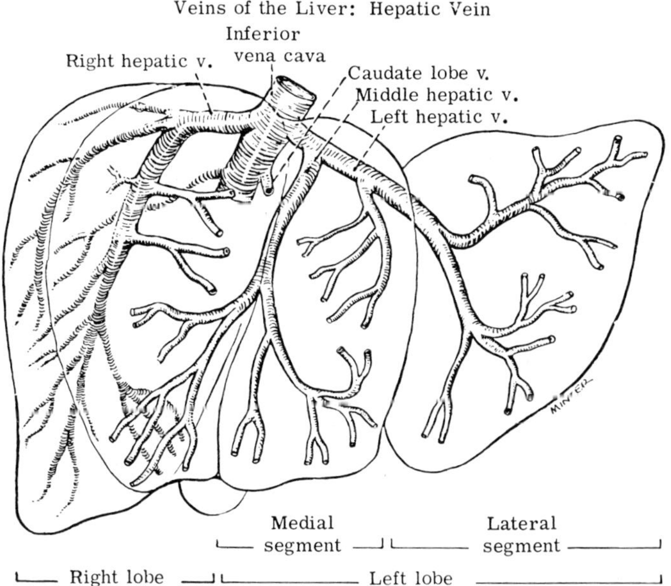

FIG. 77. Intrahepatic distribution of hepatic veins in the human liver, as demonstrable in a vinyl acetate (plastic) corrosion cast. (From Healey, 1954, reproduced by Michels, 1955, with the permission of J. B. Lippincott Co.)

portal vein. While no branches of the bile ducts, hepatic arteries or portal vein ever cross the main lobar fissure, branches of the hepatic veins do. The hepatic veins are not segmentally arranged, but lie in fissures, i.e., in intersegmental planes. Of the three main branches of the hepatic vein, the right is located in the right segmental fissure and drains the entire posterior segment, together with the superior area of the anterior segment of the right lobe. The middle hepatic vein lies in the main lobar fissure and drains the inferior area of the anterior segment of the right lobe, plus the inferior area of the medial segment. The left hepatic vein is found in the upper part of the left segmental fissure and drains the entire lateral segment of the left lobe, plus the superior area of the medial segment. In most cases, the middle and left hepatic vein unite into a common trunk before entering the inferior vena cava. One or more constant small veins drain the caudate lobe (Fig. 77).

The two systems of venous channels, the portal and hepatic veins, are interdigitated in their branches, giving patterns comparable with those obtained when, palm to palm, the fingers of each hand are made to cross each other in midline (Glisson, 1654; Melnikoff, 1924; Elias and Petty, 1952; Healey and Schroy, 1953; Couinaud, 1957).

Bergstrand (1957) has shown that the normal and pathologic x-ray anatomy of the portal ramifications can clearly be seen in portal venograms by lieno-portal venography. However, the contrast filling of the hepatic veins is usually insufficient for adequate determination of topography. From a surgical point of view, there is an urgent need for a statistical study of the variable patterns of the origin and distribution of the hepatic veins, for, such information is essential in resection of the liver. At present, the terrain of drainage of these vessels remains a phase of liver anatomy which is a major unsolved problem.

C. MICROSCOPIC AND SUBMICROSCOPIC ANATOMY

By H. Elias;[2] E. E. Selkurt[3]

1. Structure of the Liver

Before describing the circulation in the liver, it is necessary first to present briefly its general microscopic structure. This organ (Fig. 78) can be compared with a large building containing many narrow, crooked rooms, the *lacunae hepatis,* which collectively form a continuous labyrinth. The *lacunae* are separated from each other by con-

[2] Author of Subsections 1–8.

[3] Author of Subsection 9.

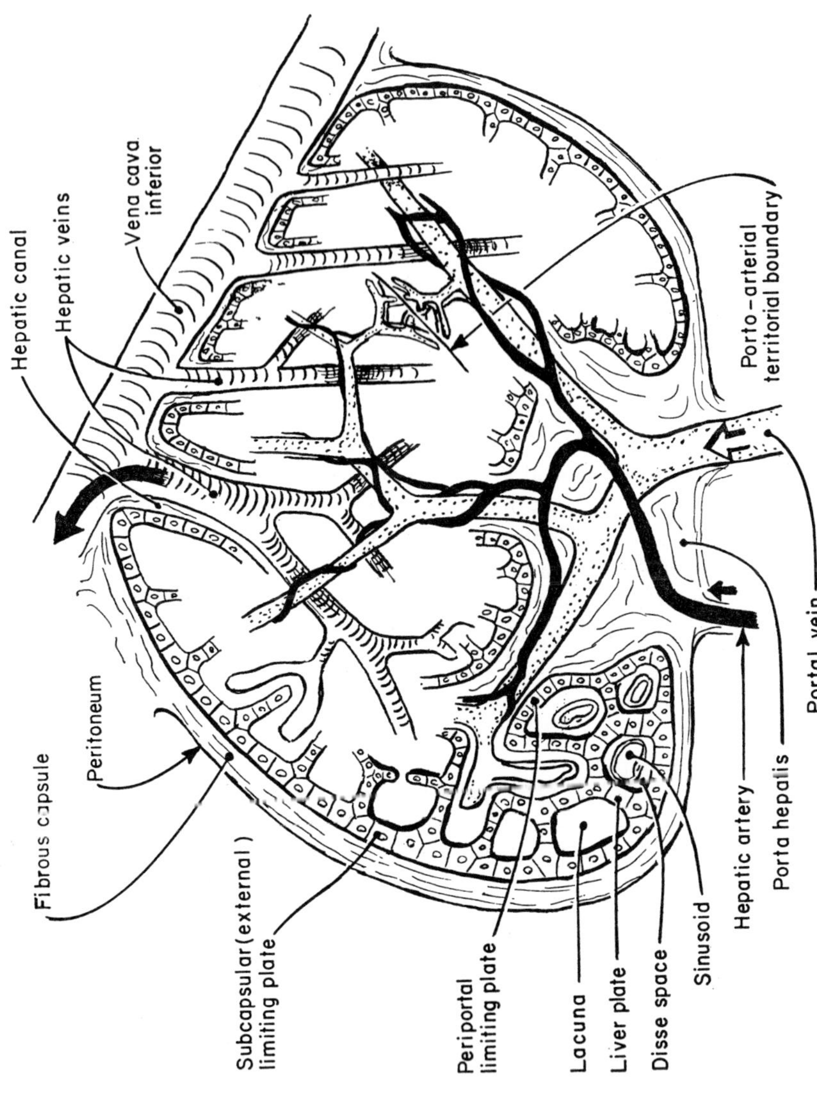

FIG. 78. Diagram of the relationship of the blood vessels of the liver to the stroma, to the parenchyma and to each other.

nected walls, one cell thick, the *laminae hepatis* or liver plates (Elias, 1949a), which also form a continuous system of walls (Elias, 1953b). Through the building run branching corridors, the portal and hepatic canals, lined by the limiting plate (Elias, 1949b), consisting of one layer of "dark" liver cells. This limiting plate is continuous with the external or subcapsular limiting plate (Elias, 1953a), as well as with the internal *muralium*.

The liver is covered by a capsule which is continuous with a large mass of connective tissue in the *porta hepatis*. Cylindrical continuations of connective tissue extend into the hepatic canals from the external capsule and into the portal canals from the *porta hepatis*. These structures carry the blood vessels, lymph vessels, nerves and ducts into the interior of the organ (Glisson, 1654).

2. Portal and Hepatic Territories

Each division of the portal vein has its own, sharply defined territory (Fig. 78), and, normally, there are no anastomoses between those of individual portal vein branches, such as the two large trunks (McIndoe and Counseller, 1927; Gruenwald, 1949) or those of "interlobular" veins (Elias and Petty, 1952). Similarly, the hepatic veins and their tributaries drain sharply delimited territories, although occasional anastomoses occur between them (Elias and Petty, 1952).

To compare the portal venous territories with the bronchopulmonary segments, as some anatomists have done (Healey *et al.*, 1953), is a mistake, because the hepatic veins and their tributaries cross portal territorial boundaries (Glisson, 1654). A system of surgical segments must therefore consider both the portal canals and the hepatic veins (Reifferscheid, 1957; Banner and Brasfield, 1958).

3. Portal Vein Branches

The portal vein branches exhibit a pattern intermediate between the monopodial and the dichotomous type. This is to say that at points of division the two branches are usually of unequal size. The larger vessel undergoes a slight bend, while the smaller one deviates more markedly from the direction of the branch of origin.

Since the portal canals are separated from the liver parenchyma by the limiting plate, the portal blood, to reach the sinusoids, must pass through inlet venules (Elias, 1949b); very short conduits that arise perpendicularly from distributing portal veins pierce the limiting plate and empty into paraportal sinusoids (Figs. 79 and 80 iv). The distributing

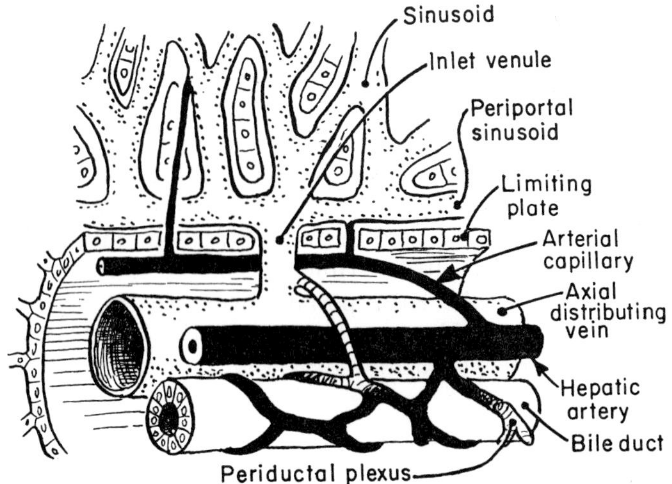

FIG. 79. Blood vessels in and near a small portal canal.

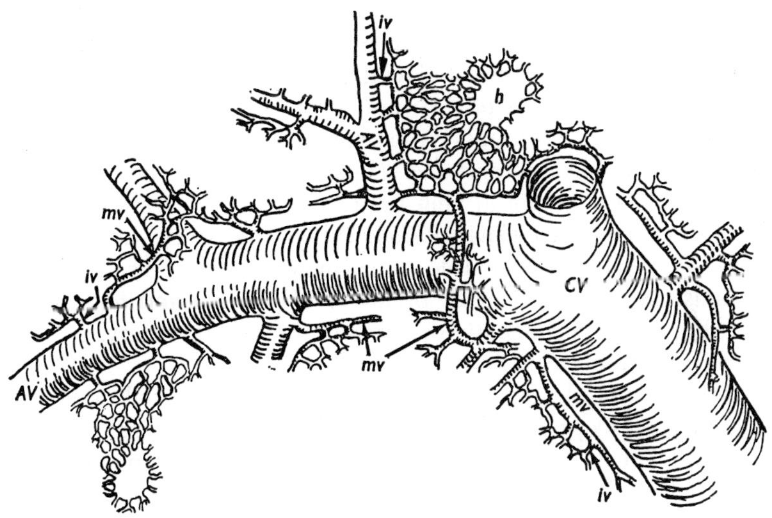

FIG. 80. Categories of portal veins in man. iv—paraportal sinusoid; AV—axial distributing veins; CV—conducting veins; mv—marginal distributing veins. (From Elias, 1955; reproduced with the permission of Biological Reviews.)

portal veins have a diameter up to 280 μ. Those which occupy an axial or slightly eccentric position in a small portal canal are called axial distributing veins (Figs. 79 and 80 AV). Portal vein branches with a diameter of 300 μ or larger do not give rise to inlet venules and are called conducting veins (Fig. 80 CV). The territory immediately surrounding large portal canals is supplied with portal blood by those inlet venules which arise from marginal distributing veins (Fig. 80 mV). These are small branches of the conducting veins that run in the same portal canals, often parallel to the conducting veins, in a forward or retrograde direction, and sometimes spirally around them.

Since the liver of man is richly supplied with marginal distributing veins, the periportal sinusoids receive a good supply of portal blood; the same being true in the dog, the rabbit, and the mouse (Elias and Popper, 1955; Lee *et al.*, 1960). However, in the rat, the regions immediately surrounding the large portal canals are so poorly supplied with portal blood (Gershbein and Elias, 1954) that they have been called "nonportal areas" (Hartroft, 1953).

4. Hepatic Arterial Branches

While the gross anatomic distribution of the branches of the hepatic artery and the occurrence of "accessory hepatic arteries" are subject to great variations (see Section B.-1b and c, this chapter), as soon as the vessels enter into the substance of the liver, i.e., into portal canals, they become strictly coordinated with the rami venae portae and their branches. Their territories of distribution are identical with those of the latter, and there are no intrahepatic, macroscopic arterial anastomoses, a fact of great surgical importance (Glauser, 1953). The rami arteriae hepaticae cling to the portal vein branches much like vines would cling to the bars of a trellis (Elias, 1953a). When "accessory" hepatic arteries are present, these play the role of segmental arterial branches, which never anastomose with arterial branches of another territory (Glauser, 1953). (For gross intrahepatic distribution of hepatic artery, see Section B.-3b, this chapter.)

Unlike the portal vein branches of which there is only one in each portal canal (except for the marginal distributing veins in the very large portal canals), there are more than one artery in almost every portal canal, even in small ones; these arteries anastomose at frequent intervals with each other.

The hepatic arteries give rise to several kinds of terminal branches, as listed below (Fig. 79):

(1) Certain microscopic branches supply periductal, capillary plexuses, which, in the case of very small bile ducts, are subepithelial in location. Medium-sized and large intrahepatic bile ducts are surrounded by two capillary plexuses: one subepithelial and the other submucosal. Both are connected by radial anastomoses and drain through venules into portal inlet venules (Aunap, 1931; Elias and Petty, 1953).

(2) Long and straight arterial capillaries run along the periphery of the portal canals, giving rise to numerous secondary arterial capillaries which pierce the limiting plate to empty into sinusoids (Elias, 1949b). They resemble bile ductules so closely that they can be truly identified only by tracing them back to their origin through serial sections and by injecting them with India ink.

(3) The intraparenchymal arterial capillaries enter into sinusoids at all levels of the liver lobules. Short arterial capillaries empty into paraportal (periportal) sinusoids; long ones penetrate as far as two-thirds of the distance toward the central vein, emptying into centrolobular sinusoids. Still others open into sinusoids at intermediate levels. Thus, the hepatic lobule is supplied with arterial blood everywhere except in the central third (Elias and Petty, 1953). This description applies to man, the horse and the dog. In the rat the penetration of arterial capillaries is not as deep as in the other species mentioned (Riedel and Moravec, 1959).

The hepatic arteries are provided with sphincters at many levels. For example, side branches, as they arise from larger rami, possess sphincters at their point of origin (Märk, 1951). The arterioles and arterial capillaries which supply sinusoids directly also are provided with such structures (Elias, 1949b).

5. Hepatic Venous Radicles

The smallest radicles of the hepatic veins are the central veins. These empty into sublobular veins which, in turn, continue into collecting veins. One encounters among the smaller and smallest hepatic veins an almost strict monopodial pattern of branching, i.e., the type of ramification prevalent among coniferes.

Under normal conditions of blood pressure, the sinusoids converge radially toward the central veins. In man and in the dog, they empty only into such vessels (Elias and Popper, 1955), whereas in the rat (Elias and Popper, 1955; Gershbein and Elias, 1954) and in the mouse (Lee *et al.*, 1960), they join hepatic veins of all sizes.

In man, because of the lack of sinusoidal openings into sublobular veins, a bottleneck exists at the entrance of the central into sublobular veins, this situation being aggravated by a constriction at the point of junction of these vessels (Popper, 1931). In the dog and a number of other mammals (Arey, 1941; Varícak, 1950, 1952), as well as in the carp (Elias and Bengelsdorf, 1952), the hepatic veins are provided with spiral muscles, which contract in anaphylaxis, causing anaphylactic shock (Mautner and Pick, 1915).

6. Sinusoids

The sinusoids are suspended in the lacunae hepatis and form a three-dimensional network which is continuous throughout the human liver, as is the *labyrinthus hepatis* in which it is contained. The sinusoids are lined with littoral cells, the so-called Kupffer cells which are all potential phagocytes (Knisely *et al.*, 1948).

It has been believed for a long time that the Kupffer cells form a syncytial continuum because their boundaries cannot be impregnated

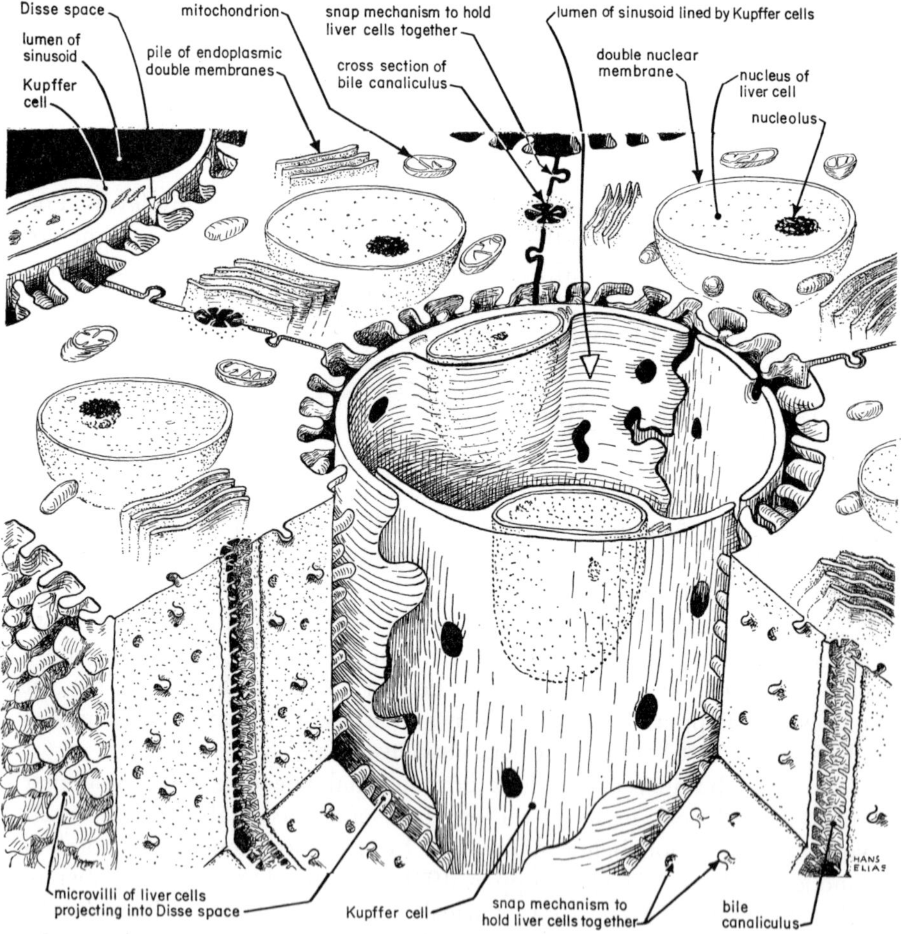

Fig. 81. Stereogram of the submicroscopic structure of the liver based on electron microscopic data from various sources. (From Elias and Pauly, 1960; reproduced with the permission of Da Vinci, Chicago.)

with silver nitrate. However, electron microscopic studies (Rüttner and Vogel, 1957) have revealed the individuality of these structures and have shown that they do not even adhere together, but instead overlap, often leaving spaces between one another (Fig. 81). Evidence has also been presented (Elias, 1949b) to indicate that the walls of the sinusoids are contractile. The exact mechanism of blockage of blood flow was not understood until electron microscopic studies (Rüttner and Vogel) revealed that the Kupffer cells can bulge into the lumen of the sinusoids, thus impeding the flow (Fig. 82).

The presence of a perisinusoidal space (the space of Disse) in life has also been debated. Its occurrence in autopsy material, in spite of its apparent absence in biopsies (by light microscopy), was attributed to

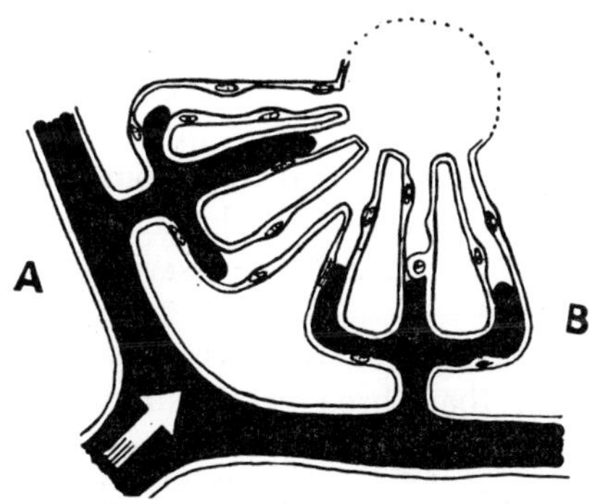

FIG. 82. Mechanism of sinusoidal mechanics, showing bulging of Kupffer cells. (From Lee *et al.*, 1960; reproduced with the permission of the Proc. Animal Care Panel.)

anoxia (Popper, 1948). Electron microscopy (Rüttner and Vogel), however, has shown that the space of Disse is a reality during life. Into it project microvilli with which the lacunar facets of the liver cells are studded.

7. Lymph Production and Lymph Vessels[4]

In man, during anoxia the walls of the sinusoids are believed to become permeable to whole plasma, which then may seep out of the sinusoidal lumen into the perisinusoidal space (Popper, 1948). As a result, the blood becomes condensed and the spaces of Disse open up. This process has also been observed in frogs *in vivo* by transillumination

[4] For general discussion of lymph formation and lymphatics, see Sections C and D, Chapter XXIII.

(Knisely *et al.*, 1948). The fluid, thus extravasated, flows along the spaces of Disse toward the portal canals where it enters into a periportal tissue space (Mall, 1906). From there, it must be filtered through the periportal connective tissue and the endothelium of the lymph vessels which form an extensive plexus in the portal canals (Elias, 1949b). A few, very fine lymph vessels enter into the parenchyma, running in the delicate adventitia of intraparenchymal arterioles and ductules. In the dog, large lymph vessels are also found around the hepatic veins. The sphincters located at the points of arterial branching (Märk, 1951) and the intimal muscle cushions in hepatic arteries (Ehrenbrandt and Burckhart, 1956) may provide the state of anoxia which appears to be necessary for lymph production.

8. Anatomic Basis for Regulation of Blood Supply

Both large and small regions of liver parenchyma are supplied with portal and arterial blood and drained of venous blood, according to the current needs of the tissues. This is brought about through several mechanisms, some of which have been described above.

a. Venous control: The portal inlet venules are spaced at such distances from each other that three to six sinusoids are between them. Thus, a candelabra-like arrangement exists at every portal inlet (Fig. 79). If the sinusoids were inactive or flaccid tubes, blood would flow along the shortest route from the inlet to the central vein (Fig. 82, left). However, it has been shown that if India ink is injected into a mesenteric vein of a living rabbit (Elias, 1949b) or of a mouse (Lee *et al.*, 1960), the material proceeds on an even front from the periportal sinusoids toward the central vein in all the sinusoids of the region, with the shortest route not necessarily being taken. Such a pattern of flow has been attributed to active contractility of the sinusoids, but now, in the light of Rüttner and Vogel's (1957) studies, it can more precisely be related to the bulging of Kupffer cells (Fig. 82).

Admission of portal blood to the parenchyma can be prevented by a sphincter mechanism at the level of the inlet venules (inlet sphincters) (Knisely *et al.*, 1948) (Fig. 83). Blood is retained in the sinusoids by bulging Kupffer cells and by "outlet" sphincters (Knisely *et al.*, 1948) at the entrance of sinusoids into central veins (Fig. 83). In the absence of demonstrable muscle fibers, it is possible that at this specific location, bulging Kupffer cells also provide the obstruction. The constriction at the mouths of the central veins may become important in pathologic states (Popper, 1931).

b. Arterial and capillary control: The flow of arterial blood can be controlled at the arterial level by sphincters (Märk, 1951) (Fig. 83) and by intimal muscle cushions (Ehrenbrandt and Burckhart, 1956). Sphincters have been described at the level of the arterioles (periportal and intraparenchymal) and of the arterial capillaries (Elias, 1949b), as

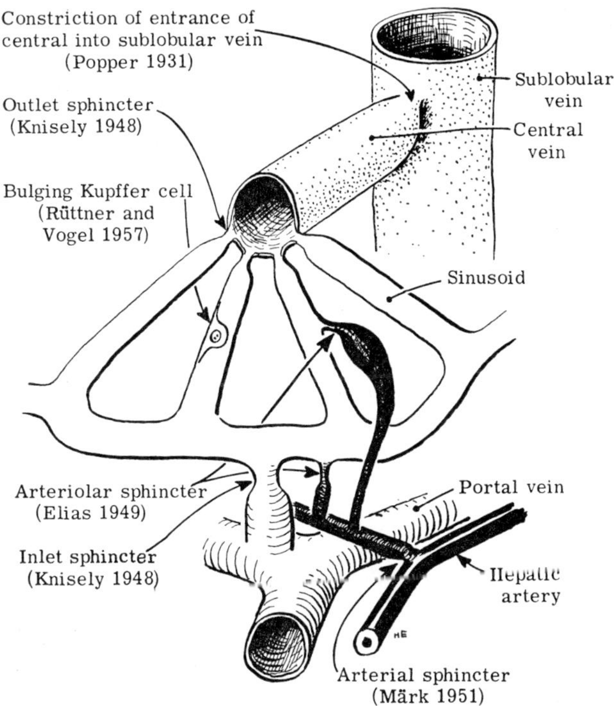

FIG. 83. Graphic summary of the microcirculation through the human liver. (From Elias and Pauly, 1960; reproduced with the permission of Da Vinci, Chicago.)

well as a narrowing of entire arterial capillaries to an external diameter of 3 μ along their intraparenchymal course.

9. SPLENIC CIRCULATION[5,6]

The splenic artery branches pass along the trabeculae of the spleen and penetrates the *white pulp*, to break up into arterioles, with husk-like

[5] By E. E. Selkurt.

[6] For gross anatomy of the splenic circulation, see Section B.-la, Chapter XI.

sheaths (penicilli). The latter, which constitute the arteriolar stopcocks, branch into arterial capillaries, which, in turn, supply the sinusoids (venous sinuses) in the red pulp of the spleen (Fig. 84). The collecting venules converge ultimately into the splenic vein, which drains into the portal vein. Sphincters have been described at the sinusoid outlets which close during storage phases. The sinusoids are composed of rod-like "barrel stave" endothelial cells, which are longitudinally arranged but not contiguous; the latter are surrounded by rings or spirals of

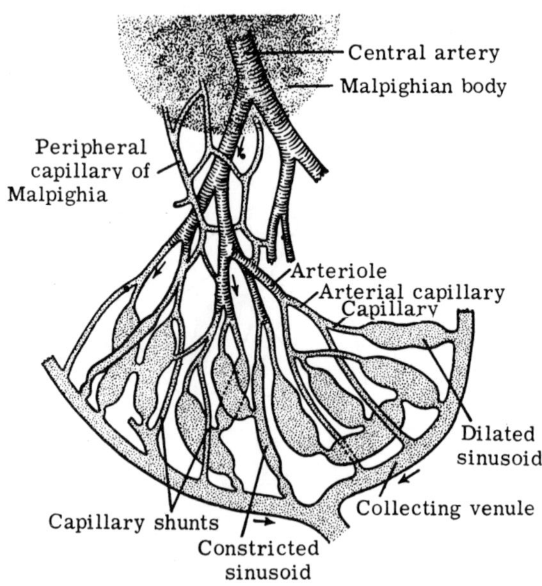

Fig. 84. Diagrammatic representation of the circulation through the spleen. (From Peck and Hoerr, 1951; reproduced with the permission of the Anatomical Record.)

reticular fibers, reminiscent of barrel hoops, which create a lattice-like structure.

a. Question of "closed" or "open" system: A commonly accepted view is that the splenic system is a closed one (Knisely, 1936; Björkman, 1947; Peck and Hoerr, 1951), but that the latticework is large enough to permit occasional erythrocytes to slip through, as observed in the transilluminated mouse spleen. Thus, the sinusoids represent a mechanism which forms a "separatory" circulation, filtering off plasma and storing red cells in high concentration in the dilated channels, and operating in a cycling manner. The separated plasma may return through smaller veins and lymphatics to the general circulation. In a spleen exhibiting a high degree of storage, constant, rapid flow is seen primarily in the capillary shunts (Fig. 84) which directly connect arterioles with venules. When the splenic capsule contracts under physiologic demand, the concentrated cells are discharged directly into the portal venous system. Filtered red cells re-enter the venous sinuses by diapedesis, which leads to destruction of fragile erythrocytes. After rupture, the released hemoglobin and stroma are ingested by the reticuloendothelial system of the spleen or elsewhere in the body. Electron microscopic examination appears to support the idea of a closed system in the rat and human spleen (Weiss, 1957).

On the other hand, MacKenzie, *et al.* (1941), have taken the position that in the mouse, the splenic circulation is an "open" one (whole blood coming in direct contact with splenic pulp cells), and not closed, as Knisely (1936) has contended. Erkoçak (1958), using a carmine-gelatin injection technique in various mammalian spleens, has also come to the same conclusion. Parpart *et al.* (1955) and Fleming and Parpart (1959) found that in the mouse the arterial and venous vessels are united by unlined pulp spaces, lacking endothelium. The terminal arterioles are said to end in flared ampullae, the last structure to show signs of endothelial walls. Blood issues forth into the pulp spaces through the ampullae and re-enters the vascular system via the open ends of the venous vessels and perforations in the venous walls. In the rat, however, nothing resembling venous sinuses or pulp spaces is observed, and instead, vascular walls are clearly discernible, with very few erythrocytes in intercellular spaces. Although differences in technique are in part basic to the disagreements, species differences may well also be contributory.

D. PHYSIOLOGY

By E. E. SELKURT

1. INTERMITTENCE OF FLOW IN THE LIVER[7]

The presumption that sphincters are located at the entry and exit of the sinusoids in the liver, permitting regulation of flow and storage of blood, has given rise to the concept of intermittence of flow. This phenomenon has been observed in the frog (Knisely, 1939), and in the rat (Wakim and Mann, 1942). Spontaneous local variations in blood flow have also been noted in the human liver, using a Hensel calorimetry needle (Graf *et al.*, 1957). It has been suggested that in the frog the number of sinusoids which are open vary greatly, with frequent shifts of activity. Although intermittency seems characteristic of the rat, more than 75 per cent of the regions studied showed inactivity when neither excitatory nor inhibitory factors were operative in the intact liver. More recently, transillumination studies in the mouse, rat, and guinea pig have demonstrated no periodic changes in flow (Maegraith, 1958), except in what were believed to be traumatized organs. Of interest is the finding that total hepatic blood flow in the dog and man is reasonably steady. This could result because of factors which tend to buffer the total hepatic circulation, including the compensatory relation between the hepatic arterial and portal venous flow (Cull *et al.*, 1956; Grayson and Mendel, 1957), and also because of the fact that the liver contributes only 15 per cent of the total splanchnic resistance.

2. NERVOUS REGULATION OF LIVER BLOOD FLOW

Due to the complexity of the vascular bed of the liver, an analysis of nervous regulation is difficult. By direct observation of liver flow, sympathetic stimulation has been seen to cause constriction of arterioles, sinusoids, and, indeed, all intrahepatic vessels. It must be kept in mind, however, that some of these changes, i.e., sinusoidal constriction, may have been a "passive recoil" effect, as the result of arteriolar constriction. Based on X-ray opacity techniques, the constrictor influence upon smaller vessels seems to vary in different regions of the liver. Little is known about the control of the arteriovenous shunts in this organ. Dilatation of the liver vessels apparently cannot be induced by vagal stimulation. It is therefore doubtful that specific vasodilator fibers run

[7] For a discussion of lymph flow in the liver, see Section D.-2d, Chapter XXIII.

to this organ. Finally, it should be kept in mind that the extrinsic control of the liver circulation adds further complexity, since vasomotor changes in the intestine and spleen will modify portal vein inflow, aside from any direct neurogenic or humoral influences on the liver.

3. Measurement of Liver Blood Flow

Numerous techniques, direct and indirect, have been used for the measurement of blood flow. The indirect methods are based on the Fick principle, the most widely used being the BSP (bromsulphalein) dye removal method devised by Bradley *et al.* (1945).[8]

A summary of numerous results in dogs by direct methods has been compiled which shows that total flow averages 403 ml/min (29.4 ml/Kg/min and 82.5 ml/100 gm liver/min). The hepatic artery contributes about 30 per cent of the total flow. Such figures are in contrast with indirect measurements which averaged a total of 570 ml/min (35.5 ml/kg/min). By a K^{42} uptake technique Sapirstein (1958) has deduced that portal flow in the dog fractionates as follows: stomach, 12.5 per cent; intestine, 73 per cent; pancreas, 8 per cent; and spleen, 6.5 per cent.

The finding that, on a per kg body weight basis, the indirect measurements yielded the higher total flow figures requires further discussion. Granting that operative procedures may have fostered the lower direct flows, yet another argument has been evoked to explain the higher dye method figures, namely that extrahepatic dye removal may have accounted for the somewhat higher apparent blood flows. This argument is based on comparison of dye removal (BSP) in intact, as compared to hepatectomized dogs (Horvath *et al.*, 1953). But the protagonists of the method insist that the rate of removal of the dye by extrahepatic

[8] Fick Principle (as applied to man): Infuse 3 mgm/min per sq M S A after 150 mgm (priming dose).

$$\text{E.H.B.F. (effective hepatic blood flow)} = \frac{R}{.01(P - H)} \times \frac{1}{1 - \text{hemat.}}$$

R is rate of dye removal (mgm/min); P is peripheral venous dye concentration in mgm/%; H is hepatic venous dye concentration in mgm/%; *R* is determined by adjusting rate of infusion (I) of dye so that blood levels do not change.

Validity of the method depends on the following conditions: (a) BSP must remain in the plasma and be removed exclusively by the liver; (b) concentration of dye in the peripheral venous blood (P) must be the same as that in blood supplying the liver (i.e., hepatic artery and portal vein); (c) with catheter technique, concentration in any one hepatic vein should be representative of total hepatic venous outflow.

If plasma level is changing, a correction can be made by measuring or estimating plasma volume (PV). If rising: $R = I - (P \times PV)$; if falling: $R = I + (P \times PV)$.

tissue is negligible, compared to hepatic removal (Casselman and Rappaport, 1954; Selkurt, 1953, 1954a; Bradley *et al.*, 1945). Enterohepatic recirculation of the dye does not seem to be an important source of error (Pratt *et al.*, 1952). Evidence has been presented that sampling from different hepatic veins does not influence the BSP extraction figures (*ca* 34–40%) so long as the catheter does not reflux inferior vena cava blood (Heinemann *et al.*, 1953; Pratt *et al.*, 1952; Casselman and Rappaport, 1954). One report, however, is in disagreement on this point (Sapirstein, 1958). In dogs with especially prepared "common hepatic veins" (occluded inferior vena cava below liver), higher hepatic blood flow figures (Rose Bengal extraction) were obtained from the common hepatic vein than from deep catherization. It was contended that the catheter caused ischemia of the portion of the liver surrounding it, thus resulting in higher extraction and lower hepatic flow determinations. A new value of 50 ml/kg/min (30% of the cardiac output) has been thought to be more nearly representative of true hepatic blood flow in the dogs (Sapirstein, 1958).

The BSP method has been applied to man by a number of investigators, and the results are consistent at a figure of about 1470 ml/min (850 ml/sq M/min) (Bearn *et al.*, 1951a,b; Bradley *et al.*, 1945; Myers, 1947). Myers (1947) compared the BSP method with the urea excretion method and found 800 and 1000 ml/sq M/min respectively.

Dobson and Jones (1952) have studied liver flow in several mammalian species, using the chromic–PO_4 disappearance technique. Their figures, given in volumes of blood per volume of tissue per minute, are: dog (anesthetized), 1.2 ± 0.3; rabbit (normal), 0.74 ± 0.12; rabbit (anesthetized), 1.12 ± 0.13; mouse (normal), 1.40 ± 0.1; mouse (anesthetized, 1.70 ± 0.1; and rat (normal), 1.20 ± 0.1.

4. Hemodynamic Considerations of Liver Blood Flow

Although about one-third of the liver blood supply in the dog is delivered by the hepatic artery at approximately the pressure gradient existing in the systemic arterial system, the greater bulk of the flow enters the sinusoids at portal venous pressure, about 8 mm Hg. Hepatic venous pressure averages 2 mm Hg, so that the perfusion pressure gradient is 6 mm (Selkurt and Brecher, 1956). Myers (1954) has estimated this gradient to be about 10 mm Hg in man (13 mm Hg in the portal vein and 3 mm Hg in the hepatic vein).

Brauer *et al.* (1956) have made an extensive study of the hemodynamics of portal vein flow in the isolated rat liver. They obtained a mean flow of 2.79 ± 0.28 ml/gm/min at a mean perfusion pressure of

13.5 cm. When oxygenated plasma was used for perfusion, the figure increased to 4.99 ± 0.17, illustrating the effect of lowered viscosity on flow. The pressure-flow curves appeared to be sigmoid in shape (Fig. 85). Flow began at *ca* 4 cm for blood (critical closing pressure?) and increased more rapidly at first than did pressure (i.e., convex to the pressure axis). At about 13 cm, flow was approximately proportional to pressure, increasing to 3.6 ml/gm/min at about 17 cm. The plasma curve began at 2 cm and reached about 6.5 ml/gm/min at 18 cm and was more sigmoid in character. The augmented flow as pressure was elevated

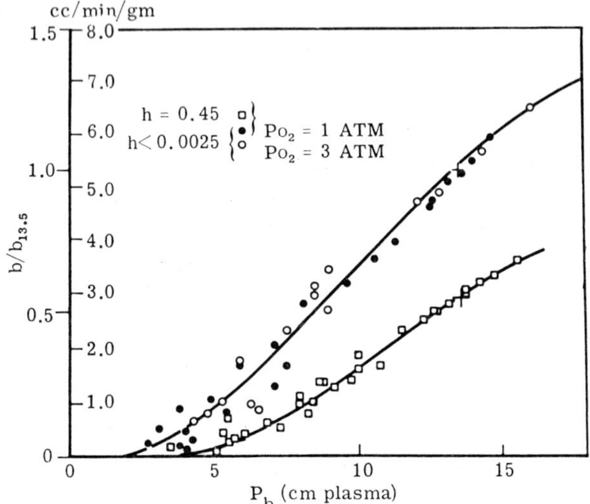

FIG. 85 Pressure flow relationship of the isolated rat liver preparation perfused with rat plasma (hematocrit < 0.0025) or with whole rat blood (hematocrit, 0.45). Flow is given in absolute values, as well as a ratio of experimental flow to the flow observed at a standard perfusion pressure of 13.5 cm plasma ($b/b_{13.5}$). (From Brauer *et al.*, 1956; reproduced with the permission of the American Journal of Physiology.)

was concluded to be the result of increase in mean effective vessel radius in the approximate range of 4–12 cm perfusion pressure.

Arising from these considerations, it is apparent that the level of perfusion pressure in the portal vein will be an important determinant of total liver flow. Thus, any modification of vasomotor tone (neurogenic or humoral) in the intestinal vascular bed could significantly influence liver flow, as would also splenic discharge. Finally, since the portal-hepatic vein gradient is a small one, it is conceivable that changes in inferior vena cava pressure transmitted to the hepatic veins could significantly alter flow. Brauer *et al.* (1959) found that increments of venous pressure as little as + 1–5 cm of water above the hilum of the liver caused visible congestion and formation of transudate on the liver surface.

5. Factors Which Modify Hepatic Blood Flow

a. Hemorrhage: Loss of blood curtails hepatic blood flow in proportion to the degree of depletion of the blood volume (Heinemann *et al.*, 1953; Werner *et al.*, 1952; Myers, 1954; Hamrick and Myers, 1955; Selkurt and Brecher, 1956; Reynell *et al.*, 1955; Sapirstein *et al.*, 1955). One of the surprising observations noted in most of this work, however, is that total splanchnic vascular resistance does not increase to any marked degree and that flow decreases in approximate proportion to blood pressure decrement. Indeed, it has been observed that after sustained hemorrhage, hepatic blood flow actually makes a certain degree of recovery within 23–70 minutes (Heinemann *et al.*, 1953). No explanation for this behavior of the splanchnic bed is forthcoming, although one may speculate that hypoxic states promote the accumulation of metabolites or humoral substances that favor splanchnic hyperemia. With restoration of blood pressure after transfusion, a marked overshooting of particularly the portal flow occurs during the early phase of normovolemic shock, the result of "opening up" of the intestinal bed during hypotension (Selkurt and Brecher). (For discussion of hepatic blood flow in shock, see (2) Shock, Section C.-4, Chapter XX.)

b. Other substances or conditions: Histamine causes an increase in hepatic blood flow in man, concomitant with a decrease in blood pressure, thus suggesting active dilatation of the splanchnic bed (Bradley, 1953). Insulin injection results in a somewhat delayed hepatic hyperemia, probably on a metabolic basis. It is suggestive of an epinephrine response (see Section E.-1, below) or other compensatory reactions to insulin-induced hypoglycemia (Shoemaker *et al.*, 1959). Ethyl alcohol, in narcotizing doses, given by gavage tube to dogs, has no significant effect on hepatic blood flow (Smythe *et al.*, 1953), while pyrogenic substances have a marked hyperemic action on the splanchnic circulation (Heinemann *et al.*, 1952; Bradley, 1949). Abdominal compression, creating an increase of *ca* 20 mm Hg in intra-abdominal pressure, results in reduction of hepatic blood flow, probably by raising portal pressure and restricting sinusoidal perfusion (Bradley, 1953). Hepatic ischemia, created by 1–2 hours of occlusion of the portal vein and hepatic artery, is followed by a period of elevated portal pressure and reduced effective hepatic blood flow (Selkurt, 1954b). During the phase of developing syncope induced by occlusion cuffs to the thighs, splanchnic blood flow is diminished, with an increase in calculated vascular resistance. At the time of fainting, hepatic blood flow decreases as blood pressure falls, but the calculated vascular resistance actually diminishes, due to an apparent "opening up" of the splanchnic bed at the time of the faint (Bearn *et al.*, 1951b). Muscular exercise reduces splanchnic circulation, thus shunting more blood to the muscles and brain (Bradley, 1949; Wade *et al.*, 1956).

6. Splanchnic Oxygen Utilization

The classic studies of Blalock and Mason (1936) have revealed that hepatic oxygen utilization in unanesthetized dogs averages 23.8 ml/min (range, 13.2–32.6), or an average of .045 ml/gm/min (.025–.065). Oxygen contents are: hepatic artery, average of 16.13 volumes per cent; portal vein, average of 10.51; and hepatic vein, average of 6.42. Mesenteric utilization (intestine, spleen, pancreas) has been calculated to be 20.4 m/min or a total splanchnic utilization of 44.2 ml/min. More recently, this has been confirmed at 38 ml/min (Werner *et al.*, 1952) and 43.15 (Selkurt and Brecher, 1956), or approximately 2.0 ml/kg/min (Hamrick and Myers, 1955).

In man, splanchnic oxygen consumption has been calculated in separate studies to be 41 and 34 ml/min sq M (Myers, 1947, 1954). From these data, oxygen utilization by the liver of man has been computed to be .05 ml/gm/min, a figure close to that of the dog liver.

The hepatic artery supplies *ca* 35 per cent of the liver oxygen supply, but this can vary up to 70 per cent (Selkurt and Brecher, 1956). Following hemorrhage, increase arteriovenous oxygen difference across the splanchnic bed provides a reasonably constant supply of oxygen despite reduction in blood flow (Werner *et al.*, 1952; Myers, 1954; Hamrick and Myers, 1955). With severe and prolonged hypotension, oxygen utilization of both the liver and intestine is depressed (Selkurt and Brecher). Under these circumstances, there is a trend for greater contribution from the hepatic artery, as portal oxygen tension drops to low levels.

7. Reservoir Function

In addition to hemorrhage and anoxia, reservoir function is called in by exercise, psychic influences, raised external temperature, certain types of anesthesia (chloroform), certain humoral agents and drugs (see Section E., below), and, subacutely, in estrus, pregnancy, and lactation. Under pentobarbital anesthesia, which apparently promotes maximal storage, severe hypoxia causes the spleen of the dog to discharge 16–20 per cent of the normal circulating blood volume. Such a finding is in line with older estimates of splenic discharge in dogs in response to exercise and hemorrhage (Barcroft *et al.*, 1927, 1929), under which circumstances an increase of 10.5 per cent in the systemic hemoglobin occurred. These findings have led to the estimate that blood stored in the spleen must carry approximately 40 gm per cent of hemoglobin, which corresponds to 95 per cent cell volume in the splenic stores. Although this seems unduly high, in lightly morphinized dogs the splenic hematocrit has been found to be 80 per cent, compared to the systemic average of 45 per cent. In line with this is the finding that the oxygen capacity of the splenic vein blood can be 80 per cent higher than arterial blood. During hypoxia, this may serve as a small emergency store of oxygen, for about 30–60 ml of oxygen are released with the 100–200 ml of stored splenic blood.

The volume of blood per gm of tissue is indeed high in the spleen, compared to other organs: in the dog, 0.420 per gm, as compared to 0.200 in the liver and 0.060 in the bowel. In the rat, 0.481 per gm is found in the spleen, as compared to 0.178 in the liver.

8. Splenic Blood Flow

Grindlay *et al.* (1939a), studying splenic inflow and outflow in intact dogs using thermostromuhrs, found that at rest and during sleep, flow averaged 97 ml/min in the splenic artery. During feeding, flow in both vein and artery increased by 26 to 100 per cent and lasted for 3–5 hours. With exercise on a treadmill, a rise in both arterial and venous flow was noted, but the change was delayed in the artery, signifying initial emptying of the spleen. Shivering brought on by sojourn in a cold room also elevated splenic flow, probably as a result of increased skeletal muscle activity. Sudden noise caused splenic discharge, as evidenced by a 38–154 per cent rise in venous flow for a minute, with no change in arterial flow. With hemorrhage, arterial flow fell sharply, while venous flow showed an initial transitory increase, again signifying discharge. Breathing 30 per cent CO_2 caused splenic contraction, acting via the splenic nerve and adrenal medulla discharge (Ramlo and Brown, 1959). (For the effect of drugs on splenic blood flow, see Section E.-4 of this chapter.)

9. Nervous Regulation of Splenic Circulation

Nervous control of the splenic circulation is achieved by sympathetic outflow in splanchnic nerve branches accompanying the artery into the spleen, presumably mediated by norepinephrine. Regulation is both by way of the smooth muscle of the capsule and trabeculae and by direct vasomotor action on the arterioles and sphincters. Pressures as high as 100 mm Hg can be developed by splenic contraction, if the vein is occluded. This response, combined with constriction of arterioles, forces the blood out. Simultaneous relaxation of the efferent sphincters aids discharge, which state is evidently initiated by reflex mechanisms, e.g., via chemoreceptor stimulation during hypoxia or baroreceptor stimulation during hemorrhage.

Since dilator fibers to the spleen have not been demonstrated, filling is taken to a passive process. This is achieved by relaxation of arterioles, through decrease in splanchnic vasomotor tone, and/or by constriction of efferent sphincters and possibly the splenic veins, by action of venomotor fibers.

10. Splenic Rhythm

Barcroft *et al.* (1932) have noted rhythmic changes in splenic volume in cats at the rate of 25–50 seconds. The alterations could be initiated by sudden elevation of blood pressure, by restoration of respiration after stoppage, by temporary interruption of the splenic circulation, and by histamine injection, the latter being accompanied by a decrease in blood pressure. The rhythm could also be induced in excised and perfused spleens. Small undulations in systemic arterial pressure accompanied this rhythmic change in volume, a small rise in blood pressure being associated with discharge of the spleen, and return to a normal level of pressure, with restoration of volume.

The significance of splenic rhythm is not known, beyond being associated with instability of the circulatory homeostatic mechanisms. In this regard, it has been found that strips of cat spleen suspended in a bath show no rhythmic activity until epinephrine or norepinephrine is added; then rhythmic activity is superimposed on a tonic contraction (DeGeus *et al.*, 1956). Grindlay *et al.* (1939b) believe that the splenic rhythm is not caused by rhythmic contractions of the splenic musculature but by rhythmic flow in the splenic circulation. Their evidence points to an arterial or arteriolar origin. Denervation does not abolish the rhythm.

E. PHARMACOLOGY

By K. I. Melville and H. E. Shister;[9] E. E. Selkurt[10]

1. Effect of Adrenergic Drugs on Liver Circulation

Injection of epinephrine or norepinephrine into either the hepatic artery or the portal vein of experimental animals leads to vasoconstriction, i.e., a reduction in inflow in the vessel into which the injection is made, and, to a lesser degree, in the other vascular bed also. Similar changes have been reported from studies both with the liver *in situ* and with saline-perfused isolated livers of dogs, cats, and rabbits.

A form of sluice valves has been described in the sublobular veins and sinusoidal capillaries of cats and rabbits, which can shunt blood into the central veins of the liver lobules (see Section C.-8 of this chapter). Small doses of epinephrine open these valves and thus lead to a diminution of liver volume. The effect has been ascribed to a constriction of all vessels of the liver, as well as to a distention of the hepatic veins. In considering the mechanism of action of the two amines, it seems probable that adrenergic substances injected into the portal vein diffuse from the sinusoids to the hepatic arterioles, the contriction of the latter leading to an increased resistance to flow from the hepatic artery to the liver sinuses.

In cats infusion of epinephrine into a systemic vein produces increased portal vein pressure, while in dogs there is a transitory increase followed by diminished flow of hepatic blood. Norepinephrine exerts a more striking constrictor effect. The initial dilatory effect in dogs is regarded as a reflex phenomenon, the constriction, as a direct action of the catecholamines. However, the problem requires further study.

In man, the intravenous infusion of epinephrine causes an increased hepatic blood flow (with a concurrent rise in hepatic output of glucose and in splanchnic oxygen consumption); norepinephrine leads to a slight decrease in flow (and no influence in hepatic glucose or splanchnic oxygen consumption). It has been estimated that in normal humans, intravenous epinephrine (0.1 μg/kg/min) increases the hepatic flow from the average normal of 1500 ml to 3000 ml. Demonstration of the effects of both of these amines on inflow in the liver of man is difficult because of reflex and associated hepatic metabolic responses, induced by intravenous injections.

[9] Authors of Subsections 1–3.
[10] Author of Subsection 4.

In recent years the technique of measuring the wedge pressure in hepatic veins (by catheterization) has been utilized as an index of portal pressure changes in man. Administration of epinephrine (0.2–0.3 ml of 1:1000 sol., subcutaneously or intravenously) to patients with normal livers results in an increase in the wedge-hepatic-vein pressure, accompanied by a parallel rise in the free-hepatic-vein pressure. Posterior pituitary extract (0.5 ml or 10 units), given intramuscularly, produces similar results. Kohn *et al.* (1959), have found that in patients with cirrhosis of the liver a paradoxical effect is obtained, i.e., no elevation of the wedge-hepatic-vein pressure following the use of both drugs. These workers have concluded that the hepatic artery is not a significant factor in the portal vein hypertension observed in cirrhosis.

The effects of other adrenergic drugs have not been systematically studied. Isoproterenol, in doses of 10 μg when injected into the portal vein or hepatic artery in dogs produces no effect (Green and Kepchar, 1959).

2. Effect of Blocking Agents on Hepatic Blood Flow

It has been shown that azapetine (Ilidar) can reduce the constrictor effect of epinephrine or norepinephrine when infused into the hepatic artery or portal vein. No vasodilator response is elicited by the use of Isuprel or epinephrine during adrenergic blockade.

3. Effect of Other Agents on Hepatic Blood Flow

a. Cholinergic agents: Acetylcholine (Ach) in small doses has no effect on the hepatic circulation of the dog; with larger amounts there is slight contraction of the portal venules and dilatation of the hepatic arterioles. No change in the hepatic or portal venules is observed with small doses in cats, while with large amounts constriction of the portal vein results.

b. Histamine: This drug shows distinct species variations. Thus, in dogs it constricts the hepatic vascular bed, whereas no such effect can be obtained in cat, rabbit, or monkey. In the presence of epinephrine, histamine appears to cause arterial dilatation rather than constriction in all species. The mechanism is unknown.

c. Nitrites and nitrates: The hepatic vessels evidently do not participate in the splanchnic dilatation following the use of amyl nitrite; both the portal pressure and the liver volume diminish as a result of the associated fall in systemic blood pressure (Melville, 1958).

4. Effect of Drugs on Splenic Blood Flow[11]

Ether anesthesia has no effect on splenic blood flow, while pentobarbital causes it to rise by an average of 32.6 per cent. Pitressin and acetylcholine decreases flow; histamine produces the same effect on arterial flow, but venous flow increases at first (discharge) and then becomes less. Epinephrine and ephedrine result in a temporary reduction in flow, preceded by a transitory rise (Grindlay *et al.*, 1939a). The findings of Otis *et al.* (1957), obtained with electromagnetic flowmeters, are more compatible with the conclusion that epinephrine and norepinephrine cause splenic discharge, since venous outflow and venous pressure rose concurrently with reduction in inflow. Such alterations could result either from arteriolar constriction, elastic recoil of walls of sinusoids and splenic tissue, or active contraction of the smooth muscle of the substance of the spleen.

The changes in flow correlate with vessel changes by microscopic observation. Epinephrine, norepinephrine, acetylcholine, and histamine induce vascular constriction in the spleens of mice (Fleming and Parpart, 1958), with epinephrine being the most potent and histamine, the weakest. Epinephrine brings about constriction of the penicilli (pulp arteries), as well as terminal arterioles. Those arteriovenous anastomoses of the same order of magnitude as the terminal arterioles are equally sensitive. Veins are never seen to react. Although epinephrine and acetylcholine are usually antagonists, the splenic vessels respond similarly to both, indicating that the constriction is probably a direct response of the vascular smooth muscle (muscarinic action).

References

Arey, L. B. (1941). Throttling veins in the livers of certain mammals. *Anat. Record* **81,** 21.

Aunap, E. (1931). Über den Verlauf der vena hepatica in der Leber. *Z. mikroskop.-anat. Forsch.* **25,** 238.

Banner, R. L., and Brasfield, R. D. (1958). Surgical anatomy of the hepatic vein. *Cancer* **11,** 22.

Barcroft, J., and Florey, H. (1929). The effects of exercise on the vascular conditions in the spleen and colon. *J. Physiol. (London)* **68,** 181.

Barcroft, J., and Stephens, J. F. (1927). Observations on the size of the spleen. *J. Physiol. (London)* **64,** 1.

Barcroft, J., Khanna, L. C., and Nisimaru, Y. (1932). Rhythmical contractions of the spleen. *J. Physiol. (London)* **74,** 294.

Bearn, A. G., Billing, B., and Sherlock, S. (1951a). The effect of adrenaline and

[11] By E. E. Selkurt.

noradrenaline on hepatic blood flow and splanchnic carbohydrate metabolism in man. *J. Physiol.* (*London*) **115,** 430.

Bearn, A. G., Billing, B., Edholm, O. G., and Sherlock, S. (1951b). Hepatic blood flow and carbohydrate changes in man during fainting. *J. Physiol.* (*London*) **115,** 442.

Bergstrand, I. (1957). Liver morphology in percutaneous lienoportal venography. *Kgl. Fysiograf. Sällskap. Lund, Förh.* **27,** 105.

Bilbey, D. L., and Rappaport, A. M. (1960). The segmental anatomy of the human liver. *Anat. Record* **136,** 165.

Björkman, S. E. (1947). The splenic circulation with special reference to the function of the spleen sinus wall. *Acta. Med. Scand.* Suppl. 191, 1.

Blalock, A., and Mason, M. F. (1936). Observations on the blood flow and gaseous metabolism of the liver in unanesthetized dogs. *Am. J. Physiol.* **117,** 328.

Bradley, S. E. (1949). Variations in hepatic blood flow in man during health and disease. *New Engl. J. Med.* **240,** 456.

Bradley, S. E. (1953). Determinants of hepatic haemodynamics. *In* "Visceral Circulation," Ciba Foundation Symposium (G. E. W. Wolstenholme, ed.), Part III, p. 219. Little, Brown, Boston, Mass.

Bradley, S. E., Ingelfinger, F. J., Bradley, G. R., and Curry, J. J. (1945). The estimation of hepatic blood flow in man. *J. Clin. Invest.* **24,** 890.

Brauer, R. W., Leong, G. F., McElroy, R. F., and Holloway, R. J. (1956). Hemodynamics of the vascular tree of the isolated rat liver preparation. *Am. J. Physiol.* **186,** 537.

Brauer, R. W., Holloway, R. J., and Leong, G. (1959). Changes in liver function and structure due to experimental passive congestion under controlled hepatic vein pressures. *Am. J. Physiol.* **197,** 681.

Brunschwig, A. (1953). Surgery of hepatic neoplasms. *Cancer* **6,** 725.

Casselman, W. G. B., and Rappaport, A. M. (1954). "Guided" catherization of hepatic veins and estimation of hepatic blood flow by bromsulphalein method in normal dogs. *J. Physiol.* (*London*) **124,** 173.

Couinaud, C. (1957). "Le foie. Études anatomiques et chirurgicales," p. 283. Masson & Cie, Paris.

Cull, T. E., Scibetta, M. P., and Selkurt, E. E. (1956). Arterial inflow into the mesenteric and hepatic vascular circuits during hemorrhagic shock. *Am. J. Physiol.* **185,** 365.

DeGeus, J. P., Bernards, J. A., and Verduyn, W. H. (1956). On rhythmic activity of the cat spleen. *Arch. intern. pharmacodynamic* **106,** 113.

Dickson, A. D. (1957). Development of the ductus venosus in man and the goat. *J. Anat.* **91,** 358.

Dobson, E. L., and Jones, H. B. (1952). The behavior of intravenously injected particulate material. *Acta Med. Scand.* **144,** Suppl. 273.

Douglass, B. E., Baggenstoss, A. H., and Hollinshead, W. H. (1950). The anatomy of the portal vein and its tributaries. *Surg., Gynecol. Obstet.* **91,** 562.

Ehrenbrandt, F., and Burckhart, T. (1956). Über Sperrarterien in der menschlichen Leber. *Acta Hepatol.* **4,** 3.

Elias, H. (1949a). A re-examination of the structure of the mammalian liver. I. Parenchymal architecture. *Am. J. Anat.* **84,** 311.

Elias, H. (1949b). A re-examination of the structure of the mammalian liver. II. The hepatic lobule and its relation to the vascular and biliary systems. *Am. J. Anat.* **85,** 379.

Elias, H. (1953a). Observations on the general and regional anatomy of the human liver. *Anat. Record* **117**, 377.

Elias, H. (1953b). Embryonic diversity leading to adult identity, as expressed in the development of the livers of vertebrates. *Anat. Record* **117**, 556.

Elias, H. (1955). Liver Morphology. *Biol. Revs. Cambridge Phil. Soc.* **30**, 263.

Elias, H., and Bengelsdorf, H. (1952). The structure of the liver of vertebrates. *Acta Anat.* **14**, 297.

Elias, H., and Pauly, J. E. (1960). "Human Microanatomy." Da Vinci Publ. Co., Chicago, Ill.

Elias, H., and Petty, D. (1952). Gross anatomy of the blood vessels and ducts within the human liver. *Am. J. Anat.* **90**, 59.

Elias, H., and Petty, D. (1953). Terminal distribution of the hepatic artery. *Anat. Record* **116**, 9.

Elias, H., and Popper, H. (1955). Venous distribution in livers: comparison of man and experimental animals and application to morphogenesis of cirrhosis. *A. M. A. Arch. Pathol.* **59**, 332.

Erkoçak, A. (1958). Recherches comparatives sur la circulation sanguine dans la rate et sur les rapports due sang avee la pulpe rouge. *Acta Anat.* **34**, 249.

Evans, H. M. (1912). The development of the vascular system. *In* "Manual of Human Embryology" (F. Keibel and F. P. Mall, eds.), Vol. II. p. 570. Lippincott, Philadelphia, Pa.

Fleming, W. W., and Parpart, A. K. (1958). Effects of topically applied epinephrine, norepinephrine, acetylcholine and histamine on the intermediate circulation of the mouse spleen, *Angiology* **9**, 294.

Fleming, W. W., and Parpart, A. K. (1959). Structure of the intermediate circulation of the rat spleen. *Angiology* **10**, 28.

Fraser, J., and Brown, A. K. (1944). A clinical syndrome associated with a rare anomaly of the vena portae system. *Surg., Gynecol. Obstet.* **78**, 520.

Gershbein, L., and Elias, H. (1954). Observations on the anatomy of the rat liver. *Anat. Record* **120**, 85.

Gilfillan, R. S. (1950). Anatomic study of the portal vein and its main branches, *A. M. A. Arch. Surg.* **61**, 449.

Glauser, F. (1953). Studies on intrahepatic arterial circulation. *Surgery* **33**, 333.

Glisson, F. (1654). "Anatomia Hepatis." O. Pullein, London.

Gordon-Taylor, G. (1943). A rare case of severe gastrointestinal haemorrhage; with a note on aneurysm of the hepatic artery. *Brit. Med. J.* **i**, 504.

Graf, K., Graf, W., Rosell, S., and Allgoth, A. M. (1957). Spontane Schwankungen der Leberdurchblutung des Menschen. *Pflügers Arch. ges. Physiol.* **266**, 36.

Grayson, J., and Mendel, D. (1957). Observations on the intrahepatic flow interactions of the hepatic artery and portal vein. *J. Physiol.* (*London*) **139**, 167.

Green, H. D., and Kepchar, J. H. (1959). Control of peripheral resistance in major systemic vascular beds. *Physiol. Revs.* **39**, 617.

Grindlay, J. H., Herrick, J. F., and Baldes, E. J. (1939a). Measurement of blood flow of the spleen. *Am. J. Physiol.* **127**, 106.

Grindlay, J. H., Herrick, J. F., and Baldes, E. J. (1939b). Rhythmicity of spleen in relation to blood flow. *Am. J. Physiol.* **127**, 119.

Gruenwald, P. (1949). Degenerative changes in the right half of the liver resulting from intrauterine anoxia. *Am. J. Clin. Pathol.* **19**, 801.

Hamilton, W. J., Boyd, J. D., and Mossman, H. W. (1952). p. 163. 2nd rev. ed., "Human Embryology," Williams & Wilkins, Baltimore, Md.

Hamrick, L. W., and Myers, J. D. (1955). The effect of hemorrhage in hepatic blood flow and splanchnic oxygen consumption of the dog. *Circulation Research* **3,** 65.

Hartroft, W. S. (1953). Cardiovascular lesions in choline-deficient rats. *In* "Symposium on Liver Injury." Josiah Macy, Jr., Foundation, New York.

Heinemann, H. O., Smythe, C. M., and Marks, P. A. (1952). The effect of hemorrhage and the pyrogenic reaction of hepatic circulation in dogs. *J. Clin. Invest.* **31,** 637.

Heinemann, H. O., Smythe, C. M., and Marks, P. A. (1953). Effect of hemorrhage on estimated hepatic blood flow and renal blood flow in dogs. *Am. J. Physiol.* **174,** 352.

Healey, J. E. (1954). Clinical and anatomic aspects of radical hepatic surgery. *J. Intern. Coll. Surgeons* **22,** 542.

Healey, J. E., and Schroy, P. C. (1953). Anatomy of the biliary ducts within the human liver; analysis of the prevailing pattern of branching and the major variations of the biliary ducts. *A. M. A. Arch. Surg.* **66,** 599.

Healey, J. E., Schroy, P. C., and Sorensen, R. J. (1953). The intrahepatic distribution of the hepatic artery in man. *J. Intern. Coll. Surgeons* **22,** 133.

His, W. (1885). "Anatomie menschlicher Embryonen." Vogel, Leipzig.

Hjortsjö, C. H. (1948). Die Anatomie der intrahepatischen Gallengänge beim Menschen, mittels Roentgen- und Injektionstechnik studiert, nebst Beiträgen zur Kenntnis der inneren Lebertopographie. *Kgl. Fysiograf. Sällskap. Lund, Handl.* [N.F.] **59,** 1.

Hjortsjö, C. H. (1951). The topography of the intrahepatic duct system. *Acta Anat.* **11,** 599.

Hochstetter, F. (1886). Anomalien der Pfortader und der Nabelvene in Verbindung mit Defect oder Linkslage der Gallenblase: Anomalien der vena coronaria ventriculi. *Arch. Anat. u. Physiol., Anat. Abt.* p. 369.

Horvath, S. M., Hutt, B. K., Knapp, D. W., and Werner, A. Y. (1953). Studies in bromsulphalein and cardiac output in the hepatectomized dog. *J. Physiol.* (*London*) **119,** 129.

Knisely, M. H. (1936). Microscopic observations of the circulatory system of living unstimulated mammalian spleens. *Anat. Record* **65,** 23.

Knisely, M. H. (1939). Microscopic observations on circulatory conditions in living frog liver lobules. *Anat. Record* **73,** Suppl. 2, 69.

Knisely, M. H., Bloch, E. H., and Warner, L. (1948). Selective phagocytosis. I. Microscopic observations concerning the regulation of the blood flow through the liver and other organs and the mechanism and rate of phagocytic removal of particles from the blood. *Kgl. Danske Videnskab. Selskab Biol. Skrifter* **4,** No. 7, 1.

Kohn, P. M., Charns, B. L., and Brofman, B. L. (1959). Effects of epinephrine and posterior pituitary extract on the wedged hepatic vein pressure in normal patients and in those with liver disease. *New Engl. J. Med.* **261,** 323.

Lee, Y. B., Elias, H., and Davidsohn, I. (1960). Vascular pattern in the liver of the mouse. *Proc. Animal Care Panel* **10,** 25.

Lewis, F. T. (1912). The development of the liver. *In* "Manual of Human Embryology" (F. Keibel and F. P. Mall, eds.), Vol. II, p. 403. Lippincott, Philadelphia, Pa.

McIndoe, A. H., and Counseller, V. (1927). Bilaterality of the liver. *A. M. A. Arch. Surg.* **15,** 589.

MacKenzie, D. W., Jr., Whipple, A. O., and Wintersteiner, M. P. (1941). Studies of the microscopic anatomy and physiology of living transilluminated mammalian spleens. *Am. J. Anat.* **68,** 394.

Maegraith, B. (1958). Sinusoids and sinusoidal flow. *In* "Liver Function" (R. W. Brauer, ed.), Chapter 2, p. 135. Am. Inst. Biol. Sci., Washington, D. C.

Märk, W. (1951). Zur Kenntnis der sog. Arterienwülste beim Menschen und bei einigen Säugern. *Anat. Nachr.* **I,** 305.

Mall, F. P. (1906). A study of the structural unit of the liver. *Am. J. Anat.* **5,** 227.

Mautner, H., and Pick, E. P. (1915). Über die durch Schockgifte erzengten Zirkulationsstörungen. *Münch. med. Wochschr.* **62,** 1141.

Melnikoff, A. (1924). Architecktur der intrahepatischen Gefässe und der Gallenwege des Menschen. *Z. Anat. Entwicklungsgeschichte* **70,** 411.

Melville, K. I. (1958). *In* "Pharmacology in Medicine" (V. A. Drill), 2nd ed., Part 9, Section 32, p. 482.

Michels, N. A. (1953). Collateral arterial pathways to the liver after ligation of the hepatic artery and removal of the celiac axis. *Cancer* **6,** 708.

Michels, N. A. (1955). "Blood Supply and Anatomy of the Upper Abdominal Organs, with a Descriptive Atlas." Lippincott, Philadelphia, Pa.

Myers, J. D. (1947). The hepatic blood flow and splanchnic oxygen consumption of man—their estimation from urea production or bromsulphalein excretion during catheterization of the hepatic veins. *J. Clin. Invest.* **26,** 1130.

Myers, J. D. (1954). The circulation in the splanchnic area. *In* "Shock and Circulatory Homeostasis" (H. D. Green, ed.), Vol. 4, Chapter 5, p. 121. Josiah Macy, Jr., Foundation, New York.

Otis, K., Davis, J. E., Jr., and Green, H. D. (1957). Effects of adrenergic and cholinergic drugs on splenic inflow and outflow before and during adrenergic blockade. *Am. J. Physiol.* **189,** 599.

Pack, G. T., Miller, T. R., and Brasfield, R. D. (1955). Total right hepatic lobectomy for carcinoma of the gallbladder; 3 cases. *Ann. Surg.* **142,** 6.

Parpart, A. K., Whipple, A. O., and Chang, J. (1955). The microcirculation of the spleen of the mouse. *Angiology* **6,** 350.

Peck, H. M., and Hoerr, N. L. (1951). The intermediary circulation in the red pulp of the mouse spleen. *Anat. Record* **109,** 447.

Pernkopf, E. (1932). Eine seltene Anomalie im Pfortader stammes zugleich Beitrag zur Entwicklungsgeschichte der Pfortader beim Menschen. *Z. Anat. Entwicklungsgeschichte* **97,** 293.

Popper, H. (1931). Über Drosselvorrichtungen an Lebervenen. *Klin. Wochschr.* **10,** 1693 and 2129.

Popper, H. (1948). Significance of agonal changes in the human liver. *A. M. A. Arch. Pathol.* **46,** 132.

Pratt, E. B., Burdick, F. D., and Holmes, J. H. (1952). Measurement of liver blood flow in the unanesthetized dog using the bromsulphalein method. *Am. J. Physiol.* **171,** 471.

Ramlo, J. H., and Brown, E. B., Jr. (1959). Mechanism of splenic contraction produced by severe hypercapnia. *Am. J. Physiol.* **197,** 1079.

Reifferscheid, M. (1957). "Chirurgie der Leber." Thieme, Stuttgart.

Reynell, P. C., Marks, P. A., Chidsey, C., and Bradley, S. E. (1955). Changes in blood volume and splanchnic blood flow in dogs after hemorrhage. *Clin. Sci.* **14,** 407.

Riedel, J., and Moravec, R. (1959). Das System der Pfortader und der Leberarterie in der Leber der Ratte. *Anat. Anz.* **107,** 99.

Rüttner, J. R., and Vogel, A. (1957). Elektronenmikroskopische Untersuchungen an der Lebersinusoidwand. *Verhandl. deut. Ges. Pathol.* **41,** 314.

Sapirstein, L. A. (1958). Indicator dilution methods in the measurement of the splanchnic blood flow of normal dogs. *In* "Liver Function" (R. W. Brauer, ed.), Chapter 2, p. 93. Am. Inst. Biol. Sci., Washington, D. C.

Sapirstein, L. A., Buckley, N. M., and Ogden, E. (1955). Splanchnic blood flow after hemorrhage. *Science* **122,** 1138.

Selkurt, E. E. (1953). Validity of the bromsulphalein (BSP) method for estimating hepatic blood flow. *Am. J. Physiol.* **175,** 461.

Selkurt, E. E., and Brecher, G. A. (1956). Splanchnic hemodynamics and oxygen utilization during hemorrhagic shock. *Circulation Research* **4,** 693.

Selkurt, E. E. (1954a). Comparison of the bromsulphalein method with simultaneous direct hepatic blood flow. *Circulation Research* **2,** 155.

Selkurt, E. E. (1954b). Effect of acute hepatic ischemia on splanchnic hemodynamics and on BSP removal by liver. *Proc. Soc. Exptl. Biol. Med.* **87,** 307.

Shoemaker, W. C., Mahler, R., Ashmore, J., and Pugh, D. E. (1959). Effect of insulin on hepatic blood flow in the unanesthetized dog. *Am. J. Physiol.* **196,** 1250.

Symthe, C. M., Heinemann, H. O., and Bradley, S. E. (1953). Estimated hepatic blood flow in the dog. *Am. J. Physiol.* **172,** 737.

Streeter, G. L. (1942). Developmental horizons in human embryos, description of age group XI (13 to 20 somites) and age group XII (21–29 somites). *Contribs. Embryol.* **30,** 211.

Tiedemann, F. (1822). "Tabulae arteriarum corporis humani." Müller, Karlsruhe, Germany.

Varićak, T. (1950). Die intrahepatischen multiplen Sperreinrichtungen im Bereiche der Lebervenen bei einigen Arctoidea. *Soc. Sci. Nat. Croat., Glasnik, biol. sekcije, period. biol., Ser. II/B* **2/3,** 106.

Varićak, T. (1952). Die Drosseleinrichtungen in den zu- und abführenden Leberblutbahnen beim Hermelin. *Acta Anat.* **16,** 203.

von Haller, A. (1756). "Icones Anatomicae." Vandenhoek, Göttingen.

von Schmidt, H., and Guttman, E. (1954). Röentgenuntersuchungen über die Verzweigung der grossen intrahepatischen Gallenwege. *Anat. Anz.* **100,** 277.

Wade, O. L., Combes, B., Childs, A. W., Wheeler, H. O., Cournand, A., and Bradley, S. E. (1956). The effect of exercise on splanchnic blood flow and splanchnic blood volume in normal man. *Clin. Sci.* **15,** 457.

Wakim, K. G., and Mann, F. C. (1942). The intrahepatic circulation of blood. *Anat. Record* **82,** 233.

Weiss, L. (1957). A study of the structure of the splenic sinuses in man and the rat with the light microscope and the electron microscope. *J. Biophys. Biochem. Cytol.* **3,** 599.

Werner, A. Y., MacCanon, D. M., and Horvath, S. M. (1952). Fractional distribution of total blood flow to and oxygen consumption of the liver as influenced by mild hemorrhage. *Am. J. Physiol.* **170,** 624.

Chapter XIII. RENAL CIRCULATION

A. EMBRYOLOGY

By R. Licata

1. Establishment of the Renal Circulation

a. Early development: The embryonic kidney or mesonephros and its related lateral segmental system of arteries provide the foundation for development of a renal vasculature (Keibel and Mall, 1912). The distribution of the lateral segmental (viscerorenal) arteries closely follows the histogenetic change involving the nephrogenic organ. Approximately thirty pairs of lateral segmental vessels supply the mesonephric mass and related structures. The organs receiving this supply, namely, the developing urogenital fold, provisional mesonephros, and dorsally situated permanent kidney, are crowded laterally to form a structural angle in relation to the aorta. Within the confines of the paravertebral gutter, the distal portions of the segmental arteries fuse into a vascular network, the rete arteriosum urogenitale. As a result, limited variability can be found in the postnatal pattern of the arterial rootlets derived from this mesh and in the proximal connection of these vessels with the aorta.

The appearance of the primordium of the permanent kidney or metanephros is accompanied by the timely regression of the mesonephros. In appropriating part of the mesonephric tubular system, the accompanying vasculature is incorporated into the framework of the definitive kidney. As a result, the primary divisions of the renal artery, which roughly include the interlobar vessels just beyond the region of the arcuate branchings, originate from vessels previously supplying the territory of the mesonephric tubular system. The cortical vessels, on the other hand, generally represent newer acquisitions. The remainder of the persisting lateral segmentals, not annexed by the permanent kidney, supplies the adnexa lodged in the field of the lateral body wall.

Departing from its original site deep within the pelvosacral cavity, the permanent kidney gradually ascends to lumbar levels. During this migration it appropriates successive pairs of lateral-stem arteries situated at each level, and rejects its connection with the vessels at hand, in anticipation of the favored segmental arteries lying at higher definitive points. From the latter are derived the main renal trunks.

b. Later development: Prenatally, the primordial kidney has a characteristic form, due to subgrouping of the developing nephrons into lobules. Although this lobulation is lost postnatally, it provides a developmental basis for a discrete segmental supply of the adult kidney. This is reflected in the fact that the major divisions of the renal artery develop into a number of interlobar branches that reach beyond the pelvis. Because of their multiple origin, the renal arteries normally consist of more than one stem. Distally, at the hilus of the kidney, the enlarging renal artery encounters a centrally located renal pelvis, around which it separates into a ventral and dorsal division. The subdivisions of the ventral ramus apparently develop more complexly than those of its dorsal mate.

Internally, the kidney may be described as undergoing a zonal histologic differentiation, which proceeds crescentically in a peripheral direction. For example, inner zonal layers develop and fade out in favor of outer glomerular zones. This event is documented postnatally in the architecture of the intrinsic blood supply of the adult organ. In accordance with the zonal maturation, lesser and greater arcs of circulation are developed. The first two proximal or hilar layers consist of vestigial and temporary tubular zones, respectively. Because of the degeneration of glomeruli, aglomerular arterioles (*arteriolae rectae verae*) of the adult kidney develop in the vestigial zone. From the temporary layer is differentiated the juxtamedullary zone, in which direct postglomerular arterioles (*arteriolae rectae spuriae* or *vasa recta*) arise, similar to the pattern normally developed from the cortex. The first vascular circuit thus derived represents one example of a number of arteriovenous bypass shunts, which constitute the lesser circulation in the postnatal kidney. The greater circulation is comprised of interlobular vessels and their pre- and postglomerular subbranchings, all of which fan out into the cortex where they are distributed in a segmental fashion without appreciably crossing intersegmental boundaries.

c. Renal vein: During the early stages of development, paired intricately interconnected venous lines of drainage are laid down in the abdomen (Keibel and Mall, 1912; Arey, 1946). From these channels are carved principally the visceral collecting circuits of circulation. Developmentally the rudiments of the renal veins are intimately associated with formation of the intermediate segment of the inferior vena cava. Along the mesial borders of the mesonephros develop two wide venous channels, called the subcardinal veins. Because of an increase in size of the provisional kidneys, with bulging toward the midline, the subcardinal veins approach each other centrally where they become confluent and produce an intersubcardinal anastomosis. Subsequently a drastic rearrangement of the primary arcs of drainage results in the rerouting of blood into a main-line channel. In this process the left subcardial vein drops out in favor of the right channel, which is retained and differentiated into part of the abdominal inferior vena cava. An unpaired main line system of veins is thus developed which absorbs the contiguous segment of the intersubcardinal anastomosis into the developing cava and converts the distal segment into the left renal vein. The asymmetrical structure of the renal veins postnatally reflects this growth. The developing renal trunks split into anterior and posterior divisions at the renal hilus. Within the substance of the kidney the secondary divisions of the renal vein follows less discretely the pattern of primary segmentation of the organ than is observed in the case of the arterial supply.

B. GROSS ANATOMY

By N. A. Michels

1. Arterial Blood Supply

a. Renal arteries: In most cases (72%), the kidney is supplied by one renal artery, derived from the aorta at the vertebral level of L 1 or between L 1 and L 2. Dual renal arteries from the aorta to the hilus occur in 10 per cent of cases (Fig. 86), while triple hilar renal arteries from the aorta are very rare (1–2%) (Merklin and Michels, 1958). The length of the right renal artery to its point of division varies from 0.5–8 cm; that of the left varies from 0.5–6 cm (Levi, 1909). The caliber of the right and left renal artery is usually the same, the average diameter being 5.5 mm, with variations from 4–7 mm (Hou-Jensen, 1930). The caliber of supernumerary renal arteries varies from 1–6 mm, the largest being the renal polars.

In most instances, the renal artery divides into a ventral and dorsal trunk near the hilus or midway between the latter and the aorta. Its division may be early, occurring immediately after its aortic origin and thus giving the impression of the presence of dual renal arteries. The ventral trunk gives rise to the anterior superior, anterior medial, and anterior inferior rami. The dorsal trunk courses posteriorly to the pelvis, being previously known as the retropyelic ramus (Levi, 1909). As a sequence to the discovery of Graves (1954) that the kidney, like the lung and liver, is a segmental organ composed of five portions (apical, upper, middle, lower, and posterior), the three terminal branches of the ventral trunk are now known as the upper, middle, and lower segmental

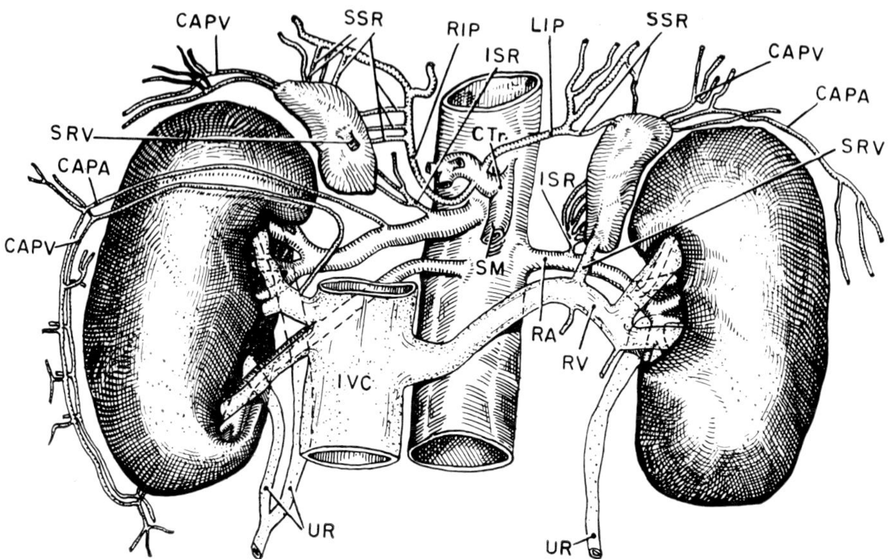

FIG. 86. Renal blood supply. Dual right renals, single left renal. Arteries: CAPA—capsular; CTr—celiac trunk; ISR—inferior suprarenal; LIP—left inferior phrenic; RA—renal; RIP—right inferior phrenic; SM—superior mesenteric; SSR—superior suprarenal. Veins: CAPV—capsular; IVC—inferior vena cava; SRV—suprarenal; RV—renal; UR—ureter. (From Merklin and Michels, 1958; reproduced with the permission of the Journal of the International College of Surgeons.)

branches, while the dorsal trunk, a continuation of the renal artery, constitutes the posterior segmental branch (Fig. 87).

When single, the renal artery gives off no other renal branches than those entering the hilus or its borders. Constant branches of the single renal are the inferior suprarenals (1 to 3), while an inconstant, yet relatively frequent, division is the testicular or ovarian artery. The latter is a component of the renal pedicle in 16 per cent of cases on the right and 34 per cent on the left (Notkovich, 1956). Atypically, parenchymatous branches may arise from the single renal trunk, from its anterior or posterior division, or from one of its terminal rami. Such branches become distributed to the upper or lower pole of the kidney as renal polars,

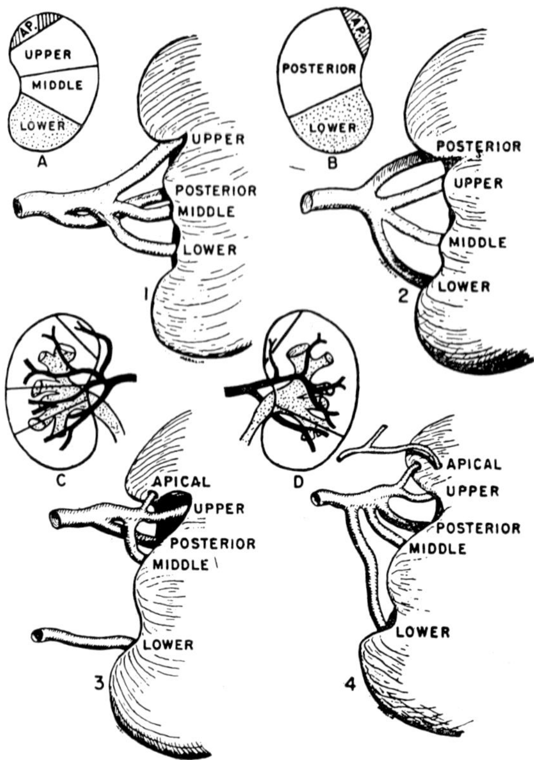

FIG. 87. Correlation of terminal renal branches with segments of kidney. A, anterior view of left kidney; B, posterior view. Kidney segments comprise: apical, upper, middle, lower and posterior; only the apical and lower being represented on both anterior and posterior planes. C and D show intrarenal segmental distribution of renal branches, each segment having its own artery with no collaterals. Dissection sketches: 1. Typical pattern showing 4 segmental branches; the fifth (apical) arises inside kidney, most likely from upper segment. 2. Four segmental branches arising from same point, apical again being intrarenal in origin. 3. Four segmental branches from main renal, apical, a branch of upper segmental, as most commonly the case. Second hilar branch from aorta becomes lower segmental, usually so with double hilar arteries. 4. Late origin of posterior segmental branch and a separate origin from aorta of the apical branch. (From Merklin and Michels, 1958; sketches A, B, C and D, modified after Graves, 1956; reproduced with the permission of the Journal of the International College of Surgeons.)

to the anterior and posterior lip of the hilus, or to any extrahilar surface of the kidney or its fatty capsule.

b. Supernumerary renal arteries: These vessels occur in about 30 per cent of kidneys, being present in equal frequency on the right and left sides. A single renal artery may be present on the right side, with multiple vessels (2 to 4) on the left, and vice versa (Fig. 86). Upper supernumerary renals in their course to the kidney may be retrocaval, while the lower ones may be precaval. The vertebral levels of the origin of the supernumerary renals may vary from T 11 to L 4, a fact to be taken into consideration in translumbar arteriography, in order to assure complete filling of all renal arteries with contrast medium. Supernumerary renals may be given off by the aorta from the level of the superior mesenteric to that of the inferior mesenteric, as high as the aortic diaphragm, and as far inferiorly as the hypogastric artery.

c. Superior polar arteries: Usually single, these vessels arise separately from the aorta (7%) or as a branch of the renal artery (12%) and supply the apical segment of the kidney (Graves, 1956) (Fig. 87). They may be double or triple with varied aortic or renal origin. Frequently, they arise from the inferior suprarenal or from its adipose capsular branch to the kidney. Conversely, the superior polar may give off several inferior suprarenals, a fact to be remembered in the removal of the adrenal gland. Cryptically concealed in connective tissue, superior polar arteries may be mistaken for strands of this tissue and torn in operative procedures, thus producing massive hemorrhage, at times immediately fatal (Eisendrath, 1920). The ligation or severance of superior polar arteries produces anemic infarcts, with resultant necrosis, since renal vessels are end arteries, having no anastomosis inside the kidney.

d. Inferior polar arteries: Generally single, these vessels are derived either from the aorta (5.5%) or from a branch of the renal artery (1.4%). When double, one may arise from the aorta, the other from the renal, or they may originate from either source. As ascertained by Graves (1956) using plastic casts, the inferior polars supply the lower segment of the kidney (Fig. 87). Clinically and surgically, these vessels are extremely important, since large series of cases have shown that they are a fundamental anatomic factor in the etiology of hydronephrosis. By constricting the upper ureter or the ureteropelvic junction, they cause renopelvic distention, with resultant intermittent pain as a sequel to retained urine. Surgical severance of the obstructing inferior polar has almost always resulted in a cure of the condition.

e. Perirenal arterial network: Adipose capsular branches from the renal, suprarenal, gonadal and lumbar arteries form a perirenal arterial circle in the subperitoneal tissue (Fig. 86). The nature and physiologic character of this long-forgotten subperitoneal plexus deserves extensive investigation as a physiologically unexplored, yet very possible route of collateral circulation to the kidney. The first anatomic description of this network was made by Sir William Turner (1863), while the most extensive one was presented by Albarran (1909).

2. Venous Drainage

a. Renal veins: In contrast to the diversity and complexity of the renal arteries, the renal veins have relatively simple patterns (Fig. 86), except for the left one, which, in many instances departs considerably from the normal pattern. (For the most extensive description of the variations in the tributaries and communications of the left renal vein, see Fagarasanu, 1938; Anson *et al.*, 1947, 1948; Anson and Kurth, 1955; for an account of the collateral venous drainage from the kidney, see Hollinshead and McFarlane, 1953.)

As with the renal arteries, the renal veins are predominantly single. In an investigation by Merklin and Michels (1958), the right kidney was found to have one vein in 79.3 per cent; two, in 16.3 per cent; three, in 3.3 per cent; and four in 1.1 per cent. The left kidney had one vein in 96.9 per cent and two in 3.1 per cent. In nearly all instances the renal veins are ventral to the aorta and renal arteries, but at a variable distance from the kidney hilus, their branches intermingle with those of the arteries. A tributary to the renal vein from the dorsal aspect of the kidney occurs frequently (30%). Being retropelvic, it is of surgical import (Brödel, 1901).

The right renal vein, which is much shorter than the left, has no tributaries. As already mentioned, the left renal vein can be amazingly varied in its tributaries and communications. Commonly, it receives the inferior phrenic and suprarenal veins as a common trunk on its cranial border and the testicular or ovarian vein on its caudal border. Aberrantly, the left renal may split, leaving the gonadal vein, to pass through the slit; in other cases, it arches over the renal vein (Notkovich, 1956). In varied manner, the left renal communicates with the second and third lumbar veins, with the ascending lumbar vein and, ultimately, with the hemiazygous root, thereby affording a communication with the vertebral system. More and more, the latter system is being accepted as a common route of cancer metastasis to the brain from various regions of the body. (For detailed description of the vertebral venous system, see Batson, 1940, 1957.)

C. MICROSCOPIC AND SUBMICROSCOPIC ANATOMY[1]

By D. C. PEASE

1. FINE STRUCTURE OF GLOMERULAR CAPILLARIES

a. Pattern of glomerular circulation: Many features of the glomerular circulation, as we understand it today, were established by 19th century morphologists. Thus, it was realized early that an afferent vessel entered the renal corpuscle and immediately subdivided into a small number of major trunks. These then distributed to capillary territories which were almost isolated as lobules. Efferent vessels assembled in reverse order, combining to produce a single efferent arteriole which distributed to the neighboring peritubular capillary bed. It was realized that the afferent arteriole was a substantially larger vessel than the efferent one. An excellent historical account of the development of this knowledge is to be found in the review of von Möllendorff (1930). Unfortunately Vintrup (1928) introduced the apparently erroneous idea that the capillary bed consists of simple loops between afferent and efferent arterioles. This has received widespread acceptance. It is the recent work of Hall (1955) which forces us to return to the original concept that the capillary bed anastomoses freely, at least within the territory of a single "lobule."

b. Histologic constituents: The fine structure of the glomerular capillary bed has proven to be close to or beyond the resolving power of the light microscope, so that very important discoveries necessarily had to await the development of electron microscopy. However, two important facts have been known for many years. First, epithelial cells are associated with the glomerular capillaries. These can be observed and studied without difficulty during the histogenesis of the renal corpuscle. In the mature corpuscle, however, conventional microscopists have had difficulty distinguishing epithelial cells from endothelial cells unless their position could be ascertained by seeing them in profile. Thus, there was no reliable way of determining the proportion of one cell type to the other, and it was not possible to decide whether or not the epithelial layer was complete over the surface of each capillary.

The second important fact established by conventional microscopy was the obvious presence of an unusually thick basement membrane

[1] For a general discussion of the submicroscopic structure of blood vessels, see Section C., Chapter I.

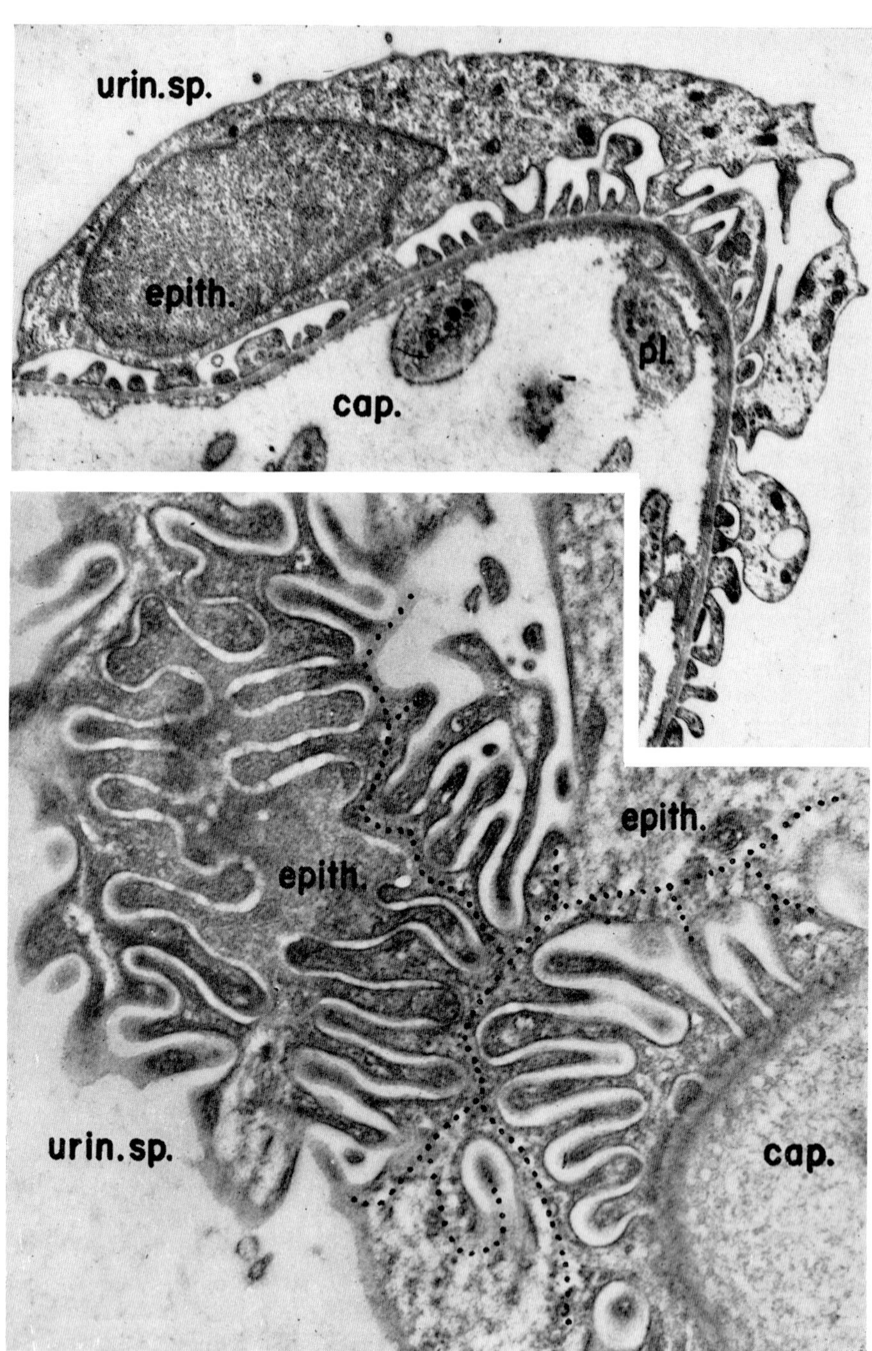

FIG. 88. Epithelial cell (epith.) within the urinary space (urin. sp.), making

between endothelium and epithelium. Its mucopolysaccharide nature has been repeatedly demonstrated by the histochemical PAS-reaction.

Conventional microscopists also have concerned themselves with the question as to whether or not a "mesangium" exists, that is to say, whether or not connective tissue cells are found between adjacent capillaries. In some submammalian forms, particularly birds, there is no question but that aggregates of morphologically unspecialized cells are to be found in this position in the vicinity of the vascular pole (Pak Poy and Robertson, 1957), and this has strengthened the arguments of those who would sponsor the concept of a mesangium.

Electron microscopy has importantly altered our concept of the epithelial cells of the glomerulus and has disclosed that the endothelium is fenestrated. It also has become quite apparent that there is no mesangium in most places. Instead, endothelium and epithelium share a common basement membrane. Other extracellular connective tissue elements are missing in the mammalian glomerulus, except insofar as they intrude at the vascular pole.

c. Fine structure of epithelial cells: The most astonishing morphologic specialization of the glomerulus is that of the epithelial cells. These do not form an epithelium in any ordinary sense of the word. They make contact with the basement membrane only by way of their foot processes, as Fig. 88 indicates. The principal mass of the cell body is suspended in the urinary space. The cells may be described as stellate, with major arms extending considerable distances over the capillary surface and between adjacent capillaries. These major arms interdigitate with those of neighboring cells. Each of these major arms breaks up into great numbers of small foot processes which make contact with the basement membrane. The foot processes are elongate structures (Fig. 89) which quite exactly interdigitate with foot processes from neighboring arms. No doubt a neighboring system often belongs to a different arm of the same cell, but, of course, it might also belong to an adjacent cell. Frequently a precise regularity of alternation can be observed, as in Fig. 89 and in the lower part of Fig. 91. The entire surface of the glomerular capillaries is effectively covered by these interdigitating foot processes. Details of this extraordinary morphology were first seen by Gautier *et*

---

contact with the capillary surface only by means of its specialized foot processes. Blood platelets (pl.) can be seen within the capillary lumen (cap.).

FIG. 89. Section nearly tangential to the surface of a capillary. Several systems of epithelial feet interdigitate one with the other in this view. The ramifications of a single arm of an epithelial cell are indicated by the dotted line. Over 20 individual feet belonging to this one system can be counted.

al. (1950), Oberling *et al.* (1951), Dalton (1951), Hall *et al.* (1953a,b), and Hall (1954a,b). Techniques of electron microscopy were relatively primitive when these original investigations were undertaken, and subsequently Pease (1955a,b) and Rhodin (1955) confirmed and extended the findings. The former investigator suggested that a slight swelling of the foot processes might effectively create an epithelial barrier to the passage of glomerular filtrate, while, conversely, a small shrinkage would open a free passageway so that the epithelial layer would cease to be a barrier of any sort. Hall (1957) has come to believe that the space between neighboring foot processes ordinarily is approximately 100 Å wide and acts essentially as a "slit-pore" controlling the movement of glomerular filtrate. It is obvious from the pathologic studies of Farquhar *et al.* (1957), Spiro (1959) and numerous others that the epithelial cells are sensitive to noxious influences and easily lose their complexity, so that pathologically they may often become simple flattened sheets of protoplasm without the specialized foot processes of the normal glomerulus. In retrospect, it is apparent that suggestions of epithelial feet were observed by conventional microscopy by Zimmermann (1915) and by von Möllendorff (1927). With foreknowledge of their existence, they can be demonstrated in well-preserved material today with a light microscope.

d. Fine structure of endothelium: The fenestrated endothelial sheet was first seen and correctly understood by Gautier *et al.* (1950). Hall *et al.* (1953a,b) and Hall (1954a,b) impressively demonstrated the pores through the attenuated endothelium following formalin fixation and extraction of the embedding medium. Although this would not be regarded as a refined technique today, it sufficed to emphasize in unequivocal terms that the endothelium was, in fact, fenestrated. This has been repeatedly confirmed since by Pease (1955a,b), Rhodin (1955, 1958), Bergstrand (1957), and others. The pores are best seen in a section tangential to the surface of a glomerular capillary, as in Fig. 91. Under these favorable circumstances, the endothelial sheet is viewed in the horizontal plane, and it is quite apparent that the pores are of reasonably uniform size (slightly more than 1000 Å in diameter) and are quite regularly spaced. In transverse section (Fig. 90) the pores are easily apparent, but their uniformity is not obvious. Although the diameter of the pores in such preparations may have been altered by preservation artifact, it is evident that the order of size is such that the endothelium cannot be regarded as a molecular barrier, for even the largest proteins easily could pass through the pores.

Although it is obvious in electron micrographs of most glomerular material that the endothelium is truly porous as it is observed, some doubts have been raised as to whether or not this situation exists in life. Rinehart (1955) felt that an extremely thin cell membrane bridged the regions of the pores. An osmiophilic line can, in fact, sometimes be seen here. Miller (1958) has published a high magnification micrograph which shows this clearly (his Fig. 16). Such a line looks exactly like those which may be present between the epithelial feet at the interface between the "outer cement substance" and the urinary space (visible in Fig. 90). In the latter case, it seems evident that this represents nothing more than a condensation of osmiophilic material on the surface of the basement membrane, and that the explanation applies equally well to the interface between basement membrane and serum. However, a final decision will have to await micrographs of very great resolution of material embedded in one of the newer media which will minimize artifact. Then, presumably, the limits of the endothelial plasma membrane will be traced unequivocably, and the matter finally settled.

e. Fine structure of basement membrane: In view of the likelihood of openings through both endothelium and epithelium, it would seem that the basement membrane must be the principal "filter" between blood and glomerular filtrate. It is therefore particularly important to consider its structure. When it is observed in transverse section, it can be said to consist of three layers (Fig. 90). In the middle is the "lamina densa" which presumably corresponds to the structural part of the total membrane. On either side of this are layers which are relatively electron transparent. The innermost of these layers provides a bed upon which the attenuated endothelial sheet spreads out. On the outer side the epithelial feet make contact with the corresponding layer and are slightly embedded in it. On the outer side one often can see its external limit, demarcated by a thin line of condensed material at its surface. This is visible in Fig. 90 between a number of the epithelial feet. Pease (1955a,b) spoke of these relatively translucent layers as consisting of "cement substance," and he (1960) emphasized that this sort of layer was common to all basement membranes. Its extension is to be found between adjacent cells in all epithelia, as well as constituting the "gap substance" of infolded plasma membranes as these are seen, for example, in the basal region of kidney tubule cells.

From histochemical studies it is clear that the basement membrane consists largely of mucopolysaccharide (PAS-positive). It can be regarded as a gel. In electron micrographs it ordinarily appears homogenous even with great resolution (Fig. 91). However, under some circumstances (particularly after phosphotungstic acid staining), fibrillar

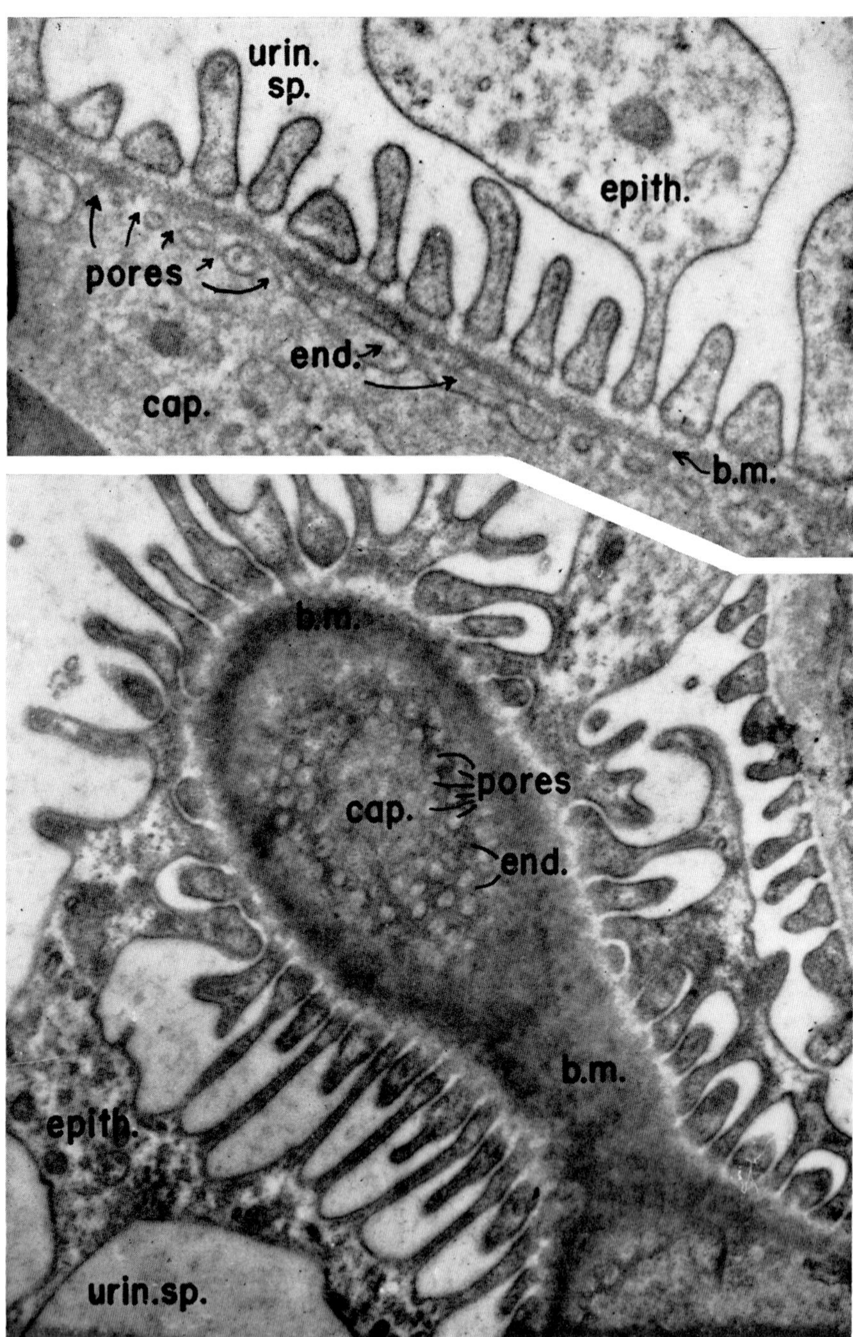
urin.
sp.
epith.
pores
end.
cap.
b.m.
b.m.
pores
cap.
end.
b.m.
epith.
urin.sp.

components can be seen, although it remains a moot question as to whether these may be fixation artifacts (Pease, 1960). There are no visible pores through the basement membrane in any organized sense, although some early work with unrefined techniques indicated this. It appears certain, however, that the basement membrane presents a diffusion barrier to at least very large molecules, even though it may be completely porous to water, salts and small organic molecules. Diffusion of the latter may possibly be controlled by the spacing of the epithelial feet in accordance with the "slit-pore hypothesis" of Hall (1957, 1960).

In the light of our present knowledge of the fine structure of the glomerular capillaries, it seems fair to call these channels the most remarkable blood vessels in the body. Their wall structure, no doubt, is partly related to the need for special reinforcement, for they derive nothing more than hydrostatic support from surrounding tissues. The internal blood pressure is regarded as relatively high for vessels of this size. No doubt part of the strength of the vessel wall comes from the relatively massive basement membrane. Indeed, in pathologic conditions glomerular capillaries can be seen denuded of epithelium which none the less had not ruptured (Pease, 1960). It seems eminently reasonable, however, that the system of epithelial feet over the surface of these vessels contributes additional support.

f. Question of "mesangium": Electron microscopy has demonstrated that mesangium does not exist in most parts of a glomerulus. It is evident that the ordinary capillary wall consists only of endothelium, epithelium, and intervening basement membrane. The occasional authors in the recent literature (Yamada, 1955; Pak Poy, 1958) who have described mesangial cells at least would agree that they have limited dis-

---

FIG. 90. High magnification micrograph of a transverse section of a glomerular capillary wall. Epithelial feet are conspicuous, slightly embedded in a "cement layer" overlying the dense portion of the basement membrane (b. m.). Very attenuated endothelium (end.) underlies the basement membrane, again separated from it by a thin "cement layer." The endothelium is fenestrated (pores).

FIG. 91. Tangential section of the surface of a glomerular capillary. Alternating interdigitations of different systems of epithelial feet (epith.) show in the urinary space (urin. sp.). Just inside the line of epithelial feet is a pale zone representing the "outer cement layer" of the basement membrane. The dense portion of the basement membrane is the homogenous dark material (b. m.). Passing through this, the "inner cement layer" is not conspicuous, but then comes the sheet of attenuated and fenestrated endothelium (end.). It can be observed that the pores are of nearly uniform diameter, approximately 0.1 μ, and are closely spaced. Finally, in the center, the plane of section breaks through into the capillary lumen (cap.) filled with plasma.

tribution. Certainly some of them have been misled by curious planes of section through regions of complex geometry. Furthermore, a number of the cells they have seen are suspect and probably endothelial. However, it is quite possible that there may be, under some circumstances, relatively undifferentiated cells inside the renal corpuscle. Clusters of such cells may sometimes be seen in the vicinity of the vascular pole, associated with the primary branchings of the afferent and efferent vessels, but records are inadequate to indicate whether or not such undifferentiated cells are confined to the glomeruli of young animals, or perhaps represent only an occasional embryonic rest in adults. They are remindful of the knot of undifferentiated cells regularly seen in this position in the Avian glomerulus (Pak Poy and Robertson, 1957). Pak Poy (1958) suggests that some may be related to fibroblasts but he avoids comment on whether or not he regards them as fully differentiated.

2. Fine Structure of Peritubular Capillary Bed

a. Pattern of blood distribution: It is generally agreed at present, that all of the blood supply for the tubular portions of a normal kidney passes first through glomeruli and then reaches peritubular areas by way of efferent arterioles. Thus, even the medulla receives its blood from the somewhat oversized juxtamedullary glomeruli by way of the *arteriolae rectae* (*spuriae*). MacCallum (1939) has demonstrated that when there seem to be direct vessels which bypass the glomeruli, they are always the result of pathologic degeneration of pre-existing glomeruli and consequent alteration of the blood channels. Such vessels (*a. rectae verae*) do not exist in young normal kidneys.

The capillary bed of the cortex occupies most of the space between the convoluted tubules, there being scant connective tissue. Drainage is into interlobular veins which parallel the interlobular arteries to reach the arcuate veins. In the medulla the pattern is somewhat more complicated, in that descending arterioles are aggregated in more or less conspicuous bundles, along with ascending venules which return blood to the arcuate veins. Particularly in the outer zone of the medulla, arteries and veins are very closely associated with each other, to form a true *rete mirabile*, as indicated by Longley *et al.* (1960). The excellent review of von Möllendorff (1930) should be consulted for anatomic details, particularly those which relate to changing patterns in different zones of the medulla.

b. Fine structure of cortical capillaries: The fine structure of the peritubular capillary bed has been a neglected subject. Pease (1955a,b) studied the kidney cortex and emphasized that the endothelium was extremely attenuated, exhibiting a system of pores similar to those found in the glomerulus, except that they were smaller (approximately 0.05 microns in diameter) and more widely dispersed. Figure 92 shows at high magnification a transverse section of such a capillary wall, including several of the pores. Figure 93, at lower magnification, is a section tangential to the surface of a cortical capillary in which a portion of the endothelial layer is viewed in the horizontal plane. The uniform size of the pores and their distribution are readily apparent. The regularity of the system is such that it would be difficult, indeed, to account for such an arrangement upon any artifactual basis. Historically this was the second fenestrated endothelium to be described, following that of the glomerulus as the type example. Occasional authors have published micrographs which incidentally have shown the fenestrated endothelium, but usually without comment. Miller (1958) however, does point these out in his Fig. 15. It seems evident that such capillaries are specialized to present a well nigh minimal barrier to diffusion processes.

c. Fine structure of medullary capillaries: In the medulla, venous capillaries reach a very large size, yet continue to have the extremely attenuated, fenestrated endothelium that one observes in the cortex. As Pease (1955b) indicated, however, the attenuated and fenestrated portions of the endothelial sheet were not as widespread or as regularly organized as in the cortex, but existed as patches. Thus in Fig. 93, thin regions can be seen at the top and left, but at the right the endothelial sheet is quite irregular in thickness, with scattered fenestrations. These relationships are quite apparent in the micrographs of venous capillaries published by Longley *et al.* (1960), in relation to their study of the *rete mirabile.*

d. Fine structure of basement membrane: The peritubular capillaries all have their own basement membrane. Ordinarily the dense component of this is very thin, much thinner than the membranes underlying the tubules as can be seen in Figs. 92 and 94. The capillary basement membranes do not fuse with those of adjacent tubules, so that ordinarily a patent connective tissue space is visible. This, however, may be quite thin as in Fig. 92.

Between the dense portion of the basement membrane and the endothelial sheet, there is a relatively electron transparent zone that has been spoken of as "cement substance." This layer no doubt provides a bed for the endothelial sheet to rest upon. Layers of this sort are invariably found associated with basement membranes elsewhere. Sometimes a thin osmiophilic layer can be observed spanning endothelial

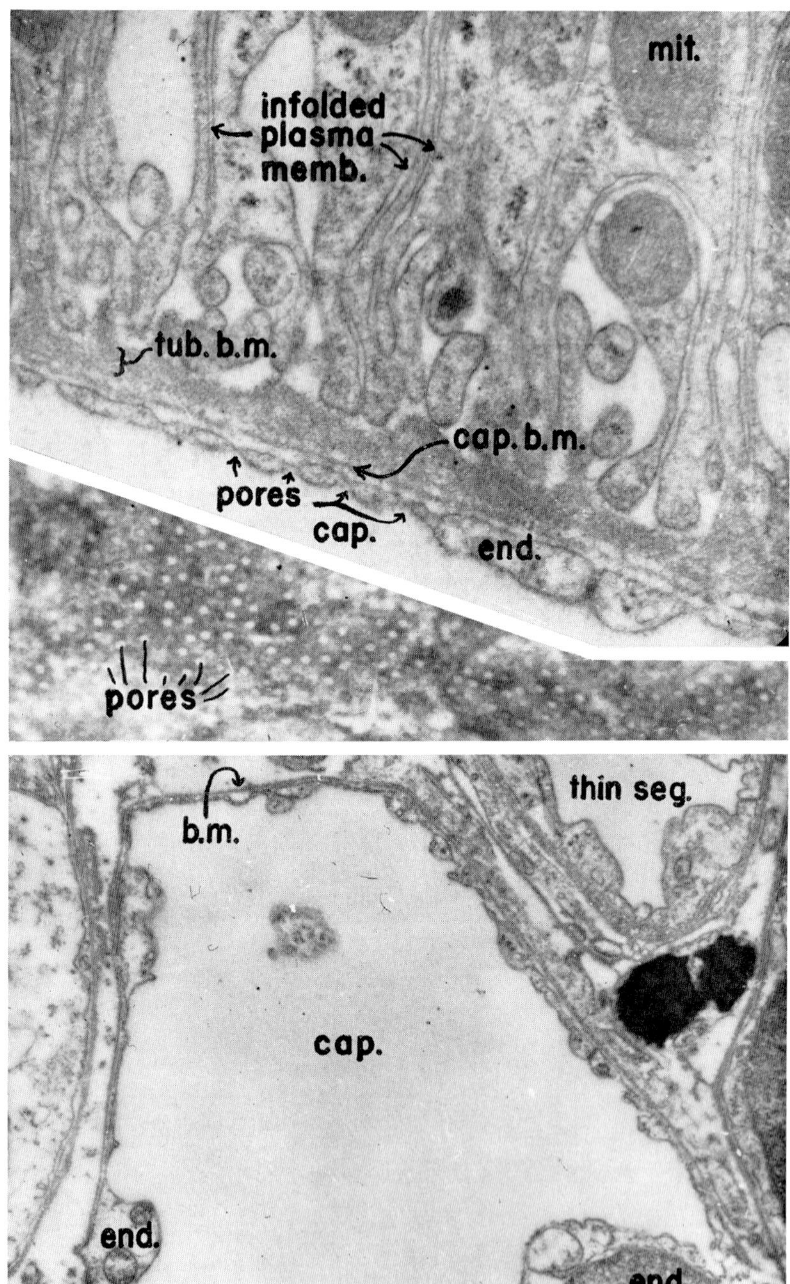
mit.
infolded
plasma
memb.
tub. b.m.
cap. b.m.
pores
cap.
end.
pores
b.m.
thin seg.
cap.
end.
end.

pores at the surface of the cement layer. Such lines are apparent in Fig. 92, particularly at the far left. Longley *et al.* (1960) believe that this represents the fused plasma membranes of the two sides of the endothelial sheet, without intervening cytoplasm. Thus they would deny that real pores exist through the endothelium. However, their micrographs lack the resolution necessary to demonstrate this. One is more inclined to think of such lines as simply representing the condensation of osmiophilic material at the interface between cement layer and blood serum. It will be necessary to employ great resolution, with well preserved material mounted in one of the newer embedding media, to settle this question. (For further discussion of the problem, see Section C.-1 of this chapter.)

e. Rete mirabile: The afferent arterial capillaries of the *rete mirabile,* as described by Longley *et al.* (1960), are curious in several respects. The endothelium of these vessels is relatively thick and lacks any suggestion of fenestration. Individual cells frequently overlap broadly, and they are pleomorphic. Pericytes are commonly present but do not form a complete layer. The endothelium shows histochemically an intense esterase activity. Although, as they descend into the medulla, the afferent arterial capillaries course close to ascending venous capillaries, the two types of vessels are not in direct contact. Thus there always seems to be an intervening connective tissue space. The anatomic pattern is such, however, that any given vessel of one type tends to be surrounded by vessels of the other sort. Because of the arrangement, Longley *et al.* believe that this region of the kidney fulfills the requirements Scholander (1951) has discussed for countercurrent exchange, as found, for example, in the swimbladders of fish. As such, it would help maintain the hyperosmolarity of the renal papilla.

FIG. 92. High magnification micrograph of a portion of capillary wall in close relationship to a distal tubule from the kidney cortex. The fenestrated nature of the endothelial sheet is apparent (pores). The very thin capillary basement membrane (cap. b.m.) is distinct from the much thicker basement membrane belonging to the adjacent tubule (tub. b.m.). Infolded plasma membranes are a conspicuous feature of the basal ends of these tubular cells. Mitochondria (mit.) are evident.

FIG. 93. Tangential section of the surface of a peritubular capillary showing fenestrations of the endothelial sheet (pores). These pores are approximately 0.05 μ in diameter (half the diameter of the corresponding pores of glomerular capillaries) and are much more widely spaced than those of the glomerulus.

FIG. 94. Low magnification micrograph of a venous capillary (cap.) in the medulla of a rat kidney. The great attenuation of the endothelial sheet (end.) is apparent. The thin basement membrane (b.m.), which underlies all such capillaries, is visible.

D. PHYSIOLOGY

By F. J. Haddy

1. Introduction

The purpose of the kidney is to help maintain homeostasis by removing waste products and by retaining useful substances. This organ is ideally suited to such a purpose by virtue of an unusual circulatory system, consisting of numerous short vessels and two sets of arterioles and capillaries in series. The capillary hydrostatic pressure in the first set is well in excess of colloid osmotic pressure, whereas that in the second is below this level. These unusual features provide a low resistance to the movement of blood (and thus a greater rate of flow than exists in most other organs), a very rapid rate of filtration at one site (glomerular capillary), and an almost equally rapid rate of reabsorption at another (peritubular capillary). Further, in the interest of homeostasis, the circulation is able to regulate the rates of blood flow, filtration and reabsorption as the need arises.

2. Blood Flow

The flow of blood through both kidneys in normal man is estimated to be slightly in excess of 1000 ml/min. It depends, as in other organs, upon the pressure gradient (the difference in pressure between the renal artery and renal vein) and the resistance to flow.

a. Pressure gradient: The flow of blood through the kidney can be varied by changing the pressure gradient. A rise of arterial pressure or a fall of venous pressure tends to increase flow, whereas a fall of arterial pressure or a rise of venous pressure has the opposite effect. Hence, in conditions characterized by low arterial pressure (hemorrhagic shock, cardiogenic shock) or high venous pressure (right heart failure, renal vein thrombosis), the rate of blood flow through the kidneys is low, in part, because of a reduction in the driving force.

b. Resistance to blood flow: Blood flow through the kidney may also be altered by a change in resistance. As in other organs, this factor has both geometric and viscous components (for further discussion of these points, see Sections D.-2g and E.-1a, Chapter II). The geometric component is a function of the length and radius of blood vessels. The renal vascular bed is short and has a large cross sectional area, both contributing to the normally low resistance to flow through this organ.

The radius is the most variable geometric dimension. It may change actively, that is, through vasoconstriction or vasodilation initiated by an increase or a decrease in the state of contraction of the smooth muscle within the walls of the vessels. Either may result from nervous, chemical or physical stimuli. Under physiologic conditions the sympathetic nerves do not appear to exert as much tonic activity in the kidney as in the limb, as indicated by the finding that resistance to blood flow is normally low, and acute denervation of kidney produces only a small and irregular fall of resistance (Sartorius and Burlington, 1956). That there is some mild tonic activity exerted by the sympathetic nerves is supported by the observations that stretching the carotid sinus and intravenous injection of pressor drugs produce a small fall of renal resistance, presumably from stimulation of the carotid baroreceptors with inhibition of neurogenic outflow (Page and McCubbin, 1953). It is of interest that direct or reflex stimulation of the sympathetic nerves can produce severe vasoconstriction. Faradic stimulation of the nerves in the hilus, carotid occlusion, section of the carotid sinus and aortic depressor nerves, and depression of blood pressure with some vasodilator drugs result in a large increase of renal resistance (Page and McCubbin, 1953). There is some evidence to indicate that the high renal resistance with resulting low flow rates, noted in hemorrhagic shock and congestive heart failure may, in part, be mediated reflexly through the carotid sinus (Barger, 1960).

Numerous chemical substances cause active change in the radius of renal vessels. The direct effect of noradrenaline, adrenaline, angiotensin, serotonin and pitressin is vasoconstriction (Page and McCubbin, 1953, Emanuel *et al.*, 1959), while that of acetylcholine and histamine is vasodilation (Page and McCubbin, 1953). Following systemic administration, these changes are, in general, antagonized by changes in the activity of the sympathicoadrenal system resulting from variation of arterial pressure in the carotid sinus. Reduction of the hydrogen ion and carbon dioxide concentrations by hyperventilation results in vasoconstriction, which is partially antagonized by activity of the sympathetic nerves (Emanuel *et al.*, 1957). A fall of oxygen tension, as by breathing 10 per cent oxygen, apparently results in vasodilatation (Fishman *et al.*, 1951). (For further discussion of the responses of renal vessels to pharmacologic agents, see Section E., this chapter.)

Among the physical agents which actively influence radius is pressure. Blood vessels resist distention by pressure with active contraction (Folkow and Lofving, 1956; Johnson *et al.*, 1960). This response may be involved in the unique capability of the kidney to keep its blood flow fairly constant despite large changes in systemic arterial pressure (Winton, 1956). The latter phenomenon, termed autoregulation of blood flow, is most apparent over a range of arterial pressure from 80–200 mm Hg. Such a progressive rise in resistance is mediated through some local mechanism, since it persists following denervation and even in the totally isolated kidney. The mechanism has been variously ascribed to changes in the geometric component of resistance, both active and passive, and to alterations in the viscous component of resistance, both static (that referable to glomerular filtration) and dynamic (that referable to axial streaming of red cells). More recent evidence tends to support the hypothesis that the resistance rise results from active constriction of preglomerular arterioles (Waugh, 1958; Haddy *et al.*, 1958a). Whatever the mechanism, the effect is to stabilize renal blood flow, glomerular capillary pressure and, hence, glomerular filtration rate, in the face of wide variations in arterial pressure. Such a response may be important to water and salt homeostasis, for it would indeed be unfortunate if glomerular filtration rate changed with each normal or pathologic variation in the level of systemic arterial pressure. Elevation of venous pressure also causes an increase of renal vascular resistance (Haddy *et al.*, 1958b). This may, in part, result from passive constriction of arterioles due to rise of interstitial pressure, but there is some evidence to indicate that these vessels also actively constrict. Such a response, together with that mediated via the carotid sinus, may account for the elevated renal vascular resistance in congestive heart failure. Hence, the low renal blood flow rate in this condition results both from a rise of resistance and a decrease of pressure gradient.

Radius may likewise change through passive mechanisms. A rise of transmural pressure (the difference in pressure across the vessel wall) tends to distend vessels, whereas a fall tends to collapse them. Transmural pressure may change either because of alterations in intraluminal or extraluminal (interstitial or tissue) pressure. Hence, elevation of arterial pressure over the range 0–70 mm Hg results in a passive increase in vessel radius (Haddy *et al.*, 1958a). The role of extraluminal pressure in determining vessel radius is also relatively important in the kidney. Renal tissue pressures may rise to high levels subsequent to increase of tissue fluid, vascular volume or tubular volume (Winton, 1956; Gottschalk, 1952; Gottschalk and Mylle, 1956). This is because both the stroma of the kidney and the renal capsule are relatively indistensible. Therefore, elevation of venous pressure may increase renal vascular resistance, in part, because the greater filling of veins raises tissue pressure, thus partially collapsing high resistance vessels (Haddy *et al.*, 1958b).

The resistance to flow may also change because of an alteration in blood viscosity. The viscosity of static blood may be high (polycythemia) or low (anemia), resulting in a slow or rapid flow rate, respectively, for a given pressure gradient. Further, a change of the filtration fraction varies the viscosity of blood perfusing post-glomerular vessels. The viscosity of dynamic blood may be affected by a change in velocity, an increase, as from elevation of blood pressure, causing axial accumulation of red cells and a resulting reduction of viscosity. A theory based on axial accumulation of red cells has been advanced to account for the apparent low hematocrit of intrarenal blood, as well as for autoregulation of blood flow and glomerular filtration rates (Kinter and Pappenheimer, 1956; Pappenheimer and Kinter, 1956). However, recent evidence indicates that axial accumulation is probably not an important factor in controlling resistance under usual conditions (Haynes and Burton, 1959).

3. Filtration

It is estimated that the glomerular filtration rate of both kidneys in normal men is slightly greater than 120 ml/min. The rate of filtration into the tubules depends upon the effective filtration pressure across the glomerular capillary membrane, as well as upon the nature of the capillary membrane.

Effective filtration pressure is equal to the sum of capillary hydrostatic pressure and filtrate colloid osmotic pressure minus the sum of capsular hydrostatic pressure and plasma colloid osmotic pressure. Because of the presence of a downstream arteriole, the pressure in the glomerular capillary is higher than in any other capillary of the body and has been estimated to be about 75 mm Hg (Winton, 1956). Capsular hydrostatic pressure is about 20 mm Hg and average colloid osmotic pressure is about 32 mm Hg (23 in the afferent arteriole and 40 in the efferent arteriole). In the normal state, filtrate colloid osmotic pressure is zero. The various pressures, therefore, result in a sizable driving force for filtration. (For further discussion of exchange mechanisms, see Section E.-2, Chapter VI.)

Some idea of the permeability of the glomerular capillary membrane may be obtained from the following estimates: filtration coefficient, 1.9–4.5 ml/min/mm Hg/100 g kidney; effective pore radius, 3.5–4.0 Å; pore surface, 500–1000 cm^2/100 g kidney, and ratio of pore surface to total surface of membrane, 5–10 per cent (Pappenheimer, 1955).

a. Effective filtration pressure: The circulation exerts control over the rate of filtration by varying capillary hydrostatic pressure. This may result from a change of arterial pressure, of upstream vessel caliber, of downstream vessel caliber, or of venous pressure. Considering only one variable at a time, capillary hydrostatic pressure tends to rise with elevation of arterial pressure, dilatation of upstream vessels (for example, afferent arterioles), constriction of downstream vessels (for example, efferent arterioles) or elevation of venous pressure. These factors, however, rarely operate alone. Hence, filtration rate rises with increasing arterial pressure over the range 40–80 mm Hg, but remains relatively constant over the range 80–200 mm Hg, probably because of progressive afferent arteriolar constriction (autoregulation of glomerular filtration rate). In the case of efferent arteriolar constriction, the rise in glomerular filtration rate is limited by an increase of colloid osmotic pressure. Elevation of venous pressure, contrary to expectation, decreases glomerular filtration rate, possibly because of a concomitant passive and active reduction of the calibers of preglomerular vessels (Haddy *et al.*, 1958b). As indicated above in Subsection 2, the afferent and efferent arterioles are also sensitive to nervous, humoral and physical stimuli. The details of their relative responsiveness to such stimuli and the resulting effect upon the rate of glomerular filtration are a fertile field for further investigation.

b. Capillary membrane: There is little evidence to indicate that the circulation can directly influence permeability of the capillary membrane. However, it may do so indirectly through changes in pressure. Hence, a rise in pressure (afferent dilatation or efferent constriction) may enlarge the area for filtration by opening closed capillaries and distending those already open and enhance permeability by increasing pore size.

4. Reabsorption

Of the tremendous amount of water filtered per day (180 liters), about 99 per cent is reabsorbed into the peritubular capillaries. This results, in part, because peritubular capillary pressure is well below colloid osmotic pressure. It is estimated that the blood that enters the peritubular capillaries has a colloid osmotic pressure of about 40 mm Hg (Winton, 1956). Direct measurements reveal average hydrostatic pressures in small and large peritubular capillaries of 14.0 and 18.7 mm Hg, respectively (Gottschalk and Mylle, 1956; Wirz, 1955). The circulation probably exerts control over the rate of reabsorption by varying capillary pressure and colloid osmotic pressure. A high filtration fraction likely tends to promote reabsorption by elevating colloid osmotic pressure, whereas a low rate of blood flow accomplishes the same by lowering hydrostatic pressure. Elevation of venous pressure may not effect reabsorption because it causes equivalent changes in intratubular pressure and peritubular capillary pressure (Gottschalk and Mylle, 1956).

5. Production of Vasoactive Materials

The most notable vasoactive substance produced in the kidney is renin. This substance has been implicated in acute renal hypertension, but its role in chronic renal hypertension is still obscure. There is some evidence to suggest that renin may be produced in the juxtaglomerular apparatus (Tobian *et al.*, 1958; Janecek *et al.*, 1960) (see (1) Arterial Hypertension, Section B.-3b, Chapter XX). However, other evidence suggests that the cells in this apparatus are mast-cell-like and, hence, may contain heparin, serotonin, and histamine (Corbascio, 1960).

E. PHARMACOLOGY

By K. I. Melville and H. E. Shister

1. Adrenergic Drugs

a. Epinephrine and norepinephrine: Both these amines increase the total resistance to flow through the kidney by a direct action. Comparative studies have shown that epinephrine reduces the normal renal blood flow in man from 1500 ml/min to 900 ml/min, while norepinephrine shows somewhat less activity, the reduction being from 1500 to 1200 ml/min (Barcroft and Swan, 1953). It is interesting to note that after intravenous injection of these adrenergic drugs, the reduction of blood flow through the kidney takes place in spite of the concomitant elevation of systemic blood pressure.

Denervation of the dog kidney increases its sensitivity to adrenergic drugs, and a more marked increase in afferent arteriolar resistance is observed than in normal animals. Infusion of epinephrine into the isolated perfused dog kidney or the intravenous injection in intact animals causes constriction of the efferent arterioles of the glomeruli, but with increased dosage, afferent arteriolar constriction also becomes apparent. In man, intravenous or intramuscular injections of epinephrine, norepinephrine, ephedrine, and paredrinol lead to a decrease in *p*-aminohippurate clearance, i.e., to a reduction of renal plasma flow. This is also true of the frog and dog (see reviews of Folkow, 1955, and Green and Kepchar, 1959).

In patients with renal disease, both epinephrine and norepinephrine cause renal vasoconstriction and increased clearance of plasma protein, presumably due to greater filtration of protein as a result of a slowed glomerular blood flow. In normal persons and dogs, these amines decrease urinary secretion of both sodium and potassium, probably also due to increased tubular reabsorption. (For complete review see Green and Kepchar, 1959.)

b. Isoproterenol (Isuprel) and other depressor amines: In the dog, intravenous administration of isoproterenol can cause a slight decrease of renal vascular resistance, but the response is less marked when the drug is directly injected into the renal artery; this is also true of other vasodepressor amines, e.g. ethyl-arterenol, suggesting that these agents exert little direct vasomotor effect in the kidney (Green and Kepchar, 1959).

2. Blocking Agents

a. Adrenergic blocking drugs: The constrictor action of epinephrine can be blocked by injections of piperoxan, tolazoline, and azapetine into the renal artery. Norepinephrine responses can also be blocked, but larger doses are required. No "reversal" of the epinephrine effect can be elicited. Normally, azapetine by itself causes dilatation, while piperoxan and tolazoline produce only slight constriction. The adrenergic blocking agents have no effect on renal vascular responses to angiotonin, serotonin, or tryptamine.

b. Ganglionic blocking drugs: In general, these drugs reduce the renal blood flow. If the simultaneous fall in systemic blood pressure is very marked, renal impairment can ensue. In a normal individual injections of tetraethylammonium leads to a decrease in renal resistance and arteriolar pressure. If epinephrine is then given, the pressor effect of the latter on renal vascular resistance is augmented rather than decreased. With norepinephrine there is an increase of urinary output, which is not influenced by hexamethonium.

3. Miscellaneous Agents

Amyl nitrite increases renal blood flow, while intravenous infusion of pitressin produces a variable response on calculated blood flow of the kidney, despite its intense antidiuretic action (Corcoran and Page, 1939).

F. PATHOLOGY

By C. L. PIRANI

1. INTRODUCTION

In considering the pathologic anatomy of the renal circulation, it is necessary to call attention to the fact that the renal vascular system is so constructed anatomically as to play an essential role in the filtering, excretory and resorptive functions of the kidneys. At the same time, this system, because of its particular anatomic arrangement and the large volume of blood which circulates through it under relatively high blood pressure, is particularly susceptible to a variety of disease processes which can affect it either directly or indirectly. Furthermore, the unique position of the glomerular capillaries, inserted in the renal arteriolar system, would appear to predispose them to disorders which usually do not involve arteriovenous capillaries elsewhere. All the different disorders affecting the renal vasculature may, therefore, lead to various degrees of renal insufficiency.

Besides such vascular diseases as benign and malignant nephrosclerosis (see (1) Arterial Hypertension, Section C.-3a,c, Chapter XX), diabetic nephropathy, pre-eclampsia and renal vein thrombosis, it is also apparent that the glomerulonephritides, no matter what the type or cause, are primarily diseases of capillaries and, hence, should also be considered vascular in nature. Furthermore, it is necessary to point out that even such conditions as chronic pyelonephritis and polycystic renal disease, which do not affect primarily the vascular system of the kidneys, are invariably associated with severe vascular changes which largely contribute to their clinical picture.

2. DIABETIC NEPHROPATHY

Diabetic nephropathy, which very frequently develops in patients with diabetes mellitus of many years duration, is characterized by the deposition of hyaline material in the glomerular tufts, in the wall of the small arteries, and along the basement membrane of the tubules (Kimmelstiel and Wilson, 1936; Allen, 1941; Muirhead *et al.*, 1956; Gellman *et al.*, 1959). In the glomeruli, this type of hyaline substance is at first deposited on the inner aspect of the capillary basement membrane, more abundantly in the axial region between capillary lumina, the so-called intercapillary space. It eventually forms nodular masses, the "Kimmel-

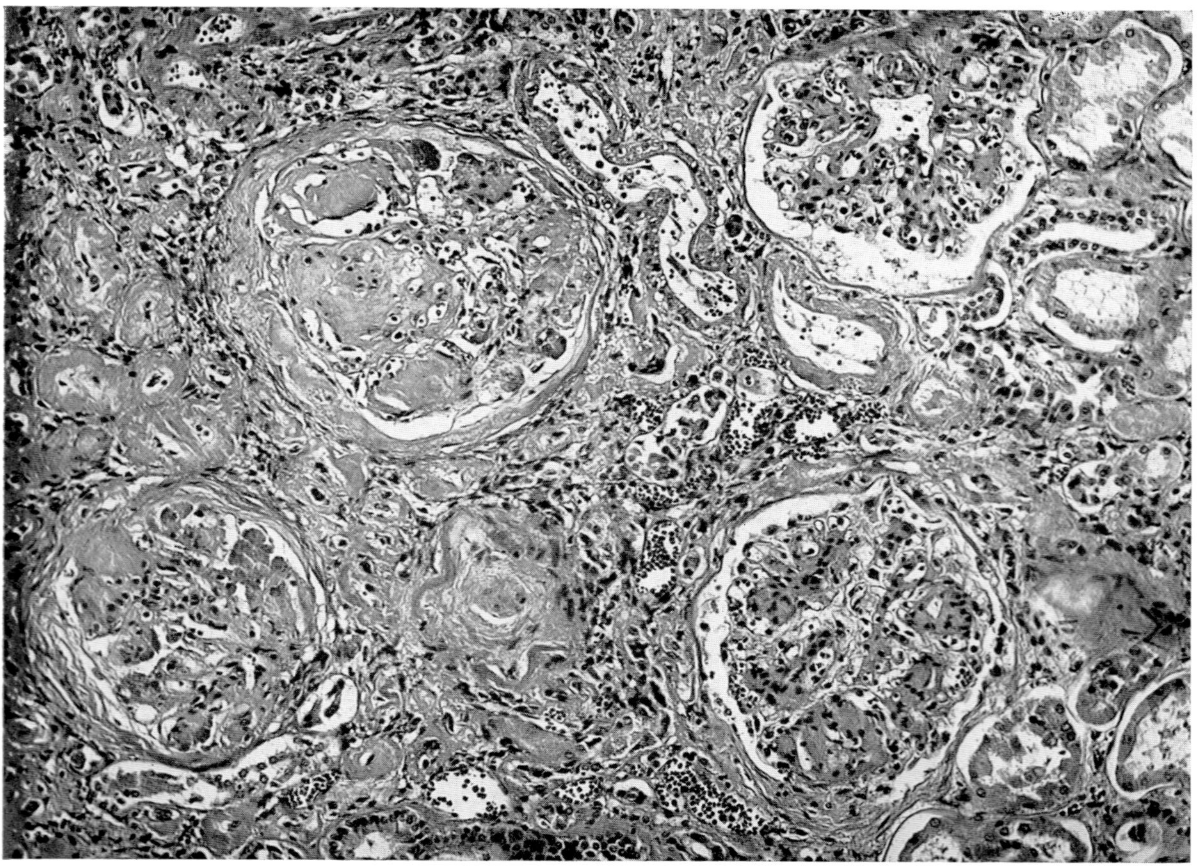

Fig. 95. Diabetic nephropathy. In the glomeruli, diffuse and nodular hyaline thickening of the capillary walls is noted. Hyaline thickening of Bowman's capsule and of tubular basement membranes is also present. H. and E.; ×190.

stiel-Wilson lesions," which project into the capillary lumen and are often associated with an aneurysmatic dilatation of the capillaries (Fig. 95). Recent electron microscopic studies of renal biopsies have led to a more precise localization of the hyaline substance within the wall of glomerular capillaries (Bergstrand and Bucht, 1959). In addition to hyaline arteriolar sclerosis, atherosclerosis of larger renal arteries is particularly prominent, involving the interlobar, arcuate and even interlobular arteries of the kidneys which, as a rule, are not affected by atherosclerosis in the absence of diabetes mellitus (Hall, 1952). (For changes in other organs produced by diabetes mellitus, see Section G.-3c, Chapter VI.)

The particularly severe and frequently observed hyaline vascular disease associated with diabetes is generally considered as an aggravation of the usual type of arteriosclerosis and is attributed to the longer life of diabetic patients, made possible by insulin and other hypoglycemic agents. The fact that even in the most severe cases of arterial and arteriolar nephrosclerosis, lesions of the diabetic type are not noted unless diabetes mellitus is present, makes these lesions specific for this disease and not for arteriosclerosis. It must be postulated, therefore, that certain metabolic abnormalities characteristic of the diabetic state, such as lipo- and glycoprotein serum changes, are important in determining the character of diabetic nephropathy. The morphologic similarities between the site of deposition of hyaline in diabetic nephropathy and that of amyloid in renal amyloidosis also support this point of view (Gellman *et al.*, 1959). It is of interest that while hypertension is a common feature of diabetic nephropathy, it is usually a relatively late manifestation of the disease and is not very marked. Further, malignant hypertension is extremely rare. From such findings, it seems likely that the histogenesis and physical character of the diabetic hyaline in the arterioles of the kidneys and elsewhere are entirely different from that of arteriosclerotic hyaline, and that, as a result, resistance to renal blood flow might be less impaired.

3. Toxemia of Pregnancy

In toxemia of pregnancy, a condition characterized by proteinuria, hypertension, and edema, many different renal diseases may be present. One, however, appears to be characteristic of pre-eclampsia-eclampsia, which is that type of toxemia developing in the last trimester of pregnancy. In the kidneys of pre-eclamptic women, the glomeruli are typically large and bloodless, as a result of swelling of endothelial and epithelial cells (Fig. 96), while the capillary lumens are extremely narrow (Sheehan, 1950). These changes are apparently due to edema which, to a lesser extent, also involves the wall of small arteries and the

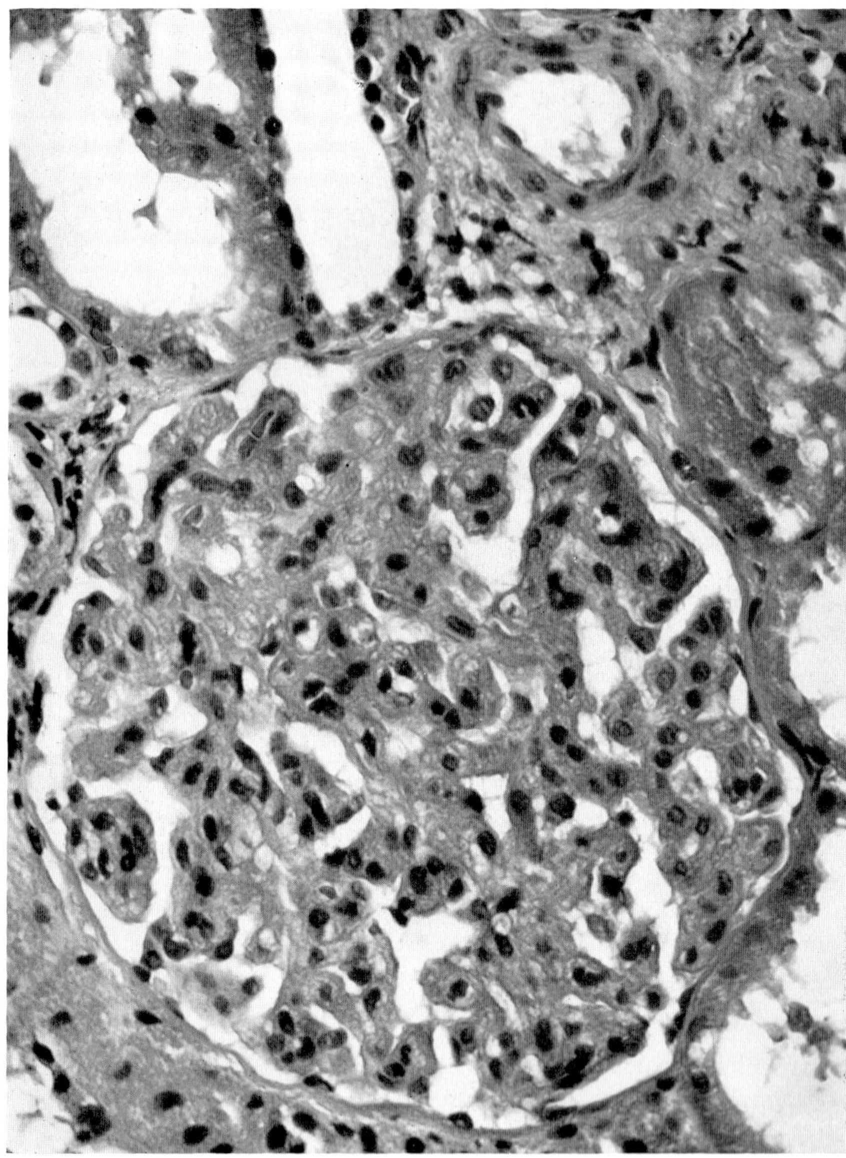

FIG. 96. Typical appearance of a glomerulus in pre-eclampsia. The glomerular tuft is swollen and bloodless, filling Bowman's space completely. The capillary walls are thick and their lumen is almost completely obliterated. Renal biopsy. H. and E.; $\times 450$.

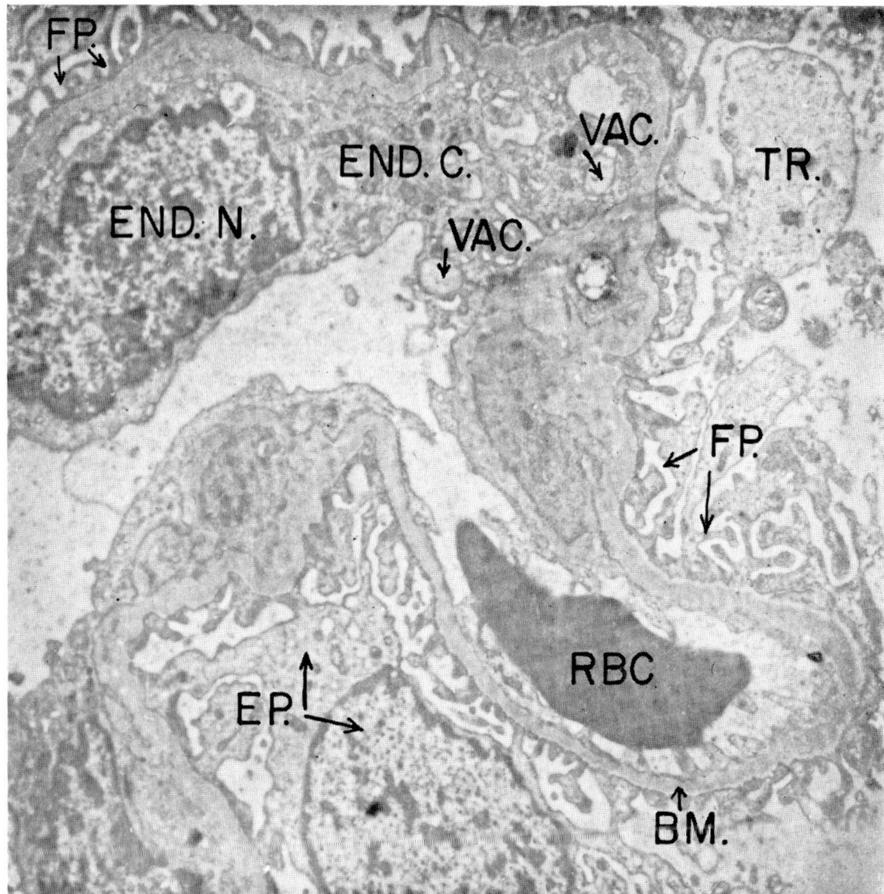

FIG. 97. Electron microphotograph of a pre-eclamptic glomerulus. The capillary containing a red blood cell (RBC) is narrow. The endothelial nucleus (END. N.) is essentially normal but the endothelial cytoplasm in one area (END. C.) is swollen and vacuolated (VAC.). The epithelial nucleus and cytoplasm (EP.) as well as the trabeculae (TR.) are swollen. The basement membrane (BM.) and the epithelial foot processes (F.P.) are normal. Renal biopsy. ×20,000.

interstitial tissue (Pollak *et al.*, 1956a). The glomerular capillary lesions have been shown by electron microscopic observations to be predominantly located in the endothelial cells, which cytoplasm is swollen, vacuolated and contains lipid droplets (Fig. 97), while the basement membrane is essentially normal (Spargo *et al.*, 1959; Pirani *et al.*, 1961). In particularly severe cases, fibrinoid changes have also been observed in the glomerular capillaries. In addition, patchy or diffuse cortical necrosis, on an ischemic basis, is not uncommon. Although the renal lesions are responsible for the major manifestations of pre-eclampsia, this disease cannot be considered primarily renal in origin. A relationship between essential hypertension and pre-eclampsia has been postulated, since the development of benign and/or malignant hypertension and of renal arteriolar sclerosis is not uncommon in young women with previous pre-eclamptic or eclamptic episodes.

4. Glomerulonephritis

Glomerulonephritis, no matter what the type, is at present considered an inflammatory disease of glomerular capillaries, due to hypersensi-

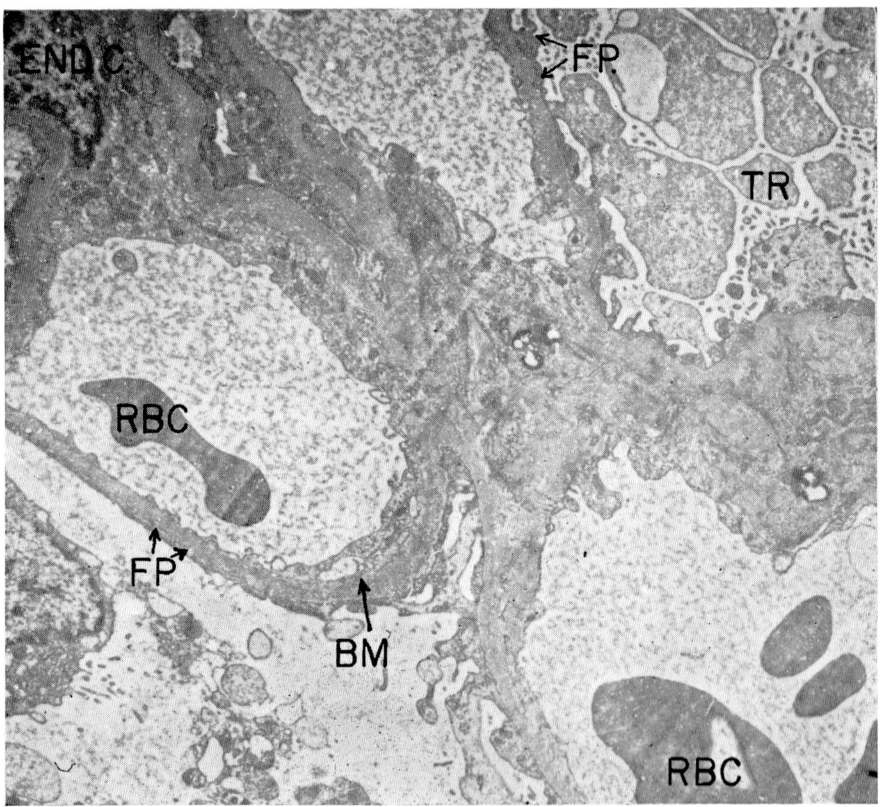

Fig. 98. Electron microphotograph of a glomerulus in a patient with "lipoid nephrosis" and the nephrotic syndrome. The lumen of the capillaries is wide and contains many red blood cells (RBC). The endothelial nucleus (END.C.) and adjacent cytoplasm are normal. The epithelial cytoplasmic trabeculae (TR.) are somewhat swollen and vacuolated. The epithelial foot processes (FP.) cannot be recognized as distinct structures in most areas and are replaced by a continuous osmiophilic smudgy layer. Renal biopsy. ×21,300.

tivity. (For discussion of changes in hypersensitive states, see Section G.-3b, Chapter VI.) Whether the predominant acute morphologic features in the glomeruli are proliferative, exudative, necrotizing, hemorrhagic or membranous, the progression of the disease in many cases will

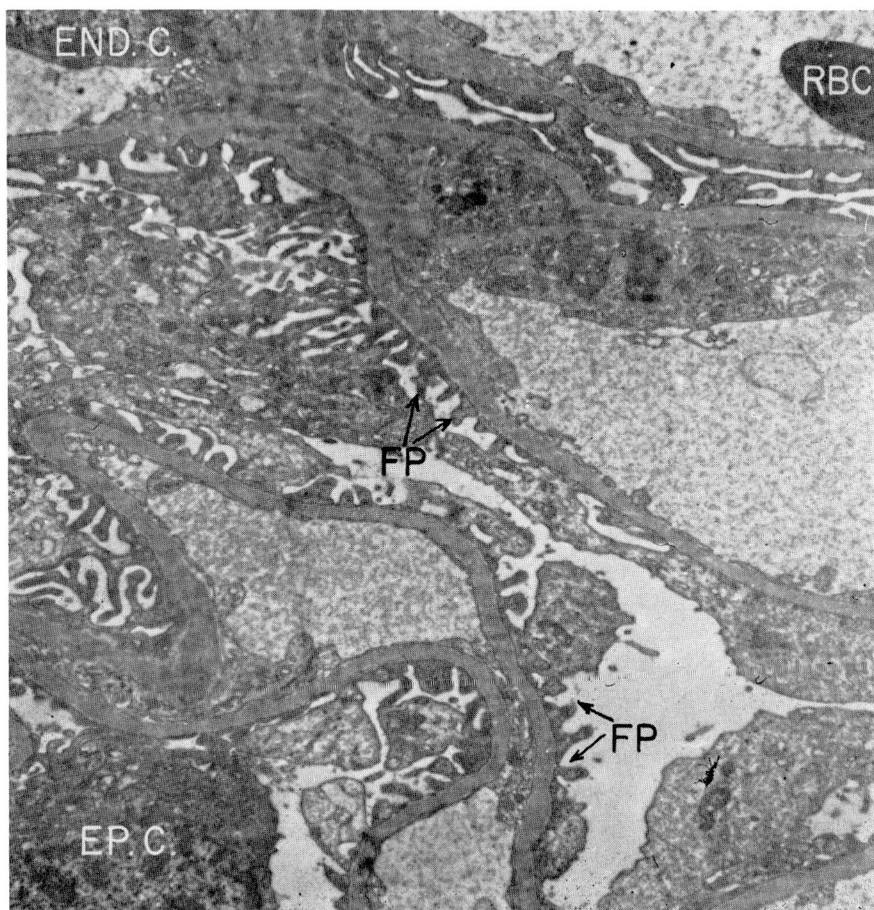

FIG. 99. Electron microphotograph of a glomerulus from the same patient of Fig. 98 after proteinuria had markedly decreased and he was no longer "nephrotic." The foot processes (FP.) are now discrete in most areas and have an essentially normal appearance. Renal biopsy. ×21,300.

lead to permanent capillary damage and eventually to obliteration of glomerular tufts (Jones, 1953). In the acute phase of the disease, the functional result of these glomerular changes will be increased glomerular capillary resistance and permeability, decreased glomerular filtration and hypertension. Changes in the renal tubules, interstitium and blood vessels will become morphologically apparent from a few days and weeks to several months and years following the acute phase. Alterations in the arterioles, in particular, usually develop in the subacute, and become prominent in the chronic, stage of the disease. They are of the arteriolar sclerotic type and less commonly may be those of malignant hypertension. The arteriolar lesions are to a considerable extent the result of high blood pressure resulting from glomerular damage and subsequent renal ischemia. However, glomerular damage and obliteration *per se* are known to cause "disuse sclerosis" of afferent and efferent arterioles.

Atherosclerosis is more severe in patients with glomerulonephritis. This can be attributed, in part, to hypertension and, in part, to the abnormalities of protein and particularly lipid metabolism, which often occur in the subacute phase of the disease and may lead to the development of a full-blown nephrotic syndrome.

In lipoid nephrosis, a form of glomerulonephritis associated with the nephrotic syndrome and particularly common in childhood, light microscopic studies have often failed to reveal distinct glomerular lesions (Bell, 1950; Allen, 1951; Ehrich *et al.*, 1952; Vernier *et al.*, 1958). By electron microscopy, however, it has been possible to determine the presence of definite changes in the foot processes of the epithelial cells lining the outer aspect of glomerular capillaries (Farquhar *et al.*, 1957). By serial renal biopsy studies, these changes have been shown to be reversible (Figs. 98 and 99), either spontaneously or following steroid therapy (Folli *et al.*, 1958).

5. So-called Collagen Diseases

In the so-called collagen diseases, renal involvement is often prominent. In systemic lupus erythematosus, proliferative, membranous and necrotizing changes are present in the glomeruli. When the thickening of the glomerular capillary basement membrane is marked and is associated with fibrinoid changes, the fairly characteristic wire-loop appearance is noted (Klemperer *et al.*, 1941). In this disease, fibrinoid changes can also occur, in the absence of hypertension, in the arterioles and, less commonly, in the interlobular arteries. As the disease progresses, more severe glomerular changes develop. However, a significant degree of renal arteriosclerosis is usually not encountered, and hypertension is relatively uncommon (Muehrcke *et al.*, 1957) (for further discussion, see (3) So-called Collagen or Systemic Connective Tissue Diseases, Section B.-3d, Chapter XXI).

In polyarteritis nodosa, typical arterial and periarterial inflammation can involve arterioles and interlobular and arcuate arteries. Interlobar arteries and other major renal artery branches are usually not affected. Fibrinoid changes are common in polyarteritis nodosa, and, as in the other "collagen diseases," are not related to hypertension. The latter, however, may be present in the chronic phase of the disease as arterial and arteriolar sclerosis develops. Steroid treatment would appear to reduce the activity of the inflammatory process but in so doing facilitates the development of sclerotic changes in the arteries (Zeek, 1952). (For a discussion of the renal changes in scleroderma, see (3) So-called Collagen or Systemic Connective Tissue Diseases, Section D.-3c, Chapter XXI.)

6. Occlusion of Renal Vessels

a. Thrombosis of renal arteries and branches: One form of arterial thrombosis, which predominantly involves arterioles and capillaries and is often associated with fibrinoid necrosis, constitutes the main feature of thrombotic thrombocytopenic purpura. This disease is a widespread vascular condition with prominent renal involvement and appears to be pathogenetically related to the so-called collagen diseases. The thrombi, often occlusive in nature, consist of fibrin and platelets (Moschcowitz, 1925; Gore, 1950). Similarly constituted thrombi are not an uncommon feature of other renal conditions and can be particularly prominent in shock (Jones and Loring, 1951).

b. Thrombosis of renal vein: This is a rare condition which may develop, often bilaterally, either in association with, or independently from, thrombotic manifestations elsewhere. In this disorder, the kidneys are usually enlarged as a result of congestion and especially of interstitial edema. In the glomeruli, thickening of the capillary basement is often prominent and is apparently responsible for the marked proteinuria and the nephrotic syndrome which is usually present (Pollak *et al.*, 1956b).

c. Renal thromboembolism: This condition is a common complication of such cardiac diseases as mitral stenosis and myocardial infarction, in which mural thrombi, either of the left atrium or ventricle, are the source of the emboli. Septic emboli originating from the vegetations of mitral or aortic bacterial endocarditis can lead to either abscess formation or septic infarcts of the kidneys. So-called embolic glomerulonephritis, usually associated with subacute bacterial endocarditis and characterized by focal necrotizing and proliferative changes in the glomeruli, is at present generally recognized as being due to streptococcal hypersensitivity rather than to true embolization of glomerular capillaries (Allen, 1951). Other types of emboli which may involve the renal vascular system include fat, as a result of fractures, and atheromatous material from aortic atheroma, more commonly following surgical resection of an aortic aneurysm (Thurlbeck and Castleman, 1957).

7. Pyelonephritis

Among the primary renal diseases which secondarily cause pronounced changes in the renal arteries, a prominent position is occupied by pyelonephritis. Vascular lesions usually become apparent in the chronic stage of the disease, apparently as a result of interstitial inflammation and fibrosis. These arteriosclerotic lesions are aggravated by ensuing hypertension, and in areas of scarring, "disuse sclerosis" is also a factor. If malignant hypertension develops, the changes of malignant nephrosclerosis may become superimposed on those of pyelonephritis. There is no definite way to differentiate the vascular changes of pyelonephritis from those of arterial and arteriolar sclerosis. However, in pyelonephritis the alterations are often more focal, intimal changes of an endoarteritic type are prominent, and great tortuosity of the intra- and interlobular arteries may be present, apparently as a result of the predominant interstitial fibrosis and scarring (Weiss and Parker, 1939). The difficulties involved in the differential histologic diagnosis between chronic pyelonephritis and arterial and arteriolar nephrosclerosis has been responsible for the reported differences in the incidence of pyelonephritis as the renal disease responsible for hypertension (Jackson *et al.*, 1957; Saphir and Cohen, 1959). The problem is even further complicated by the frequent superimposition of pyelonephritis on benign nephrosclerosis. There is no question, however, that pyelonephritis is the most common cause of hypertension resulting from a primary renal disease (Kincaid-Smith *et al.*, 1958).

References

Albarran, J. (1909). "Médicine operatoire des voies urinaires." Masson, Paris.

Allen, A. C. (1941). So-called intercapillary glomerulosclerosis—a lesion associated with diabetes mellitus: Morphogenesis and significance. *A. M. A. Arch. Pathol.* **32**, 33.

Allen, A. C. (1951). "The Kidney." Grune & Stratton, New York.

Anson, B. J., and Kurth, L. E. (1955). Common variations in the renal blood supply. *Surg., Gynecol. Obstet.* **100**, 156.

Anson, B. J., Cauldwell, E. W., Pick, J. W., and Beaton, L. E. (1947). The blood supply of the kidney, suprarenal gland and associated structures. *Surg., Gynecol. Obstet.* **84**, 313.

Anson, B. J., Cauldwell, E. W., Pick, J. W., and Beaton, L. E. (1948). The anatomy of the pararenal system of veins, with comments on the renal arteries. *J. Urol.* **60**, 714.

Arey, L. B. (1946). "Developmental Anatomy," 5th ed. Saunders, Philadelphia, Pa.

Barcroft, H., and Swan, H. J. C. (1953). "Sympathetic Control of Human Blood Vessels." Edward Arnold, London.

Barger, A. C. (1960). The kidney in congestive heart failure. *Circulation* **21**, 124.

Batson, O. V. (1940). The function of the vertebral veins and their role in the spread of metastases. *Ann. Surg.* **112**, 138.

Batson, O. V. (1957). The vertebral vein system. Caldwell Lecture. *Am. J. Roentgenol.* **78**, 195.

Bell, E. T. (1950). "Renal Diseases." Lea & Febiger, Philadelphia, Pa.

Bergstrand, A. (1957). Electron microscopic investigations of the renal glomeruli. *Lab. Invest.* **6**, 191.

Bergstrand, A., and Bucht, H. (1959). The glomerular lesions of diabetes mellitus and their electron microscope appearances. *J. Pathol. Bacteriol.* **77**, 231.

Brödel, M. (1901). The intrinsic blood vessels of the kidney and their significance in nephrotomy. *Bull. Johns Hopkins Hosp.* **12**, 10.

Corbascio, A. W. (1960). Action of long-chain polymers on kidney juxtaglomerular cells and connective tissue mast cells. *Circulation Research* **8**, 390.

Corcoran, A. C., and Page, I. H. (1939). Effects of renin, pitressin, and pitressin and atropine on renal blood flow and clearance. *Am. J. Physiol.* **126**, 354.

Dalton, A. J. (1951). Structural details of some epithelial cell types in the kidney of the mouse as revealed by the electron microscope. *J. Natl. Cancer Inst.* **11**, 1163.

Ehrich, W. E., Forman, C. W., and Seifer, J. (1952). Diffuse glomerular nephritis and lipid nephroses. *A. M. A. Arch. Pathol.* **54**, 463.

Eisendrath, D. N. (1920). The relation of variations in the renal vessels to pyelotomy and nephrectomy. *Ann. Surg.* **71**, 726.

Emanuel, D. A., Fleishman, M., and Haddy, F. J. (1957). Effect of pH change upon renal vascular resistance and urine flow rate. *Circulation Research* **5**, 607.

Emanuel, D. A., Scott, J., Collins, R., and Haddy, F. J. (1959). The local effect of serotonin upon renal vascular resistance and urine flow rate. *Am. J. Physiol.* **196**, 1122.

Fagarasanu, I. (1938). Recherches anatomique sur la veine rénale gauche et ses collatérales. Leurs rapports avec la pathogénie du varicocèle essentiel et des varices du ligament large. *Ann. anat. pathol. et anat. normale méd.-chir.* **15**, 9.

Farquhar, M. G., Vernier, R. L., and Good, R. A. (1957). An electron microscope study of the glomerulus in nephrosis, glomerulonephritis, and lupus erythematosus. *J. Exptl. Med.* **106**, 649.

Fishman, A. P., Maxwell, M. H., Crowder, C. H., and Morales, P. (1951). Kidney function in cor pulmonale—Particular consideration of changes in renal hemodynamics and sodium excretion during variation in level oxygenation. *Circulation* **3**, 703.

Folkow, B. (1955). Nervous control of blood vessels. *Physiol. Revs.* **35**, 629.

Folkow, B., and Löfving, B. (1956). The distensibility of the systemic resistance blood vessels. *Acta Physiol. Scand.* **38**, 37.

Folli, G., Pollak, V. E., Reid, R. T. W., Pirani, C. L., and Kark, R. M. (1958). Electronmicroscopic studies of reversible glomerular lesions in the adult nephrotic syndrome. *Ann. Internal Med.* **49,** 775.

Gautier, A., Bernhard, W., and Oberling, C. (1950). Sur l'existence d'un appareil lacunaire péri-capillaire du glomérule de Malpighi, révélée par le microscope électronique. *Compt. rend. soc. biol.* **144,** 1605.

Gellman, D. D., Pirani, C. L., Soothill, J. F., Muehrcke, R. C., and Kark, R. M. (1959). Diabetic nephropathy: A clinical and pathologic study based on renal biopsies. *Medicine* **38,** 321.

Gore, I. (1950). Disseminated arteriolar and capillary thrombosis. A morphologic study of its histogenesis. *Am. J. Pathol.* **26,** 155.

Gottschalk, C. W. (1952). A comparative study of renal interstitial pressure. *Am. J. Physiol.* **169,** 180.

Gottschalk, C. W., and Mylle, M. (1956). Micropuncture study of pressures in proximal tubules and peritubular capillaries of the rat kidney and their relation to ureter and renal venous pressures. *Am. J. Physiol.* **185,** 430.

Graves, F. T. (1954). The anatomy of the intrarenal arteries and their application to segmental resection of the kidney. *Brit. J. Surg.* **42,** 132.

Graves, F. T. (1956). The aberrant renal artery. *J. Anat.* **90,** 553.

Green, H. D., and Kepchar, J. H. (1959). Control of peripheral resistance in major systemic vascular beds. *Physiol. Revs.* **39,** 617.

Haddy, F. J., Scott, J., Fleishman, M., and Emanuel, D. (1958a). Effect of change in flow rate upon renal vascular resistance. *Am. J. Physiol.* **195,** 111.

Haddy, F. J., Scott, J., Fleishman, M., and Emanuel, D. (1958b). Effect of change in renal venous pressure upon renal vascular resistance, urine and lymph flow rates. *Am. J. Physiol.* **195,** 97.

Hall, B. V. (1954a). Studies of normal glomerular structures by electron microscopy. *Proc. 5th Ann. Conf. on the Nephrotic Syndrome (Natl. Nephrosis Found.), New York* p. 1.

Hall, B. V. (1954b). Observations on the number, size, and structure of rat glomerular capillaries (Abstract). *Anat. Record* **118,** 425.

Hall, B. V. (1955). Further studies of the normal structure of the renal glomerulus. *Proc. 6th Ann. Conf. on the Nephrotic Syndrome (Natl. Nephrosis Found.), New York* p. 1.

Hall, B. V. (1957). The protoplasmic basis of glomerular ultrafiltration. *Am. Heart J.* **54,** 1.

Hall, B. V. (1960). A slit-pore theory of glomerular filtration based on dimensional data provided by electron micrographs of the glomerular capillary wall (Abstract). *Anat. Record* **136,** 205.

Hall, B. V., Roth, E., and Johnson, V. (1953a). The ultramicroscopic structure and minute functional anatomy of the glomerulus (Abstract). *Anat. Record* **115,** 315.

Hall, B. V., Roth, E., and Johnson, V. (1953b). Electron microscopy of the rat kidney (Abstract). *Anat. Record* **115,** 427.

Hall, G. F. M. (1952). The significance of atheroma of the renal arteries in Kimmelstiel-Wilson's syndrome. *J. Pathol. Bacteriol.* **64,** 103.

Haynes, R. H., and Burton, A. C. (1959). Role of the non-Newtonian behavior of blood in hemodynamics. *Am. J. Physiol.* **197,** 943.

Hollinshead, W. H., and McFarlane, J. A. (1953). The collateral venous drainage from the kidney following occlusion of the renal vein in the dog. *Surg., Gynecol. Obstet.* **97**, 213.

Hou-Jensen, H. M. (1930). Die Verästelung der arteria renalis in der Niere des Menschen. *Z. Anat. Entwicklungsgeschichte* **91**, 1.

Jackson, G. G., Poirier, K. P., and Grieble, H. G. (1957). Concepts of pyelonephritis: Experience with renal biopsies and long-term clinical observations. *Ann. Internal Med.* **47**, 1165.

Janecek, J., Tobian, L., Otis, M., and Huneke, J. (1960). Studies on renin and juxtaglomerular cells, components of a regulating system in the kidney. *Federation Proc.* **19**, 100.

Johnson, P. C., Knockel, W. L., and Sherman, T. L. (1960). Pressure volume relationships of small arterial segments. *Federation Proc.* **19**, 91.

Jones, D. B. (1953). Glomerulonephritis. *Am. J. Pathol.* **29**, 33.

Jones, D. B., and Loring, W. E. (1951). Glomerular thrombosis. *Am. J. Pathol.* **27**, 841.

Keibel, F., and Mall, F. P. (1912). "Manual of Human Embryology," Vol. II. Lippincott, Philadelphia, Pa.

Kimmelstiel, R., and Wilson, C. (1936). Intercapillary glomerulosclerosis. *Am. J. Pathol.* **12**, 83.

Kincaid-Smith, P., McMichael, J., and Murphy, E. A. (1958). The clinical course and pathology of hypertension with papilloedema (malignant hypertension). *Quart. J. Med.* **27**, 117.

Kinter, W. B., and Pappenheimer, J. R. (1956). Role of red blood corpuscles in regulation of renal blood flow and glomerular filtration rate. *Am. J. Physiol.* **185**, 399.

Klemperer, P., Pollack, A. D., and Baehr, G. (1941). Pathology of lupus erythematosus disseminatus. *A. M. A. Arch. Pathol.* **32**, 569.

Levi, G. (1909). Le variazioni delle arterie surrenali e renali studiate col metodo statistico seriale. *Arch. ital. anat. e embriol.* **8**, 35.

Longley, J. B., Banfield, W. G., and Brindley, D. C. (1960). Structure of the rete mirabile in the kidney of the rat as seen with the electron microscope. *J. Biophys. Biochem. Cytol.* **7**, 103.

MacCallum, D. B. (1939). The bearing of degenerating glomeruli on the problem of the vascular supply of the mammalian kidney. *Am. J. Anat.* **65**, 69.

Merklin, R. J., and Michels, N. A. (1958). The variant renal and suprarenal blood supply with data on the inferior phrenic, ureteral and gonadal arteries. *J. Intern. Coll. Surgeons* **29**, 41.

Muehrcke, R. C., Kark, R. M., Pirani, C. L., and Pollak, V. E. (1957). Lupus nephritis. *Medicine* **36**, 1.

Muirhead, E. E., Montgomery, P. O., and Booth, E. (1956). The glomerular lesions of diabetes mellitus. *A. M. A. Arch. Internal Med.* **98**, 146.

Miller, F. (1958). Orthologie und Pathologie der Zelle im electronenmikroskopischen Bild. *Verhandl. deut. Ges. Pathol.* **42**, 261.

Moschcowitz, E. (1925). An acute febrile pleiochromic anemia with hyaline thrombosis of the terminal arterioles and capillaries. *Arch. Internal Med.* **36**, 89.

Notkovich, H. (1956). Variations of the testicular and ovarian arteries in relation to the renal pedicle. *Surg., Gynecol. Obstet.* **103**, 497.

Oberling, C., Gautier, A., and Bernhard, W. (1951). La structure des capillaires glomérulaires vue au microscope électronique. *Presse méd.* **59**, 938.

Page, I. H., and McCubbin, J. W. (1953). Renal vascular and systemic arterial pressure responses to nervous and chemical stimulation of the kidney. *Am. J. Physiol.* **173,** 411.

Pak Poy, R. K. F. (1958). Electron microscopy of the mammalian renal glomerulus. The problems of intercapillary tissue and the capillary loop basement membrane. *Am. J. Pathol.* **34,** 885.

Pak Poy, R. K. F., and Robertson, J. S. (1957). Electron microscopy of the avian renal glomerulus. *J. Biophys. Biochem. Cytol.* **3,** 183.

Pappenheimer, J. R. (1955). Über die Permeabilität der Glomerulammembranen in der Niere. *Klin. Wochschr.* **33,** 362.

Pappenheimer, J. R., and Kinter, W. K. (1956). Hematocrit ratio of blood within mammalian kidney and its significance for hemodynamics. *Am. J. Physiol.* **185,** 377.

Pease, D. C. (1955a). Electron microscopy of the vascular bed of the kidney cortex. *Anat. Record* **121,** 701.

Pease, D. C. (1955b). Fine structures of the kidney seen by electron microscopy. *J. Histochem. Cytochem.* **3,** 295.

Pease, D. C. (1960). The basement membrane: substratum of histological order and complexity. *Proc. 4th Intern. Conf. on Electron Microscopy, Berlin, 1959* **2,** 139.

Pirani, C. L., Pollak, V. E., Lannigan, R., Nettles, J. B., and Stein, P. (1961). Light and electronmicroscopic studies of the renal lesions in toxemias of pregnancy with observations on some clinico-pathologic relationships. *Proc. 7th Intern. Conf. Soc. Geogr. Pathol., London, 1960 Path. Microbiol.* **24,** 586.

Pollak, V. E., Kark, R. M., Pirani, C. L., Shafter, H. A., and Muehrcke, R. C. (1956a). Renal vein thrombosis and the nephrotic syndrome. *Am. J. Med.* **21,** 496.

Pollak, V. E., Pirani, C. L., Kark, R. M., Muehrcke, R. C., Freda, V. C., and Nettles, J. B. (1956b). Reversible glomerular lesions in toxaemia of pregnancy. *Lancet* **ii,** 59.

Rhodin, J. (1955). Electron microscopy of the glomerular capillary wall. *Exptl. Cell Research* **8,** 572.

Rhodin, J. (1958). Electron microscopy of the kidney. *Am. J. Med.* **24,** 661.

Rinehart, J. F. (1955). Fine structure of renal glomerulus as revealed by electron microscopy. *A. M. A. Arch. Pathol.* **59,** 439.

Saphir, O., and Cohen, N. A. (1959). Chronic pyelonephritis lenta and the "malignant phase of hypertension." *A. M. A. Arch. Internal Med.* **104,** 748.

Sartorius, O. W., and Burlington, H. (1956). Acute effects of denervation on kidney function in the dog. *Am. J. Physiol.* **185,** 407.

Scholander, P. F. (1954). Secretion of gasses against high pressures in the swim-bladder of deep sea fishes. II. The rete mirabile. *Biol. Bull.* **107,** 260.

Sheehan, H. L. (1950). *In* "Toxaemias of Pregnancy" (J. Hammond, F. J. Browne, and G. E. W. Wolstenholme, eds.). Churchill, London.

Spargo, B., McCartney, C. P., and Winemiller, R. (1959). Glomerular capillary endotheliosis in toxemia of pregnancy. *A. M. A. Arch. Pathol.* **68,** 593.

Spiro, D. (1959). The structural basis of proteinuria in man. *Am. J. Pathol.* **35,** 47.

Thurlbeck, W. M., and Castleman, B. (1957). Atheromatous emboli to kidneys after aortic surgery. *New Engl. J. Med.* **257,** 442.

Tobian, L., Thompson, J., Twedt, R., and Janecek, J. (1958). The granulation of juxtaglomerular cells in renal hypertension, desoxycorticosterone and postdesoxycorticosterone hypertension, adrenal regeneration hypertension, and adrenal insufficiency. *J. Clin. Invest.* **37**, 660.

Turner, W. (Sir) (1863). Cited by Hughes, A. (1892). Abnormal arrangement of arteries in the region of the kidney and suprarenal body. *J. Anat. and Physiol.* **26**, 305.

Vernier, R. L., Farquhar, M. G., Brunson, J. G., and Good, R. A. (1958). Chronic renal disease in children. *A. M. A. J. Diseases Children* **96**, 306.

Vintrup, B. (1928). On the number, shape, structure, and surface area of the glomeruli in the kidneys of man and mammals. *Am. J. Anat.* **41**, 123.

von Möllendorff, W. (1927). Einige Beobachtungen über den Aufbau des Nierenglomerulus. *Z. Zellforsch.* **6**, 441.

von Möllendorff, W. (1930). Der Exkretionsapparat und weibliche Genitalorgane. *In* "Handbuch der mikroskopischen Anatomie des Menschen" (W. von Möllendorff, ed.), Vol. VII, Part 1. Springer, Berlin.

Waugh, W. H. (1958). Myogenic nature of autoregulation of renal flow in the absence of blood corpuscles. *Circulation Research* **6**, 363.

Weiss, S., and Parker, F., Jr. (1939). Pyelonephritis: Its relation to vascular lesions and to arterial hypertension. *Medicine* **18**, 221.

Winton, F. R. (1956). "Modern Views on the Secretion of Urine." Little, Brown, Boston, Mass.

Wirz, H. (1955). Druckmessung in Kapillaren und Tubuli der Niere durch Mikropunktion. *Helv. Physiol. et Pharmacol. Acta* **13**, 42.

Yamada, E. (1955). The fine structure of the renal glomerulus of the mouse. *J. Biophys. Biochem. Cytol.* **1**, 551.

Zeek, P. M. (1952). Periarteritis nodosa: A critical review. *Am. J. Clin. Pathol.* **22**, 777.

Zimmermann, K. W. (1915). Über das Epithel des glomerularen Endkammerblattes der Säugerniere. *Anat. Anz.* **48**, 335.

Chapter XIV. THE BLOOD VESSELS OF THE PITUITARY AND THE THYROID[1]

By W. C. Worthington, Jr.

1. The Pituitary

A. INTRODUCTION

Although there were a number of early studies of the blood vessels of the pituitary, this work was primarily academic until the beginning of the series of discoveries concerning pituitary function which forms the basic core of endocrine knowledge. Since then, the pituitary vessels have become a subject of fundamental importance. A thorough knowledge of the vasculature of any endocrine gland is essential to our understanding of its functioning by the very definition of the word, endocrine. In the case of the pituitary, even more than the other endocrine glands, insight into the structure and physiology of the vessels has been necessary before major forward steps in our knowledge of the gland and its functions could be made. This is due, in most part, to the complexity and unique arrangement of the pituitary vessels.

This arrangement consists in its fundamentals of a portal system of vessels, apparently phylogenetically constant from the tailless amphibia to man, joining those parts of the neurohypophysis most closely related to the brain with the anterior pituitary. In all forms studied thus far save one (the rabbit), the blood supply of the anterior lobe is exclusively by way of these portal vessels. This vascular link between the median eminence and pituitary stalk, on the one hand, and the anterior lobe, on the other, is apparently necessary for adequate functioning of the gland. Since the experimental and review literature on these points is extensive, it will be discussed in its relation to vascular architecture only.

A significant segment of the study of the special portal system of vessels has pertained to the direction of blood flow. This point is fundamental to the concept of humoral mediators between the brain and anterior lobe by way of these vessels. The matter was academic when these vessels were discovered by Popa and Fielding (1930) and even when the correct direction of blood flow (from the hypothalamus to the pituitary) was observed by Houssay *et al.* (1935) and deduced by

[1] This chapter was written during the author's tenure of a grant from The Commonwealth Fund for the study of pituitary microcirculation.

Wislocki and King (1936). However, shortly thereafter, the concept of a humoral mediator linking the brain with the anterior lobe of the pituitary (suggested earlier by Hinsey and Markee, 1933) became the subject of wide interest and intense study. It developed very rapidly through the particular efforts of G. W. Harris and J. D. Green, and many workers have made major contributions to the field (among others, Markee *et al.*, 1946; Hume, 1949; Fortier, 1951; DeGroot, 1952; Long, 1952; Porter, 1953; Slusher and Roberts, 1954; Guillemin, 1955; Saffran *et al.* 1955; Nikitovitch-Wiener and Everett, 1958). This hypothesis, in brief, states that a humoral mediator or mediators, released at hypothalamic nerve endings in the median eminence and stalk, are absorbed into a primary capillary system and transported by way of portal trunks to the sinusoids of the anterior lobe. Despite the importance of the direction of blood flow for this concept, it has only been in recent years that contention concerning this point has ceased to be significant in discussions of pituitary function.

Another important reason for detailed analysis of the vascular structure of the pituitary is its susceptibility under certain circumstances to infarction, due presumably to vascular insufficiency (Sheehan, 1937). This situation is probably attributable to the specialized nature of the vascular structure of the pituitary, certain peculiarities in the physiology of its vascular bed, its position in the brain case and its enlargement during pregnancy.

Still another reason for an analysis of the vascular structure of the pituitary is the necessity for determining to what extent variations in the vascular physiology may modify pituitary function and how such modifications fit in with neurohumoral and humoral mechanisms of control (see Worthington, 1960).

In this section as much emphasis as possible will be placed on descriptions of human material, particularly when dealing with the larger vessels. However, much of our knowledge concerning both the structure and the physiology of these vessels and their relation to the functioning of the gland has been derived from animal experimentation. Such work will be described when it is deemed necessary.

B. EMBRYOLOGY[2]

1. Development in the Human

A detailed and dynamic account of the development of the blood supply of the hypophysis has been given by Wislocki (1937) on the

[2] For a general discussion of the embryology of the blood vessels to the brain and its membranes, see Section A., Chapter VIII.

basis of studies in a series of human embryos. His report will be described in some detail.

a. Relationships of Rathke's pouch, brain, and pial-hypophyseal plexus: At the 18 mm and 21 mm stages there may be observed, covering the base of the diencephalon in the region of the pituitary, a plexus of vessels in a layer of undifferentiated mesenchyme. This occurs between the brain and the developing epithelial portion of the pituitary (Rathke's pouch) and is called the pial-hypophyseal plexus. Beneath and lateral to the epithelial pituitary there are capillaries and venules which communicate with the anterior cardinal veins. The plexus between the pituitary and the brain is supplied by arterial twigs arising from the internal carotid or posterior communicating arteries. At these stages the venous blood is drained from this pial plexus by specific pial venules, while the area beneath the pituitary which is destined to become the sella turcica is drained separately by venules associated with the primordia of the dural plexuses.

At the 25 mm stage and later, the arrangements just described become modified as changes occur in the primordium of the epithelial pituitary (Rathke's pouch). That portion of Rathke's pouch which is to become the anterior pituitary forms a cup-like depression, the tuberal processes grow upward, and that portion which is to form the intermediate lobe begins to enfold the posterior lobe. In many places the mesenchyma lying between the pituitary and brain becomes imbedded in hollows within the developing epithelial pituitary and begins to undergo intercrescence with its epithelium. As the parenchyma becomes subdivided by the stroma, the stroma carries in with it a rich plexus of capillaries which is derived by extension from the plexus which was originally present between the brain and the pituitary. These vessels eventually break through the tissue of the epithelial pituitary and connect with the cavernous plexuses in the developing sella turcica. At this point the epithelial portion of the pituitary has a dual venous drainage, one by primary pial venules associated originally with the pial-hypophyseal plexus in the region of the pituitary above and one by the secondary connections established with the cavernous plexuses below. Some of these connections with cavernous plexuses will become a portion of the definitive venous drainage of the hypophysis. The primary venous drainage of the pial-hypophyseal plexus eventually disappears.

b. Arteries: The arterioles from the internal carotid or posterior communicating arteries to the pial-hypophyseal plexus develop into superior hypophyseal arteries. Wislocki pointed out that if the growth of the pituitary had not modified the development of the meninges in this region, one would have anticipated that the arterioles which are destined to become the superior hypophyseal arteries would have entered the brain as pial arteries ordinarily do. He regarded these arterioles as vessels which become deflected from their original destination by the ingrowth of the pial-hypophyseal mesenchyma into the pituitary.

c. Vessels of posterior lobe: The blood supply to the infundibular process or posterior lobe is derived primarily from the mesenchyma in the region of the future sella turcica which will form the dura. While some of the vessels in the region between the two parts of the gland may be derived from a thin lamina of vascular pia occurring between intermediate and posterior lobe, the major portion of the blood vessels of the posterior lobe probably comes from its free posterior pole. Here the posterior lobe is unencumbered by intermediate lobe and a delicate condensation of pia mater which was originally present has disappeared. This leaves the free pole to become encapsulated by dura like the remainder of the pituitary. It is by virtue of the foregoing facts that the posterior lobe does not receive its definitive vessels until relatively late in development. Further, Wislocki pointed out that even in the latest stages which he examined, the embryonic infundibular stalk, while ensheathed by the pial-hypophyseal plexus, is relatively avascular in the interior when compared with the extensive ingrowth of pial vessels to the brain proper. He stated that "vessels are apparently slow in penetrating the embryonic neural stalk and may remain small and inconspicuous during the early fetal period." However, as the tuberal processes grow upward, fuse, and become the pars tuberalis which surrounds the stalk in the adult, that portion of the pial-hypophyseal plexus which occurs between these two parts of the gland develops, creating a rich, widecalibered capillary network, and forms at least a part of the primary capillary plexus at this stage. The point regarding the ingrowth of the distinct capillary tufts of the adult described below is an extremely important one; this ingrowth apparently does not occur until much later in the development of these vessels.

d. Portal trunks: The material studied by Wislocki did not contain fetal stages of late enough development for observations to be made on the portal trunks themselves. The oldest well-injected fetus which he had at his disposal was 60 mm. He did surmise, however, in a most convincing way, that portal vessels are derived from the original primary venules of the pial-hypophyseal plexus after they have lost their connections with the systemic drainage.

A considerably less detailed paper on this subject has been written by Niemineva (1950). He found that, while the hypophyseal portal system is fully developed in human fetuses at term, in the middle of the fetal period capillary tufts penetrating into the median eminence and into the neural stalk cannot be demonstrated. He concluded that hypothalamic regulation by way of portal vessels cannot begin until late in the fetal period.

2. Development in the Rat

A much more recent study of the development of the hypophyseal portal vessels has been made in the rat by Glydon (1957). His material covered an age range of 14 days of gestation to 50 days after birth. It is not inconsistent in some of its essentials with earlier accounts of human material. Glydon described a perihypophyseal capillary plexus, part of which is trapped beneath the median eminence by the upgrowth of the pars tuberalis. Subsequently, with the development of the pituitary stalk, the vessels between this trapped portion and the capillary plexus of the pars distalis become "stretched" and give rise to the portal trunks. He made the points that portal vessels may be first seen three days before birth and that "the first signs of the primary capillary plexus are not evident until five days after birth."

3. Physiologic Significance

Glydon quoted a series of workers whose experiments indicate that there is adenohypophyseal function before birth and stated that this, coupled with the late development of the portal trunks and still later development of the primary capillary plexus in the rat, would indicate the ability of the anterior pituitary to function in the absence of the distinctive vascular arrangement which is characteristic of the adult median eminence.

Several considerations would suggest a more cautious interpretation of this embryologic data. First, it would seem that those vessels lying within the pars tuberalis or between the pars tuberalis and median eminence (pial-hypophyseal plexus of Wislocki or perihypophyseal plexus of Glydon) constitute at least the beginnings of the primary capillary plexus. According to Wislocki's description, they are supplied by the superior hypophyseal arteries. Second, although portal trunks do not appear until late in the rat, the presence of capillary connections at earlier stages is implicit in the descriptions. Third, little is known of the chemical nature and physical properties of the hypothalamic humoral mediators which, current theory holds, are released at nerve endings in the median eminence and stalk and pass by way of the primary capillaries and portal trunks to the anterior lobe. The smaller size of the stalk and median eminence and less dense tissues in the fetus might make the characteristic capillary formations of the adult unnecessary.

C. ARTERIAL SUPPLY AND SYSTEMIC VENOUS DRAINAGE

Of the three most recent descriptions of the gross (or semigross) arterial supply to the human hypophysis (McConnell, 1953; Xuereb *et al.*, 1954a; and Stanfield, 1960), that of Xuereb *et al.* seems to be the most detailed and complete. Their description will be followed in this work and any variations or additions which have been presented by others will be discussed subsequently.

1. Principal Arterial Supply

The principal arterial supply of the human hypophysis is derived from paired right and left superior hypophyseal arteries and from paired right and left inferior hypophyseal arteries.

a. Superior hypophyseal arteries: These vessels arise from the internal carotid arteries on their medial sides just after the carotids have passed through the cavernous sinuses and dura mater. Each superior hypophyseal artery divides into two branches, one of which passes to the anterior, and the other to the posterior, aspect of the pituitary stalk. These divisions may arise separately from the internal carotid. The anterior branch of the superior hypophyseal artery sends vessels to the optic nerve and chiasma and to the supraoptic and infundibular portions of the hypothalamus, as well as to the pituitary. The posterior branches send some branches to the optic tract and to the tuber cinereum. There is a complex system of anastomoses between anterior and posterior branches of the superior hypophyseal arteries. The communications occur both anterior and posterior to the stalk between corresponding vessels from the opposite side, and on the lateral aspects of the stalk between anterior and posterior branches of the superior hypophyseal arteries of the same side. This complex of arterial vessels surrounding the stalk sends smaller branches both superiorly and inferiorly into the stalk itself, as well as to some other structures. The papers cited above, especially Xuereb *et al.*, should be consulted for details.

Another important division of the superior hypophyseal arterial system is that branch of the anterior superior hypophyseal artery on each side which is called by McConnell and by Stanfield, the "loral artery" and by Xuereb *et al.*, the "artery of the trabecula." This vessel passes downward on each side, either immediately adjacent to the stalk or some distance from it, anteriorly in the subarachnoid space. It may enter the anterior lobe directly alongside the stalk or at a position somewhat forward and laterally. The further ramifications of this vessel will be discussed below (Section D. of this chapter).

b. Inferior hypophyseal arteries: These paired vessels arise from the intracavernous portions of the internal carotids and follow a tortuous course (giving off several branches) to the inferior lateral aspect of the pituitary. As each artery approaches the gland it sends a vessel to the dural covering of the anterior lobe and then forms a medial and a lateral branch. The medial branch passes toward the midline in the subhypophyseal venous sinus to anastomose with the corresponding branch from the opposite side. The lateral branch passes upward along the line of separation between anterior and posterior lobes and, turning medially, anastomoses with the lateral branch of its fellow behind the hypophyseal stalk. This produces an arterial ring around the posterior lobe. Branches from this arterial ring pass to the posterior lobe and into the lower or intraglandular portion of the pituitary stalk, as well as to the dura covering both the posterior and anterior lobes. Further details concerning the distribution of the branches of this ring will be given below.

c. Arterioarterial anastomoses: Xuereb *et al.* have identified a number of major arterioarterial anastomoses which they consider characteristic of the vascular system of the pituitary. Several of these anastomoses within the superior hypophyseal and inferior hypophyseal arterial systems, respectively, have been described above. Anastomoses between the superior and inferior arterial systems are apparently variable, but Xuereb *et al.* (1954a) give three major arterioarterial connections which are most frequently found: (1) the artery of the trabecula with the lateral branch of the inferior hypophyseal artery on the same side; (2) the artery of the trabecula with the lateral branch of the inferior hypophyseal artery on the opposite side; and (3) the artery of the trabecula with a vessel which they termed "the inferior artery of the lower infundibular stem," a single artery arising variably from some portion of the arterial loop of the inferior hypophyseal arterial system and supplying the intraglandular portion of the stalk.

2. Venous Drainage of the Hypophysis

The veins which drain the human pituitary have been discussed by a number of workers, including Xuereb *et al.*, and have been the subject of a recent study by H. T. Green (1957). Green described not only details and groupings of small veins and venules draining various parts of the pituitary, but also the detailed anatomy of the cavernous and intercavernous sinuses.

a. Intercavernous sinuses: Like the arteries, this venous drainage is somewhat complex. Green found that the cavernous sinuses are joined by three vessels which traverse the sella turcica in a transverse direction. One of these lies between the anterior margin of the anterior lobe and the anterior portion of the sella turcica; another lies in front of the posterior clinoid plate; while a third passes across the floor of the sella just anterior to the groove between the anterior and posterior lobes. These three vessels were designated, respectively, anterior, posterior, and inferior intercavernous sinuses.

b. Plexiform sinuses and posterior venous network: In addition to these channels, Green found two meshworks of vessels closely applied to the pituitary surface. One, the superior plexiform sinus, occurs on both anterior and posterior lobes superiorly and the other, the anteroinferior plexiform sinus, occurs on the anteroinferior surface of the anterior lobe. Green also noted a plexus of small veins about the posterior and lateral aspects of the posterior lobe which he called the posterior venous network.

c. Veins draining the anterior and posterior lobes: The veins leaving the anterior lobe were described as falling into three principal groups. A lateral group drains into the superior and the anteroinferior plexiform sinuses, into the cavernous and the inferior intercavernous sinuses, and into the posterior venous network. A superior group drains into the superior plexiform sinus. An anteroinferior group of hypophyseal veins drains into the anteroinferior plexiform and the inferior intercavernous sinuses. The small venous vessels of the posterior lobe run peripherally toward the anterior part of the lateral surface, emerging close to branches of the inferior hypophyseal artery, and drain either into the posterior intercavernous sinus and into a vein accompanying the inferior hypophyseal artery or into the posterior venous network. The drainage of blood from the median eminence and upper and lower portions of the pituitary stalk by way of the special and highly significant hypophyseal portal system of vessels will be described below (Section

D. of this chapter). Green, in confirmation of earlier workers, was unable to find any systemic venous drainage from these regions.

D. HYPOPHYSEAL PORTAL SYSTEM

1. Scope of Previous Anatomic Studies

Discussion of the detailed ramifications of the blood vessels within the pituitary centers of necessity deals with the hypophyseal portal system of vessels. This third portal venous system of vessels in vertebrates was discovered by Popa and Fielding (1930) in human material. It has been the subject of detailed comparative anatomic studies by J. D. Green (1951a; 1952) and of anatomic studies in specific animals. Among the latter are the following: frog (Green, 1947); rat (Green and Harris, 1949; Morin and Bottner, 1941; Landsmeer, 1951; Daniel and Prichard, 1956); cat (Morato, 1939); rabbit and dog (Green and Harris, 1947); mouse, guinea pig and bat (Morin and Bottner, 1941); sheep (Daniel and Prichard, 1957); goat (Prichard and Daniel, 1958); and human (Fumagalli, 1942; McConnell, 1953; Xuereb *et al.*, 1954b; Stanfield, 1960).

2. General Characteristics

There are certain essential facts concerning this portal system which should be kept clearly in mind. First, in all of the animals examined it involves not one but a series of portal trunks (the term "vein" is avoided here purposely). Second, these portal trunks join a system of capillary-sized vessels (generally referred to as the primary capillary network) which occurs in the neurohypophysis, with capillary-sized vessels (sinusoids) in the anterior lobe. Third, these portal vessels are the sole significant source of blood supply to the anterior lobe in most animals. Fourth, despite the rather wide range of variation in the general morphology of the pituitary gland in different animals and in details of vascular structure (configuration of capillaries, specific relationships of capillaries to surrounding tissues and variation in length and comparative numbers of portal vessels), the essentials of this system are constant in all animals thus far examined, from the tailless amphibia to man.

3. The Vessels of the Portal System

The following descriptions of the portal vessels, of the finer ramifications of the arterial supply, and of capillary patterns will again be drawn heavily from studies of human material made by Xuereb *et al.* (1954a,b).

a. Vessels supplying primary capillaries: The primary capillaries in man are derived from ascending and descending branches of the complex of superior hypophyseal arterial branches which encircles the hypophyseal stalk, from the ascending branches of the artery of the trabecula ("loral artery") and from small ascending vessels which arise from the anastomoses between the superior and inferior hypophyseal arterial systems. Ascending branches from the artery of the trabecula and descending branches from the network about the stalk superiorly anastomose within the stalk. Many of these vessels run within the pars tuberalis and between the pars tuberalis and the neurohypophysis and finally penetrate the neurohypophysis, where they subdivide into capillaries.

b. Distribution of primary capillaries: The arrangement of the capillaries in the primary capillary plexus seems to be more complex in man than in most common laboratory animals. Xuereb *et al.*, McConnell, Fumagalli and others have described the complexity and the variability of the capillaries which occur in the neurohypophyseal tissue. Xuereb *et al.* (1954b) described them as "simple loops," "compact tufts of convoluted loops," and "intricate formations in which the convoluted loops are arranged in the form of a long 'spike.'" These authors stated further that "each of the more complex tufts of capillaries is supplied by one or more vessels derived from an infundibular artery, and drains through one or more efferent vessels into a hypophyseal portal vessel." These heavy concentrations of complex capillaries are characteristic of the neurohypophyseal tissue and are quite distinct from the capillary beds of the surrounding neural tissue at the point of attachment of the stalk. Xuereb *et al.* found very little connection between the capillary systems of the two types of tissue, an observation which had been made by Wislocki and King in 1936.

The capillary beds of the lower infundibular stem (intraglandular portion of the stalk) are derived from the artery of the trabecula, arterial branches of the inferior hypophyseal arterial system and those anastomotic vessels connecting superior and inferior hypophyseal arterial systems. Capillary loops similar to those elsewhere in the pituitary stalk may be observed here but they seem to be considerably less complex.

c. The portal trunks: The capillary beds drain into the hypophyseal portal vessels which, in most instances, both in man and other animals, seem to be formed by the confluence of two or more efferent vessels from the capillary bed in the neurohypophyseal tissue. In man they may receive further tributaries during their course downward through the pituitary stalk. Anastomoses between hypophyseal portal trunks proper have not been reported in man but have been seen in some lower animals.

The hypophyseal portal trunks are of varying lengths in the human. Many of the longer ones run within the pars tuberalis; however, some are located deeper within the neurohypophyseal tissue. There are some hypophyseal portal vessels which may be seen on the stalk not only on its anterolateral aspect but also, more rarely, on its posterior surface. The capillaries of the lower infundibular stem (intraglandular portion of the stalk) are drained by quite short portal vessels running parallel to each other deep in the stalk. These vessels supply specifically the sinusoidal bed of the posterior portion of the anterior pituitary, while the longer portal trunks described above supply the anterior and lateral portions.

d. Anterior lobe sinusoids (secondary capillaries): Sinusoids which are derived from the branching and breakup of hypophyseal portal vessels were described by Xuereb *et al.* and others as anastomosing freely with one another and forming a dense network. However, Daniel and Prichard (1956) cauterized surface hypophyseal portal vessels in the rat and found that infarction resulting from such treatment developed in fairly definite territories. They noted that the lateral, dorsal, and caudal regions of the anterior lobe remained apparently healthy and that those regions which were unaffected derived blood from hypophyseal portal vessels which had not been interfered with by the cautery. These data, coupled with the comment of J. D. Green that the portal vessels are virtually end arteries in the pituitary, would lead one to question whether or not the sinusoidal network of the anterior pituitary in man is truly freely anastomotic or at least whether or not it is effectively so. It may be that while there is anatomic continuity between the sinusoids, the pressure relationships are such as to make these anastomotic connections ineffective in maintaining an adequate blood supply in the presence of occlusion of principal vessels to a given area. On the basis of data derived from studies in living pituitaries, it would appear that the effective forward pressure of blood in the sinusoids of the anterior lobe must be very low (Worthington, unpublished observations).

e. Extent of direct arterial supply to the anterior lobe: The papers of Xuereb *et al.* and subsequently of Prichard and Daniel emphasize very strongly that in man, rat, goat, and sheep the anterior pituitary tissue receives no direct arterial supply. This is an extremely significant point. The absence of a direct arterial supply may render the gland parenchyma susceptible to infarction after local reductions in rate of blood supply. Capillary anastomoses may well be unable to compensate for the relatively large reductions in blood flow which must follow occlusion of a large portal trunk. In a recent paper Stanfield (1960), besides supplying some additional detail and confirming much of the previous work on the distribution of blood vessels within the pituitary, has attempted to clarify several points of remaining disagreement, including the degree of anastomosis between the inferior and superior hypophyseal arterial systems. He apparently feels that it is unsafe to generalize regarding the degree of overlap of the territory supplied by the two systems. While Xuereb *et al.* and Prichard and Daniel were quite emphatic regarding the lack of end arteries within the anterior lobe proper, Stanfield stated that a variable number of small end arteries supply the anterior lobe directly. These vessels come from the trabecular artery, from the capsule and from the region between the lobes. Stanfield believes that such terminal arterial twigs can be distinguished from anterior lobe capillaries or sinusoids only by fairly fine elastic tissue staining. One suspects from the description that these smaller arterial vessels would be incapable in themselves of markedly altering the vascular physiology of the anterior lobe and are probably of comparatively little significance.

In this regard, a question arises concerning the extent of the direct arterial blood supply which supposedly is present in the rabbit pituitary. It is described as arising from the internal carotid artery and passing directly to the anterior lobe (Harris, 1947; Daniel and Prichard, 1960). The descriptions by these authors are not elaborate, however. Furthermore it is difficult to develop any concept from them concerning the extent of direct arteriolar connections in the anterior pituitary, whether these connections supply their own capillary network or join the sinusoids and, if they do, in precisely what manner they join them. It would appear that in considering a system so constant phylogenetically in its essential detail, great care should be exercised in interpreting morphologic data which imply a radically different vascular arrangement in a single species.

E. MICROSCOPIC AND SUBMICROSCOPIC ANATOMY

1. Histologic Studies

Much of the information available on the anatomy of the blood vessels of the pituitary gland concerns either their distribution and configuration or (from more recent studies) their ultrastructure. However, J. D. Green (1948), in his study of the microscopic structure of the hypophyseal stalk in man, gave some excellent descriptions of histologic structure of the large or intermediate sized vessels and some of the finer ramifications.

a. Arterial vessels: In the superior hypophyseal arteries Green observed a readily discernible smooth muscle coat as well as an elastic lamina. He pointed out that some of these arteries subdivide into arterioles in the pars tuberalis and that the smooth muscle coat is still prominent in this location. Arterial branches in the pars tuberalis and in the field of distribution of the inferior hypophyseal arterial system, with readily discernible uniform smooth muscle coats, have been observed in routine sections (Worthington, unpublished observations).

b. Primary capillaries and portal trunks: Green gave some interesting data concerning the diameter of the capillary vessels supplied by these arterial vessels. He described "curious tufted vessels" and elsewhere in the same paper he referred to "skeins" of vessels. His Fig. 1 gives the impression of tufts or even glomerular formations as well as of long "skeins" of vessels. Green found that these vessels were of large caliber for capillaries and variable in size. For example, in measurements taken in one specimen he found an extreme range of 5.8–39.7 micra in 119 vessels. This wide variation in the diameters of the primary capillaries of the hypophyseal portal system is also a characteristic of species other than man and has been observed in living preparations (Worthington, 1955).

The capillaries, as well as the venules draining them, have been described as having a connective tissue sheath which stains in a manner similar to collagen. Green found that the cells in this sheath were somewhat varied, some resembling smooth muscle, while others were possibly derived from the reticuloendothelial system. None of these cells resembled glial cells. Green also described the collecting venules leaving the tufts of vessels as tending to run together, particularly into the pars tuberalis, to form portal channels of large caliber. Their courses have already been described (Section D. of this chapter).

In a number of specimens of human pituitaries (stained with hematoxylin and eosin or the periodic acid Schiff reaction), Worthington (unpublished observations) found, in confirmation of the observations of other workers, that these wide hypophyseal portal trunks are extremely thin walled, consisting of a very thin endothelium and comparatively small amounts of collagenous or reticular connective tissue. There is, however, a very definite suggestion of the continuation of the connective tissue sheath, as described by Green, along the courses of the large hypophyseal portal vessels. Within these thin walls, there are cells, all of which do not appear to be ordinary connective tissue cells. Some have a definite longitudinal or transverse orientation and a few scattered cells could not be distinguished from smooth muscle. However, there is no direct evidence which indicates that they are smooth muscle cells. Data from the literature supporting the thesis that there are prominent contractile elements in the walls of the portal trunks or that there are valve-like structures obstructing blood flow or in some way preventing the backflow of blood in the hypophyseal portal vessels have not been numerous or convincing. Contraction of the walls of portal trunks has never been observed in living animals (Worthington, 1960 and unpublished observations).

c. Neurovascular zone: Green described a morphologically highly specialized neurovascular zone on the posterior aspect of the human pituitary stalk. This zone apparently escaped attention as a specialized area prior to his study because of its continuity with the pars tuberalis on either side of the stalk and below it. It may extend posteriorly almost to the mammillary bodies. It is described as having a sharp line of demarcation between it and the tuber cinereum. It contains collagen and islands of glandular cells but apparently consists primarily of blood vessels and nerve fibers. The nerve fibers are arranged in extremely complex and highly individual perivascular plexuses which are unusually thick. As attractive as the finding of very definite connections between this neurovascular zone and the tuberohypophyseal tract might be, in the light of current hypotheses on the neurovascular link between the hypothalamus and the anterior pituitary, these connections were not noted.

d. Perivascular nerve endings of hypophyseal stalk: In the hypophyseal stalk Green found nerve endings to be common in the vessel sheaths previously described and these took a variety of forms. He also described nerve fibers in the pars tuberalis. Regarding the origin of these nerve fibers, he made the cautious statement that "many seem to come from the tractus hypophyseus proper." He believed that some of the neurovascular zone fibers gave rise to vasomotor type endings which were perhaps supplying some of the vessels in the neural stalk.

An interpretation that at least some of the fibers ending around blood vessels in the primary capillaries and portal trunks in the human neurohypophysis are derived from the tractus hypophyseus is consistent with the findings in other animals and is the only view in accord with the uniform application of the neurohumoral hypothesis. The ever-present problem of whether silver-stained fibers are neural or reticular might give us considerably more concern here if it were not for Green's extremely judicious handling of material of this type, both in previous and subsequent work. He demonstrated, by the painstaking use of silver stains and auxiliary techniques, that while there may be some vasomotor fibers accompanying the blood vessels which enter the anterior lobe, there are no nerve fibers ending directly as secretomotor fibers on cells of the anterior lobe. This work (J. D. Green, 1951b) is quite convincing and lends credence to the interpretations concerning the presence and sources of origin of nerve fibers in the vicinity of the blood vessels in the neurohypophysis and special neurovascular zone.

e. Organization of neurohypophyseal perivascular structures: Liss (1958) has given an even more detailed description of the perivascular structures in the human neurohypophysis. In brief, those structures which immediately surround the neurohypophyseal capillaries are a nerve plexus, "pericytes," perivascular glia, pituicytes and connective tissue. The nerve plexus is described as lying directly on the vessel wall, and, on the basis of previous work, Liss felt that these fibers are derived partially from the hypothalamic-hypophyseal tracts and partially from autonomic plexuses. "Pericytes" lie in the same plane as these nerve fibers. The cells, presumably connective tissue cells of mesodermal origin, are applied directly to the vessel wall. Liss further described perivascular glial cells lying close to the "pericytes" but having their processes oriented in two separate planes, one adjacent to the vessel wall and the other somewhat farther away from it. There are pituicytes lying quite near the perivascular glial tissue, and their processes extend toward, as well as away from, the blood vessels, where they come into contact with the processes of other pituicytes deeper in the neurohypophysis. The processes of these pituicytes are arranged in a very complicated way in reference both to each other and to the connective tissue. Some of the pituicyte processes end on blood vessels in a type of sucker foot. The perivascular connective tissue has no direct contact with the wall of the vessel but rather forms a dense sheet in which all of the other perivascular structures are contained. It seems to serve chiefly as a supporting structure. The arrangement is quite complicated and the reader is referred to Liss' paper for details. One significant suggestion which Liss made is that the pituicytes are a type of modified astrocyte.

2. Ultrastructure

a. Neurohypophysis: The capillaries of the neurohypophysis of the rat have been studied with the electron microscope by Hartmann (1958) before and after the administration of histamine, a substance which increases the titer of posterior lobe hormone in the circulating blood.

Posterior lobe capillaries were studied previously by Bargmann and Knoop (1957) in the cat and dog and by Palay (1957) in the rat.

According to Hartmann's description, the posterior lobe capillaries are surrounded by two layers of basement membrane which appear homogeneous at moderate magnification and show only a moderate electron density. There is no real suggestion of fibrillar structure within the basement membrane. The inner of these membranes occurs immediately adjacent to the capillary endothelium and the outer one borders pituicyte processes and nerve terminals. This outer membrane, which is separated from the inner by a space of variable width, is described by Hartmann as "complexly folded and compartmented." The space between these two basement membranes appears to be, for the most part, optically empty, but sometimes fibers of connective tissue and an occasional fibroblast may be seen. Fenestrations in the endothelium of the capillaries of the neurohypophysis of the rat were also described by Hartmann. These had been observed previously in the neurohypophysis of the rat by Palay (1957). The fenestrations are bridged by a single very thin membrane and resemble fenestrations described by Ekholm (1957) in thyroid capillaries.

Hartmann held some reservations concerning Palay's interpretation of the thin membrane bridging these discontinuities in the endothelium as a derivative of the plasma membrane. He considered the possibility that there is an absorption of electron-dense material on the inner portion of the endothelial basement membrane at those points where the endothelial discontinuities occur. Ekholm (1957), in the case of the thyroid, simply stated that the plasma membranes of the endothelial cells meet at the borders of the discontinuities.

In the posterior pituitary capillaries, as in the case of the thyroid, Hartmann found that the discontinuities do not coincide with the borders between separate endothelial cells. Also, as in the thyroid, there are no discontinuities observed in the basement membrane applied to the endothelial cells. In addition to these observations, Hartmann noted thickenings of endothelial cytoplasm at the points of junction between adjacent endothelial cells. He also reported that the endothelial cells are fairly thick in the region of the nucleus. In this site they contain mitochondria, golgi material, and some elements of endoplasmic reticulum. Farther away from the nuclei the endothelial cytoplasm is very thin, measuring as little as 150 ångstroms in places. Hartmann also observed small vesicles in the endothelial cytoplasm which are increased in number in rats that have been treated with histamine. The capillary endothelium in the histamine-treated rats is somewhat swollen, but no other distinct changes in the capillaries are noted following the histamine treatment. Quite significantly, however, no neurosecretory vesicles, or granules which can be associated with the electron-dense centers of neurosecretory vesicles, are found in pericapillary spaces or in the capillary lumina.

The whole arrangement of endothelial cells, endothelial discontinuities and basement membranes and the relationships of these structures to the parenchyma show a marked similarity in posterior pituitary and thyroid. Apparently this similarity extends to other endocrine organs as well, and it has been suggested by several workers that the endocrine glands have this structural arrangement in common and that it has some physiologic significance. (For general discussion of submicroscopic structure of blood vessels, see Section C., Chapter I.)

b. Adenohypophysis: The ultrastructural organization of the sinusoids of the anterior pituitary in the rat has been described by Rinehart and Farquhar (1955). They noted the endothelial lining of the sinusoids and a successive series of layers from the capillary lumen outward, including the series of structures mentioned above for the thyroid and posterior pituitary. There is present an inner and outer endothelial cell border, a thin basement membrane closely applied to the endothelial cells, a connective tissue space, and a second basement membrane more closely applied to the cells of the parenchyma. Rinehart and Farquhar believed that the endothelium formed a continuous lining but pointed out also that the endothelial cytoplasm of the sinusoids of the anterior pituitary may be extremely thin in some areas, probably below the level of resolution for light microscopy. The cytoplasm of the endothelial cells contains mitochondria, small vesicles, very fine particles, a golgi complex and an endoplasmic reticulum. The space between the two basement membranes is defined as a perisinusoidal space of varying width and completely surrounding the sinusoids. Most of the perisinusoidal space is of fairly low electron density, and Rinehart and Farquhar suggested that it contains a substance relatively high in water content. Fibrils showing the definite periodicity characteristic of connective tissue fibrils are also present. Further, Rinehart and Farquhar reported that "granules and small segments of granule-containing cytoplasm derived from the anterior pituitary parenchymal cells are frequently found lying free in the perisinusoidal and intersinusoidal spaces." Of further significance is the fact that in rats previously treated with trypan blue, very little of this dye is observed in the cytoplasm of the endothelial cells, although it is frequently present in cells in the perisinusoidal spaces. Rinehart and Farquhar felt that under the stimulus of the administration of the dye, these cells in the perisinusoidal space, which are evidently macrophages, can project pseudopodia through the basement membrane and endothelium into the lumen of the sinusoid. With these changes there is little accumulation of trypan blue in the endothelial cells, but the cells enlarge and there is an increase in the number of small vesicles in the cytoplasm. The account of Rinehart and Farquhar clearly shows that the phagocytic cells in the anterior pituitary lie outside the endothelium and that the sinusoidal endothelium of the anterior lobe of the pituitary does not constitute part of the "reticulo-endothelium" and is not lined by highly phagocytic cells, such as occur in liver, spleen, and bone marrow. This observation, so beautifully and clearly demonstrated with the advanced techniques of electron microscopy, makes the same finding in 1929 by Cappell, who used only vital dyes and light microscopy, an even more significant accomplishment.

More recent studies on the ultrastructural organization of the sinusoids of the anterior pituitary by Farquhar (1961) have revealed additional details which make it apparent that the basic pattern of organization of the sinusoidal wall and of the relationships of the sinusoid to the surrounding parenchyma is quite similar to that found in thyroid and posterior pituitary, as well as parathyroids and adrenals. In this more recent work Farquhar described fenestrations in the endothelium of the sinusoids of the anterior pituitary of roughly the same order of magnitude as those found in other endocrines. These fenestrations are bridged by the same very narrow membrane and have the same relationship to the underlying basement membrane and to the points of junction between endothelial cells as do the other organs mentioned. In other words, there are no discontinuities in the basement membranes apposed to the fenestrations, and the fenestrations are quite separate and distinct from the points of junction between neighboring endothelial cells.

In this later paper Farquhar described secretion granules, with their membranes in continuity with the cell membrane at the vascular poles of parenchymal cells, parenchymal cell pseudopodia projecting into the perisinusoidal space, and elaborate foldings and plications of parenchymal cell membranes facing sinusoids. She hypothesized that there is rapid solubilization of granules in the connective tissue space after discharge and that the surface arrangements of the parenchymal cells facilitate metabolic exchange and granule release.

F. MICROCIRCULATION[3]

1. Observations in Control Animals

a. Direction of portal flow: The first observations of blood flow in the living pituitary were made by Houssay *et al.* in 1935. Despite their

[3] For general discussion of the microcirculation, see Chapters V and VI.

description of the correct direction of blood flow in the portal vessels (hypothalamic-hypophyseal), a long controversy on this point ensued which has only in recent years begun to subside. The correct direction of flow was subsequently observed by J. D. Green (1947) in the frog and by Green and Harris (1949) and Barrnett and Greep (1951) in the rat. Green and Harris described a uniform nonpulsatile flow in the hypophyseal portal vessels in the rat.

b. Vasomotion, degree of vascularity, capillary flow: More extensive microcirculatory studies have been conducted in the pituitary stalk of the mouse by Worthington (1955, 1960). Although this work was carried out primarily in a single species and the observations were made following (of necessity) a highly traumatic operative procedure, these investigations are the only fairly extensive ones on small blood vessels in the living gland. High magnification, water immersion studies have also been made on the frog by Worthington *et al.* (in press). An account of some of this work on pituitary vascular responses and microcirculation in the mouse follows.

Blood flow in hypophyseal portal trunks has been downward into the pars distalis in every one of a long series of cases. Under a variety of different experimental conditions, reversals in the direction of blood flow are found only in primary capillaries and some of the arterial anastomoses around the stalk. The very fine arterioles or arterial vessels supplying the primary capillary system of the median eminence and stalk are highly contractile. Some intermittency of blood flow may be observed in the primary capillary system proper, in the sense that under the comparatively low magnifications which are possible, a number of primary capillaries may be lost to view for varying periods of time. The degree of vascularity of different regions of the stalk and of the whole stalk varies from animal to animal. It is possible under some conditions to make tentative correlations between the type of treatment to which the animal has been subjected and the degree of vascularity. For example, there are usually fewer open capillaries in the stalk following superior cervical ganglionectomy (Blakely and Worthington, unpublished data). No suggestion of active contractility of the hypophyseal portal trunks proper has ever been observed.

2. Effects of Experimental Procedures

For several reasons, discussed in detail elsewhere (Worthington, 1960), it was concluded that linear velocity of blood flow in the portal trunks is an accurate index of changes of blood flow into the anterior lobe of the pituitary in the mouse. Such changes in linear velocity were used as an indicator of alterations in total blood flow into the anterior pituitary.

a. Drugs: Sodium pentobarbital and morphine sulphate, both previously reported as blocking agents for the adrenocorticotrophic response to stress, are capable of markedly reducing the rate of blood flow in the hypophyseal portal vessels under some circumstances. They also bring about a vasodilatation of the superior hypophyseal arteries and arterioles.

Epinephrine applied locally to the region of the pituitary stalk in dilutions as great as 1:5,000,000 results in constriction of superior hypophyseal arteries and arterioles and a reduction of the rate of portal blood flow. Single doses given parenterally are followed by gradual acceleration of portal flow rates and transient narrowing of the arterial vessels, and then by a return to the original or slightly larger diameters. Both local and peripheral administration of *l*-norepinephrine bitartrate in fairly small doses gives inconsistent results. Low concentrations may be extremely active, in that they may produce marked arteriolar constriction and reduction of portal flow rates; however, this is a result which cannot always be repeated. Large unphysiologic doses given intravenously elicit considerably more consistent results in the form of respiratory embarrassment, temporary cessation of blood flow in the portal trunks and primary capillaries, and extreme slowing of the circulation in the superior hypophyseal arteries. All of this is followed by intense arteriolar constriction and extremely rapid portal blood flow. Mecholyl or acetylcholine applied locally to the stalk in dilutions of 1:1000–1:100,000 always gives arterial and arteriolar dilatation and acceleration of portal blood flow.

b. Physical stimuli: Intense painful stimuli induce arterial and arteriolar vasoconstriction around the median eminence and stalk and subsequent acceleration of hypophyseal portal blood flow. Total temporary occlusion of the tracheotomy opening, resulting in hypoxia and hypercapnia, produces extremely marked arterial vasodilation and extremely rapid portal blood flow. The vasomotor responses to hemorrhage of the portal system are varied, depending on the degree of blood loss. A small hemorrhage is usually followed by a short period of slowing of the portal blood flow and then acceleration, while a sudden large one results in almost immediate and long-lasting reduction of the rate of portal blood flow and arterial dilatation. This type of change occurs also with numerous small hemorrhages over a prolonged period of time.

Another interesting finding is that bilateral common carotid artery ligation results in very marked slowing of blood flow in the portal vessels and primary capillaries but not complete cessation of blood flow in these vessels. In experiments in which the carotids are ligated while the pituitary vessels are under observation, damage sufficient to result in the death of the animal does not occur before termination of the experiment. However, in a number of attempts to maintain an animal after bilateral common carotid ligation, the animals have died, usually in about 12 hours. Apparently the vertebral circulation in the mouse is sufficient to support some continued circulation within the hypophyseal portal system for a while but it is not enough to maintain life for any length of time.

c. Superior cervical ganglionectomy: Recent experiments by Blakely and Worthington (1960) have suggested that the hypophyseal arterial vessels receive vasomotor fibers by way of the superior cervical ganglia. When the superior cervical ganglia were removed bilaterally from mice, dilatation of the arterial vessels of the stalk and acceleration of portal vessel flow resulted. The arterial vessels were much more sensitive to epinephrine following this procedure.

3. Significance of Microcirculatory Studies

One general aim of this work has been to test the hypothesis that the many experimental procedures which have been applied to the pituitary and to the animal in general, with the view of elucidating the relationship between the hypothalamus and anterior pituitary, all involve the possibility of vascular interference. Sufficient attention has not been paid to the physiology of the blood vessels. While the neurohumoral hypothesis appears to be very strong currently, it still seems fairly apparent that the possibility that vasomotion controls some of the trophic functions of the anterior pituitary has not been satisfactorily ruled out. This matter has been reviewed by Worthington (1960).

While it is apparent that the blood vessels of the pituitary stalk have an extremely active vasomotor system and respond quite clearly and, on occasions, dramatically to a number of drugs and peripheral stimuli, the available microcirculatory data are not sufficient to prove that vasomotor effects are primarily responsible for any trophic function. It can only be hoped that refinements in technique will make it possible to advance our knowledge of the physiology of the pituitary blood vessels beyond its present status. Thus far, microcirculatory techniques have offered the only hope of getting any insight into the physiology of these vessels, and the studies which have thus far been made on them constitute a satisfactory beginning.

2. The Thyroid

A. INTRODUCTION

Much of our information concerning the gross and microscopic anatomy of the thyroid vessels falls into that realm of anatomic knowledge which we would like to consider relatively stable. However, even a cursory examination of the sources of such information makes it apparent that there are areas in which some of our generally accepted information can be clarified by the available literature. In addition, there are aspects of thyroid structure which urgently need further investigation.

In the following survey it was necessary to select references rather than attempt a detailed review of the entire literature. Despite this deficiency, however, it should be possible from the bibliography of this section to locate either primarily or secondarily most of the significant literature relating to the anatomy, both gross and microscopic, of the thyroid blood vessels. In addition to these specific references, the paper on thyroid blood flow by Söderberg (1958) contains an excellent bibliography.

B. GROSS ANATOMY

1. Principal Arterial Supply

The extraglandular course and relations of the thyroid arteries are described in detail in textbooks of gross anatomy. However, certain important and highly variable relations require comment.

a. Superior and inferior thyroid arteries: The superior thyroid artery comes into close relationship (parallel and in close proximity) with the superior laryngeal nerve on the outer surface of the inferior constrictor of the larynx, and there is some surgical risk of ligating the nerve when the artery is ligated (Nordland, 1930).

The inferior thyroid artery on each side, usually described as the larger of the two, has important surgical relationships to the laryngeal nerve which were described and illustrated in beautiful plates by Nordland (1930) and were described more recently by Reed (1943). According to Nordland there are no constant differences in relationship between the right and the left sides. The findings in his series of thirty-one cadavers are listed in Table XV. Reed, in his longer series (253

TABLE XV[a]
RELATIONS OF INFERIOR THYROID ARTERY TO RECURRENT LARYNGEAL NERVE IN 31 CADAVERS

	Right Side	Left Side
Nerve anterior to artery	15 (2 posterior to a small branch)	11 (2 posterior to a small branch)
Nerve posterior to artery	8 (3 anterior to a small branch)	11 (1 anterior to a small branch)
Nerve between equal branches of artery	7	7
Nerve branched around artery		2
Artery not present	1	

[a] From Nordland, 1930 (abridged).

cadavers), gave a more detailed breakdown of these relationships, listing some twenty-eight anatomic variants of the relation of inferior laryngeal nerve to inferior thyroid artery. The more commonly occurring of these are presented in Table XVI.

TABLE XVI[a]
RELATIONS OF INFERIOR THYROID ARTERY TO RECURRENT LARYNGEAL NERVE IN 253 CADAVERS

	Right Side	Left Side
Nerve anterior to artery or branches	65	29
Nerve posterior to artery or branches	68	130
Nerve anterior to ½ of the branches and posterior to ½ of the branches	56	55
Nerve anterior to ⅓ of the branches and posterior to ⅔ of the branches	33	20
Nerve divided into 2 branches while in direct relationship to artery	17	4

[a] From Reed, 1943 (abridged).

Reed (1943) reviewed the dissections of Berlin and Lahey (1929), Ziegelman (1933), and Fowler and Hanson (1929). Of these the largest series was that of Fowler and Hanson (400 nerves) which showed 65.5 per cent of the nerves passing posterior to the artery, 26 per cent, anterior and 8.5 per cent passing between branches of the artery. In this study there was no breakdown as to right and left side. Reed stated that this is in general agreement with his finding that the nerve more frequently passes deep to the artery. The difference in total number, however, is greater in Fowler and Hanson's series. It is evident from the foregoing that although there is a readily discernible pattern of relations between inferior thyroid artery and recurrent laryngeal nerve, the variation is marked and in any single case at least twenty-eight possible variants may be encountered. Of course there may be others unreported.

b. Arteria thyroidia ima: A fifth thyroid arterial vessel, the *arteria thyroidia ima,* is generally said to be present in about 10 per cent of cases. This figure is given by Wangensteen (1929) in his review of the subject and is not inconsistent with the data reviewed by Cayotte and Sommelet (1952). According to Cayotte and Sommelet, it is sometimes difficult to evaluate data on the frequency of occurrence of this vessel since one cannot always be assured that, for any given number of cadavers reported, a systematic search has been exercised in each case. They also reported that this vessel occurs more frequently in men than in women and that there appears to be no racial variation.

The origin of the *arteria thyroidia ima* is usually given in textbooks of anatomy as the brachio-cephalic trunk and the aortic arch. However, Cayotte and Sommelet pointed out that it may arise from the common carotid or internal mammary artery. The *arteria thyroidia ima* arises most frequently from the right; rarely it is paired. Its course in relation to the trachea is given as almost transverse, oblique, or vertical and median, depending on its site of origin. It seems to have a more significant role in the vascularization of the left lobe of the thyroid gland than the right. The caliber of this vessel is highly variable; it may be an extremely narrow vessel, sometimes described as an arteriole, or it may be as large as 3 or 4 mm in diameter and even larger. Cayotte and Sommelet stated that the difference in size is conditioned by the means of vascularization of the thyroid and that when one of the inferior thyroid arteries is lacking, the thyroidia ima replaces it and thereby acquires a greater size. The importance of these data for emergency tracheotomy is obvious.

c. Collateral circulation: The presence of a significant collateral circulation for the thyroid has been recently studied by Faller (1957), who made corrosion preparations of glands injected with radio-opaque substances. He pointed out that collateral vessels reach the thyroid by way of vessels from the esophagus and larynx. Wangensteen (1929) described the work of Petenkofer in 1914 and Erderlen and Hotz in 1918 in which they demonstrated that after ligation of the four thyroid arteries, material injected by way of the ascending aorta reached the thyroid vessels. Further, Curtis (1930) demonstrated filling of parathyroid vessels after quadruple ligation of the major thyroid arteries.

d. Arterioarterial anastomoses: Descriptions of anastomoses between the vessels on the surface of the thyroid vary slightly from one gross anatomy text to the other. According to Gardner *et al.* (1960), the superior thyroid artery has an anterior branch which anastomoses with its fellow along the upper border of the isthmus and a posterior branch which anastomoses with the inferior thyroid. The inferior thyroid artery has an ascending branch which anastomoses with the superior thyroid and a descending branch which anastomoses with its fellow from the opposite side along the lower border of the isthmus. Gudernatsch (1953) gave a similar description and reported that the anastomosis of all four or five of the major vessels may occur in the region of the isthmus. Johnson (1953; 1954) stated unequivocally that surface anastomoses between the thyroid arterial vessels are numerous. This point seems to have been clear quite early, in that it was made by Major (1909), who referred to it as a confirmation of the work of earlier authors. It should be pointed out that Major stated that the general scheme of these anastomoses varies considerably in different human thyroids. In general, however, the pattern seems to be that which is described above.

C. MICROSCOPIC AND SUBMICROSCOPIC ANATOMY

1. Intraglandular Distribution

Information on the ramifications of the smaller branches of vessels on the surface of the thyroid and deep in the gland is considerably more diffuse and difficult to obtain than that concerning the gross blood supply. Differences of opinion exist about the general histologic structure of the thyroid which are not adequately reflected in general works on histology, particularly in the case of the question of the lobular structure of the thyroid gland and the details of the distribution of its vessels.

a. Structural organization and relationship of the thyroid to its vessels: Major's paper (1909), although by no means the first on the blood supply of the thyroid, gave such a clear and concise description of the distribution of the thyroid vessels, including the finer ramifications (so well confirmed and complimented almost fifty years later by the work of Johnson) that it seems basic to our understanding of the distribution of these vessels. Major divided the vasculature of the thyroid into a series of units, the largest of which, called the prime unit, is the gland proper with its major blood supply. He then described arteries of the first order, which ramify on the surface, supplying various regions of the thyroid, and penetrate to give rise to branches of the second order. The arteries of the second order supply structural lobes of the thyroid, which are collections of smaller units, the lobules, and which are separated by somewhat thicker connective tissue septa than the lobules. The arteries of the second order surround collections of lobules and give rise to branches of the third order. The branches of the third order pass between the lobules, giving rise to arteries of the fourth order which run over the clumps of follicles and have, as their direct branches, the follicular arteries. The arteries of the fourth order constitute the arterial supply to the lobular unit, while those of the smallest order supply the follicles and are the vascular-follicular units. Thus there is a series of histologic structural units of decreasing size, which is supplied by an accompanying series of vascular units of decreasing size. The acceptance of this scheme is dependent upon the assumption that the lobule is one of the units of structure in the thyroid. However, there were differences of opinion about this at the time Major wrote and this situation still exists.

A recent account of the distribution of the smaller vessels of the thyroid is that of Johnson (1953), which also is based in part on the concept of a distinct lobular structure of the thyroid. According to this author, there are three orders of structure within the adult thyroid gland: the lobule, consisting of 20–40 follicles bound together in loose connective tissue; a structural lobe, consisting of a group of lobules bound together in a common connective tissue sheath; and an anatomic lobe, made up of structural lobes surrounded by a considerably heavier layer of connective tissue which forms the capsule of the gland. Within this general framework the ramifications of the arterial system are as follows:

Branches from the major trunks penetrate the surface of the gland where they run in fibrous septa between structural lobes, with some subdivision occurring during their courses between the lobes. Branches of these interlobar vessels pass into the structural lobe where they divide into branches supplying a lobule. Johnson stated rather emphatically that each lobule receives an individual twig which he designated the lobular artery. Occasionally a lobule has two small vessels entering it, but this is the exception. The lobular artery then breaks up into a series of smaller twigs which pass to the individual follicles, forming a capillary plexus on their surfaces. The density and arrangement of this plexus vary from one follicle to the next according to differences in the activity of the follicle. Both Johnson and Major indicated that the venous drainage follows essentially the same pattern as does the arterial supply.

While the descriptions of these two workers are mutually confirmatory in most respects, they do show some minor differences. For example, Johnson stated that having more than one vessel supplying a lobule is the exception rather than the rule, while Major pointed out that, depending on the size of the lobule, 2–5 small arteries may supply it.

Rienhoff (1929), on the basis of his wax plate reconstructions from thick sections of the gland as well as on maceration and microdissection studies, stated that there is no evidence for a lobular structure of the thyroid. He felt that the thyroid consists of continuous bars, columns, and plates of tissue containing discontinuous individual follicles. However, Johnson pointed out that some of the connective tissue septa separating what he considers lobules are very thin and may conceivably have been missed in the wax plate reconstructions. He found that, in his own work, glands rendered edematous by saline injections showed very definite planes of cleavage between separable lobules and that in injected glands, lobules could be dissected out with their blood supply intact. It seems clear from descriptions of the blood supply by both Major and Johnson, as well as others, that something very closely resembling lobular units within the thyroid can be distinguished on the basis of the vascular structure even without considering the other evidence which has been presented here. It would seem, then, that the anatomic basis for Johnson's theories (1955a) on the evolution of nodular goiter is a reasonably secure one.

b. Intraglandular anastomoses: The problem of anastomoses, both arteriovenous and arterioarterial, may well be introduced by discussion of another major contribution to the distribution of the small blood vessels within the thyroid, that of Modell (1933), who performed his examinations in the dog. This author observed anastomoses within the depths of the gland between arterial trunks, between branches of the larger trunks, and between two branches of the same trunk. His description of this, however, does not lead one to believe that these anastomoses are a frequent occurrence. Johnson (1953) also demonstrated this phenomenon. Modell found some capillaries passing from the arterioles directly to veins which seemed to be in a position to supply only a few follicles. This arrangement was noted fairly frequently. Such capillaries arose by the rather abrupt loss of the muscular coat of an arteriole. Larger arteriovenous anastomoses were present but were not found very often. Major made no mention of such arteriovenous connections, nor have other references to these structures been found thus far. Johnson (1953) in his discussion seemed to believe that such connections must be present but reached his conclusion only on the basis of Modell's work in the dog, by clinical inference, and by indirect evidence.

Veins have been observed to anastomose with great frequency both on the surface and deeper in the gland.

2. Histology

Comparatively little information is available concerning the specific histology of the thyroid vessels, as compared with vessels of similar order of size in other organs. Johnson (1955b) found that even in goiters there is little change in the larger arteries either on the surface of the gland or deeper, with the possible exception of occasional examples of hypertrophy of the media. Modell (1933) reported some rather unusual areas of dilatation and abrupt thinning and thickening of the muscular coats in arterial vessels in the dog. Johnson (1955b) called attention to the occurrence in nodules of hyperplastic thyroid tissue of very wide and thin-walled sinusoidal vessels, which may be found individually or in groups of several such vessels.

One major structural detail of thyroid vessels which has been investigated and discussed at length is the presence of "buds," "cushions," or "sphincters" at points of arterial branching within the thyroid. There seems to have been no early general agreement on the nature of these structures. Thickening of the intima by proliferation of connective tissue, localized proliferation of endothelial cells, involvement of both intima and media, and various arrangements of smooth muscle have been described. The subject has been reviewed by Modell (1933) and more recently by Sato (1955). Modell found that the back extension of the muscle coat of a branch into the wall of the parent trunk gave additional smooth muscle in the wall of the larger vessel at the point of branching. Sato described in the hamster complete and incomplete thickenings of the circular smooth muscle at the points of branching of the arteries.

3. Ultrastructure

The ultrastructure of the thyroid capillaries has been studied by Monroe (1953), Dempsey and Wislocki (1955), Dempsey and Peterson (1955), Lever (1956), and Ekholm (1957). This excellent series of papers, in addition to its inherent scientific interest, makes an interesting study in the technical progress of the preparation of tissues for electron microscopy. Each paper gives new information, frequently not only on the basis of further investigation and interpretation of the material but also on the basis of better tissue preparation and greater resolution. Of the papers in this group, only that of Ekholm (1957) dealt exclusively with the fine structure of thyroid vessels. The others were concerned either with other morphologic and/or physiologic aspects of thyroid or with other organs as well as the thyroid.

Monroe (1953) believed that portions of the walls of the thyroid capillaries consist of a basement membrane only and that, while endothelial nuclei and cytoplasm are plentiful, areas exist in which there is not even a highly attenuated endothelial cytoplasmic layer present. In larger vessels in the same material, a continuous endothelial lining was always observed. Dempsey and Peterson (1955), in a more complicated study involving variation in physiologic state induced by hypophysectomy, cold exposure, or thiouracil administration, noted very little change in the ultrastructure of the thyroid capillaries under these different conditions. Apposition of capillaries to the thyroid epithelial cells was observed in the hypophysectomized animal rather than the indentation which was present in normal animals. In addition to a fine epithelial basement membrane (quite separate from the plasma membrane of the cell itself), Dempsey and Peterson described a basement membrane, about 200 Å in thickness, for the endothelium of the capillaries. Between these membranes a space occurred which was described as variable in thickness. Occasional collagenous fibers were noted in the space. Studies made about the same time by Dempsey and Wislocki (1955) have shown that both of these membranes are capable of reducing silver salts ingested by an animal subjected to chronic argyria, a property held in common with many other basement membranes elsewhere in the body.

Dempsey and Peterson considered Monroe's finding of discontinuities in the endothelium, leaving exposed basement membrane in contact with the lumen, unlikely and explained their occurrence on the basis of low resolution or tissue distortion during preparation. An amorphous layer applied directly to the epithelial and endothelial cells, with an intervening space, has also been described in the thyroid gland of the rat by Lever (1956).

A more recent paper on the electron microscopy of thyroid capillaries is that of Ekholm, who directed his attention to the fine structure of the small blood vessels alone. He described the area between epithelial and endothelial cells as consisting of three layers, an osmiophilic layer immediately adjacent to epithelial cells, an osmiophilic layer immediately adjacent to endothelial cells, and an osmium deficient layer or space between the two. Each of the dense layers varies from 400 to 500 Å in thickness, and each runs uninterruptedly over cell boundaries. Ekholm was able, in some areas, to discern a lamellated structure in the dense layers which had not been seen in previous studies. He also described material bridging the osmium deficient layer between the two dense layers. The osmium deficient layer in his material was between 500 and 1000 Å thick and contained no structures other than the bridges between the two dense layers. He did describe, however, widenings in the osmium deficient layer which sometimes contained aggregates of homogenous material of moderate density or circular oval bodies bordered by what appeared to be a surface membrane. While the endothelial cells themselves were variable in thickness, most of the endothelium was very thin, varying between 200 and 600 Å. The endothelial cells showed a distinct plasma membrane which measured about 70 Å.

Ekholm again raised the question of discontinuities in the endothelial cytoplasm, which he described as having one dimension of at least 400 Å. In these areas the endothelial cells were replaced by very thin membrane having a thickness of about 50 Å. The plasma membranes of the endothelial cells met at the borders of the discontinuities. These discontinuities did not coincide with endothelial cell boundaries, nor were there any in the dense layer applied to the epithelium.

Ekholm discussed the various possibilities of interpretation derived from his morphologic data and suggested that the round bodies bordered by surface membrane might be derived from cells located outside of the level of section. While he did not note fibrillar structures showing periodicity in the narrowest portions of the osmium-deficient layer, he did see such fibrils in osmium-deficient areas separating the endothelium of neighboring capillaries and separating endothelium from connective tissue cells, thus confirming the previous observation by Dempsey and Peterson of the presence of collagenous fibers. While recognizing the fact that methods of preparation may have had something to do with the appearance of discontinuities in the endothelial lining, Ekholm seemed to be convinced that they were real. He believed that he had excellent preservation of cytoplasmic structure in the endothelial and epithelial cells and that the clear junction between the inner and outer portions of the plasma membrane of the endothelial cells, at the margins of the discontinuities, did not appear to be artifactual. In fact, Ekholm's electron micrographs are quite convincing. Bennett *et al.* (1959) pointed out that the work of Luft and Hechter (1957) on the bovine adrenal may have some bearing on the question of fenestrations in endothelium. However, the differences between Luft and Hechter's oxygenated perfused and control adrenal sinusoids suggest an entirely different situation from that described by Ekholm. In Luft and Hechter's study the control glands show irregular and apparently complete (i.e., no intervening membrane) fenestrations, whereas Ekholm's sections of thyroid capillaries seem to demonstrate considerably greater regularity of the discontinuities and the presence of a very thin membrane bridging the discontinuity. Recently Wissig (1960) suggested that the fine single-layered membranes which bridge the fenestrations in rat thyroid endothelium are artifacts.

Thus, there are differences of interpretation of electron micrographs of thyroid capillaries and marked variations in structure of adrenal sinusoidal endothelium with physiologic state, as reported by Luft and Hechter. Bennett *et al.* have suggested that anoxia or variation in physiologic state may be responsible for some details of ultrastructure which have been observed in capillaries in many different areas. This cautious attitude is well taken. More data obtained from lines of attack similar to that of Luft and Hechter should prove most profitable.

The fixation of tissues for electron microscopy by the method of exposing them *in situ,* with the circulation intact, and fixing superficial portions of organs and tissues after the normality of the circulation is established, would seem to show some promise as a method for maintaining a state as close as possible to that existing before the application of the fixative. This would tend to eliminate, or at least alleviate to some extent, the problem of what degree of anoxic change is taking place in an organ or tissue in even the very short period of time between sacrifice and immersion of the tissues in the fixative. Such methods have been attempted on the pituitary stalk of the mouse, and it has been found that the circulation in the capillaries of the stalk stops instantaneously upon application of buffered osmium tetroxide (Worthington, unpublished observations).

D. MICROCIRCULATION[4]

Studies on the microcirculation of living thyroid tissue seem thus far to have been somewhat limited in both number and scope. Williams (1944) commented on the microcirculation in the thyroid in a paper which did not deal primarily with microcirculation per se, but with a number of other aspects of living thyroid structure and function. He stated that the capillaries at the periphery of the follicles in the rat were very numerous and of large caliber and the flow in them extremely rapid. Under the conditions of his experiments the rate of flow did not fluctuate, and he was unable to see any empty vessels. In his studies on microdissection he found that these thyroid capillaries were quite adherent to the epithelial cells of the follicle, and it was very difficult if not impossible to remove them from their attachment to the thyroid cells. These observations of Williams were made with the thyroid *in situ.*

The microcirculation of the thyroid was also investigated by Thomas

[4] For general discussion of the microcirculation, see Chapters V and VI.

(1945), who performed injection experiments as well. He described two sets of capillaries in their relations to the follicle. One consisted of large interfollicular capillaries, which outlined the follicle in transilluminated specimens, and in which the blood flow was very rapid. The other was a system of very narrow capillary vessels, which much more closely enveloped the follicles and through which individual red cells passed in single columns and in "a somewhat jerky fashion." Thomas further stated that the corpuscles were frequently distorted by the narrow lumen. This observation of two sets of capillaries of quite different caliber was likewise supported by his injected material. Thomas also carried out some experimental work by treating animals with thiourea for a period of several months and examining the glands histologically after both routine and perfusion fixation. Under these conditions he noted enlargement of what he called the interfollicular or larger set of capillaries into very large sinusoidal vessels. Enlargement of the smaller and more closely adherent capillaries also occurred. No observations in living animals after treatment with thiourea were reported. Johnson (1953) was unable to find evidence from his injection experiments on human material in favor of the concept of two distinct sets of capillaries of different sizes in relation to the thyroid follicle.

The close relationship between the follicular capillaries and follicular cells proper seems to be very well established, and the indentation of the thyroid follicles by very fine capillaries is an observation which has been made frequently in the course of the routine study of the histology of the thyroid. Thomas calls the more intimately related set of capillaries "intrinsic capillaries," which is a perfectly satisfactory term, but probably their description as intraepithelial is an overstatement. An earlier paper of Williams (1937) described the vascularization of transplants of thyroid tissue into a Clark rabbit ear chamber. He found that the blood vessels were restricted to the interfollicular connective tissue, forming a capillary bed which surrounded the follicles. This description seems to be consistent with the findings of Merwin and Wollman in their grafts of normal thyroid tissue into special chambers in the backs of mice (reported by Algire and Merwin, 1955). Merwin and Wollman observed that one follicle is frequently supplied by capillaries arising from more than one arteriole and that a given capillary may pass by at least two follicles. Their statement that this is in contradiction to earlier work, indicating that each follicle has an individual arteriole and capillary network, does not seem to be warranted by the information published thus far. It would seem entirely possible that the revascularization of a graft in a site foreign to the normal location might proceed in such a manner as to give ultimately a vascularization which is somewhat modified from that which one would find in the gland *in situ*. Merwin and Wollman were unable to find evidence of arteriovenous anastomoses.

The number of studies on the vasculature of the thyroid using microcirculatory techniques has been insufficient. However, those few workers who have investigated this subject have shown how materially such techniques can contribute to our understanding of the anatomy and physiology of the gland; the excellent motion pictures by Wollman and Merwin, shown at the Seventh Microcirculatory Conference in 1959 at Bethesda, Maryland, are a case in point. Further, *in vivo* studies might now be profitably directed toward analysis of those factors which modify or control blood flow through the thyroid capillaries in different phases of activity. These techniques should be coupled with such experiments as the use of stimulating and goitrogenic drugs in different species of animals. The lack of a sufficient number of studies by independent workers, using the more classical techniques and microcirculatory techniques, has left us with a number of unresolved differences in both observation and interpretation which could have long since been clarified. For example, Johnson's failure to find two separate morphologic types of capillaries in man, whereas these were described in some detail by Thomas in the rat, may be due to species differences. Furthermore, differences which have been described between the capillary bed of the follicles and tissues *in situ* and those in thyroid transplants to distant sites might also be easily resolved by workers in independent laboratories doing correlative studies in the same species of animal.

References

Algire, G. H., and Merwin, R. M. (1955). Vascular patterns in tissues and grafts within transparent chambers in mice. *Angiology* **6**, 311.

Bargmann, W., and Knoop, A. (1957). Electron microkopische Beobachtungen an der Neurohypophyze. *Z. Zellforsch. u. mikroskop. Anat.* **46**, 242.

Barrnett, R. J., and Greep, R. O. (1951). The direction of flow in the blood vessels of the infundibular stalk. *Science* **113**, 185.

Bennett, H. S., Luft, J. H., and Hampton, J. C. (1959). Morphological classifications of vertebrate blood capillaries. *Am. J. Physiol.* **196**, 381.

Berlin, D. D., and Lahey, F. H. (1929). Dissections of the recurrent and superior laryngeal nerves. *Surg., Gynecol. Obstet.* **49**, 102.

Blakely, J. L., and Worthington, W. C., Jr. (1960). Vascular changes in the pituitary stalk following superior cervical ganglianectomy. *Anat. Record* **136**, 180.

Cappell, D. F. (1929). Intravitam and supravital staining: II. Blood and organs. *J. Pathol. Bacteriol.* **32**, 629.

Cayotte, J., and Sommelet, J. (1952). A propos de deux observations d'artère thyroidienne moyenne. *Arch. anat. histol. embryol.* **35**, 227.

Curtis, G. M. (1930). The blood supply of the human parathyroids. *Surg., Gynecol. Obstet.* **51**, 805.

Daniel, P. M., and Prichard, M. M. L. (1956). Anterior pituitary necrosis. Infarction of the pars distalis produced experimentally in the rat. *Quart. J. Exptl. Physiol.* **41**, 215.

Daniel, P. M., and Prichard, M. M. L. (1957). The vascular arrangements of the pituitary gland of the sheep. *Quart. J. Exptl. Physiol.* **42**, 237.

Daniel, P. M., and Prichard, M. M. L. (1960). The blood supply of the pituitary gland. *Anat. Record* **136**, 180.

DeGroot, J. (1952). "The Significance of the Hypophyseal Portal Circulation." Van Gorcum, Assen, The Netherlands.

Dempsey, E. W., and Peterson, R. R. (1955). Electron microscopic observations on the thyroid glands of normal, hypophysectomized and cold exposed and thioracil-treated rats. *Endocrinology* **56**, 46.

Dempsey, E. W., and Wislocki, G. B. (1955). The use of silver nitrate as a vital stain and its distribution in several mammalian tissues as studied with the electron microscope. *J. Biophys. Biochem. Cytol.* **1**, 111.

Ekholm, R. (1957). The ultra-structure of the blood capillaries in the mouse thyroid gland. *Z. Zellforsch. u. mikroskop. Anat.* **46**, 139.

Faller, A. (1957). Le comportement des artères thyroidiennes dans la capsule et dans le parenchyme de la glande thyroide. *Helv. Med. Acta* **24**, 220.

Farquhar, M. G. (1961). Fine structure and function in capillaries of the anterior pituitary gland. *Angiology* **12**, 270.

Fortier, C. (1951). Dual control of adrenocorticotrophic release. *Endocrinology* **49**, 782.

Fowler, C. H., and Hanson, W. A. (1929). Surgical anatomy of the thyroid gland with special reference to the relations of the recurrent laryngeal nerve. *Surg., Gynecol. Obstet.* **49**, 59.

Fumagalli, Z. (1942). La vascolarizzazione dell'ipofisi umana. *Z. Anat. Entwicklungsgeschichte* **111**, 266.

Gardner, E., Gray, D. J., and O'Rahilly, R. (1960). "Anatomy: A Regional Study of Human Structure." Saunders, Philadelphia, Pa.

Glydon, R. St. J. (1957). The development of the blood supply of the pituitary in the albino rat with special reference to the portal vessels. *J. Anat.* **91**, 237.

Green, H. T. (1957). The venous drainage of the human hypophysis cerebri. *Am. J. Anat.* **100**, 435.

Green, J. D. (1947). Vessels and nerves of amphibian hypophyses. A study of the living circulation and of the histology of the hypophyseal vessels and nerves. *Anat. Record* **99**, 21.

Green, J. D. (1948). The histology of the hypophyseal stalk and median eminence in man with special reference to blood vessels, nerve fibers and a special neurovascular zone in this region. *Anat. Record* **100**, 273.

Green, J. D. (1951a). The comparative anatomy of the hypophysis, with special reference to its blood supply and innervation. *Am. J. Anat.* **88**, 225.

Green, J. D. (1951b). Innervation of the pars distalis of the adenohypophysis studied by phase microscopy. *Anat. Record* **109**, 99.

Green, J. D. (1952). Comparative aspects of the hypophysis, especially of blood supply and innervation. *In* "Ciba Foundation Colloquia on Endocrinol." Vol. IV, Book I. "Anterior Pituitary Secretion," pp. 72–86. McGraw-Hill (Blakiston), New York.

Green, J. D., and Harris, G. W. (1947). The neurovascular link between the neurohypophysis and adenohypophysis. *J. Endocrinol.* **5,** 136.

Green, J. D., and Harris, G. W. (1949). Observation of the hypophysio-portal vessels of the living rat. *J. Appl. Physiol.* **108,** 359.

Gudernatsch, F. (1953). The endocrine glands. *In* "Morris' Human Anatomy," Section 13, pp. 1579–1580. McGraw-Hill (Blakiston), New York.

Guillemin, R. (1955). A re-evaluation of acetylcholine, adrenaline, noradrenaline, and histamine as possible mediators of the pituitary-adreno-corticotrophic activation by stress. *Endocrinology* **56,** 248.

Harris, G. W. (1947). The blood vessels of the rabbit's pituitary gland, and the significance of the pars and zona tuberalis. *J. Anat.* **81,** 343.

Hartmann, J. F. (1958). Electron microscopy of the neurohypophysis in normal and histamine-treated rats. *Z. Zellforsch. u. mikroskop. Anat.* **48,** 291.

Hinsey, J. C., and Markee, J. E. (1933). Pregnancy following bilateral section of the cervical sympathetic trunks in the rabbit. *Proc. Soc. Exptl. Biol. Med.* **31,** 270.

Houssay, B. A., Biasotti, A., and Sammartino, R. (1935). Modifications fonctionnelles de l'hypophyse après les lesions infundibulo-tubériennes chez le crapaud. *Compt. rend. soc. biol.* **120,** 725.

Hume, D. M. (1949). The role of the hypothalamus in the pituitary adreno-cortical response to stress. *J. Clin. Invest.* **28,** 790.

Johnson, N. (1953). The blood supply of the thyroid gland. I. The normal gland. *Australian New Zealand J. Surg.* **23,** 95.

Johnson, N. (1954). The blood supply of the thyroid gland. II. The nodular gland. *Australian New Zealand J. Surg.* **23,** 241.

Johnson, N. (1955a). The blood supply of the human thyroid gland under normal and abnormal conditions. *Brit. J. Surg.* **42,** 587.

Johnson, N. (1955b). The blood supply of the thyroid gland. III. The histology of the thyroid vessels. *Australian New Zealand J. Surg.* **24,** 303.

Landsmeer, J. M. F. (1951). Vessels of the rat's hypophysis. *Acta Anat.* **12,** 82.

Lever, J. D. (1956). The subendothelial space in certain endocrine tissues. *J. Biophys. Biochem. Cytol.* Suppl. 2, 293.

Liss, L. (1958). Die perivascularen structuren der menschlichen Neurohypophyse. *Z. Zellforsch. u. mikroskop. Anat.* **48,** 288.

Long, C. N. H. (1952). Regulation of ACTH secretion. *Recent Progr. in Hormone Research* **7,** 75.

Luft, J., and Hechter, O. (1957). An electron microscopic correlation of structure with function in the isolated perfused cow adrenal, preliminary observation. *J. Biophys. Biochem. Cytol.* **3,** 615.

McConnell, E. M. (1953). The arterial supply of the human hypophysis cerebri. *Anat. Record* **115,** 175.

Major, R. H. (1909). Studies on the vascular system of the thyroid gland. *Am. J. Anat.* **9,** 475.

Markee, J. E., Sawyer, C. H., and Hollinshead, W. H. (1946). Activation of the anterior hypophysis by electrical stimulation in the rabbit. *Endocrinology* **38,** 345.

Modell, W. (1933). Observations on the structure of the blood vessels within the thyroid gland of the dog. *Anat. Record* **55,** 251.

Monroe, B. G. (1953). Electron microscopy of the thyroid. *Anat. Record* **116,** 345.

Morato, M. J. X. (1939). The blood supply of the hypophysis. *Anat. Record* **74,** 297.

Morin, F., and Bottner, V. (1941). Contributi alla conoscenza della irrorazione sanguina dell 'ipofisi e dell 'ipotalama di alcuni mammiferi. *Morphol. Jahrb.* **85**, 470.

Niemineva, K. (1950). Observations on the development of the hypophyseal-portal system. *Acta Paediat.* **39**, 366.

Nikitovitch-Wiener, M., and Everett, J. W. (1958). Functional restitution of pituitary grafts retransplanted from kidney to median eminence. *Endocrinology* **63**, 916.

Nordland, M. (1930). The larynx as related to surgery of the thyroid based on an anatomical study. *Surg., Gynecol. Obstet.* **51**, 449.

Palay, S. L. (1957). The fine structure of the neurohypophysis. *In* "Ultrastructure and Cellular Chemistry of Neural Tissue" (H. Waelsch, ed.), Chapter II. Hoeber-Harper, New York.

Popa, G. T., and Fielding, U. (1930). A portal circulation from the pituitary to the hypothalamic region. *J. Anat.* **65**, 88.

Porter, R. W. (1953). Hypothalamic involvement in the adreno-cortical pituitary response to stress stimuli. *Am. J. Physiol.* **175**, 515.

Prichard, M. M. L., and Daniel, P. M. (1958). The effects of pituitary stalk section in the goat. *Am. J. Pathol.* **34**, 433.

Reed, A. F. (1943). The relation of the inferior laryngeal nerve to the inferior thyroid artery. *Anat. Record* **85**, 17.

Rienhoff, W. F., Jr. (1929). Gross and microscopic structure of the thyroid gland in man. *A. M. A. Arch. Surg.* **19**, 986.

Rinehart, J. A., and Farquhar, M. G. (1955). The fine vascular organization of the anterior pituitary gland. An electron microscopic study with histochemical correlations. *Anat. Record* **121**, 207.

Saffran, M., Schally, A. V., and Benfey, B. C. (1955). Stimulation of the release of corticotropin from the adenohypophysis by a neurohypophyseal factor. *Endocrinology* **57**, 439.

Sato, T. (1955). On the sphincter apparatus in the artery of the thyroid gland of the hamster. *Okajimas Folia Anat. Japon.* **27**, 3[illegible].

Sheehan, H. L. (1937). Post-partum necrosis of the anterior pituitary. *J. Pathol. Bacteriol.* **45**, 189.

Slusher, M. A., and Roberts, S. (1954). Fractionation of hypothalamic tissue for pituitary stimulating activity. *Endocrinology* **55**, 245.

Söderberg, U. (1958). Short term reactions in the thyroid gland revealed by continuous measurement of blood flow, rate of uptake of radio-active iodine and rate of release of labelled hormones. *Acta Physiol. Scand.* **42**, Suppl. 147.

Stanfield, J. P. (1960). The blood supply of the human pituitary gland. *J. Anat.* **94**, 257.

Thomas, O. L. (1945). The vascular bed in normal and thiourea activated thyroid glands of the rat. *Anat. Record* **93**, 23.

Wagensteen, O. H. (1929). The blood supply of the thyroid gland with special reference to the vascular system of the Cretin. *Surg., Gynecol. Obstet.* **48**, 613.

Williams, R. G. (1937). Microscopic studies of living thyroid follicles implanted in transparent chambers installed in the rabbit's ear. *Am. J. Anat.* **62**, 1.

Williams, R. G. (1944). Some properties of living thyroid cells and follicles. *Am. J. Anat.* **75**, 95.

Wislocki, G. B. (1937). The meningeal relations of the hypophysis cerebri. II. An embryological study of the meninges and blood vessels of the human hypophysis. *Am. J. Anat.* **61**, 95.

Wislocki, G. B., and King, L. S. (1936). The permeability of the hypophysis and hypothalamus to vital dyes with a study of the hypophyseal vascular supply. *Am. J. Anat.* **58**, 421.

Wissig, S. L. (1960). The anatomy of secretion in the follicular cells of the thyroid gland: I. The fine structure of the gland in the normal rat. *J. Biophys. Biochem. Cytol.* **7**, 419.

Worthington, W. C., Jr. (1955). Some observations on the hypophyseal portal system in the living mouse. *Bull. Johns Hopkins Hosp.* **97**, 343.

Worthington, W. C., Jr. (1960). Vascular responses in the pituitary stalk. *Endocrinology* **66**, 19.

Worthington, W. C., Jr., Shealy, J. R., and Blakely, J. L. (1961). Histologic studies on the neurohypophysis of the living frog. *Z. Zellforsch. u. mikroskop. Anat.*, in press.

Xuereb, G. P., Prichard, M. M. L., and Daniel, P. M. (1954a). The arterial supply and venous drainage of the human hypophysis cerebri. *Quart. J. Exptl. Physiol.* **39**, 199.

Xuereb, G. P., Prichard, M. M. L., and Daniel, P. M. (1954b). The hypophyseal portal system of vessels in man. *Quart. J. Exptl. Physiol.* **39**, 219.

Ziegelman, E. F. (1933). Laryngeal nerves. Surgical importance in relation to the thyroid arteries, thyroid gland and larynx. *A. M. A. Arch. Otolaryngol.* **18**, 793.

Chapter XV. PLACENTAL CIRCULATION

By E. M. Ramsey

A. INTRODUCTION

From the circulatory standpoint, the placenta may be defined as the organ in which maternal and fetal blood vessels come into close propinquity and in which physiologic exchange takes place between the two blood streams. It is thus an organ of dual nature. Its maternal and fetal components are intimately intertwined but essentially independent in origin and in behavior. It follows that in considering any attribute of the placenta, it is necessary to describe both the maternal and fetal aspects, as well as their interrelationship, and that in referring to "the placenta," it must be specified whether a single aspect is under consideration or the composite organ as a whole.

An additional factor must be taken into account in the case of the placenta of menstruating primates (humans, Old World monkeys and the higher apes), namely, the manner in which cyclic endometrial changes provide the basis for formation of the maternal placenta. This is particularly true of the endometrial blood vessels in which a smooth, unbroken sequence of changes proceeds from day one of the menstrual cycle through to parturition.

The following account of current information and opinion on uterine and placental vasculature and circulation is necessarily an abbreviated one. The bibliographic references direct the reader to sources of detailed data on specific points.

B. GROSS ANATOMY

1. Maternal

a. Uterine vasculature during the menstrual cycle: Arterial blood is carried to the uterus by the uterine arteries and by branches of the ovarian arteries which traverse the broad ligaments bilaterally. The vessels pass directly to the middle layer of the myometrium and break up to form a circumferential wreath of arcuate arteries. Radial branches of the wreath run at right angles into the endometrium where they promptly divide into basal and spiral endometrial arteries (Fig. 100). The former supply the basal layer of the endometrium and, like it, are relatively insensitive to hormonal stimulation, showing no striking changes during the menstrual cycle or in pregnancy. In contrast, the spiral arteries are extremely sensitive. Figure 101a and b shows how they grow continually closer to the surface of the endometrium as the cycle progresses and become increasingly coiled. In the absence of pregnancy, the withdrawal of hormonal support about day 24 occasions decrease in endometrial thickness and resultant increase in arterial coiling. There is a concomitant but unrelated vasoconstriction. Thereafter, the endometrium becomes ischemic, necrotic and hemorrhagic, and finally the functionalis sloughs away, carrying the superficial two thirds of the spiral arterioles with it. Regeneration proceeds from the

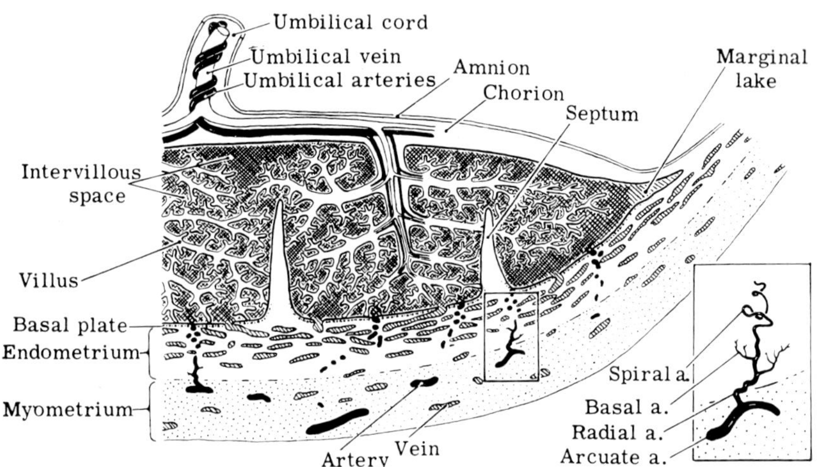

Fig. 100. Diagrammatic representation of a portion of the primate (hemochorial) placenta attached to the uterine wall. The inset shows, at higher magnification, the components of a single arterial stem. (From Ramsey, 1955a; reproduced with the permission of M & R Laboratories.)

uninjured nidus of the basalis (Daron, 1936; Markee, 1940; Phelps, 1958).

The venous pattern of the endometrium is characterized by a deep and a superficial chain of dilated lakes connected with each other and with the venous plexus in the myometrium by numerous slender radial venules. The large collecting veins roughly parallel the arteries with which anastomoses are common. A rich capillary network occupies the layer of endometrium between the tips of the arterioles and the surface epithelium (Daron, 1937; Bartelmez, 1957) (Fig. 101a and b).

Lymphatics are abundant in the myometrium, particularly at the myoendometrial junction. Little is known of the endometrial lymphatics except that Wislocki and Dempsey (1939) observed them only in the basalis and deepest layer of the functionalis of the rhesus monkey. In pregnancy all the uterine lymphatic channels become dilated, though there is no evidence of increase in their number. (For general discussion of lymphatics, see Chapter XXIII.)

b. Vascular adaptations of the uterus to pregnancy: Within a matter of hours after implantation of a fertilized ovum, invading trophoblast opens up the walls of regional capillaries, permitting maternal blood to seep into the lacunar spaces which form within the trophoblastic shell. During the succeeding week, in consequence of progressive growth of spiral arterioles and increasing invasion of the endometrium by the trophoblast, the tips of maternal arterioles are opened. At the same time, the lacunae enlarge and become confluent, and maternal blood, entering from the arterioles under higher pressure, commences to circulate. The trophoblastic columns intervening between the lacunae now differentiate into characteristic chorionic villi with a mesodermal core and a border of cytotrophoblast and syncytium. The latter forms the outer covering of the villi, thus lining the lacunae and coming in contact with the maternal blood. With further elaboration of what may now be designated "the placenta," the confluent lacunae come to form a single, amorphous intervillous space in which the villous branches and twigs hang free (Wislocki and Streeter, 1938).

Meanwhile the changes in endometrial spiral arterioles which were initiated during the early phases of the cycle continue. During the first one third of pregnancy there is progressive growth of the arteries, combined with decreasing width of the endometrium, the latter occasioned by trophoblastic erosion and pressure of the overlying conceptus. As a result, not only increased coiling of the arterioles but also back-and-forth and lateral looping become necessary, so that the vessels can be accommodated within the endometrium (Fig. 102). Then at mid-term, when growth of the component parts of the uterus is superseded by simple stretching as the effective factor in uterine enlargement (the epoch named "conversion" by Reynolds, 1947), the coils of the spiral arteries are paid out and their subsequent course is a gently undulating one (Ramsey, 1949).

The morphology of the connections between the spiral arterioles and the intervillous space is somewhat different in monkey and man. In the former, the terminal segment of the arteriole dilates to form a small sac. Two or more adjacent arterioles occasionally enter the placenta through a common terminal sac (Ramsey, (1949). An opposite configuration prevails in the human where single arterioles are believed to discharge blood into the intervillous space through multiple narrow mouthpieces (Spanner, 1935).

The pregnancy changes in the endometrial veins are essentially

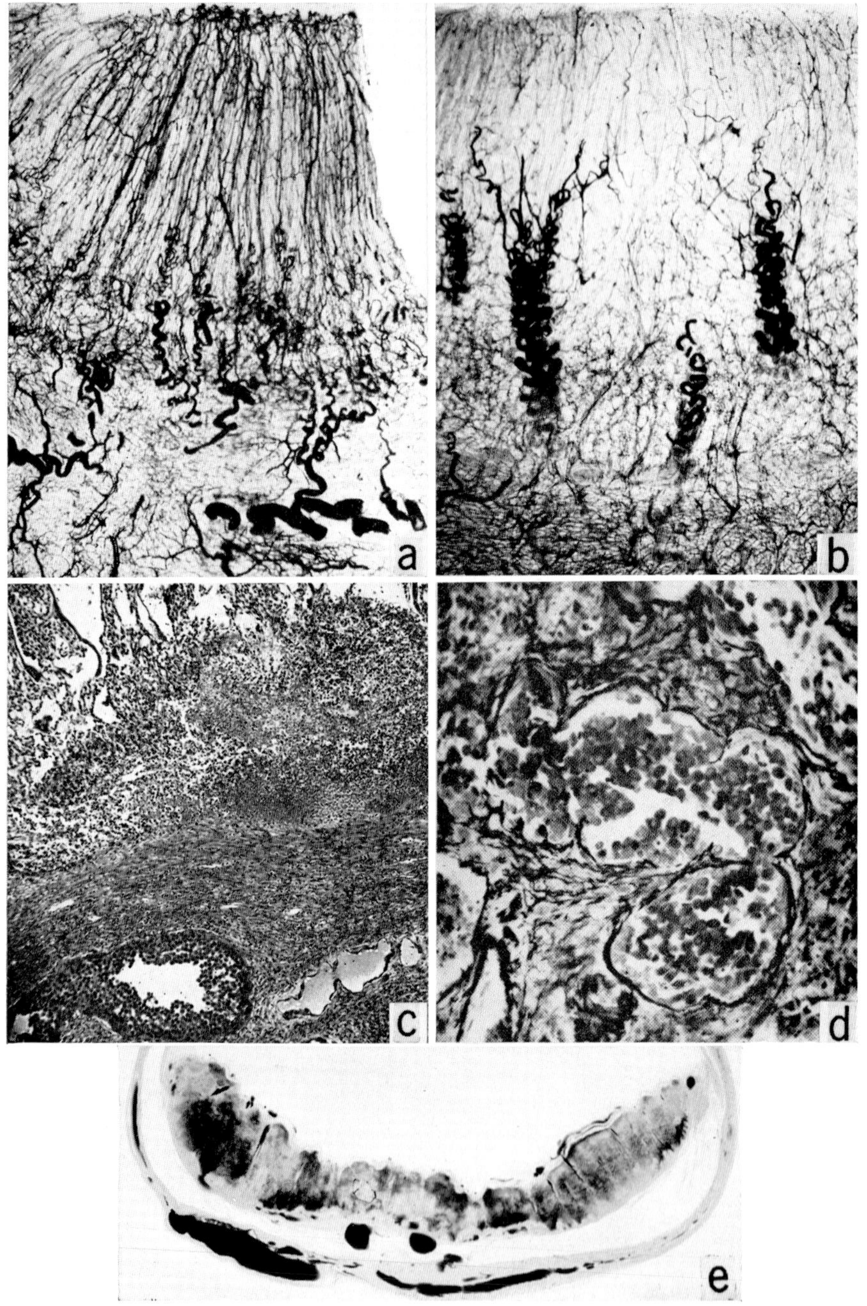
a
b
c
d
e

passive ones, namely, stretching and dilatation (Ramsey, 1954) (Fig. 103). In the myometrium, a shift in dominance between the uterine and the ovarian drainage systems has been reported, depending upon the level of the placental attachment, low implantation favoring the uterine veins and high implantation, the ovarian (Bieniarz, 1958).

Characteristic pregnancy changes occur only in the endometrial vessels connecting directly with the intervillous space. The remaining vessels, which are concerned with mural circulation, remain small and unchanged. True uteroplacental vessels number some twenty arteries and forty veins in the mature rhesus monkey (Ramsey, 1958) and, in the human, arterial entries are estimated to be 100–150 at four months and 200–300 at term (Boyd, 1956). The distribution of arteries and veins is totally haphazard and reflects the distribution in the progravid endometrium of channels which the trophoblast opens indiscriminately. External compression of the thin-walled veins by the endometrial stroma effects reduction in their number as the placenta matures. Patent channels remain, however, in all regions of the placental base. Those draining the central position tend, especially in the human, to be slender and to run an oblique course. More prominent ones frequently drain the periphery of the placenta, the so-called "marginal sinus" (Spanner, 1935). The existence of this structure has been denied in recent years by some investigators (Hamilton and Boyd, 1951) and reaffirmed by others who base their theory of circulation in the maternal placenta upon it

---

FIG. 101. a. Cross section of the uterine wall of a monkey on day 12 of the menstrual cycle. A spiral arteriole displays minimal coiling in the lower third of the endometrium. Its straight distal end reaches half way to the surface epithelium. India ink and starch injection. B 338. ×15. (From Ramsey, 1955b; courtesy of G. W. Bartelmez; reproduced with the permission of *Angiology.*)

b. Cross section of the uterine wall of a monkey on day 17 of the menstrual cycle. The abundant coils of several arterioles are seen in the middle third of the endometrium. The distal ends reach three fourths of the way to the surface epithelium. India ink and starch injection. B 309. ×15. (From Ramsey, 1955b; courtesy of G. W. Bartelmez; reproduced with the permission of *Angiology.*)

c. Implantation site in a monkey on the 29th day of pregnancy. Above are seen the cytotrophoblastic tips of anchoring villi. A cross section of a spiral arteriole is noted in the endometrium below at the left. This vessel enters the intervillous space in an adjoining section. The intense proliferation of its endothelial lining contrasts with the relatively inert venule at the right. ×55.

d. High power view of cells in the lumen of a human spiral artery from an early 4th month pregnancy. Periodic acid-Schiff stain. (From Boyd, 1956; reproduced with the permission of the author and the Josiah Macy, Jr. Foundation.)

e. A section of the entire placenta and the uterine wall in a monkey 123 days pregnant. A "spurt" of arterial blood into the intervillous space is seen at the far left. Secondary placenta. ×2.

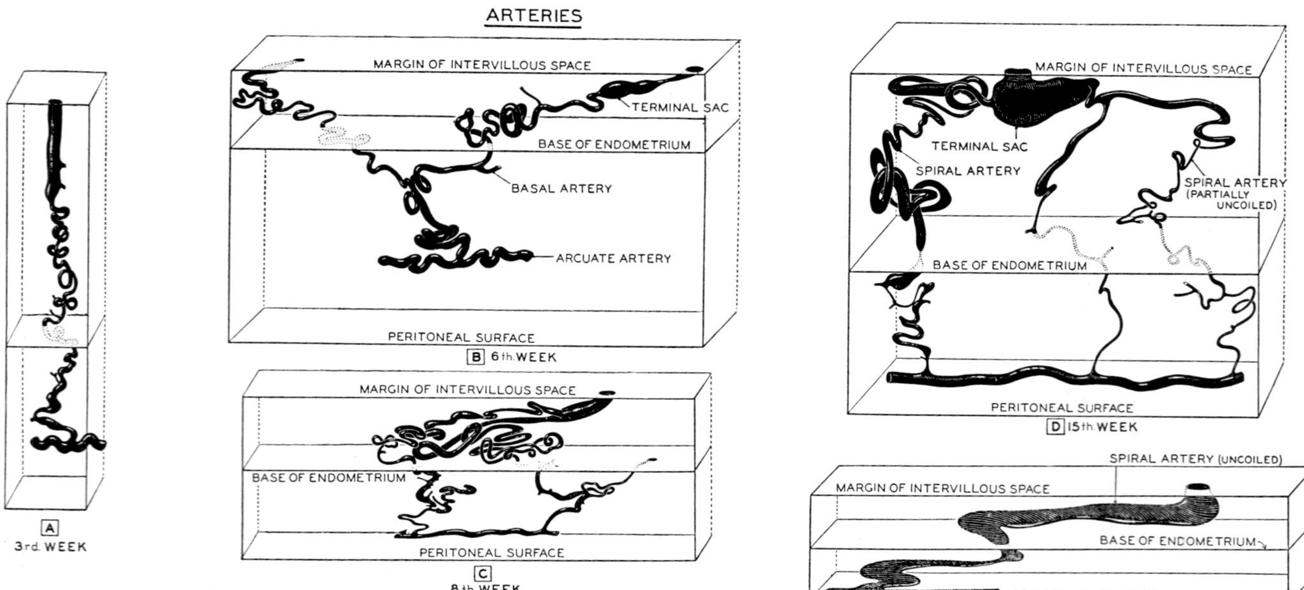

FIG. 102. Diagrammatic representations of the course and configuration of the spiral arteries of the maternal placenta of the monkey at various stages of pregnancy, based upon three-dimensional reconstructions of serial sections. (From Ramsey, 1955b; reproduced with the permission of Carnegie Institution of Washington.)

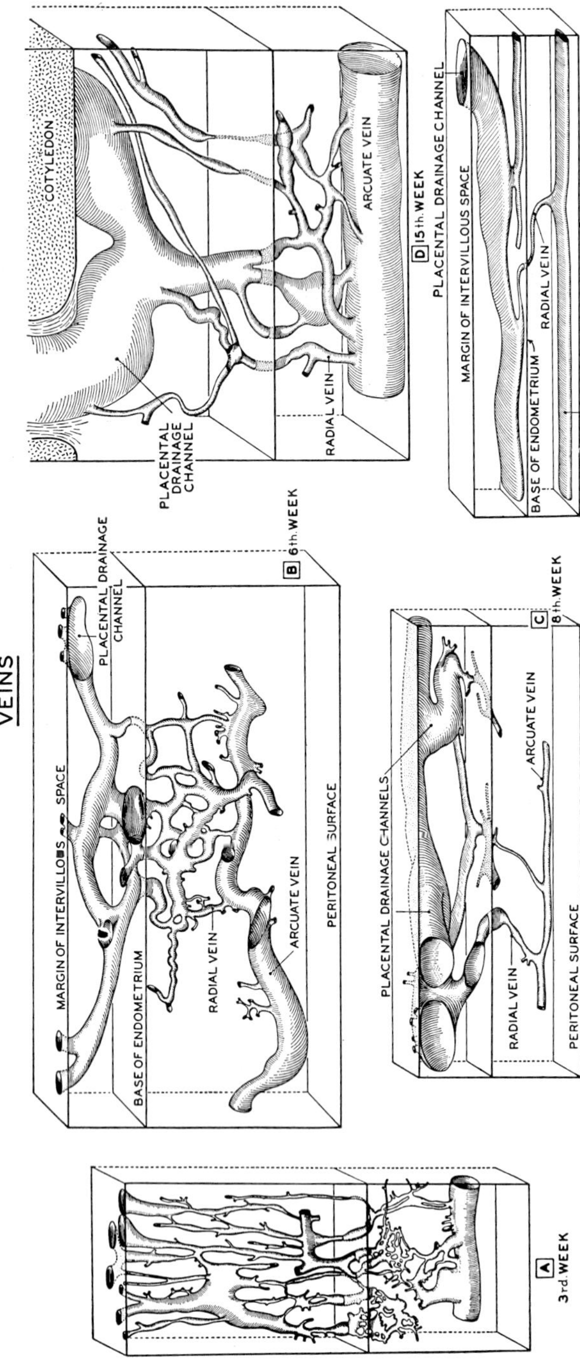

FIG. 103. Diagrammatic representations of the course and configuration of the veins of the maternal placenta of the monkey at various stages of pregnancy, based upon three-dimensional reconstructions of serial sections. (From Ramsey, 1955b; reproduced with the permission of Carnegie Institution of Washington.)

(Earn and Nicholson, 1952; Schneider, 1952; Ferguson, 1955; Fish, 1955). In any event the term "sinus" is a misnomer for, in fact, the structure is merely the peripheral portion of the intervillous space itself. By analogy with the "subchorial lake" (with which it is continuous, Fig. 100), it is more accurate to call the peripheral portion of the intervillous space the "marginal lake" and, since it is composed of an interrupted series of pools, the plural form *marginal lakes* is the designation to be preferred (Ramsey, 1956). It is important that the intraplacental marginal lakes be clearly differentiated from the maternal or venous "sinusoids" which regularly form a circumferential wreath within the endometrium at the periphery of the placenta. These are dilated maternal veins and not necessarily in immediate connection with the placental circulation.

c. The mature maternal placenta: In summary of the developmental sequence described in the foregoing, attention is directed to Fig. 100 which presents in schematic form the components of the mature primate placenta and their relationships. Also to be noted is the manner in which the placenta is irregularly subdivided into a variable number of maternal cotyledons[1] (15–20 in the human) by slender septa of connective tissue in which endometrial and trophoblastic elements mingle (Ramsey, 1960). The septa run from the base of the placenta nearly to the chorionic plate; a few actually connect with the plate. Communication via the subchorial lake and perforations in the septa prevent the cotyledons from being entirely self-contained circulatory units (Boyd, 1956; Ramsey, 1954).

2. Fetal

The differentiation of the primitive trophoblastic columns into true chorionic villi, as described above, is an achievement of the first two weeks after implantation. Subsequent growth and development of the villi result in the attainment of a final form, which may be likened to that of a tree, with roots in the chorionic plate and trunk, limbs, branches, and twigs hanging down into the intervillous space (Fig. 104). Many of the terminal twigs adhere to the endometrium as anchoring villi.

Some two hundred fetal cotyledons[1] comprise the average human

[1] Two entirely separate uses of the term "cotyledon" are current. It is necessary, therefore, to differentiate carefully betwen (1) the *maternal* cotyledon, which is the portion of the placenta lying between two septa, and (2) the *fetal* cotyledon, which is the portion of placenta related to the ramifications of a single main stem villus.

placenta. Their number, once attained, is stationary, but the weight and length of individual cotyledons increase progressively up to term, by virtue of a continuous process of proliferation of syncytial trophoblast. This is the means whereby the placenta keeps pace with the requirements of the growing fetus, rather than by spread of the area of attachment to the endometrium (Crawford, 1959).

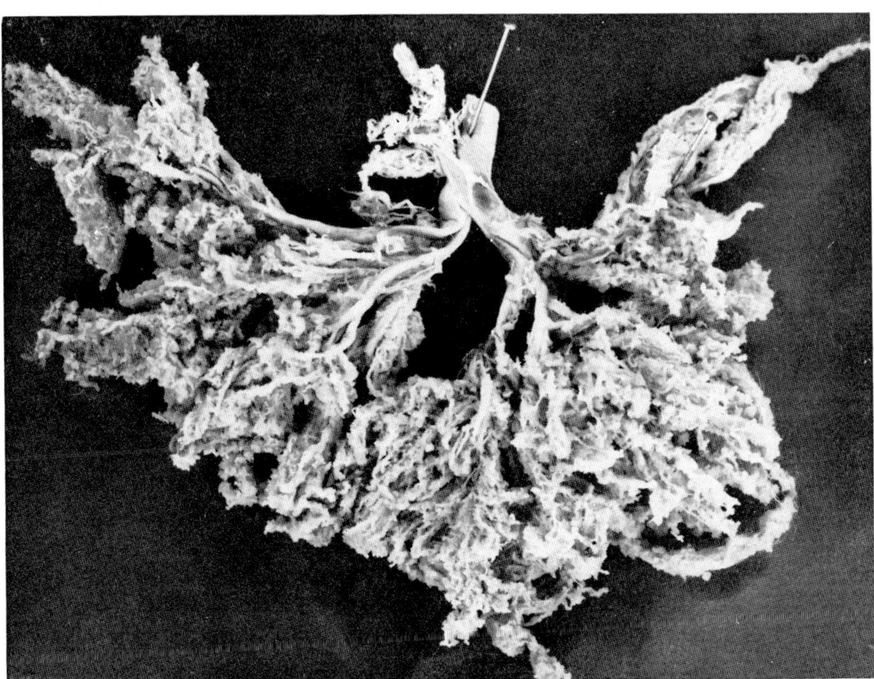

FIG. 104. A cotyledon has been digested and the subcotyledons sufficiently separated to show their origin from the main trunk. (From Crawford, 1956; reproduced with the permission of the Journal of Obstetrics and Gynaecology of the British Empire.)

The treelike configuration of the villus is repeated by the pattern of the fetal vessels which they enclose (Romney and Reid, 1951; Crawford, 1956). The latter are derived from the umbilical vessels which, anatomically, consist of two arteries and one vein. It must be remembered, however, that the arteries carry venous blood from the fetus to the placenta and that the vein returns oxygenated blood to the fetus. The umbilical arteries divide into branches which course laterally in all directions through the chorionic plate and dip sharply to enter the villous trunks. Long branches of relatively wide caliber, often coiled, travel toward the terminal villi.

The absorptive units of the vascular tree lie in the twigs of the villi. The precapillary arteriolar bed is intricately ramified and connects with two types of richly anastomosing capillary complexes: a superficial network, and a paravascular system, providing extravillous shunts which act as safety valves to prevent overloading of the villous circulation (Bøe, 1953) (Fig. 105). Throughout gestation newly formed capillaries project into syncytial buds on the growing ends of the villi. As a result, the vascularity of the fetal placenta becomes increasingly rich and elaborate (Crawford, 1959).

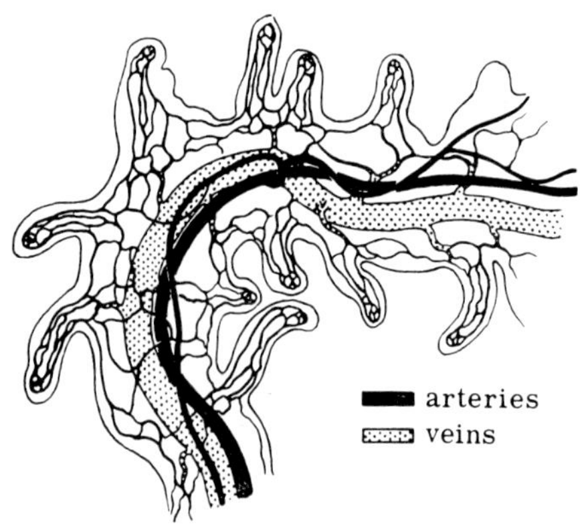

FIG. 105. Semischematic drawings to show the arterio-capillary-venous system in a terminal villus from an 18-week human pregnancy. (From Bøe, 1953; reproduced with the permission of the *Acta Obstet. Gynecol. Scand.*)

Spanner (1935), Wilkin (1954, 1958), and others describe the path of venous return as a "weeping willow" or "chandelier" pattern whereby the villi and their contained vessels recurve from the trophoblastic to the chorionic plate. Others consider that the veins, in returning to the chorionic plate, parallel the afferent arteries and arterioles (Romney and Reid, 1951; Bøe, 1953, 1954). Crawford (1959) finds the recurving only in peripheral cotyledons. Stieve's (1941) "labyrinthine network" is now regarded as illusory.

C. MICROSCOPIC ANATOMY

1. MATERNAL

a. In the nonpregnant uterus: The myometrial blood vessels present the histologic pattern characteristic for arteries and veins supplying smooth muscle elsewhere. In functional activity the local contractility of the radial arteries is of significance (Bartelmez, 1957). This activity appears to be hormonally controlled and is elicited by diminution in blood concentration of either estrogen or progesterone (Daron, 1936; Phelps, 1958).

The basal and the spiral branches of the radials differ markedly. The former are characterized, throughout the whole reproductive cycle, by paucity of elastic fibrils and predominance of a myoepithelioid type of muscle cell, displaying abundant cytoplasm and large nuclei (Okkels and Engle, 1938). They grow but little and remain intact at the time of menstruation. The spiral arterioles, on the contrary, have a well-developed muscle coat, containing a rich network of elastic fibrils which extends into the precapillary area of the subepithelial zone. As the cycle advances and the spiral arterioles lengthen, the elastic and muscle constituents progress distally. In the composition of their walls the spiral arterioles closely resemble the radial arteries, a similarity heightened by the intermittent contractility which they display. It is this phenomenon of vasoconstriction, rather than the increased coiling of the arterioles during the progestational phase of the cycle, which creates stromal ischemia, leading to necrosis, sloughing, and menstrual hemorrhage (Bartelmez, 1957).

In contractile capacity the individual spiral arterioles appear to be independent units, and constriction is not necessarily synchronous. Furthermore, the arterioles supplying the paramedian region of the uterine corpus and fundus, the potential implantation sites, exhibit greater sensitivity to hormonal withdrawal and earlier vasoconstriction than do those to the rest of the endometrium. This is of importance from the standpoint of pregnancy, since it allows for formation of a pablum suitable for nourishment of the implanting ovum (Markee, 1940). There is evidence that persistence of this vasoconstrictive activity during pregnancy may cause the flow of blood from individual arterioles into the intervillous space to be pulsatile (Ramsey, 1959).

Although hormonal control of uterine vessels is a distinctive and highly significant feature, the distribution of nonmyelinated nerves indicates that the nervous mechanism of control is also operative. Extensive myometrial plexuses have been demonstrated, roughly paralleling the course of the myometrial arteries and following both the basal (Hirsch and Martin, 1943) and the spiral (Okkels, 1950) arterioles into the endometrium. Vater-Pacinian sensory corpuscles are present in the adventitia of the myometrial branches of the uterine arteries. One study of the human uterus at term has shown abundant nonmyelinated fibers following the spiral arteries to within a few cells' breadth of the surface epithelium (Pribor, 1951).

b. Changes during implantation and the formation of the placenta: Studies in rhesus monkeys have demonstrated that when the trophoblast taps the spiral arterioles during implantation, a wave of proliferative activity is initiated in the cells lining the vessels. Seen first at the point of entry of the arteriole into the intervillous space, the process gradually extends proximally almost to the muscularis. It is most marked in the immediate vicinity of the placental base, where the proliferated cells are heaped up into several layers and replace the tissues of the media and adventitia (Fig. 101c). Special stains confirm this appearance, demonstrating in particular the disappearance of elastic fibrils. The resultant dilatation is the origin of the terminal sac, previously noted as characteristic of the monkey.

Similar proliferation of lining cells of uteroplacental arteries has been reported in the human (Fig. 101d), but no final decision has been reached regarding their origin: endothelial or trophoblastic (Ramsey, 1956; Boyd, 1956), though some preference is shown for the endothelial.

By midpregnancy the lining cell proliferation abates and in both humans and monkeys, epithelioid transformation of the media accompanies intimal fibrosis. Actual occlusion resembling arteriosclerosis may involve some endometrial stems, and in arcuate arteries partial occlusion is frequently seen. Reduplication of elastic lamellae appears in the vessels so involved.

Uteroplacental veins partake to a very minor degree in the early proliferative phenomena. They show none of the intimal and medial changes characteristic of the arteries later on.

c. Postpartum changes: The changes in the myometrial arteries tend to be permanent, so that it is usually possible to state whether a given uterus has been pregnant, especially if the organ is examined before the onset of vascular changes related to advancing years (Boyd, 1956). The number of antecedant pregnancies cannot, however, be deduced, as the degree of permanent change is highly variable.

2. Fetal

The blood vessels of the chorionic villi originate, in the second week after implantation, from angioblasts in the mesodermal core of the primitive villi. By progressive coalescence the independent units form a continuous system. This connects with the vessels of the body stalk, the latter eventually forming the umbilical cord with its vessels.

The early angioblasts mature into typical vascular endothelium, and condensations of mesoderm around the endothelial tubes form media and adventitia.

In the umbilical vessels characteristic musculoelastic walls appear.

The fetal capillaries are composed of endothelial tubes embedded in the loose connective tissue core of the villi. Early in pregnancy their location is central, but as the placenta ages they gradually come to lie just beneath the surface. At the same time the syncytial covering of the villus thins markedly (Paine, 1957). Further aging phenomena in all the branches of the fetal vascular tree appear in the form of elastoid degeneration and thickening of the basement membrane beneath the endothelium.

The evidence for the presence of lymphatic channels and vasomotor nerves in the umbilical cord and the chorionic villi (reviewed by Krantz, 1958) is equivocal.

D. PHYSIOLOGY

1. Maternal

a. Theories of mechanism: Several successive theories of placental circulation have been formulated (see Ramsey, 1960). Bumm (1893), believing that endometrial arteries coil upward through the septa to orifices at the septal summits, stated that arterial blood bathes the villi as it flows back to venous drainage openings in the trophoblastic plate. Spanner (1935), with refined histologic techniques, showed that arterial orifices are located in the trophoblastic plate itself. His injection techniques led him to believe that the venous exits occur exclusively at the periphery of the placenta, so that arterial blood rises to the subchorial lake between fluid-tight septal dividers and from the subchorial lake passes laterally to the "marginal sinus," whence alone drainage takes place (Spanner, 1935).

Studies in the past 25 years, combining histologic techniques and the injection of dyes which pass through the drainage channels rather than hardening within the intervillous space (Daron, 1936; Bartelmez, 1957), have shown arterial and venous orifices to be indiscriminately distributed along the trophoblastic plate, basally as well as at the margin (Ramsey, 1956). In anatomic preparations (Fig. 101e), as well as *in vivo* radio-angiographic studies (Fernström, 1955; Borell, 1958; Ramsey *et al.*, 1960), arterial blood has been seen to enter the intervillous space in characteristic "spurts" or fountain-like "jets" (Ramsey, 1959). This observation has led to formulation of a "physiologic concept" of placental circulation, which envisages the *vis a tergo* of maternal blood pressure as the controlling factor rather than morphologic arrangements. This pressure is sufficiently higher than that in the intervillous space, so that arterial blood is driven well up toward the chorion before lateral diffusion occurs. Thereafter the blood falls back upon the orifices in the basal plate which connect with maternal veins, and, since there is an additional fall in pressure between the intervillous space and the endometrial veins, drainage is accomplished. The pressure differential is enhanced by the intermittent myometrial contractions throughout pregnancy (Braxton Hicks).

Studies of the intra-amnionic and intervillous space pressures in humans (Alvarez and Caldeyro-Barcia, 1954; Caldeyro-Barcia, 1957; Hendricks, 1958) and macaques (Ramsey *et al.*, 1959) have confirmed the pressure gradients assumed in this working hypothesis and have established the character and effect of myometrial contractions.

Supporters of the "physiologic" concept do not deny the existence of marginal lakes and marginal drainage but regard them as only part of the total drainage system (Ramsey, 1956, 1959).

b. Blood volume and rate of flow: Determination of the volume of blood in the maternal placenta and its rate of flow through the intervillous space is made difficult by the necessity of separating the placental and uterine components in values obtained for the pregnant uterus as a whole. Both the Kety (NO_2) modification of the Fick principle and miniature electromagnetic flowmeters have been employed by Assali and his associates (1960), who have shown total uterine flow values rising from 51.7 ml/min at 10 weeks to 185 ml/min at 28 weeks. At term it was found to be 500–700 ml/min (Assali *et al.*, 1953; Metcalfe *et al.*, 1959). The disappearance time of intraplacentally injected radioactive sodium has been monitored, and the blood flow in the placenta alone has been established as 600 ml/min (Browne and Veall, 1953; Browne, 1954, 1957). No actual measurement of the volume of the placental pool has been made though Browne estimates it to be about 250 ml.

2. Fetal

The role of the *vis a tergo* of maternal pressure in control of the circulation on the maternal side of the placenta has been noted. Similarly, the head of fetal blood pressure dominates the circulation on the fetal side. Data on fetal circulation emanate for the most part from studies in animals, particularly in ungulates (sheep and goats) (Barcroft and Barron, 1945; Barcroft, 1946; Reynolds, 1956; Dawes, 1957). Normal pressures of 20–35 mm Hg have been recorded in the umbilical vein and 50–70 mm, in the umbilical arteries (Reynolds and Paul, 1958). The high venous pressure is maintained by action of a special sphincter between the umbilical vein and the ductus venosus, and in the umbilical arteries pressure remains high because the unusual width of the vessels prevents loss of pressure by friction. The resultant capillary pressure is considerably above that in other capillary beds and appreciably exceeds the resting pressure in the maternal intervillous space.

Assali and his associates (1953, 1960) have been able to determine fetal circulation in the same human cases utilized for study of uterine blood flow (see above) and found that blood flow increased from 8.5 ml/min at 12 weeks to 80.0 ml/min at 28 weeks.

3. Exchange

Grosser (1927) formulated a morphologic classification of the placentas of various animals, based on the number of layers of tissue separating the fetal and maternal blood streams. Thus the placenta of the horse, in which maternal and fetal blood travels in closed blood vessels embedded, respectively, in epithelium-covered endometrial folds and trophoblast-covered chorionic villi, is described as epitheliochorial. In it substances must cross six layers of tissue to pass from one blood stream to the other. The primates have a hemochorial type of placenta, with maternal blood flowing freely around the chorionic villi, separated from the fetal blood by only three tissue layers: the trophoblast covering the villi, the mesodermal core of the villi, and the endothelium of the fetal capillaries.

The Grosser classification, around the finer details of which much controversy has raged (Amoroso, 1955), has demonstrated physiologic as well as anatomic usefulness. Flexner and Gellhorn (1942), in particular, have interpreted variations in rate of placental transfer in different species upon the basis of the grouping. Furthermore, variations in these values in a single animal at different stages of pregnancy or under pathologic conditions may be referred to changes in the thickness of any of the tissue layers intervening between the two circulations.

In lower animals the blood in maternal and fetal placental vessels generally flows in opposite directions (Mossman, 1926), reduced fetal blood making its first contact with maternal blood in the venous state and gradually coming into contact with increasingly pure arterial blood. It has been shown that maximum crossing of a permeable membrane takes place under such circumstances (Noer, 1946), such an arrangement being of wide occurrence in biologic systems (e.g. the gills of fish). Whether this principle of "counterflow" is honored by the hemochorial placenta remains open to question (Ramsey, 1956).

E. PHARMACOLOGY

It is assumed that uterine blood vessels react, as do vessels elsewhere, to vasopressor and vasodilator agents, but no drugs are known to have a selective action upon them. On the other hand, as is described in some detail above, the uterine arteries have a special sensitivity to the intrinsic ovarian hormones (Bartelmez, 1957; Markee, 1940). Incomplete as is present understanding of these relationships, therapeutic administration of natural and synthetic endocrine preparations has long been practiced on an empirical basis for the control of pathologic bleeding from the nonpregnant endometrium and from the placental site. Variable success attends these forms of therapy, failures stemming in large part from the wide gaps in present knowledge of the metabolism of the hormones and of their sites and modes of action, coupled with the difficulty of achieving early and accurate diagnosis.

It is to be noted that both ovarian and anterior pituitary hormones act upon other uterine tissues besides blood vessels. The resulting responses, in turn, have a secondary effect upon the vascular channels. Thus, progesterone, by relaxing the myometrium, enhances flow, and the oxytocics, such as pituitrin, cause the contracting muscle fibers to occlude at least the thin-walled veins.

Despite the fundamental work of such investigators as Csapo (1959) in the field of progesterone metabolism and the studies of a host of biochemists and clinicians, the pharmacology of the uterus and placenta constitutes an "area of ignorance" which urgently requires exploration and promises profoundly significant and clinically useful rewards to the explorer.

F. PATHOLOGY

1. Premature Separation of the Placenta

The high incidence of hemorrhage as a major clinical sign of placental pathology is an index of the importance of the vascular system in this regard. It is probable that an appreciable percentage of extremely early abortions is associated with anomalies of implantation in which abnormally early opening of arteriolar stems, before the lacunar spaces can withstand the force of maternal blood pressure, leads to hemorrhage into the decidua and necrosis of the area surrounding the implantation site. Occasionally, in specimens of later abortion, healed clots and scars in this region suggest that the early ovum may survive such episodes with or without harmful effects upon development. Information and statistics on this very early stage are obviously difficult to obtain. Hertig's study of young ova (Hertig *et al.*, 1956) remains the classic in the field.

Pathologic separation of the normally implanted placenta (*placenta praevia* is excluded from this category) is important as a clinical problem and as a cause of abortion. It may arise from (1) rupture of uteroplacental arteries; (2) rupture of uteroplacental veins, especially the "venous sinusoids" around the periphery of the placenta (see above); and (3) rupture of a marginal lake. Relatively little is yet known about such *abruptio placentae,* but it is clear that each of these types is a special entity with separate etiology and consequences.

Hertig (1953) has shown that the walls of spiral arterioles leading to the hemorrhagic areas at the base of the placenta in central separation frequently exhibit degenerative changes, which may well account for their disintegration or rupture under arterial pressure. Although this explains the hemorrhage which, under the same pressure, dissects through the decidua and separates the placental attachment, the etiology of the vascular lesions, in their turn, must be sought. The results of this type of abruptio are particularly grave, since the separation tends to extend and the flow of arterial blood to the intervillous space is interrupted, with serious consequences for the fetus.

That rupture of maternal venous sinuses or of a marginal lake may be precipitated by acute elevations of venous pressure is suggested by experimental observations of Mengert *et al.* (1953). Clots in these venous channels may also play a part. Hemorrhages dissect less widely and have less marked effect upon fetal well-being, because the blood extravasated has already completed the placental circuit. Clinical statistics confirm such reasoning. Fish (1955), for example, has noted 32.5 per cent fetal mortality in central abruptio against 4.1 per cent in rupture of a marginal lake (see also Hellman, 1953; Hertig, 1945).

A serious condition which Schneider (1959) regards as a sequel of massive abruptio is afibrinogenemia (fibrinopenia; hypofibrinogenemia). In this condition the circulating blood is defibrinated, as the result, in Schneider's opinion, of "autoextraction" of tissue-thromboplastin, i.e., the expanding retroplacental hematoma formed in abruptio ruptures into the intervillous space, carrying with it an extract of necrotic endometrial tissue which eventually reaches the general circulation. Other theories of etiology include: amniotic fluid emboli; obstetric shock; retention of a dead fetus; fibrinolysis. Reid (1959) wisely states: "How fibrinogenemia develops remains in the realm of theory." Associated with the fibrinopenia of the circulating blood may be disseminated fibrin clots which produce characteristic pathologic and clinical symptoms.

Placental separation and the poorly defined condition of placental infarction have both been invoked in connection with toxemia of pregnancy, sometimes as cause and sometimes as effect. Fibrinoid degeneration of arteriolar walls (Hertig, 1945) appears in both pathologic states, associated with degeneration of the trophoblastic covering of the villi, deposition of fibrin in the intervillous space and thrombosis. The villous changes may involve large segments of the placenta, causing adhesions between the villi and rendering the villous core hemorrhagic and necrotic (red infarcts), with eventual fibrosis (white infarcts). The part played in these changes, and especially in the degeneration of arteriolar walls, by hypertension, abnormal blood concentrations of adrenal or other hormones, specific toxins or exogenous factors has been widely investigated without attainment of any concensus (Hertig, 1945). Bieniarz's (1958) conclusion that site of placental location is decisive should be noted: a high location leading to preponderant draining through the ovarian veins, initiating toxemia; a low site, favoring drainage through the uterine veins and provoking hemorrhage.

References[2]

Alvarez, H., and Caldeyro-Barcia, R. (1954). Fisiopatologia de la contraccion uterina y sus aplicaciones en la clinica obstetrica. Presented at *2nd Congr. Latino Amer. Obstet. Ginecol.* and at *4th Congr. Brasil. Obstet. Ginecol., San Pablo, Brasil.*

*Amoroso, E. C. (1955). The comparative anatomy and histology of the placental barrier. *In* "Gestation," Trans. 1st Conf. (L. B. Flexner, ed.), Josiah Macy, Jr., Foundation, New York.

Assali, N. S., Douglass, R. A., Jr., Baird, W. W., Nicholson, D. B., and Suyemoto, R. (1953). Measurement of uterine blood flow and uterine metabolism. IV. Results in normal pregnancy. *Am. J. Obstet. Gynecol.* **66**, 248.

Assali, N. S., Rauramo, L., and Peltonen, T. (1960). Measurement of uterine blood flow and uterine metabolism. VIII. Uterine and fetal blood flow and oxygen consumption in early human pregnancy. *Am. J. Obstet. Gynecol.* **79**, 86.

Barcroft, J. (1946). "Researches on Pre-Natal Life." Blackwell, Oxford.

Barcroft, J., and Barron, D. H. (1945). Blood pressure and pulse rate in the fetal sheep. *J. Exptl. Biol.* **22**, 63.

*Bartelmez, G. W. (1957). The form and the functions of the uterine blood vessels in the rhesus monkey. *Carnegie Inst. Wash. Publ. No.* 611 (*Contribs. Embryol.* **36**, 153).

Bieniarz, J. (1958). The patho-mechanism of late pregnancy toxemia and obstetrical hemorrhages. I. Contradiction in the clinical pictures of eclampsia and placenta previa depending upon the placental site. *Am. J. Obstet. Gynecol.* **75**, 444.

Bøe, F. (1953). Studies on the vascularization of the human placenta. *Acta Obstet. Gynecol. Scand.* **32**, Suppl. 5, 1.

Bøe, F. (1954). The mammalian fetus: physiological aspects of development. Vascular morphology of the human placenta. *Cold Spring Harbor Symposia Quant. Biol.* **19**, 29.

Borell, U. (1958). Eine arteriographische Studie des Plazentarkreislaufs. *Geburtshilfe Frauenheilk.* **18**, 1.

[2] Asterisks before certain references indicate particularly useful and comprehensive reviews of general aspects of the subject treated.

*Boyd, J. D. (1956). Morphology and physiology of the uteroplacental circulation. *In* "Gestation," Trans. 2nd Conf. (C. A. Villee, ed.), p. 132. Josiah Macy, Jr., Foundation, New York.

Browne, J. C. M. (1954). Utero-placental circulation. *Cold Spring Harbor Symposia Quant. Biol.* **19**, 60.

Browne, J. C. M. (1957). Measurement and significance of utero-placental and uterine muscle flow. *In* "Oxygen Supply to the Human Fetus," Macy Foundation C.I.O.M.S. Conf. (J. Walker and A. C. Turnbull, ed.), p. 113. Charles C Thomas, Springfield, Ill.

Browne, J. C. M., and Veall, N. (1953). The maternal placental blood flow in normotensive and hypertensive women. *J. Obstet. Gynaecol. Brit. Empire* [N.S.] **60**, 141.

Bumm, E. (1893). Über die Entwicklung des mütterlichen Blutkreislaufes in der menschlichen Placenta. *Arch. Gynäkol.* **43**, 181.

*Caldeyro-Barcia, R. (1957). *In* "Physiology of Prematurity," Trans. 1st Conf. (J. T. Lanman, ed.), p. 128. Josiah Macy, Jr., Foundation, New York.

Crawford, J. M. (1956). The foetal placental circulation. Part IV. The anatomy of the villus and its capillary structure. *J. Obstet. Gynaecol. Brit. Empire* **63**, 548.

Crawford, J. M. (1959). A study of human placental growth with observations on the placenta in erythroblastosis foetalis. *J. Obstet. Gynaecol. Brit. Empire* **66**, 885.

Csapo, A. (1959). Function and regulation of the myometrium. *Ann. N. Y. Acad. Sci.* **75**, 790.

Daron, G. H. (1936). The arterial pattern of the tunica mucosa of the uterus in *Macacus rhesus*. *Am. J. Anat.* **58**, 349.

Daron, G. H. (1937). The veins of the endometrium (*Macacus rhesus*) as a source of the menstrual blood. *Anat. Record* **67**, Suppl., 13.

Dawes, G. S. (1957). The fetal and placental circulation in late pregnancy. *In* "Physiology of Prematurity," Trans. 1st Conf. (J. T. Lanman, ed.), pp. 81–139. Josiah Macy, Jr., Foundation, New York.

Earn, A. A., and Nicholson, D. (1952). The placental circulation, maternal and fetal. *Am. J. Obstet. Gynecol.* **63**, 1.

Ferguson, J. H. (1955). Rupture of the marginal sinus of the placenta. *Am. J. Obstet. Gynecol.* **69**, 995.

Fernström, I. (1955). Arteriography of the uterine artery; its value in the diagnosis of uterine fibromyoma, tubal pregnancy, adnexal tumour, and placental site localization in cases of intra-uterine pregnancy. *Acta Radiol.* Suppl. 122, 1.

Fish, J. S. (1955). "Hemorrhage of Late Pregnancy," Am. Lect. Gynecol. Obstet. No. 225, pp. 819 and 832. Charles C Thomas, Springfield, Ill.

Flexner, L. B., and Gellhorn, A. (1942). The comparative physiology of placental transfer. *Am. J. Obstet. Gynecol.* **43**, 985.

Grosser, O. (1927). "Frühentwicklung, Eihautbildung und Placentation des Menschen und der Säugetiere." Bergmann, Munich.

*Hamilton, W. J., and Boyd, J. D. (1951). Observations on the human placenta. *Proc. Roy. Soc. Med.* **44**, 489–496.

Hellman, L. M. (1953). Clinical impact of late pregnancy hemorrhage. *In* "Prematurity, Congenital Malformation and Birth Injury," Conf. Assoc. Aid Crippled Children, 1952, pp. 198–202. N. Y. Acad. Med., New York.

Hendricks, C. H. (1958). The hemodynamics of a uterine contraction. *Am. J. Obstet. Gynecol.* **76**, 969–982.

Hertig, A. T. (1945). Vascular lesions in the hypertensive albuminuric toxemias of pregnancy. *Clinics* **4,** 602.

Hertig, A. T. (1953). The pathology of late pregnancy hemorrhage. *In* "Prematurity, Congenital Malformation and Birth Injury," Conf. Assoc. Aid Crippled Children, 1952, pp. 202–215. N. Y. Acad. Med., New York.

*Hertig, A. T. (1960). Pathological aspects. *In* "The Placenta and Fetal Membranes" (C. A. Villee, ed.), p. 109. Williams & Wilkins, Baltimore, Md.

Hertig, A. T., Rock, J., and Adams, E. C. (1956). A description of 34 human ova within the first 17 days of development. *Am. J. Anat.* **98,** 435.

Hirsch, E. F., and Martin, M. (1943). The distribution of the nerves in the adult myometrium. *Surg., Gynecol. Obstet.* **76,** 697.

Krantz, K. E. (1958). The anatomy and endocrinology of the human placenta. *In* "Essentials of Human Reproduction" (J. T. Velardo, ed.), pp. 101–130. Oxford Univ. Press, London and New York.

*Markee, J. E. (1940). Menstruation in intraocular endometrial transplants in the rhesus monkey. *Carnegie Inst. Wash. Publ. No.* 518 (*Contribs. Embryol.* **28,** 219).

Mengert, W. F., Goodson, J. H., Campbell, R. G., and Haynes, D. H. (1953). Observations on the pathogenesis of premature separation of the normally implanted placenta. *Am. J. Obstet. Gynecol.* **66,** 1104.

Metcalfe, J., Romney, S. L., Swartwout, J. R., Pitcairn, D. M., Lethin, Anton N., Jr., and Barron, D. H. (1959). Uterine blood flow and oxygen consumption in pregnant sheep and goats. *Am. J. Physiol.* **197,** 929.

Mossman, H. W. (1926). The rabbit placenta and the problem of placental transmission. *Am. J. Anat.* **37,** 433.

Noer, R. (1946). A study of the effect of flow direction on the placental transmission, using artificial placentas. *Anat. Record* **96,** 383.

Okkels, H. (1950). Histophysiology of the human endometrium. *In* "Menstruation and Its Disorders" (E. T. Engle, ed.), pp. 139–163. Charles C Thomas, Springfield, Ill.

Okkels, H., and Engle, E. T. (1938). Studies on the finer structure of the uterine blood vessels of the macacus monkey. *Acta Pathol. Microbiol. Scand.* **15,** 150.

Paine, C. G. (1957). Observations on placental histology in normal and abnormal pregnancy. *J. Obstet. Gynaecol. Brit. Empire* **64,** 668.

*Phelps, D. (1958). Menstruation. *In* "Essentials of Human Reproduction" (J. T. Velardo, ed.), pp. 55–87. Oxford Univ. Press, London and New York.

Pribor, H. C. (1951). Innervation of the Uterus. *Anat. Record* **109,** 339.

Ramsey, E. M. (1949). The vascular pattern of the endometrium of the pregnant rhesus monkey (*Macaca mulatta*). *Carnegie Inst. Wash. Publ. No.* 583 (*Contribs. Embryol.* **33,** 113).

Ramsey, E. M. (1954). Venous drainage of the placenta of the rhesus monkey (*Macaca mulatta*). *Carnegie Inst. Wash. Publ. No.* 603 (*Contribs. Embryol.* **35,** 151).

Ramsey, E. M. (1955a). Uterine and placental circulation. *In* "Respiratory Problems in the Premature Infant," Report 15th M & R Pediat. Research Conf., pp. 23–34.

Ramsey, E. M. (1955b). Vascular patterns in the endometrium and the placenta. *Angiology* **6,** 321.

Ramsey, E. M. (1956). Distribution of arteries and veins in the mammalian placenta. *In* "Gestation," Trans. 2nd Conf. (C. A. Villee, ed.), pp. 229–251. Josiah Macy, Jr., Foundation, New York.

Ramsey, E. M. (1956). Circulation in the maternal placenta of the rhesus monkey and man, with observations on the marginal lakes. *Am. J. Anat.* **98,** 159.

Ramsey, E. M. (1958). Vascular anatomy of the utero-placental and fetal circulation. *In* "Oxygen Supply to the Human Fetus," Macy Foundation C.I.O.M.S. Conf. p. 67. Charles C Thomas, Springfield, Ill.

*Ramsey, E. M. (1959). Vascular adaptations of the uterus to pregnancy. *Ann. N. Y. Acad. Sci.* **75,** 726.

*Ramsey, E. M. (1960). The placental circulation. *In* "The Placenta and Fetal Membranes" (C. A. Villee, ed.), p. 36. Williams & Wilkins, Baltimore, Md.

Ramsey, E. M., Corner, G. W., Jr., Long, W. N., and Stran, H. M. (1959). Studies of amniotic fluid and intervillous space pressures in the rhesus monkey. *Am. J. Obstet. Gynecol.* **77,** 1016.

Ramsey, E. M., Corner, G. W., Jr., Donner, M. W., and Stran, H. M. (1960). Radioangiographic studies of circulation in the maternal placenta of the rhesus monkey. *Proc. Natl. Acad. Sci. U. S.* **46,** 1003.

Reid, D. E. (1959). Clinical considerations in hypofibrinogenemia. *Ann. N. Y. Acad. Sci.* **75,** 685.

Reynolds, S. R. M. (1947). Relation of maternal blood-flow within the uterus to change in shape and size of the conceptus during pregnancy; physiological basis of uterine accommodation. *Am. J. Physiol.* **148,** 77.

Reynolds, S. R. M. (1956). Pressures in the fetal circulatory system of the sheep. *In* "Gestation," Trans. 2nd Conf. (C. A. Villee, ed.), pp. 219–228. Josiah Macy, Jr., Foundation, New York.

Reynolds, S. R. M., and Paul, W. M. (1958). Pressures in umbilical arteries and veins of the fetal lamb *in utero*. *Am. J. Physiol.* **193,** 257.

Romney, S. L., and Reid, D. E. (1951). Observations on the fetal aspects of placental circulation. *Am. J. Obstet. Gynecol.* **61,** 83.

Schneider, C. L. (1952). Rupture of the basal (decidual) plate in abruptio placentae: A pathway of autoextraction from the decidua into the maternal circulation. *Am. J. Obstet. Gynecol.* **63,** 1078.

Schneider, C. L. (1959). Etiology of fibrinopenia: fibrination defibrination. *Ann. N. Y. Acad. Sci.* **75,** 634–675.

Spanner, R. (1935). Mütterlicher und kindlicher Kreislauf der menschlichen Placenta und seine Strombahnen. *Z. Anat. Entwicklungsgeschichte* **105,** 163.

Stieve, H. (1941). Die Entwicklung und der Bau der menschlichen Placenta. 2. Zotten, Zottenraumgitter und Gefässe in der zweiten Hälfte der Schwangerschaft. *Z. mikroskop.-anat. Forsch.* **50,** 1.

Wilkin, P. (1954). Contribution à l'étude de la circulation placentaire d'origine foetale. *Gynécol. et obstét.* **53,** 239.

*Wilkin, P. (1958). *In* "Le Placenta humain—Aspects morphologiques et fonctionels" (J. Snoeck, ed.), p. 60. Masson, Paris.

Wislocki, G. B., and Dempsey, E. W. (1939). Remarks on the lymphatics of the reproductive tract of the female rhesus monkey (*Macaca mulatta*). *Anat. Record* **75,** 341.

*Wislocki, G. B., and Streeter, G. L. (1938). On the placentation of the macaque (*Macaca mulatta*), from the time of implantation until the formation of the definitive placenta. *Carnegie Inst. Wash. Publ.* No. 496 (*Contribs. Embryol.* **27,** 1).

Chapter XVI. CUTANEOUS CIRCULATION

A. EMBRYOLOGY

By K. CHRISTENSEN

1. EARLY VASCULOGENESIS

Early vasculogenesis of significance in the cutaneous areas begins with the formation of peripheral vascular networks in the embryonic subcutaneous tissue.[1] These structures have been demonstrated by the injection method in various parts of the body in the embryonic and early fetal stages of the human and other mammalian species. Presumably such vascular networks, at least in part, have an *in situ* origin in the somatic mesoderm. The cutaneous vascular network develops out of the subcutaneous networks, and later definitive cutaneous blood vessels differentiate. The plexuses contain both arterial and venous vessels even though the arterial connections are the more frequently noted. At the 12 mm stage (pig embryo), the edges of the superficial vascular plexus have already become modified in the upper pair of limb buds, to establish the superficial veins and their anastomoses, typical of the adult pattern.

2. PRIMORDIAL VESSELS

a. Arteries: The intersegmental arteries, which connect the aorta with the subcutaneous vascular plexuses in the trunk, appear about the 2 mm stage (human). When the embryo is about 4.5 mm long, over thirty pairs of intersegmental arteries have already been formed. Portions of these intersegmental vessels may be plexiform, and anastomoses between them is of significance in final differentiation. In the trunk the intersegmental arteries divide into posterior and anterior branches, from which posterior, lateral, and anterior cutaneous vascular rami are given off. Cervical intersegmental arteries and a subcutaneous plexus on the sides and vault of the head contribute to the cutaneous vascular rami in this region and also in the neck. In the upper extremity the seventh cervical intersegmental artery is continuous with the axial artery. In the lower extremity, after participation of several intersegmental arteries, the fifth lumbar

[1] For a summary of the early studies on this subject, see Evans (1912).

intersegmental artery joins the main axial vessel. Early in the investigation of this subject it was believed that the cutaneous vascular rami developed equally and symmetrically throughout the body, but it became apparent in subsequent studies that this was not the case. In the adult, evidence for segmental symmetry has been almost completely lost. It would be of interest, however, to plot diagrammatically through the utilization of early developmental stages the vascular "dermatomes."

b. Veins: Venous development in relation to the skin has merely been mentioned, but it follows a plan similar to that for the arteries. The cardinal veins are the first important venous channels to receive blood from the developing subcutaneous and cutaneous plexuses.

c. Histogenesis: The histogenesis of the peripheral vascular system, as it extends from the subcutaneous vascular channels, has been described by Finley (1922). Her observations in the 20 mm embryo, although limited to the head, help to explain the origin of the cutaneous vascular plexuses in other parts of the body as well. According to this author, the inner zone, which is the most differentiated, has a vascular net of open blood vessels, while the second zone, which is just external, appears as a solid vascular network whose vessels await canalization. The third zone illustrates the means for vascular extension because at this level there is active endothelial sprouting from the solid primordia of the second zone. In it blood islands also form in the embryonic connective tissue and contribute to the advancing endothelium by attachment to nearby sprouts. The fourth and most external zone is completely avascular and awaits vascular invasion. The sequence of zonal changes accounts for the manner of cutaneous vascular development.

3. Cutaneous Vascular Pattern

The cutaneous vascular system was first described in detail by Spalteholz (1893). Cutaneous arteries from the subcutaneous tissue enter the dermis and form a deep plexus; branches are given off from the latter, that ascend to the upper part of the dermis where a second plexus is formed. Arterial twigs from the superficial plexus, without further anastomoses, join the capillaries in the connective tissue papillae. The veins returning blood from the dermis also form superficial and deep plexuses, and in some localities, as in the plantar skin, two additional venous plexuses are present between these. According to Zweifach (1957), recent investigations have shown that the terminal ramifications (either of cutaneous arteries or veins), instead of branching repeatedly to form successively smaller arborizations, show a regular pattern of interconnecting links, to form, respectively, arterial and venous arcades. Adjacent arterial arcades may be joined by interconnecting vessels, and venous arcades likewise may be interconnected. Most of the precapillaries and capillaries which nourish the tissues are branches of the arterial arcades. The capillaries, in turn, enter the companion series of venous arcades. In addition, a series of arteries which join the veins directly establish the arteriovenous anastomoses of the skin. Early papers on cutaneous arteriovenous anastomoses are listed by Boyd (1953). (For structure and function of these structures, see Sections B.-1e, 2g and C.-3 of this chapter.)

4. Nature of Vascular Differentiation in the Skin

Cutaneous vascular differentiation from the vascular plexuses might be more complicated were it not for the special plexiform arrangement of cutaneous blood vessels. Spalteholz's (1893) classical study revealed an organization of anastomotic blood vessels whose complexities, besides size and number of vessels, are determined by their relative position in the various cutaneous areas of the body. According to Spalteholz, the structural characteristics of the cutaneous blood supply are present and fully differentiated at the time of birth. Recent studies have both revised and added to the observations of Spalteholz. Developmentally, the revisions in the circulatory pattern of the skin (Zweifach, 1957) have only modified the complexity of the vascular networks that are retained in cutaneous vascular differentiation.

5. Clinical Implications

The relation of developmental stages, as well as differentiated patterns of cutaneous vascular system, to clinical practice is often only suggested. Pillsbury *et al.* (1956) have discussed diseases of blood vessels of the skin from an anatomic standpoint, and a somewhat similar plan has been followed in pediatric dermatology (McKee and Cipollaro, 1946). However, in many instances the relationship of cutaneous diseases to cutaneous vascular patterns are not mentioned, although attention may at times be called to an abnormal vascular development and the resulting congenital lesion (McKee and Cipollaro, 1946).

B. ANATOMY

By A. P. Moreci and E. M. Farber

The human integument consists of fairly well-defined strata, the most superficial of which is the epidermis. This is an avascular, stratified squamous epithelium which depends for its nutrition upon the blood supply conveyed to the subjacent capillaries in the dermis. The latter stratum consists mainly of dense, irregularly arranged bundles of collagen and elastic fibers embedded in an amorphous matrix or ground substance. Two insensibly merging parts comprise the dermis, a thin superficial papillary layer and a deep reticular layer. The appendages of the skin—hair follicles, sebaceous and sweat glands—are situated in the dermis. Loose bundles of collagen fibers secure the skin to the underlying layer, which consists of areolar and adipose tissue (telea subcutanea or superficial fascia).

1. Distribution of Blood Vessels in the Dermis

a. Arterial blood supply: The arterial supply of the skin is derived from a network of vessels in the subcutaneous tissue. The perforating arteries enter the skin and form an extensively anastomosing system, the cutaneous plexus (Fig. 106), which lies between the deep reticular portion of the dermis and the underlying tissue (Spalteholz, 1893, 1927; Horstmann, 1957). Some branches from this plexus are given off to the subcutaneous layer, where they supply the adipose tissue and sweat glands. Other vessels, referred to as "Candelaberarterien" because of the manner in which they branch, ascend through the reticular portion of the dermis. At various levels within this area, adjacent candelabra vessels communicate with one another through arched anastomoses, to form an irregular plexus. The more superficial arcades are more numerous and of smaller caliber, and constitute the subpapillary arteriolar plexus of Spalteholz (1893). Terminal branches from this network do not anastomose but give rise to the capillary loops of the papillae. Each papilla may contain one or several capillaries. Further studies of the blood supply to the skin have recently been made by Moretti *et al.* (1959), who presented an extensive description of the distribution in selected areas of the face.

b. The cutaneous veins: The venous system is arranged in four networks which are disposed parallel to the skin surface and at different

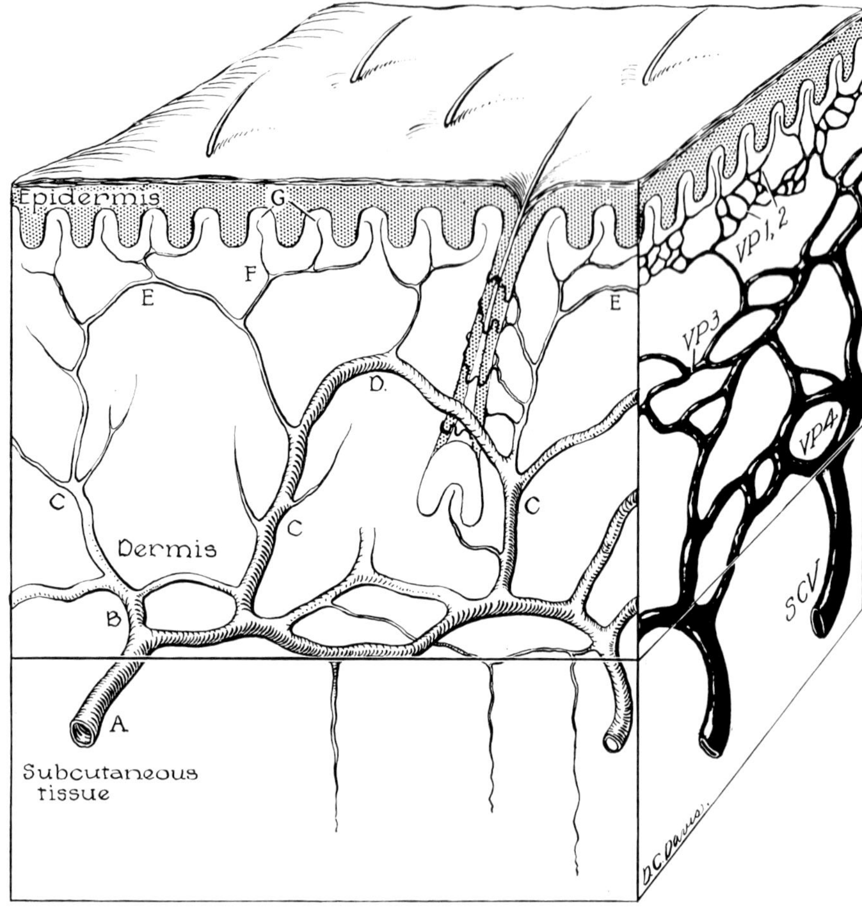

FIG. 106. Diagrammatic representation of the arterial and venous systems of the skin. A. Perforating artery. B. Cutaneous plexus. C. "Candelaberarterie" (arteriole). D and E. Arched anastomoses between candelabra vessels. The more superficial arcades (E) are of smaller caliber and more numerous than the deeper ones (D). The superficial arcades form the subpapillary arteriolar plexus. F. Terminal arteriole. G. Arterial limb of capillary in a papilla. VP 1 and VP 2 form the subpapillary venous plexus. VP 3 and VP 4 form the deeper venous plexuses. SCV, subcutaneous vein.

levels within the dermis (Fig. 106). Small venules which drain the papillary loops merge to form the most superficial plexus just below the papillae. Another plexus lies a little deeper than the first and is in close association with the subpapillary arteriolar network. Petersen (1935) and Horstmann (1957) are of the opinion that the first and second venous plexuses should not be regarded as separate entities but as parts of a single plexus ("the venous main-plexus"). Blood from the first and second networks is collected into a third plexiform system, which is situated approximately in the middle portion of the dermis. The fourth or deepest venous network occupies a position between the dermis and subcutaneous tissue, in close relation to the cutaneous arterial plexus. Many of the branches which collect blood from the sweat glands and adipose tissue empty into this plexus. The latter is served by the large subcutaneous veins and also by the deep venous system which accompanies the arteries.

c. Capillary circulation: Microscopic studies of the superficial vessels of the skin *in vivo* have been made by many investigators (Davis and Lawler, 1958; Goldman, 1957; Lovett Doust, 1955; Gilje *et al.*, 1954). Most investigations have been limited to the capillary bed of the nailfolds, since the vessels in these sites lie parallel to the dorsal surface of the fingers and may be visualized through the intact skin by microscopy. The capillaries and other vessels lying close to the surface in other regions may be seen to best advantage after removing the keratin layer by successive applications of cellulose tape (Davis and Lorincz, 1957; Pinkus, 1952). Davis and Lawler (1958) have used this method to study the end capillaries of the dermal papillae in several hundred normal and diseased individuals. They found that each normal capillary formed a loop which was oriented in a plane perpendicular to the skin surface and consisted of an ascending arterial limb of an average diameter of 0.010 mm and a descending venous limb of an average diameter of 0.015 mm. The apical portion of the loop was considered to be a part of the venous limb on the basis of its caliber. Diameters of the nailfold capillaries ranged from 0.010 to 0.013 mm on the arterial side and from 0.013 to 0.020 mm on the venous side. The longest capillaries in the dorsal skin of the hands and feet varied in length from 0.15 to 0.20 mm; the average length of these in the nailfold was 0.30 mm. The number of capillaries per square millimeter was about 20 in the nailfold and between 60–70 in the dorsal skin of the hands and feet. It has been reported that the connective tissue of the dermis contains no capillaries of its own (Horstmann, 1957), this tissue being served by the capillary nets which accompany the larger arteries and veins, nerves and dermal appendages.

d. Blood supply of the dermal appendages: The capillary networks of the hair follicles receive several side branches from the candelabra arteries. These branches may be given off from any section of the parent vessel, as illustrated in Fig. 106. Durward and Rudall (1949) and Montagna and Ellis (1957) have provided detailed descriptions of the distribution of capillaries to the hair follicles. They have also discussed the changes in blood supply which occur in relation to the growth activity of these structures.

The blood supply of the sweat glands is derived from vessels of the cutaneous arterial plexus and from branches of the candelabra arteries. Ellis *et al.* (1958) studied the distribution of blood vessels to the sebaceous and sweat glands by the Azo-dye method of Gomori (1952) for alkaline phosphatase and have reported that the capillary networks of sebaceous glands generally spring from a single parent arteriole which arises from vessels of the hair follicle. These workers also described the arrangements of capillary networks of eccrine and apocrine sweat glands.

e. Arteriovenous anastomoses: Arteriovenous anastomoses (AVA) form direct communications between the arterial and venous systems and hence may serve to bypass the capillary circulation. Existence of these vessels in human skin has been reported (Spalteholz, 1893, 1927; Grant and Bland, 1931; Clark, 1938; Horstmann, 1957). Mescon *et al.* (1956) and Hale and Burch (1960) have recently reviewed the morphology and some functional aspects of arteriovenous anastomoses in digital skin. Some of these form structures called glomi. A typical glomus consists of an afferent arteriole (the AVA proper), which may give rise to smaller branches, and a reservoir-like primary collecting venule. (For submicroscopic structure of arteriovenous shunt, see Section B.-2g of this chapter; for function, see Section C.-3.)

2. Microscopic and Submicroscopic Structure

a. Distributing arteries: These vessels are composed of three closely adherent layers of varying histologic character, a tunica intima, a tunica media and a tunica adventitia. The innermost of these, the tunica intima consists of a single layer of endothelial cells and a well-developed internal elastic membrane. In some specimens, a thin layer of connective tissue may be distinguished between the endothelial and elastic elements. The chief component of the middle layer is smooth muscle which is closely bound together by connective tissue. The external coat, the tunica adventitia, consists of loosely-arranged collagen and elastic fibers. The latter are more abundant and form in most cases an external elastic lamina, which is in close relation to the tunica media (Hamm, 1953; Bailey, 1953).

b. Arterioles: The most prominent feature of these vessels is a well-developed muscular media. It constitutes the largest proportion of the vessel wall thickness. The tunica intima is composed of endothelium and an internal elastic membrane, similar to that seen in the distributing arteries. A scant amount of delicate connective tissue may sometimes be seen between the two layers of the intima. Loosely arranged connective tissue comprises the tunica adventitia.

On the basis of morphologic differences, Hibbs *et al.* (1958) have divided the arterioles of the dermis and subcutis of the finger tip and abdomen into two types. Those located mainly in the deep subcutaneous tissue are indistinguishable from arterioles present in most other organs. The second type is found in the dermis and subcutis. The chief features of the arterioles in this group are an endothelial lining that is thicker than that ordinarily seen in other vessels; the presence of a large number of endothelial nuclei with respect to the vessel size; the presence of bundles of filaments, rod-shaped granules and vesicular bodies in the endothelial cytoplasm; and the absence of an elastic lamina. Many of these vessels are associated with sweat glands.

The ratio of wall thickness to lumen diameter is an important factor in the classification of a vessel as either a distributing artery or an arteriole. In a study of cutaneous arterioles by Farber (1947), it was noted that the average wall to lumen ratio in individuals with normal blood pressure was approximately 1:2, while that in patients with essential hypertension was about 1:1. In the case of normal muscle arterioles, Kernohan *et al.* (1929) found an average ratio of 1:2 (range of 1:1.7 to 1:2.7), while in patients with hypertension, it was greater, 1:1.4.

c. Metarterioles: These vessels have been described as consisting of an intimal layer of endothelium, a discontinuous tunic of smooth muscle cells in the media and a scant amount of connective tissue which forms the adventitia. No elastic laminae are present.

d. Capillaries:[2] These vessels consist essentially of an endothelial membrane one cell thick. A scant amount of delicate connective tissue, in which pericytes and various other types of cells are found, forms an outer layer which is sometimes referred to as perithelium. Hibbs *et al.* (1958) have described two types of capillaries in digital and abdominal skin. In one type, found predominantly in the adipose tissue of the superficial subcutis and deep subcutis, the vessels are similar to those generally encountered in other tissues. When widened, these capillaries present very thin walls and a smooth luminal surface, except for the presence of endothelial nuclei. Electron microscopy has revealed numerous vesicles, 500 Å–700 Å in diameter, in the cytoplasm. The endothelium of the second type of capillaries found in the dermis and subcutis dis-

[2] For a general discussion of the submicroscopic structure of vessels, see Section C., Chapter I.

tinguishes them from those already discussed. The endothelial cells are less flattened when the vessel is widened and they assume a columnar or pyramidal contour when it is narrowed. In the latter state the capillary lumen may be diminished or completely occluded. More nuclei are encountered in cross section. The most striking feature of these cells is the large bundles of filaments found running parallel to the nuclear membrane within the cytoplasm. It has been suggested that these filaments are actively contractile (Hibbs *et al.*, 1958).

e. Venules: These vessels consist of a single layer of endothelial cells adherent to a thin layer of connective tissue which forms the outer wall. Smooth muscle cells are generally absent in the smallest venules but are present in the larger ones. It has been reported that the endothelial cells of most cutaneous venules examined by electron microscopy contain thick filaments similar to those seen in the second type of capillary already discussed (Hibbs *et al.*, 1958).

f. Veins: Three layers similar to those present in arteries occur in veins of medium and large size. The great morphologic difference between veins and arteries, however, is that the walls of the former are thinner and contain less muscle than those of the latter. The smooth muscle found in the tunica media of veins is more interspersed with connective tissue than that found in arteries.

g. Arteriovenous anastomoses: Mescon *et al.* (1956) have discussed the histology of the arteriovenous anastomosis proper in a glomus body of human digital skin. In fixed sections this structure has been described as a thick-walled vessel with a barely perceptible lumen. The wall (20–40 micra thick) consists of endothelium, a subendothelial reticular layer, and a tunica media containing an outer loosely arranged collagenous reticulum. No internal elastic membrane is identified. (For function of arteriovenous shunt, see Section C.-3 of this chapter; for discussion of glomus tumor, see (1) Vascular Anomalies Due to Developmental Defects, Section B., Chapter XXI.)

C. PHYSIOLOGY

By A. P. Moreci and E. M. Farber;[3] L. A. Sapirstein[4]

1. Methods for Studying the Cutaneous Circulation

Diverse methods have been described for the study of cutaneous circulation. Among these are colorimetry (Lewis, 1927), thermometry (Beekman, 1959; Hensel and Bender, 1956; Winsor, 1954; Burton, 1948), plethysmography (Burch, 1954; Winsor, 1953; Hertzman, 1937), calorimetry (Mendlowitz, 1950; Greenfield and Scarborough, 1949), stroboscopic techniques (Lovett Doust and Salna, 1955), clearance of radioactive isotopes (Kety, 1948, 1949), and microscopy (Davis *et al.*, 1960; Roth, 1946). The relative merits and disadvantages of some of these methods have been discussed by Burton (1954).

2. Role of the Cutaneous Circulation

Skin nutrition depends not only upon an adequate minute-volume flow of blood but also upon its distribution between true capillaries and the extensive system of arteriovenous anastomoses. Since shunt flow has been shown not to participate in transvascular exchanges (Kety, 1949), only blood which reaches the vessels of the capillary bed can serve in the processes of tissue nutrition. It has been estimated that a blood flow rate of 0.8 cc/min/100 cc of skin is required to meet normal metabolic needs. However, the circulation in digital skin of resting individuals is normally 15–40 cc/min/100 cc and may even be as high as 90 cc/min/100 cc under conditions which produce full vasodilation (Burton, 1939). Most of this flow rate increase is mediated by arteriovenous shunts. The wide fluctuations in local circulation above the minimum requirements for skin metabolism reflect the multiple role of the cutaneous circulation in the physiologic processes of the body as a whole. In addition to its nutritive role, therefore, the cutaneous circulation serves the thermoregulatory system and may also act, through the many vascular plexuses, shunts and other vessels, as a blood reservoir. Furthermore, it has been suggested that the arteriovenous anastomoses of the skin may serve to moderate blood pressure (Burton, 1959). Hence, caution must be observed in any attempt to correlate total skin blood flow rates with the nutritional state of the tissues.

[3] Authors of Subsections 1, 2, 4 and 5.

[4] Author of Subsection 3.

3. Role of the Arteriovenous Shunts[5]

a. Physiologic considerations: Microscopic communications between the arterial and venous systems, through which blood may pass without exchanging its contents with the extravascular fluids (arteriovenous shunts), were considered of little physiologic significance until the early nineteen thirties. The regularity of their occurrence in the acral portions of the body and the fact that their development in the rabbit ear is accelerated when the animal is kept in a warm environment and reduced at a low temperature (Clark and Clark, 1934) suggest that these shunts are of functional significance. (For anatomic considerations, see Section B.-1e and 2g of this chapter.)

The most striking reaction of arteriovenous anastomoses is to temperature. Dilation occurs below 15° C and also with slight warming of the whole body. The shunts are constricted by epinephrine and dilated by acetylcholine and histamine (Grant, 1930). The innervation is of sympathetic origin, the fibrils containing a specific cholinesterase, which suggests that they are cholinergic. In the rabbit ear, the response to local cooling is delayed but not abolished by denervation (Grant and Bland, 1931).

Despite the rich cholinergic innervation of the arteriovenous shunts, the vasodilator effect of body heating is not abolished in the human hand by atropine (Roddie *et al.*, 1957). On the other hand, forearm vasodilation can no longer be elicited. Since there are few, if any, arteriovenous anastomoses in the forearm, the question must be raised whether the increased skin blood flow and temperature are as dependent on the opening of arteriovenous shunts in man as they seem to be in the rabbit.

Though the opening of a shunt may be expected to reduce the arteriovenous pressure gradient slightly, it seems improbable that there is diversion of blood from the smaller vessels to the shunted area. Rather, it would be anticipated that total flow would increase by the amount of shunt flow. It has, however, been suggested that the capillary bed in the vicinity of open shunts may be perfused by cell-poor blood in consequence of plasma skimming (Grant *et al.*, 1932).

The amount of blood carried by the shunts may be very considerable in certain portions of the body. Studies with microspheres (Rondell *et al.*, 1955) suggest that in certain areas shunt flow may range from 5–70 per cent of the total flow. It must, however, be emphasized that these areas are small. In this regard, studies on the pentobarbital anesthetized rat with K^{42} indicate that rather less than 3 per cent of the cardiac output passes through nonexchanging channels (Sapirstein,

[5] By L. A. Sapirstein.

1958). This value is increased by sympathomimetic drugs, but precise quantitation has not been accomplished.

b. Role of arteriovenous shunts in heat conservation and dissipation: The fact that the major avenue of heat loss from the body is the skin is so well known that it requires no documentation. Heat losses through this structure, occurring by radiation, convection, and evaporation, are dependent on the difference between cutaneous and ambient temperature. The exposed portions of the skin, which are ordinarily much cooler than the interior of the body, afford a regulatory mechanism for the fine adjustment of body temperature. This is ordinarily accomplished by increasing cutaneous blood flow, so that the tissues come more nearly into temperature equilibrium with the body.

The magnitude of the regulatory effect may be appreciated by considering the heat losses which occur through the hand at various environmental temperatures. It has been shown (Forster *et al.*, 1946) that at room temperatures of 15.2° C, the heat loss through one hand is of the order of 0.5 Cal/hr. When the ambient temperature is raised to 37.4°, 2.45 Cal/hr are lost. Considering both hands as organs for heat loss, and neglecting other tissues which may serve in the same capacity, this represents a "thermostatic ability" of the order of 100–120 Cal/day.

The increased effectiveness of the hand as an organ for heat loss at high temperatures is associated with an increase in blood flow. For example, at 15.2° C, the flow rate is of the order of 0.75 ml/100 cc/min, while at 37.4° C, it is 10–30 ml/100 cc/min.

It seems improbable that these increases in local circulation should occur through the true capillaries, though no positive evidence to this effect is available. It is more probable that the excessive blood flows are associated with the opening of the arteriovenous anastomoses which appear to be peculiarly suited to serve the function of heat dissipation and conservation. Their location at the tips of the exposed portions of the body can accomplish local warming on dilation; furthermore, by increasing the return of blood through those veins which lie in close proximity to the arteries, the cooling of arterial blood in its passage to the extremities is diminished by a counter current mechanism. The whole extremity can, therefore, be warmed by dilation of a peripheral arteriovenous shunt. In addition, the arteriovenous shunts increase heat loss from the extremities by augmenting venous return. Finally they seem ideally suited, anatomically and physiologically, to prevent, by their dilation, overcooling of those portions of the body which by position and shape are most likely to suffer from exposure to cold, e.g., the fingertips and ears. However, before accepting this attractive notion uncritically, it would be well to study the following statement made by Clark (1938) in his review on arteriovenous anastomoses: " . . . altogether too many erroneous notions have crept into physiology as the result of unbased teleological explanations. Far healthier . . . is the attitude that definite physical and chemical factors will be found to account for structure."

4. Sex Differences in Blood Flow[6]

A recent plethysmographic study of 49 healthy individuals has indicated that digital blood flow rates in men are significantly higher than those in women (Farber *et al.*, 1959). The median blood flow rate, calculated for a group of 20 men, was 30.3 cu mm/5 cu cm tissue/sec, while that for a group of 29 women was 18.0 cu mm. It was suggested that the difference might be related to the lower basal metabolic rate in women, compared to men (Fulton, 1955). However, many other factors which could affect the degree of humoral and nervous control of the circulation might also be involved, such as differences in activity of the endocrine or sympathetic nervous systems. In connection with the observation that digital blood flow rates are lower in normal females than in normal males, it is of interest to note that vasospastic disorders, such as acrocyanosis, Raynaud's disease and livedo reticularis, occur predominantly in women.

5. Effect of Rubefacients on the Microcirculation[6]

Various nicotinic acid esters, oil of cloves, and mustard oil are examples of rubefacients which have been used for many years in the treatment of muscle and joint pain. These substances have been reported to produce erythema, hyperemia and local increase of temperature when rubbed into the skin (Lange and Winer, 1949; Gross and Merz, 1948); however, the mechanism of action has not been well established. To cast some light upon this subject, a study was performed on the effects of rubefacients on the microcirculation of the hamster cheek pouch (Fulton *et al.*, 1958, 1959), the results of which are presented elsewhere (Section D.-4c, Chapter V). In brief, it was found that the predominant response of arterioles and muscular venules was dilation, while no change in caliber of capillaries or small venules could be detected by ocular micrometry. The dilation produced in muscular vessels was prompt and extended over a larger portion of the vessel than that in direct contact with the drug. All segments of the reacting vessel appeared to dilate simultaneously (Fig. 107), suggesting that the response was mediated by

[6] By A. P. Moreci and E. M. Farber.

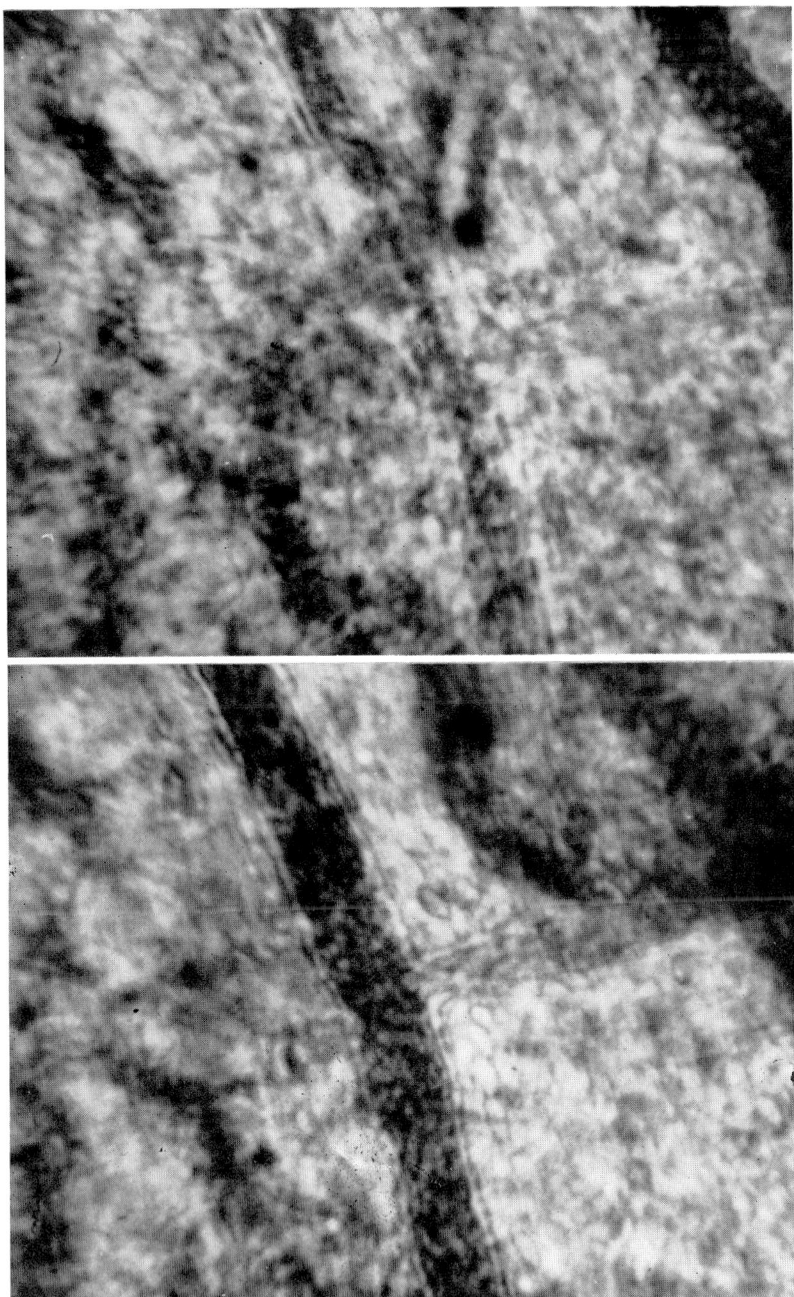

FIG. 107. Dilation of arterioles produced by topical application of mustard oil. Top: Constricted arterioles of hamster cheek pouch. Note pipette in upper right quadrant of photograph. Bottom: Mustard oil (1% in mineral oil) produced prompt, extensive vasodilation. Note that the dilation extended beyond the site of rubefacient deposition, upper right quadrant of photograph.

a conduction system with nerve-like properties. The widely distributed, nonmyelinated perivascular nerve plexus demonstrated by Lutz *et al.* (1950) provides a morpholgic basis for this interpretation. Evidence which indicates that the response to rubefacients in human skin does not depend upon the integrity of peripheral nerves has been presented by Crismon *et al.* (1958). These authors found that after the cutaneous nerves in the area had been blocked by a local anesthetic, a benzyl ester of nicotinic acid applied to the forearm still produced normal vasodilation.

D. PHARMACOLOGY

By A. P. Moreci and E. M. Farber

1. Vasoconstrictor and Vasodilator Agents

Various pharmacologic agents which affect the blood vessels may be generally classified as either vasoconstrictors or vasodilators. The latter are of greater clinical interest, since many peripheral vascular diseases involve vasospasm or mechanical obstruction of vessels, with consequent diminution of blood flow to the affected parts. Exceptions to this are diseases such as erythermalgia (erythromelalgia), which are characterized by vasodilation, increased peripheral blood flow and pain in the affected areas (Allen *et al.*, 1946). Since cutaneous vasomotor tone is affected to a great extent by sympathetic nerve impulses and by circulating catecholamines, such as epinephrine and norepinephrine, therapeutic agents have been sought which either produce adrenergic block, antagonize circulating epinephrine, or inhibit the contraction of vascular smooth muscle itself.

a. Locus of action: Vasoconstriction induced by the stimulation of adrenergic fibers can be blocked by a variety of drugs (Goodman and Gilman, 1956; Patton, 1960). Among these are the imidazoline compounds, such as tolazoline (Priscoline) and phentolamine (regitine); the ergot alkaloids (ergotoxin, ergotamine and their derivatives); and the beta-haloalkylimines (Dibenamine and Dibenzyline). It is believed that these compounds act by competitive inhibition at the receptor cell level. Agents which are of value in antagonizing the effects of circulating epinephrine and norepinephrine include regitine and benzodioxane. Other drugs diminish peripheral vasomotor tone by blocking synaptic transmission in autonomic ganglia. This group includes the salts of tetraethylammonium, pentamethonium and hexomethonium. Still other drugs, such as the barbiturates, veratrum alkaloids and the so-called tranquilizers, affect higher centers in the central nervous system. The effects of adrenergic stimulation may also be blocked by drugs which have a direct action on vascular smooth muscle. Examples of this type of compound include the xanthine derivatives (caffeine, theophylline and theobromine); the kinins; the benzylisoquinoline group of alkaloids, of which papaverine is representative; and the nitrites.

b. Clinical application: The value of drugs in the clinical management of peripheral vascular diseases is, however, limited because of the undesirable side effects produced by many of them. In addition they are inactivated by the tissues or excreted by the kidneys, or the patient may already have, or may develop, resistance to them. Thus, permanent results are not always assured. For example, Lippmann (1952) treated patients with impaired arterial circulation by intra-arterial injection of Priscoline and found that while improved circulation was obtained initially, the effects were transient. Papaverine hydrochloride has also been used extensively by many investigators, but it too does not have a lasting effect.

In many instances surgical sympathectomy is a better means of reducing vasospasm, but this likewise may produce only transient results. Initially, blood flow may be increased appreciably by such a procedure, since the neurogenic vasoconstrictor influences are diminished. However, a high degree of tone in the vessels may return later. This has been well illustrated by many investigators, including Mendlowitz and Touroff (1952), who studied the effects of sympathectomy on the hallucal circulation in essential hypertension. In every subject it was possible to demonstrate a return of sympathetic nerve function of varying degree. It was also found that non-neurogenic or intrinsic resistance had increased due to sympathectomy. Since the heightened resistance could not be reversed by benzodioxane, it was concluded that circulating epinephrine was probably not responsible for the effect. The change in resistance was ascribed to shortening of the circular smooth muscle of arterioles through relative disuse atrophy.

2. Some Biologically Elaborated Substances

a. Epinephrine: It is well known that injection of epinephrine into the skin results in local blanching. This effect is due to strong constriction of all blood vessels in the area containing smooth muscle. The statement has often been made that epinephrine also constricts the capillaries, but evidence for this is lacking. Fulton *et al.* (1958, 1959) obtained constriction of the arterioles, precapillary sphincters and muscular venules in the hamster cheek pouch by topical application of epinephrine or norepinephrine, but neither of these amines produced any detectable change in the caliber of true capillaries.

b. Norepinephrine: This compound, which has been shown to be the mediator in sympathetic nerves (von Euler, 1956), also constricts muscle-containing blood vessels in the skin, but the effect is not so marked as that produced by epinephrine.

c. Serotonin: A review of this subject has been made by Page (1958). The effects of intra-arterial infusion of serotonin on blood flow in the human hand and forearm have been described by Roddie *et al.* (1955). These workers found a marked erythema and a decrease in the blood flow rate, attributed to constriction of resistance vessels and dilation of the minute cutaneous vessels. Reid (1952) and Lorincz and Pearson (1959) have reported that intradermal injection of 5–hydroxytryptamine (serotonin) produced constriction of large subcutaneous veins. However, the latter investigators found that this response did not occur if a tourniquet was placed between the vessels and the site injected with serotonin. This suggested that serotonin did not act as a mediator in axon reflexes but had a direct venoconstrictor effect.

The effects of intradermal injection of serotonin have also been described by Clendenning *et al.* (1959). The reactions obtained in the forearm skin of 34 normal subjects consisted of erythema at the site of injection, a flare within thirty seconds, and formation of a wheal, 5–8 mm in diameter. No effect on the adjacent subcutaneous veins was noted. This is in contrast to the findings of other investigators and may possibly have been due either to differences in dosage of serotonin used, the manner in which it was administered, or failure of the drug to reach the veins.

A more recent study of the cutaneous effects of serotonin on 50 healthy subjects has been made by Demis *et al.* (1960). The tests were carried out on intact forearm skin and also in a similar site from which the keratin layer had been removed by successive applications of cellulose tape. Intradermal injection of intact skin produced erythema, flare and constriction of large cutaneous veins in all subjects. Wheal formation was observed in only three individuals, despite the fact that serotonin concentrations were as high as 200 μ/ml. The same concentrations of serotonin applied directly to the surface of skin from which the keratin had been removed produced a slowing of the blood flow rate within ten to fifteen minutes, but no morphologic changes in the capillaries or subpapillary venules could be demonstrated. An increase in the dosage had in general no greater effect. There was, however, some evidence of edema formation.

E. PATHOLOGY

By A. P. Moreci and E. M. Farber

Pathologic changes in the blood vessels of the skin may be due to diverse causes. Among these are hereditary defects, internal disease, or external factors, such as extremes of temperature, exposure to X-rays or ultraviolet light, other physical trauma, and chemical injury. In some instances structural and functional changes may occur in several types of cutaneous blood vessels simultaneously, while in other cases they may involve predominantly one type of vessel. The following disorders have been selected to exemplify this.

1. Telangiectasia

This vascular lesion appears on the skin or mucous membranes and consists of widened capillaries, venules or arterioles. A simple telangiectasis is a discrete, erythematous macule in which dilated capillaries occur. These fade in color when pressure is applied to the surface. Telangiectases appear following excessive exposure to X-rays, sunburn or application of excessive heat to one part of the skin. These lesions may also appear in patients with lupus erythematosus, or they may be found encircling and covering the surface of an epithelioma.

Cutaneous "spiders" are telangiectases composed of thin-walled dilated arterioles. They may be hereditary or acquired. The morphology of spiders varies widely (Bean, 1958). However, the lesions have some characteristics in common: a body, legs and surrounding erythema. A central papule forms the body, while the legs are formed by a varying number of arterioles which radiate from it. Vascular spiders frequently occur in individuals with hepatic cirrhosis; however, they may also appear in the absence of organic disease (idiopathic spiders). In Rendu-Osler-Weber disease (hereditary hemorrhagic telangiectasia), the vascular spiders in the skin and other organs rupture easily, so that patients with this disorder periodically bleed profusely (for further discussion of this entity, see (1) Vascular Anomalies Due to Developmental Defects, Section D., Chapter XXI).

2. Purpura

The term, purpura, relates to the escape of blood from capillaries into the skin or mucous membranes (Fig. 108). This response may occur in many unrelated diseases and is therefore a symptom rather than a specific disease entity. It has been suggested that in purpura associated with scurvy, the endothelial lining of the capillaries may develop defects

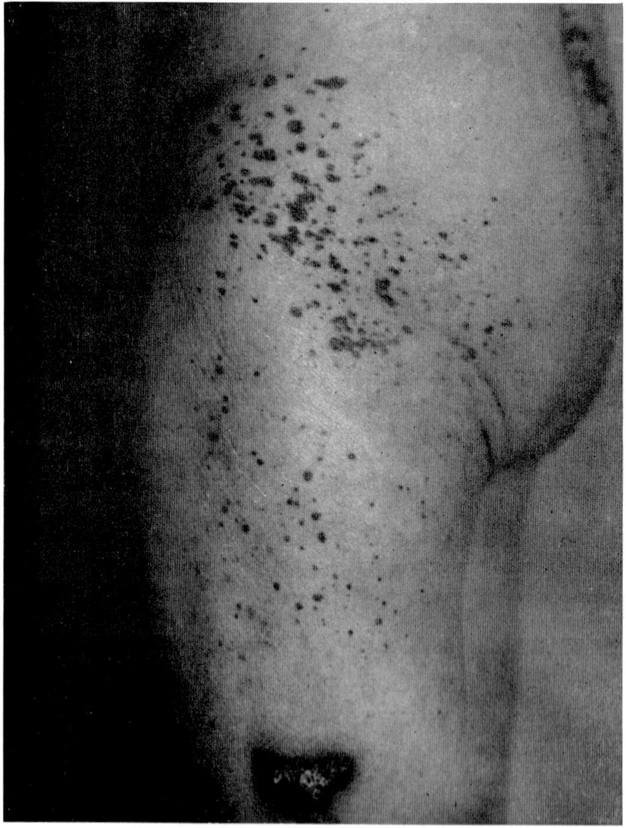

Fig. 108. Petechiae and ecchymoses seen in a patient with purpura.

and become susceptible to hemorrhage. Thrombosis or bacterial emboli may also be responsible for the rupture of capillaries. A large number of hematologic disorders, characterized by low levels of prothrombin and fibrinogen and low platelet count, may be associated with purpura. These deficiencies may result in failure to maintain the normal repair of the capillary wall, which could account for the bleeding.

3. Hypertensive Ischemic Ulcer of the Leg

Widespread cutaneous arteriolar sclerosis is present in some patients with essential hypertension (Farber, 1947). The walls of arterioles are thickened and their lumen size is diminished. In some individuals in whom the disease has been present for a long time, an indolent, painful, ischemic ulcer may develop on the ankle (Fig. 109). The pre-

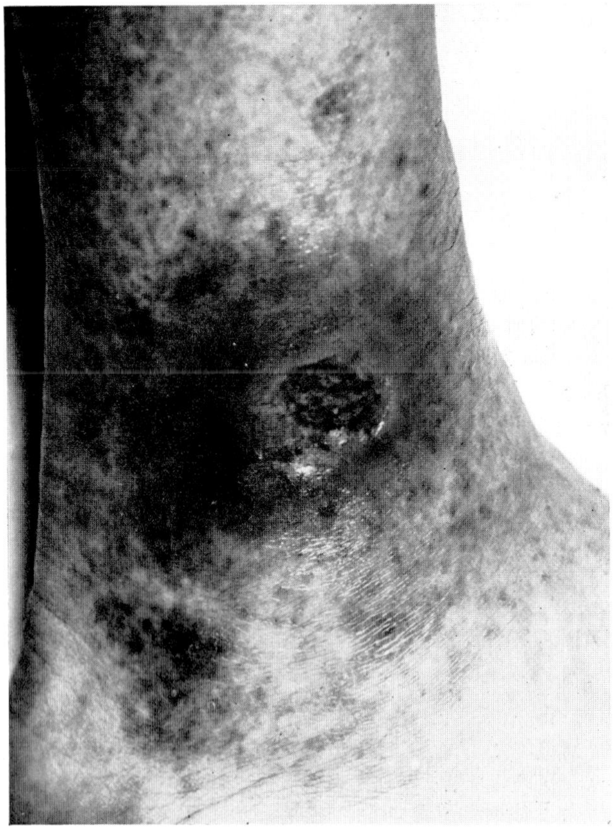

Fig. 109. Hypertensive ischemic ulceration of the ankle.

cipitating factor may be trauma to the diseased arterioles, producing thrombosis. The incidence of hypertensive ischemic leg ulcers is much higher in women than in men (Farber and Schmidt, 1950).

4. Livedo Reticularis

This disease is characterized by a cyanotic discoloration of the skin in a mottled, netlike pattern. The pathologic changes include arteriolar narrowing and widening of the capillaries and venules (Farber and Barnes, 1955). Arterioles with thickened walls are found in the deep cutis and upper subcutaneous tissue. The lumina of these vessels may be occluded in some instances as the result of endothelial proliferation. Widened capillaries are seen in the upper dermis.

The clinical categories of this disease are: cutis marmorata, which is precipitated by exposure to cold but disappears on rewarming; livedo reticularis idiopathica, which is also precipitated by cold but does not disappear on rewarming; and livedo reticularis symptomatica, which may be present independently of the environmental temperature and may be associated with other diseases, such as lupus erythematosus and polyarteritis nodosa.

5. Acrocyanosis

Acrocyanosis is a condition in which the skin of the hands and feet are persistently cold and assume a diffusely cyanotic hue at moderate environmental temperatures. The discoloration deepens as the ambient temperature is lowered and is less marked in a warm environment. Elevation of the extremity or sleep may diminish or abolish the symptoms. The etiology of acrocyanosis is not known, although over-activity of the sympathetic nervous system and hormonal dysfunction have been suggested as causative factors. It is generally conceded that the immediate abnormality lies in excessive constriction of the arterioles of the affected extremities.

6. Dermatologic Conditions

Reports in the literature have suggested that a vascular dysfunction exists at some distance from visible lesions in patients with lupus erythematosus (Huff *et al.*, 1950) and also in those with psoriasis (Huff, 1955; Huff and Taylor, 1953; Milberg, 1947). However, plethysmographic studies of blood flow rates and pulse wave contours in digits free of visible lesions have not substantiated this, since the results were not significantly different from those observed in normal individuals of the same sex (Farber *et al.*, 1959; Moreci *et al.*, 1959; Herdenstam, 1959).

The question of vascular dysfunction in the apparently uninvolved skin of patients with psoriasis has been further investigated by thermometric means (copper-constantan thermocouples). Vasomotor responses to rapid changes of environmental temperature were studied in the skin of normal individuals, in psoriatic plaques and in the surrounding normal-appearing skin (Moreci *et al.*, 1960). At an environmental temperature of 25° C, psoriatic plaques were found to be warmer than the surrounding uninvolved skin, the difference being about one degree centigrade in most cases. Following a rapid decrease or increase in ambient temperature, the rate of temperature change of plaques was slower than that of the normal-appearing skin, the responses in the latter being similar to those obtained in normal subjects. These results indicated that resistance blood vessels in apparently uninvolved skin of patients with psoriasis responded to environmental temperature changes in a manner similar to that found in the skin of subjects without disease. On the other hand, these vessels in psoriatic plaques were chronically dilated and failed to respond normally to a rapid change in environmental temperature.

References

Allen, E. V., Barker, N. W., and Hines, E. A., Jr. (1946). "Peripheral Vascular Diseases." Saunders, Philadelphia, Pa.

Bailey, F. R. (1953). "Bailey's Textbook of Histology," 13th ed. Williams & Wilkins, Baltimore, Md.

Bean, W. B. (1958). "Vascular Spiders and Related Lesions of the Skin." Charles C Thomas, Springfield, Ill.

Beekman, Z. M. (1959). The value of skin thermometry in peripheral vascular disease of the lower extremities. *Angiology* **10**, 70.

Boyd, J. D. (1953). General survey of visceral vascular structures. *In* "Visceral Circulation," Ciba Foundation Symposium (G. E. W. Wolstenholme, ed.), pp. 3–20. Little, Brown, Boston, Mass.

Burch, G. E. (1954). "Digital Plethysmography," Modern Med. Monographs, Vol. 11. Grune & Stratton, New York.

Burton, A. C. (1939). The range and variability of the blood flow in the human fingers and the vasomotor regulation of body temperature. *Am. J. Physiol.* **127**, 437.

Burton, A. C. (1948). *In* "Methods in Medical Research" (V. R. Potter, ed.), pp. 146–166. Yearbook Publrs., Chicago, Ill.

Burton, A. C. (1954). *In* "Peripheral Circulation in Man," Ciba Foundation Symposium (G. E. W. Wolstenholme, J. S. Freeman, and J. Etherington, eds.), pp. 3–22. Little, Brown, Boston, Mass.

Burton, A. C. (1959). Physiclogy of cutaneous circulation, thermoregulatory functions. *In* "The Human Integument" Publ. No. 54 (S. Rothman, ed.), p. 77. Am. Assoc. Advance. Sci., Washington, D. C.

Clark, E. R. (1938). Arterio-venous anastomoses. *Physiol. Revs.* **18**, 229.

Clark, E. R., and Clark, E. L. (1934). The new formation of arterio-venous anastomoses in the rabbit's ear. *Am. J. Anat.* **55**, 407.

Clendenning, W. E., De Oreo, G. A., and Stoughton, R. B. (1959). Serotonin, its effect on normal and atopic skin. *A. M. A. Arch. Dermatol.* **79**, 503.

Crismon, J. M., Fox, R. H., Goldsmith, R., and MacPherson, R. K. (1958). Forearm blood flow after inunction of rubefacient substances. *J. Physiol.* (*London*) **145**, 47.

Davis, M. J., and Lawler, J. C. (1958). The capillary circulation of the skin. *A. M. A. Arch. Dermatol.* **77**, 690.

Davis, M. J., and Lorincz, A. L. (1957). An improved technique for capillary microscopy of the skin. *J. Invest. Dermatol.* **28**, 283.

Davis, M. J., Demis, D. J., and Lawler, J. C. (1960). The microcirculation of the skin. *J. Invest. Dermatol.* **34**, 31.

Demis, D. J., Davis, M. J., and Lawler, J. C. (1960). A study of the cutaneous effects of serotonin. *J. Invest. Dermatol.* **34**, 43.

Durward, A., and Rudall, K. M. (1949). Studies on hair growth in the rat. *J. Anat.* **83**, 325.

Ellis, R. A., Montagna, W., and Fanger, H. (1958). Histology and cytochemistry of human skin. XIV. The blood supply of the cutaneous glands. *J. Invest. Dermatol.* **30**, 137.

Evans, H. M. (1912). The development of the vascular system. *In* "Manual of Human Embryology" (F. Keibel and F. P. Mall, eds.), Vol. II, pp. 570–709. Lippincott, Philadelphia, Pa.

Farber, E. M. (1947). The Arteriole of the Skin in Essential Hypertension. M.S. Thesis, Mayo Clinic, Rochester, Minn.

Farber, E. M., and Barnes, V. R. (1955). Livedo reticularis. *Stanford Med. Bull.* **13**, 183.

Farber, E. M., and Schmidt, O. E. L. (1950). Hypertensive-ischemic leg ulcers. *Calif. Med.* **72**, 4.

Farber, E. M., Moreci, A. P., and Sage, R. D. (1959). Digital blood flow rates in lupus erythematosus. *A. M. A. Arch. Dermatol.* **79**, 340.

Finley, E. B. (1922). The development of the subcutaneous vascular plexus in the head of the human embryo. *Contribs. Embryol.* **7**, 157.

Forster, R. E., Ferris, B. G., and Day, R. (1946). The relationship between total heat exchange and blood flow in the hand at various ambient temperatures. *Am. J. Physiol.* **146**, 600.

Fulton, G. P., Farber, E. M., and Moreci, A. P. (1958). Motion picture: "The Responses of the Microcirculation to Rubefacients." Filmed and produced by the Department of Dermatology, Stanford Univ. School of Med., Stanford, California.

Fulton, G. P., Farber, E. M., and Moreci, A. P. (1959). The mechanism of action of rubefacients. *J. Invest. Dermatol.* **33**, 317.

Fulton, J. F. (1955). "Textbook of Physiology," 17th ed., p. 1049. Saunders, Philadelphia, Pa.

Gilje, O., Kierland, R., and Baldes, E. J. (1954). Capillary microscopy in the diagnosis of dermatologic diseases. *J. Invest. Dermatol.* **22**, 199.

Goldman, L. (1957). Clinical studies in microscopy of the skin at moderate magnification. *A. M. A. Arch. Dermatol.* **75**, 345.

Gomori, G. (1952). "Microscopic Histochemistry, Principles and Practice." Univ. of Chicago Press, Chicago, Ill.

Goodman, L. S., and Gilman, A. (1956). "The Pharmacological Basis of Therapeutics," 2nd ed. Macmillan, New York.

Grant, R. T. (1930). Observations on direct communications between arteries and veins in the rabbit's ear. *Heart* **15**, 281.

Grant, R. T., and Bland, E. F. (1931). Observations on arteriovenous anastomoses in human skin and in the bird's foot with special reference to the reaction to cold. *Heart* **15**, 385.

Grant, R. T., Bland, E. F., and Camp, P. D. (1932). Observations on the vessels and nerves of the rabbit's ear with special reference to the reaction to cold. *Heart* **16**, 69.

Greenfield, A. D. M., and Scarborough, H. (1949). An improved calorimeter for the hand. *Clin. Sci.* **8**, 211.

Gross, F., and Merz, E. (1948). The pharmacology of Trafuril with hyperemization of skin. *Schweiz. med. Wochschr.* **78**, 1.

Hale, A. R., and Burch, G. E. (1960). The arteriovenous anastomosis and blood vessels of the human finger. *Medicine* **39**, 191.

Hamm, A. W. (1953). "Histology," 2nd ed. Lippincott, Philadelphia, Pa.

Hensel, H., and Bender, F. (1956). Fortlaufende Bestimmung der Hautdurchblutung am Menschen mit einem elektrischen Warmleitmesser. *Pflügers Arch. ges. Physiol.* **263**, 603.

Herdenstam, C. G. (1959). Digital plethysmography in psoriasis. *Acta Dermato-Venereol.* **39**, 41.

Hertzman, A. B. (1937). Photoelectric plethysmography of fingers and toes in man. *Proc. Soc. Exptl. Biol. Med.* **37**, 529.

Hibbs, R. G., Burch, G. E., and Phillips, J. H. (1958). The fine structure of the small blood vessels of normal human dermis and subcutis. *Am. Heart J.* **56**, 662.

Horstmann, E. (1957). Die Haut. Die Milchdrüse. *In* "Handbuch der mikroskopischen Anatomie des Menschen" (W. von Möllendorff, ed.), Vol. III, Part 3, p. 198. Springer, Berlin.

Huff, S. E. (1955). Observations on peripheral circulation in various dermatoses. *A. M. A. Arch. Dermatol.* **71**, 575.

Huff, S. E., and Taylor, H. L. (1953). Observations on peripheral circulation in psoriasis. *Arch. Dermatol. Syphilol.* **68**, 385.

Huff, S. E., Taylor, H. L., and Keys, A. (1950). Observations on peripheral blood flow in chronic lupus erythematosus. *J. Invest. Dermatol.* **14**, 21.

Kernohan, J. W., Anderson, E. W., and Keith, N. M. (1929). Arterioles in cases of hypertension. *Arch. Internal Med.* **44**, 395.

Kety, S. S. (1948). Quantitative measurement of regional circulation by the clearance of radioactive sodium. *Am. J. Med. Sci.* **215**, 352.

Kety, S. S. (1949). Measurement of regional circulation by the local clearance of radioactive sodium. *Am. Heart J.* **38**, 321.

Lange, K., and Winer, D. (1949). Effect of certain hyperkinemics on blood flow through the skin (evaluation of counterirritants). *J. Invest. Dermatol.* **12**, 263.

Lewis, T. (1927). "The Blood Vessels of the Human Skin and Their Responses." Shaw & Sons, London.

Lippmann, H. I. (1952). Intra-arterial priscoline therapy for peripheral vascular disturbances. *Angiology* **3**, 69.

Lorincz, A. L., and Pearson, R. W. (1959). Studies on axon reflex vasodilatation and cholinergic urticaria. *A. M. A. Arch. Dermatol.* **32**, 429.

Lovett Doust, J. W. (1955). The capillary system in patients with psychiatric disorders. *A. M. A. Arch. Neurol. Psychiat.* **74**, 137.

Lovett Doust, J. W., and Salna, M. E. (1955). A stroboscopic method for estimating nailfold capillary blood flow in the skin of man. *J. Nervous Mental Disease* **121**, 511.

Lutz, B. R., Fulton, G. P., and Aker, R. B. (1950). The neuromotor mechanism of the small blood vessels in membranes of the frog. (Rana pipiens) and the hamster (Mesocricetus auratus) with reference to the normal and pathological conditions of blood flow. *Exptl. Med. Surg.* **8**, 258.

McKee, G. M., and Cipollaro, A. C. (1946). Vascular diseases. *In* "Skin Diseases in Children," Chapter VIII. Hoeber-Harper, New York.

Mendlowitz, M. (1950). Observations on the calorimetric method for measuring digital blood flow. *Angiology* **1**, 247.

Mendlowitz, M., and Touroff, A. S. W. (1952). The effect of sympathectomy for essential hypertension on the hallucal circulation. *Circulation* **5**, 577.

Mescon, H., Hurley, H. J., and Moretti, G. (1956). The anatomy and histochemistry of the arteriovenous anastomosis in human digital skin. *J. Invest. Dermatol.* **27**, 133.

Milberg, I. L. (1947). Response of the uninvolved skin of patients with psoriasis. *J. Invest. Dermatol.* **9**, 31.

Montagna, W., and Ellis, R. A. (1957). Histology and cytochemistry of the human skin. XIII. The blood supply of the hair follicle. *J. Natl. Cancer Inst.* **19**, 451.

Moreci, A. P., Farber, E. M., and Sage, R. D. (1959). Digital blood flow rates in psoriasis under normal conditions and in response to local mild ischemia. *J. Invest. Dermatol.* **33**, 113.

Moreci, A. P., Farber, E. M., and Heinz, M. (1960). Vasomotor responses in plaques of psoriasis and apparently uninvolved skin to rapid changes in environmental temperature. Presented to the *12th Autumn Meeting Am. Physiol. Soc., Stanford Univ. School of Med., California. The Physiologist* **3**, 117, 1960.

Moretti, G., Ellis, R. A., and Mescon, H. (1959). Vascular patterns in the skin of the face. *J. Invest. Dermatol.* **33**, 103.

Page, I. H. (1958). Serotonin (5-hydroxytryptamine); the last four years. *Physiol. Revs.* **38**, 277.

Patton, W. D. M. (1960). Pharmacology of vasodilator drugs with special reference to the skin. *In* "Progress in the Biological Sciences in Relation to Dermatology" (A. Rook, ed.), pp. 429–446. Cambridge Univ. Press, London and New York.

Petersen, H. (1935). "Histologie und Mikroskopische Anatomie des Menschen." Munich. (Publisher not available.)

Pillsbury, D. M., Shelley, W. B., and Kligman, A. M. (1956). The blood vessels and nerves of the skin. *In* "Dermatology," Chapter 10. Saunders, Philadelphia, Pa.

Pinkus, H. (1952). Examination of the epidermis by the strip method. II. Biometric data on regeneration of the human epidermis. *J. Invest. Dermatol.* **19**, 431.

Reid, G. (1952). Circulatory effects of 5-hydroxytryptamine. *J. Physiol. (London)* **118**, 435.

Roddie, I. C., Shepherd, J. T., and Whelan, R. F. (1955). The action of 5-hydroxytryptamine on the blood vessels of the human hand and forearm. *Brit. J. Pharmacol.* **10**, 445.

Roddie, I. C., Shepherd, J. T., and Whelan, R. F. (1957). The contribution of constrictor and dilator nerves to the skin vasodilation during body heating. *J. Physiol.* (*London*) **136**, 489.

Rondell, P. A., Keitzer, W. F., and Bohr, D. F. (1955). Distribution of flow through capillaries and arterio-venous anastomoses in the rabbit's ear. *Am. J. Physiol.* **183**, 523.

Roth, G. M. (1946). *In* "Peripheral Vascular Diseases" (E. V. Allen, N. W. Barker, and E. K. Hines, Jr.), p. 148. Saunders, Philadelphia, Pa.

Sapirstein, L. A. (1958). Regional blood flow by fractional distribution of indicators. *Am. J. Physiol.* **193**, 161.

Spalteholz, W. (1893). Die Vertheilung der Blutgefässe in der Haut. *Arch. Anat. u. Physiol.* **1.**

Spalteholz, W. (1927). Blutgefässe der Haut. *In* "Handbuch der Haut- und Geschlechtskrankheiten" Vol. I, Part 1, 379. Springer, Berlin.

von Euler, U. S. (1956). "Noradrenaline." Charles C Thomas, Springfield, Ill.

Winsor, T. (1953). Clinical plethysmography. Part I: An improved direct writing plethysmograph. *Angiology* **4**, 134.

Winsor, T. (1954). Skin temperatures in peripheral vascular disease. A description of the thermistor thermometer. *J. Am. Med. Assoc.* 1404.

Zweifach, B. W. (1957). General principle governing the behavior of the microcirculation. *Am. J. Med.* **23**, 684.

Chapter XVII. CIRCULATION IN SKELETAL MUSCLE

A. INTRODUCTION

By C. M. Pearson

Skeletal muscle has a rich blood supply which can be adjusted to a remarkable degree, to accommodate the variable requirements of this tissue during rest or active contraction. The vascular bed in muscles has so great a total capacity that potentially it can contain the entire blood volume. Hence, it is essential that vascular adjustments be constantly made in this part of the circulatory system in order to prevent complete sequestration of blood.

The vascular patterns within muscle vary widely between muscles, depending to some extent upon their functional requirements or upon the availability of collateral circulation from other arterial sources. The factors controlling the flow of blood through muscle are not completely delineated, although it is reasonably clear that the capillaries, as well as the arterioles and some veins, show active contraction and relaxation under the influence of exercise and certain neurohumoral agents.

B. EMBRYOLOGY

By K. Christensen

1. General Considerations

a. Early vasculogenesis significant to muscular vascularization: The literature concerning development of the circulation to and in skeletal muscles is meager. Only scattered references are given in the review by Evans (1912), and subsequent studies are limited.

As for other tissues, the primordia of blood vessels to skeletal muscles are represented by indifferent vascular networks which are present throughout somatic regions of the embryo. Out of these networks, definitive vessels are derived. By injection methods or by reconstructions from serial sections, portions of these networks have been demonstrated in various parts of the embryonic or early fetal body.

b. Primordial vessels (especially dorsal segmental vessels) for muscular circulation: In the head, Finley (1922) observed two vascular plexuses, the meningeal, which first appears in 4 mm embryos, and the subcutaneous, which appears about the 20 mm stage. From the subcutaneous plexus, the vessels of the skin and of the head musculature develop. The two cephalic plexuses unite in common with vessels of the neck, but on the sides and vault of the head they are completely separated. For the trunk musculature, the main arterial connections are provided by the muscular rami given off from the posterior and anterior divisions of the intersegmental arteries. The intersegmental arteries, branches of the dorsal aortae which early show more or less of a plexiform organization, begin to appear in the 6–somite embryo and by the 40–somite stage are practically complete; i.e., as far as the fourth sacral segment. The vascular plexuses of the extremities are also supplied by intersegmental vessels. Originally four or five intersegmental arteries are present in the vicinity of each limb bud, but a single vessel finally provides the axial connection. This is the seventh cervical intersegmental artery for the upper extremity and fifth lumbar intersegmental artery for the lower.

The developing posterior branches of intersegmental arteries at first establish a vascular plexus about the central nervous system, at a stage when neighboring premuscle masses are practically avascular. Later this vascular plexus, which is medial to the somite, grows laterally and then invades the somite.

Mall (1898) observed that as the intercostal arteries differentiate, the myotomes are among the first tissues to be supplied with blood vessels. Muscular buds are derived from these myotomes, and as they wander away to definitive positions, they have associated with them not only the branches of segmental nerves but also intersegmental blood vessels. More specific data concerning the development of the vascular system in relation to the differentiation of myotomes or of muscle masses developing in situ are practically unknown.

2. Specific Changes

a. Nature of vascular differentiation in the development of the circulation in the muscles: Detailed studies of the vascularity of differentiating or differentiated individual skeletal muscles have been few and sporadic, although it is known that vascular patterns may be quite variable. One of the earliest is the classical study of Spalteholz (1888). Subsequent ones include those of Campbell and Pennefather (1919) and Le Gros Clark and Blomfield (1945). Campbell and Pennefather classified muscles into three groups based on the number of sources of vascular rami and the extent to which anastomoses between these rami might be expected (for detailed discussion of this classification, see Section C.-1 of this chapter). Muscles representative of these three groups have been selected from the lower extremity because some of the most complete studies of the developing vascular system which provides for the developing muscles have been made by Senior (1919, 1924, 1925) in this limb, and details of muscular development in the same locality have been described in the comprehensive descriptions by Bardeen (1906–07).[1] According to the latter investigator, three typical muscles, the vastus externus (Class 1), the semitendinosus (Class 2), and the gracilis (Class 3) have definite primordia by the 14 mm stage, and differentiation is well advanced by the 20 mm stage. At the 14 mm stage, the semitendinosus has its two primordia. At the 20 mm stage, the muscle is as well developed, and the tendinous inscription between its two parts is as well marked, as in later life. The embryonic blood supply of these structures is provided by muscular rami of the femoral artery or of branches of the femoral and popliteal arteries and of corresponding veins developing in the vicinity. The femoral artery is a secondary formation of the vascular plexus in the hind limb bud, and its union with the popliteal artery is an even later development. Out of neighboring portions of the vascular plexus the veins originate. Senior states that the femoral arterial plexus participates in the blood supply of the leg at the 14 mm stage, and branches that spring from its side at these early stages (14–17.8 mm) provide the muscular branches; later on, these branches assume their differentiated status, some by joining the deep femoral artery. It is obvious from the separate studies of Bardeen and of Senior that muscular differentiation proceeds along with the provision for an adequate developing blood supply.

b. Clinical implications of developmental studies: Clinically, the type of differentiated vascular pattern of a muscle has practical importance. Some knowledge of this pattern would be requisite for proper treatment of muscular injuries or for surgical intervention, but roentgenological studies of individual muscles are few and incomplete, and this information is not readily available. Any muscular devascularization,

[1] See also review by Lewis (1910).

either complete or only partial, may set up the problem of anaerobic infection, as well as of muscle degeneration. If the main vascular supply of a muscle is interrupted, especially when a single main trunk is involved, there is always the possibility of vascularization by small vessels often present at the sites by muscular attachment. On the other hand, the extent of vascular anastomoses in a muscle supplied by several vessels is of special importance when one or more of these cease functioning. The end results may not be anticipated until the efficiency of the patent channels is known. Spalteholz pointed out another problem, namely, that even though the arterial anastomoses are abundant, if, at the same time, they are small, they might be incapable of providing an adequate circulation if the occlusion of some of the vessels occurred suddenly. Finally, the possibilities of anomalous vascular patterns are always present and they likewise must be taken into account (Le Gros Clark and Blomfield, 1945).

C. GROSS ANATOMY

By C. M. Pearson

1. Types of Vascular Patterns

In general, one or more arteries enter the muscle substance and then branch repeatedly before reaching capillary size. The actual intramuscular vascular patterns, however, have been given very little attention. From the few studies that have been made it is apparent that the arterial supply and the pattern of anastomoses vary widely from one muscle to another, rendering some very susceptible to extensive ischemic necrosis, whereas others are remarkably resistant to such a change. Information of this type is of great value to the surgeon, especially when amputation or the surgery of trauma is undertaken. Wollenberg (1905) was one of the first to investigate the vasculature of different human muscles, but his preparations were imperfect and only the roughest idea of intramuscular vascular patterns can be obtained from this study. Campbell and Pennefather (1919), under the stimulus of traumatic wound surgery and numerous cases of gas gangrene in World War I, published a brief paper on the arterial blood supply of various human muscles after injection of cadavers with a radiopaque material, followed by radiographic analysis of dissected whole muscle specimens. They demonstrated that there were numerous anastomotic communications in the arterial tree, which were similar to those found in the brain, heart or small intestine, although less prominent than the anastomotic loops in the mesentery or subcutaneous tissue. They concluded that the circulation to muscles could be separated into three general patterns:

Class I—Those in which the blood supply is derived from many different sources and the potential for collateral circulation is great. In this group are included the deltoid, pectoralis major and minor, the scapular muscles, triceps, adductor magnus, vastus internus and externus, soleus and gluteus medius and minimus.

Class II—Those in which there are two or three different sources of arterial supply but in which fewer anastomotic intercommunications exist. These include the gluteus maximus, rectus femoris, hamstring muscles, sartorius and tibialis anterior.

Class III—Those in which, for all practical purposes, the blood supply is derived from only one source and when it is occluded, almost the entire muscle becomes necrotic. These include the crureus, gracilis, long head of the biceps brachii, and both heads of the gastrocnemius. It is obvious that muscles of this type are much more liable to the development of necrosis and hence to the propagation of gas bacillus infection which thrives on devitalized tissue. Clinical experience has verified this point of view (for further discussion, see Section B.-2b of this chapter).

The only other study on the intramuscular vascular patterns in man is a short communication by Blomfield (1945). He confirmed the vulnerability of the gastrocnemius to necrosis and showed that its blood supply is derived entirely from a single main artery, which enters the proximal end of the muscle, divides repeatedly and supplies this structure throughout its entire length. The underlying soleus, in contrast, is supplied with blood from at least five separate main vessels which enter the muscle in succession from above downwards. Branches from these arteries form a detailed anastomotic complex.

A further example of the variability in vascular pattern of muscular tissue is found in the muscles of the anterior tibial compartment. Both the tibialis anterior and the extensor digitorum longus contain, in their entire length, an elaborate series of anastomotic loops which are formed by a sequence of entering vessels. In turn, smaller loops arise from the major ones, thus forming a complex series of anastomosing arcades. However, while the tibialis anterior derives its entire supply from the anterior tibial artery, the extensor digitorum longus receives in addition a plentiful supply from perforating arteries which penetrate the interosseous membrane from the posterior compartment of the leg. The extensor hallicus longus stands midway between these two extremes, for although it is supplied from both the anterior tibial artery and perforating arteries, the amount of blood from the latter source is relatively meager.

From his concise study, Blomfield was able to recognize the following five main types of intramuscular vascular patterns:

A pattern of longitudinal anastomoses arising from a number of separate arteries which enter the muscle throughout its length, e.g., soleus and peroneus longus.

A longitudinal pattern of arcades which originate from a single artery entering the upper end of the muscle, e.g., gastrocnemius.

A radiating pattern of collaterals which arise from a single main vessel that enters the midportion of the muscle, e.g., biceps brachii.

A pattern of anastomosing arcades formed by a series of vessels which enter the muscle throughout its length, e.g., tibialis anterior, extensor digitorum longus and long extensors of the leg.

An open rectangular pattern with sparse anastomotic connections, e.g., extensor hallicus longus.

2. Formation of Collateral Circulation

The establishment of collateral circulation in muscle following occlusion of a nutrient artery to muscle is remarkably variable and dependent upon the factors mentioned previously. In some muscles, such as the biceps brachii in man, Power (1945) has shown that interruption of a small supplying vessel causes focal necrosis of small portions of muscle. In other muscles, such as the soleus, collateral circulation is readily established after occlusion of one or several nutrient arteries. Experimentally LeGros Clark and Blomfield (1945) noted that, in the rabbit, ligation of one of the chief supplying vessels of the tibialis anterior results in devascularization and necrosis of a large proportion of the muscle. In contrast, the gracilis muscle is not readily devascularized because of the multiple sources for its blood supply.

D. MICROSCOPIC ANATOMY

By C. M. PEARSON

Upon entry of an artery into the muscle substance, it promptly branches freely and forms many anastomoses which comprise the primary arterial network of the muscle. From the meshes of this net arise smaller arteries which also freely anastomose to create the secondary network. From the threads of the latter the arterioles branch off and usually cross at right angles to the muscle fibers (Fig. 110). They are

FIG. 110. Diagram of typical distribution pattern of the small arteries (A), capillaries, and veins (B) in striated muscle. (Adapted from Spalteholz, 1888.)

spaced at fairy regular intervals of about 1 mm. The capillaries, in turn, branch from the arterioles and run parallel to, and between, the muscle fibers. There are numerous capillary anastomoses which form long narrow meshes about the fibers. The capillaries empty into venules which then unite into larger veins, and these, in general, follow the course of the arterioles and arteries. Krogh (1929) points out the interesting fact that all veins down to the smallest branches are supplied with valves which allow blood to flow only in the direction of the heart. The great practical significance of this device is that when muscle contracts, capillary and venous blood is driven forward toward the heart, the muscle thus acting as an effective pump which aids venous return. During the relaxation phase, the empty capillaries are filled only from the arteriolar end.

1. Variability in Size of Capillary Bed

The capillaries, when viewed in transverse section, are distributed very regularly between the muscle fibers. The number of visible ones varies greatly, depending upon (a) the animal species, (b) the muscle examined and (c) the state of rest or activity immediately prior to examination.

a. Species variations: Krogh (1929) has pointed out that there are about six times more capillaries per mm^2 in dog muscle than in that from a cold-blooded species (frog or cod), in which the metabolic rate is much lower. It seems likely that the density of capillaries roughly parallels the metabolic rate of the animal species and is greatest in small warmblooded animals.

Lee (1958) noted that in the rabbit the red muscle, which contracts more slowly but has greater endurance, contains many tortuous capillaries, and sac-like dilatations are commonly found at the branching points of arterioles into capillaries. These dilatations may serve as emergency reservoirs of blood. In white muscle, on the other hand, the arteriolar branching is regular and the capillaries follow a straight course. Complete capillary patterns in muscle have not been clearly worked out in man. So far as is known, the arterioles, venules and capillaries in muscle are histologically identical to those in other tissues and organs of the body.

Although no systematic measurement has been made of the spacing of capillaries in various muscles of a single species, Krogh has estimated the distance from the centers of capillaries in a few fresh guinea pig muscles and found the following: abdominal wall, superficial layer, 83 μ; deeper layer, 29.6 μ; pretracheal muscle, 43 μ; and diaphragm, 17 μ.

b. Capillary bed at rest and with exercise: The number of capillaries that are patent in resting muscle are few, in contrast to the change that occurs after exercise or infusion of a vasodilating substance, such as histamine. For example, when an isotonic solution of India ink particles is injected into a narcotized guinea pig, the leg muscles will become only slightly darkened. However, after several tetanic contractions, the color will now turn a deep gray to black color. Shown in Fig. 111 is the capillary distribution before and after activity. In some muscles a twenty- to fortyfold increase in number of opened capillaries has been observed.

The capillary bed in muscle is potentially very great. Krogh has calculated that there is a maximum of 1,350 capillaries per mm^2 in the transverse section of horse gastrocnemius. He has further estimated that in a large man, whose muscle mass weighs 50 kg, the total length of

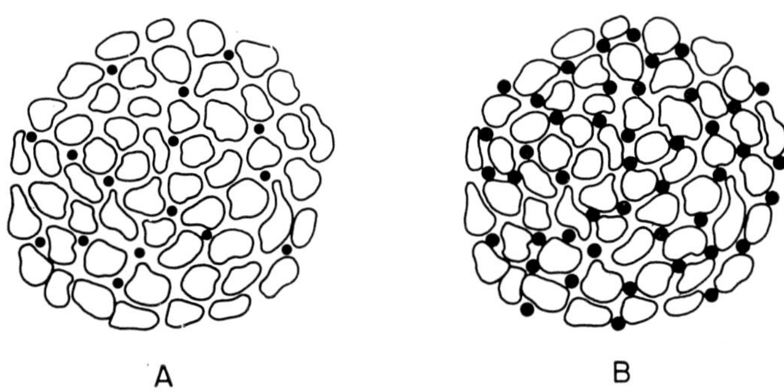

FIG. 111. Transverse section of guinea pig muscle at rest (A), and immediately after exercise (B). Note increase in both number and diameter of capillaries in muscle after work. (Adapted from data of Krogh, 1929.)

these vessels would be 62,000 miles, or 2.5 times around the earth. Furthermore, they would have a total surface area of 6,300 square meters.

c. *Effect of training on capillary bed:* In the guinea pig Petrén *et al.* (1937) have shown that training by repeated exercise over a period of weeks raised the number of capillaries in the gastrocnemius by 40 per cent. Moreover, the increase in these vessels was proportional to the severity of training. Untrained muscle (masseter) in the same animal showed no change in vascular supply.

2. Lymphatic System

Although there is an abundant supply of lymphatic vessels in connective tissue sheaths and tendons, as far as is known there are few if any within the skeletal muscle proper. The vessels that are present in the connective tissue planes collect and convey lymph from the local areas. The manner in which the lymph reaches these vessels is still a matter of debate among anatomists. Clearly the fluid must travel a considerable distance in some areas, and it has been suggested that it moves through intercommunicating spaces along facial planes between muscle fibers. Aagard (1912) has demonstrated intrafasicular lymphatics in muscle by injection techniques, and Shdanow (1936) has elaborated upon this study. (For general discussion of distribution of lymphatic capillaries, see Section C.-2, Chapter XXIII.)

E. PHYSIOLOGY

By L. A. SAPIRSTEIN

1. EFFECTS OF VARIOUS FACTORS ON BLOOD FLOW

The blood flow through resting muscle, estimated by a variety of methods, is of the order of .01–.03 ml/gm/min. In exercise (single stimuli, one/sec) the blood flow of the rat gastrocnemius increases between five- and twentyfold (Sapirstein, unpublished observations). This extraordinary range of local circulation exceeds that observed in any other organ excepting portions of the skin. Since skin blood flow may occur in part through arteriovenous shunts, and since the latter structures are rare, if indeed they exist at all, in skeletal muscle, the range indicated for muscle is the largest so far observed through a true capillary network. The changes in muscle blood flow brought about by exercise are due, in part, to local influences and, in part, to central effects.

a. Local physical effects: Unlike the blood vessels of the skin, those of muscles do not appear to be dilated by heat (Edholm *et al.*, 1956). The effects of micromechanical stimulation have not been studied exhaustively. It has been well established (see review by Barcroft and Swan, 1953) that during sustained muscular contraction, blood flow to muscle increases less than it does during intermittent contraction. This is attributable to the greater external pressure exerted on the vessels during muscular contraction. The resulting effect is to counteract, in part, the vasodilation produced by other factors.

b. Local chemical effects: Little information is available concerning alterations of muscle blood flow produced by local changes in oxygen and carbon dioxide tension. The vasodilator effect of lactic has been studied extensively (see review by Lundholm, 1956), but the results are equivocal. It is teleologically attractive to assume that the products of muscle activity (increased carbon dioxide tension and diminished oxygen tension and increased lactic acid) act to increase muscle blood flow. This subject deserves further investigation.

c. Central chemical effects: Epinephrine and norepinephrine have been most intensively studied with regard to their influence upon muscle circulation, a subject which is discussed in detail in Section F.-1a, in this chapter. All that can be stated at present, however, is that the question of local and central chemical control of skeletal muscular blood flow requires further investigation.

d. Central nervous effects: The existence of sympathetic vasoconstrictors to forearm muscle blood vessels is well established (Barcroft and Swan, 1953). The presence of sympathetic vasodilators in man has also been demonstrated in studies made in posthemorrhagic fainting (Barcroft *et al.*, 1944). In these circumstances, the forearm blood flow increases while arterial pressure falls, signifying vasodilation. This effect is greatly reduced by nerve block, indicating the existence of sympathetic vasodilator fibers. Such a response has no known animal counterpart.

Roddie *et al.* (1958) noted three- to fourfold increases in forearm blood flow associated with elevated intrathoracic pressure changes resulting from obstructed breathing. They attributed this effect to a reflex due to stimulation of receptors in the low pressure area of the circulation within the thorax.

Centers in the cerebral cortex potentially capable of modifying muscular blood flow through reflex mechanisms have been demonstrated by Green and Hoff (1937). Stimulation of the motor and premotor areas in cats and monkeys resulted in increased limb volume, an effect which persisted despite curarization. More recently Abrahams and Hilton (1958) have demonstrated muscular vasodilation after stimulation of an area of the brain stem between the anterior commisure and the supraoptic nucleus.

Another point requiring further study is concerned with the possibility of a dual circulation through the minute vessels of muscle. Increasing muscle blood flow by body warming (decreasing sympathetic vasoconstriction) does not reduce the hyperemia of exercise (Barcroft and Swan, 1953). Neither does sympathetic block increase the "tissue clearance" of Na^{24} (Rapaport *et al.*, 1952). These findings suggest that two types of blood vessels are present in muscle. Local metabolic requirements may dictate the behavior of one group, while central stimuli may control the other via the sympathetic outflow. Barcroft and Swan suggest that the central stimuli act primarily on anastomotic vessels (such as the "thoroughfare channels" of Chambers and Zweifach, 1946), while metabolic requirements influence the flow of blood through the true capillary bed which subserves the nutrition of muscle. It is, however, by no means certain that "thoroughfare channels" are a regular part of the microcirculation. Whether or not they occur in most muscles will have to be determined by further studies. (For a discussion of "thoroughfare channels," see Section C.-1c, Chapter V.)

Venous return is facilitated through muscular activity. The "muscle pump" (Smirk, 1936) reduces the venous pressure in the dorsum of the foot during the exercise of rhythmic raising and lowering of the heels. This mechanism can return blood to the heart even against pressures of 90 mm Hg (Barcroft and Swan, 1953). Thus, the muscular exercise, which makes demands on the circulation for increased blood flow to the active muscles, satisfies these demands, in part, by making available to the heart more blood for delivery.

F. PHARMACOLOGY

By K. I. Melville and H. E. Shister

1. Adrenergic Drugs

a. Epinephrine and norepinephrine: Following intra-arterial injections in cats and dogs, small doses of epinephrine (0.1 μg) produce vasodilatation in the limbs; with larger amounts (1–10 μg), there is constriction followed by dilatation. The same is true of man: 0.002–0.05 μg increases blood flow in the forearm and calf, while 1–2 μg reduces it, without change in arterial blood pressure. On the other hand, norepinephrine, in similar dose ranges, initially causes only vasoconstriction (cats, dogs, man), without changes in blood pressure, and then secondary vasodilatation, which is also less pronounced than with epinephrine (Barcroft and Swan, 1953). However, Youmans *et al.* (1955) have reported similar constrictor effects with both amines.

Following intravenous infusions in cats and dogs, epinephrine (0.1

μg) produces a variable response, depending on the existing state of the vascular bed. If the latter is constricted, dilatation occurs, probably due to a reflex response; if the vessels are dilated, constriction takes place. Intravenous infusion of norepinephrine in cats produces a slight initial increase, followed by a diminution in flow. In dogs, the same responses are obtained as with epinephrine.

In man, intravenous infusion of epinephrine (10 μg/min) results in a marked increase in flow in the forearm (180–200% of the control value); this is soon followed by a return to the control level and then a secondary slow dilatation. A 10 μg dose of norepinephrine causes an elevation of blood pressure (systolic and diastolic), a slowing of the heart, and an initial increase, followed by a decrease, in muscle blood flow in the forearm and calf.

b. Other adrenergic compounds: Neosynephrine has been shown to produce vasoconstriction in the hind limb of the anesthetized dog without secondary dilatation (Johnson *et al.*, 1953). Under similar conditions, isoproterenol (Isuprel), 1–10 μg, produces marked vasodilatation (Green *et al.*, 1954).

2. Blocking Agents

a. Adrenergic blocking agents: Phenoxybenzamine (Dibenzyline), azapetine (Ilidar), tolazoline, and phentolamine show more or less the same effects. Thus, epinephrine constriction in dogs, produced by 1 μg administered intravenously, is easily diminished; with increasing doses of the blocking agents a "reversal" effect is observed, i.e., dilatation. With still higher amounts, this response can also be blocked. Norepinephrine requires much larger doses of the blocking agent to abolish its constrictor effect, and only a very mild, if any, dilator effect ensues. Isuprel reacts to adrenergic blockade in the same way as epinephrine. Finally, the ergot derivatives can block the constrictor but not the dilator phase of epinephrine action.

In dogs adrenergic blockade can abolish the vasoconstriction in the skeletal muscle of the hind limb, elicited by stimulation of the lumbar sympathetic chain (and similar to that obtained by the administration of epinephrine or norepinephrine). At the same time, a dilating effect is unmasked, which can be readily abolished by atropine and augmented by eserine.

Folkow and Uvnäs (1948), in connection with studies of adrenaline blockade, showed that vasodilator fibers are distributed only to the skeletal muscle; furthermore, they maintained that there is a cholinergic substance which is probably the mediator of the vasodilatation induced by adrenergic nerve stimulation. It has also been postulated that the adrenergic-mediated vasoconstriction is under the control of the vasomotor centers in the medulla, while the hypothalamus and cortex control the cholinergic dilatation. (For review of this problem, see Folkow, 1955; also, Section A.-2, Chapter IV.)

b. Ganglionic blocking agents: These substances cause vasodilatation and increased blood flow in the human arms and legs. Since the vascular change takes place in the presence of a fall in blood pressure, it is presumably due to a decreased resistance in the limb vessels. The mode of action is not a direct one on the blood vessels but is consequent to a block at the ganglia, thus interfering with the postganglionic sympathetic discharge (constrictor effect). Hence, in a sympathectomized limb the ganglion blocking agents do not produce vasodilatation.

3. Cholinergic Drugs

Acetylcholine (Ach) and related compounds produce vasodilatation in the dog; in man, injection of 0.25 μg of Ach into the brachial artery leads to increased forearm blood flow (Duff *et al.*, 1953).

4. Histamine

The response to intra-arterial histamine is rather similar to that of Ach. In man, if epinephrine is added to the histamine, its action is antagonized, while in experimental animals, after a preliminary injection of histamine, the epinephrine response is increased. (For complete summary of present concepts of skeletal muscle circulation and its pharmacologic responses, see Green and Kepchar, 1959.)

G. PATHOLOGY

By C. M. Pearson

1. Intermittent Claudication

An intermittent deficiency of the blood supply to muscle gives rise to the characteristic symptom of *claudication*, which is not a disease but a manifestation of an inadequate supply of arterial blood to contracting muscle. Claudication occurs most commonly in the leg muscles in obstructive arterial disease and is characterized by (1) the absence of pain or discomfort in the limb during rest; (2) the onset of pain, tension or weakness after walking is begun; (3) intensification of the symptoms until walking is impossible; and (4) the disappearance of symptoms after a period of rest. Pickering and Wayne (1934) have pointed out that claudication is the skeletal muscle counterpart of angina pectoris. They have also noted that anemia may initiate or greatly accentuate it in an individual with only mild vascular disease, and furthermore, that, although a rare occurrence, very severe anemia may induce claudication in exercising muscles with apparently normal vasculature.

The cause of pain in claudication or muscular ischemia has been a matter for considerable debate and is incompletely settled at the present time. Lewis (1932) has stated that it is not primarily due to an oxygen lack but rather is the result of an accumulation of "stable chemical agents" which amass in the interstitial tissues and stimulate sensory nerve endings. The nature of these chemical substances has not been determined.

2. Vascular Disorders

a. Atherosclerosis: In severe peripheral arteriosclerosis obliterans the cardinal symptom is claudication upon exercise. Nevertheless, muscular atrophy secondary to vascular disease or disuse is much less common than are atrophic changes in the adjacent skin and subcutaneous tissues. In fact, it is not unusual to see advanced cutaneous wasting and a well-preserved musculature.

Atherosclerosis seldom extends to the intramuscular arteries, and when it does, only the larger ones are involved. Wartman (1933) studied the distribution of atherosclerosis in different organs, and he rarely observed atheromatous deposits within arteries of skeletal muscles. This is explained by the fact that most of the intramuscular arteries are small and it is, of course, a common observation that atheromatous plaques are rarely seen in blood vessels less than 250 μ in diameter. Hence, the use of muscle biopsy to aid in the diagnosis of atherosclerosis is to be discouraged. In fact, in a recent review of 930 muscle biopsy specimens taken at autopsy from 110 unselected cases, chiefly elderly men (Pearson, 1959), no significant vascular changes were observed except for some thickening of the walls of arteries or arterioles in a few muscles. (For general discussion of histologic changes in atherosclerosis, see (1) Atherosclerosis, Section F., Chapter XIX.)

b. Generalized arteriosclerosis and hypertension: Both of these conditions may affect the intramuscular arteries, especially the larger and medium-sized ones. However, necrotizing arteriosclerosis seldom occurs even in the most malignant phases of hypertension and nephrosclerosis. The most common change in hypertension, as pointed out by Kernohan *et al.* (1929), is a thickening of the arteriolar walls, due chiefly to hypertrophy and hyperplasia of the muscularis coat and, to a lesser extent, of the intima and adventitia. (For general discussion of changes in muscular arteries in hypertension, see (1) Arterial Hypertension, Section C.-2b, Chapter XX.)

c. Arterial occlusion: Muscles undergo gangrene following complete arterial occlusions, the type of accompanying change in the musculature depending upon (a) the rapidity with which the obstruction occurs, (b) the status of the collateral circulation, and (c) the venous drainage, whether patent or occluded. It has already been pointed out (Section C.-1, above) that the severity of the ischemia is highly variable from one muscle to another and is related to the individual anatomic features of the vascular supply to the muscle. The various types of ischemic necrosis that muscle may undergo are described in detail elsewhere (Adams *et al.*, 1954).

d. "Collagen disorders": In most of the systemic connective tissue ("collagen") diseases there is involvement of the blood vessels, often extending to the intramuscular vessels. In polyarteritis the pan-necrosis of the medium-sized arteries, with accompanying inflammatory infiltrate and often thrombosis, is similar to the changes observed elsewhere in the body. Frequently small hemorrhages are noted in the muscle, but foci of infarct necrosis are rare. Although muscle biopsy is commonly utilized in an attempt to diagnose a suspected case, Maxeiner *et al.* (1952) have reported that the chances of finding an arteritic lesion in muscle are directly related to the size and multiplicity of the biopsy. They observed that in 26 proven cases of polyarteritis a random muscle biopsy gave a positive result in only 30 per cent. From 136 biopsies, made in 106 patients in whom the disease was suspected clinically, a positive biopsy for polyarteritis occurred in only 13 cases. Hence, although the presence of typical arteritis provides an indisputable diagnosis, a randomly-taken muscle biopsy provides no guarantee that the lesion will be found, even in a case that is ultimately proved to be polyarteritis. (For general discussion of histologic changes in polyarteritis nodosa, see (3) So-called Collagen or Systemic Connective Tissue Diseases, Section C.-3, Chapter XXI.)

In all of the other "collagen" diseases a vasculitis or perivasculitis may occasionally be noted, although in systemic lupus erythematosus, it is much more common. The lesion consists of a loose collection of mononuclear cells, resembling lymphocytes, which surround the wall of a small intramuscular artery, arteriole, venule or vein. There is no evidence that the vessel wall is involved in the process. For the present, perivasculitis must be considered to be a nonspecific pathologic reaction which exists with much greater frequency, but not exclusively, in the "collagen" diseases. (For general discussion of histologic changes in systemic lupus erythematosus, see (3) So-called Collagen or Systemic Connective Diseases, Section B.-3, Chapter XXI.)

There is no known disorder which affects the intramuscular capillaries or lymphatics. As mentioned previously, the smaller arteries, arterioles and veins show no unique pathologic changes, but they may be involved incidentally in the course of a systemic disease of the blood vessels.

References

Aagard, O. C. (1912). Über die Lymphgefässe der Zunge, des quergestreiften Muskelgewebes und der Speicheldrüsen des Menschen. *Anat. Hefte* **47,** 493.

Abrahams, V. C., and Hilton, S. M. (1958). Active muscle vasodilation and its relation to the "flight" and "fight" reaction in the conscious animal. *J. Physiol.* **140,** 16.

Adams, R. D., Denny-Brown, D., and Pearson, C. M. (1954). "Diseases of Muscle. A Study in Pathology." Hoeber-Harper, New York.

Bardeen, C. R. (1906–1907). Development and variation of the nerves and the musculature of the inferior extremity and the neighboring regions of the trunk in man. *Am. J. Anat.* **6,** 259.

Barcroft, H., and Swan, H. J. C. (1953). "Sympathetic Control of Human Blood Vessels." Edward Arnold, London.

Barcroft, H., Edholm, O. G., McMichael, J., and Sharpey-Schafer, E. P. (1944). Post hemorrhagic fainting: Studies on cardiac output and forearm blood flow. *Lancet* **i,** 489.

Blomfield, L. B. (1945). Intramuscular vascular patterns in man. *Proc. Roy. Soc. Med.* **38,** 617.

Campbell, J., and Pennefather, C. M. (1919). The blood supply of muscles; with special reference to war surgery. *Lancet* **i**, 294.

Chambers, R., and Zweifach, B. W. (1946). Functional activity of the blood capillary bed with special reference to visceral tissue. *Ann. N. Y. Acad. Sci.* **46**, 683.

Duff, F., Greenfield, A. D. M., Shepherd, J. T., and Thompson, I. D. (1953). A quantitative study of the response to acetylcholine and histamine of blood vessels of human hand and forearm. *J. Physiol. (London)* **120**, 160.

Edholm, O. G., Fox, R. H., and Macpherson, R. K. (1956). The effect of body heating on the circulation in skin and muscle. *J. Physiol. (London)* **134**, 612.

Evans, H. M. (1912). The development of the vascular system. *In* "Manual of Human Embryology" (F. Keibel and F. P. Mall, eds.), Vol. II, p. 570. Lippincott, Philadelphia, Pennsylvania.

Finley, E. B. (1922). The development of the subcutaneous vascular plexus in the head of the human embryo. *Contribs. Embryol.* **7**, 157.

Folkow, B. (1955). Nervous control of blood vessels. *Physiol. Revs.* **35**, 629.

Folkow, B., and Uvnäs, B. (1948). Distribution and functional significance of sympathetic vasodilators to hind limbs of cat. *Acta Physiol. Scand.* **15**, 327.

Green, H. D., and Hoff, E. C. (1937). Effects of faradic stimulation of the cerebral cortex on limb and renal volumes in the cat and monkey. *Am. J. Physiol.* **118**, 641.

Green, H. D., and Kepchar, J. H. (1959). Control of peripheral resistance in major systemic vascular beds. *Physiol. Revs.* **39**, 617.

Green, H. D., Shearin, W. T., Jr., Jackson, T. W., Keach, L. M., and Denison, A. B., Jr. (1954). Isopropylnorepinephrine blockade of epinephrine reversal. *Am. J. Physiol.* **179**, 287.

Johnson, H. D., Green, H. D., and Lanier, J. T. (1953). Comparison of adrenergic blocking action of Ilidar (Ro 2-3248), Regitine (C-7337) and Priscoline in innervated saphenous arterial bed and femoral arterial bed of anesthetized dog. *J. Pharmacol. Exptl. Therap.* **108**, 144.

Kernohan, J. W., Anderson, E. W., and Keith, N. M. (1929). The arterioles in cases of hypertension. *A. M. A. Arch. Internal Med.* **44**, 395.

Krogh, A. (1929). "The Anatomy and Physiology of Capillaries." Yale Univ. Press, New Haven, Connecticut.

Lee, J. C. (1958). Vascular patterns in the red and white muscles of the rabbit. *Anat. Record* **132**, 597.

Le Gros Clark, W. E., and Blomfield, L. B. (1945). The efficiency of intramuscular anastomoses, with observations on the regeneration of devascularized muscle. *J. Anat.* **79**, 15.

Lewis, T. (1932). Pain in muscular ischemia *A. M. A. Arch. Internal Med.* **49**, 713.

Lewis, W. H. (1910). The development of the muscular system. *In* "Manual of Human Embryology" (F. Keibel and F. P. Mall, eds.), Vol. I, p. 454. Lippincott, Philadelphia, Pennsylvania.

Lundholm, L. (1956). The mechanism of the vasodilator effect of adrenaline. I. Effect on skeletal muscle vessels. *Acta Physiol. Scand.* **39**, Suppl. 133.

Mall, F. P. (1898). Development of the ventral abdominal wall in man. *J. Morphol.* **14**, 347.

Maxeiner, S. R., McDonald, J. R., and Kirklin, J. W. (1952). Muscle biopsy in the diagnosis of periarteritis nodosa; An evaluation. *Surg. Clin. North Am.* **32**, 1225.

Pearson, C. M. (1959). Incidence and type of pathologic alterations observed in muscle in a routine autopsy survey. *Neurology* **9**, 757.

Petrén, T., Sjöstrand, T., and Sylvén, B. (1937). Der Einfluss des Trainings auf die Häufigkeit der Capillaren in Herz- und Skeletmuskulatur. *Arbeitsphysiol.* **9**, 376.

Pickering, G. W., and Wayne, E. J. (1934). Observations on angina pectoris and intermittent claudication in anemia. *Clin. Sci.* **1**, 305.

Power, R. W. (1945). Gas gangrene. With special reference to vascularization of muscles. *Brit. Med. J.* **i**, 656.

Rapaport, S. I., Saul, A., Hyman, A., and Morton, M. E. (1952). Tissue clearance as a measure of nutritive blood flow and the effect of lumbar sympathectomy upon such measures in calf muscle. *Circulation* **5**, 594.

Roddie, I. C., Shepherd, J. T., and Whelan, R. F. (1958). Reflex changes in human skeletal muscular blood flow associated with intrathoracic pressure changes. *Circulation Research* **6**, 232.

Senior, H. D. (1919). The development of the arteries of the human lower extremity. *Am. J. Anat.* **25**, 55.

Senior, H. D. (1924). The description of the larger direct or indirect muscular branches of the human femoral artery; a morphogenetic study. *Am. J. Anat.* **33**, 243.

Senior, H. D. (1925). An interpretation of the recorded arterial anomalies of the human pelvis and thigh. *Am. J. Anat.* **36**, 1.

Shdanow, D. A. (1936). Die Kollaterollymphwege der Brusthohle des Menschen. *Anat. Anz.* **82**, 417.

Spalteholz, W. (1888). Die Vertheilung der Blutgefässe im Muskel. *Abhandl. math.-phys. Kl. sächs. Ges. Wiss.* **14**, 509.

Smirk, F. H. (1936). Observations on the causes of oedema in congestive heart failure. *Clin. Sci.* **2**, 317.

Wartman, W. B. (1933). The incidence and severity of arteriosclerosis in organs from 500 autopsies. *Am. J. Med. Sci.* **186**, 27.

Wollenberg, G. A. (1905). Die Arterienversargung von Muskeln und Sehnen. *Z. Orthop. Chir.* **14**, 312.

Youmans, P. L., Green, H. D., and Denison, A. B., Jr. (1955). Nature of vasodilator and vasoconstrictor receptors in skeletal muscle of the dog. *Circulation Research* **3**, 171.

Chapter XVIII. CIRCULATION OF BONE

Bone is a connective tissue unique because of its mineralization. As a result, its vascular system does not lend itself easily to observation.

Studies of blood supply to bone were made by many of the early anatomists, Langer's presentation (1876) being considered the classic anatomic description of this vascular bed. Other investigations have been reported by Lexer *et al.* (1904) and Lexer (1904), Doan (1922a), Drinker *et al.* (1922) and Johnson (1927). Renewed interest in the elucidation of this problem has been apparent during the past decade with the introduction of new techniques by Barclay (1951) and Trueta and Harrison (1953).

A. EMBRYOLOGY

By P. J. Kelly and L. F. A. Peterson

Two types of formation of bone are generally recognized, which differ in origin but not in ultimate mechanism. Intramembranous bone formation occurs in a condensation in the mesenchyme, while endochondral bone formation is laid down on and within a pre-existing cartilage model (Maximow and Bloom, 1957).

a. Intramembranous bone: At about the mid-portion of the embryonic period, intramembranous bone is formed by specialized mesenchymal cells, the "osteoblasts," that develop and produce a specialized fibrillar matrix called "osteoid," which is capable of calcification. Small trabeculae are formed which calcify in the original mesenchymal condensation and become bone. As the intramembranous bone grows, a layer of osteoblasts becomes evident on the surface, and the open spaces between the trabeculae are filled with an invading tissue derived from the mesenchyme, which is to form the primary marrow. The angioblasts present form endothelial tubes which become incorporated into the vascular system. Those in the primary marrow develop rapidly into capillaries and wider endothelial-lined cavities which appear as sinusoids. These are supplied by the periosteal vessels, some of which enlarge on further development to carry the major portion of the blood supply to the cancellous portion of this type of bone.

b. Cartilaginous bone: The vessels which form in cartilagenous bone appear in a similar manner. The cartilage model of a bone develops as a condensation in the mesenchyme and is transformed into tissue recognizable as cartilage. A cuff of thin bone is laid down around the midportion of the shaft of the cartilage model, either before or after calcification has occurred in the cartilage. Invasive tissue, developing from the inner layers of the periosteum, passes through small pores in the primary bony cuff; this tissue, consisting of marrow elements and angioblasts, then proliferates and branches to become capillaries and sinusoids. The vessels grow rapidly, with the remainder of the cartilage in the cuff being destroyed and transformed into the primary marrow cavity. The epiphyseal cartilage region may now be identified. Vessels then grow into the area which is later destined to become the epiphysis. Streeter (1949) has documented these changes for the humerus of the human.

B. ANATOMY

By P. J. Kelly and L. F. A. Peterson;[1] D. C. Pease and L. Zamboni[2]

1. Techniques Used in Study of the Blood Supply

a. Injection techniques: The Spalteholz technique (1911) for clearing bone is a reliable method. It depends on the injection of an opaque media into the vascular system and subsequent placement of the bone in a material with a refractive index similar to collagen, to render it transparent. Barclay (1951) and Trueta and Harrison (1953) utilized microangiography which depends on injection of a radiopaque medium into the vascular system. Three dimensional studies may be carried out by means of stereoscopic microangiograms (Bellman, 1953; Peterson *et al.*, 1959).

[1] Authors of all sections (except 4c) in "B. ANATOMY."
[2] Authors of Section 4c.

b. Microscopy: Standard microscopy techniques are necessary to confirm the types of vessels found by injection techniques. It is desirable to use both thick and thin sections which have previously been studied by the above methods (Trueta and Harrison, 1953; Kelly *et al.*, 1959a).

c. Suppression of a portion of the extraosseous blood supply: This procedure depends on the destruction of a particular portion of the extraosseous blood supply to bone. Although crude, if correlated with the previously mentioned techniques, it provides additional information (Kistler, 1934; Foster *et al.*, 1951; Trueta and Cavadias, 1955; Johnson, 1927; Brookes, 1957; Larson, 1959).

d. Radioisotope studies: This method depends on clearance of radioactive isotopes after they have been deposited in a particular region of bone and is promising since nondestructive determinations can be made. Radioactive phosphorus, iodine, and sodium have been used with this technique (Tucker, 1950; Boyd *et al.*, 1955; Laing and Ferguson, 1958, 1959; Pauporte *et al.*, 1958).

e. Determinations of intramedullary pressure: Stein *et al.* (1958) described a method whereby intramedullary pressure can be measured and the effect of various agents can be determined by this parameter.

f. Vital microscopy: Brånemark (1959) in his excellent description of vital microscopy of the bone marrow of rabbit's fibulas related the blood flow in the marrow cavity to flow in the haversian canals. Kinosita *et al.* (1956) also described circulatory dynamics.

2. Vascular Anatomy of Cancellous Bone

The talus is an excellent example of a bone that is entirely cancellous. Arteries, entering cancellous bone from all sides, branch until they become oriented to each marrow space formed by the individual trabeculae. The smaller subdivisions thus formed are arterioles with definite muscular walls. Finally, thin-walled capillaries or sinuses are seen about the individual trabeculae or abutting against them, with nutritive exchange apparently occurring at this level. It is difficult to tell where the sinuses or capillaries end and venous drainage begins. Intraosseous veins are extremely thin-walled structures, which enlarge and become spacially arranged, generally to parallel small arteries, and which finally exit at the points where the arterial supply enters cancellous bone. Immediately on gaining the exterior of the bone, the vein has a discernible but thin wall and is easily differentiated from its accompanying artery (Kelly *et al.*, 1959b; Haliburton *et al.*, 1958).

The epiphyseal and subchondral circulation has been described by Trueta and Harrison (1953) for the femoral head of the human. Peterson *et al.* (1957) and Morgan (1959) noted the same picture in the femur of the dog and the tibia of the rabbit. Arterial arcades form tiers of smaller arcades on the larger ones as the vessels come closer to the

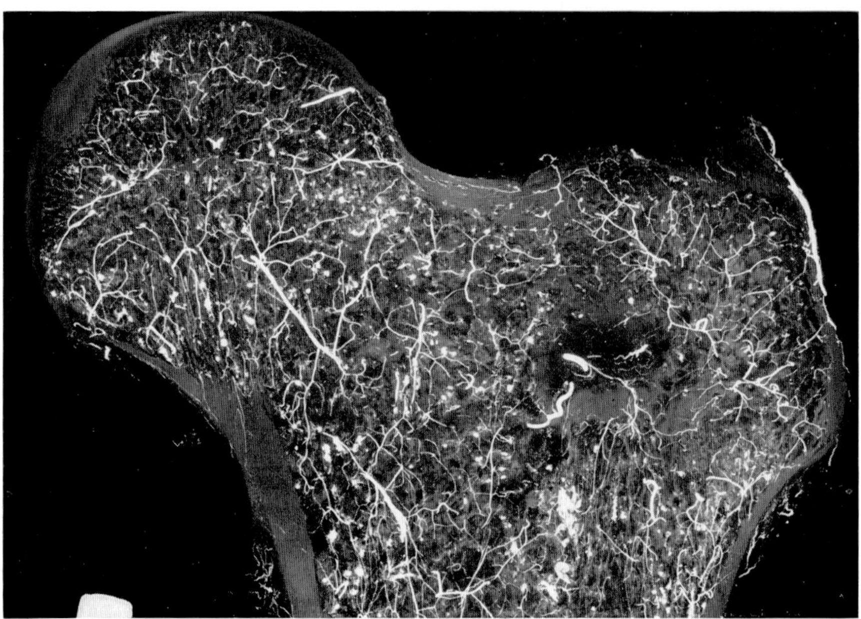

FIG. 112. Microangiogram of femoral head of the dog. The arterial arcades in the subchondral region are seen. The pattern of vessels is typical for cancellous bone. ($\times$ 2¾).

articular surface (Fig. 112). Finally, small terminal capillaries project vertically toward the subchondral surface.

3. Vascular Anatomy of Compact Bone

Cortical bone is organized into a complex arrangement of longitudinal structures known as "haversian systems." This arrangement has been described in detail by Pritchard (1956) and by Cohen and Harris (1958). As pointed out by Ham (1952), such a plan allows for an intimate relationship between bone cells and vessels of the haversian canals.

The reported differences in the number of vessels in the canals seem to depend on the species of animal and the particular bone studied. In the rabbit (Brookes and Harrison, 1957) and in the rat (Brookes, 1958a), the haversian canal contains a single vessel which is a simple endothelial tube. In human material the average number is two vessels per haversian canal (Langer, 1876). Nelson *et al.* (1961) observed that the haversian canals in the tibia from adult patients without evidence of vascular disease each contained one to four vessels (Fig. 113, inset), the com-

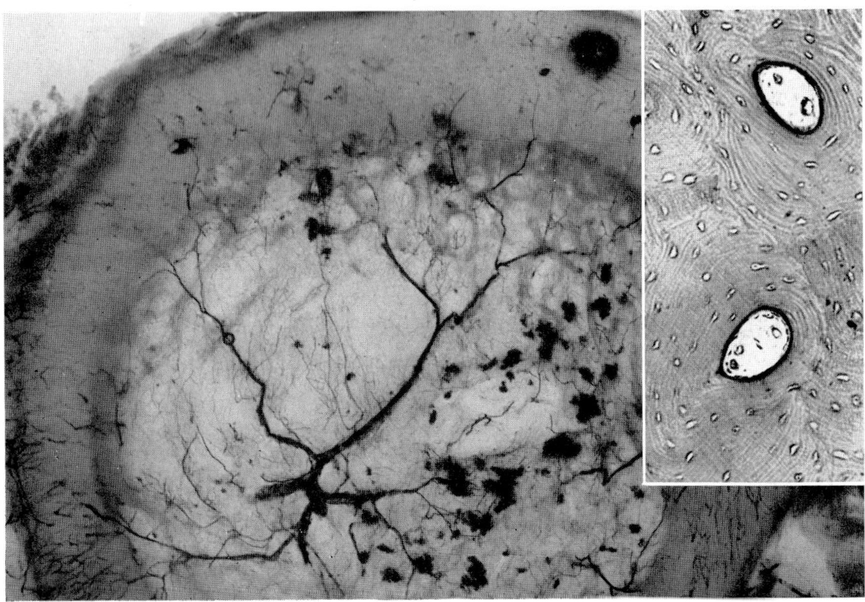

FIG. 113. Nutrient artery with lateral branches going centrifugally toward the cortex. In the cortex further subdivision occurs to allow supply of the vessels in the haversian canals.

Inset. Capillary vessels of the haversian canals supplied by lateral branches of nutrient artery. Note two capillaries in superior haversian canal and three in inferior canal.

monest number being two. In most haversian canals, capillary-like vessels with simple endothelial walls are distinguished. In the dog usually a single vessel of capillary type is noted (Peterson *et al.*, 1959). In a few canals of the human tibia an arteriole is seen, with an accompanying venule (Nelson *et al.*, 1961).

4. Vascular Anatomy of Long Bones

The general architecture of the vasculature of long bones will be presented without reference to any particular bone of any single species, although where differences of opinion exist, these will be noted. In this regard, Kelly *et al.* (1959a,b), Peterson *et al.* (1957, 1959) and Nelson *et al.* (1961) have had most of their experience with the vascular supply of the dog femur and human tibia, while Brookes (1958a,b), Brookes and Harrison (1957) and Morgan (1959) have described the blood supply of long bones of other species. The general pattern of vascular anatomy of a long bone will now be presented.

a. Epiphyseal-metaphyseal vessels: These two groups of vessels will be considered as a single functional unit although divided during growth by the epiphyseal plate. They usually stem from the same extracortical source and enter the bone either in the epiphysis or the metaphysis. The epiphyseal vessels lie close to the epiphyseal plate, anastomose in vascular arcades toward the articular cartilage, and send vertically oriented capillary loops to perforate the subchondral bone plate (Holmdahl and Ingelmark, 1951). The latter channels bend back on themselves to form an aneurysmal dilatation immediately beneath the cartilage. Trueta and Morgan (1960) demonstrated that vessels (crossing the bone plate) on the epiphyseal side of the epiphyseal plate perforate the bone plate and expand before returning to the epiphysis, frequently through a different passage. Metaphyseal vessels penetrate the metaphyseal cortex, to branch, divide and send capillaries into the growing epiphyseal plate. These minute channels also anastomose with the branches of the nutrient artery. In the human tibia the metaphyseal-epiphyseal arteries are arranged radially and are accompanied by a vein (Nelson *et al.,* 1961).

Conflicting evidence has been presented concerning the existence of a circulation across the epiphyseal plate (see Tilling, 1958, for review of the literature on this subject). However, the reported differences of opinion may be due primarily to the age of the species in question when the examinations were carried out. In this regard Trueta (1957) described vessels crossing the epiphyseal area in the extremely young child but ceasing after the epiphyseal plate becomes well-developed. No further vessels cross the plate until epiphyseal fusion has occurred, creating anastomoses by means of spiral arteries between the epiphysis and metaphysis. Trueta and Morgan (1960) described the intimate anatomy of the growing epiphyseal plate. The opinion is generally held that when the epiphyseal plate is well developed, no vessels directly cross the plate. However, some may persist around the periphery. A very small artery was described by Morgan (1959) as circumscribing the epiphyseal plate of the tibia of the growing rabbit and connecting both the epiphyseal and metaphyseal circulations.

b. Nutrient system: All long bones have one or more nutrient arteries. Each supplies the medullary cavity and the major portion of the diaphysis, and its branches anastomose with the metaphyseal vasculature at the ends of the long bone. The nutrient artery, accompanied by the nutrient vein or veins and often by myelinated nerves, enters the bone through a diagonal nutrient canal, usually directed away from the growing end of the bone (Nelson *et al.*, 1961). After penetrating the marrow cavity, the nutrient artery divides, to send one or several main branches toward each end of the long bone. From the main channels smaller arterial branches, given off radially, pass in a centrifugal direction to the cortical bone (Nelson *et al.*, 1961) (Fig. 113). In the human tibia a vein accompanies these arterial branches. Further subdivision of the vessels occurs as the haversian canals are entered.

c. Vascular bed of red bone marrow:[3] Towards the end of the nineteenth century it was already recognized that the vascular bed of red bone marrow was unusual. Not only were the vessels atypically large, but their walls were greatly attenuated and often seemed incomplete, suggesting that there might be an "open" circulation. Early investigators perfused the vascular bed with suspensions or colloids, and more often than not observed the materials extravascularly. Two groups of workers who have influenced our thoughts greatly (Maximow, 1910; Drinker *et al.*, 1922) subsequently concluded that the circulation was mostly closed, but that gaps appeared as local clusters of normoblasts completed erythropiesis, thus allowing the erythrocytes to be washed into the blood stream. However, it is fair to say that the question of an open or closed vascular bed has not been settled to everyone's satisfaction to this day.

Doan (1922b) introduced a new concept of the vascular pattern of marrow. In birds which had received perfusions with India ink, he believed he saw a system of "intersinusoidal capillaries," in addition to the vascular structures already noted. He allied himself with Cunningham and Sabin, and subsequently they wrote a series of influential papers which developed this theme and which provided the morphologic basis for their belief that the erythrocytic cell line developed intravascularly from endothelium within these vessels. In the absence of erythrogenic activity, the vessels were thought to be largely collapsed and therefore difficult to identify (see the review of Sabin, 1932). Admittedly the presence of intersinusoidal capillaries was hard to establish at best, and particularly so in normal mammalian marrow (Doan *et al.*, 1925). The great resolving power of the electron microscope seemed necessary finally to decide whether or not such structures really existed.

Electron microscopists have tended to avoid bone marrow as an organ, although isolated cells have been studied repeatedly. Pease (1955,

[3] By D. C. Pease and L. Zamboni.

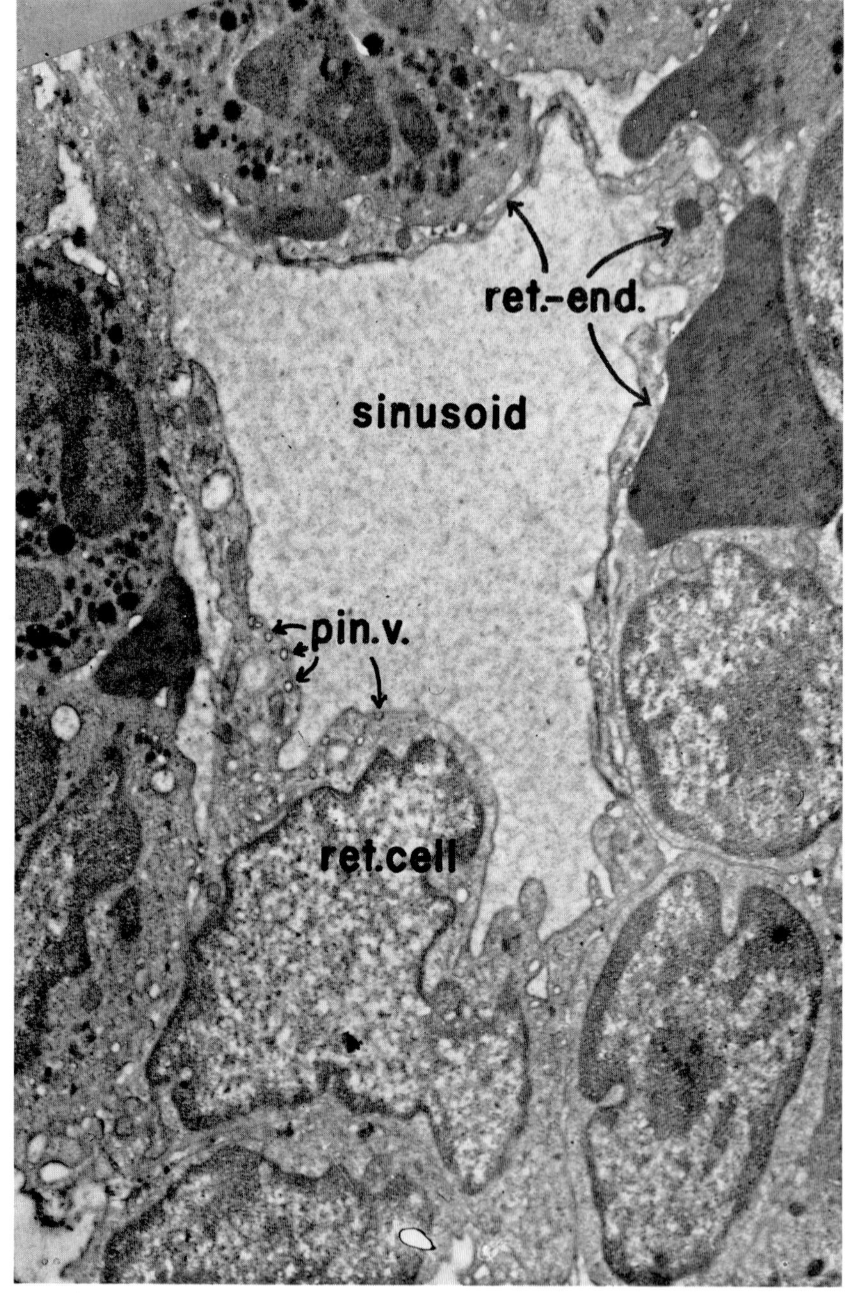
ret.-end.
sinusoid
pin.v.
ret.cell

1956), however, removed blocks of tissue in a relatively undisturbed fashion and observed that the sinusoidal linings were quite incomplete.[4] Frequently metamyelocytes could be observed passing into the vascular lumen, thus interrupting the wall structure. He concluded that the circulation was fundamentally of the "open" type. There remained the possibility, however, that some of the gaps seen were fixation artifact, and hence it has seemed wise recently to reopen the investigation. The following account is based mainly upon this effort (Zamboni and Pease, 1961).

Doan (1922b) has given perhaps the most detailed presentation of the general pattern of vascularization of a long bone, the radius of the pigeon. This description applies with only minor variations to the long bones of the rat (femur, tibia and humerus) (Zamboni and Pease, 1961). The nutrient artery bifurcates and penetrates deeply into the center of the marrow. Long branches then turn towards both ends of the bone. Relatively rarely small arteriolar branches are given off and turn towards the surface. In the pigeon, just before reaching the surface, these channels are said to break up into a tuft of short arteriolar capillaries which soon dilate and branch further as venous sinusoids. In the rat the terminal arterioles proceed all of the way to the endosteal surface and there abruptly terminate in a very rich, anastomosing, endosteal, sinusoidal bed. This bed freely anastomoses with capillaries which penetrate the substance of the adjacent bone and so supply haversian systems. From this bed, also, arises a rich supply of sinusoids which penetrate the marrow substance, eventually to drain into the central vein. Thus, the penetrating sinusoids are arranged radially.

In electron micrographs of the arterial system, nerve bundles have been seen repeatedly in the adventitia. Presumably these exert a vasomotor control which has been indicated by Drinker *et al.* (1922) as functionally important in regulating erythrocytic levels in blood.

It is quite possible to preserve marrow *in situ* in such a way that

[4] For general discussion of ultramicroscopic structure of blood vessels, see Section C.-1 and 2, Chapter I.

FIG. 114. Small sinusoid in red marrow, fixed *in situ*, and so mechanically undisturbed. Irregularly thin sheets of reticuloendothelial cell cytoplasm (ret.-end.) outline the vascular channel. Cytoplasmic sheets of neighboring cells make haphazard and probably unstable contacts with each other. The characteristic nucleus of reticular cells (ret. cell) is shown. This is relatively pale and watery with condensed chromatin around its periphery. Ordinarily it is without a definitive nucleolus, although occasionally it may show one. Pinocytotic vesicles (pin. v.) are a common feature of the reticular cell cytoplasm. Note the paucity of supporting structures other than the parenchymal cells of marrow.

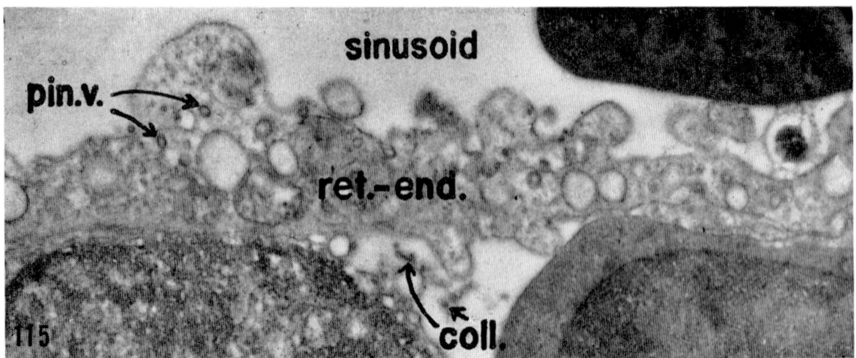
sinusoid
pin.v.
ret.-end.
coll.
115

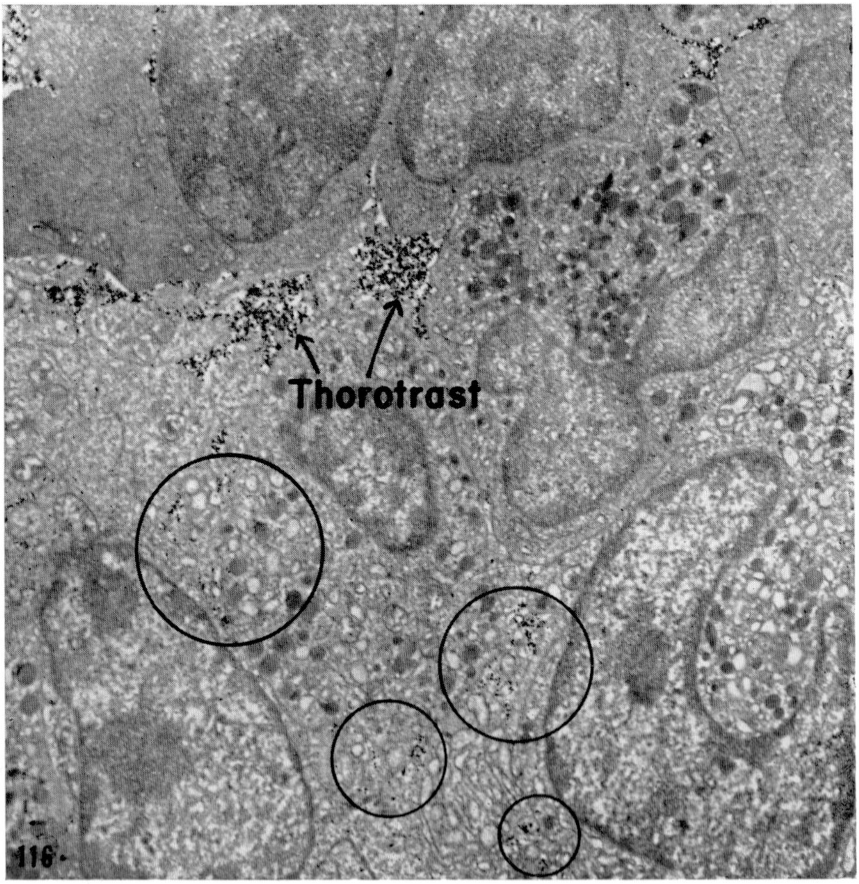

there is no danger of important mechanical artifact in removing tissue blocks for electron microscopy. Under these circumstances it is a fair presumption that the sinusoidal bed retains its normal relationships. Reticuloendothelial cells possess distinctive morphologic characteristics, and transitional forms with the other cell types are not normally seen. These cells are quite flattened but irregularly so (Fig. 114). They have long processes which tend to make very simple contact with the processes of neighboring cells. Most commonly, two processes simply butt against one another without any further morphologic specialization. Often gaps in the wall are visible.

No basement membrane is to be found underlying the reticuloendothelium. The only supporting structures that might be regarded as being of a semipermanent nature are small loose aggregates of collagenous fibers (Fig. 115). These are so sparsely distributed that they can hardly be regarded as an adventitial layer and, indeed, this is the only stroma to be observed in red marrow, away from the larger blood vessels. Often metamyelocytes and red cells can be found in all stages of passing from the marrow parenchyma into the sinusoidal lumen. Thus, on the basis of simple observation of carefully preserved material, it would appear that the marrow circulation is fundamentally open.

Weiss (1959, 1960) has also begun an extended study of marrow, the results of which are so far available only in abstract form. However, his findings are in close agreement on all major points with those presented above (Zamboni and Pease, 1961).

It seemed desirable to test the patency of the sinusoidal lining experimentally. With this in mind the hind limbs of rats were perfused with Thorotrast (Zamboni and Pease, 1961). This was administered by way of the aorta in living animals, the hind limbs usually being left intact until the injection was completed. Marrow samples were collected from 5 to 15 minutes after the perfusion. Thorotrast particles in-

---

FIG. 115. Transverse section of sinusoidal wall at fairly high magnification. It is to be noted particularly that there is no basement membrane underlying the cytoplasmic sheet. Connective tissue elements are only represented in the form of sparse and scattered groups of collagenous fibers (coll.) along with a little nearly amorphous material that is probably mucopolysaccharide. The sinusoidal surface of the reticuloendothelium is apt to be irregular, as here, suggesting protoplasmic mobility.

FIG. 116. Marrow fixed *in situ* 15 min after administration of Thorotrast by way of the abdominal aorta. Thorotrast particles had reached almost all interstitial spaces, even very narrow slits between cells, as noted in the uppermost part of this figure. The encircled areas show regions where Thorotrast particles have been phagocytized by early myelocytes. Such particles are seen to be mainly in cisterns of the endoplasmic reticulum.

variably were found throughout the interstitial spaces, even when these were but narrow slits between cells, as demonstrated in Fig. 116. Even within 5 minutes there had been time for phagocytic activity, so that Thorotrast particles were found deeply within certain cell types (encircled areas, Fig. 116). Phagocytosis was particularly evident in early myelocytes and megakaryocytes. The particles seemed usually, if not always, to be within compartments of the endoplasmic reticulum. Thus, these experiments have reinforced the belief that the reticuloendothelial lining of sinusoids is functionally incomplete. Furthermore, the rapidity with which Thorotrast became widely dispersed within the parenchymal spaces hardly seems possible on the assumptions of passive diffusion and Brownian movement. Rather the parenchymal cells must have been actively moving about, churning the small quantities of tissue fluid (Zamboni and Pease, 1961). Probably, too, the contacts between neighboring reticuloendothelial cells were constantly in a state of dynamic flux and readjustment.

The ease with which suspended particles of suitable size became distributed in the parenchyma suggests an explanation for Doan's (1922b) descriptions of "intersinusoidal capillaries." His strings of carbon particles were best seen in yellow marrow between neighboring fat cells. Thorotrast easily gets into such loci. With the resolution available with an electron microscope, there was no possibility that intersinusoidal capillaries could have been overlooked had they existed in red marrow. However, it must be pointed out that yellow marrow or physiologically hypoplastic marrow was not systematically studied. It may also be categorically stated that erythroblasts, normoblasts and juvenile red cells were all present simply as free clusters of cells in parenchymal tissue. They were not surrounded by an endothelium, which would have indicated an intravascular origin. Reticular cells were almost entirely confined to the walls of vascular channels. They were not a part of the stroma penetrating the parenchymal tissue, for such a stroma does not exist beyond the presence of a few scattered collagenous fibers. Thus, there was no possibility of confusing sinusoid with capillary or reticular tissue.

d. Periosteal blood supply:[5] The classic concept of the blood supply to long bones has been that cortical bone receives a definite portion of its arterial blood from the periosteum. More recently, Brookes and Harrison (1957) and Brookes (1958a,b) minimized this possibility on the basis of studies on the rabbit, rat and human fetus. MacNab (1957) reached similar conclusions concerning the human tibia, and these were summarized by Jackson and MacNab (1959). Nelson *et al.* (1961) ob-

[5] By P. J. Kelly and L. F. A. Peterson.

served in the human tibia that the anterior tibial artery gives rise to transverse periosteal branches at regular intervals as it courses down the interosseous membrane near the posterolateral border of the tibia. Each transverse artery is accompanied by a vein on either side. These branches are seen on the lateral and posterior surface of the tibia. The anteromedial surface presents a more irregular arrangement of anastomoses of the periosteal vessels over the superficial layer of the periosteum. Although arteries are not observed entering the cortical bone, *per se,* small capillaries, originating from nests of arterioles in the periosteum, are seen penetrating the cortex. Larger-calibered vessels with thin walls representing venules leave the cortex and empty into small periosteal veins (Nelson *et al.,* 1961). In contrast to this, Morgan (1959) has described the periosteal circulation in the tibiofibula of the rabbit as being represented by six or seven longitudinally-arranged arteries forming a network over the periosteum. Fine branches enter the cortex to supply its outer layers. The periosteal system anastomoses with the metaphyseal-epiphyseal systems and contributes its major supply to the cortex in the region of the metaphysis. Nelson *et al.* (1961) and Morgan (1959) have each observed capillaries extending from the periosteal to the endosteal surface.

In summary it appears that the periosteal blood supply plays the following roles: It provides the minimal portion of the arterial supply to the major portion of the shaft of the long bone, becoming increasingly important as the ends of the bone are approached. Superficial haversian vessels are provided by appositional new bone forming over periosteal capillaries. In addition, the periosteal blood supply represents a reserve circulation, capable of enlargement if the usual sources are disrupted.

e. Venous drainage.[6] The veins in bone are thin-walled structures and even the largest ones accompanying the nutrient artery have walls only two or three cells in thickness (Nelson *et al.,* 1961). The centrally placed nutrient vein or sinus in the human tibia accompanies the nutrient artery. In the tibia of the rabbit, differences in the venous pattern exist, due to fusion of the tibia and fibula and also to variation in distribution of red marrow. The main nutrient vein has an ascending portion and a descending limb; the latter empties through the nutrient canal into the general venous circulation and it is distensible (Morgan, 1959). The venous drainage of the epiphysis is by veins paralleling the arterial supply. That of the metaphysis is through a series of longitudinally-arranged sinusoids, into which the metaphyseal supply empties after it has coursed into the vessels growing up the degener-

[6] By P. J. Kelly and L. F. A. Peterson.

ating cartilage cell columns. These vessels drain primarily to the emerging metaphyseal veins (Trueta and Morgan, 1960). The remainder of the venous drainage is from the sinusoids and capillaries of the medullary cavity directly to the nutrient vein; in addition, some venous drainage is to the periosteal system which is more numerous in the metaphyseal region and decreases as the midportion of the diaphysis is approached. (For detailed description of these anatomic points, see Morgan, 1959, and Nelson *et al.*, 1961.)

5. Lymphatics of Bone[7]

At present one general statement can be made concerning the lymphatics of bone, namely, that it has not been conclusively shown that they exist. Kolodney (1925) injected India ink into the medullary cavity of long bones and later observed the appearance of the carbon particles in the regional lymph glands. He concluded that this indicated the presence of lymph channels. This only proved, however, the existence of a transport system for pigments but not necessarily the presence of lymphatics in bone. Kallius (1932) reported the demonstration of a channel system in bone, but other investigators have not confirmed his findings. Finally, Anderson (1960) was unable to visualize an organized lymphatic collecting system by means of instillation of India ink into the medullary cavities of long bones.

C. PHYSIOLOGY

By P. J. Kelly and L. F. A. Peterson

1. Blood Flow in Bone

Information on direct measurement and observation of blood flow in bone is limited. As already mentioned, the blood enters the epiphysis by means of circumferentially located, radially oriented epiphyseal arteries and is distributed to the subchondral region and epiphyseal plate by means of capillaries, vertically oriented to the cartilage. It returns by

[7] By P. J. Kelly and L. F. A. Peterson.

means of a paralleling venous system (Trueta and Morgan, 1960). Stein *et al.* (1958), in their studies on the intramedullary pressures in the epiphysis, found that the pulse pressure in this location was not as great as that in the metaphysis. This observation indicates that the epiphyseal arterial supply is made up of smaller caliber vessels than the diaphyseal arterial supply.

Johnson (1927), Trueta and Morgan (1960), Brookes (1957), Brookes and Harrison (1957), and Nelson *et al.* (1961) stated that the nutrient system supplies most of the blood to the diaphysis. Nonetheless, the epiphyseal-metaphyseal vascular system is able to support the diaphysis if the periosteal and nutrient circulations are destroyed (Johnson, 1927, and Larson, 1959).

Brånemark (1959), using vital microscopy, described the flow in the marrow space as it relates to the circulation in the cortex of the bone of the fibula of the rabbit. He stated that a bone-marrow arteriole divides into capillaries, which then enter sinusoids or continue directly into a venule. The sinusoids are hexagonal or spindle-shaped and unite to form sinusoidal systems. These, in turn, are drained by collecting venules. The sinusoids vary rhythmically in degree of dilatation and thus alter the velocity of flow within them. Capillaries enter haversian canals from the endosteal portion of the diaphyseal bone and then return to the marrow cavity, emptying into venules. The velocity of flow in the haversian vessels is higher than that in the capillaries of the bone marrow, while the velocity in the sinusoids is slower than in either of the other two sites. Brånemark (1959) stated that whenever two vessels are seen in the same haversian system, the velocity in one is always considerably higher than that in the other and flow is in an opposite direction. The velocity of flow in the cortical bone is rapid and steady. Brånemark found no evidence of a fluctuating type of flow from within outward or from without inward, as described by Brookes and Harrison (1957).

Kinosita *et al.* (1956), in direct observations on regenerating bone marrow, reported that the arterioles divide into several capillaries which then merge into three or four hexagonal sinusoids. The blood flow in the sinusoids is variable, with these vessels, in turn, emptying into the collecting venules. The latter vessels are contractile and therefore can regulate the flow in various portions of the sinusoidal network. Drainage is by means of the nutrient vein.

The arterial supply to the metaphysis consists of branches from the nutrient system and the metaphyseal vessels. These two groups of vessels may be considered as specialized forms of periosteal vessels. However, in reality they perforate the cortex without branching to supply the interior of the bone. Long slender arteries approach the metaphysis, to divide into numerous capillaries which ascend into each degenerating cell column in the epiphyseal plate. The latter vessels turn abruptly and enter venous sinusoids which then drain into the nutrient and metaphyseal veins.

The medullary cavity has as its principal supply the nutrient system which anastomoses at either end of the bone with the metaphyseal systems. There are no direct observations on the amount of blood which is contributed by each system. The marrow receives its circulation through direct capillaries and by means of sinusoids.

Original workers, such as Langer (1876) and Lexer (1904; Lexer *et al.*, 1904), stated that the periosteal system is of considerable importance. More recent workers, however, have given it relatively limited significance in normal bone.

It may be said that the major arterial supply to the diaphysis of long bones is via the nutrient system. The arterial circulation in the medullary cavity joins the venous return by means of sinusoids and capillaries. The velocity of flow in the capillaries is higher than that in the sinusoids. The latter are contractile, so that a variable flow occurs through them. The major portion of the cortical blood supply is from capillaries, derived from the nutrient system, that enter from the medullary cavity. The circulation in the cortex has a relatively high velocity and the major venous runoff from this structure is to the medullary cavity and thence to the general circulation by means of the nutrient venous system.

2. Effects of Aging

Jaffe and Pomeranz (1934) and Sherman and Selakovich (1957) reported small areas of necrosis in bone with advancing age and related this to vascular insufficiency. Frost (1960) also recorded an increasing percentage of osteocytes dying with age.

3. Effects of Alteration in Blood Supply on Growth

Bier (1905) annotated many early observations on the growth and healing of bone, as related to circulation in this tissue. Recently, work by Doumanian (1960) has cast doubt on vein-ligation techniques, such as those of Pearse and Morton (1930, 1937). These workers ligated the popliteal vein of a dog's hind limb and concluded that this hastened healing of surgical defects created in the fibula. Doumanian noted a persistent elevation of venous pressure only if the inferior vena cava and ipsilateral common femoral vein were ligated. Ligations at levels lower in the venous tree, such as the femoral vein, did not have a similar effect except for a short period.

Trueta (1953a) postulated the increase in bone length in children following fracture or osteomyelitis as due to blockage of the nutrient artery, with resultant increased arterialization in the metaphyseal region. However, Brookes (1957) was unable to increase the length of the femur in rabbits by suppressing the circulation in nutrient vessels, although, for the middle period of growth, the femur affected by suppression appeared to grow faster than the one on the control side. De Sapio (1953) increased the length of dogs' femurs by stripping the periosteum. Hutchison and Burdeaux (1954) used tourniquets to increase the length and width of bones in dogs and attributed the changes observed to venous stasis.

Kelly *et al.* (1959b) attempted to elucidate the problem by determining the effect of an arteriovenous fistula on bone growth in the dog. Three changes were seen with femoral or iliac arteriovenous fistulas: (1) increased length of the femur and tibia, (2) elevated temperature of the bone, and (3) hypervascularity of the affected bones, especially about the epiphysis, as demonstrated by microangiography. More recently Stein *et al.* (1959), in a series of six dogs, noted a significant increase in intramedullary blood pressure in epiphyseal bone shortly after institution of a femoral arteriovenous fistula; intramedullary pressure in the diaphyseal bone, however, was not altered significantly. Foster *et al.* (1951) caused infarction of metaphyseal marrow and found temporary thickening of the epiphyseal plate until revascularization occurred. The thickening was considered to indicate continued growth of the cartilagenous plate with cessation of remodeling into bone.

D. PHARMACOLOGY

By P. J. Kelly and L. F. A. Peterson

The blood vessels which enter bone are affected by various pressor and depressor drugs. Evidence for these statements depends on studies by Bloomenthal *et al.* (1952) and Stein *et al.* (1958). The latter investigators found that pharmacologic agents alter the intramedullary pressure by two mechanisms: (1) an effect on systemic pressure, and

(2) a local effect on the small nutrient vessels carrying blood to bone. They presented rather convincing arguments for this thesis, such as the cessation of bleeding from a drill hole in the tibia noted after systemic injection of epinephrine. This observation indicates that the effect of this drug in lowering intramedullary pressure is on the arteries entering bone and is not due to a dilation of marrow arterioles, as stated by Bloomenthal *et al.* (1952).

E. PATHOGENESIS AND PATHOLOGY

By P. J. Kelly and L. F. A. Peterson

1. Inflammatory Changes

Lexer (1904) has discussed the relationship between the vasculature of bone and osteomyelitis, and Kistler (1935) and Trueta (1953b) have presented the pathogenesis of this disorder. The two latter workers have stated that the response of the vasculature to infection is essentially the same as elsewhere in the body, with the exception that the usual inflammatory reaction is encased in a rigid nonyielding bone. This makes the effect of pressure noticeable earlier and isolates the infected area from the resistance of the host by reducing blood flow. The usual sites of infection are those which contain sinusoids in which the blood flow is slower.

2. Avascular Necrosis

Avascular necrosis of bone is caused by caisson disease (Kahlstrom *et al.*, 1939), sickle cell anemia (Golding *et al.*, 1959) and Gaucher's disease (Laufer, 1957). Trauma as a cause of avascular necrosis of the femoral head (Boyd, 1957), the proximal fragment of the navicular (Watson-Jones, 1952) and the talus is well-known. Another large group, idiopathic avascular necrosis (osteochondrosis), typified by Legg-Calvé-Perthe disease, is of less certain origin. In this regard Goff *et al.* (1954) have stated that "the fundamental pathogenesis in osteochondrosis is an occlusive vascular phenomenon resulting in bone necrosis of the fortuitously selected center." Luck (1950) took a less positive attitude, saying that while the pathogenesis of all the osteochondroses is fundamentally the same, the cause is not known. Ponseti (1956) surmised that in Legg-Calvé-Perthe disease the interruption of the blood supply is secondary to obstruction of the retinacular vessels as they cross near the edge of the disrupted epiphyseal plate. Lemoine (1957) caused changes similar to those observed in this disease in the femoral head of the rabbit by ligation of the main capsular artery.

3. Vascular Changes During Fracture Healing

The vascular changes after fracture of a bone have been reviewed only incidentally during histologic and biophysical investigations of fracture healing (Ham, 1930, 1952; Ham and Harris, 1956; McLean and Urist, 1955; Enneking, 1948; Nilsonne, 1959; Koskinen, 1959). The most commonly-held view of fracture healing in a long bone is that the ends are rather quickly united by a periosteal cuff of callus surrounding the fracture hematoma. The hematoma is primarily passive in nature and is not required for healing. Blood vessels which proliferate from the periosteum occur most numerously at some distance from the site of fracture, to grow into the area of new bone formation from the periosteum. Necrosis occurs in the cortical bone a short distance from the fracture site, to a point at which there still remains a useful circulation in the haversian vessels. Little vascular proliferation from the haversian vessels is seen in the early stages of fracture healing. The medullary cavity contributes to the vascular response, but this again is not as marked as that of the periosteum. Vessels rapidly grow into the metaplastic cartilage, removing it and laying down new bone. Gradually over the course of the first several months, remodeling takes place and the vessels become more stabilized in their position. In the case of compact bone, marked resorption eventually occurs in the haversian canals and for a short distance back from the site of fracture (Nilsonne, 1959). By the time healing has eventually taken place, beyond which no new bone forms, the resorbed areas in the haversian canals have again filled in.

Intramedullary nailing in the radius of the rabbit destroys the nutrient vessel, and, as a result, the periosteal vessels revascularize the cortex (Trueta and Cavadias, 1955). Some necrotic cortical bone is present as long as 8 months after the surgical procedure. As stated before, the epiphyseal-metaphyseal vessels can also help support diaphyseal blood supply.

Holmstrand (1957) has observed that vessels tend to align themselves in their position prior to fracture, and that such a response is influenced by the degree of difficulty in removing compact bone. He demonstrated that the vascular channels in a cortical bone graft maintain their orientation to the graft and not to the host. In cancellous bone treated similarly, the vascular channels are oriented in the axis of the long bone.

4. Effect of Irradiation on the Blood Vessels

The effect of irradiation, in particular, therapeutic irradiation, on the blood supply of bone has not been conclusively determined. Ewing (1926) suggested that changes seen in three amputated limbs following heavy therapeutic irradiation were secondary to impairment of blood supply. Gratzek *et al.* (1945) attributed the spontaneous fractures of the femoral neck observed in females receiving pelvic irradiation to an effect on vascular supply. Bonfiglio (1953) presented convincing histopathologic and clinical evidence to the contrary.

Hinkel (1943), using rats and India ink injection, found, after doses of 950 r had been given to hind limbs, that filling of vessels decreased by the third to the seventh day. Further work by Carlson *et al.* (1960), using microangiography and histologic sections, did not demonstrate an alteration in the number of filled vessels or any morphologic change of the blood vessels in bone following irradiation.

References

Anderson, D. W. (1960). Studies of the lymphatic pathways of bone and bone marrow. *J. Bone and Joint Surg.* **42A,** 716.

Barclay, A. E. (1951). "Micro-Arteriography and Other Radiological Techniques Employed in Biological Research." Blackwell, Oxford.

Bellman, S. (1953). Microangiography. *Acta Radiol.* Suppl. 102, 7.

Bier, A. K. G. (1905). "Hyperemia as a Therapeutic Agent." Robertson, Chicago.

Bloomenthal, E. D., Olson, W. H., and Necheles, H. (1952). Studies on the bone marrow cavity of the dog: Fat embolism and marrow pressure. *Surg., Gynecol. Obstet.* **94,** 215.

Bonfiglio, M. (1953). The pathology of fractures of the femoral neck following irradiation. *Am. J. Roentgenol., Radium Therapy Nuclear Med.* **70,** 449.

Boyd, H. B. (1957). Avascular necrosis of the head of the femur. *In* "American Academy of Orthopedic Surgeons Instructional Course Lectures" (R. B. Raney, ed.), Vol. 14, pp. 196–204. Edwards, Ann Arbor, Michigan.

Boyd, H. B., Zilversmit, D. B., and Calandruccio, R. A. (1955). The use of radioactive phosphorus (P^{32}) to determine the viability of the head of the femur. *J. Bone and Joint Surg.* **37A,** 260.

Brånemark, P. I. (1959). Vital microscopy of bone marrow in rabbit. *Scand. J. Clin. & Lab. Invest.* **11**, Suppl. 38.

Brookes, M. (1957). Femoral growth after occlusion of the principal nutrient in day-old rabbits. *J. Bone and Joint Surg.* **39B**, 563.

Brookes, M. (1958a). The vascular architecture of tubular bone in the rat. *Anat. Record* **132**, 25.

Brookes, M. (1958b). The vascularization of long bones in the human fetus. *J. Anat.* **92**, 261.

Brookes, M., and Harrison, R. G. (1957). The vascularization of the rabbit femur and tibiofibula. *J. Anat.* **91**, 61.

Carlson, H. C., Williams, M. M. D., Childs, D. S., Jr., Dockerty, M. B., and Janes, J. M. (1960). Microangiography of bone in the study of radiation changes. *Radiology* **74**, 113.

Cohen, J., and Harris, W. H. (1958). The three-dimensional anatomy of haversian systems. *J. Bone and Joint Surg.* **40A**, 419.

de Sapio, F. S. (1953). Scollamento del periostio ed allungamento degli arti. *Ortoped. e traumatol. app. motore* **21**, 339.

Doan, C. A. (1922a). The capillaries of the bone marrow of the adult pigeon. *Bull. Johns Hopkins Hosp.* **33**, 222.

Doan, C. A. (1922b). The circulation of the bone marrow. *Contribs. Embryol.* **14**, 29.

Doan, C. A., Cunningham, R. S., and Sabin, F. R. (1925). Experimental studies on the origin and maturation of avian and mammalian red blood cells. *Contribs. Embryol.* **16**, 163.

Doumanian, A. V. (1960). Experimental Deep Venous Insufficiency in the Dog. Thesis, Graduate School, University of Minnesota, Minneapolis, Minnesota.

Drinker, C. K., Drinker, K. R., and Lund, C. C. (1922). The circulation in the mammalian bone-marrow. *Am. J. Physiol.* **62**, 1.

Enneking, W. F. (1948). The repair of complete fractures of rat tibias. *Anat. Record* **101**, 515.

Ewing, J. (1926). Radiation osteitis. *Acta Radiol.* **6**, 399.

Foster, L. N., Kelly, R. P., and Watts, W. M. (1951). Experimental infarction of bone and bone marrow: Sequelae of severance of the nutrient artery and striping of periosteum. *J. Bone and Joint Surg.* **33A**, 396.

Frost, H. M. (1960). In vivo osteocyte death. *J. Bone and Joint Surg.* **42A**, 138.

Goff, C. W., Shutkin, N. M., and Hersey, M. R. (1954). "Legg-Calvé-Perthes Syndrome and Related Osteochondroses of Youth." Charles C Thomas, Springfield, Illinois.

Golding, J. S. R., MacIver, J. E., and Went, L. N. (1959). The bone changes in sickle cell anaemia and its genetic variants. *J. Bone and Joint Surg.* **41B**, 711.

Gratzek, F. R., Holmstrom, E. G., and Rigler, L. G. (1945). Post-irradiation bone changes. *Am. J. Roentgenol. Radium Therapy* **53**, 62.

Haliburton, R. A., Sullivan, C. R., Kelly, P. J., and Peterson, L. F. A. (1958). The extra-osseous and intra-osseous blood supply of the talus. *J. Bone and Joint Surg.* **40A**, 1115.

Ham, A. W. (1930). A histologic study of the early phases of bone repair. *J. Bone and Joint Surg.* [N.S.] **12**, 827.

Ham, A. W. (1952). Some histophysiological problems peculiar to calcified tissues. *J. Bone and Joint Surg.* **34A**, 701.

Ham, A. W., and Harris, W. R. (1956). *In* "The Biochemistry and Physiology of Bone" (G. H. Bourne, ed.), Chapter 16. Academic Press, New York.

Hinkel, C. L. (1943). The effect of irradiation upon the composition and vascularity of growing rat bones. *Am. J. Roentgenol. Radium Therapy* **50,** 516.

Holmdahl, D. E., and Ingelmark, B. E. (1951). The contact between the articular cartilage and the medullary cavities of the bones. *Acta Anat.* **12,** 341.

Holmstrand, K. (1957). Biophysical investigations of bone transplants and bone implants. *Acta Orthopaed. Scand.* Suppl. 26, 7.

Hutchison, W. J., and Burdeaux, B. D., Jr. (1954). The influence of stasis on bone growth. *Surg., Gynecol. Obstet.* **99,** 413.

Jackson, R. W., and MacNab, I. (1959). Fractures of the shaft of the tibia: A clinical and experimental study. *Am. J. Surg.* **97,** 543.

Jaffe, H. L., and Pomeranz, M. M. (1934). Changes in the bones of extremities amputated because of arteriovascular disease. *A. M. A. Arch. Surg.* **29,** 566.

Johnson, R. W., Jr. (1927). A physiological study of the blood supply of the diaphysis. *J. Bone and Joint Surg.* [N.S.] **9,** 153.

Kahlstrom, S. C., Burton, C. C., and Phemister, D. B. (1939). Aseptic necrosis of bone: Infarction of bones in caisson disease resulting in encapsulated and calcified areas in the diaphysis and in arthritis deformans. *Surg., Gynecol. Obstet.* **68,** 129.

Kallius, H. U. (1932). Experimentelle Untersuchungen über die Lymphgefässe der Röhrenknochen. *Beitr. klin. Chir.* **155,** 109.

Kelly, P. J., Peterson, L. F. A., and Janes, J. M. (1959a). A method of using sections of bone prepared for microangiography for subsequent histologic study. *Proc. Staff Meetings, Mayo Clin.* **34,** 274.

Kelly, P. J., Janes, J. M., and Peterson, L. F. A. (1959b). The effect of arteriovenous fistulae on the vascular pattern of the femora of immature dogs: A micro-angiographic study. *J. Bone and Joint Surg.* **41A,** 1101.

Kinosita, R., Ohno, S., and Bierman, H. R. (1956). Observations on regenerating bone marrow tissue in situ (Abstract). *Proc. Am. Assoc. Cancer Research* **2,** 125.

Kistler, G. H. (1934). Sequences of experimental infarction of the femur in rabbits. *A. M. A. Arch. Surg.* **29,** 589.

Kistler, G. H. (1935). Sequences of experimental bacterial infarction of the femur in rabbits. *Surg., Gynecol. Obstet.* **60,** 913.

Kolodney, A. (1925). The relation of the bone marrow to the lymphatic system: Its role in the spreading of carcinomatous metastases throughout the skeleton. *A. M. A. Arch. Surg.* **11,** 690.

Koskinen, E. V. S. (1959). The repair of experimental fractures under the action of growth hormone, thyrotropin, and cortisone: A tissue analytic, roentgenologic and autoradiographic study. *Ann. Chir. et Gynaecol. Fenniae* **48,** Suppl. 90, 1.

Laing, P. G., and Ferguson, A. B., Jr. (1958). Sodium-24 as an indicator of the blood supply of bone. *Nature* **182,** 1442.

Laing, P. G., and Ferguson, A. B., Jr. (1959). Iodine-131 clearance-rates as an indication of the blood supply of bone. *Nature* **183,** 1595.

Langer, K. (1876). Über das Gefässsystem der Röhrenknochen, mit Beiträgen zur Kenntniss des Baues und der Entwicklung des Knochengewebes. *Kgl. Akad. Wiss. Math.-Naturw. Cl.* **36,** 1.

Larson, R. L. (1959). The Effect of Suppression of the Periosteal and Nutrient Arterial Blood Supply of the Femora of Immature and Mature Dogs: A Roentgenographic, Histologic and Microangiographic Study. Thesis, Graduate School, University of Minnesota, Minneapolis, Minnesota.

Laufer, A. (1957). Aseptic necrosis of the femoral head. *J. Mount Sinai Hosp.* **24**, 957.

Lemoine, A. (1957). Vascular changes after interference with the blood flow of the femoral head of the rabbit. *J. Bone and Joint Surg.* **39B, 763.**

Lexer, E. (1904). Weitere Untersuchungen über Knochenarterien und ihre Bedeutung für krankhafte Vorgänge. *Arch. klin. Chir.* **73**, 481.

Lexer, E., Kuliga, and Türk, W. (1904). "Untersuchungen über Knochenarterien mittelst Röntgenaufnahmen injizierter Knochen und ihre Bedeutung für einzelne pathologische Vorgänge am Knochensysteme." Hirschwald, Berlin.

Luck, J. V. (1950). "Bone and Joint Diseases: Pathology Correlated with Roentgenological and Clinical Features." Charles C Thomas, Springfield, Illinois.

McLean, F. C., and Urist, M. R. (1955). "Bone: An Introduction to the Physiology of Skeletal Tissue." Univ. of Chicago Press, Chicago, Illinois.

MacNab, I. (1957). Blood supply of tibia (Abstract). *J. Bone and Joint Surg.* **39B**, 799.

Maximow, A. A. (1910). Untersuchungen über Blut und Bindegewebe. III. Die embryonale Histogenese des Knochenmarks der Säugetiere. *Arch. mikroskop. Anat. u. Entwicklungsmech.* **76**, 1.

Maximow, A. A., and Bloom, W. (1957). "A Textbook on Histology," 7th ed. Saunders, Philadelphia, Pennsylvania.

Morgan, J. D. (1959). Blood supply of growing rabbit's tibia. *J. Bone and Joint Surg.* **41B**, 185.

Nelson, G. G., Jr., Kelly, P. J., Peterson, L. F. A., and Janes, J. M. (1961). Blood supply of the human tibia. *J. Bone and Joint Surg.* **42A**, 625.

Nilsonne, U. (1959). Biophysical investigations of the mineral phase in healing fractures. *Acta Orthopaed. Scand.* Suppl. 37, 5.

Pauporte, J., Lowenstein, J. M., Richards, V., and Davison, R. (1958). Blood turnover rates distal to an arteriovenous fistula. *Surgery* **43**, 828.

Pearse, H. E., Jr., and Morton, J. J. (1930). The stimulation of bone growth by venous stasis. *J. Bone and Joint Surg.* [N.S.] **12**, 97.

Pearse, H. E., Jr., and Morton, J. J. (1937). Venous stasis accelerates bone repair. *Surgery* **1**, 106.

Pease, D. C. (1955). Marrow cells seen with the electron microscope after ultrathin sectioning. *Rev. hématol.* **10**, 300.

Pease, D. C. (1956). An electron microscopic study of red bone marrow. *Blood* **11**, 501.

Peterson, L. F. A., Kelly, P. J., and James, J. M. (1957). Ultrastructure of bone: Technic of microangiography as applied to the study of bone: Preliminary report. *Proc. Staff Meetings Mayo Clinic* **32**, 681.

Peterson, L. F. A., Neher, M., Janes, J. M., and Kelly, P. J. (1959). A stereoscopic microradiographic camera with vacuum filmholder and a stereomicroscope. *Proc. Staff Meeting Mayo Clinic* **34**, 283.

Ponseti, I. V. (1956). Legg-Perthes disease: Observations on pathological changes in 2 cases. *J. Bone and Joint Surg.* **38A**, 739.

Pritchard, J. J. (1956). *In* "The Biochemistry and Physiology of Bone" (G. H. Bourne, ed.), Chapter 1. Academic Press, New York.

Sabin, F. R. (1932). Bone marrow. *In* "Special Cytology" (E. V. Cowdry, ed.), 2nd ed., p. 505. Hoeber-Harper, New York.

Sherman, M. S., and Selakovich, W. G. (1957). Bone changes in chronic circulatory insufficiency. *J. Bone and Joint Surg.* **39A**, 892.

Spalteholz, W. (1911). "A Method for Clearing of Human and Animal Specimens. With an Appendix: Bone Stains." Hirzel, Leipzig.

Streeter, G. L. (1949). Developmental horizons in human embryos (fourth issue): A review of the histogenesis of cartilage and bone. *Contribs. Embryol.* **33** (Nos. 213–221), 149.

Stein, A. H., Jr., Morgan, H. C., and Porras, R. F. (1958). The effect of pressor and depressor drugs on intramedullary bone-marrow pressure. *J. Bone and Joint Surg.* **40A,** 1103.

Stein, A. H., Jr., Morgan, H. C., and Porras, R. (1959). The effect of arteriovenous fistula on intramedullary bone pressure. *Surg., Gynecol. Obstet.* **109**, 287.

Tilling, G. (1958). The vascular anatomy of long bones: A radiological and histological study. *Acta Radiol. Suppl.* 161, 5.

Trueta, J. (1953a). The influence of the blood supply in controlling bone growth. *Bull. Hosp. Joint Diseases* **14,** 147.

Trueta, J. (1953b). Acute haematogenous osteomyelitis: Its pathology and treatment. *Bull. Hosp. Joint Diseases* **14,** 5.

Trueta, J. (1957). The normal vascular anatomy of the human femoral head during growth. *J. Bone and Joint Surg.* **39B**, 358.

Trueta, J., and Cavadias, A. X. (1955). Vascular changes caused by the Küntscher type of nailing. *J. Bone and Joint Surg.* **37B**, 492.

Trueta, J., and Harrison, M. H. M. (1953). The normal vascular anatomy of the femoral head in the adult man. *J. Bone and Joint Surg.* **35B,** 442.

Trueta, J., and Morgan, J. D. (1960). The vascular contribution to osteogenesis. I. Studies by the injection method. *J. Bone and Joint Surg.* **42B,** 97.

Tucker, F. R. (1950). The use of radioactive phosphorus in the diagnosis of avascular necrosis of the femoral head. *J. Bone and Joint Surg.* **32B,** 100.

Watson-Jones, R. (1952). "Fractures and Joint Injuries," Vol. I. Williams & Wilkins, Baltimore, Maryland.

Weiss, L. (1959). The organization of the connective tissue elements in bone marrow. I. Fat cells (Abstract). *Anat. Record* **133**, 439.

Weiss, L. (1960). The organization of the connective tissue elements in bone marrow. II. Reticular cells (Abstract). *Anat. Record* **136,** 300.

Zamboni, L., and Pease, D. C. (1961). The vascular bed of red bone marrow. *J. Ultrastract Research* **5,** 65.

PART THREE

Disorders Affecting the Arterial or Venous Circulation

Chapter XIX. ARTERIOSCLEROSIS: 1. Atherosclerosis; 2. Mönckeberg's Sclerosis

1. Atherosclerosis

A. INTRODUCTION

By J. STAMLER

1. DEFINITION OF TERMS

In any discussion of atherosclerosis, it is important that clarity prevail from the outset concerning definition of terms—particularly the difference that exists between the terms, arteriosclerosis and atherosclerosis, at least, on the basis of pathologic changes. Many years ago arteriosclerosis was designated as a generic term, encompassing all the vascular scleroses or pathologic abnormalities resulting in thickening and hardening of the vessels. Several different processes (almost certainly with different etiologies) produce abnormalities in the vessel wall of the sclerotic variety; hence, in order to point up the generic nature of the designation, it would perhaps be better to use the term, the arterioscleroses.

Atherosclerosis is a specific disease process, one among the arterioscleroses. In terms of production of human morbidity and mortality, it is by far the most important entity in this category. For example, it is the underlying pathologic lesion in at least 90 per cent of cases of myocardial infarction.

Atherosclerosis is primarily a lesion of the intima of the wall of large and small arteries. The morphologic hallmark of this lesion, recognized over 100 years ago, in the early days of scientific cellular pathology, is its content of fatty materials, neutral fats, phospholipids, and, particularly, cholesterol. These are very often distinguishable grossly, and almost invariably, microscopically.

Considerable evidence is available indicating that deposition of fatty materials is the initial process in atherogenesis. As the lesion develops, these lipids accumulate both intracellularly and extracellularly, in globules and amorphous pools. Concomitantly, necrotic and proliferative processes supervene, resulting in fibrosis, scarification and calcification. The term, atherosclerosis, derived from the Greek athere, meaning mush, and skleros, meaning hard, therefore characterizes the lesion well.

B. EXPERIMENTAL ATHEROSCLEROSIS

By J. STAMLER

1. EARLY ATTEMPTS

Once the atherosclerotic lesion was clearly identified by the early cellular pathologists, attempts began to reproduce it in laboratory animals. Throughout the second half of the nineteenth century, up until 1908, considerable experimental work was done, involving a variety of approaches, depending upon the hypotheses of the various investigators (Anitschkow, 1933; Hueper, 1944, 1945; Katz and Stamler, 1953).[1] For example, efforts were made to induce atherosclerosis with toxins, bacteria, pressor and depressor substances, hormones, mechanical injury to vessels, and through many other means. Not unexpectedly, some of these procedures did result in pathologic changes in arteries which resembled virtually every type of arteriosclerosis seen in man, except atherosclerosis.

This historic experience is very important because it constitutes substantial evidence, of a negative type, ruling out several processes as decisive or primary or sole factors in the causation of atherosclerosis. However, it must be emphasized that they cannot be eliminated as possible factors in its pathogenesis or etiology, since, as it will be shown, subsequent experimental work has yielded substantial evidence indicating an adjuvant role for several of these agents in the multifactorial causation of atherosclerosis.

2. RECENT WORK

Atherosclerosis was first successfully produced in animals in the period 1908–10 by a group of Russian scientists led by Anitschkow (1933). In the course of a study on the relationship of diet to kidney function and high blood pressure, these workers fed whole animal tissues to rabbits. After a considerable period of time on this regimen, the animals were observed at postmortem to have developed atherosclerosis.

In subsequent experiments, Anitschkow vigorously followed up this lead. He recognized that the whole animal tissues fed to the rabbits contained both animal fats and proteins. He also noted that these

[1] Here and elsewhere in Sections B and C reference is made to comprehensive reviews with extensive bibliographies.

animals were markedly hyperlipemic and that the atherosclerotic lesions were laden with lipids. He therefore set out to determine whether the lipids (neutral fats and cholesterol) or the proteins in the rabbits' diet were the atherogenic factors. He was soon able to demonstrate that hypercholesterolemic hyperlipemia and atherosclerosis could be produced by feeding rabbits pure cholesterol and fat. This achievement stimulated a tremendous amount of research during the second and third decades of this century, principally in Russia, Germany, and the United States.

During the years since World War II, a tremendous further surge of animal experimentation has occurred. Atherosclerosis has been successfully induced in virtually all species used in the laboratory: mammalian and avian, omnivorous, carnivorous, and herbivorous, including rat, dog, monkey, guinea pig, hamster, pig, chick, duck, and goose (Katz and Stamler, 1953; Katz *et al.*, 1958).

It is of decisive theoretical importance to recognize the common feature which characterizes the experiments on different species. Invariably (with a few "exceptions," which are not really exceptions at all) in all species studied the *sine qua non* for inducing atherosclerotic lesions proved to be a dietary alteration, involving increased intake of fat and cholesterol, with a resulting sustained hypercholesterolemic hyperlipemia. In the dog, for example, hypercholesterolemia and atherosclerosis were induced by feeding a high-fat, high-cholesterol diet and rendering the animal hypothyroid. In the rat and the monkey, hypercholesterolemia and atherosclerosis were initially produced by feeding a diet high in fat and cholesterol and deficient in the sulfur-containing amino acid, methionine.

In susceptible species (e.g. chick and rabbit), it proved possible, by appropriate manipulation of the dietary regimen, to induce atherosclerosis in the presence of only minimal hypercholesterolemia and organ lipidosis (Anitschkow, 1933; Katz and Stamler, 1953). Thus, cockerels fed mash, supplemented with small amounts of cholesterol (1/4%) and triglycerides (5% fat) over long periods of time, amounts quite similar to those in prevailing American diets, developed moderate hypercholesterolemia, with little or no organ cholesterosis (Katz and Stamler, 1953; Stamler, 1959a). After about nine months on this diet, gross atherosclerosis supervened (Fig. 117). Anitschkow (1933) has reported similar findings in rabbits.

In the last few years, an additional basic advance has been achieved. All the "complications" of the severe atherosclerotic plaque seen in man, not only scarification, calcification, cartilaginization, and ossification, but also hemorrhage, ulceration, thrombosis, and myocardial infarction, have been induced in laboratory animals, including primates, by dietary means. (For the role of lipids in the production of atherosclerosis in man, see Section D, below.)

The successful production of atherosclerosis in a host of animal species stimulated a fresh wave of intensive research on the parameters of the phenomenon. Two main lines of work have generally been pursued: the study of the interplay (1) among various nutrients, and (2) between exogenous (dietary) and endogenous (e.g., hormonal) factors.

In the area of dietary inter-relationships, recent studies have focussed on the effects of different triglycerides (saturated, monounsaturated and

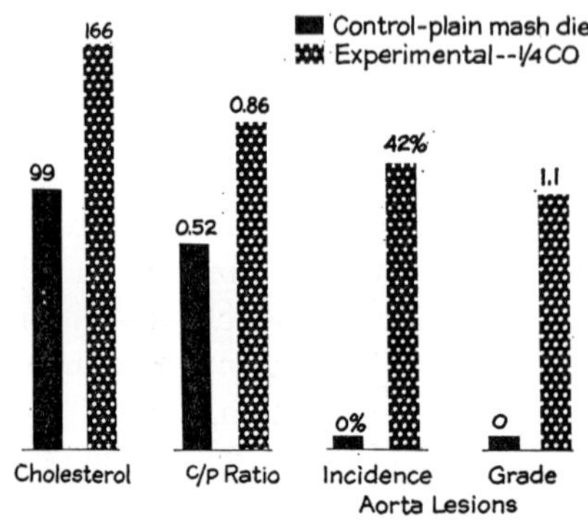

FIG. 117. Atherosclerosis in cockerels with minimal hypercholesterolemia due to feeding mash supplemented with ¼% cholesterol + 5% cottonseed oil. Age period: Weeks 5–40.

C/P ratio—the ratio of serum total cholesterol to total phospholipids. ¼ CO—mash supplemented with ¼% cholesterol and 5% cottonseed oil. Incidence of lesions is per cent of birds with gross aortic lesions at the end of the experiment. The grade is the mean grade of lesions, on an arbitrary scale 0–4.

polyunsaturated), carbohydrates, proteins and amino acids, vitamins and trace minerals in animals on high-cholesterol diets (Katz *et al.*, 1958; Stamler, 1958, 1959a; Portman and Stare, 1959). An analysis of the results of the many experiments is beyond the scope of this brief review. Suffice it to note that many studies have yielded positive results, demonstrating that several constituents of the diet influence hypercholesterolemia and atherogenesis.

It should be emphasized that all the experiments on nutritional interrelationships involve analyses of the effects of specific nutrients in animals on high-fat, high-cholesterol diets. By demonstrating significant influences of some of these nutrients (e.g., proteins), the studies lend support to the concept that the overall pattern of the diet and nutritional imbalance plays a key role in the causation of hypercholesterolemia and atherogenesis. At the same time, the results reinforce the basic theoretical concept that high-fat, high-cholesterol intake is the decisive nutritional aberration in the etiology of hypercholesterolemia and atherogenesis. (For an hypothesis which differs somewhat from this point of view, see Section D., below.)

3. Reversibility of Lesion

Another major contribution of experimental atherosclerosis has been the demonstration that the pathologic plaque is, within limits, a reversible lesion (Anitschkow, 1933; Katz and Stamler, 1953; Stamler,

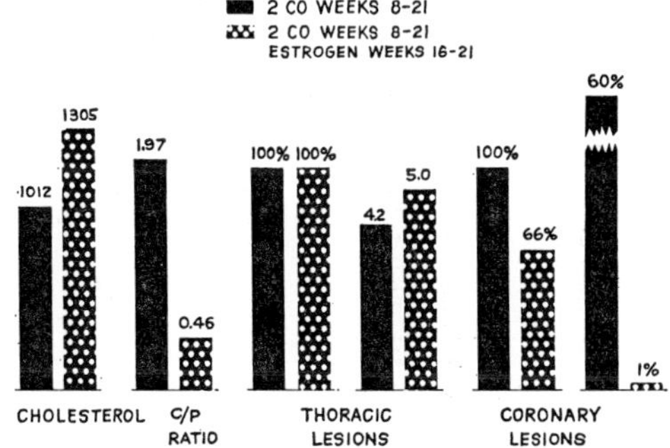

Fig. 118. Estrogen-induced reversal of coronary atherosclerosis in cholesterol-fed cockerels. Same symbols for dietary supplements used as in Fig. 117.

Coronary lesions are those noted in the coronary vessels microscopically. The first set of bars is the per cent of birds with lesions. The second is per cent involvement of the coronary vessels seen microscopically, and is, therefore, an index of severity (in addition to incidence) of coronary lesions. Both control and treated groups consumed mash supplemented with 2% cholesterol and 5% cottonseed oil throughout the experiments. Beginning with the 16th week, the second group was given estrogens, and this was continued for the remainder of the experiment, with resultant disappearance of coronary but not aortic lesions. (A third group on a similar diet, not illustrated on graph, was sacrificed at 16 weeks of age, before initiation of estrogen therapy, to ascertain extent of atherosclerosis at this juncture. Severe aortic and coronary lesions were observed.)

1959a). This has been shown in rabbits, chicks and dogs. Experiments on this aspect of the problem involved producing lesions by feeding an atherogenic diet. Upon discontinuation of this diet and substitution of a low-fat, low-cholesterol ration, hypercholesterolemia and atherosclerosis gradually regressed.

Reversal of atherosclerosis in the coronary vessels of cockerels has also been achieved by another means: estrogen administration (Fig. 118) (Stamler *et al.*, 1956; Pincus, 1959; Stamler *et al.*, 1961). Under the influence of this ovarian hormone, lesions regress even when the birds remain on the high-fat, high-cholesterol diet.

The cited effect of estrogen, an endogenously secreted, physiologic hormone in the mature female, is an example (one among many from experimental atherosclerosis) of a fundamental truth with respect to atherogenesis, namely, that a complex interplay occurs between exogenous and endogenous factors (Katz and Stamler, 1953; Katz *et al.*, 1958; Stamler, 1959b; Pincus, 1959). Given the long-term ingestion of a potentially atherogenic diet, the nature of the organism (its genetics, heredity, sex, endocrinology, disease history, etc.) exerts an important influence. In addition to the effects of estrogens, it has been demonstrated that thyroid hormone, adrenal medullary and cortical hormones, blood pressure, previous vascular damage, and exercise, among other variables, may affect hypercholesterolemia and atherosclerosis in animals on high-fat, high-cholesterol diets.

Thus, extensive data from experimental animals indicate that atherosclerosis has a multifactorial causation, with habitual diet, a key factor, interacting with others.

4. Conclusions

Four fundamental generalizations may be validly drawn from the vast experimental evidence: (1) Atherosclerosis is a nutritional-metabolic disease. Longterm intake of diets high in fat and cholesterol, leading to a sustained hypercholesterolemic hyperlipemia, is a key factor in its pathogenesis and etiology. (2) Atherosclerosis is a disease of multifactorial causation. A large number of factors, exogenous and endogenous, interact with diet to influence the development of the lesion. (3) Atherosclerosis is a potentially reversible lesion, in whole or part, depending upon its severity and extent. (4) The wealth of research data amassed by animal experimentation and the basic conclusions flowing from these data are of great theoretical and practical importance for science's effort to understand and master the problem of atherosclerosis in man (Stamler, 1960b).

C. PATHOGENETIC FACTORS

By J. STAMLER

1. FACTORS ASSOCIATED WITH INCREASED RISK OF ATHEROSCLEROTIC DISEASE

One of the most important recent advances in the field of atherosclerosis is the definitive identification of several abnormalities associated with increased risk of atherosclerotic disease. Knowledge concerning these factors has not arisen *de novo*. The findings of clinical investigation during the 1920's and 1930's originally led to the implication of several of them in atherogenesis. Incontrovertible evidence, however, has come only recently, from the data of longterm prospective epidemiologic studies in representative samples of the U. S. population.

The abnormalities associated with increased risk of atherosclerotic coronary heart disease in middle-aged men include hypercholesterolemia, hypertension, obesity, diabetes mellitus, hypothyroidism, kidney disease, heavy smoking, and a family history of premature vascular disease. Sedentary living and psychological stress-tension-frustration have also been implicated, but clearcut evidence on such variables is not yet available.

a. Hypercholesterolemia: This single abnormality is associated with a three- to sixfold increase in risk of coronary disease (Stamler, 1959b; Katz *et al.*, 1958; Symposium, 1957; Symposium, 1960; Stamler, 1960b). In the Framingham study (Symposium, 1957), for example, a coronary disease incidence rate of 13/1000/4 years was found in men aged 45–62 with cholesterol levels under 225 mg per cent, in contrast with a rate of 80/1000/4 years for the group with cholesterol levels of 260 mg per cent and above. These data have additional significance because they illuminate a longstanding problem, i.e., proper standards for normal serum cholesterol. It is certainly sound to regard 225 mg per cent as the upper limit of normal; further work indicates the validity of accepting an even lower level.

b. Hypertension: This disorder, as a single abnormality, is also associated with a three- to sixfold increase in risk of coronary disease. Thus, coronary disease incidence rate was 17/1000/4 years in middle-aged men with diastolic blood pressures less than 90 mm Hg, in contrast with 100/1000/4 years for patients with diastolic pressures of 95 mm Hg and greater. Other data indicate that diastolic pressures of 90–94 mm Hg are likewise associated with significantly increased coronary proneness. (For discussion of role of hypertension in atherosclerosis, see (1) Arterial Hypertension, Section B.-9a, Chapter XX.) (Stamler, 1959b; Symposium, 1957; Symposium, 1960; Stamler *et al.*, 1960.)

c. Obesity: This state, particularly gross obesity, is associated with a two- to threefold increase in susceptibility to coronary disease. Thus, incidence rates were 43/1000/4 years in nonobese or slightly obese men aged 50–59 and 87/1000/4 years in the frankly obese. Similar prospective epidemiologic data are available on the influences of diabetes, hypothyroidism, renal disease, heavy cigarette smoking and positive family history of premature coronary disease.

d. Combination of factors: The effects of the various abnormalities in combination have by now been quantitated fairly precisely (Stamler, 1959b; Katz *et al.*, 1958; Symposium, 1957). Thus, middle-aged men with normal values for serum cholesterol, blood pressure and weight had a coronary disease incidence rate of only 10/1000/4 years in the Framingham study, in contrast with the group with two or three abnormalities, who had a rate of 143/1000/4 years, a fourteenfold difference in susceptibility. Expressed differently, these data mean that a 45 year old American male, free of clinical evidence of coronary disease and of all the foregoing abnormalities, has only one chance in twenty or thirty of developing clinical coronary disease before the age of 65. His counterpart with two or more defects has one chance in two or two chances in three.

Such findings are not only of theoretical significance for investigators studying the natural history, pathogenesis and etiology of atherosclerotic disease, but they are also of great practical importance for clinicians. They illuminate the way to the identification of high-risk, susceptible, coronary-prone individuals before they are victimized by the disease. This information is most important, since identification of susceptibles has always been a key aspect of the effort to achieve effective prevention of disease.

2. Prevention of Atherosclerotic Disease

With respect to coronary-prone men, the approach to prevention is quite straightforward, since most of the abnormalities associated with coronary susceptibility are amenable to effective medical-nutritional-hygienic correction and/or control (Stamler, 1960b; Symposium, 1960). This has long since been true for obesity, diabetes mellitus, hypothyroidism, heavy smoking, and lack of exercise. Powerful, safe pharmacologic and dietary tools are now available for the treatment of hypertension, and in recent years it has been conclusively demonstrated that in most cases hypercholesterolemia responds to such nutritional therapy as low-total-fat, low-cholesterol diets, or low-saturated-fat, low-cholesterol diets, containing unsaturated vegetable and marine oils in moderate amounts. These advances constitute the basis and rationale for modern attempts to prevent and treat coronary heart disease and atherosclerotic disease generally, namely, by dietary-hygienic-medical means. For it follows logically from these observations that prevention and correction of the abnormalities making for coronary proneness might be effective in preventing and treating the disease. Both primary and secondary prevention (treatment) of atherosclerotic disease might be achieved in this way.

This is not merely a logical conclusion, since considerable evidence is available suggesting that such prevention can, in fact, be accomplished. Thus, three investigations have yielded data indicating that low-fat, low-cholesterol diets are effective in the secondary prevention (treatment) of coronary heart disease. Life insurance findings indicate that overweight men convert from high to standard mortality experience when obesity is corrected. A significant effect of altered diet is also suggested by the mortality experience during World War II, when a decline in death rates from arteriosclerotic heart disease apparently occurred in the Scandinavian and Low Countries and in Leningrad. Considerable evidence is also available indicating that good control in diabetes mellitus, involving close adherence to prescribed diet, is associated with a lower occurrence rate of vascular complications. Similar findings are also beginning to appear concerning the positive long-term results of the pharmacologic control of hypertension. Thus, there are good reasons, both practical and theoretical, to infer that successful large-scale prevention of atherosclerotic disease has become feasible, as a result of recent research advances.

Atherosclerotic disease, particularly coronary heart disease, is the major present-day public health problem in the United States and in other economically developed countries. It exacts a terrible toll, not only among the elderly, but also among the middle-aged. The incidence rate among middle-aged American men is about 1500/100,000/year, with 25–35 per cent of first attacks being fatal during the acute stage. Furthermore, of the survivors of an acute attack, 20–40 per cent (depending upon clinical circumstances) are dead in five years. The over-all mortality rate from arteriosclerotic heart disease is 330/100,000/year in men aged 45–54, thus accounting for one-third of all deaths in this middle-aged group. Truly this is our Number One epidemic disease at mid-century.

There is no reason to believe that a policy of "watchful waiting," "judicious neglect," "therapeutic nihilism," "symptomatic treatment," or "general supportive measures" will in any way alter this grim picture. Specifics are indicated, and the measures proposed for rational long-term prevention, control and treatment, including the nutritional, hygienic and medical measures, are safe, moderate, sound and free of danger. A wealth of evidence points to the likelihood of their value. Their widespread prophylactic utilization by public health and internal medicine, in dealing with well men, even as pediatrics and obstetrics deal with well babies and well expectant mothers, would therefore seem to be the order of the day.

D. ETIOLOGY AND PATHOGENESIS

By G. M. Hass

The cause of atherosclerosis is unknown. There are reasons for believing that more than one factor is responsible, either as a basic change or in the acceleration of its development in a generalized, systemic, segmental or local sense. Though it begins in childhood, the condition usually progresses more rapidly in middle life and affects men more adversely than women. Often it has a hereditary background. It is especially severe in most patients with diabetes mellitus and tends to be more serious in patients with either pulmonary or systemic (particularly diastolic) hypertension than in normotensive individuals (see (1) Arterial Hypertension, Section C.-2a, Chapter XX). Not only is the blood pressure important, but the rate or volume of blood flow also seems to have an effect. This is manifest principally where the peripheral demand of tissues supplied by a given arterial system decreases or where the volume of flow is reduced peripherally by narrowing of the proximal channel. In these instances the arterial muscle undergoes atrophic changes, while the intima demonstrates fibroelastic proliferative reactions which reduce the caliber of the lumen. In a sense, therefore, the arterial wall undergoes adaptive modifications in response to the demands of the tissues which it supplies.

Inasmuch as lipids invariably accumulate in the intima of many parts of the aged arterial system, the idea has arisen that some metabolic abnormality enhances the tendency for these substances to accumulate through either a mural mechanism, a change in the composition of plasma lipids or an increase in their level. There is little doubt that the level and composition of blood lipids, especially cholesterol and its esters, are factors in development of the disease, especially when the total blood cholesterol is long sustained at high levels. But there is no proof that cholesterol or lipids levels, customary in man, are in themselves responsible for the disease. The reasons for this are as follows: The disease occurs eventually in everyone, irrespective of the magnitude or the duration of these levels. There is no consistent correspondence between the rate of progression or severity of the disease and the level of blood cholesterol or lipids. Patients with high levels of long duration may have less disease than those with lower levels of similar duration. Furthermore, the distribution of xanthomatous deposits, if governed exclusively by blood levels, follows no consistent pattern. Finally, it has not been possible to produce in animals an arterial disease, which resembles the usual human disease, by prolonged maintenance of blood cholesterol or lipids at levels comparable to those regarded as normal for man. When the levels in animals are maintained at much higher values than those common in man, arterial xanthomatosis develops, but its resemblance to the usual human disease in its pattern of distribution or microscopic characteristics is meager indeed. Medial degeneration, with associated calcification and intimal proliferative reactions with conspicuous fibrogenesis, so prominent in human disease, is inhibited, while generalized xanthomatosis, rarely encountered in human disease, acquires prominence. (For a somewhat different point of view on this controversial subject, see Section B.-2, above.)

These matters lead, therefore, to a concept of human disease, involving events which characterize the susceptibility and reactions of arterial walls to injury, in the presence of blood of variable lipid and other compositions, circulating under various pressures at variable rates of flow. It is convenient to discuss this concept in terms of arterial potentials because each part of the arterial system has its own characteristic susceptibility to injury and its own reactive capacities.

1. Calcific-degeneration Potential

All vessels do not undergo medial degeneration, of the types commonly found in human disease, at the same rate or to the same degree. Even when medial degenerative changes in different arterial systems do have a microscopic resemblance, these abnormalities do not necessarily attract calcium to a similar extent. Nor is there any indication that the alterations, with or without calcification, occur at the same time or to the same degree in a given part of the arterial system of different people. However, certain parts of the vascular system and certain elements of these parts have predictable forms of medial degeneration and of distributions of calcium deposits, while others have little or no tendency toward such changes. Furthermore, there seems to be no absolute relation between the occurrence of the degenerative and the calcific changes. Thus, it is necessary to deal with local and systemic calcific-degeneration potentials characterizing the susceptibilities of each part of the arterial system to deleterious factors which may be multiple and not necessarily identical in all systems or in all people.

2. Mesenchymal Reaction Potential

As a consequence of the medial calcific-degenerative changes, each part of the arterial system displays a characteristic structural reactive capacity—its mesenchymal reaction potential. This is manifest principally by proliferation and differentiation of arterial mesenchyme, governed, in part, by composition of the blood, circulatory dynamics and factors which regulate mesenchymal reactions in general. For instance, in the course of medial degeneration, the deteriorative changes may activate mesenchyme in the media, adjacent intima and adventitia. A common early reaction in both the media and adventitia leads to an increase in collagen; in the adventitia there is also proliferation of immature vascularized mesenchyme which, in the presence of appropriate medial disease, tends to penetrate the media. Such a change assists in the reparative and resorptive sequences in the degenerate media, occasionally even extending into the intima. An important early mesenchymal reaction in the intima is a proliferation of intimal endothelium, a reaction which does not necessarily require for its excitation any recognizable disturbance in medial structure. It may be elicited, in part, by factors which activate certain reticuloendothelial cellular responses. In this connection there is indication that some xanthomatous plaques may arise through activation of endothelial and associated mesenchymal cells to absorb, phagocytose, or otherwise entrap neighboring lipid droplets. This activation is not restricted to arterial mesenchyme.

An equally important early mesenchymal reaction in the intima is attributable to disturbances in structural integrity of subendothelial tissues. The first evidence of this reaction is activation or proliferation of mesenchymal cells beneath the endothelium and production of collagen.

Continuation and differentiation of this reaction, whether mediated by local medial disease, composition of the blood, or circulatory dynamics, are governed principally by the mesenchymal reaction potential peculiar to the part of the arterial system affected. This reaction is often so specific that it can be used to define the location from which a microscopic section of some part of the arterial system was taken. Under special conditions in certain arteries, the mesenchymal reaction potential is such that a new arterial wall may be constructed within the framework of the proliferated intima.

3. Lipoidosis Potential

Although there is often a close relation between the medial calcific-degeneration potential and the mesenchymal reaction potential (with the former serving to elicit the latter), and although this may be sufficient to explain the sequences in certain arterial systems, it does not account for the pattern of lipid accumulation, which is a reflection of the tendency toward lipoidosis—the lipoidosis potential. All arterial systems, irrespective of the correspondence of medial calcific-degeneration and mesenchymal activation potentials, do not equally acquire lipid deposits. It is convenient, therefore, to regard each arterial segment and system as having its own lipoidosis potential. Such potentials are easily defined in animals with dietary or metabolic hyperlipemic hypercholesterolemia. In this experimental condition, patterns of distribution of intimal lipoidosis are similar among animals of the same species and closely resemble one another among animals of different species, when blood cholesterol levels are similar in magnitude and duration. Therefore, there is reason for assuming that the lipoidosis potentials of normal human arteries may not be greatly different from those of experimental animals.

However, the potentials of normal animal arteries are such that no significant atheromatous disease occurs unless the blood cholesterol is maintained at levels much higher than those regularly encountered in man. In order to produce significant atheromatous localization in animals with blood cholesterol levels in the average normal human range, the lipoidosis potential of arterial walls must be increased. Experimentally, this can be accomplished by various forms of injury which lead to activation of arterial mesenchyme, especially in or near the intima. This suggests that the average person would not acquire atheromatous intimal deposits if it were not for factors which increase the probability of lipid localization through an increase in arterial lipoidosis potentials. Though these factors may be obscure, it is clear that they exist in man and are manifest principally by the production of medial calcific-degenerative changes and simultaneous or subsequent mesenchymal activation. Inasmuch as these events are essentially focal initially and not necessarily confined to a specific pattern, the variable distribution, rate of development and magnitude of atheroarteriosclerosis in different people and in different arterial systems become understandable in theory.

The aspect which is most difficult to understand is the difference in lipoidosis potentials of normal arteries and the fact that these potentials persist in a relative sense even when increased by disease productive of mesenchymal activation. Thus, arteries with similar dimensions, structures and mesenchymal activation potentials, conveying the same blood under similar pressures, may have greatly different lipoidosis potentials. This is best illustrated by the conspicuous atheromatous deposits in some arteries and the scarcity or absence of such deposits in their large branches. One good example of this is the common severity of atheroarteriosclerosis in the extramyocardial coronary arterial system and the insignificant nature of the disease in the immediate primary intramyocardial branches (for further discussion of coronary atherosclerosis, see Section G., Chapter IX). The same applies, as a rule, to the cerebral, renal, hepatic, splenic, and many other important organ systems. Though this may be due to local metabolic differences, it seems desirable to seek for explanations based on differences in regulation of the functions of these organs. It is equally difficult to understand the refractoriness of the venous system to lipoidosis despite the frequency of medial degenerative and intimal proliferative changes in this portion of the vascular tree. One distinction between systems with high lipoidosis potentials and those with negligible potentials is the tendency of the former to acquire considerable calcium in the medial degenerative lesions and often in the newly-formed intimal mesenchyme as well. This is commonplace throughout the arterial and venous systems.

4. Thrombosis Potential

The thrombosis potential, the tendency to favor local thrombosis, varies throughout the arterial system. The probability of thrombosis is great in some locations and negligible in others. The level of this potential depends principally upon the three other potentials. Medial degeneration alone is not nearly as important as calcific medial degeneration; mesenchymal activation by itself is also not especially important. Nor is lipoidosis, whether independent or associated with avascular mesenchymal activation, a very significant factor. However, the excitation of vascularized intimal mesenchymal reactions does play a role, occurring principally as a consequence of vascularization of medial degenerative disease. Medial vascularizing reaction alone is relatively innocuous unless it penetrates the intima. As a rule, such a response is not stimulated by an uncomplicated intimal proliferation but is elicited by a combination of lipoidosis and secondary calcific-degenerative changes in the newly-formed intima. Apparently, lipid accumulations interfere with the production of a suitable fibrous support for the intimal system of new capillaries and, as interstitial degenerative changes occur, the capillaries leak fluid, plasma or whole blood, a type of alteration which enhances the likelihood of intraluminal mural thrombosis. Such a series of events accounts for most of the facts in man and is in accord with observations on experimental atheroarteriosclerosis in animals. Also, it explains the infrequency of spontaneous nonembolic mural thrombosis under the following conditions in man and animal: severe avascular medial calcific degeneration, severe occlusive avascular intimal proliferation with degeneration, and severe occlusive avascular intimal lipoidosis with degeneration. It also accounts for the frequency of mural thrombosis when various combinations of these processes undergo vascularization. In addition to these inter-related events, there are mysterious mechanisms which seem to be independent of factors which affect the potentials of arterial walls. There are reasons for believing that they may be related to saturation of systems with the highest potentials, so that events normally occurring in such systems under constant conditions tend to proceed elsewhere in systems with the next highest potentials. There is some indication experimentally that the spread and distribution of xanthomatous, as well as medial calcific-degenerative, disease may reflect the operation of some such mechanism.

In general, for the solution of the problem, it is necessary to search for factors which lead to calcific medial degeneration. Once this process attains sufficient activity in certain arterial channels, the other mechanisms are set in motion. It is likely that if calcific medial degeneration could be prevented, other factors which enhance progression of the most undesirable atheroarteriosclerotic changes would be tolerated well by most people. If calcific medial degeneration can not be prevented, the chain of reactions arising therefrom could best be minimized or interrupted by measures for keeping the blood cholesterol level below 180 milligrams per cent, for preventing unduly increased blood pressure and, when indicated, for reducing the clotting tendency of the blood.

E. GROSS CHANGES[2]

By C. B. TAYLOR

Atherosclerosis is a degenerative arteriopathy which primarily affects large and medium-sized arteries and is characterized by a patchy distribution of focal intimal thickenings (plaques) made up largely of stainable lipid and fibrous tissue. Because of their marked pleomorphism, the lesions may best be described by separately considering the several typical types (which may often be but different stages in the natural history of atherosclerotic plaques) and their secondary changes.

1. TYPES OF LESIONS

a. Early plaques: The earliest lesions are generally believed to be the thin yellow spots or streaks ("fatty flecks or streaks") that may be found at any age and are common in young people, even children. In fact, gross fat stains show lesions in the aorta of almost everyone over 3 years of age. They increase in number and size with the progression of time and typically show a beginning distribution around and between the orifices of intercostal arteries, along the posterior wall of the thoracic aorta. The intimal location of early plaques is easily demonstrated by stripping the intima from a not overly diseased aorta; the yellow plaques also strip off, appearing to be suspended within the thin, almost transparent intima. Early plaques either may change little for long periods, may regress with only a minute fibrous scar remaining, or may show progressive lipid deposition and/or fibrous tissue proliferation.

b. Fatty plaques: If lipid accumulation predominates, plaques may become quite large, up to 0.5 cm or more in thickness and 1–2 cm in cross section diameter. Contiguous lesions frequently coalesce. Many plaques are yellow, often a bright carrot-yellow, but sometimes residual pigment from previous hemorrhage gives a gray or dark brown color. Yellow always indicates a high lipid content, although the yellow substance itself is not a lipid but a mixture of xanthophyll and carotene pigments. Fatty plaques are usually particularly abundant, large, and bright yellow in cases of extreme hypercholesterolemia, e.g., in nephrosis, xanthomatosis, and xanthomatous biliary cirrhosis. A fibrous cap

[2] For a discussion of the gross changes in coronary atherosclerosis, see Section G.-3, Chapter IX.

over the luminal surface often diminishes the prominence of the yellow color of older plaques (Fig. 119). (For a discussion of the genesis of atheroma, see Section D.-4c, Chapter II.)

c. Fibrous plaques: Other early plaques, or possibly later stages of fatty plaques, develop into firm, dense intimal "buttons" that have a milky-white to pearly-gray surface. Arteries of adults usually contain both fatty and fibrous plaques; however, occasional cases, particularly in the older age groups, show chiefly fibrous lesions. The cut surface of fibrous plaques often is stratified and gives the impression of a layer of dense tissue stuck on the underlying intima; these lesions also may frequently be stripped off with the intima. Many plaques appear to be all fibrous tissue when judged by their intimal surface, but on cross section are seen to contain one or more yellow layers of lipid. In this respect most plaques are really "mixed," but the fibrous component may be the predominant component and not exceptionally it is the sole one, as determined by gross examination (Fig. 119). Characteristically then, every plaque contains both elements: lipid and fibrous tissue, since the atherosclerotic lesion is not purely degenerative, but also, in large measure, proliferative.

2. Secondary Changes

a. Central necrosis: Progressive widening and thickening, especially of fatty plaques, may be associated with degeneration and necrosis of the center, which becomes a pool of soft, often greasy, necrotic material that frequently literally sparkles from the presence of many small flat shiny white crystals of cholesterol. Dissection may reveal extension of the "softening" into the underlying media, and the lesion can now no longer be cleanly removed with the intima. With larger, more advanced, plaques, one can often squeeze the margins and express—through central soft spots or ulcerations in the luminal surface—the grumous contents which ooze out in semiliquid consistency (Figs. 119 and 120).

b. Surface ulcerations: Irregular surface defects of variable size, from minute erosions to ragged ulcerations of 1 cm or larger, are common in the more severe cases of atherosclerosis, especially in the aorta. Fatty plaques, particularly advanced lesions with central necrosis, usually underlie large surface defects. It is not unusual to find thrombi attached to such rough spots; what is surprising, however, is the frequency of ulcerations, even large ones, that do not have associated thrombi (Fig. 120).

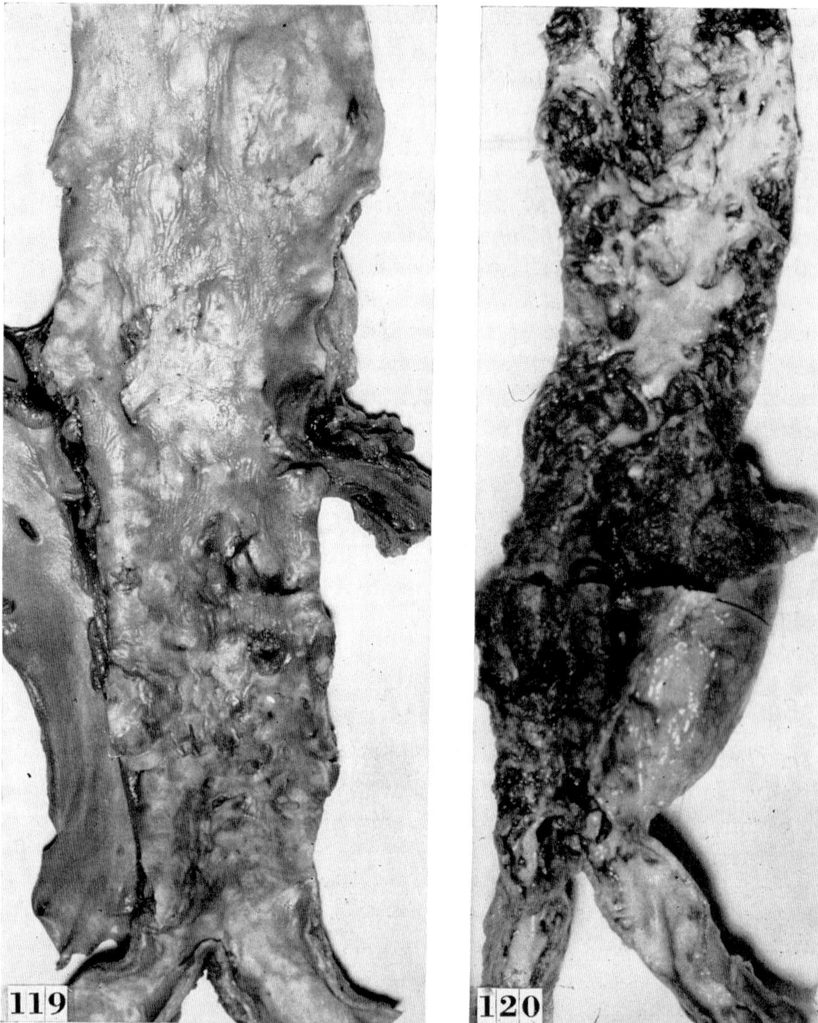

FIG. 119. Abdominal aorta showing moderately far-advanced atherosclerosis, with large and coalesced plaques, in most instances covered with a "pearly" fibrous cap. Several plaques in the lower half of figure show central necrosis and ulceration. Circumference of vessel increased due to medial degeneration and scarring.

FIG. 120. Far advanced aortic atherosclerosis, with a large aneurysm of the lower abdominal aorta which has been opened. The majority of the large, coalesced atheromata in this vessel have undergone extensive necrosis and ulceration, with the surface of each ulcer being covered with a light-brown greasy material which actually sparkles from the many small crystals of cholesterol it contains. Some of the ulcers have loosely attached mural thrombi attached to their surfaces. Aneurysm is nearly occluded by a thick, laminated mural thrombus attached to the degenerated, ulcerated wall.

c. Hemorrhage within plaques: Grossly visible hemorrhage is more frequently found in degenerated fatty plaques. Single or repeated hemorrhages may be a significant causative factor in the size and softness of some plaques, as well as associated surface ulcerations and thrombi. However, some workers believe that the luminal thrombus may frequently be primary, with the subjacent intramural hemorrhages being part of the local reaction to the thrombus.

d. Calcification: Calcium salts frequently deposit within atherosclerotic plaques, forming thin, hard flakes that are brittle and resist bending but can be cracked fairly easily. These flakes or plates vary in size up to 1–2 cm in cross section diameter and are generally anchored in dense fibrous tissue which completely surrounds them. As with dystrophic calcification elsewhere, this type is noted in sites of previous cellular injury or degeneration (atherosclerotic plaques). However, the more florid fatty lesions often are associated with minimal gross calcification, while extensive calcification may occur with minimal lipid deposition, as seen in the lower abdominal aorta and common iliac arteries of middle aged or older people. Coronary arteries in the latter group frequently are so severely calcified that many segments simply shatter when serial section is attempted with a sharp knife.

e. Involvement of media: Medial involvement is a common consequence of progressive development of the intimal lesions and is a frequent feature in the wall of atherosclerotic aneurysms, present in the lower aorta and iliac arteries (Fig. 120). Fibrosis and calcification also cause tortuosity and loss of elasticity of large arteries.

3. Vagaries of Atherosclerosis

The distribution of lesions is often so patchy and so variable from one artery to the next, that no single principle appears to explain it. This also applies to the severity of the condition. For example, marked coronary atherosclerosis may accompany a much lesser degree of aortic atherosclerosis, or vice versa. The renal arteries usually show very little involvement, despite the often advanced disease of the abdominal aorta from which they arise. Arteries of the head and neck may be prominently affected with yellow fatty plaques, but those tend to develop later in life than do the changes in the aorta or coronary arteries. Pulmonary atherosclerosis is usually much less severe than aortic or coronary disease; however, its development is considerably stimulated by pulmonary hypertension. Arteries of the upper extremities show minimal atherosclerosis, with rarely a completely obstructive lesion, whereas severe atherosclerosis, with obstruction, is not at all unusual in the lower extremities. Atherosclerosis usually stops sharply at the point of entry of vessels into firm visceral tissue.

F. MICROSCOPIC AND SUBMICROSCOPIC CHANGES[3]

By C. B. TAYLOR

1. MICROSCOPIC CHANGES

a. Genesis of lesions: Different investigators have variously interpreted atherosclerotic lesions as being initiated by either prior degenerative changes in the media or internal elastic lamina (Gillman, 1959), fibrin clots adherent to endothelium (Duguid, 1959), or degenerative changes in the ground substance of the intima (Moon, 1959).

It would also appear that more emphasis should be given to the reaction of subendothelial cells to excesses of lipids in their environment. The intima of young healthy arteries consists of a layer of endothelial cells and a very thin subendothelial layer which contains scattered primitive mesenchymal cells, with multipotential regenerative powers; the latter play a major role in the repair of vascular tissue and in the genesis of atherosclerosis (Taylor, 1955). The intima and inner media depend upon diffusion from luminal blood for nutrition, while this is supplied to the adventitia and the outer media by vasa vasorum.

The diffusion of luminal plasma into and through the intima and superficial media is of major importance in the genesis of atherosclerosis, for there is a large body of clinical and experimental evidence which indicates that plasma lipids precipitate out into subendothelial spaces of the intima during diffusion of plasma lipoproteins through the intima and superficial media. In this regard, recent studies on monkeys have indicated a critical level of total serum cholesterol of about 250 mg/100 ml; at this level or higher, lipids precipitate in the subendothelium and are phagocytized by the multipotential mesenchymal cells in the subendothelium (Taylor *et al.*, 1957). Phagocytosis of lipids in the subendothelium represents an early form of atherosclerosis (Fig. 121).

[3] For a discussion of the microscopic changes in coronary atherosclerosis, see Section G.-4, Chapter IX.

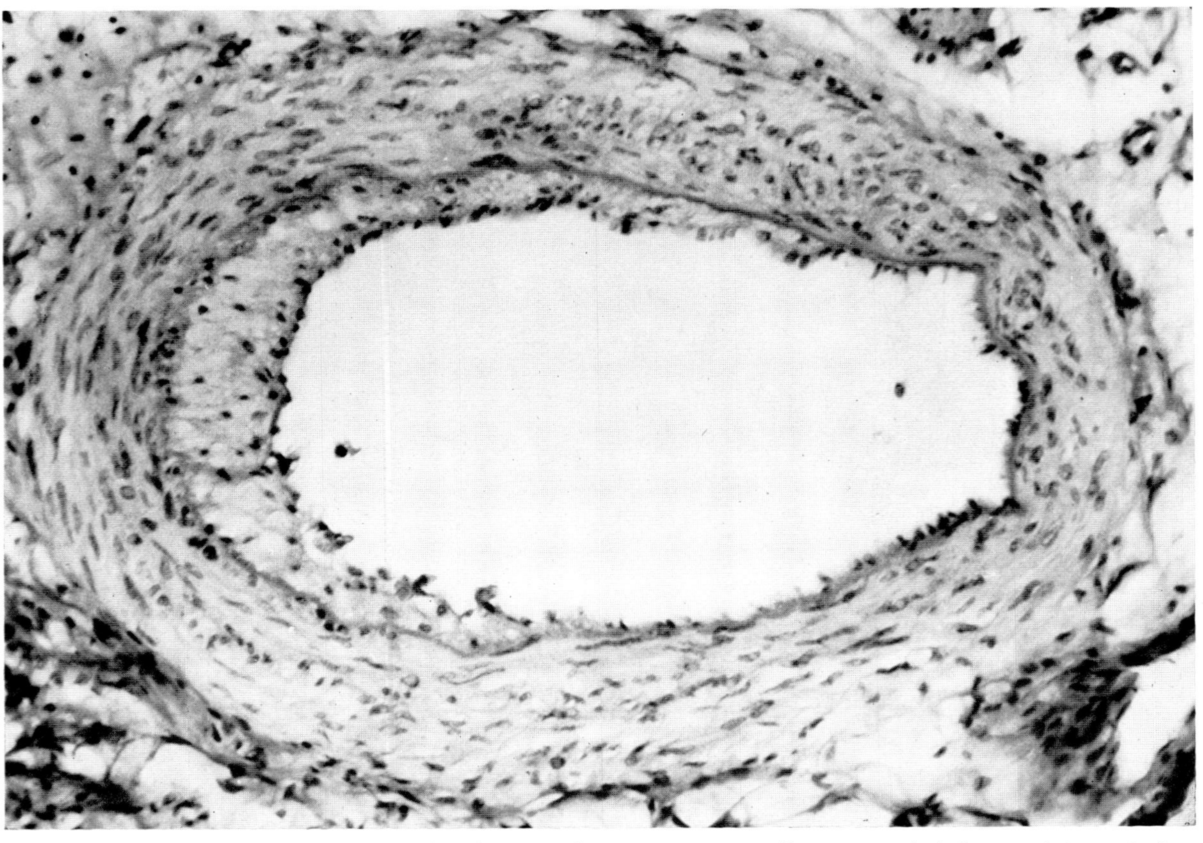

Fig. 121. Photomicrograph of an early atheroma of a coronary artery (between endothelium and internal elastic membrane, on left side), made up entirely of macrophages filled with lipid ("foam cells"). No fibrous cap or medial change noted at this stage.

While the propensity of lipids to accumulate within the intima is striking, it would appear that the internal elastic membrane usually delimits early deposits of lipids and atheromas in the media of arteries. Flow of interstitial fluids in the media via vasa vasorum is accomplished under much more stable conditions, since lipids do not precipitate out in this layer unless the structure of the media has been markedly altered. It has been proposed that the presence of lipids in the intima also stimulates proliferation of primitive mesenchymal cells into phagocytes, resulting in a mixed intimal lesion composed of lipophages and connective tissue (Duff, 1936). Fibrous elastic scars may become quite thick and the intima vascularized with vasa vasorum originating in the media and from the endothelial surface. They lie in the poorly formed, often degenerating, tissue of the plaques and frequently rupture with subsequent hemorrhage into the plaques and subsequent thrombosis (Patterson, 1955) (Fig. 122).

b. Changes in intima: Early lesions show fine droplets of lipids scattered through the ground substance and also accumulated within macrophages (foam cells). There is considerable evidence that these lipids originate from the plasma by diffusion through the endothelium. However, the endothelial cells rarely show detectable changes. Fibrous tissue begins to proliferate in response to the presence of the lipids.

As atheromatous lesions develop, fibrous tissue increases in amount, and often the new tissue is of poor quality, showing varying degrees of hyalinization and sometimes fibrinoid necrosis. With increasing concentration of lipids within the deeply-placed foam cells, many of these cells disintegrate and an acellular mass of lipid and necrotic debris forms. In such areas, there are numerous crystals of precipitated cholesterol (Fig. 122). As this necrotic, atheromatous tissue encroaches upon the elastic membranes, fraying, fragmentation, and rupture occur and the media may be invaded by the atheromatous process (Fig. 122). Intact but lipid-filled macrophages usually form the outer margin of these plaques, and beyond this, varying thicknesses of fibrous tissue surround the lesions (Fig. 122). A few fibrocytes and lymphocytes can usually be found. Indeed, in far advanced lesions, foci of lymphocytes may be quite prominent.

In the presence of the degeneration and necrosis, calcium salts frequently deposit; initially calcium stains demonstrate the presence of fine granules of this material around elastic and collagen fibers. Calcification may proceed to form large, solid aggregations.

c. Changes in media: Changes in the media are primarily degenerative, with minimal proliferative reaction, and are usually associated with

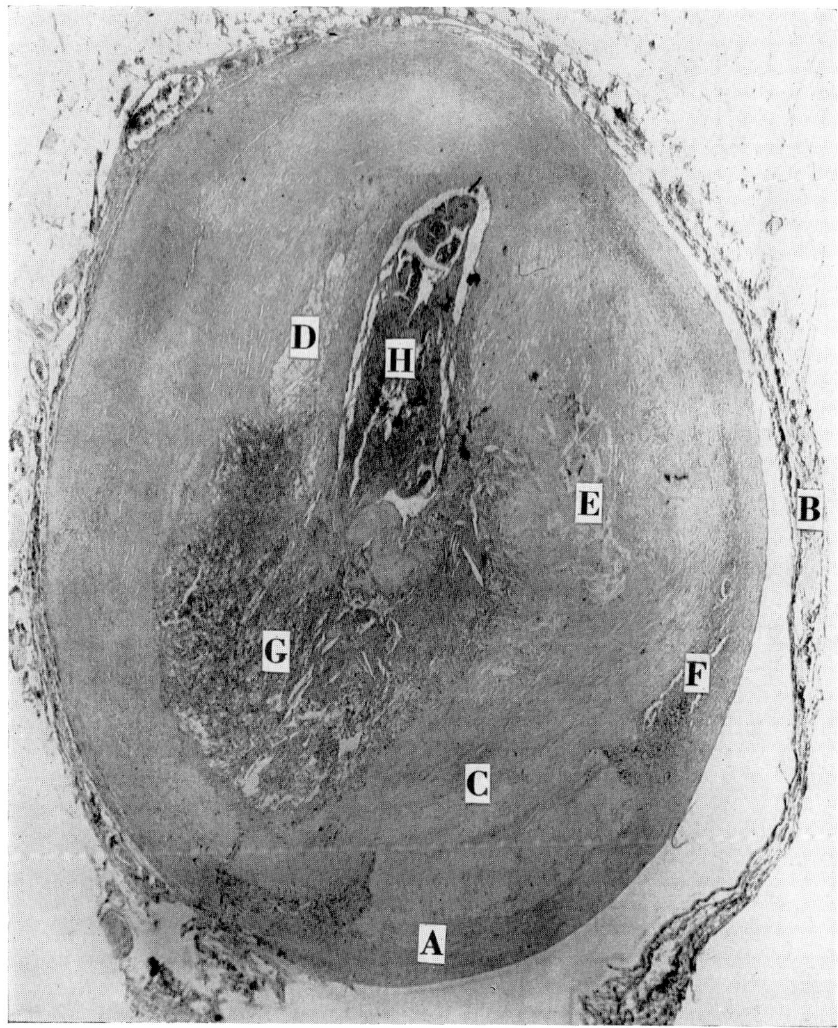

FIG. 122. Photomicrograph of severe coronary atherosclerosis with intramural hemorrhage and thrombosis. (A), media, the darker, more cellular, thin rim of tissue lying just inside the loose-textured adventitia (B). (C), intima, enlarged to three to five times the thickness of the media and consisting of grumous, fatty plaques (lower right and right of central thrombus). (D), new groups of fat-laden macrophages. (E), dense hyaline scar containing degenerated atheromatous deposit, with fat and cholesterol crystals lying in its center. (F), vasa vasorum at junction of media with thickened intima. (G), large intramural hemorrhage resulting from rupture of a vasa vasorum which had proliferated into a partially degenerated atheromatous deposit. (H), dark blood clot in lumen, attached to the site of maximum disruption and inflammation of the endothelium caused by reaction to intramural hemorrhage.

overlying intimal atheromas. As the degeneration of the intima causes a breakdown of the internal and deeper elastic membranes, the process extends into the media (Fig. 122). The first changes may be abnormal staining reactions, followed by fragmentation with structural discontinuities of the elastic membranes. Smooth muscle cells may then undergo degeneration and be completely absent from relatively large areas. These alterations may be unassociated with foam cells, or lipid infiltration may be prominent, as in the "collar button" lesions of cerebral arteries (Duff, 1955) or in advanced central necrosis of fatty plaques anywhere. In long-standing lesions, calcification becomes prominent. There appears to be minimal reaction surrounding the calcium deposits, with some of the larger ones becoming converted into bone.

d. Role of other arteriopathies: Atherosclerosis can develop secondarily in arteriopathies due to other causes. Almost all of the regenerative and reparative capacity of arterial tissue stems from the subendothelial primitive mesenchymal cells mentioned earlier. As demonstrated in the experimental animal, these multipotential cells have a capacity to form essentially a new vascular wall in the framework of a proliferated intima, as contrasted with the media in which very little regeneration of the vascular tissue occurs (Taylor, 1955) (Fig. 123). In hypercholesterolemic animals and apparently in moderately hypercholesterolemic man, intimal proliferation following medial degeneration consists of a mixed response, with some primitive mesenchymal cells being diverted to phagocytosis of subendothelial lipid, and, as a result, the more desirable response of regeneration of new functional vascular tissue being adversely affected (Fig. 123). Thus, it appears that in the moderately hypercholesterolemic man, arteriopathies due to other causes enhance the rate of development of atherosclerosis.

Ideally, intimal scars should contain well oriented, new elastic membranes, fibrocytes, smooth muscle cells and a moderate number of well oriented collagen bundles. In hypercholesterolemic animals and apparently in moderately hypercholesterolemic man, intimal scars are composed of an abundance of fat-laden macrophages, poorly formed and often hyalinized collagen, very meager amounts of poorly oriented elastic tissue, no smooth muscle cells, and a few scattered fibrocytes (Fig. 123). With partial loss of elastic and contractile characteristics, the arterial bed is left with a decreased quantity of functional tissue available to dilate and contract with each cardiac cycle. Consequently, the potential for development of new arterial lesions, with subsequent development of more atherosclerosis and poor intimal scars, is enhanced. This vicious

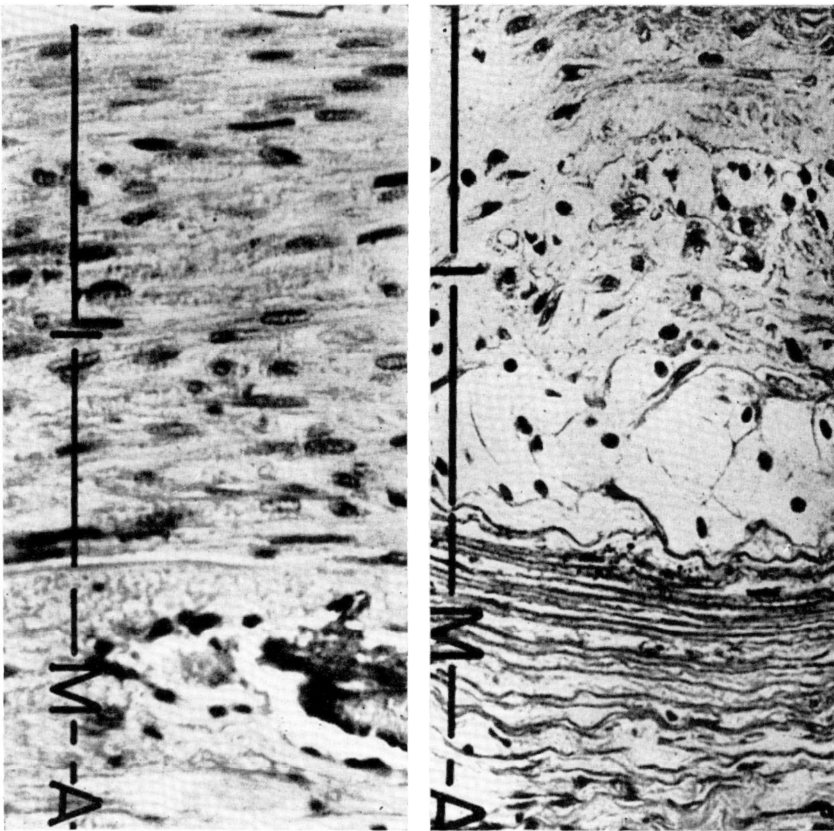

FIG. 123. Photomicrographs demonstrating profound effect of hypercholesterolemia on vascular repair at sites of injury and degeneration. *Left:* Cross section of a healed arterial lesion produced by freezing the aorta of a normocholesterolemic rabbit on a low-fat, cholesterol-free diet. (A), adventitia, composed of dense bundles of collagen. (M), damaged, degenerated and partially calcified media, showing few viable cells and practically no reparative response. (I), intima, containing multipotent mesenchymal cells which have proliferated, differentiated, and formed an essentially new arterial wall lying within the degenerated shell of the old injured media (M). *Right:* Cross section of inadequate healing process occurring at site of injury produced by freezing the aorta of a hypercholesterolemic rabbit (average serum cholesterol level, 1,090 mg/100 ml). (A), adventitia; (M), degenerated media; (I), proliferated intima. Comparison with the section on the left demonstrates the marked alteration in intimal repair. Lipophages occupy the lower one-half of the proliferated intima. The upper half of the intimal scar shows only a few scattered fibroblasts lying in partially hyalinized scar tissue, with "mucinous" degeneration noted in some areas.

self-accelerating cycle of events explains, in part, why atherosclerosis usually develops so rapidly during the later decades of life.

e. Involvement of other vascular channels: Other vascular channels, such as lymph vessels and veins, rarely if ever develop atherosclerosis, a finding which is probably related to their low luminal pressure. However, lymphatic drainage of arterial walls may be an important local factor in the genesis of atherosclerosis; further studies in this area are critically needed.

2. Submicroscopic Changes[4]

Studies using the electron microscope have confirmed many findings observed by light microscopy and have demonstrated several new features that are interesting and probably significant. Among these is the presence of numerous pockets or recesses in the surface of the endothelial cell which communicate with the outside of the cell and extend inward (Buck, 1957). These pockets, lined with extensions of the cell membrane which can be pinched off to form free vesicles within the cytoplasm, have been called "caveola intracellularis." Electron microscopy, therefore, gives a picture of the luminal surface of endothelial cells as not being smooth but highly irregular, in the form of round or elongated indentations which may become filled with plasma constituents, then be pinched off and transported through the cell, thereby conveying plasma constituents to the subendothelial tissue.

In atherosclerotic lesions of cholesterol-fed rabbits, dilated cisternas have been seen in the endoplasmic reticulum of endothelial cells, over the surface of lesions (Buck, 1959), filled with material interpreted to be lipoprotein. The interiors of lesions contain highly vacuolated cells (foam cells), considered to be macrophages that have phagocytized lipid freed from its protein complex. More recently Parker (1960) has described in rabbits dense material, probably lipoprotein, adherent to the luminal surface of endothelial cells, after 24 hours to 108 days of fat and cholesterol feeding. This lipoprotein complex has also been found in numerous "caveola intracellularis," apparently representing transport through endothelial cells. Parker has likewise described "initial elastic lesions," both in acute (24 hours) and chronic lipemia, which consisted of discrete focal areas in the internal elastic lamina showing a decrease in density and a swelling of the lamina. These lesions have been interpreted to represent lipid substances which entered into the elastic

[4] For general discussion of submicroscopic anatomy of blood vessels, see Section C., Chapter I.

protein to form a solid solution of lipid in elastic tissue. The electron microscope has not demonstrated pores in the endothelial surface (excepting in renal glomeruli, see Section C.-1c, Chapter XIII) or "cement substance" between endothelial cells (Buck, 1957). Instead, in cholesterol-fed rabbits there is a minute space between what is interpreted to be the cytoplasmic membranes of adjacent endothelial cells, presumably filled with some sort of interstitial fluid and enlarged irregularly.

G. BIOCHEMICAL CHANGES

By J. E. Kirk

1. Composition of Arteriosclerotic Tissue

Several systematic studies have been conducted on the composition of arteriosclerotic tissue, the majority having been made on human aortic tissue. The results, for the most part, show fairly good agreement, although some aspects of the subject remain controversial. It seems likely that the differences in the reported findings are due to the various analytical procedures used and to the employed classifications of the arteriosclerotic changes.

a. Changes in connective tissue: Connective tissue analyses of intima-media samples of the human thoracic aorta with varying degrees of arteriosclerosis have been performed by Buddecke (1958). These studies showed that with the development of arteriosclerosis there was a marked decrease in the elastin concentration of the tissue, but no significant change in the arterial collagen content. In the case of both proteins, however, considerable variations were found for the values reported for different arteriosclerotic samples. The mean elastin concentrations, expressed as per cent of dry, lipid-free tissue, noted for normal aortas and for samples with moderate and severe arteriosclerosis were, respectively, 28.6, 22.9, and 16.8. The corresponding average concentrations found for collagen were 22.4, 21.8, and 23.0.

b. Changes in lipid content: It is generally agreed that, in connection with the development of arteriosclerosis, the major changes in the composition of the arterial wall occur in the lipid composition of the intima. It has been shown by several authors (Weinhouse and Hirsch, 1940; Buck and Rossiter, 1951; Hevelke, 1956) that the absolute tissue concentrations of total lipids, free cholesterol, ester cholesterol, lecithin, sphingomyelin, and neutral fat usually increase with the severity of the disease. Since a rise in lecithin and neutral fat has not been observed in connection with the aging process of the aorta, it has been concluded by Buck and Rossiter (1951) that an increase in the tissue concentrations of these compounds is directly related to the arteriosclerotic disease process.

Although the concentrations of both free and ester cholesterol are elevated in atherosclerotic tissue portions, the values reported by various

TABLE XVII

PERCENTAGE COMPOSITION OF LIPID MATERIAL EXTRACTED FROM HUMAN AORTIC TISSUE PORTIONS EXHIBITING VARIOUS DEGREES OF ATHEROSCLEROSIS

	Weinhouse and Hirsch (1940) (Intima)			Buck and Rossiter (1951) (Intima-media)						Hevelke (1956) (Intima-media)		
	Free cholesterol	Total cholesterol	Total phosphatides	Free cholesterol	Total cholesterol	Total phosphatides	Lecithin	Sphingomyelin	Cephalin	Free cholesterol	Total cholesterol	Total phosphatides
Atherosclerosis, Grade 1	15.5	37.5	21.1	18.3	31.9	28.2	6.1	12.2	9.9	15.3	21.6	25.8
Atherosclerosis, Grade 2	17.4	44.4	14.1	25.1	40.4	19.1	4.8	10.1	4.3	20.2	25.8	23.1
Atherosclerosis, Grade 3	21.1	48.1	12.9	29.5	46.8	15.3	4.4	9.2	1.8	24.1	29.8	14.9

authors for the ratio of free sterol/ester sterol in such lesions of the aorta show great differences, the calculated ratios ranging from about 0.7 (Schönheimer, 1926; McArthur, 1942) to 5.0 (Hevelke, 1956). A tendency for the ratio to decrease with the development of severe atherosclerosis has been noted by Schönheimer (1926). In contrast to this, observations by Meeker and Jobling (1934), Zeek (1936) and Buck and Rossiter (1951) have indicated that the relative proportion of free to ester cholesterol is higher in late than in early lesions of the aorta.

Analyses of the percentage composition of the lipid material extracted from atheromatous aortic tissue samples have been reported by various authors. Although somewhat different criteria were used for the grading of the lesions, the results of three main studies, which have been summarized in Table XVII, show good agreement, denoting an increase in the cholesterol fraction and a decrease in the phosphatide fraction in relation to the aggravation of the atherosclerosis. The values listed in the table are essentially of the same magnitude as those previously reported by Lehnherr (1933), Meeker and Jobling (1934), and Page (1941).

The absolute values observed by Weinhouse and Hirsch (1940) for fatty, fibrous, calcified, and ulcerated intimal aortic plaques are presented in Table XVIII. The data reveal that considerable differences

TABLE XVIII
COMPOSITION OF AORTIC INTIMA IN VARIOUS TYPES OF ARTERIOSCLEROSIS[a]

Type of arteriosclerosis	Dry matter	Total lipids	Free cholesterol	Sterol ester	Total cholesterol	Total phosphatides	Total ash	Calcium
Fatty plaques	32.7	7.71	1.22	1.70	2.92	1.55	1.29	0.27
Fibrous plaques	33.5	9.05	1.67	2.76	4.43	1.20	3.85	1.35
Calcified plaques	61.6	7.69	1.62	2.05	3.67	0.99	39.50	15.40
Atheromatous ulcer	39.1	14.15	3.67	3.14	6.81	2.15	11.05	3.94

[a] Values are expressed in percentage of wet tissue weight and are calculated on the basis of data reported by Weinhouse and Hirsch (1940).

exist in the composition of the various types of arteriosclerotic aortic lesions, the most notable findings being the high percentages of dry matter, total ash, and calcium in the calcified plaques and the high ash and lipid contents of the ulcerated tissue.

c. Changes in inorganic constituents: Analyses of the inorganic constituents of the intima-media layers of aortas with varying degrees of arteriosclerosis have been reported by Buck (1951). The studies showed an increase in nonlipid phosphorus from 0.116% of wet tissue weight in samples with mild arteriosclerosis to 1.238% in aortas with severe arteriosclerosis. The corresponding values for calcium were 0.212% and 3.093%, and for magnesium 0.015% and 0.046%. It has been shown by various investigators (Baldauf, 1906; Schönheimer, 1928) that the ratio of calcium to nonlipid phosphorus in arteriosclerotic tissue was similar to that found in bone. In the study by Buck (1951) on aortic tissue, the increments of calcium were accompanied by proportional increments of phosphorus in the ratio of 1:0.36. Such a relationship was found to be constant despite great variations in the degree of arteriosclerosis, and it remained unchanged even in samples exhibiting a thousandfold increase in the concentration of calcium. The ratio observed by Buck (1951) agreed fairly well with that of 1:0.40 found by Schönheimer (1928) in two calcified samples of the aorta. The relative proportion in which the minerals are deposited may, according to Buck, indicate that they are in chemical combination.

An increase in the magnesium concentration of aortic tissue in connection with the development of arteriosclerosis has been reported both by Buck (1951) and by Rechenberger and Hevelke (1955).

d. Changes in components of mucopolysaccharides: Studies by Kirk and Dyrbye (1956) have revealed no significant correlation between the degree of arteriosclerosis and the concentrations of hexosamine and acid-hydrolyzable sulfate in the intima-media layers of human aortic tissue. Similar findings have been made by Noble *et al.* (1957) for the hexosamine content of normal and atherosclerotic portions of the aortic intima. In contrast to this, analyses performed by Buddecke (1958) on lipid-free, dry aortic tissue have shown notable increases in both the hexosamine and sulfate concentrations of the aorta in connection with the development of arteriosclerosis. The fact that the accumulation of nonmucopolysaccharide material in severely arteriosclerotic tissue will affect the hexosamine and sulfate values when these are expressed on the basis of wet, lipid-containing tissue weight may to some extent explain the difference between the reported findings.

e. Changes in internal carotid and cerebral arteries: A study on the chemical composition of the human internal carotid artery and cerebral arteries in relation to arteriosclerosis was made by Buck *et al.* (1954). They found that although the concentrations of free and ester cholesterol in both vessels were significantly correlated with the severity of the arteriosclerosis, the ester cholesterol concentration was markedly lower in the cerebral arteries than in the carotid artery tissue. In contrast to the observations made by Buck and Rossiter (1951) on the human aorta, the ratio of free sterol/ester sterol was found to decrease in both the internal carotid and cerebral arteries with increasing severity of the arteriosclerosis. While the analyses of the carotid artery showed a considerable increase in the calcium and nonlipid phosphorus contents of the arteriosclerotic tissue, the values for these compounds remained remarkably low in the cerebral arteries, even in samples exhibiting a severe degree of pathologic change. It has been suggested by Buck *et al.* (1954) that the fact that the cerebral arteries differ morphologically from the muscular-walled arteries may reflect itself in the chemical manifestations of the disease processes.

2. Enzyme Activities of Arteriosclerotic Tissue

Comparisons of the mean enzyme activities of arteriosclerotic tissue with the values observed for normal portions of the same blood vessels (Section G.-2, Chapter III) have, in general, revealed lower values for the former when the results were expressed on the basis of wet tissue weight. The investigations have further shown a rather high degree of variation in the activities recorded for individual arteriosclerotic samples. In the case of most of the enzymes, however, the average differences between arteriosclerotic and normal tissue areas were not significant when the activities were calculated on the basis of the tissue nitrogen content. Nevertheless, in the case of some enzymes, as for example, aldolase and glucose-6-phosphate dehydrogenase, distinctly lower values have been recorded for the arteriosclerotic tissue. It should be pointed out in this connection that *in vitro* studies on human aortic samples performed by Dyrbye (1959) have revealed a markedly lower incorporation rate of radioactive-labeled sulfate into arteriosclerotic tissue than into normal tissue. Since the permeability of arteriosclerotic aortic tissue has been found to be distinctly higher than that of normal tissue (Kirk and Laursen, 1955), these observations by Dyrbye suggest a reduced activity in the arteriosclerotic tissue of the enzymes associated with the exchange of the sulfate groups of the acid mucopolysaccharides. It should be mentioned in this connection, as previously pointed out by Kirk and Laursen, that the increased permeability of the arteriosclerotic tissue may be responsible for the loss from the blood vessel wall of compounds of significance for the metabolism of the tissue. It remains uncertain, however, whether the increased permeability should be considered as a cause or a result of the arteriosclerotic changes.

2. Mönckeberg's Sclerosis

By G. M. Hass

A. DEFINITION

Mönckeberg's sclerosis is a term used to classify a form of arteriosclerosis characterized principally by medial calcific-degenerative changes in the predominantly muscular arteries which supply the extremities. It begins in early life and gradually progresses with advancing age in both sexes of all races. It is more conspicuous in the arteries of the legs than of the arms and tends to be more severe in patients with diabetes mellitus. The cause is unknown but it seems to be a variant of the generalized arteriosclerotic process and not an independent disorder. The disease follows a well-defined course, more rapidly and more severe in some than in others. It is first most easily recognized proximally in the large muscular and musculoelastic branches of the aorta. With the passage of time, it progresses distally affecting first the larger and later the smaller branches of these vessels. It tends to diminish in severity with the decreasing size of arteries, more so in the upper than the lower extremities. It is seldom conspicuous in vessels which serve as the immediate intrinsic supply of highly cellular functional tissues, and it is characterized initially by medial degeneration and its consequences.

B. PATHOGENESIS

The medial degeneration, which is ordinarily of unknown pathogenesis, leads to a decreased ability of the vessel to withstand internal blood pressure, so that there is a tendency toward elongation, tortuosity, and dilatation. This is preceded and accompanied by a decrease of muscular tone and efficiency, with an increase in firmness and inelasticity. These changes vary in severity from place to place and are principally a reflection of the distribution or magnitude of degenerative and reparative sequences. The degenerative changes are characterized principally by accumulation of calcium and lipids, while the reparative ones primarily involve mesenchymal reactions, often of complex nature. The accumulation of calcium tends to occur in the media as rather uniformly distributed segmental foci which progressively enlarge in the circumferential and longitudinal directions, fusing with one another to form circular and spiral bands; the latter slowly spread and unite to make a cylindrical calcified shell within the media. As this occurs, there is a reactive fibrosis in the media and intima, with secondary xanthomatous deposition occurring principally in the reactive intimal mesenchyme. The variations in these gross changes from place to place lead to caliber changes, varying from considerable enlargement to total obstruction, the latter being noted principally in the anterior and posterior tibial arteries.

C. GROSS CHANGES

Though the dilatation of rigid calcified arterial channels decreases the circulatory efficiency, the encroachment upon the lumen by newly-formed intimal mesenchyme undergoing degenerative and reparative changes, with lipid accumulation, calcification and vascularization, is most important. This leads to several secondary changes: (1) the formation of a collateral circulation by which arteries proximal and distal to the point of obstruction become enlarged to accommodate the tissue demand for blood supply; (2) the insidious disuse atrophy by which the lumen of the artery distal to a point of obstruction is progressively decreased in size in rough proportion to the reduction in blood flow; (3) enhancement of the likelihood of significant obstruction by emboli from proximal thrombi or atheromatous lesions, so common in the aorta and iliac arteries; and (4) the increase in probability of significant local thrombosis, initiated principally by hemorrhage into degenerate, vascularized, partly-obstructive fibroatheromatous intimal plaques. In general, though thrombi frequently form on these lesions of the intima of large arteries, they acquire their greatest significance clinically when they are dissipated as fibrinous or lipid embolic thrombi, completing the obstruction of smaller more distal important arteries. More important, however, are the thrombi which form locally in these main arteries, usually between the popliteal fossa and the ankle joint. Though gangrenous changes may make their initial and only appearance in the toes, the obstruction responsible for these alterations is usually one of several recent or old occlusive thrombotic processes engrafted upon degenerate vascularized fibroxanthomatous intimal lesions of the anterior and posterior tibial arteries. Any occlusive process occurring in the more distal branches is usually attributable to embolism from proximal thrombi or to effects of circulatory insufficiency, gangrene and infection. Spontaneous local occlusive nonembolic thrombosis seldom occurs in arteries beyond the ankle joint and is very rare in any artery to the upper extremity distal to the subclavian artery. It is much more common in the internal carotid arterial system, the vertebral arteries and their major branches.

D. MICROSCOPIC CHANGES

The microscopic changes which reflect the major histologic sequences in the evolution of Mönckeberg's sclerosis occur first in the media. There is some doubt as to whether these alterations are primarily in smooth muscle or the interstitial tissues. Both are affected early and almost simultaneously. Usually these changes are first recognizable in the internal elastic membrane or the tissues upon which it rests. The first alteration in the internal elastic membrane is a loss of undulation and pliability, associated with a change in its optical and staining properties. Thereafter, the membrane becomes more discontinuous than normal and acquires calcium on its surface and in its intrinsic structure. Often, the deposition of calcium occurs just beneath the membrane and spreads from this location along fibers between smooth muscle cells, along the processes of smooth muscle cells and in the cytoplasm of the cells themselves. The muscle cells acquire atrophic and degenerative changes and the process of deterioration with or without massive calcification progresses more deeply into the media in both the circular and longitudinal axes. The tendency for this to extend in a spiral or circular manner is often pronounced and is evident long before the full thickness of the media has become involved. As this goes on, there is a mesenchymal reaction characterized principally by an increase of collagen in the media and an intimal proliferative response. The intimal proliferative reaction usually occurs first beneath the endothelium at points where the internal elastic membrane shows the most conspicuous degenerative changes, especially those characterized by mobile discontinuities. Here, there is fibroblastic proliferation and accumulation, first of a mucinous matrix and later of an increasing number of collagenous fibrils. The orientation and activity are such that the newly-formed mesenchyme tends to heal the defect in the internal elastic membrane and to splint it by what is essentially an avascular fibrous mound-like scar. Depending on local conditions, the minute mesenchymal reactions may follow several courses. The local reaction may subside; it may continue and spread in all directions; or it may establish continuity with adjacent focal reactions of similar nature. In general, it progresses as the disease in the internal elastic membrane and media increases in severity. The rate and magnitude of the intimal reaction depend upon factors which regulate mesenchymal reactions in general, as discussed in the section on the etiology and pathogenesis of atherosclerosis (1) Atherosclerosis, Section D., this chapter. This regulation becomes most apparent where mechanical conditions and demands incidental to circulatory flow lead to a differentiation of the newly-formed mesenchyme. As this differentiation occurs, the general tendency is in the direction of formation of a new vascular wall. Though this is usually abortive and obscured by interfering influences, new elastic membranes and fibers often appear in the maturing mesenchymal framework. Occasionally, these fibers and membranes are perfectly arranged as single or multiple concentric lamella. In the latter instance the lamella are usually separated by concentrically-arranged bands of collagen. Also it is not unusual for the pluripotential cells of the mesenchyme to differentiate into smooth muscle cells so that an essentially new arterial wall is constructed internal to the shell of the degenerate media of the original artery.

These mesenchymal reactions of a reparative nature do not occur equally in all vessels, each having its own particular potential, as discussed in the section dealing with the pathogenesis of atherosclerosis

((1) Atherosclerosis, Section D., this chapter). The full potential in many parts of the arterial system is seldom realized because of the following complicating factors: (1) the limitation imposed by the lack of development of a circulation in the newly-formed mesenchyme; (2) the gradual involvement of the newly-formed mesenchyme by the same type of calcific-degenerative changes which characterized the subjacent medial disease; and (3) the tendency for xanthomatous material to accumulate in the activated mesenchyme. The last factor is most significant, since it apparently inhibits fibrogenesis, as the fibrogenic cells become lipid-laden or as the intercellular matrix acquires an excess of lipids and hyalin properties. The cells, laden with lipids, and the interstitial lipid deposits deteriorate, thus only enhancing mechanical obstruction to flow and providing conditions which encourage local thrombosis.

The conditions which are conducive to local thrombosis arise through the apparent operation of three degenerative sequences. The first is the ulcerative disintegration of the degenerate lipid-laden intima and the subsequent edema or excitation of local fibrinous deposits. This is often accompanied or preceded by an inflammatory reaction which in many instances involves the entire thickness of the vessel wall. Although this may be acute, more often it is chronic and occasionally so conspicuous that a suspicion of arteritis of unknown origin arises. The second of the degenerative sequences is the stimulus to vascularization which accompanies degeneration in the intima and inflammation. The vascularization may proceed from the lumen of the artery, from nearby branches as they pass through the arterial wall, or from adventitial vessels penetrating through the degenerate media. Vascularization alone is not necessarily a source of trouble, but this process may lead to the third degenerative sequence, namely, frequent rupture of the thin-walled capillaries as they traverse a degenerate intimal plaque, even in the absence of significant inflammatory changes. The subsequent local hemorrhage, effusion of plasma and edema greatly enhance the probability of formation of fibrinous thrombi on the surface of degenerate plaques, as well as the tendency of thrombi to propagate in the lumen of the vessel. One common source of vascularization of the degenerate newly-formed intima is from newly-formed vascular channels penetrating the media from adventitial plexuses. The stimulus to penetration is usually attributed to progression of medial degeneration outward from the region of the internal elastic membrane.

As a rule, medial calcific degeneration does not elicit vascularizing sequences and other signs of mesenchymal activation in the adventitia until the wave of degeneration approaches the junction of the media and adventitia. Then, the activated adventitial mesenchyme engages the media and begins to excavate and resorb degenerate calcified structures. As this proceeds, vascularity ordinarily increases and occasionally the mesenchymal cells deposit an osteoid tissue along the margins of the resorbing calcified medial structure. This, in turn, is converted into bone and the entrapped mesenchyme differentiates into a tissue resembling bone marrow. However, as the reparative mesenchyme approaches the junction of the media and the thickened intima, a barrier against further penetration tends to persist; this is fairly effective unless degenerative changes develop in the thick intima. In such an event, further vascular penetration often occurs and anastomotic channels communicating with the lumen of the main vessel may develop. Though this may provide a useful by-pass of the circulation around a point of obstruction, it has the danger of initiating a sequence of events leading to further obstruction, by nutritive enhancement of intimal proliferative activity, by conveying lipids into the thickened intima and by leakage or rupture with subsequent mural thrombosis. It is apparent that this form of vascularization may favor xanthomatous accumulation, for occasionally this seems to be the only source for xanthomatous deposits found in the degenerate media at a distance from similar deposits in the intima.

Mönckeberg's sclerosis, therefore, is a variant of the generalized arteriosclerotic process. Its peculiarities reflect the relations existing among the composition of the blood, the dynamics of flow and the characteristic susceptibilities and reactivities of the affected arteries.

References

Anitschkow, N. (1933). Experimental arteriosclerosis in animals. *In* "Arteriosclerosis" (E. V. Cowdry, ed.), Chapter 10. Macmillan, New York.

Baldauf, L. K. (1906). The chemistry of atheroma and calcification. *J. Med. Research* **15**, 355.

Buck, R. C. (1951). Minerals of normal and atherosclerotic aortas. *A. M. A. Arch. Pathol.* **51**, 319.

Buck, R. C. (1957). Electron microscopic study of arterial endothelium. *Circulation* **16**, 484.

Buck, R. C. (1959). The fine structure of the aortic endothelial lesions in experimental cholesterol atherosclerosis of rabbits. *Am. J. Pathol.* **34**, 897.

Buck, R. C., and Rossiter, R. J. (1951). Lipids of normal and atherosclerotic aortas. *A. M. A. Arch. Pathol.* **51**, 224.

Buck, R. C., Paterson, J. C., and Rossiter, R. J. (1954). Chemical composition of cerebral arteries. The concentration of lipids and minerals compared with those in the internal carotid. *Can. J. Biochem. and Physiol.* **32**, 539.

Buddecke, E. (1958). Untersuchungen zur Chemie der Arterienwand. II. Arteriosklerotische Veränderungen am Aortenbindegewebe des Menschen. *Hoppe-Seylers Z. physiol. Chem.* **310**, 182.

Duff, G. L. (1936). The nature of experimental cholesterol arteriosclerosis in the rabbit. *A. M. A. Arch. Pathol.* **22**, 161.

Duff, G. L. (1955). Functional anatomy of the blood vessel wall; adaptive changes. Symposium on Atherosclerosis. *Natl. Acad. Sci.—Natl. Research Council Publ.* No. 338, 33.

Duguid, J. B. (1959). The role of connective tissues in arterial diseases. *In* "Connective Tissue, Thrombosis and Atherosclerosis" (I. H. Page, ed.), p. 13. Academic Press, New York.

Dyrbye, M. O. (1959). Studies on the metabolism of the mucopolysaccharides of human arterial tissue by means of S^{35}, with special reference to changes related to age. *J. Gerontol.* **14**, 32.

Gillman, T. (1959). Reduplication, remodeling, regeneration, repair and degeneration of elastic membranes. *A. M. A. Arch. Pathol.* **67**, 624.

Hevelke, G. (1956). Angiochemische Untersuchungen der Aorta zur Frage der Physiosklerose, Arteriosklerose und diabetischen Angiopathie. *Deut. Arch. klin. Med.* **203**, 528.

Hueper, W. C. (1944). Arteriosclerosis. *A. M. A. Arch. Pathol.* **38**, 162, 245, 350.

Hueper, W. C. (1945). Arteriosclerosis. *A. M. A. Arch. Pathol.* **39**, 51, 117, 187.

Katz, L. N., and Stamler, J. (1953). "Experimental Atherosclerosis." Charles C Thomas, Springfield, Illinois.

Katz, L. N., Stamler, J., and Pick, R. (1958). "Nutrition and Atherosclerosis." Lea & Febiger, Philadelphia, Pennsylvania.

Kirk, J. E., and Dyrbye, M. (1956). Hexosamine and acid-hydrolyzable sulfate concentrations of the aorta and pulmonary artery in individuals of various ages. *J. Gerontol.* **11**, 273.

Kirk, J. E., and Laursen, T. J. S. (1955). Diffusion coefficients of various solutes for human aortic tissue, with special reference to variation in tissue permeability with age. *J. Gerontol.* **10**, 288.

Lehnherr, E. R. (1933). Arteriosclerosis and diabetes mellitus. *New Engl. J. Med.* **208**, 1307.

McArthur, C. S. (1942). The acetone-soluble lipid of the atheromatous aorta. *Biochem. J.* **36**, 559.

Meeker, D. R., and Jobling, J. W. (1934). A chemical study of arteriosclerotic lesions in the human aorta. *A. M. A. Arch. Pathol.* **18**, 252.

Moon, H. D. (1959). Connective tissue reactions in the development of arteriosclerosis. *In* "Connective Tissue, Thrombosis and Atherosclerosis" (I. H. Page, ed.), p. 33. Academic Press, New York.

Noble, N. L., Boucek, R. J., and Kao, K. Y. T. (1957). Biochemical observations of human atheromatosis. Analysis of aortic intima. *Circulation* **15**, 366.

Page, I. H. (1941). Some aspects of the nature of the chemical changes occurring in atheromatosis. *Ann. Internal Med.* **14**, 1741.

Parker, F. (1960). An electron microscopic study of experimental atherosclerosis. *Am. J. Pathol.* **36**, 19.

Patterson, J. C. (1955). The reaction of the arterial wall to intramural hemorrhage. Symposium on Atherosclerosis. *Natl. Acad. Sci.—Natl. Research Council Publ.* No. 338, 65.

Pincus, G., ed. (1959). "Hormones and Atherosclerosis," Proc. Brighton Conf., 1958. Academic Press, New York.

Portman, O. W., and Stare, F. J. (1959). Dietary regulation of serum cholesterol levels. *Physiol. Revs.* **39**, 407.

Rechenberger, J., and Hevelke, G. (1955). Der Magnesiumgehalt der Aorta beim Diabetiker und seine Abhängigkeit vom Lebensalter. *Z. Altersforsch.* **9,** 309.

Schönheimer, R. (1926). Zur Chemie der gesunden und der atherosklerotischen Aorta. I. Über die quantitativen Verhältnisse des Cholesterins und der Cholesterinester. *Hoppe-Seylers Z. physiol. Chem.* **160,** 61.

Schönheimer, R. (1928). Zur Chemie der gesunden und der atherosklerotischen Aorta. II. Weitere Untersuchungen über die quantitativ-chemischen Veränderungen in der atherosklerotischen Aorta. *Hoppe-Seylers Z. physiol. Chem.* **177,** 143.

Stamler, J. (1958). Diet and atherosclerotic disease. *J. Am. Dietet. Assoc.* **34,** 701, 814, 929, 1053, 1060.

Stamler, J. (1959a). Experimental studies on dietary fats, cholesterol, and atherosclerosis. *Food Technol.* **13,** 50.

Stamler, J. (1959b). The epidemiology of atherosclerotic coronary heart disease. *Postgraduate Med.* **25,** 610, 685.

Stamler, J. (1960a). The problem of elevated blood cholesterol. *Am. J. Public Health* **50,** No. 3, Pt. 2.

Stamler, J. (1960b). Current status of the dietary prevention and treatment of atherosclerotic coronary heart disease. *Progr. in Cardiovasc. Diseases,* **3,** 56.

Stamler, J., Pick, R., and Katz, L. N. (1956). Experiences in assessing estrogen anti-atherogenesis in chick, rabbit, and man. *Ann. N. Y. Acad. Sci.* **64,** 596.

Stamler, J., Lindberg, H. A., Berkson, D. M., Shaffer, A., Miller, W., and Poindexter, A., with the assistance of Colwell, M., and Hall, Y. (1958). "Epidemiological Analysis of Hypertension and Hypertensive Disease in the Labor Force of a Chicago Utility Company," 7th Ann. Proc. of the Council for High Blood Pressure Research, p. 23. Am. Heart Assoc., New York.

Stamler, J., Lindberg, H. A., Berkson, D. M., Shaffer, A., Miller, W., and Poindexter, A., with the assistance of Colwell, M., and Hall, Y. (1960). Prevalence and incidence of coronary heart disease in strata of the labor force of a Chicago industrial corporation, *J. Chronic Diseases* **11,** 405.

Stamler, J., Katz, L. N., Pick, R., Lewis, L. A., Page, I. H., Pick, A., Kaplan, B. M., Berkson, D. M., and Century, D. (1961). Effects of long-term estrogen therapy on serum cholesterol-lipid-lipoprotein levels and on mortality in middle-aged men with previous myocardial infarction. *In* "Drugs Affecting Lipid Metabolism" (S. Garattini and R. Paoletti, eds.), p. 432. Elsevier, Houston, Texas.

Symposium (1957). Measuring the risk of coronary heart disease in adult population groups. *Am. J. Public Health* **47,** No. 4, Pt. 2.

Symposium (1960). Prevention and control of heart disease. *Am. J. Public Health* **50,** No. 3, Pt. 2.

Taylor, C. B. (1955). The reaction of arteries to injury by physical agents. With a discussion of factors influencing arterial repair. Symposium on Atherosclerosis. *Natl. Acad. Sci.—Natl. Research Council Publ.* No. 338, 74.

Taylor, C. B., Nelson-Cox, L. G., Hall-Taylor, B. J., and Cox, G. E. (1957). Atherosclerosis in monkeys with moderate hypercholesterolemia induced by dietary cholesterol. *Federation Proc.* **16,** 374.

Weinhouse, S., and Hirsch, E. F. (1940). Chemistry of atherosclerosis. I. Lipid and calcium content of the intima and of the media of the aorta with and without atherosclerosis. *A. M. A. Arch. Pathol.* **29,** 31.

Zeek, P. M. (1936). A chemical analysis of atherosclerotic lesions in human aortas. *Am. J. Pathol.* **12,** 115.

Chapter XX. CONDITIONS AFFECTING BLOOD PRESSURE

1. Arterial Hypertension

A. ETIOLOGY

By L. Tobian

1. General Factors Implicated in Elevated Blood Pressure

In hypertension the lumina of the small arterioles throughout the systemic circulation are excessively narrowed, despite which the peripheral flow to most tissues is maintained at normal levels. This is accomplished through an elevation of the central arterial pressure, hence the name, "arterial hypertension." Since the cardiac output is also normal, it is therefore necessary for the left ventricle to pump its usual amount of blood against a high head of pressure in the aorta, thus resulting in an increased work load and hypertrophy of the ventricular wall. Blood volume and blood viscosity are not sufficiently elevated to contribute significantly to the rise in arterial pressure. It seems probable that the initial change is the narrowing of the arterioles, with the arterial pressure becoming elevated as a compensatory mechanism.

2. Types of Clinical Hypertension

There are several different forms of arterial hypertension found in man, and it is by no means certain that the underlying mechanism is the same in all. In "essential" hypertension, the arterioles are narrowed, but the reason for this change is not known. Sustained arterial hypertension can also result from various forms of renal disease, such as chronic glomerulonephritis, chronic pyelonephritis, intercapillary glomerulosclerosis (as seen in diabetics), polycystic disease of the kidney, atrophy of the kidneys, and also various conditions which partially narrow one or both renal arteries. Among the latter are a thrombus, an embolus lodging in the renal artery, a congenital narrowing, an atheroma, a tumor, or a constricting band. In the usual type of coarctation of the aorta, the blood pressure in the upper part of the body is elevated, while that in the renal arteries is lowered, the reduced pressure in the renal arterial bed probably contributing to the rise in arterial pressure in the head and arms. Hypertension may be caused by hypersecretion of glucocorticoids, as in Cushing's syndrome, or by hypersecretion of mineralocorticoids, as in primary hyperaldosteronism. It may likewise result from a sustained hypersecretion of catecholamines from a pheochromocytoma. Certain brain lesions, brought on by poliomyelitis and other diseases, may also produce a sustained hypertension.

3. Experimental Hypertension

Many of the hypertensive states in man can be reproduced in other mammals. A narrowing of one renal artery, with the opposite kidney in or out, will produce hypertension in the rat. Wrapping some constricting material around a kidney to compress it will also produce experimental hypertension. Administration of desoxycorticosterone or aldosterone, along with a generous intake of sodium chloride, will likewise produce experimental hypertension (Gross *et al.*, 1957). Excessive doses of cortisone by itself will have a similar effect (Knowlton *et al.*, 1952), as will feeding large amounts of sodium chloride for a prolonged period (about 9 months) (Meneely *et al.*, 1953).

B. PATHOGENESIS

By L. Tobian

1. Type of Change Producing Narrowing of Arteriolar Lumen

a. Passive narrowing: The question arises as to what mechanism is responsible for narrowing of the lumen of arterioles. In this regard, it has been shown independently by both Folkow (1957) and Conway (1960) that in patients with essential hypertension, when all muscle tone in the arterioles has been abolished, the arteriolar lumen is still narrower than that of normotensive subjects. In other words, there is a passive narrowing of the arteriolar lumen, unrelated to contraction of the smooth muscle. Furthermore, this nonmyogenic narrowing becomes greater as the hypertension becomes more severe. Conway (1960) has shown that this type of change in the hypertensive subject is not due to a permanent structural alteration, since it can be reversed with thiazide drugs, the fall in pressure being associated with an enlargement of the vessel's lumen. It is possible that the passive narrowing may be caused by waterlogging of the walls of the arterioles (Tobian and Binion, 1952). In this regard, it has been shown that the walls of the renal artery in human hypertensives and of the aorta in hypertensive rats contain an increased content of water per unit of solids (Tobian and Binion, 1952, 1954). If a similar change existed in the arterioles as well, the increased content of water could conceivably be responsible for an encroachment on the lumen of the arterioles. Tobian *et al.* (1961b) have recently found that the arteriolar wall in rats with renal hypertension has an increased content of cations (Na + K). This would almost surely be accompanied by an increased water content in the arteriolar wall, just as the increased sodium and potassium in the aortic wall of hypertensive rats is associated with an increase in water (Tobian and Binion, 1954). Furthermore, this type of narrowing in man could be readily reversed by thiazide diuretics.

b. Increased arteriolar tonus: In addition to passive narrowing, the possibility of increased tonus in the arteriolar smooth muscle must also be considered. Of interest in this regard is the fact that the media of the arterioles appears to be thickened and that, in general, a thickening of smooth muscle often indicates increased tonus. Moreover, as vasomotion of the arterioles and precapillary sphincters is studied in the conjunctiva (Lee and Holze, 1951; Jackson, 1958) and in the nailfold (Greisman, 1954), it is apparent that the number of contractions per minute which are strong enough to stop the flow of blood is greater in hypertensive than in normotensive individuals. Furthermore, a given dose of norepinephrine administered intra-arterially will produce a greater reduction in blood flow in the hypertensive subject (Doyle *et al.*, 1959). Such data support the view that there may be an increased tonus of the arteriolar muscle. However, since, as already mentioned, nonmyogenic resistance to flow in arterioles is increased in hypertensives, it is also possible that some of the various phenomena cited above could be dependent on such a change. For example, with a vessel already passively narrowed, due to an increased amount of "waterlogging," a given degree of shortening of arteriolar smooth muscle might produce an exaggerated increase in arteriolar resistance in a hypertensive subject. In aortic strips from hypertensive rats, there is no increased sensitivity to norepinephrine (Redleaf and Tobian, 1958b). It would seem, then, that the question of increased tonus in hypertension must remain open until further information has been obtained.

2. Role of the Sympathetic Nervous System

There are a number of systemic factors which are known to influence arteriolar narrowing, and it is therefore necessary to consider them in the elucidation of the pathogenesis of hypertension. Among the first of these is the sympathetic nervous system. It has been shown that the daily excretion of norepinephrine in the urine of patients with essential hypertension is not significantly different from that of normotensive subjects (von Euler, 1956; Møller *et al.,* 1957; Griffiths and Collinson, 1957). Since the level of excretion is a rough index of the rate of sympathetic discharge, this would suggest that hypertension is not related to hyperactivity of the sympathetic nervous system. Further proof is the absence of hyperhidrosis and tachycardia, both of which are signs of sympathetic hyperactivity. Moreover, hypertensive patients are often not benefited by extensive sympathectomies, and paralyzing the sympathetic nerves does not produce a greater increase in blood flow in hypertensives, compared to normotensives. All these findings, therefore, speak against the presence of sympathetic hyperactivity in hypertensive patients. Nevertheless, it is possible that the peripheral arterioles of hypertensive subjects are hypersensitive to a normal rate of sympathetic outflow, a condition which would imitate sympathetic hyperactivity. In favor of this is the finding that a given dose of norepinephrine administered topically or intra-arterially will produce a much greater decrease in blood flow in hypertensives than in normotensives (Doyle *et al.,* 1959). However, this does not necessarily prove that the circular smooth muscle in hypertensive patients is truly hypersensitive. Moreover, this same hypersensitivity to norepinephrine can be demonstrated in some patients who do not have abnormal elevations of blood pressure. For example, Mendlowitz *et al.* (1958) have shown that cortisone can produce hypersensitivity of the small finger vessels to this amine, in the absence of hypertension. Furthermore, Merrill (1958) transplanted a normal kidney into the circulation of a girl with severe renal hypertension, following which her blood pressure fell to fairly normal levels. Despite this, however, the patient retained her hypersensitivity to norepinephrine, and only after her own diseased kidneys were removed did she lose it.

Although very little conclusive evidence has been collected in its favor, the question of hypersensitivity has always been an attractive and reasonable hypothesis, and more information must be obtained to learn whether or not it is an important pathogenetic factor. Along these lines, blockers of sympathetic impulses, such as guanethidine or pentolinium, lower elevated arterial pressures, but they do so mainly by reducing cardiac output rather than by decreasing peripheral resistance. If a heightened sympathetic effect were the main cause of the elevated blood pressure, sympathetic paralysis could be expected to produce a more physiologic cure of hypertension than it actually does. Again, these observations do not lend support to the notion that the sympathetics play an important part in hypertension.

3. Role of the Kidney

In any consideration of the pathogenesis of hypertension, the kidney has a prominent place. Many kidney diseases bring on the hypertensive state, and certain kidney manipulations in animals reliably produce arterial hypertension (see Section A.-3 of this chapter). Regardless of whether a narrowing of the renal artery decreases the pressure in the renal arterial bed distal to the constriction or whether an encapsulating material compresses the parenchyma of the kidney, the resulting stimuli are somehow sensed and a chain of events occurs which ultimately produces hypertension. Under either of these circumstances, if the sensing element were a stretch receptor in the wall of a preglomerular renal artery, it would become stretched to a lesser degree. With a clip on the renal artery, the afferent arteriole is stretched less because the pressure in it is lower. With encapsulation, the whole parenchyma, including the afferent arterioles, is compressed, thus decreasing the stretch of the walls of the latter vessels.

a. Role of juxtaglomerular cells: Of interest in regard to the above points is the finding of specialized granular cells, the juxtaglomerular cells, within the wall of the renal afferent arteriole, usually just before the vessel enters the glomerular tuft. These structures have all the appearances of secretory cells, with secretory granules in their cytoplasm. When the pressure in the renal arterial bed is raised, the granules tend to disappear; when it is lowered, the granularity of these cells increases (Tobian *et al.*, 1958, 1959a). These various responses make it quite plausible and likely that the juxtaglomerular cells act as stretch receptors which appreciate a narrowing of the renal artery or an encapsulation of the kidney. When the appropriate signal is received, they alter their rate of secretion.

b. Role of renin: There are three lines of evidence which suggest that the juxtaglomerular cells secrete renin or its precursor. First, the renin in a kidney is mainly found associated with glomeruli but not in the glomerular tuft itself (Cook and Pickering, 1959; Bing and Wiberg, 1958). From microdissection studies, all the renin in a kidney seems to be located either in the granular juxtaglomerular cells or in the macula densa (Bing and Kazimierczak, 1960; Cook, personal communication). Secondly, in the rat the quantity of extractable renin and the granularity of the juxtaglomerular cells have a very strong correlation in various experimental situations (Tobian *et al.*, 1959b). Thirdly, fluorescent antibodies to renin predominantly localize in the cytoplasm of the juxtaglomerular cells in rabbits on a sodium-deficient diet (Hartroft and Edelman, 1960). It therefore seems probable that a decreased pressure in the renal arterial system lessens the stretch on the juxtaglomerular cells and the latter, in turn, cause the secretion of greater amounts of renin either from the juxtaglomerular cells, from the macula densa, or from both.

In view of the above, it becomes important to discuss the function of renin, which is a proteolytic protein, capable of splitting a decapeptide, angiotensin, from the plasma protein, angiotensinogen. Angiotensin is a very powerful pressor agent and is present either as a decapeptide (angiotensin I) or as an octapeptide (angiotensin II) (for further discussion of these points, see Section E.-4b, Chapter II). Although renin is capable of splitting off angiotensin, this may not necessarily be its most important function. Moreover, even if it does produce some angiotensin, this may or may not be the cause of the sustained elevation of arterial pressure in hypertension. In this regard, Peart (1959), using fairly sensitive methods, has tested for renin in the renal venous blood of rabbits made hypertensive by a narrowing of the renal artery and has been unable to detect an increased amount. He obtained similar results in the case of patients with hypertension, assaying renin by measuring its ability to split off angiotensin. Blaquier *et al.* (1960) reported comparable findings in the case of rats. They noted that no rise in blood pressure occurred in a normal animal when cross-circulated with a hypertensive one, a type of preparation which has been shown to be very sensitive to pressor materials such as angiotensin. In contrast to these data is the study of Helmer and Judson (1959) who have been able to detect a pressor material in the renal venous blood of hypertensive patients, which may be similar to the "sustained pressor principle" originally described by Shipley *et al.* (1947). Moreover, animals with renal hypertension, made tachyphylactic to renin, are still able to maintain their elevated arterial pressure. The sum total of the evidence, therefore, suggests that the renin-angiotensin system, acting by itself, is not directly responsible through a simple pressor effect for the sustained elevation of blood pressure in hypertension. However, it is quite possible or even probable that renin may have a more subtle but still important role in the hypertensive process, either that of an intrarenal hormone, of an agent affecting peripheral vessels, or of both.

c. *Antihypertensive effect of normal renal tissue:* There is abundant evidence that the kidney has influence on hypertension other than through the direct pressor action of the renin-angiotensin system. First, it has been noted that the presence of a contralateral normal kidney in the dog ameliorated the renal hypertension produced by narrowing one renal artery. Moreover, Braun-Menendez and von Euler (1947) have demonstrated that bilateral nephrectomy produces hypertension in the rat, a finding which has abundantly been confirmed by others in the rat and the dog, but not necessarily in man (Merrill, 1958). It has been shown that the hypertension is not due simply to failure of excretion, since implantation of the ureters into the vena cava does not bring it on. The state of hydration influences the degree of this "renoprival" hypertension, but it still may exist in the absence of overhydration. It is a true type of arterial hypertension, due to an increase in arteriolar resistance.

The above experiments point to an antihypertensive effect of normal renal tissue, as has been strikingly confirmed by the maneuver of Merrill *et al.* (1956) of transplanting a normal kidney from one identical twin into the circulation of the other who had malignant renal hypertension. This procedure promptly brought the blood pressure down to normal levels for several months. Then it began to rise again but returned to normal when the diseased kidneys were removed, leaving only the transplanted normal kidney. This phenomenon can also be reproduced in a short period of time in dogs and rats with renal hypertension (Gomez *et al.*, 1960).

It seems very likely, therefore, that normal renal tissue has some antihypertensive activity, the lack of which brings on the elevated blood pressure and is probably the main cause of most types of arterial hypertension. In experiments in which a renal artery is narrowed or a kidney is encapsulated, the juxtaglomerular cells probably receive the stimulus initially and respond by secreting increased amounts of renin, which somehow cause an inhibition of the antihypertensive activity in the kidney, thus initiating hypertension. Such a hypothesis would explain why antibodies to renin can lower blood pressure in renal hypertension, as Wakerlin (1958) has repeatedly shown. These antibodies would tie up the excess amount of renin and thereby prevent it from inhibiting the antihypertensive mechanism in the kidney.

It is not known which cells in the kidney carry on this antihypertensive activity. Muirhead (1959) believes that they may be in the outer part of the renal medulla. In certain diseases, such as essential hypertension in man and metacorticoid or postdesoxycorticosterone hypertension in the rat, it may be that an exhaustion-atrophy of these "antihypertensive cells" produces the hypertension. It would be akin to the exhaustion atrophy of the β-cells of the islets of Langerhans which results in diabetes. In human essential hypertension, restricting the intake of sodium or inhibiting the activity of the sympathetic nervous system would be expected to take some of the strain away from these cells, and this might be responsible for their partial recovery. In such a manner the basic hypertensive diathesis may be benefited if the cells are not irrevocably damaged.

Recent experiments (Tobian *et al.*, 1961a) indicate that the antihypertensive activity of a normal kidney is strongly brought into play when this organ is exposed to high arterial pressures. On the other hand, when the kidney is exposed to low normal arterial pressures, the antihypertensive activity receives little or no stimulation. Thus, there appears to be a feed-back arrangement designed to maintain normal arterial pressures.

4. Role of Adrenal Cortex

An additional factor influencing hypertension is the steroidal secretion of the adrenal cortex. There is little question that the level of adrenal steroids in the body has an influence on the lumen caliber of arterioles. With adrenal insufficiency, arterial pressure is sometimes diminished even when the extracellular fluid volume is normal. With increased levels of aldosterone or desoxycorticosterone in man or animals, the blood pressure rises, due to an increase in arteriolar resistance. With excessive levels of glucocorticoids in man or in the rat, hypertension again may become manifest.

5. Influence of Sodium

When the adrenals have been removed, it is difficult to produce any of the various forms of experimental renal hypertension. However, when sodium, in large amounts, is given after extirpation of the adrenals, this ion seems to be able to take the place of the absent adrenal tissue, and characteristic hypertension can now be produced (Knowlton *et al.*, 1950; Floyer, 1951). An adequate sodium intake is absolutely necessary for the production of desoxycorticosterone hypertension; a very low intake of this ion will prevent it. On the other hand, cortisone hypertension can be produced in conjunction with a low intake of sodium (Knowlton *et al.*, 1952). Humans and animals with adrenal insufficiency lack the normal sensitivity to norepinephrine or renin; rats with desoxycorticosterone hypertension are hypersensitive to these two substances.

From the above observations, it is obvious that sodium is an important factor in hypertension. Patients with essential hypertension may be greatly improved if their intake of this ion is drastically curtailed, although rats that have become hypertensive from the narrowing of a renal artery are not significantly improved on a low intake of sodium (Redleaf and Tobian, 1958a). Large amounts of sodium in the diet, by itself, will produce hypertension in otherwise normal animals (Meneely *et al.*, 1953). That salt probably encourages hypertension in man is supported by the finding that Japanese farmers with a very large salt intake have an uncommonly high incidence of hypertension. Dahl and Love (1957) have also noted this correlation in the United States.

6. Influence of Potassium

An extremely low intake of potassium will lower the blood pressure of hypertensive men (Perera, 1953) and rats (Friedman *et al.*, 1951) and even of normal rats (Freed and Friedman, 1950). A very high intake of potassium will also lower the level of blood pressure in rats with renal hypertension (Bach *et al.*, 1956). A high intake of potassium likewise prevents mesenteric hemorrhages in rabbits with renal hypertension (Gordon and Drury, 1956) and decreases the hypertensive effects of a large intake of sodium in both man (McQuarrie *et al.*, 1936) and rat (Meneely and Ball, 1958).

7. Role of Thiazide Diuretics

The thiazide diuretic drugs have a strong tendency to lower blood pressure in patients with essential hypertension by actually decreasing peripheral resistance and providing a truly physiologic ameliorating effect. They produce this response without very much change of body sodium or potassium, extracellular fluid volume, blood volume, or cardiac output, although total body water is diminished (Lauwers and Conway, 1960).

The mechanism of action is probably the same as a drastic reduction of sodium in the diet. In recent studies (Tobian, "unpublished observations"), it was found that chlorothiazide causes an increase in the abundance of granular juxtaglomerular cells in the normal rat, similar to that brought on by a sodium deficient diet (Tobian *et al.,* 1958). In these same rats the chlorothiazide did not change the content of sodium and potassium in the arteriolar wall. It would appear, therefore, that with the low salt diet, it is not the "desalting" *per se* which reduces the blood pressure but rather some physiologic adaptation to the very minor degree of "desalting" and minor reduction in blood volume. As mentioned above, a low sodium intake in the rat does cause hypergranularity of the juxtaglomerular cells and an increased quantity of extractable renin (Pitcock and Hartroft, 1959). The latter response might decrease the stimulus for activity of the "antihypertensive cells" and hence somehow "rest" these cells and allow for their partial recovery.

8. Conditions Producing Permanent Hypertension

There are several examples of a temporary hypertension becoming permanent if allowed to last long enough. If one renal artery is narrowed in the rat with resulting hypertension, the blood pressure can be brought to normal if the "ischemic" kidney is removed within two months. However, if the hypertension is allowed to go on for six months, such an excision will now usually produce some drop in blood pressure but not to levels within the normal range (Wilson and Byrom, 1939; Tobian *et al.,* 1959c). Similarly, if desoxycorticosterone (DOCA) and salt are given to a rat for one month, hypertension ensues but gradually disappears after discontinuation of the DOCA. However, if the period of administration is increased to six months, the hypertension will usually be maintained indefinitely, despite elimination of the DOCA and salt. Prolonged use of high intakes of sodium chloride will do the same thing (Fregly, 1960). In man, a pheochromocytoma may produce hypertension for so long that it persists after the removal of the tumor. Similarly, if pre-eclampsia with hypertension is allowed to go on for about two months, a permanent elevation of blood pressure may become evident following delivery.

There are two possible explanations for the above phenomenona. During the temporary hypertension of various types, arteriolar lesions in the kidney may be produced, and these, in turn, may be responsible for the continuation of a permanent hypertension after the initial cause has been removed. It is also possible that the initial hypertension provoked a very strong stimulus to the renal antihypertensive mechanism, which produced an exhaustion of this mechanism and possibly some irreversible damage to it, thus bringing on a permanent hypertension.

9. Complications of Hypertension

a. Increased rate of formation of atherosclerosis: Essential hypertension in man would be fairly harmless if it were not for certain complications. In grades I and II hypertension ("benign"), the main rise in mortality rate is caused by an increased tendency to atherosclerosis of the coronary and cerebral vessels, combined with a greater strain on the left ventricle (see (1) Atherosclerosis, Section C.-1b, Chapter XIX). Of further interest in this regard is the finding that hypertension is much more poorly tolerated by men than by women, since the former have a greater natural predisposition to develop atherosclerosis, a higher pressure in the arterial system somehow accelerating the rate of growth of the lesions. The increased load on the left ventricle, superimposed upon coronary disease, frequently leads to congestive heart failure.

b. Arteriolar necrosis: In hypertension of grades III and IV, especially in the latter, the element of arteriolar necrosis is added. It is particularly severe in the kidney and usually causes death from renal failure. Encephalopathy may also be produced. There are several possible factors responsible for the arteriolar necrosis. It has been demonstrated clinically and experimentally that the very high pressure itself partly contributes to the lesion. For example, if a particular arteriolar bed can be protected from the high arterial pressure, the lesions do not develop even though the arterioles are probably exposed to the same humoral factors as the unprotected vessels in which necrosis occurs (Robert and Nezamis, 1957; Wilson and Byrom, 1939). Byrom (1954) believes that the involved arterioles are undergoing the Bayliss phenomenon, namely, contracting in response to an elevated pressure. Other studies indicate that hypertensive arterioles have an abnormal content of sodium (Tobian *et al.*, 1961b), a change in the intracellular environment which may be partially responsible for the lesions. Moreover, in malignant hypertension there is evidence that there may be an increase in renin-like substances in the circulation (Kahn *et al.*, 1952; Helmer and Judson, 1959), as well as a rise in aldosterone (Genest *et al.*, 1958; Laragh *et al.*, 1960). Such a combination will produce severe arteriolar lesions in the rat (Masson *et al.*, 1951, 1952) and may contribute to the arteriolar necrosis in human malignant hypertension.

C. GROSS AND MICROSCOPIC CHANGES

By C. L. PIRANI

1. RELATIONSHIP OF HYPERTENSION TO ARTERIOSCLEROSIS

Hypertension, no matter what the cause, will inevitably induce morphologic changes in arteries if this condition is sufficiently protracted and severe. The alterations are essentially of two types, often found in combination: fibrosis of the intima and infiltration of this layer by lipids with resulting atherosclerosis. Medial changes, consisting of hypertrophy and/or fibrosis, are usually less prominent findings.

Uncontroversial and ample human autopsy and experimental evidence indicates that not only will arterio- and arteriolarsclerosis occur, but the development of atherosclerosis or further aggravation of existing atheromatous lesions will be facilitated by hypertension. The arteriosclerotic and atherosclerotic changes in the pulmonary arteries in mitral stenosis and in certain types of congenital heart disease producing significant pulmonary hypertension (Heath and Whitaker, 1955a,b; Heath and Edwards, 1958) and in the arteries of the upper half of the body in coarctation of the aorta are proof of this role of hypertension. Experimentally, atherosclerotic lesions induced by cholesterol feeding can be aggravated in dogs and rabbits if hypertension is produced by renal ischemia (Moss *et al.*, 1951; Moses, 1954; Bronte-Stewart and Heptinstall, 1954). Not only are the level and duration of hypertension important in inducing or aggravating arterio- and atherosclerotic lesions, but evidence has been provided to indicate that the rapidity with which high blood pressure develops is also a factor (Pickering, 1952).

In essential hypertension, therefore, arterial and arteriolar sclerosis of all tissues, including the kidneys, is generally considered to be the result, and not the cause, of the elevated blood pressure. Widespread arteriolar sclerosis, in turn, maintains and further aggravates pre-existent hypertension caused by other factors. On the other hand, sclerosis or atheroma of larger arteries is usually a factor of little or no importance in the production or aggravation of hypertension, unless the secondary narrowing of the arterial lumen is strategically located in one or both major renal arteries.

In such renal conditions as pyelonephritis and polyarteritis nodosa, however, inflammation involving the wall of small arteries will induce narrowing and sclerotic changes which precede, and are apparently responsible for, the hypertension. In glomerulonephritis, destruction and obliteration of glomerular capillaries are a major factor in increasing peripheral resistance and causing the rise in pressure. In this type of renal disease, arteriolar sclerosis usually will become apparent only when hypertension has been present for some time. In diabetes mellitus and in other conditions associated with hypercholesterolemia, lipid infiltration in the wall of arteries will eventually lead to severe and widespread atherosclerosis and often to hypertension. Obviously, when hypertension is present in these and in the previously mentioned renal conditions, it cannot be properly considered of the essential type.

2. Extrarenal Arteries

The relationship of hypertension to arteriosclerosis becomes more apparent if one considers the vascular alterations occurring in the former condition in arteries of different caliber and in different organs.

a. Large arteries: In the large arteries, including the aorta, the pulsating blood stream is converted in a continuous flow, a function attributed to the preponderance of elastic fibers in the media. Degeneration of the elastic fibers, resulting from aging changes and from pathologic processes, leads to an increase of pulse pressure and of systolic pressure. There is no evidence, however, to indicate that such degenerative changes of the media or of the intima in large arteries will lead to an increase of diastolic blood pressure.

On the other hand, hypertension unquestionably aggravates already existent atherosclerosis and probably facilitates the development of atheroma in previously healthy arteries (for further discussion of this point, see Section B.-9a of this chapter). This has been well established for the aorta and some of its larger branches (Wilens, 1947) and for the major pulmonary arteries, while it is not entirely proven for the coronary arteries (Davis and Klainer, 1940a,b) and the cerebral arteries of the circle of Willis. That hemodynamic forces are important in the pathogenesis of atherosclerotic lesions is also suggested by the location of atheromatous plaques at the orifice of arterial branches, at the point of bifurcation and, in general, at sites where the blood exerts jet stream or lateral pressure effects.

The mechanism by which hypertension acts in the pathogenesis of atherosclerosis has not been clearly determined. However, it would seem that changes in the amount and character of intimal polysaccharides (Movat *et al.*, 1959) may facilitate the deposition of lipids and that endothelial cell damage may predispose to the "insudation" of lipids and other blood components in the arterial wall.

b. Muscular arteries and arterioles: These vessels are mostly responsible for peripheral resistance of the arterial system and, therefore, their functional and structural changes are closely related to diastolic pressure and essential hypertension. In this regard, medial hypertrophy has been repeatedly observed in both human and experimental hypertension, usually, but not always, in association with intimal changes (Moritz and Oldt, 1937). It has been suggested that medial hypertrophy is the result of active vasoconstriction and is a true work hypertrophy. This change, although particularly prominent in the kidneys, can also be seen in the lungs, spleen, pancreas, liver, gastrointestinal tract, and skeletal muscles (see Section G.-2b, Chapter XVII). According to some investigators, arteriolar spasm is not uncommon in hypertension and can be differentiated morphologically from muscular hypertrophy (Sommers *et al.*, 1958) (see Section B.-1 and 2, above). Other medial changes, such as mucoid degeneration, medial fibrosis and calcification, are often observed in the elastic and muscular arteries. However, they are more distinctly the result, rather than the cause, of high blood pressure. Intimal thickening, due to thickening and reduplication of elastic fibers, so-called intimal elastosis, although best seen in the kidneys, can be observed with decreasing frequency in the pancreas, suprarenals, skeletal muscle (see Section G.-2b, Chapter XVII), and gastrointestinal tract. It is not uncommon in the digital arteries and in the lungs, in association with conditions leading to pulmonary hypertension.

In the arterioles, sclerotic and hyaline changes are often associated with lipid infiltration of the wall (Baker and Kent, 1950) and involve, with particular frequency and severity, kidneys, spleen, pancreas, adrenals, gastrointestinal tract and skeletal muscles (Kernohan *et al.*, 1929; Morlock, 1939). They are particularly prominent in diabetes mellitus.

The similarity in type and severity of the arteriolar changes in essential hypertension and in hypertension due to chronic renal disease is well established and strongly suggests that arteriolar sclerosis is the result of hypertension, no matter what the cause. However, arteriolar sclerosis of a milder degree can be found rather commonly in older individuals with normal blood pressure (Bell, 1951) and is often absent even in patients with hypertension (Smithwick and Castleman, 1951; Sommers *et al.*, 1958).

The rarity with which pronounced arteriolar sclerosis is observed in the myocardium is of considerable interest. The heart muscle is supplied by blood during diastole and is protected from the impact of systolic blood pressure by the systolic contraction of the myocardium. This type of evidence would support the view that arteriolar sclerosis is the result of systolic and not of diastolic hypertension. On the other hand, biopsy studies indicate that arteriolar sclerosis is an excellent index of diastolic blood pressure (Foa *et al.*, 1943; Merriam *et al.*, 1958; Gellman *et al.*, 1959), and that sustained elevations of diastolic blood pressure are due to, and maintained by, functional and anatomic changes of the arterioles.

Arteriolar necrosis, the so-called "fibrinoid necrosis," is a true necrosis only in part, since, to a considerable extent, it is due to an infiltration of the arterial and arteriolar walls by plasma constituents, predominantly fibrin (Movat and More, 1957; Wolfgarten and Magarey, 1959). It is particularly common in the kidneys in malignant hypertension but is also seen in the spleen, gastrointestinal tract, pancreas, and in other tissues. Only rarely is this process associated with a periarterial cellular inflammatory infiltration. In these cases, the process may simulate polyarteritis nodosa and hypersensitivity angiitis. Fibrinoid necrosis has been attributed to the level of diastolic blood pressure (Goldblatt, 1948b; Byrom and Dodson, 1948) and not to such factors as uremia and cardiac failure. It is possible that arterial spasm, secondary to hypertension, plays an important role in its pathogenesis (Byrom, 1954). This view has been recently disputed (Schaffenburg and Goldblatt, 1957).

In the brain of hypertensive individuals, proliferative and degenerative changes of capillaries and arterioles have been described (Scheinker, 1943, 1948) which are somewhat different from those found in other tissues.

In the conjunctiva, the capillaries present a hypertensive pattern characterized by narrowing, elongation, tortuosity, and thickening of the wall (Lack *et al.*, 1949).

In the retina, the lesions of hypertensive retinopathy are related directly to hypertension but not necessarily to vascular lesions. Other factors, such as constriction of retinal arterioles, increased capillary pressure in the retina and elevated intracranial pressure, are also important. The main change consists of arteriolar necrosis, which may be followed by hyalinization and lipid infiltration and is often associated with marked narrowing (Keith *et al.*, 1928; Friedenwald, 1935). Medial hypertrophy and intimal proliferation have also been observed although they appear to be more pronounced in the choroid than in the retina (Manlove, 1946). The exudates and hemorrhages observed in the retina in malignant hypertension are not necessarily associated with arteriolar disease. Further, cases of malignant hypertension with minimal or absent arteriolar retinal changes have been reported (Fishberg, 1954).

Finally, it should be emphasized that most, if not all, the different arterial changes previously described can occur, at least to a mild or moderate degree, in the absence of systemic hypertension. Moreover, no single type of arterial lesion, *per se,* can be considered pathognomonic for hypertension, and only when the changes are relatively severe and occur in combination can a definitive diagnosis of hypertension be made on morphologic grounds.

3. Renal Changes in Essential Hypertension

In the absence of the specific changes of glomerulonephritis, pyelonephritis, diabetic nephropathy, and of other conditions, renal arterial and especially arteriolar sclerosis should always be considered directly related to essential hypertension, except in the milder forms. Usually the alterations in the arteries and especially arterioles of the kidneys are considerably more severe than those found in other organs.

The so-called senile arteriosclerotic kidney, so commonly found in the older age group, is of little clinical significance. Although its pathologic anatomy is similar to that of the less severe forms of benign nephrosclerosis, it is not associated with a significant degree of hypertension.

a. Benign nephrosclerosis: This form of renal disease, associated with benign essential hypertension, is more exactly an arteriolar sclerosis of the kidney. Only when significant degrees of interstitial fibrosis are present should the term "arteriolar nephrosclerosis" be used. On the basis of the work of Castleman and Smithwick (1948), who found no significant arteriolar sclerosis in a signficant proportion of renal biopsies from hypertensive patients, the concept that this change is the result of hypertension, rather than the primary cause (Goldblatt *et al.,* 1934; Goldblatt, 1948a,b), has gradually gained general acceptance.

The sclerosis of interlobular arteries and arterioles with narrowing of their lumen, which is invariably the result of persistent hypertension, will lead to a progressively more severe ischemic atrophy of the kidneys. The resulting glomerular and interstitial fibrosis usually causes bilateral scarring of the kidneys, characterized by a decrease in renal weight which, however, rarely reaches the extreme proportion often observed in chronic glomerulonephritis and pyelonephritis. Even in advanced cases, the reduction usually does not amount to more than 30 or 40 per cent of the original weight. The cortical surfaces are finely granular and pale red-brown. On section, the cortex is reduced in thickness while the medulla, which is less ischemic, is darker in color and less atrophic. Histologically, fibrosis and hyalinization of arterioles and intralobular arteries are the dominant feature (Fig. 124). These processes involve not only the intima but also the media of the involved vessels. The thickening of the wall, with corresponding narrowing of the vascular lumen, results in progressive glomerular fibrosis, which typically begins at the glomerular hilus and extends from this region in an axial and radial fashion toward the peripheral portion of the tuft. As a rule, arteriolar sclerosis must be present before glomerular scarring appears. The disease process does not necessarily affect all arterioles and to the

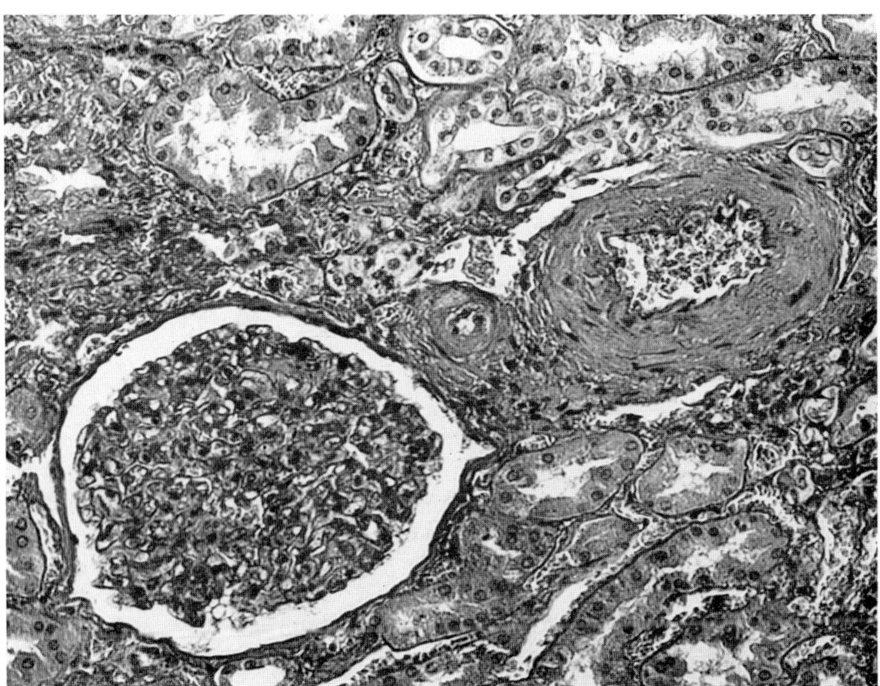

FIG. 124. Arterial and arteriolar sclerosis of the kidney in benign hypertension. In the center of the figure is the thick and hyalinized wall of an arteriole with narrowing of the lumen. To the right is an interlobular artery with moderate fibrosis of the wall and only slight narrowing of the lumen. The glomerulus is normal. Hematoxylin and eosin, ×270.

same degree. This somewhat focal involvement results in a patchy distribution of interstitial and glomerular fibrosis. Fibrosis of Bowman's capsules contributes to the eventual complete fibrous obliteration of many glomeruli. The associated tubular atrophy and dilatation are the result of both glomerular obliteration and of interstitial fibrosis. Hyaline, broad and waxy casts are often seen within the tubules, being evidence of a rather profound functional derangement (Kimmelstiel and Wilson, 1936).

In cases of benign essential hypertension of relatively short duration, especially in younger individuals, the lesions may be exclusively limited to the arterioles and intralobular arteries. However, in older patients, with progression of the disease, invariably the interlobular, arcuate and interlobar arteries will also be affected. This involvement is characterized by various degrees of intimal sclerosis, with mild to moderate narrowing of the lumen and generally with little fibrosis of the media.

When the larger renal artery branches participate in the sclerotic process, the expression "arterial and arteriolar sclerosis of the kidneys" or "arterial and arteriolar nephrosclerosis" is used. Even to a greater extent than in the arterioles, the degree of intimal sclerosis in the larger arteries tends to vary somewhat from vessel to vessel. Distinct atheromatous plaques are not commonly seen in the intraparenchymal renal arterial branches, except when diabetes mellitus and/or hypercholesterolemia is present. However, it is possible to demonstrate the presence of a fairly considerable amount of lipids in the sclerotic intima of arteries and in the wall of the arterioles (Baker and Kent, 1950).

b. Extraparenchymal renal arteriosclerosis: The focal and patchy character of sclerotic and atherosclerotic lesions in benign nephrosclerosis is more obvious in the two major renal arteries at their aortic orifice or in their primary branches before entering the renal parenchyma. At times, these lesions are of sufficient severity to narrow or stenose the lumen significantly at one or more points. Their functional importance can be evaluated accurately at autopsy only in their most severe forms. Renal arteriography often reveals marked narrowing of the blood stream column, although dissection of the artery may not disclose a stenosis of corresponding severity at the indicated point. The relationship of these findings to Goldblatt's theory (1948b) of hypertension is obvious, especially in the case of unilateral obstruction, regardless of whether the basis is arteriosclerotic, some other type of vascular involvement or extravascular. Recently, the practical importance of this type of lesion has been particularly emphasized, since it could possibly be the cause of hypertension more commonly than previously thought (Poutasse, 1959; Gellman, 1958; Hodges, 1960).

The character of the renal changes resulting from extraparenchymal renal arterial obstruction will vary with the rapidity with which the condition develops, as well as with its degree. When occlusion is rapid, infarction of the kidney will occur, the size of the lesion depending on the caliber of the artery obstructed. In occlusion of the renal artery, the entire kidney may become infarcted. However, in slowly progressing narrowing of the renal artery or major branches, a gradual atrophy and fibrosis of the kidney or corresponding renal segment will develop. Under these circumstances, hypertension usually occurs if renal ischemia affects either the entire kidney or most of it. As a consequence, the opposite organ and that part of the kidney which still receives blood under increased pressure will undergo the changes of arteriolar nephrosclerosis previously described. The kidney, or the part which was previously supplied by the obstructed artery, is "protected" from the effects of hypertension, and shows little or no arteriolar sclerosis. The involved part will be considerably smaller than normal, but its surface will be smooth, since the process of atrophy and interstitial fibrosis is diffuse and homogeneous, rather than focal, as in arteriolar nephrosclerosis.

c. Malignant nephrosclerosis: The renal changes associated with, and probably the result of, malignant hypertension are referred to as malignant nephrosclerosis. The term is somewhat improper, since sclerosis, either of the arteries or of the interstitium, is not the dominant feature (Klemperer and Otani, 1931; Weiss *et al.*, 1933; Kimmelstiel and Wilson, 1936, Kincaid-Smith *et al.*, 1958).

In the "pure" untreated form of malignant hypertension, in which the disease has appeared suddenly and has progressed rapidly without having been preceded by other renal disease or by a phase of benign hypertension, the kidneys are either normal in size or slightly enlarged. The cortical surfaces are congested and mottled, the lighter areas being related to focal ischemia. There are usually numerous petechial hemorrhages but no granularity or other evidence of scarring, similar changes being noted on the cut surface. The cortex is not reduced in thickness.

Histologically, striking changes are noted in the interlobular and intralobular arteries and the afferent arterioles. Fibroblastic proliferation, concentric lamination with elastic reduplication, and edema of the intima are associated with hypertrophy of smooth muscle cells in the media (Figs. 125 and 126). In addition, the afferent arterioles may undergo necrosis of their wall which, when particularly severe, may extend into the glomeruli (Fig. 127). If the arterioles become thrombosed, infarction of the glomerular tufts often results. This process of necrotizing arteriolitis is generally attributed to the marked elevation of blood pressure and not to uremia which is almost always present in malignant hypertension. Actually, these vascular changes, particularly arteriolar necrosis, are strikingly absent in a relatively high percentage of biopsies from patients with malignant hypertension (Castleman and Smithwick, 1948; Heptinstall, 1954). They may also not be noted or they may be of mild severity if the individuals die of nonrenal causes, such as heart failure or cerebral hemorrhage. These findings suggest that arteriolar necrosis is a late feature of the disease and develops only when hypertension is extremely severe. At this stage, uremia is present, and profound ischemia exists in the distal vascular segments. Other changes resulting from the marked vascular involvement include proliferative

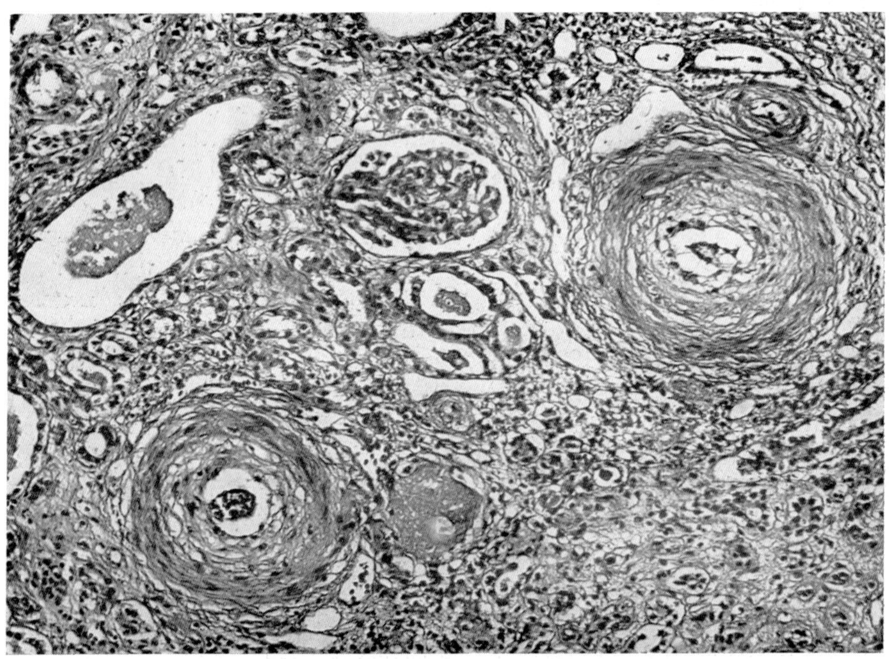

FIG. 125. Intimal swelling and lamination of interlobular renal arteries in malignant hypertension. Some hypertrophy of the muscular coat is present and the lumen is markedly narrowed. Fibrosis and edema of the interstitium are noted. The glomerulus is normal. Hematoxylin and eosin, ×190.

glomerular changes, at times simulating glomerulonephritis, severe degenerative and regenerative changes of the tubular epithelium and interstitial edema. Small cortical infarcts and interstitial hemorrhages are common.

Malignant hypertension and related vascular alterations are more commonly observed, not as a primary disease or pure form, but as a secondary condition superimposed on other renal diseases, such as benign nephrosclerosis and chronic pyelonephritis and, less frequently, glomerulonephritis. Malignant hypertension has been reported in association with the renal changes of scleroderma (see (3) So-called Collagen or Systemic Connective Tissue Diseases, Section D.-3c, Chapter XXI) and occasionally has been seen in patients with lupus nephritis, polyarteritis nodosa and other so-called collagen diseases treated with adrenocortical steroids.

d. Relationship of benign to malignant nephrosclerosis: Although pathogenetically and morphologically, benign and malignant nephrosclerosis must be considered as entirely distinct forms of renal disease,

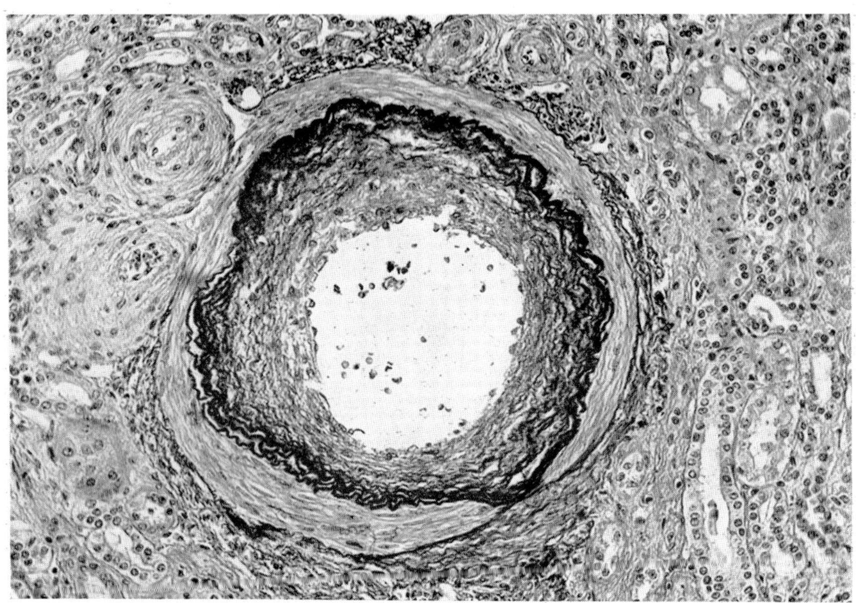

FIG. 126. Marked thickening and reduplication of the internal elastic lamella, so-called intimal elastosis, of an interlobular artery in malignant hypertension. Two adjacent arterioles on the left have marked muscular hypertrophy and some intimal thickening, with pronounced luminal narrowing. Weigert-Van Gieson, ×200.

associated with two different clinical types of hypertension, it is apparent that transitional forms are often seen. Benign essential hypertension may be slowly progressive, with gradual increase in severity over a period of many years, and with death resulting more frequently either from arteriosclerotic heart disease or from cerebrovascular accidents, rather than from renal failure. In a certain proportion of cases, however, the course may be more accelerated, the period of clinically benign hypertension lasting only one or two years and being followed by a relatively short phase of clinically malignant hypertension, with eventually death due to uremia. In these instances, it would seem possible that the "benign" phase of the disease might actually be associated with "malignant" morphologic vascular changes. This discrepancy between clinical and morphologic features of the disease is frequently observed in many renal diseases and is due to the more subjective

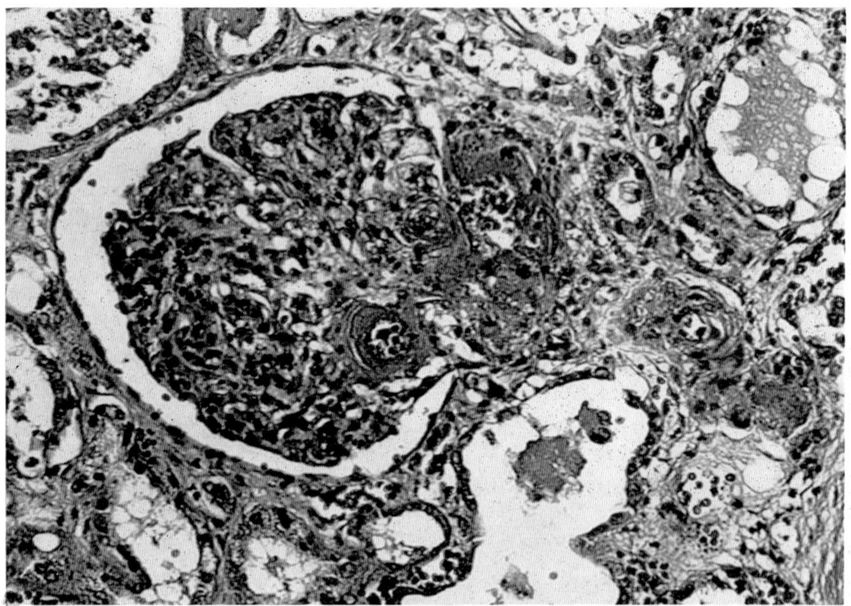

FIG. 127. Arteriolar necrosis of the kidneys in malignant hypertension. The afferent arteriole and some of its major branches extending into the glomerular tuft have a smudgy and granular appearance of their wall—so-called fibrinoid necrosis. Hematoxylin and eosin, ×270.

character of clinical criteria and the unreliability of clinical histories in many of the patients.

e. Effects of therapy: The introduction of more effective therapeutic agents for the treatment of hypertension, such as sympathomimetic and ganglion-blocking drugs, has led to considerable modifications in the clinical course and apparently also in the renal changes of malignant hypertension (Harington *et al.*, 1959; McCormack *et al.*, 1958). The most severe symptoms of either benign or malignant hypertension can be at least partially controlled by these agents, and in the malignant form, prolongation of life has been obtained in many cases. It is of interest that in the patients in whom therapy has been effective, postmortem examination of the kidneys discloses definitely less prominent necrotizing vascular changes, while intimal sclerosis of interlobular and intralobular arteries becomes more pronounced.

2. Shock

By V. V. Glaviano

A. HISTORICAL RÉSUMÉ

1. Early History

Although the complex pattern of signs and symptoms associated with shock were established approximately one hundred years ago, the term "shock" has constantly undergone changes in definition and interpretation.

In the latter half of the 19th century, the outstanding characteristic of this state was firmly established as a failure of the circulation. At the turn of the century and up to World War I, various causes were considered, such as paralysis of the vasomotor center, a toxic factor, overactivity of the sympathetic nervous system, acidosis, acapnia, fat emboli, and exhaustion of pressor hormones from the adrenal medulla. In this period, Henderson (1910) formulated the important concept that circulatory failure in shock was due to an inadequate venous return to the heart. Since serious involvement of the myocardium was not found to occur in experimental shock, the direction of research was toward the peripheral circulation (Cannon, 1923).

In the period between the two World Wars, shock was primarily thought to result from a disturbance in the capillary circulation. Moon (1938) specifically implicated the loss of capillary tonus and abnormal permeability of the capillary endothelium as causes for this state.

2. Experimental Era of World War II

The large number of battle and civilian casualties resulting from World War II served as a strong stimulus for undertaking large-scale research programs on the basic problems of shock. A contribution of major importance was the standardization of the methods of inducing this state in experimental animals by hemorrhage or trauma. A more accurate terminology was devised, with rules of application. Distinction was made between simple hypotension and hypotension of a low enough level and of long enough duration to cause irreversible shock. Methods for measuring blood flow were improved, thus making possible an accurate evaluation of the role of the liver and kidney in the pathogenesis of shock. This era was also a period when biochemical techniques were extensively utilized in the elucidation and understanding of the metabolic and electrolyte changes accompanying shock (Gregerson, 1946; Wiggers, 1950).

B. PRESENT-DAY CONCEPTS OF SHOCK

1. Forms of Shock

a. Primary shock: The term, primary shock, sometimes referred to as neurogenic shock, is frequently employed to designate a state of acute circulatory collapse. This type of shock is usually not considered to have a fatal outcome. The pulse is slowed, blood pressure is reduced, the patient appears pale, and he may or may not be unconscious; sweating and nausea are common. The syndrome is very often seen in blood donors ("common faint"), in patients with spinal injuries, or in those experiencing pain and fear. The acute circulatory failure is usually explained on the basis of a generalized arteriolar dilatation of central nervous system origin.

b. Medical shock: It is obvious that circulatory insufficiency can arise from many diseases, with the patient typically exhibiting signs and symptoms that have become the accepted criteria for the diagnosis of shock: tachycardia, vasoconstriction as recognized by cold extremities, central nervous system depression in the form of stupor and apathy, a gradual decreasing blood pressure, thirst, anuria, and depression of body temperature. Among the disorders which may be responsible for such clinical findings are pericardial tamponade, pulmonary embolus, diabetic acidosis, and Addison's disease. However, Wiggers (1950) has objected to applying the term "shock" to these conditions, because "true" shock can exist only when the venous return to the heart has declined, this, in turn, leading to the reduction in cardiac output and arterial blood pressure.

c. Secondary or "true" shock: Most forms of "true" shock involve a reduction in circulating blood volume, the etiologic factor being trauma, burns, hemorrhage or dehydration. The basic underlying defect for the development of this type of shock is an imbalance between the remaining blood volume and the vascular space. However, Gregerson and Rawson (1959) have made a special point in emphasizing that all forms of "true" shock are not necessarily associated with such a situation. For example, a reduction in blood volume is not present in experimental anaphylactic shock caused by the administration of histamine (Deyrup, 1944). Furthermore, occlusion of a major coronary artery can undoubtedly cause a state of shock, while still not producing a decline in circulating blood volume. In this condition a decrease in cardiac output brings about a reduction in peripheral blood flow, with a consequent anoxia of the tissues (Richards, 1954).

In shock caused by extensive injuries or a massive hemorrhage, the reduced blood volume produces a decline in the venous return, which, in turn, leads to an eventual decrease in blood pressure and blood flow. As a result, a condition of stagnant hypoxia develops, characterized by a high arteriovenous oxygen difference. Such hemodynamic alterations initiate a series of homeostatic compensatory mechanisms primarily designed to reduce the vascular space and maintain, to a limited extent, blood flow to "vital organs," such as the brain and heart. A selective vasoconstriction occurs in the liver and kidney, which, in the course of shock, is largely responsible for the altered metabolism and anuria. The threat to a change in homeostasis is also met by activation of the adrenal medulla and the hypothalamic-hypophyseal-adrenal cortical system. As a consequence, the secretion of catecholamines, cortical steroids and ADH (antidiuretic hormone) becomes an important adjunct to a complex pattern for adjusting and eventually overcoming the state of shock (Cournand *et al.*, 1943; Selkurt *et al.*, 1947; Lauson *et al.*, 1944; Farrell *et al.*, 1956; Pekkarinen, 1960; Ramey and Goldstein, 1957; Hume, 1953; Glaviano *et al.*, 1960).

C. REGIONAL CHANGES

1. Brain

Blalock (1944) reported that in different forms of shock produced in dogs a reduction occurred in cerebral venous blood oxygen. However, such an alteration did not permit him to draw any conclusions regarding cerebral blood flow. Davis *et al.* (1949) observed that hemorrhagic shock in unanesthetized rabbits induced wave changes in the electroencephalogram that were similar to those observed in hypoxia. Despite the fact that brain cells are highly sensitive to anoxia, brain damage does not usually occur in patients that have recovered from severe shock (Frank, 1953). Such findings may therefore imply that hypoxia, if present, is not of a sufficient degree to cause damage. More recent studies on cerebral blood flow in hemorrhagic shock indicate that the decline in local circulation is not sufficient to diminish the supply of oxygen seriously (Kovach *et al.*, 1958). It has also been found that, while the supply of oxygen during shock is adequate, its distribution in the brain may not be uniform (Erdélyi *et al.*, 1958).

2. Heart

a. Coronary flow: The role of the heart in shock was not considered to be of importance for many years, because of the fact that improving the venous return by transfusion in many cases was found to arrest the progressing circulatory failure. However, Wiggers (1947) focused attention on the heart in shock by his detailed studies of cardiac pressure and volume curves in dogs exposed to long periods of oligemic shock. Despite the presence of normal levels of venous pressure, this author reported loss of expulsive force by the heart, which he termed myocardial depression. It was his opinion that this was probably due to a decline in the compensatory factors maintaining coronary blood flow. Opdyke and Foreman (1947) have found that, although a decrease in coronary flow, to the extent of 30 to 60 per cent, does occur in oligemic shock, restoration of blood volume, with elevation of the coronary flow from 120 to 420 per cent above the control flow, still does not eliminate myocardial depression. The conclusion reached was that myocardial depression, although present in the terminal period of shock, was not due to an inadequate coronary flow. On the other hand, Saranoff *et al.* (1954) have reported that a state of coronary insufficiency does exist in dogs exposed to long periods of hemorrhagic hypotension. They observed an increase in right and left atrial pressure which occurred concurrently with cardiac dilation. Furthermore, dilatation and atrial pressures declined when coronary flow was augmented in the shocked dog by extracorporeal perfusion of its coronary arteries. (For a discussion of the use of drugs in cardiogenic shock, see Section F.-2d, Chapter IX.)

b. Pathologic changes: Striking gross and histologic changes have been observed in the hearts of humans and animals dead from hemorrhagic or traumatic shock. Among these are fatty infiltration and myocardial fiber necrosis (Mallory, 1950; Melcher and Walcott, 1951). Subendocardial hemorrhagic necrosis has been postulated to be due to the persistent tachycardia, hyperactivity of the sympathetic nervous system, and circulating adrenaline (Hackel and Goodale, 1955).

3. Intestinal Tract

Hemorrhagic necrosis of the intestinal tract in experimental and clinical shock has focused the efforts of many investigators in the last decade on the role of the "gut factor" in the pathogenesis of shock. Numerous reports have appeared in the literature confirming the damaging effect of protracted hypotension on the intestinal wall. Furthermore, besides the liver, the intestinal tract has been frequently considered to contain the common denominator for triggering the shock syndrome. Fine *et al.* (1959) believed that the fatal outcome was due to a breakdown of the intestinal barrier, thus allowing passage of bacteria and bacterial toxins into the blood stream. MacLean *et al.* (1956) observed an increase in the intestinal weight in irreversible endotoxin shock, a type change which Aust *et al.* (1957) reported was due to sequestration of plasma, probably on the basis of alteration in capillary permeability. However, Johnson and Selkurt (1958) have noted that the increase in intestinal weight in experimental hemorrhagic shock was not an invariable finding and that it resulted primarily from blood and sloughed cells in the intestinal lumen. Alexander (1955) observed that the pooling of blood in the intestinal tract in hemorrhagic hypotension was probably aided by a loss of venomotor tone, although a recent report has questioned the splanchnic pooling of blood, since the portal system in liverless animals has been found to drain against an increased resistance (Selkurt, 1958). The importance of the intestinal factor in the pathogenesis of shock has received further support from experiments in which irreversible shock was prevented by maintenance of the superior mesenteric arterial inflow through a donor dog (Lillehei, 1957).

4. Liver

Despite the decline in mesenteric resistance, hepatic blood flow in shock has generally been reported to decrease (Selkurt *et al.*, 1947; Erskine, 1958). Werner *et al.* (1952) have noted that although a reasonable supply of oxygen is delivered to the liver following hemorrhage, hepatic anoxia is present because the arteriovenous oxygen difference is not offset by the decline in hepatic blood flow. Reynell *et al.* (1955) have demonstrated hepatic involvement in shock by finding that liver extraction of BSP decreases in prolonged states of hypotension. Evidence has also been presented by Cull *et al.* (1956) to indicate that a rising portal pressure, together with a decline in mesenteric and hepatic arterial inflow, presents the liver as a region of high resistance in shock. Erskine (1958) believes that although much work has been done to clarify the role of liver in shock, its function, especially in irreversible shock, has not been sufficiently emphasized. (For further discussion of hepatic blood flow in hemorrhage, see Section D.-5a, Chapter XII.)

D. STATUS OF IRREVERSIBILITY

Many investigations have been devoted to the study of the type of shock which is irreversible despite transfusion, and countless theories have been proposed. Some have been disproved, some have been simply forgotten, while others are at present providing a dynamic direction for research. For the most part, the emphasis has been directed toward finding the mechanism responsible for irreversibility of the changes occurring in an organ. Etiologic factors which have been considered are a toxic substance and an alteration in metabolism or electrolyte distribution. While the collected data have made important contributions to the understanding of the genesis of shock, the evidence has not been sufficient to identify a single factor as the cause of death.

1. Nervous Factor

a. Afferent stimulation: Wang and Overman (1949) found striking differences between dogs subjected to standardized hemorrhagic shock and traumatic shock. With equal reduction in blood volume for both groups of animals, the survival rate was three times greater in those subjected to simple hemorrhage than in those exposed to trauma. The latter had higher heart rates and mean blood pressure and more pronounced signs of central nervous depression, all of which were considered to be due to a nervous factor. Of interest in this regard was the finding that stimulation of the central end of the sciatic nerve in hemorrhaged dogs gave all the symptoms and signs of traumatic shock. Wang and Overman concluded that "afferent impulses arising in the area of injury reflexly produce an intense and prolonged activity of the sympathetic nervous system. While such sympathetic activity operates to maintain a higher blood pressure level and may indeed be considered a compensatory mechanism early in the shock syndrome, the extreme and lasting vasoconstriction may become finally detrimental to the animal through further reduction in tissue blood flow and its sequelae—tissue hypoxia and acidosis."

b. Sympathetic nervous system: The increase in sympathetic nervous system activity that accompanies a decline in blood pressure has been implicated by Freeman *et al.* (1938) as a possible cause of the fatal outcome of shock. Such an hypothesis was supported by the finding that traumatic shock could be prevented in animals with spinal transection (Freedman and Kabat, 1940; Swingle *et al.*, 1944). Further evidence for this point of view was obtained using the adrenergic blocking compound, Dibenamine, which increased the survival rate of animals subjected to standardized lethal shock procedures (Remington *et al.*, 1948, 1950; Wiggers *et al.*, 1950).

c. Adrenal catecholamines: The recent improvements in methods for analyzing plasma catecholamines has once again aroused interest in the role of epinephrine and norepinephrine in shock. High peripheral blood levels of epinephrine and, to some extent, increases in norepinephrine, have been reported in hemorrhagic shock in dogs (Watts and Bragg, 1957; Manger *et al.*, 1957). Glaviano *et al.* (1960) believe that the increased peripheral blood levels of epinephrine in both reversible and irreversible hemorrhagic shock are due to stimulation of the adrenal medulla. These authors noted that the animals maintained their output of adrenal epinephrine above the control level, to the point where complete cardiorespiratory collapse intervened. However, it is necessary to point out that there is no evidence to indicate that adrenal medullary stimulation in shock is due entirely to a neurogenic mechanism, since anoxia or plasma levels of potassium can also be involved in the adrenal secretion of catechols (von Euler, 1958).

d. Pituitary-adrenal axis: Porter (1953), among others, has reported evidence which supports the participation of the central nervous system in pituitary-adrenal responses. He observed that the application of stressing agents to animals resulted in increased electrical activity recorded from the posterior hypothalamus. In support of such findings is the view of Hume (1953) who postulated that afferent impulses from an area of injury reach neurohumoral centers of the anterior pituitary. In addition to the direct relationship of the hypothalamus to the pituitary-adrenal axis, a relationship has also been suggested between adrenal-medullary and adrenal-cortical hormones. With regard to the latter point, Fritz and Levine (1951) observed that the responsiveness of blood vessels to the pressor effects of norepinephrine required the presence of cortical extracts of the adrenals. Ramey and Goldstein (1957) believe that adrenal cortical steroids and epinephrine appear to operate largely as a physiologic unit. The function of steroids in this unit is to maintain the responsiveness of tissues to stress, which is accomplished by their reacting to neurohumoral substances.

2. Bacterial Factor

Fine *et al.* (1959) have presented a large number of observations supporting a toxic substance of bacterial origin as the cause of irreversible shock. These investigators believe that the source of this factor is an endotoxin from bacteria in the intestinal tract which is released into the blood when normal defense mechanisms in the wall of the gut break down in the course of shock. Evidence to support such a concept is the following: (1) animals have been found to become resistant to hemorrhagic shock when pretreated with nonabsorbable antibiotics; (2) they also become resistant when small doses of endotoxins are administered over a period of time; and (3) rabbits or dogs subjected to a nonlethal hemorrhage are put into irreversible shock when given blood from an animal in irreversible shock. The toxicity of plasma from animals in shock has been found to be contained in the lipopolysaccharide fraction.

Fine *et al.* (1959) also believe that shock becomes lethal only when the neutralizing action of the reticuloendothelium system becomes impaired, a system which is believed to be constantly detoxifying endotoxins liberated in small amounts in the normal animal. These investigators explained the protective effects of Dibenamine and reserpine against endotoxin shock as being due to the blocking action of catecholamines on blood vessels. In this regard, it was previously reported that the necrotizing action of epinephrine and norepinephrine on blood vessels was potentiated manyfold by endotoxins (Thomas, 1954).

Zweifach (1959) has objected to the theory that irreversibility is due solely to bacterial toxins, since he found that the germ-free rat could develop irreversible shock as readily as the one with an intestinal flora. Furthermore, recent work has raised the possibility that tissue substances similar in chemical nature to bacterial lipopolysaccharides may also be responsible for causing terminal shock (Landy and Shear, 1957).

3. Humoral Factors

Zweifach (1958) has emphasized the importance of the effect of humoral factors on the microcirculation in the fatal outcome of shock. He theorized that the compensatory responses, particularly to hemorrhage, involve the action of vasoactive agents as mediators for regulating tissue blood flow. He considered the vasodepressor material (VDM) to be a local mediator of blood flow rather than attributing a systemic function to it, as originally believed. It was his feeling that in shock "decompensation would not necessarily develop as the consequence of a single set of humoral factors," the fatal outcome most likely being the result of many regional reactions, especially those occurring in the liver and intestine. (For further discussion of the changes in the microcirculation in shock, see Section G.-2a, Chapter III.)

4. Blood Clotting

The theory has been advanced that irreversible shock can be caused by the formation of small blood clots due to excessive coagulability of blood (Crowell and Read, 1955). In this regard, Turpini and Stefanini (1959) have reported that intravascular clots in the lungs, liver, and kidney of animals in shock are due to a clot-accelerating factor appearing in blood.

References

Alexander, R. S. (1955). Venomotor tone in hemorrhage and shock. *Circulation Research* **3**, 181.

Aust, J. B., Johnson, J. A., and Visscher, M. B. (1957). Plasma sequestration in endotoxin shock. *Surg. Forum* **8**, 8.

Bach, I., Händel, M., and Sós, J. (1956). The effect of dietary sodium-potassium relations on the experimental renal and neurogenic hypertension of the rat. *Acta Physiol. Hung.* **10**, 437.

Baker, R. D., and Kent, S. P. (1950). Nature of the lipids of hyaline arteriosclerosis. *A. M. A. Arch. Pathol.* **49**, 568.

Bell, E. T. (1951). The pathological anatomy in primary hypertension. *In* "Hypertension" (E. T. Bell, ed.), p. 183. Univ. of Minnesota Press, Minneapolis, Minnesota.

Bing, J., and Kazimierczak, J. (1960). Renin content of different parts of the periglomerular circumference. *Acta Pathol. et Microbiol. Scand.* **50**, 1.

Bing, J., and Wiberg, B. (1958). Localization of renin in the kidney. *Acta Pathol. et Microbiol. Scand.* **44**, 138.

Blalock, A. (1944). Utilization of oxygen by brain in traumatic shock. *A. M. A. Arch. Surg.* **49**, 167.

Blaquier, P., Bohr, D. F., and Hoobler, S. W. (1960). Evidence against an increase in circulating pressor material in renal hypertensive rats. *Am. J. Physiol.* **198**, 6.

Braun-Menendez, E., and von Euler, U. S. (1947). Hypertension after bilateral nephrectomy in rat. *Nature* **160**, 905.

Bronte-Stewart, B., and Heptinstall, R. H. (1954). The relationship between experimental hypertension and cholesterol-induced atheroma in rabbits. *J. Pathol. Bacteriol.* **68**, 407.

Byrom, F. B. (1954). The pathogenesis of hypertensive encephalopathy and its relation to the malignant phase of hypertension. Experimental evidence from the hypertensive rat. *Lancet* **ii**, 201.

Byrom, F. B., and Dodson, L. F. (1948). The causation of acute arterial necrosis in hypertensive disease. *J. Pathol. Bacteriol.* **60**, 357.

Cannon, W. B. (1923). "Traumatic Shock." Appleton, New York.

Castleman, B., and Smithwick, R. H. (1948). Relation of vascular disease to hypertensive state; adequacy of renal biopsy. *New Engl. J. Med.* **239**, 729.

Conway, J. (1960). Is increased vasomotor tone present in hypertension? *Physiologist* **3**, 40.

Cook, W. F., and Pickering, G. W. (1959). The location of renin in the rabbit kidney. *J. Physiol. (London)* **149**, 526.

Cournand, A., Riley, R. L., Bradley, S. E., Breed, E. S., Noble, R. P., Lauson, H. D., Gregerson, M. I., and Richards, D. W. (1943). Studies of circulation in clinical shock. *Surgery* **13**, 964.

Crowell J. W., and Read, W. L. (1955). In vivo coagulation—a probable cause of irreversible shock. *Am. J. Physiol.* **183**, 565.

Cull, T. E., Scibetta, M. P., and Selkurt, E. E. (1956). Arterial inflow into the mesenteric and hepatic vascular circuits during hemorrhagic shock. *Am. J. Physiol.* **185**, 365.

Dahl, L. K., and Love, R. A. (1957). Etiological role of sodium chloride intake in essential hypertension in humans. *J. Am. Med. Assoc.* **164**, 397.

Davis, D., and Klainer, M. J. (1940a). Studies in hypertensive heart disease. I. The incidence of coronary atherosclerosis in cases of essential hypertension. *Am. Heart J.* **19**, 185.

Davis, D., and Klainer, M. J. (1940b). Studies in hypertensive heart disease. II. The role of hypertension, per se, in the development of coronary sclerosis. *Am. Heart J.* **19**, 193.

Davis, H. A., Grant, W. R., McNeill, W. P., Wilkinson, R. F., and Marsh, C. (1949). Effects of hemorrhagic shock upon the electroencephalogram. *Proc. Soc. Exptl. Biol. Med.* **72**, 641.

Deyrup, I. J. (1944). Circulatory changes following the subcutaneous injection of histamine in dogs. *Am. J. Physiol.* **142**, 158.

Doyle, A. E., Fraser, J. R. E., and Marshall, R. J. (1959). Reactivity of forearm vessels to vasconstrictor substances in hypertensive and normotensive subjects. *Clin. Sci.* **18**, 3.

Erdélyi, A., Kovach, A. G. B., Menyhart, J., Petocz, L., Nagy, I., and Csuzi, S. (1958). Die Verteilung der breisenden Blutmenge nach Organen bei normalen und in Schockzustand befindlichen Ratten. *Acta Physiol. Hung.* **12**, Suppl. 5.

Erskine, J. M. (1958). The relation of the liver to shock. *Surg. Gynecol. Obstet.* **106**, 207.

Farrell, G. L., Rosnagle, R. S., and Rauschkolb, E. W. (1956). Increased aldosterone secretion in response to blood loss. *Circulation Research* **4**, 606.

Fine, J., Frank, E. D., Ravin, H. A., Rutenburg, S. H., and Schweinburg, F. B. (1959). The bacterial factor in traumatic shock. *New Engl. J. Med.* **260**, 214.

Fishberg, A. M. (1954). "Hypertension and Nephritis." Lea & Febiger, Philadelphia, Pennsylvania.

Floyer, M. A. (1951). The effect of nephrectomy and adrenalectomy upon the blood pressure in hypertensive and normotensive rats. *Clin. Sci.* **10**, 405.

Foa, P. P., Foa, N. L., and Peet, M. M. (1943). Arteriolar lesions in hypertension: A study of 350 consecutive cases treated surgically. An estimation of the prognostic value of muscle biopsy. *J. Clin. Invest.* **22**, 727.

Folkow, B. (1957). Structural, myogenic, humoral, and nervous factors controlling peripheral resistance. *In* "Hypotensive Drugs," Symposium, London, 1956 (M. Harington, ed.), p. 163. Pergamon, New York.

Frank, H. A. (1953). Present-day concepts of shock. *New Engl. J. Med.* **249,** 445, 486.

Freed, S. C., and Friedman, M. (1950). Hypotension in the rat following limitation of potassium intake. *Science* **112,** 788.

Freedman, A. M., and Kabat, H. (1940). The pressor response to adrenalin in the course of traumatic shock. *Am. J. Physiol.* **130,** 620.

Freeman, N. E., Shaffer, S. A., Holling, H. E., and Schecter, A. E. (1938). The effect of total sympathectomy on the occurrence of shock from hemorrhage. *J. Clin. Invest.* **17,** 359.

Fregly, M. J. (1960). Production of hypertension in adrenalectomized rats given hypertonic salt solution to drink. *Endocrinology* **66,** 240.

Friedenwald, J. S. (1935). *In* "The Kidney in Health and Disease" (H. Berglund and G. Medes, eds.), Chapter 38. Kimpton, London.

Friedman, M., Rosenman, R. H., and Freed, S. C. (1951). Depressor effect of potassium deprivation on the blood pressure of hypertensive rats. *Am. J. Physiol.* **167,** 457.

Fritz, I., and Levine, R. (1951). Action of adrenal cortical steroids and norepinephrine on vascular response to stress in adrenalectomized rats. *Am. J. Physiol.* **165,** 456.

Gellman, D. D. (1958). Reversible hypertension and unilateral renal artery disease. *Quart. J. Med.* **27,** 103.

Gellman, D. D., Pirani, C. L., Soothill, J. F., Muehrcke, R. C., and Kark, R. M. (1959). Diabetic nephropathy: A clinical and pathological study based on renal biopsies. *Medicine* **38,** 321.

Genest, J., Koiw, E., Nowaczynski, W., and Leboeuf, G. (1958). Further studies on urinary aldosterone in human arterial hypertension. *Proc. Soc. Exptl. Biol. Med.* **97,** 676.

Glaviano, V. V., Bass, N., and Nykiel, F. (1960). Adrenal medullary secretion of epinephrine and norepinephrine in dogs subjected to hemorrhagic hypotension. *Circulation Research* **8,** 564.

Goldblatt, H. (1948a). Experimental renal hypertension. *Am. J. Med.* **4,** 100.

Goldblatt, H. (1948b). "The Renal Origin of Hypertension." Charles C Thomas, Springfield, Illinois.

Goldblatt, H., Lynch, J., Hanzal, R. F., and Summerville, W. W. (1934). Studies on experimental hypertension: Production of persistent elevation of systolic blood pressure by means of renal ischemia. *J. Exptl. Med.* **59,** 347.

Gomez, A. H., Hoobler, S. W., and Blaquier, P. (1960). Effect of addition and removal of a kidney transplant in renal and adrenocortical hypertensive rats. *Circulation Research* **8,** 464.

Gordon, D., and Drury, D. (1956). The effect of potassium on the occurrence of petechial hemorrhages in renal hypertensive rabbits. *Circulation Research* **4,** 167.

Gregerson, M. I. (1946). Shock. *Ann. Rev. Physiol.* **8,** 335.

Gregerson, M. I., and Rawson, R. A. (1959). Blood volume. *Physiol. Revs.* **39,** 307.

Greisman, S. E. (1954). The reaction of the capillary bed of the nailfold to the continuous intravenous infusion of levo-norepinephrine in patients with normal blood pressure and with essential hypertension. *J. Clin. Invest.* **33,** 975.

Griffiths, W. J., and Collinson, S. (1957). The estimation of noradrenaline in tissue and its excretion in normal and hypertensive subjects. *J. Clin. Pathol.* **10,** 120.

Gross, F., Loustalot, P., and Meier, R. (1957). Production of experimental hypertension by aldosterone. *Acta Endocrinol.* **26,** 417.

Hackel, D. B., and Goodale, W. T. (1955). Effects of hemorrhagic shock on the heart and circulation of intact dogs. *Circulation* **11,** 628.

Harington, M., Kincaid-Smith, P., and McMichael, J. (1959). Results of treatment in malignant hypertension. *Brit. Med. J.* **ii,** 980.

Hartroft, P., and Edelman, R. (1960). The juxtaglomerular cells and their influence on sodium metabolism. *In* "Hahnemann Symposium on Edema," p. 63. Saunders, Philadelphia, Pennsylvania.

Heath, D., and Edwards, J. E. (1958). The pathology of hypertensive pulmonary vascular disease. *Circulation* **18,** 533.

Heath, D., and Whitaker, W. (1955a). Pulmonary vessels in patent ductus arteriosus. *J. Pathol. Bacteriol.* **70,** 285.

Heath, D., and Whitaker, W. (1955b). Pulmonary vessels in mitral stenosis. *J. Pathol. Bacteriol.* **70,** 291.

Helmer, O. M., and Judson, W. E. (1959). The presence of vasoconstrictor and vasopressor activity in renal vein plasma of patients with arterial hypertension. *Hypertension* **8,** 38.

Henderson, Y. (1910). Venopressor mechanisms. *Am. J. Physiol.* **27,** 152.

Heptinstall, R. H. (1954). Renal biopsies in hypertension. *Brit. Heart J.* **16,** 133.

Hodges, C. V. (1960). Hypertension of renal vascular origin. *Modern Concepts Cardiovasc. Disease* **29,** 595.

Hume, D. M. (1953). The neuro-endocrine response to injury: present status of problem. *Ann. Surg.* **138,** 548.

Jackson, W. B. (1958). The functional activity of the human conjuctival capillary bed in hypertensive and normotensive subjects. *Am. Heart J.* **56,** 222.

Johnson, P. C., and Selkurt, E. E. (1958). Intestinal weight changes in hemorrhagic shock. *Am. J. Physiol.* **193,** 135.

Kahn, J. R., Skeggs, L. T., Shumway, N. P., and Wisenbaugh, P. E. (1952). The assay of hypertensin from the arterial blood of normotensive and hypertensive human beings. *J. Exptl. Med.* **95,** 523.

Keith, N. M., Wagener, H. B., and Kernohan, J. W. (1928). The syndrome of malignant hypertension. *A. M. A. Arch. Internal. Med.* **41,** 141.

Kernohan, J. W., Anderson, E. W., and Keith, N. W. (1929). The arterioles in cases of hypertension. *A. M. A. Arch. Internal. Med.* **44,** 395.

Kimmelstiel, P., and Wilson, C. (1936). Benign and malignant hypertension and nephrosclerosis. *Am. J. Pathol.* **12,** 45.

Kincaid-Smith, P., McMichael, J., and Murphy, E. A. (1958). The clinical course and pathology of hypertension with papilloedema (malignant hypertension). *Quart. J. Med.* **27,** 117.

Klemperer, P., and Otani, S. (1931). Malignant nephrosclerosis. *A. M. A. Arch. Pathol.* **11,** 60.

Knowlton, A. I., Loeb, E. N., Seegal, B. C., Stoerk, H. C., and Berg, J. L. (1950). Development of hypertension in the adrenalectomized nephritic rat maintained on NaCl. *Proc. Soc. Exptl. Biol. Med.* **74,** 661.

Knowlton, A. I., Loeb, E. N., Stoerk, H. C., White, J. P., and Hefferman, J. F. (1952). Induction of arterial hypertension in normal and adrenalectomized rats given cortisone acetate. *J. Exptl. Med.* **96,** 187.

Kovach, A. G. B., Roheim, P. S., Iranyi, M., Kiss, S., and Antal, J. (1958).

Effects of isolated perfusion of the head on the development of ischemic and hemorrhagic shock. *Acta Physiol. Hung.* **14**, 231.

Lack, A., Adolph, W., Ralston, W., Leiby, G., Winsor, T., and Griffith, G. (1949). Biomicroscopy of conjunctival vessels in hypertension. *Am. Heart. J.* **38**, 654.

Landy, M., and Shear, M. J. (1957). Similarity of host responses elicited by polysaccharides of animal and plant origin and by bacterial endotoxins. *J. Exptl. Med.* **106**, 77.

Laragh, J. H., Ulick, S., Januszewicz, V., Deming, Q. B., Kelly, W. G., and Lieberman, S. (1960). Aldosterone secretion and primary and malignant hypertension. *J. Clin. Invest.* **39**, 1091.

Lauson, H. D., Bradley, S. E., and Cournand, A. (1944). The renal circulation in shock. *J. Clin. Invest.* **23**, 381.

Lauwers, P., and Conway, J. (1960). Effect of long-term treatment with chlorothiazide on body fluids, serum electrolytes, and exchangeable sodium in hypertensive patients. *J. Lab. Clin. Med.* **56**, 401.

Lee, R. E., and Holze, E. A. (1951). Peripheral vascular hemodynamics in the bulbar conjunctiva of subjects with hypertensive vascular disease. *J. Clin. Invest.* **30**, 539.

Lillehei, R. C. (1957). The intestinal factor in irreversible hemorrhagic shock. *Surgery* **42**, 1043.

McCormack, L. J., Beland, J. E., Schneckcloth, R. E., and Corcoran, A. C. (1958). Effects of antihypertensive treatment on the evolution of the renal lesions in malignant nephrosclerosis. *Am. J. Pathol.* **34**, 1011.

MacLean, L. D., Weil, M. H., Spink, W. W., and Visscher, M. B. (1956). Canine intestinal and liver weight changes induced by E. coli endotoxin. *Proc. Soc. Exptl. Biol. Med.* **92**, 602.

McQuarrie, I., Thompson, W. H., and Anderson, J. A. (1936). Effects of excessive ingestion of sodium and potassium salts on carbohydrate metabolism and blood pressure in diabetic children. *J. Nutrition* **11**, 77.

Mallory, T. B. (1950). General pathology of traumatic shock. *Surgery* **27**, 629.

Manger, W. M., Bollman, J. L., Maher, F. T., and Berkson, J. (1957). Plasma concentration of epinephrine and norepinephrine in hemorrhagic and anaphylactic shock. *Am. J. Physiol.* **190**, 310.

Manlove, F. R. (1946). Retinal and choroidal arterioles in malignant hypertension. *A. M. A. Arch. Internal. Med.* **78**, 419.

Masson, G. M. C., Corcoran, A. C., and Page, I. H. (1951). Experimental production of a syndrome resembling toxemia of pregnancy. *J. Lab. Clin. Med.* **38**, 213.

Masson, G. M. C., Corcoran, H. C., and Page, I. H. (1952). Renal and vascular lesions elicited by "renin" in rats with desoxycorticosterone hypertension. *A. M. A. Arch. Pathol.* **53**, 217.

Melcher, G. W., Jr., and Walcott, W. W. (1951). Myocardial changes following shock. *Am. J. Physiol.* **164**, 832.

Mendlowitz, M., Gitlow, S., and Naftchi, N. (1958). Work of digital vasoconstriction produced by infused norepinephrine in Cushing's syndrome. *J. Appl. Physiol.* **13**, 252.

Meneely, G. R., and Ball, C. O. T. (1958). Experimental epidemiology of chronic NaCl toxicity and the protective effect of KCl. *Am. J. Med.* **25**, 713.

Meneely, G. R., Tucker, R. G., Darby, W. J., and Auerbach, S. H. (1953). Chronic sodium chloride toxicity in the albino rat. *J. Exptl. Med.* **98**, 71.

Merriam, J. C., Sommers, S. C., and Smithwick, R. H. (1958). Clinico-pathological correlations of renal biopsies in hypertension with pyelonephritis. *Circulation* **17,** 243.

Merrill, J. P. (1958). A possible mechanism for human renal hypertension. *Trans. Am. Clin. and Climatol. Assoc.* **70,** 81.

Merrill, J. P., Murray, J. E., Harrison, J. H., and Guild, W. R. (1956). Successful homotransplantation of the human kidney between identical twins. *J. Am. Med. Assoc.* **160,** 277.

Møller, P., Buus, O., and Bierring, E. (1957). Blood pressure and urinary excretion of nor-adrenaline (NA). *Scand. J. Clin. & Lab. Invest.* **9,** 331.

Moon, V. H. (1938). "Shock and Related Capillary Phenomena." Oxford Univ. Press, London and New York.

Moritz, A. R., and Oldt, N. R. (1937). Arteriolar sclerosis in hypertensive and non-hypertensive individuals. *Am. J. Pathol.* **13,** 679.

Morlock, C. G. (1939). Arterioles of the pancreas, liver, gastrointestinal tract and spleen in hypertension. *A. M. A. Arch. Internal. Med.* **63,** 100.

Moses, G. (1954). Development of atherosclerosis in dogs with hypercholesterolemia and chronic hypertension. *Circulation Research* **2,** 243.

Moss, W. G., Kiely, J. P., Neville, J. B., Bourgue, J. E., and Wakerlin, G. E. (1951). Effect of experimental renal hypertension on experimental cholesterol atherosclerosis. *Federation Proc.* **10, 94.**

Movat, H. Z., and More, R. H. (1957). The nature and origin of fibrinoid. *Am. J. Clin. Pathol.* **28,** 331.

Movat, H. Z., Haust, M. D., and More, R. H. (1959). The morphologic elements in the early lesions of arteriosclerosis. *Am. J. Pathol.* **35,** 93.

Muirhead, E. E. (1959). The renal medulla and renoprival hypertension. *Hypertension* **8,** 21.

Opdyke, D. F., and Foreman, R. C. (1947). A study of coronary flow under conditions of hemorrhagic hypotension and shock. *Am. J. Physiol.* **148,** 726.

Peart, W. S. (1959). Renin and hypertensin. *Ergeb. Physiol.* **50,** 409.

Pekkarinen, A. (1960). *In* "The Biochemical Response to Injury" (H. B. Stoner, ed.), p. 217. Blackwell, Oxford.

Perera, G. A. (1953). Depressor effects of potassium deficient diets in hypertensive man. *J. Clin. Invest.* **32,** 633.

Pickering, G. W. (1952). The pathogenesis of malignant hypertension. *Circulation* **6,** 599.

Pitcock, J. A., and Hartroft, P. M. (1959). Pressor activity (renin) of kidneys from sodium deficient rats and correlation with granulation of juxtaglomerular cells. *Federation Proc.* **18,** 500.

Porter, R. W. (1953). Hypothalamic involvement in the pituitary-adrenocortical response to stress-stimuli. *Am. J. Physiol.* **172,** 515.

Poutasse, E. F. (1959). Surgical treatment of renal hypertension: Results in patients with occlusive lesions of renal arteries. *J. Urol.* **82,** 403.

Ramey, E. R., and Goldstein, M. S. (1957). The adrenal cortex and the sympathetic nervous system. *Physiol. Revs.* **37,** 155.

Redleaf, P. D., and Tobian, L. (1958a). Sodium restriction and reserpine administration in experimental renal hypertension. *Circulation Research.* **6,** 343.

Redleaf, P. D., and Tobian, L. (1958b). The question of vascular hyperresponsiveness in hypertension. *Circulation Research.* **6,** 185.

Remington, J. W., Wheeler, N. C., Boyd, G. H., Jr., and Caddell, H. M. (1948).

Protective action of dibenamine after hemorrhage and after muscle trauma. *Proc. Soc. Exptl. Biol. Med.* **69**, 150.

Remington, J. W., Hamilton, W. F., Boyd, G. H., Jr., Hamilton, W. F., Jr., Caddell, H. M. (1950). Role of vasoconstriction in the response of the dog to hemorrhage. *Am. J. Physiol.* **161**, 116.

Reynell, P. C., Marks, P. A., Chidsey, C., and Bradley, S. E. (1955). Changes in splanchnic blood volume and splanchnic blood flow in dogs after haemorrhage. *Clin. Sci.* **14**, 407.

Richards, D. W. (1954). Nature and treatment of shock. *Circulation* **9**, 606.

Robert, A., and Nezamis, J. E. (1957). Hypertensive eye lesions in the rat. Effect of carotid ligation. *Experientia* **13**, 457.

Saranoff, S. J., Case, R. B., Warthe, P. E., and Isaacs, J. P. (1954). Insufficient coronary flow and myocardial failure as a complicating factor in late hemorrhagic shock. *Am. J. Physiol.* **176**, 439.

Schaffenburg, C., and Goldblatt, H. (1957). Pathogenesis of arteriolar necrosis of malignant hypertension. *Proc. Soc. Exptl. Biol. Med.* **96**, 421.

Scheinker, I. M. (1943). Hypertensive disease of the brain. *A. M. A. Arch. Pathol.* **36**, 289.

Scheinker, I. M. (1948). Alterations of cerebral capillaries in the early stages of arterial hypertension. *Am. J. Pathol.* **24**, 211.

Selkurt, E. E. (1958). Mesenteric hemodynamics during hemorrhagic shock in the dog, with functional absence of the liver. *Am. J. Physiol.* **193**, 599.

Selkurt, E. E., Alexander, R. S., and Patterson, M. B. (1947). The role of the mesenteric circulation in the irreversibility of hemorrhagic shock. *Am. J. Physiol.* **149**, 732.

Shipley, R. E., Helmer, O. M., and Kohlstaedt, K. G. (1947). The presence in blood of a principle which elicits a sustained pressor response in nephrectomized animals. *Am. J. Physiol.* **149**, 708.

Smithwick, R. H., and Castleman, B. (1951). Some observations on renal vascular disease in hypertensive patients based on biopsy material obtained at operation. *In* "Hypertension" (E. T. Bell, ed.), p. 199. Univ. of Minnesota Press, Minneapolis, Minnesota.

Sommers, S. C., Relman, A. S., and Smithwick, R. H. (1958). Histologic studies of kidney biopsy specimens from patients with hypertension. *Am. J. Pathol.* **34**, 685.

Swingle, W. W., Kleinberg, W., Remington, J. W., Eversole, W. J., and Overman, R. R. (1944). Experimental analysis of the nervous factor in shock induced by muscle trauma in normal dogs. *Am. J. Physiol.* **141**, 54.

Thomas, L. (1954). Endotoxins: Physiological disturbances. *Ann. Rev. Physiol.* **16**, 467.

Tobian, L., and Binion, J. T. (1952). Tissue cations and water in arterial hypertension. *Circulation* **5**, 754.

Tobian, L., and Binion, J. (1954). Artery wall electrolytes in renal and DCA hypertension. *J. Clin. Invest.* **33**, 1407.

Tobian, L., Thompson, J., Twedt, R., and Janecek, J. (1958). The granulation of juxtaglomerular cells in renal hypertension, desoxycorticosterone and postdesoxycorticosterone hypertension, adrenal regeneration hypertension, and adrenal insufficiency. *J. Clin. Invest.* **37**, 660.

Tobian, L., Tomboulian, A., and Janecek, J. (1959a). The effect of high perfusion pressures on the granulation of juxtaglomerular cells in an isolated kidney. *J. Clin. Invest.* **38**, 605.

Tobian, L., Janecek, J., and Tomboulian, A. (1959b). Correlation between granulation of juxtaglomerular cells and extractable renin in rats with experimental hypertension. *Proc. Soc. Exptl. Biol. Med.* **100**, 94.

Tobian, L., Janecek, J., and Tomboulian, A. (1959c). Effect of a high sodium intake on the development of permanent nephrosclerotic hypertension. *J. Lab. Clin. Med.* **53**, 842.

Tobian, L., Winn, B., and Janecek, J. (1961a). The influence of arterial pressure on the antihypertensive action of a normal kidney, a biological servomechanism. *J. Clin. Invest.* **40**, 1085.

Tobian, L., Janecek, J., Tomboulian, A., and Ferreira, D. (1961b). Sodium and potassium in the walls of arterioles in experimental renal hypertension. *J. Clin. Invest.* **40**, 1922.

Turpini, R., and Stefanini, M. (1959). The nature and mechanism of the homeostatic breakdown in the course of experimental hemorrhagic shock. *J. Clin. Invest.* **38**, 53.

von Euler, U. S. (1956). "Noradrenaline." Charles C Thomas, Springfield, Illinois.

von Euler, U. S. (1958). Regulation of production and release of catechol hormones. *Acta Endocrinol.* Suppl. 38, 36.

Wakerlin, G. E. (1958). Antibodies to renin as proof of the pathogenesis of sustained renal hypertension. *Circulation* **17**, 653.

Wang, S. C., and Overman, R. R. (1949). A neurogenic factor in experimental traumatic shock. A summary of recent studies including observations on procainized and spinal dogs. *Ann. Surg.* **129**, 207.

Watts, D. T., and Bragg, A. D. (1957). Blood epinephrine levels and automatic reinfusion of blood during hemorrhagic shock in dogs. *Proc. Soc. Exptl. Biol. Med.* **96**, 609.

Weiss, S., Parker, F., Jr., and Robb, G. P. (1933). Malignant nephrosclerosis. *Ann. Internal. Med.* **6**, 1599.

Werner, A. Y., MacCanon, D. M., and Horvath, S. M. (1952). Fractional distribution of total blood flow to, and oxygen consumption of, the liver as influenced by mild hemorrhage. *Am. J. Physiol.* **170**, 624.

Wiggers, C. J. (1947). Myocardial depression in shock. *Am. Heart J.* **33**, 633.

Wiggers, C. J. (1950). "Physiology of Shock." Commonwealth Fund, New York.

Wiggers, H. C., Goldberg, H., Roemhild, F., and Ingraham, R. C. (1950). Impending hemorrhagic shock and the course of events following administration of dibenamine. *Circulation* **2**, 179.

Wilens, S. L. (1947). Bearing of general nutritional state on atherosclerosis. *A. M. A. Arch. Internal Med.* **79**, 129.

Wilson, C., and Byrom, F. B. (1939). Renal changes in malignant hypertension. *Lancet* **236**, 136.

Wolfgarten, M., and Magarey, F. R. (1959). Vascular fibrinoid necrosis in hypertension. *J. Pathol. Bacteriol.* **77**, 597.

Zweifach, B. W. (1958). Microcirculatory derangements on a basis for the lethal manifestations of experimental shock. *Brit. J. Anaesthesia* **30**, 466.

Zweifach, B. W. (1959). Hemorrhagic shock in germ free rats. *Ann. N. Y. Acad. Sci.* **78**, 315.

Chapter XXI. PERIPHERAL VASCULAR DISORDERS

1. Vascular Anomalies Due to Developmental Defects

A. CONGENITAL ARTERIOVENOUS FISTULA

By R. J. Booher

An arteriovenous fistula is a vascular anomaly, with one or more abnormal communications between an artery and vein, which by-passes the normal capillary bed (Ward and Horton, 1940). It is formed either as a result of persistence of communications of the primary anlage of the vascular bed or following trauma to anatomically normal vessels. The first group is the province of the foregoing discussion.

1. Embryology

The concept of vascular anomalies giving rise to arteriovenous fistulization is based on the investigations of Sabin (1917–1918) and Woolard (1922) who traced the origin of arteries, veins, and capillaries from the anlage of a common bed of mesenchyme. They found that the latter first differentiates into a diffuse capillary network and that this is succeeded by a retiform stage of enlarged tubes which coalesce to form vascular stems. Persistence of any vessels from this primitive capillary network leads to various vascular anomalies, dependent on the stage of development in which the aberration occurs.

2. Pathophysiology

Bertelsen and Dohn (1953), using the plethysmographic method, reported that in congenital arteriovenous fistula the rate of blood flow in the leg is about ten times greater than normal, while that for the foot is forty times greater. Similarly, there is a fourfold increase in the pulse volume in the leg and a sevenfold increase in the foot. These changes are due to arterial or arteriolar dilatation, associated with excessive filling of the venous bed and, consequently, greater blood volume of the part. While the fistulization is parasitic upon the circulation, the extensive collateral circulation and dilatation result in a larger than necessary blood supply for normal nutrition and growth, and, as a result, hypertrophy follows. Holman (1940) has suggested that when dilatation outstrips hypertrophy, decompensation of the circulation in the part follows, as it does in the heart itself.

There may be an increase of pulse pressure due to elevation of the systolic and lowering of the diastolic pressure, although systolic pressure is reduced distal to the fistula. Absent or decreased peripheral pulses are noted in a small percentage of cases, associated with increased oscillometric readings above, and lower readings below, the fistula. Oxygen saturation of peripheral venous blood has been found to show an average increase of 24.6 per cent over the control level (Leonard and Vassos, 1951). With the lessening of peripheral resistance in venous side of the fistula, there is an increased flow rate to the heart. Murphy *et al.* (1956) found that in five patients with abnormal arteriovenous communications, the range of circulation time from femoral artery to heart was from 1.0 to 8.8 seconds, as compared with an average reading of 21 seconds for normal subjects.

3. Clinical Syndromes

a. Classification of entities: Lewis (1930) observed that it was frequently impossible to differentiate among the various types of syndromes of congenital vascular anomalies, since characteristics of each type merge in varying proportions, thus confusing attempts at concise classification. Nevertheless, the classifications of Thompson and Shafer (1951), presented below, are useful clinically.

(1) Anomalies of the large vessels conducting blood to a part (arterial, racemose, or cirsoid aneurysms): In this type a proximal nutrient artery is dilated, or if placed distally, only a short segment is involved. The veins are distended, and communications between artery and veins develop.

(2) Anomalies of large vessels conducting blood away from a part [phlebectasia or diffuse systemic angioma (Allen *et al.*, 1955)]: The importance of phlebectasia in opening communications is posed by Malan (1958) who, in studying vascular syndromes produced by dilatation of arteriovenous communications in the sole of the foot, came to the conclusion that these changes are secondary to the varicose state itself and that the dilatation occurs whenever valves are incompetent and the weight of the column of blood is on the small postcapillary venous branches. It is possible that there is a relationship between this condition and the syndrome of arterial varices (Pratt, 1949).

(3) Anomalies of minute vessels: This class includes the port wine stain, the spider angioma, and the cavernous angioma. Reid (1925) has emphasized that these are the same fundamental type of lesion as the more complicated systemic angiomatous states.

(4) Anomalies of obvious communications between arteries and veins (true arteriovenous fistulas) (Fig. 128): The most common lesion in this group is the process of diffuse phlebariectasia. The condition may

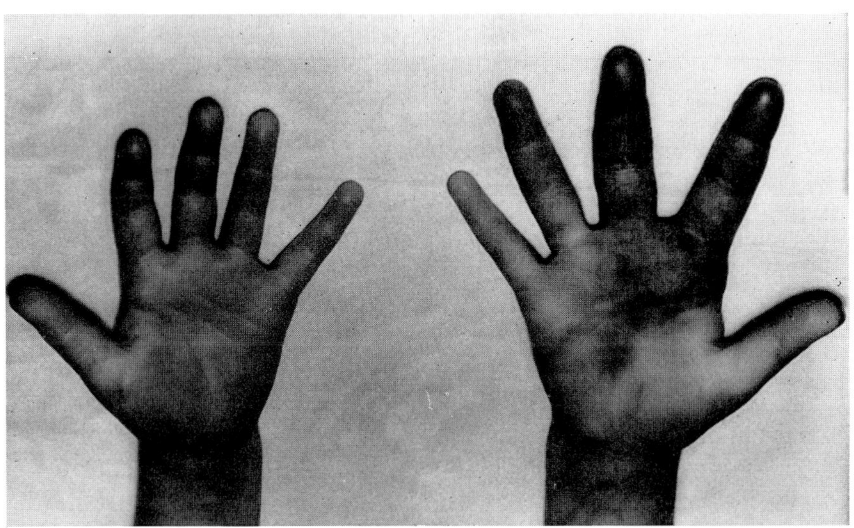

FIG. 128. Multiple congenital arteriovenous fistulae in right hand of an eight year old child. Redness of involved areas existing since birth; progressive dilatation of veins noted since age of four years, coincident with the onset of pain in hand and wrist. Hypertrophy of the second, the third and, to a lesser extent, the fourth fingers, with some degree of compressibility present.

Physical findings and tests: (1) brownish-red color of skin over lesion; (2) thrill felt over involved area; (3) bruit heard over entire extremity; (4) blood pressure, right arm, 114/80 mm Hg; left arm 98/72; (5) normal carpal bones by X-ray; (6) increase of plasma volume (Evans blue dye method) by 15 per cent and increase of total blood volume by 9 per cent.

not present the direct artery and vein anastomosis shown in the first group (Blumgart and Ernstene, 1932).

(5) Anomalies presenting various manifestations noted in previous groups but with organoid developmental defects: The latter may take the form of angiolipomatosis or of multiple sites of nodular masses simulating glomus tumors.

b. Sites of predilection: Cross *et al.* (1958) have estimated that about 200 cases of congenital arteriovenous fistula have been reported.

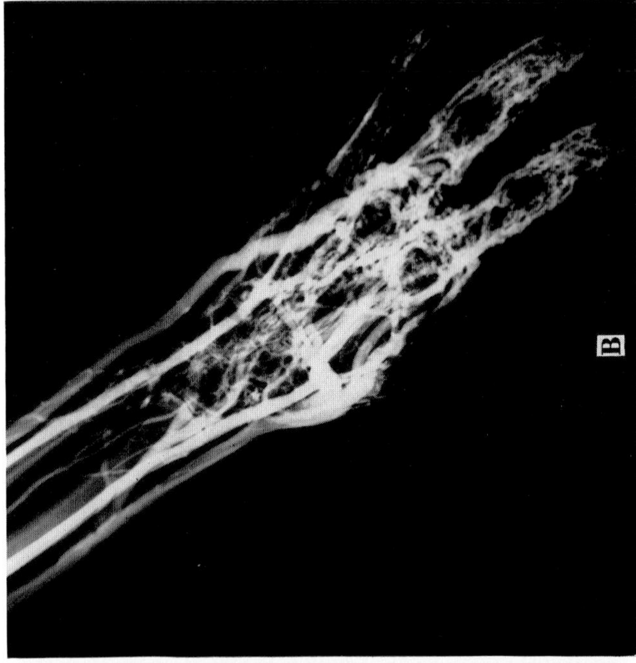

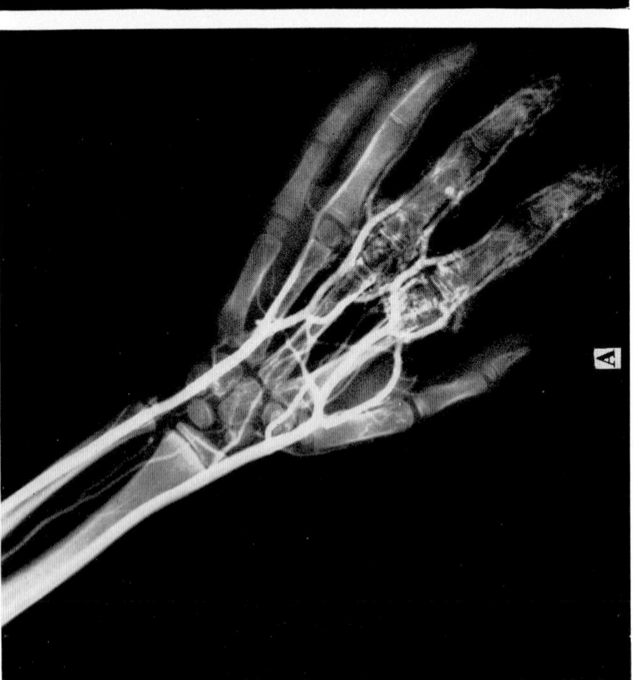

Fig. 129. Arteriograms obtained during a serial study. A. Exposure made one second after injection of contrast medium. (Same patient as in Fig. 128.) Complex arteriovenous fistulae, involving the right second, third and, to a lesser degree, fourth fingers, visualized, as contrasted with the normal vasculature of the other fingers. B. Exposure made three seconds after injection. Marked dilatation of veins now observed. (Courtesy of C. T. Dotter.)

While this type of lesion has been found in many sites, such as the coronary system (Fell *et al.*, 1958), the lung (Taber and Ehrenhaft, 1956), the ischioanal fossae (Smith *et al.*, 1957), the brain (Paillas *et al.*, 1958), to mention only a few, space precludes consideration of only the problem in the extremities, in which the condition is most common.

c. Clinical picture: Although congenital arteriovenous fistula may be found in any age group, in general, it occurs in the first three decades of life, with no difference being noted in sex distribution. The most common complaint is asymmetry of the affected extremity, the lower being involved more frequently than the upper one; this may be associated with pain. In some cases simple dilatation of veins in the area or a vascular mass may be the presenting complaint.

Increased length or girth of the part has been noted in 80 per cent of the patients (Coursley *et al.*, 1956), with the average increase being 3.2 cm in 29 of 51 patients studied for this finding by Leonard and Vassos (1951). Elevated temperature of the skin, up to 6° C, is almost as frequent a finding, followed by varices, bruit, vascular mass, port wine stain, stasis changes, and increased sweating, with or without greater hair growth. The presence of a thrill is about half as frequent as an audible bruit, but with multiple small fistulae distally, such findings are apparent less than half the time. The bradycardiac reaction is frequently present (Coursley *et al.*, 1956), and reductions of four to fourteen beats per minute have been reported (Leonard and Vassos, 1951).

d. Arteriography: Murphy and Margulis (1956) have indicated that the demonstration of small congenital arteriovenous communications is difficult, and unless there is massive involvement, the radiographic visualization of these lesions often consists simply of an increased prominence of findings normally observed. Coursley *et al.* (1956) reported that in 28 patients studied by arteriography, definite arteriovenous shunts were visualized in only four on the first trial. In eight others presumptive evidence was based on nonfilling of arteries distal to a suspected fistulous site, with early filling of proximal veins and abrupt termination of filling of arteries. There may be dilatation of the arteries leading to the fistula and pooling of the contrast medium in the region of the lesion. Serial arteriography with exact chronometry, however, gives the greatest help in the delineation of a fistula (Fig. 129).

B. GLOMUS TUMOR

By M. Riveros

1. Definition

A glomus tumor is a benign organoid tumor which develops at the expense of a neuro-myo-arterial glomus. (For discussion of the structure and function of the normal glomus, see Sections B.-1e, 2g. and C.-3, Chapter XVI.) The tumor is most frequently located on the fingers, but may also be observed in the arm, forearm and leg, among other sites, as well as in such noncutaneous tissues as muscle (Thomas, 1933), uterus (Durante, 1933), bone (Iglesias de la Torre *et al.*, 1939; Riveros and Pack, 1951, 1953) and stomach (Stout, 1935).

At first the glomus tumor was considered to be merely a painful subcutaneous nodule, but later, through histologic studies, its vascular origin was recognized. Some of the early investigators believed it to be a malignant tumor, resembling an angiosarcoma (Kolaczek, 1878), a tubular epithelioma (Chandelux, 1882), or a perithelioma (Müller, 1901). In 1924 Masson properly described it as a glomic neuro-myo-arterial tumor. Of great importance, too, in the elucidation of the problem were the reports of Sucquet (1862) and Hoyer (1877) on the anatomy and physiology of the normal glomus structure. The tissue culture studies of Murray and Stout (1942) and Stout and Murray (1942) show the close relationship between the epithelioid cell of Masson and the pericyte of Zimmermann (1923), which could be the basis for the presence of glomus tumors in other tissues.

The glomus tumor has no predilection for sex or race; it appears more frequently in individuals over forty years of age. It has two main clinical manifestations: pain and the presence of a mass. It is usually single although it may be multiple. The color of the tumor varies from a deep red to purple or blue; typically there is a striking and characteristic variation in color with changes in environmental temperature. Generally the mass is sharply demarcated from the surrounding structures and it does not ulcerate.

The pain associated with the glomus tumor is characteristic. It occurs spontaneously and without pressure or trauma, although it may develop when the tumor is touched lightly, as by a glove, clothing or bed clothes. It may be intermittent or continuous and may radiate up an entire extremity in the form of a stabbing, lancinating, and agonizing sensation. The pain is usually exaggerated by cold weather, while at times immersion of the part in warm or hot water may cause partial relief. Associated with the appearance of the pain may be signs of vasomotor imbalance, such as localized sweating or Horner's syndrome.

The fear of precipitating symptoms causes the patient to adopt a protective attitude toward the involved limb. As a result, actual disuse atrophy of an entire arm and shoulder may result.

2. Diagnosis

The pin test (Love, 1944) is useful when a glomus tumor is suspected but no obvious lesion is noted. The head of a straight pin is rubbed gently over the skin, which on coming in contact with the invisible tumor will elicit a trigger point of sensitivity and hence pain. Also of use as a diagnostic tool is the X-ray, since destruction of the cortex of the terminal phalanx is generally observed by this means (Rypins, 1941). Such findings may likewise be useful in localizing the lesion and in its surgical excision.

a. Differential diagnosis: The glomus tumor may be confused with fibromas, neuromas, neurofibromas, nevus, and angiomas. However, none of these masses produces the marked pain associated with a glomus tumor. Typically the latter is felt but not seen, while the others are seen but generally not felt. The melanoma, because of its frequent subungual location, may present a problem of differential diagnosis, but it is painless and rarely affects bone. The Boeck sarcoid, a tumor of dark red coloration, is vascular but painless.

The one tumor which offers considerable difficulty in differentiation is the hemangioerictoma which is a large mass, 8 to 15 centimeters in diameter, located anywhere in the body. In contrast to the glomus tumor, it is often malignant and metastasizes. Histopathologic examination is characteristic.

C. HEMANGIOMA

By G. T. Pack and T. R. Miller

1. Definition and Etiology

An angioma, as defined by Ewing, is a true neoplastic process involving vascular and lymphatic tissue. In this manner it differs distinctly from simple, self-limiting vessel hypertrophy, such as occurs in granulation tissue, and it bears no relation to the ordinary dilatation of previously formed vessels, such as is noted in varices or arteriovenous shunts (Ribbert, 1898). It is generally believed that simple angiomas are congenital in nature and have their beginning in embryonic sequestrations of mesodermal tissue. Also it seems well established that the three types of benign blood and lymph vessel tumors (hemangiomas, lymphangiomas, and hygroma) not only have a common origin, but grow in an identical fashion by the projection of buds of endothelial tissue.

Strongly supporting the theory that hemangiomas are congenital in origin is the fact that 73 per cent of the patients with this difficulty present evidence of tumor at birth. More than 85 per cent of the lesions develop before the end of the first year, and undoubtedly many of these are present at birth but are unnoticed because of their small size (Pack and Anglem, 1939).

The skin must be considered to be an organ which is anatomically supplied by end arteries. Of interest is the finding that the cavernous hemangiomas frequently appear at the site of the termination of a terminal artery. Various combinations of capillary angioma, port wine stain, and cavernous hemangioma may be present.

The rarity with which these tumors persist after the age of two years indicates that they are developmental anomalies which may be influenced by some error in the clotting mechanism. This is further emphasized by the fact that in the large hypertrophic angiomas there is a marked deficiency in the platelet count, the disappearance of this type of tumor often coinciding with an increase in the platelet count. Females are more frequently affected than males, the ratio being 65 to 35 per cent, a relationship which is a consistent feature in the world literature. In fact, in some clinics the proportion runs as high as 3 to 1. Although no scientific reason for such a sex differentiation has ever been offered, it is suggested that hemangiomas may in some fashion be related to the female sex hormone. In this connection it is of interest to note that a hemangioma that has been quiescent may often increase rapidly in size with the onset of menses or at the beginning of a pregnancy. For example, hemangiomas of the liver enlarge rapidly in the early months of pregnancy, as do peripheral lesions.

In the vascular system of the adult most arteries are accompanied by vena comites. However, in the development of the vascular system, the veins and arteries are always separated by a capillary bed. Such a situation, which may be interpreted as primary, persists in the brain and the lung. In general, however, the adult state can only be reached by a modification of primary conditions and by addition of new veins and arteries. The positional relationships of most veins in adults are therefore secondary. With changes that occur before birth and at birth, it is not to be wondered at that the hemangioma makes up the largest group of neoplasms occurring in childhood (Pack and Anglem, 1939). Such alterations in the blood vascular system may account for the congenital neurocutaneous syndromes associated with the angiomatoses.

Much has been written regarding the inheritance of nevi and moles, as well as of birthmarks (Gates, 1948), the latter, in fact, having been regarded from the earliest times as an indication of relationship. A study of this subject shows that there is a greater similarity in monozygos twins than in dizygos twins. The nevus flammens nuchi, "Unna's nevus," is an example of this hereditary relationship. The syndromes of Bourneville, von Recklinghausen, and Lindau-von Hippel (discussed below) are three related congenital hereditary diseases. A fourth form, encephalotrigeminal angiomatosis (Sturge-Weber's disease), has also been described. All four of these may show tumors in the retina and may be accompanied by telangiectasia, buphthalmos, and calcifications in the cerebrum. Wide variability is noted in hereditary syndromes of this sort. A peculiarity of human genetics is that the same abnormal character may be dominant in one pedigree, recessive in another, and sex linked in the third. This triple method of inheritance has been reported in many human abnormalities.

2. PATHOGENESIS

In addition to the hereditary aspects of the various forms of hemangiomata and the related congenital hereditary syndromes, the finding of spider angiomas in pregnancy (Bean *et al.*, 1949) and liver disease (Patek *et al.*, 1940), as well as in patients who have been subjected to liver toxins, suggests the relationship of the female sex hormone to the development of such tumors. Of further interest in this regard is the enlargement of these lesions during pregnancy or at the time of puberty.

The nevus flammens, described by De Morgan, is relatively more common in the elderly and is found in slightly greater number in patients with malignant tumors than in normal individuals. Here the relationship appears to be with age rather than with malignant disease.

The hemangioma inherits from its embryonic parent a certain power of irregular and unrestrained growth. The average tumor pursues a benign course, and if untreated, grows slowly up to adult life. All angiomas, however, do not follow this general rule. Occasionally one will suddenly develop rapid acceleration of growth, become locally destructive, and may even cause death by hemorrhage or sepsis if not treated properly. In all but the racemose type, tumor growth generally stops with full body growth, and in the capillary, cavernous, and hypertrophic types, spontaneous regression is the rule (Shaw, 1928).

3. Pathology

The hemangioma develops and enlarges by extending solid buds of endothelium into adjacent tissues. The process is no different from that seen in the repair of tissues. The solid cords of endothelial tissue become canalized and establish communication with a parent vessel. They do not appear to invade or communicate with surrounding normal vessels. Some histologic studies tend to confirm Ribbert's original theory that a hemangioma, with the exception of the racemose type, has no anastomotic connection with the surrounding vessels and possesses only one afferent and one efferent vessel. The occasional rapid disappearance of one of these lesions after a fortunate injection of sclerosing agent would tend to confirm this theory. However, such a response could also be related to the fact that the hemangioma in the skin arises from a terminal arteriole. The fortuitous occlusion of the terminal vessel or the tenuous blood supply from a tumor, such as Ribbert (1898) describes, could therefore explain the phenomena in either case.

Many of the systemic angiomas involving a limb bud will also present multiple arteriovenous fistulae. These are intermittently opened and closed and can present a problem if not completely dissected out. (For further discussion of congenital arteriovenous fistula, see Section A., above.)

The decrease in platelets noted in the hypertrophic angiomas, and occasionally in the cavernous and capillary angioma, may be due to the tumor or to some common factor influencing its growth. With the disappearance of the tumor, the platelet count is restored to normal almost simultaneously.

In general the blood vessels making up hypertrophic angiomas are capillary in nature. The vessels are lined with a single layer of endothelial cells and surrounded by a reticulated fibrous sheath. The lumen of the vessels comprising the tumor are either empty or contain a few immature or degenerate corpuscles.

4. Clinical Types

A rigid and all-inclusive classification of the hemangiomas is difficult. For the purpose of therapy, the following simplified classification is suggested:

A. Capillary hemangioma (strawberry mark).
B. Cavernous hemangioma.
C. Mixed cavernous and capillary angioma.
D. Angioblastic or hypertrophic hemangioma.
E. Racemose hemangioma.
F. Nevus vinosus or port wine stain.
G. Spider angioma (nevus araneus).
H. Nevus flammens (De Morgan's spot).
I. Systemic hemangiomatosis or hemangioma unis lateris, infectious hemangioma, pyogenic granuloma.
J. Special regional hemangiomas of the brain, tongue, gastrointestinal tract, liver, skeletal muscle and bone.
K. Congenital neurocutaneous syndromes associated with angiomatosis.
   1) von Recklinghausen's neurofibromatosis in angiomas of the skin.
   2) Bourneville's syndrome with tuberous sclerosis.
   3) Sturge-Weber's disease (encephalofacial angiomatosis).
   4) Lindau-von Hippel's disease (hemangiomatosis of retina and cerebellum).

a. Capillary hemangioma: This is the most common of all hemangiomas. It occurs in the skin and mucous membrane, and typically it is a circumscribed, sessile, lobulated, bright red tumor.

b. Nevus vinosus or port wine stain: This is a flat, purplish-pink lesion, made up of very fine vessels in the form of telangiectasia in the derma. No proliferative masses, as noted in the other types of angioma, are found. The purplish patch blanches on pressure and tends to become darker with age. The face is the most common site, and the mucosa of the lip, cheek, and oral cavity may be involved in continuity.

c. Nevus araneus or spider angioma: This is a tiny, red angioma resembling a small spider. It has a central point from which radiate many fine, hairlike, wavy strands, 0.5–1 cm in length. The central point or body of the lesion may be so small as to be seen only by magnification, or it may be a millimeter in diameter.

d. Sclerosing angioma: This tumor may well represent the regressed, cavernous angioma. It is composed of such connective tissue proliferations that the perivascular endovascular thickenings form a solid mass in which the lumens of the majority of the vessels are often invisible. However, some of the capillaries may be seen, as well as phagocytes containing hemosiderin and lipids. The tumor is firm, noncompressible, and resembles a benign dermatofibrosarcoma protuberans.

e. Angioblastic or hypertrophic hemangioma: This is the benign analog of the malignant hemangioendothelioma. It is a solid, noncompressible tumor of variable hue, although usually purplish-red. The overgrowth of the endothelial cells in the tumor tends to obliterate the lumen of the blood vessels. The lesion is not well localized and is often locally aggressive, tending to recur after operation.

f. Systemic hemangiomatosis: This is a diffuse vascular tumor which generally occupies the entire extremity or a portion of the head or trunk. The anlage of this tumor probably began at the time of limb budding, so that as the various structures formed, all of the tissues of the extremity became infiltrated by the abnormal, tortuous vessels. Because of the local increase in circulation, the involved tissues become hypertrophied, particularly the bone. As a result, the affected leg or arm may become so heavy and cumbersome that the patient is functionally handicapped in addition to his disfigurement. The systemic hemangioma is actually a congenital arteriovenous aneurysmal anomaly, with multiple communications between arteries and veins (Janes and Musgrove, 1949). The tumor may be partly capillary in structure, or it may consist of a mixed hemangiolymphangioma. Beside the involvement of the limbs, there may be hemangiomatous lesions in other organs, such as the liver, kidney, lung, or brain.

Of interest is the finding that the systemic hemangioma is constantly associated with hypertonicity of the sympathetic nervous system, manifesting itself in the form of marked hyperhydrosis and vasoconstriction limited to the involved limb. Cyanosis, which is a common finding, may also be due, in part, to delayed venous return. The vasoconstriction of the arteries may be a manifestation of an attempt to maintain adequate local blood pressure, in the face of loss of capillary resistance due to the substitution of wider arteriovenous channels.

5. Neuroectodermal Defects Associated with Hemangioma

Certain neuroectodermal defects involving the central nervous system and peripheral nerves are also frequently found in association with hemangiomatous lesions; these may be intracranial, dermal, and visceral in distribution. With regard to the brain, the middle cerebral artery is more likely to be involved by hemangiomatous lesions because of the wide distribution of its vessels (Cushing and Bailey, 1928). Both neuroectodermal mesenchyme and mesodermal angioblastic tissue participate in the various diseases falling into this category (Yakovlev and Guthrie, 1931).

a. von Recklinghausen's neurofibromatosis: This disorder is associated not only with multiple neurofibromas but also with angiomatous lesions of the skin and mucous membranes. Diffuse port wine stains (nevus vinosus), capillary hemangiomas, and rarely, systemic angiomas are found in patients with the classical stigmata of peripheral nerve involvement.

b. Bourneville's syndrome with tuberous sclerosis: Of the various neuroectodermal syndromes, this syndrome is most constant in its manifestations. The great majority of the patients are idiots or imbeciles. Numerous nodules, varying from 1 mm to 1 cm in size, are distributed in a butterfly fashion over the skin of the nose, nasolabial folds, cheeks, and skin. The lesions are angiofibromas, analogous in many ways to von Recklinghausen's neurofibromas. The condition is associated with various types of congenital malformations, as well as systemic angiomas of the viscera and extremities.

c. Encephalotrigeminal angiomatosis (Sturge-Weber's disease): This is a disorder characterized by the facial distribution of angioma, various central nervous system phenomena, ocular lesions, and distinctive roentgenographic findings (Bielawski and Tatelman, 1949). The facial angioma consists of a vascular port wine stain usually limited to the distribution of the trigeminal nerve, particularly the ophthalmic division. The mucosa, as well as the skin, is affected. The cerebral manifestations may be due not only to involvement of the brain substance by the tumor but also by hemorrhage within it. Generalized convulsions, unilateral epileptiform seizures, spastic hemiplegia, and feeble-mindedness are expressions of the cerebral damage. The ocular manifestations may consist of buphthalmos (infantile glaucoma), unilateral exophthalmos, ocular atrophy, nystagmus, hemianopsia, and ocular palsies. Typical X-ray findings are sinuous linear parallel streaks of calcium deposition and a distinctive diffuse calcific density outlining the contours of a part of, or the whole of, the cortex of one cerebral hemisphere (Bielawski and Tatelman, 1949).

d. Hemangiomatosis of the retina and cerebellum (Lindau-von Hippel's disease): In this disorder there are cystic changes in the cerebellum, angiomatosis of the retina and pancreatic, hepatic, adrenal, and renal cysts, with angiomas (Levy, 1930). With destruction of the retinal vessels, due to angioma formation, subretinal hemorrhages are produced. Besides the involvement in the cerebellum, cysts and hemangiomas are less frequently found in the medulla and spinal cord and rarely, if ever, in the cerebrum.

D. TELANGIECTASIA[1]

By G. T. PACK AND T. R. MILLER

1. ETIOLOGY

Telangiectasia may be either congenital or acquired. Hereditary hemorrhagic telangiectasia, described by Osler (1907), Weber and Rendu, is the best known congenital type.

a. Congenital types: Elliptocytosis and telangiectasia have been noted in several families as a dominant character which skips. Cleidocranialis dysostosis digitalis has been recorded in monozygous twins (Gates, 1948). This may also be accompanied by telangiectasia which may appear in the same or different position, or it may be discordant. Congenital hemihypertrophy is a condition in which one side of the body is larger than the other. In many of these cases a combination with telangiectasia is seen.

Over 1000 cases of familial telangiectasia have now been recorded in about 150 families (Gates, 1948). From a hereditary point of view, telangiectasia is a dominant character and may involve the mucous membranes as well as the skin. The coagulation and bleeding times and the prothrombin time are all normal. The tourniquet test is negative. Platelet count and platelet resistance are also normal.

b. Acquired types: These are noted to follow exposure to radiation therapy, sun, and heavy weather, or may appear as the spider-like nevi in pregnancy, liver disease, and in the skin of persons exposed to liver toxins.

[1] This subject is also discussed in Section E.-1, Chapter XVI.

2. Pathogenesis

The symptom complex of Osler-Weber-Rendu's disease is an inherited maldevelopment of minute blood vessels in localized areas predisposing them to injury and serious bleeding (Goldstein, 1931). The requisite triad fulfilling the definition of this disease is:

(1) Mucosal hemorrhages, usually epistaxis.
(2) Familial occurrence.
(3) The presence of telangiectatic small angiomas of the skin and mucosa, commonly in the oral and nasal cavities.

The disease is transmitted, as a simple dominant affecting both sexes, either through the male or the female, with atavistic skipping of individuals or generations. Although the anlagen of the angiomas may be congenital, the minute tumors appear relatively late in the development of the individual. The onset of the disease may be heralded in children by nosebleeds, but the severe hemorrhages in clinically evident telangiectasias start usually in the fourth or fifth decade of life.

Hemorrhages have also been known to occur from the upper respiratory passages, the gastrointestinal tract, and even the kidney or bladder (Goldstein, 1947). Epistaxis, however, is the most frequent expression and can follow simple sneezing. Recurrent exsanguinating hemorrhages can occur from the mucous membrane of the septum, even after both external carotid arteries have been tied. There may also be almost intractable bleeding from the vermilion border of the lips. The intermittency and severity of the bleeding may necessitate repeated transfusions. The disease is a serious and crippling one. It is never relieved by spontaneous remission or cure as are some of the hemangiomas (Quick, 1942). The differential diagnosis between hereditary hemorrhagic telangiectasia, pseudohemophilia, thrombocytopenic purpura, and hemophilia should not be difficult. The blood coagulating mechanism is normal in the telangiectasias. Splenomegaly has been described as appearing late in the course of a very severe case.

3. Pathology

The angiomas are usually tiny, telangiectatic, reddish or purplish dots, seldom larger than 3 mm in diameter. They are punctate, round, macular, sessile, or spider-like.

The telangiectatic areas in Osler-Weber-Rendu's disease appear grossly as tiny reddish-purplish spots, 1–3 cm in diameter. Under microscopic examination, they consist of extremely thin, dilated venules and capillaries, lined with a delicate layer of endothelial cells. The telangiectasias may occur anywhere along the gastrointestinal tract.

It is interesting to note that 3 per cent of the beef livers in the United States show telangiectasia (Gates, 1948). The origin of this condition is unknown, but such livers have a high vitamin A content, as well as much iron and copper. The highest incidence of telangiectatic liver is in steers fed on highly concentrated fattening rations.

Telangiectasia and capillary fragility are often linked and appear to follow a dominant sex link form.

4. Clinical Picture

In Osler-Weber-Rendu's disease, the onset may be heralded by nasal hemorrhages. The most manifest evidence of the disease rarely appears before the fourth or fifth decade in life (Koch *et al.*, 1952). At this time hemorrhages occur from the mucous membranes of the nose and the mouth in increasing severity. They may proceed to fatal termination despite the use of various surgical methods of controlling the bleeding. Even ligation of the external carotid artery is ineffective, since the middle ethmoidal branch is derived from the internal carotid.

Because estrogenic therapy restores the normal texture of the nasal mucous membranes, it has been used to protect the telangiectases from trauma and subsequent bleeding. Such a program has been found of value in both male and female patients. The fact that the exsanguinating hemorrhages appear in the fourth and fifth decades of life suggests that the endocrine changes concomitant with the menopause are responsible for the increase in symptoms. Atrophy of the nasal mucous membrane, such as appears in ozena in the menopausal patient, probably increases the susceptibility of the membrane to trauma.

2. Thromboangiitis Obliterans

By G. C. McMillan

A. INTRODUCTION

Pathologically, thromboangiitis obliterans is an entity of inflammatory nature that causes segmental lesions of arteries and veins, with thrombosis and obliteration of their lumens. It affects arteries more markedly than veins. The disease involves the vessels of the legs most frequently, although it may affect those of the arms as well. Very rarely, it may produce typical lesions in the blood vessels of viscera.

Although it had been described previously, the disease was first clearly delineated from other forms of occlusive peripheral vascular disease by Buerger in 1908. In particular, his work served to distinguish the disease from arteriosclerosis and its sequelae. The distinction is not always easy, and the clarity that Buerger (1924) brought to the literature is not always maintained to-day. In fact, recently Wessler *et al.* (1960) have reviewed the matter both clinically and pathologically and have concluded that thromboangiitis obliterans cannot be considered an entity in either the clinical or pathological sense. On the other hand, the concept has also been extended beyond that held by Buerger to encompass a form of coronary artery disease designated as stenosing arteritis (Zak *et al.*, 1952) or juvenile occluding coronary sclerosis (von Albertini, 1943). It would appear at present that Buerger's disease should be considered an entity, since there is evidence that it begins as an acute exudative angiitis, often with a granulomatous component in both arteries and veins, and that, while secondary thrombosis is very common, it is not invariable in either arteries or veins (von Albertini, 1943; Jäger, 1932).

B. ETIOLOGY

No specific etiologic factor or factors have been isolated as the cause of thromboangiitis obliterans. Nevertheless, a discussion of this subject reveals several facts of interest. Age and sex are both important. The disease usually develops between the ages of 20 and 40 years and is rare before the age of 20 or after 60. It is extremely rare among females. It probably occurs in all parts of the world and among all peoples. It may possibly be more common among Jews, though this has not been shown clearly. Occasional examples of thromboangiitis appearing in members of the same family have been reported.

Various suggestions have been put forward from time to time as to the etiologic importance of local or distant infections with bacteria, viruses, or fungi, or of such factors as hypersensitivity, ergotism, cold, adrenal cortical hyperplasia, thrombosing tendencies, and tobacco. None of these agents has been established as the cause of the disease or as predominant factors in its causation. Cold makes the disease worse, but thromboangiitis obliterans is found in tropical countries. Adrenalectomy has disclosed adrenal cortical hyperplasia in some patients, but this is a subtle change that is a rather common incidental finding in routine autopsies. Unless it is shown to be of greater degree or more common than in control cases, it must be regarded as merely coincidental. There is some slight evidence that patients with Buerger's disease may show hypercoagulability of the blood (Theis and Freeland, 1939; Hagedorn and Barker, 1948), but this does not necessarily imply an increased tendency to thrombosis. In any case, the disease seems to begin with inflammatory and proliferative lesions of vessels, and not as a primary thrombosis.

Tobacco smoking is more clearly related; it is most rare to observe the disease in nonsmokers. Smoking makes the disease worse, while abstinence usually produces improvement. Some investigators have found that the intradermal hypersensitivity reactions to tobacco extracts are positive in the majority of patients with Buerger's disease but others have failed to confirm this (Allen *et al.*, 1955). Nevertheless, the disease does occur in nonsmokers and in those with negative intradermal reactions to tobacco, and some patients with thromboangiitis continue to smoke without apparent detriment.

Thus, a single pre-eminent etiologic factor has not been isolated in thromboangiitis obliterans, even though the pathologic lesions and course are such as to suggest that a specific cause exists. The possibility remains that different agents may produce the disease in different individuals. All that can be said now is that the etiologic factors that have been identified most clearly as of relevance are an age of 20–40 years, male sex, and the smoking of tobacco.

C. PATHOLOGY

Thromboangiitis is a segmental disease that causes occlusion of arteries. It may affect many vessels or only a few. It is episodic and may have a course of only a few attacks or of many. Occasionally it may be fulminant and progressive. Not uncommonly it is self-limiting. Because it results in segmental vascular occlusion of variable distribution, its effects will depend not only on the site and extent of the occlusion, but on the state of anastomotic circulation at the time of occlusion and on the subsequent development of further anastomotic blood flow. The clinical presentation and course are diverse because of the variable location and development of lesions.

1. Gross Changes

Grossly the small and medium-sized arteries of the lower limb are most often affected. Small and medium-sized veins are less commonly the site of typical lesions, though an acute superficial thrombophlebitis is frequent, especially in the lower extremities. Similar arterial and venous lesions may occur in the upper limbs. Involvement of large vessels, such as the femoral, iliac, brachial, or subclavian arteries, only occurs late in the disease. The presence of the process in visceral vessels is most rare. Not uncommonly, the pathologic changes noted in the large arteries or in visceral vessels after the disorder has been present for many years prove ultimately to be arteriosclerotic and secondary in nature and not characteristic of thromboangiitis obliterans.

At the site of the lesion the vessel is firm and somewhat retracted or narrowed. On section, it may be occupied by a recently-formed red thrombus or an older, gray, organized one. The organized thrombus may possess small recanalized passages from which a drop of blood can be squeezed, but, nevertheless, it is firm, fibrous and occlusive. Where obstructions have developed recently, propagated clot may be found. Acute lesions may show adventitial swelling with inflammatory edema, and older lesions may have adventitial fibrosis that occasionally binds together the associated vessels and nerves into a bundle.

The affected limb may show ischemic and trophic changes, such as distal edema, gangrene, which is usually dry, atrophy of nails or muscle, and osteoporosis.

2. Microscopic Changes

a. Early lesions: Microscopically the lesions are of characteristic appearance and different from other types of vasculitis. The acute, early lesion is often markedly exudative, being more obvious in veins than in arteries. Frequently, the reaction affects only one vessel in a given section of a neurovascular bundle, all three coats being involved. The vessel wall is edematous, with an infiltration of small round cells and a variable number of polymorphonuclear leucocytes. The cellular infiltration is, however, rather variable in amount and nature. It is often rather scanty and focal, but occasionally it may be prominent and contain numerous polymorphonuclear leucocytes. Cellular necrosis is minimal. A slight fibrinous exudate may be found in the wall, but a prominent fibrinoid change is absent. There are endothelial swelling and proliferation, both in the main lumen and in the vasa vasorum. Thrombosis apparently usually occurs promptly and the thrombus itself often contains numerous inflammatory cells. Indeed, so many of these may be present that the lesion resembles a septic thrombus. Thrombosis, however, is not always present.

Within a short time, perhaps two or three weeks, the reaction diminishes somewhat and, instead of the small round cells and polymorphonuclear leucocytes, the lesion now contains large numbers of somewhat swollen macrophages. Fibroblastic proliferation takes place in the intima and adventitia and, to a lesser degree, in the media. It is possible that this fibroblastic activity occurs earlier in less fulminant lesions. Organization of the thrombus begins from the periphery of the process, but, frequently, it seems to be impeded by the cellular exudate. Not uncommonly a small focus of polymorphonuclear inflammatory cells,

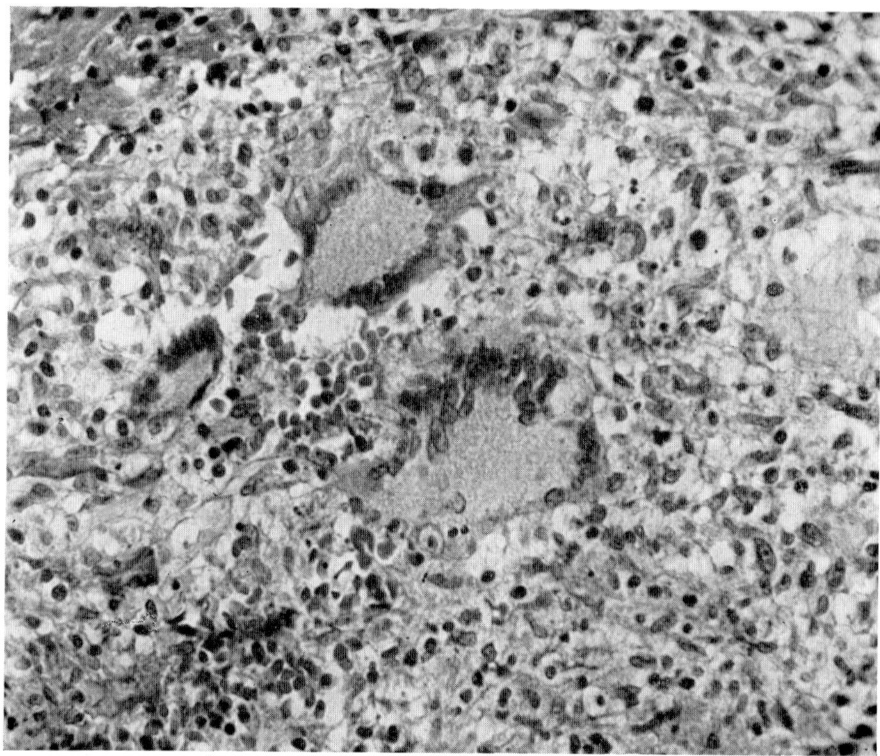

FIG. 130. Inflammatory exudate in a miliary focus in a recently formed thrombus in a vein.

often containing a multinucleated giant cell but otherwise resembling a miliary abscess, is found in the thrombus adjacent to the intima. Similar lesions, consisting chiefly of macrophages with one or more giant cells and resembling pseudotubercles, may also be noted. The giant cells may at times be an abortive ingrowth of endothelium during the earliest stages of organization and vascularization of the thrombus, or they may be due to fusion of macrophages (Fig. 130). When such foci are sectioned longitudinally, they may extend for a short distance along the thrombus.

b. Later lesions: Later in the disease, the inflammatory reaction in the thrombus, in the vessel and in the adjacent tissue subsides further, and the cellular exudate comes to consist almost exclusively of lymphocytes. Fibroblastic organization and collagen formation replace the

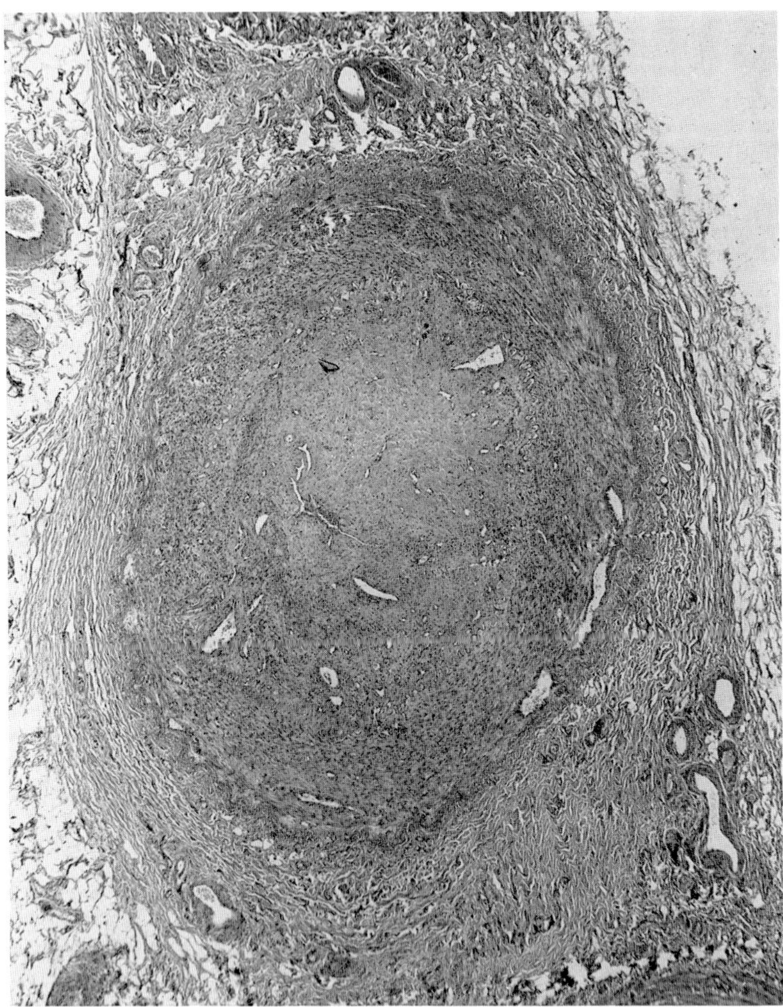

FIG. 131. An old lesion of thromboangiitis obliterans in an artery.

thrombus, with only slight canalization. The elastic interna is usually remarkably intact. The media is slightly and diffusely altered by fibroblastic proliferation. Numerous prominent vasa develop in it, some being surrounded by a few lymphocytes. In some lesions these newly formed vasa may be very numerous and may connect both with channels in the adventitia and in the thrombus. It has been suggested that they may shunt blood past the occlusion. The adventitia is increased by fibroblastic proliferation, and the whole of the neurovascular bundle may at times be encased in an excess of fibrous tissue.

c. Old lesions: The obsolete lesion is characterized by a fibrous, even hyaline connective tissue mass, which occludes the lumen of the affected vessel or vessels, an intact elastic membrane, a slightly fibrous media, often with prominent vasa vasorum, and circumferential fibrosis of the adventitia. A few lymphocytes may remain in the lesion for a long time (Fig. 131).

The nerves may show ischemic changes with demyelinization, loss of fibers and fibrosis. The alterations are considered to be due to vascular occlusion with ischemia rather than to perineural fibrosis. Occasionally the vasa of the nerves may show the lesions of thromboangiitis obliterans. The sympathetic ganglion cells also may demonstrate nonspecific degenerative changes. It should be remembered that similar nerve and ganglion lesions can be found in other forms of vascular disease, for example, in arteriosclerotic gangrene of the foot.

3. So-called Collagen or Systemic Connective Tissue Diseases

By E. R. Fisher

A. GENERAL CONSIDERATIONS

1. Entities

The terms collagen disease and systemic connective tissue disease refer to those disorders purported to develop from alterations of the mesenchymal or connective tissue of the body. The conceptual nature of this terminology is apparent, yet it has been frequently misinterpreted to imply a common pathogenesis, notably hypersensitivity. Although there is some evidence to suggest such a possibility in a number of these diseases, in others it is conspicuously lacking.

It is obvious that a host of recognizable entities might qualify for consideration as a collagen disease. Indeed, absolute criteria and more precise definition of the term are elusive. On the other hand, clinical, as well as pathologic, experience has revealed significant similarities among certain disorders which warrant their collective consideration. These include polyarteritis nodosa, hypersensitivity angiitis, lupus erythemato-

TABLE XIX

Pathologic Features of Some Necrotizing Angiitides

	Polyarteritis nodosa	Hypersensitivity angiitis	Thrombotic thrombocytopenic purpura	Rheumatic arteritis	Rheumatoid arteritis
Sites of predilection	Any site; common in G.I. tract; usually not lungs, spleen	Any site; common in lungs, spleen; usually not in G.I. tract	Any site; common in brain, spleen, myocardium, kidney	In lung, heart, G.I. tract, gonads, kidneys	In skel. muscle, nerve, synovia; rarely in heart or G.I. tract
Size of vessels	Medium and small arteries	Small arteries, arterioles, capillaries, venules	Small arteries, arterioles	Large and small arteries, arterioles	Small arteries
Layers involved	All	All	Intima, media	All	Adventitia; occ. all
Necrosis	Marked	Marked		Moderate	
Endothelium	Swollen, with proliferation	Swollen	Swollen or normal	Slight proliferation	Slight proliferation
Lumen	Eccentric	Central	Eccentric	Central	Central
Thrombosis	Frequent	Rare	Characteristic	Rare	Rare
Int. elastic memb.	Damaged	Damaged	Damaged	Damaged	Damaged
Aneurysms	Yes	No	Yes, microscopic	No	No
Lesions in various stages	Yes	No	Yes	Yes	Yes
Infarct	Common	No	Occasional, small	Rare	No
Infiltrate	Polys, plasma cells, lymphs. eos. (Marked)	Polys, lymphs, eos. plasma cells (Moderate)	Absent	Polys, lymphs monos. (Marked)	Polys, lymphs monos (Moderate)
Healed stage	Scar, marked in adventitia	Resolution	Rare; intimal fibrosis	Scar, slight in adventitia	Resolution or slight scar

sus, dermatomyositis, scleroderma (or more appropriately designated, progressive systemic sclerosis (PSS)), and thrombotic thrombocytopenic purpura (TTP). A detailed discussion of the pathologic features of rheumatic fever and rheumatoid arthritis, which also might qualify for inclusion in such a group, is beyond the scope of this presentation, although the morphologic features of the vascular alterations encountered in these disorders are tabulated in Table XIX. All of the diseases in this category are associated with vascular alterations, which may frequently represent their salient pathologic manifestation (Table XIX). Because of this they have not infrequently been considered under the general term of necrotizing angiitis. Multiple system involvement and cardiac and joint manifestations are common.

2. Fibrinoid Degeneration

a. Definition of term: An outstanding morphologic alteration encountered in some, but not all instances, of the disorders noted above is fibrinoid degeneration or necrosis. The term, fibrinoid, designates a type of tissue injury in connective tissue and blood vessels which exhibits some of the tinctorial and morphologic features of fibrin, particularly its affinity for acid dyes, such as eosin. Unfortunately, this change, as in the case of the term, collagen disease, has been taken to indicate a specific response peculiar to a particular group of diseases and to imply their common pathogenesis. However, the presence of fibrinoid in the base of a peptic ulcer, in the placenta and in the vascular changes of accelerated or malignant hypertension indicates the diverse pathogenetic mechanisms which may be responsible for its appearance. Furthermore, alteration of collagen may occur without evidence of fibrinoid degeneration. The typical basophilic degeneration of the collagen of the corium in lupus erythematosus, as the term indicates, reveals an affinity for basic rather than acid dyes, although admittedly the basophilia is probably, in large part, the result of damaged elastic, as well as collagen, fibers. In addition to the inconsistencies concerning the tinctorial properties of fibrinoid, there is no unanimity concerning its morphologic structure, being variously described as hyaline, granular, or fibrillar.

b. Origin of fibrinoid: The preceding comments strongly suggest that fibrinoid is not a singular substance, but that there are various types which may differ in composition. The structural peculiarities of the different tissues which may be the site of fibrinoid change, notably connective tissue and blood vessels, are in keeping with this concept. It is, therefore, not surprising to note that some investigators have considered vascular fibrinoid to be, at least in part, derived from altered smooth muscle (Muirhead *et al.*, 1957) or from a deposition of ground substance and protein (Altshuler and Augevine, 1949); in fact, it has also been considered to consist of fibrin *per se*. It is readily conceivable that all of the above may be involved in the formation of fibrinoid.

c. Composition of fibrinoid: More recently, immunohistochemical studies of the lesions encountered in some of the collagen diseases have revealed dissimilarities between the composition of the fibrinoid encountered, particularly with reference to various plasma protein components. For example, the vascular lesions of thrombotic thrombocytopenic purpura have been demonstrated to consist exclusively of fibrinogen (Craig and Gitlin, 1957), whereas those of polyarteritis nodosa (Mellors and Ortega, 1956), systemic lupus erythematosus, rheumatic fever, and rheumatoid arthritis contain an excess of gamma globulin without evidence of fibrin or albumin moieties (Vazquez and Dixon, 1957). In the case of progressive systemic sclerosis, no specific localization of any particular plasma protein component in the cutaneous lesions has been noted (Fisher and Rodnan, 1960); the same applies to dermatomyositis, although only a few examples of the latter have been available for study. Information derived from the utilization of the immunohistochemical technique not only provides evidence attesting to the heterogeneity of fibrinoid but also may have pathogenetic significance. It becomes apparent, therefore, that the terms, fibrinoid, collagen, and systemic connective tissue disease, possess inherent vagueries, and their subsequent utilization in this text shall be made with this realization.

3. Discrepancy between Clinical and Pathologic Findings

Another general consideration which appears worthy of note is the frequent practice of conveniently ascribing many of the unusual and often widespread clinical manifestations encountered in the "collagen" diseases to vascular involvement of specific sites, despite failure to disclose such an alteration through careful necropsy examination. This is particularly so in the case of systemic lupus erythematosus in which, in some instances, pathologic alterations may be exceedingly sparse. Further, the pathologist is often left to explain numerous clinical manifestations of this disorder, with only evidence of pathologic alteration being lamination of splenic periarteriolar collagen. It, therefore, becomes apparent that undue attention to a purely anatomic basis for the clinical manifestations of the disorders in this category may result in the neglect of other intravitam investigations which may provide significant information in this regard. With regard to the ultrastructure of the tissues of these disorders, a few studies have been made.

4. Similarity of Clinical Entities

The overlapping clinical and pathologic findings in some of the "collagen" diseases also attest to their interrelationship. For example, positive L. E. cell tests have been observed in some patients with disseminated lupus erythematosus, in whom clinical, as well as pathologic, evidence of one of the other disorders of this group has been manifest. Nevertheless, there are sufficient anatomic and pathologic dissimilarities to warrant the consideration of the "collagen" disorders as distinct and separate entities. Although the differentiation of some of them principally on the basis of differences in vascular alterations may appear somewhat artificial, nevertheless, their individual presentation focuses attention away from the more general designation of "collagen" disease. Such an approach may also have definite therapeutic advantages.

B. SYSTEMIC LUPUS ERYTHEMATOSUS

1. Etiology and Pathogenesis

a. Immunologic mechanism: Evidence is rapidly accumulating which strongly implicates an immunologic mechanism in disseminated lupus erythematosus. The identification of the serum factor responsible for the L. E. phenomenon as a gamma globulin (Haserick *et al.*, 1950) and the subsequent demonstration of the latter by immunohistochemical techniques within the phagocytized L. E. body (Holman and Kunkel, 1957) (Fig. 132), as well as within the lesions of lupus erythematosus (Vazquez and Dixon, 1957; Mellors *et al.*, 1957), might be considered as strong evidence in this regard. Such information supersedes previous concepts relating the L. E. body to depolymerized DNA. Indeed, Godman *et al.* (1958) have indicated that DNA is fully polymerized but altered by the entrance of other protein into the nucleus. That the development of hematoxylin bodies within tissues (a pathognomonic histologic feature of this disorder) may represent an analogous phenomenon gains support from the studies of Robbins *et al.* (1957) and Bardawil *et al.* (1958) who demonstrated a reaction between the sera of patients with L. E. and homologous and heterologous nuclei. More recently Deicher *et al.* (1960) observed anticytoplasmic factors in the sera of patients with L. E. cells.

That other factors may also be significant for the development of L. E. cells *in vitro* and their tissue prototype, the hematoxylin body, is suggested by the uniqueness of their derivation almost exclusively from mesenchymal cells; yet the immunohistochemical reactions noted above appear in nuclei of epithelia. It should be noted, however, that Pollack (1959) has observed hematoxylin bodies within the islets of Langerhans which he attributed to be of epithelial origin. Although such structures have been observed in as high as 85 per cent of necropsied cases of lupus erythematosus (Worthington *et al.*, 1959), it is common knowledge that their presence is extremely variable and unrelated to treatment with steroids.

Further correlation of the pathologic and serological phenomena in systemic lupus erythematosus is suggested by the finding of delayed hypersensitivity to autologous, as well as heterologous, leucocytes in

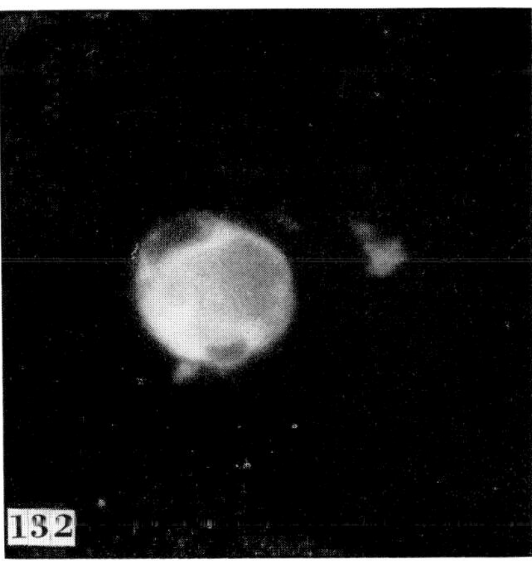

FIG. 132. Specific fluorescence of gamma globulin in L. E. cell, as demonstrated by the fluorescent antibody technic. (Courtesy of Dr. J. J. Vazquez.)

patients with this disorder, suggesting that a cellular or humoral antibody may be responsible for such a reaction (Friedman *et al.*, 1960). However, it is of interest that the clinical manifestations of L. E. have been observed to develop in a patient with agammaglobulinemia (Good *et al.*, 1957). Further, it has not been conclusively demonstrated that the circulating globulin on which the reactions cited above are based is also responsible for the clinical manifestations of systemic lupus erythematosus. Similar considerations appear warranted at this time in regard to the autoimmune nature of this disorder. (For a discussion of changes produced in hypersensitive states, see Section G.-3b, Chapter VI.)

2. Gross Changes

As indicated previously, it is not unusual to encounter few or no significant visceral macroscopic alterations in patients succumbing from lupus erythematosus. Occasionally, scant pleural or pericardial effusions may be present. Involvement of the valves of the heart has been observed in a variable number of instances. Such a lesion, commonly designated as atypical verrucous endocarditis (Libman and Sacks, 1924), appears as small, solitary or confluent excrescences of the valve, measuring no more than several millimeters in diameter. It exhibits a distinct predilection for the valve pockets, as well as other sites away from the lines of closure. Nonspecific hepatomegaly, splenomegaly and lymphadenopathy may also be evident.

3. Microscopic Changes[2]

a. Hematoxylin bodies: Although it is frequently stated that there are no pathognomonic microscopic features of systemic lupus erythematosus, nevertheless, certain alterations appear significantly unique to allow for such a diagnosis. Hematoxylin bodies, which may be noted in practically every tissue (especially, cardiac valves and adjacent endocardium, kidneys, serosal surfaces, lymph nodes, ovary, and spleen), appear to be pathognomonic. Their presence in clusters and in various sites in any particular case should allow for their differentiation from karyolytic nuclei observed in foci of nonspecific necrosis associated with other disorders.

b. Adventitial lamination of collagen: A distinctive adventitial lamination of collagen about the central arteries of the spleen has been observed in 95 per cent of cases (Klemperer *et al.*, 1941). Although a similar alteration may be evident in a variety of other disorders, it is not of the same magnitude as in lupus erythematosus, in which 7 or more lamellae may be evident.

c. Cutaneous manifestations: The cutaneous alterations often allow for a definitive diagnosis of lupus erythematosus when characterized by alternating areas of epidermal atrophy and hyperkeratosis, follicular plugging, liquefaction degeneration of basal cells, degeneration of collagen (which frequently appears basophilic in sections stained with hematoxylin and eosin), and a nonspecific inflammatory infiltrate about skin appendages and vessels. Fibrinoid degeneration of vessels is infrequent. Occasionally, a striking condensation of connective tissue at the dermal-epidermal junction may be evident.

d. Changes in kidney: The renal alterations, aside from the possible presence of hematoxylin bodies, do not appear pathognomonic of the

[2] For microscopic changes in muscle vessels, see Section G.-2d, Chapter XVII.

disorder. The so-called wire-loop thickening of basement membranes of glomerular capillaries, as well as the occasional proliferative glomerular changes, may be observed in a variety of unrelated diseases (for further discussion, see Section F.-5, Chapter XIII).

e. Valvular changes: The valvular lesions appear microscopically to be comprised of herniated masses of altered, brightly eosinophilic and argyrophilic masses of collagen, with superimposed deposits of fibrin. A similar appearance is evident in the valvular lesion designated as degenerative verrucal endocardiosis (indeterminate, nonbacterial endocarditis, terminal endocarditis) (Allen and Sirotta, 1944; Fisher and Baird, 1956), although the vegetations in this disorder classically appear at lines of valvular closure.

f. Changes in vessels: Lesions affecting small arteries and arterioles, aside from those previously referred to, have been observed in practically all tissues. Their occurrence, as well as apparent intensity, however, may be extremely variable. Absolute vascular necrosis has been noted (Klemperer *et al.*, 1941).

g. Reactive granuloma: A rare lesion observed in lupus erythematosus is the reactive granuloma, characterized by epitheloid and multinucleated giant cells, and frequently, but not invariably, by foci of fibrinoid necrosis and hematoxylin bodies. This reaction has been observed within serous membranes, lungs, lymph nodes, mediastinal and esophageal connective tissues and the liver. Its presence has been considered by some as a possible expression of hypersensitivity (Pollack, 1959).

C. POLYARTERITIS NODOSA

1. Etiology and Pathogenesis

a. Role of hypersensitivity: Although the precise mechanism concerned with the pathogenesis of polyarteritis nodosa remains to be elucidated, there is some evidence to suggest that this disorder is also immunoallergic in nature. The exclusive localization of an excess of gamma globulin, as demonstrated by the fluorescent antibody technique, provides strong but not unequivocal, evidence in support of this concept (Mellors and Ortega, 1956). Vascular lesions similar to those observed in polyarteritis nodosa are often found in rabbits subjected to experimental serum sickness (Rich and Gregory, 1943). It is worthy of note that similar vascular lesions have also been described following the experimental induction of hypertension (Smith *et al.*, 1944), and after the administration of adrenal cortical steroids and renin (Masson *et al.*,

1953). However, it seems highly unlikely that these latter play a role in the pathogenesis of polyarteritis nodosa in man. The recognition of examples of this disease in patients receiving sulfonamides, foreign serum (Rich, 1942), and a variety of other drugs also provides some suggestive evidence concerning the etiologic role of hypersensitivity. (For discussion of changes produced in hypersensitive states, see Section G.-3b, Chapter VI.)

2. Gross Changes

The macroscopic features of polyarteritis nodosa consist of a segmental thickening and frequently aneurysmal dilatation of medium-sized and small arteries, which exhibit a predilection for sites of arterial bifurcation. All viscera and organ systems have been observed to be involved, although pulmonary involvement is exceedingly rare, occurring only in the presence of pre-existing pulmonary hypertension. Similarly, involvement of the spleen is unusual. Visceral infarcts may be evident, as a result of thrombosis of involved vessels or marked luminal narrowing following administration of cortisone or related agents or spontaneous healing. Digital gangrene may be noted subsequent to involvement of peripheral vessels. Although vascular changes exist in a variety of sites in the usual case of polyarteritis nodosa, it is noteworthy that in a few instances localized involvement of the gallbladder, appendix and uterus has been observed (Talbott and Ferrandis, 1956). Whether such a finding represents a *forme fruste* of polyarteritis or a disease *sui generis* remains to be demonstrated.

3. Microscopic Changes[3]

Microscopically, the lesions of polyarteritis nodosa appear confined to medium and small arteries. Arteriolar involvement may also be evident, but veins and venules appear spared. Four stages of the disease have been recognized (Arkin, 1930), although characteristically, all phases or several may be apparent in the same case. The first is the degenerative phase, characterized by vascular edema, fibrinous exudate, endothelial proliferation, and fibrinoid degeneration of the media. The second or inflammatory stage consists of marked infiltration of all vascular coats by neutrophils, lymphocytes, and plasma cells (Fig. 133). Variable numbers of eosinophils may also be present. The elastic lamina may be disrupted resulting in aneurysmal dilation, and thrombosis may supervene. In the third or reparative stage, fibroblastic proliferation

[3] For microscopic changes in muscle vessels, see Section G.-2d, Chapter XVII.

occurs, with the appearance of reparation of the damaged media and adventitia. The fourth or healed stage is characterized by marked adventitial fibrosis (Fig. 134). The media is also, at least in part, replaced by fibrous connective tissue. The intima consists of loose connective

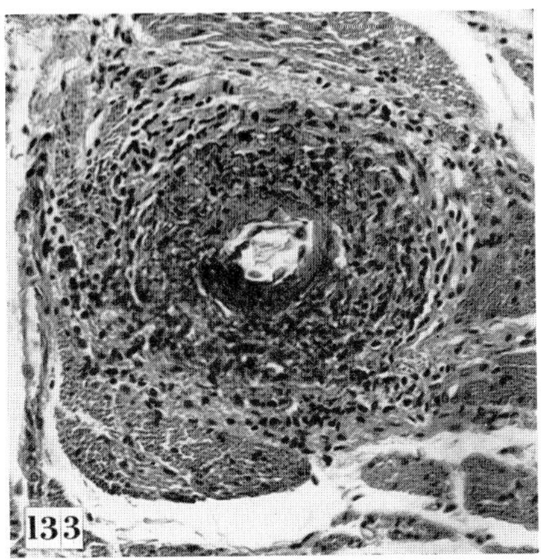

FIG. 133. Appearance of small artery in active inflammatory stage of polyarteritis nodosa. Subintimal fibrin deposition, medial degeneration and a marked inflammatory infiltrate are evident.

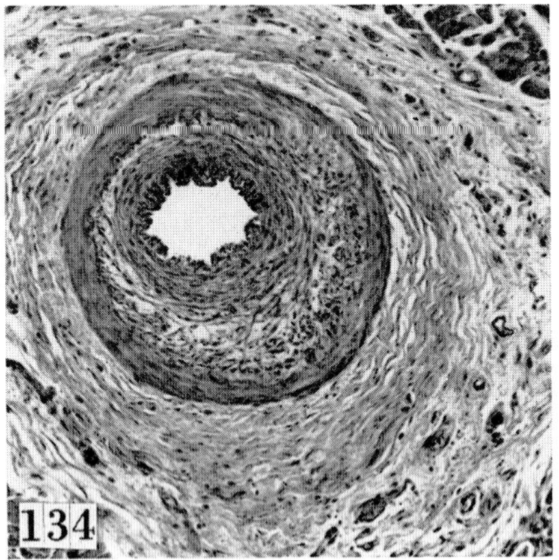

FIG. 134. Marked adventitial, as well as medial fibrosis, characteristic of "healed" lesion of polyarteritis nodosa.

tissue containing variable amounts of histochemically demonstrable ground substance. The lumen is small and may often appear centrally placed. The appearance of the healed lesion of polyarteritis is distinctive from other forms of healed vasculitis.

In addition to the alterations of blood vessels noted, affected organs may reveal classic infarcts as a result of ischemia. In the kidney the latter may produce the so-called "endocrine" appearance, in which portions of renal tubules appear indistinguishable. Proliferation of the cellular elements of the glomerular capillaries and the alterations of hypertension (if the latter represented a manifestation of the clinical course of the patient's illness) may also be frequently observed (for further discussion, see Section F.-5, Chapter XIII).

4. Entities Resembling Polyarteritis Nodosa

There are several clinical and pathologic syndromes which exhibit similar vascular alterations to that observed in classical examples of polyarteritis nodosa. These are designated as Cogan's syndrome, Wegner's granulomatosis, allergic granulomatous angiitis and hypersensitivity angiitis.

a. Cogan's syndrome: This condition consists of nonsyphilitic interstitial keratitis with vestibuloauditory symptoms (Cogan, 1945). In addition, a large number of patients will exhibit a necrotizing angiitis, resembling polyarteritis nodosa, at the time these clinical manifestations are evident or subsequently in the course of their disease.

Pathologic examination of the eyes from a patient with this disorder who received large doses of adrenal cortical steroids revealed nonspecific interstitial keratitis but no vascular lesions (Fisher, unpublished observations). The inner ear also failed to disclose alterations which might be related to vascular involvement, although lesions indistinguishable from healed polyarteritis nodosa were apparent in other sites. Furthermore, a biopsy obtained *intra vitam* prior to steroid therapy disclosed the characteristic features of active polyarteritis nodosa in small arteries.

b. Wegner's granulomatosis: This condition (Wegner, 1936) also might be considered a variant of polyarteritis nodosa. Although the vascular alterations are similar to the latter, they are principally restricted to the pulmonary and renal vessels. In addition, necrotizing granulomatous inflammation of the upper respiratory passages and lungs is present.

c. Allergic granulomatous angiitis: In 1951, Churg and Strauss described vascular changes similar to those observed in polyarteritis nodosa in 13 patients with severe asthma, fever, and eosinophilia. In addition, granulomatous foci were observed in vascular walls, as well as in extravascular sites, particularly the epicardium. An eosinophilic infiltrate of extravascular tissues was also apparent. In some cases the vascular and connective tissue lesions were widespread, whereas in others they affected only a few organs.

d. Hypersensitivity angiitis: Zeek and associates (1948) called attention to a form of necrotizing angiitis, with pathologic and some clinical features which were distinct from classical polyarteritis nodosa, and which they designated as hypersensitivity angiitis. All of the patients in this group had clinical evidence of hypersensitivity to some therapeutic agent, particularly sulfonamides. Although the vascular lesions resembled those of polyarteritis nodosa, notable differences were apparent in regard to the distribution of the lesions, caliber of vessels involved and lack of evidence of a "healed" stage (Table XIX).

D. PROGRESSIVE SYSTEMIC SCLEROSIS (PSS)

1. Etiology and Pathogenesis

The etiology and pathogenesis of progressive systemic sclerosis is unknown. As with the other diseases described in this section, hypersensitivity has been considered. However, no convincing evidence in support of such a contention has been presented. The significance of the presence of a factor in the serum of patients with PSS which reacts with autologous, as well as homologous and heterologous nuclei (Bardawil *et al.*, 1958), has not been elucidated. Similarly, no convincing support of hypotheses indicating an infectious or endocrine etiology has been presented. The association of Raynaud's phenomenon in some patients with this disorder has led to speculation that it might be the result of aberrations of the sympathetic nervous system. However, therapy based upon this consideration has been unrewarding.

The emphasis placed upon the vascular alterations of PSS has led some to consider them to be of primary pathogenetic significance (Ingram, 1958). It appears noteworthy that much of the information concerning the latter concept has been derived from studies performed on the localized or limited forms of scleroderma, viz. sclerodactylia, acrosclerosis, and morphea. However, their precise relationship to the systemic disease is at present uncertain. Further, exploration of histochemically demonstrable enzymes of skin, including those of blood vessels, has failed to disclose significant alterations in their activity, even in advanced lesions (Fisher and Rodnan, 1960).

2. Gross Changes

Aside from the macroscopically prominent cutaneous alterations in progressive systemic sclerosis, variable involvement of the gastrointestinal tract, heart, and lungs may be encountered. Portions of the gastrointestinal tract, and particularly the esophagus, may be thickened and appear shortened. Mucosal ulceration is not infrequent. The heart may disclose varying-sized areas of myocardial fibrosis which appear unrelated to the distribution of the coronary arteries. Cystic change and fibrosis of the pulmonary parenchyma may be macroscopically evident.

3. Microscopic Changes

a. Cutaneous alterations: Microscopic examination of the skin in minimally involved sclerodermatous areas may fail to disclose distinctive histologic alteration from normal. Diagnosis is less difficult in classic cases, being characterized by epidermal atrophy, with pronounced flattening of the rete pegs. The dermal collagen appears increased in thickness and homogenized, and skin appendages, as well as vessels, disclose varying degrees of hyalinization. A nonspecific inflammatory infiltrate may be present at the dermal subcutaneous junction. Elastic fibers may appear normal, decreased, or altered similarly to that noted in senile elastosis of skin.

Histochemical examination of the skin in PSS has failed to disclose significant changes from normal, and edematous foci of skin appear devoid of acid mucopolysaccharide (Fisher and Rodnan, 1960). Similarly, electron microscopy of the dermis discloses normal appearing collagen fibers (Fisher and Rodnan, 1960). Although Bahr (1956) has noted angular bodies adjacent to the longitudinal axis of the latter in two examples studied by this technique, this observation has not been confirmed. The lack of apparent histochemical and ultrastructural changes suggests that the alteration observed in sclerodermatous skin may, at least in large part, be quantitative rather than qualitative.

b. Alterations in heart, gastrointestinal tract, and lungs: Hyalinization of the type seen in the skin is also apparent in other structures, such as the heart. The presence of well-preserved myocardial fibers within islands of fibrosis is unlike the appearance of myocardial fibrosis due to insufficiency of coronary blood flow. The lamina propria, submucosa, and muscularis propria of the intestinal tract also may reveal bland fibrosis. Interstitial fibrosis and cystic change of the lung reflect the macroscopic alterations in this organ. The synovial membrane may reveal similar hyalinization.

c. Alterations in kidney: The renal changes of PSS are observed only in those patients succumbing from severe hypertension and/or renal failure. In other instances in which these manifestations are not present, the kidneys may appear normal or manifest signs of other unrelated renal disease, such as pyelonephritis or mild arteriolar nephrosclerosis.

The microscopic appearance of the kidneys in patients succumbing from severe hypertension and renal failure is indistinguishable from that of malignant hypertension (Fisher and Rodnan, 1958). Furthermore, mucinous alteration of interlobular branches of the renal artery (Fig.

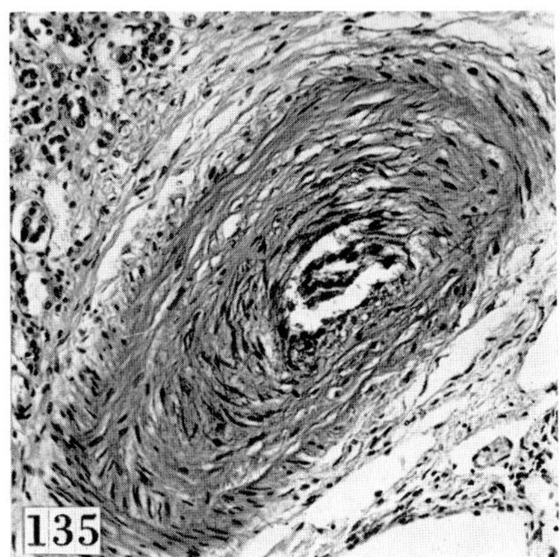

FIG. 135. Interlobular branch of renal artery in patient with PSS. "Mucinous edema" (hyaluronic acid appearing gray) of intimal-medial zone is evident. This change also observed in kidneys from patients with malignant hypertension.

135), considered pathognomonic in PSS, may often be observed in the kidneys from patients with malignant hypertension alone. Histochemical features of the kidneys in both diseases are identical disclosing similar deposits of fibrinogen but no preferential localization of gamma globulin (Fennell and Vazquez, 1960). The possibility that the renal changes noted in PSS may be the result of a renal form of Raynaud's phenomenon, producing severe hypertension and its renal sequelae, has been suggested (Sokoloff, 1956; Fisher and Rodnan, 1958). Renal vasospasm has also been considered as a possible mechanism in the pathogenesis of malignant hypertension. It is of interest that fibrinoid alteration of blood vessels in the kidney and other sites has only been observed in those examples of PSS dying from hypertension (Fisher and Rodnan, 1958).

E. DERMATOMYOSITIS

1. Etiology

The etiology of dermatomyositis is unknown. Both bacterial infection and hypersensitivity have been implicated, despite the absence of conclusive evidence in support of such contentions. The association of dermatomyositis with malignant visceral neoplasms has been observed more frequently than that of chance, although the precise relationship between the two disorders is unknown. In this regard, Grace and Dao (1959) described a case of proven carcinoma of the breast in which the clinical manifestations of dermatomyositis developed. Since the patient exhibited an immediate skin reaction to extracts of the breast tumor, as well as a positive passive transfer test, they considered this suggestive evidence in support of an immunoallergic pathogenesis for dermatomyositis.

2. Pathologic Changes

The pathologic alterations encountered in patients with dermatomyositis are for the most part nonspecific. Definitive diagnosis on the basis of cutaneous alterations is attendant with difficulty; most frequently the skin discloses dermal edema and a banal inflammatory infiltrate. Vascular lesions are not specific in this site or elsewhere, and fibrinoid change of vessels is the exception. A panniculitis of subcutaneous tissue may also be present.

The most pronounced pathologic alterations are to be observed in skeletal muscles of the extremities or oropharynx, as well as in the myocardium and the muscular layer of the gastrointestinal tract. The muscle fibers disclose evidence of degeneration, as well as active necrosis, with an inflammatory infiltrate of neutrophils, lymphocytes, and histiocytes. The degree of muscular alteration may approximate that observed in trichinosis. Other visceral changes attributable to dermatomyositis *per se* are lacking.

Examination of a limited number of skin and skeletal muscle specimens from patients with dermatomyositis by the fluorescent antibody technique has failed to disclose preferential localization of gamma globulin or other plasma protein components within the lesions (Fisher, unpublished observations). However, antinuclear antibodies have been observed in the serum of patients with this disorder (Bardawil *et al.*, 1958).

F. OTHER LESS COMMON DISORDERS

1. Thrombotic Thrombocytopenic Purpura (TTP)

Little difficulty is encountered in establishing the pathologic diagnosis of thrombotic thrombocytopenic purpura, the characteristic lesions appearing as hyaline masses within the lumina of small arteries and arterioles (Fig. 136). However, the nature and derivation of these

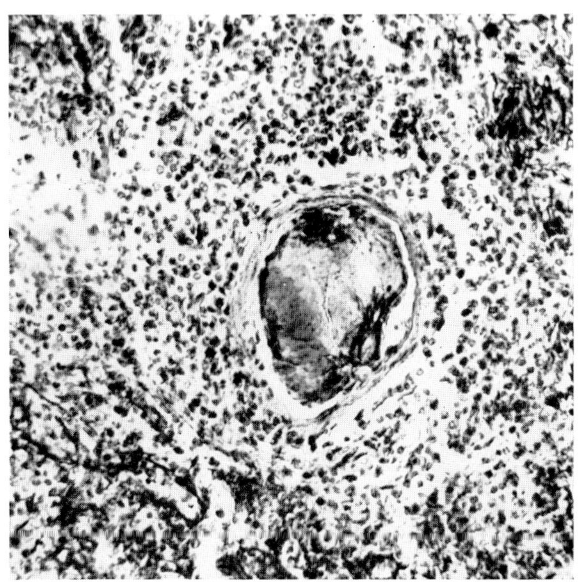

Fig. 136. Intraluminal "hyaline" mass in arteriole of spleen in patient with TTP. Phosphotungstic acid hematoxylin stain discloses the "thrombus" to be comprised of two components: a dark fibrillary material and a more homogeneous pale-staining substance.

"thrombi" have been the subject of much debate. There have been repeated observations indicating that the lesions arise primarily as the result of alteration within the walls of affected vessels; and it has been suggested that the thrombus-like intraluminal masses may represent herniation of this material (Orbison, 1952; Gore, 1950) or be secondarily imposed upon this defect (Fisher and Creed, 1955). Such lesions may be observed in a variety of tissues, particularly spleen, brain, lymph nodes, and kidney. The presence of a fibrin moiety has been demonstrated by the fluorescent antibody technique (Craig and Gitlin, 1957). It should be emphasized, however, that this finding does not eliminate from consideration the presence of other substances within the lesion. The similarity of the latter to altered smooth muscle has also been shown (Muirhead *et al.,* 1957), and some of the tinctorial properties appear different from those observed in other fibrin clots.

It is noteworthy that TTP does not represent a disease *sui generis* at least in so far as its pathologic features are concerned. Identical lesions, as well as some of its clinical manifestations, have been observed in patients with mucinous carcinoma (Fisher and Baird, 1956), obstetrical accidents (such as abruptio placenta), infantile enteritis, and hemolytic reactions to transfusions (McKay *et al.,* 1953; McKay and Wahle, 1955). The similarity of these, as well as TTP, to the generalized Shwartzman reaction in rabbits has also been cited (McKay *et al.,* 1953). Since there is some evidence to indicate the latter may result from a defect in the coagulability of the blood (McKay and Shapiro, 1958), the possibility arises that TTP may be the result of such a derangement.

It would appear that the demonstration of a relationship between the pathologic observations of altered vascular walls and a possible defect in the coagulation mechanism would be essential for a clear understanding of the pathogenesis of TTP.

4. Raynaud's Disease[4]

By G. C. McMillan;[5] K. I. Melville and H. E. Shister[6]

A. INTRODUCTION

Several diseases may produce transient or intermittent changes in the circulation of the extremities, resulting in blanching, mottling, pallor, cyanosis, or redness of the digits. Such effects are due to aberrations in local vasomotor activity. They may be secondary to some organic disease of the vessels or may occasionally occur in the absence of demonstrable occlusive vascular disease and be functional in nature. The group of signs and symptoms indicative of such intermittent vascular disturbances is designated "Raynaud's phenomenon" (or syndrome).

Raynaud's phenomenon can thus be secondary to some associated disease (viz.: scalenus anticus syndrome, thromboangiitis obliterans, ergot poisoning, disseminated lupus erythematosus, lymphosarcoma, etc.) or be idiopathic, when it is called Raynaud' disease.

[4] For discussion of other vasospastic disorders, see Section E.-4 (livedo reticularis), 5 (acrocyanosis), Chapter XVI.

[5] Author of Sections A., B., C., and E.

[6] Authors of Section D.

B. ETIOLOGY

By definition, the cause of Raynaud's disease is not known. Its onset is usually in youth; it does not often begin after the age of 40, and the diagnosis is only to be made with reservations if the onset is after the age of 50. The great majority of patients are female, and it is not uncommon for the condition to be associated with menstruation or with the menopause. It may become less marked during gestation. Migraine is not uncommonly associated. The condition is usually initiated by exposure to mild or moderate cold or by an emotional upset.

C. PATHOGENESIS

Raynaud's disease is a functional disorder and without specific anatomic lesions. The color changes in the digits depend on arteriolar spasm and reduced capillary blood flow. Pallor is due to ischemia; cyanosis occurs because of capillary relaxation or dilatation, in the presence of arteriolar constriction, resulting in a relatively static hyperemia. Although usually pallor precedes cyanosis, with capillary dilatation probably resulting from hypoxia, occasionally mottling or cyanosis may be the first color change that is noted. This may be due to a patchy vasoconstriction, with the development of small areas of capillary dilatation that gradually spread and merge with one another; it has also been suggested that spasm of veins may occur (Naide and Sagen, 1946). Following a period of restricted blood flow, the vascular spasm disappears, and active hyperemia occurs. The capillary bed, as in any area of relative ischemia, will remain dilated for a time, and so the affected member will continue to be unusually red.

D. PHARMACOLOGIC RESPONSES[7]

Peacock (1959) has reported increased concentrations of both epinephrine and norepinephrine in the venous blood collected at the wrist after an attack of digital spasm. During cold stimulation the observed increase was proportional to the severity of the response and was due mainly to norepinephrine. It has been suggested, therefore, that some metabolic disturbance, probably incomplete destruction of the catecholamines by amine oxidase, might be involved. In this regard it has also been shown that cold increases the sensitivity of isolated vascular strips to epinephrine and norepinephrine (Smith, 1952).

[7] By K. I. Melville and H. E. Shister.

a. Vasodilator drugs: Phenoxybenzamine (Dibenzyline), tolazoline (Priscoline), azapetine (Ilidar), nylidrin (Arlidin), and other similarly acting agents have been widely used in an attempt to eliminate or diminish the attacks of digital spasm. All exert their actions by blocking the effects of circulating catecholamines at adrenergic receptor sites; at the same time, the cardiac excitatory responses to epinephrine and norepinephrine are not affected.

Among other vasodilator drugs is ethyl alcohol in the form of whiskey. This medication has been reported to give good relief during attacks, probably because of its transient vasodilator action. This effect may be due, in part, to central nervous system depression. The local application of a 2 per cent ointment of glyceryl trinitrate has also been reported to be effective in some cases. Overdosage may lead to systemic absorption, with toxicity, headache, and syncope. Other vasodilator agents include cholinergic drugs (methacholine by mouth or by ion transfer) and ganglionic-blocking agents (tetraethylammonium, hexamethonium).

b. Estrogens: Beneficial effects from the use of estrogens have been reported, but their use is restricted to those cases related to menopausal disturbances. Of interest in this regard is that during pregnancy vasospastic conditions may show improvement.

E. PATHOLOGY[8]

1. Gross Changes

Raynaud's disease is almost always bilateral and symmetrical, involving the distal parts of the fingers. Subsequently it may affect the proximal parts. It may also be noted in the toes.

The ischemic episodes are accompanied by mild local edema, and in time, there may be a slight permanent change, giving rise to swelling and induration of the affected region. The texture of the tissue may even come to resemble that of scleroderma. The frequent repetition of ischemic epidodes may lead also to minor trophic changes or even to gangrene of the fingertips.

2. Microscopic Changes

Examination of the vessels of fingers amputated because of gangrene occurring in Raynaud's disease shows stenosis of the lumen, due to

[8] By G. C. McMillan.

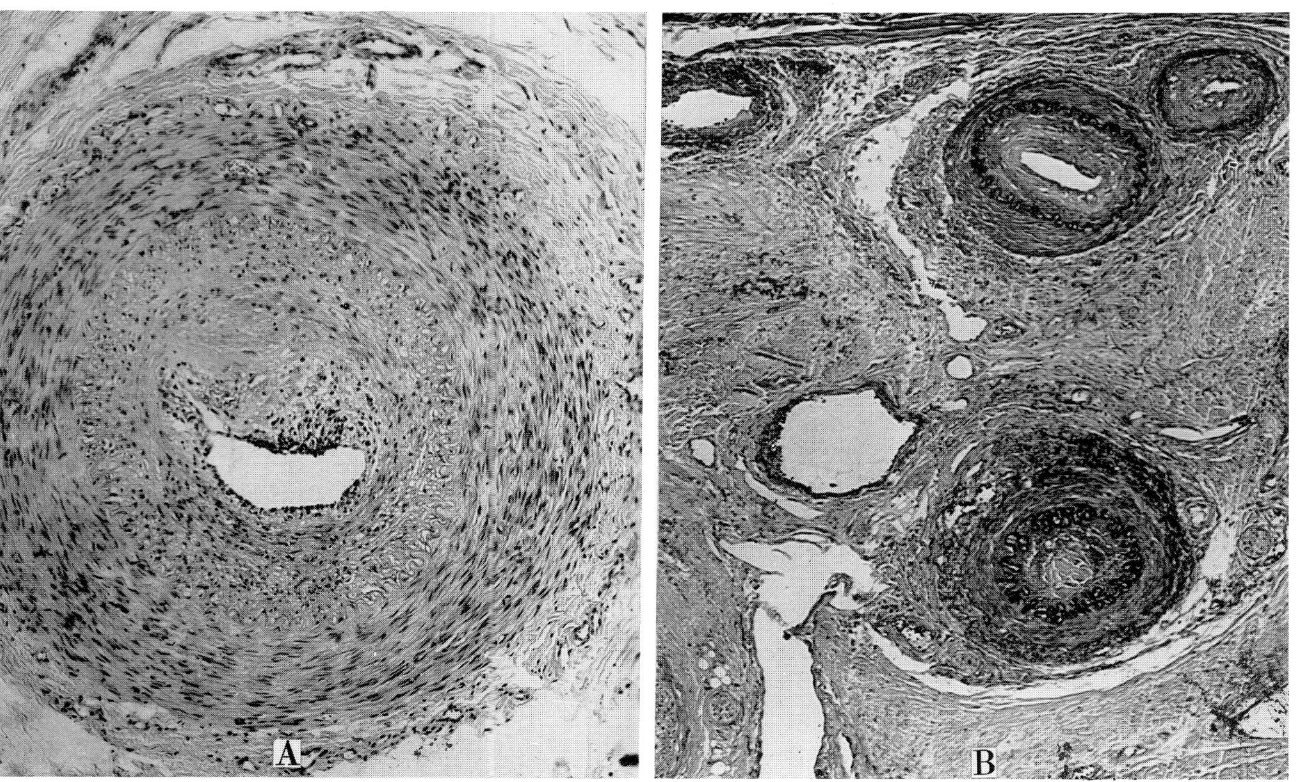

Fig. 137A. Digital artery from a finger amputated because of Raynaud's disease, showing advanced intimal arteriosclerosis. B. Divisions of the digital artery from a finger amputated because of Raynaud's disease, showing intimal arteriosclerosis. One vessel is occluded by a thrombus.

thickening of the intima of small arteries by fibrosis (Lewis, 1937). Actual occlusion by thrombus may be found (Fig. 137). This type of vascular lesion is more prominent than the intimal arteriosclerosis commonly noted in the digital vessels of individuals of the same age who do not have Raynaud's disease. Its pathogenesis is not known, but it is likely that it is secondary to reduced blood flow and that it is to be regarded as a form of arteriosclerosis. The vessels of digits less severely injured by Raynaud's disease do not show specific lesions. They manifest about the same degree of intimal arteriosclerosis as those in digits from normal individuals. The lesions illustrated in Fig. 137 are, therefore, not regarded as a cause of Raynaud's disease but rather as secondary to it. They may, however, be the cause of severe ischemic changes and gangrene.

References

Allen, A. C., and Sirotta, J. H. (1944). The morphogenesis and significance of degenerative verrucal endocardiosis; terminal endocarditis; endocarditis simplex; nonbacterial thrombotic endocarditis. *Am. J. Pathol.* **20,** 1025.

Allen, E. V., Barker, N. W., and Hines, E. A., Jr. (1955). "Peripheral Vascular Diseases," 2nd ed. Saunders, Philadelphia, Pennsylvania.

Altshuler, C. H., and Angevine, D. M. (1949). Histochemical studies on the pathogenesis of fibrinoid. *Am. J. Pathol.* **25,** 1061.

Arkin, A. (1930). A clinical and pathological study of periarteritis nodosa; a report of five cases, one histologically healed. *Am. J. Pathol.* **6,** 401.

Bahr, G. F. (1956). Elektronenoptische Untersuchungen bei Sklerodermie. Zugleich ein Beitrag zur Kenntnis des Bindegewebes der Haut. *Ärztl. Forsch.* **10,** 6.

Bardawil, W. A., Toy, B. L., Galins, N., and Bayles, T. B. (1958). Disseminated lupus erythematosus, scleroderma, and dermatomyositis as manifestations of sensitization to DNA-protein. I. An immunohistochemical approach. *Am. J. Pathol.* **34,** 607.

Bean, W. B., Cogswell, R., Dexter, M., and Embick, J. F. (1949). Vascular changes in the skin in pregnancy. Vascular spiders and palmar erythema. *Surg., Gynecol. Obstet.* **88,** 739.

Bertelsen, A., and Dohn, K. (1953). Congenital arteriovenous communications of the extremities. Clinical and patho-physiological investigations. *Acta Chir. Scand.* **105,** 244.

Bielawski, J. G., and Tatelman, M. (1949). Intracranial calcification in encephalotrigeminal angiomatosis. *Am. J. Roentgenol., Radium Therapy* **62,** 247.

Blumgart, H. L., and Ernstene, A. C. (1932). Hemangiectatic hypertrophy and congenital phlebarteriectasis, with particular reference to the diagnostic importance of peripheral vascular phenomena. *A. M. A. Arch. Internal Med.* **49,** 599.

Buerger, L. (1908). Thromboangiitis obliterans: A study of the vascular lesions leading to presenile spontaneous gangrene. *Am. J. Med. Sci.* **136,** 567.

Buerger, L. (1924). "The Circulatory Disturbances of the Extremities: Including Gangrene, Vasomotor and Trophic Disorders." Saunders, Philadelphia, Pennsylvania.

Ceelen, W. (1932). Über Extremitätenbrand. *Arch. klin. Chir.* **173**, 742.

Chandelux, A. (1882). Récherches histologiques sur les tubercules sous-cutanés douloureux. *Arch. physiol. norm. et pathol.* [2] **9**, 639.

Churg, J., and Strauss, L. (1951). Allergic granulomatosis, allergic angiitis and periarteritis nodosa. *Am. J. Pathol.* **27**, 277.

Cogan, D. G. (1945). Syndrome of nonsyphilitic interstitial keratitis with vestibulo-auditory symptoms. *A. M. A. Arch. Ophthalmol.* **33**, 144.

Coursley, G., Ivins, J. C., and Barker, N. W. (1956). Congenital arteriovenous fistulas in the extremities. An analysis of sixty-nine cases. *Angiology* **7**, 201.

Craig, J. M., and Gitlin, D. (1957). The nature of hyaline thrombi in thrombotic thrombocytopenic purpura. *Am. J. Pathol.* **33**, 251.

Cross, F. S., Glover, D. M., Simeone, F. A., and Oldenburg, F. A. (1958). Congenital arteriovenous aneurysms. *Ann. Surg.* **148**, 649.

Cushing, H., and Bailey, P. (1928). "Tumors Arising from the Blood Vessels of the Brain: Angiomatous Malformations and Hemangioblastomas." Charles C Thomas, Springfield, Illinois.

Deicher, H. R. G., Holman, H. R., and Kunkel, H. G. (1960). Anti-cytoplasmic factors in the sera of patients with systemic lupus erythematosus and certain other diseases. *Arthritis Rheumat.* **3**, 1.

Durante, G. (1933). Myomètre et vaisseaux utérins (glomus utérin). *Bull. soc. obstet. gynécol.* **22**, 463.

Fell, E. H., Weinberg, M., Gordon, A. S., Gasul, B. M., and Johnson, F. R. (1958). Surgery for congenital coronary artery arteriovenous fistulae. *A. M. A. Arch. Surg.* **77**, 331.

Fennell, H. R., and Vazquez, J. J. (1960). Personal communication.

Fisher, E. R., and Baird, W. F. (1956). The nature of arteriolar and capillary occlusion in patients with carcinoma. *Am. J. Pathol.* **32**, 1185.

Fisher, E. R., and Creed, D. L. (1955). Thrombotic thrombocytopenic purpura: Report of a case with discussion of its tinctorial features. *Am. J. Clin. Pathol.* **25**, 620.

Fisher, E. R., and Rodnan, G. P. (1958). Pathologic observations concerning the kidney in progressive systemic sclerosis. *A. M. A. Arch. Pathol.* **65**, 29.

Fisher, E. R., and Rodnan, G. P. (1960). Pathologic observations concerning the cutaneous lesion of progressive systemic sclerosis: An electron microscopic, histochemical and immunohistochemical study. *Arthritis Rheumat.* **3**, 536.

Friedman, E. A., Bardawil, W. A., Merrill, J. P., and Hanau, C. (1960). "Delayed" cutaneous hypersensitivity to leukocytes in disseminated lupus erythematosus. *New Engl. J. Med.* **262**, 486.

Gates, R. R. (1948). "Human Genetics," Vols. I and II. Macmillan, New York.

Godman, G. C., Deitch, A. D., and Klemperer, P. (1958). The composition of the L. E. and hematoxylin bodies of systemic lupus erythematosus. *Am. J. Pathol.* **34**, 1.

Goldstein, H. I. (1931). Heredofamilial angiomatosis with recurring familial hemorrhages. Rendu-Osler-Weber's disease. *A. M. A. Arch. Internal Med.* **48**, 836.

Goldstein, H. I. (1947). Unusual (benign) hematemesis; gastric telangiectatic dysplasia; Rendu-Osler-Weber's disease (Goldstein's hematemesis). *Rev. Gastroenterol.* **14**, 258.

Good, R. A., Varco, R. L., Aust, J. B., and Zak, S. J. (1957). Transplantation studies in patients with agammaglobulinemia. *Ann. N. Y. Acad. Sci.* **64**, 882.

Gore, I. (1950). Disseminated arteriolar and platelet thrombosis; morphologic study of its histogenesis. *Am. J. Pathol.* **26**, 155.

Grace, J. T., Jr., and Dao, T. L. (1959). Dermatomyositis in cancer: A possible etiological mechanism. *Cancer* **12**, 648.

Hagedorn, A. B., and Barker, N. W. (1948). Response of persons with and without intravascular thrombosis to a heparin tolerance test. *Am. Heart J.* **35**, 603.

Haserick, J. R., Lewis, L. A., and Bortz, D. W. (1950). Blood factor in acute disseminated lupus erythematosus. I. Determination of gamma globulin as specific plasma fraction. *Am. J. Med. Sci.* **219**, 660.

Holman, E. (1940). The anatomical and physiological effects of an arteriovenous fistula. *Surgery* **8**, 362.

Holman, H. R., and Kunkel, H. G. (1957). Affinity between the lupus erythematosus serum factor and cell nuclei and nucleoprotein. *Science* **126**, 162.

Hoyer, H. (1877). Über unmittelbare Einmündung kleinster Arterien in Gefässäste venösen Charakters. *Arch. mikrskop. Anat. u. Entwicklungsmech.* **13**, 603.

Iglesias de la Torre, L., Gomez Camejo, M., and Palacios, G. (1939). Consideraciónes clinicas, anatómicas, radiológicas y quirúrgicas del glomus tumoral de Masson. *Cir. ortop. y traumatol. (Habana)* **7**, 11.

Ingram, J. T. (1958). Acrosclerosis and systemic sclerosis. *A. M. A. Arch. Dermatol.* **77**, 79.

Jäger, E. (1932). Zur pathologischen Anatomie der Thromboangiitis Obliterans bei juveniler Extremitätengangrän. Parts 1 and 2. *Virchows Arch. pathol. Anat. u. Physiol.* **284**, 526 and 584.

Janes, J. M., and Musgrove, J. E. (1949). Effect of arteriovenous fistula on growth of bone. Preliminary report. *Proc. Staff Meetings Mayo Clinic* **24**, 405.

Klemperer, P., Pollack, A. D., and Baehr, G. (1941). Pathology of disseminated lupus erythematosus. *A. M. A. Arch. Pathol.* **32**, 569.

Koch, H. J., Jr., Escher, G. C., and Lewis, J. S. (1952). Hormonal management of hereditary hemorrhagic telangiectasia. *J. Am. Med. Assoc.* **149**, 1376.

Kolaczek, J. (1878). Über das Angio-Sarkom. *Deut. Z. Chir.* **9**, 1.

Leonard, F. C., and Vassos, G. A. (1951). Congenital arteriovenous fistulization of the lower limb. Report of a case successfully treated by total excision. *New Engl. J. Med.* **245**, 885.

Levy, G. (1930). L'angiomatose du systeme nerveux central. *Presse méd.* **38**, 37.

Lewis, D. D. (1930). Congenital arteriovenous fistulae. *Lancet* **219**, 621.

Lewis, T. (1937). The pathological changes in the arteries supplying the fingers in warm-handed people and in cases of so-called Raynaud's disease. *Clin. Sci.* **3**, 287.

Libman, E., and Sacks, B. (1924). A hitherto undescribed form of valvular and mural endocarditis. *A. M. A. Arch. Internal Med.* **33**, 701.

Love, J. G. (1944). Glomus tumors; diagnosis and treatment. *Proc. Staff Meetings Mayo Clinic* **19**, 113.

McKay, D. G., and Shapiro, S. S. (1958). Alteration in the blood coagulation system induced by bacterial endotoxin. I. In vivo (generalized Shwartzman reaction). *J. Exptl. Med.* **107**, 353.

McKay, D. G., and Wahle, G. H., Jr. (1955). Epidermic gastroenteritis due to Escherichia coli O 111 B_4. II. Pathologic anatomy (with special reference to the presence of the local and generalized Shwartzman phenomenon). *A. M. A. Arch. Pathol.* **60**, 679.

McKay, D. G., Merrill, S. J., Weiner, A. E., Hertig, A. T., and Reid, D. E. (1953). The pathologic anatomy of eclampsia, bilateral renal cortical necrosis, pituitary necrosis, and other acute fatal complications of pregnancy, and its possible relationship to the generalized Shwartzman phenomenon. *Am. J. Obstet. Gynecol.* **66**, 501.

Malan, E. (1958). Vascular syndromes from dilatation of arteriovenous communications of the sole of the foot. *A. M. A. Arch. Surg.* **77**, 783.

Masson, G. M. C., del Greco, C. F., Corcoran, A. C., and Page, I. H. (1953). Acute diffuse vascular disease elicited by renin in rats pretreated with cortisone. *A. M. A. Arch. Pathol.* **56**, 23.

Masson, P. (1924). Le glomus neuromyo-artériel des régions tactiles et ses tumeurs. *Lyon chir.* **21**, 257.

Mellors, R. C., and Ortega, L. G. (1956). Analytical pathology: III. New observations on the pathogenesis of glomerulonephritis, lipid nephrosis, periarteritis nodosa and secondary amyloidosis in man. *Am. J. Pathol.* **32**, 455.

Mellors, R. C., Ortega, L. G., Noyes, W. F., and Holman, H. R. (1957). Further pathogenetic studies of disease of unknown etiology, with particular reference to disseminated lupus erythematosus and Boeck's sarcoid. *Am. J. Pathol.* **33**, 613.

Müller, R. F. (1901). Zur Kenntniss der Fingergeschwülste. *Arch. klin. Chir.* **63**, 348.

Muirhead, E. E., Booth, E., and Montgomery, P. O'B. (1957). Derivation of certain forms of fibrinoid from smooth muscle. *A. M. A. Arch. Pathol.* **63**, 213.

Murphy, T. O., and Margulis, A. R. (1956). Roentgenographic manifestations of congenital peripheral arteriovenous communications. *Radiology* **67**, 26.

Murphy, T. O., Sandhaus, S., and Ryan, J. M. (1956). Congenital peripheral arteriovenous communications: Use of femoral artery to heart circulation time in diagnosis. *Minn. Med.* **39**, 389.

Murray, M. R., and Stout, A. P. (1942). The glomus tumor: Investigation of its distribution and behavior, and the identity of its "epitheliod" cell. *Am. J. Pathol.* **18**, 183.

Naide, M., and Sagen, A. (1946). Venospasm. Its part in producing the clinical picture of Raynaud's disease. *A. M. A. Arch. Internal Med.* **77**, 16.

Orbison, J. L. (1952). Morphology of thrombotic thrombocytopenic purpura with demonstration of aneurysms. *Am. J. Pathol.* **28**, 129.

Osler, W. (1907). On telegiectasis circumscripta universalis. *Bull. Johns Hopkins Hosp.* **18**, 401.

Pack, G. T., and Anglem, T. J. (1939). Tumors of the soft somatic tissues of infancy and childhood. *J. Pediat.* **15**, 372.

Paillas, J. E., Bonnal, J., Berard-Badier, M., and Serratrice, G. (1958). Arteriovenous angiomas of the brain in children. *Presse méd.* **66**, 525.

Patek, A. J., Jr., Post, J., and Victor, J. C. (1940). The vascular spider associated with cirrhosis of the liver. *Am. J. Med. Sci.* **200**, 341.

Peacock, J. R. (1959). Peripheral venous blood concentrations of epinephrine and norepinephrine in primary Raynaud's disease. *Circulation Research* **7**, No. 6, 821.

Pollack, A. D. (1959). Some observations on the pathology of systemic lupus erythematosus. *J. Mt. Sinai Hosp.* **26**, 224.

Pratt, G. H. (1949). Arterial varices: Syndrome. *Am. J. Surg.* **77**, 456.

Quick, A. J. (1942). Hereditary hemorrhagic diathesis, pseudohemophilia and telangiectasia. "The Hemorrhagic Diseases and the Physiology of Hemostasis," pp. 167–183. Charles C Thomas, Springfield, Illinois.

Reid, M. R. (1925). Abnormal arteriovenous communications, acquired and congenital. II. The origin of arteriovenous aneurysms, cirsoid aneurysms, and simple angiomas. *A. M. A. Arch. Surg.* **10**, 996.

Ribbert, V. A. (1898). Über Bau, Wachstum, und Genese der Angiome, nebst Bemerkungen über Cystenbildung. *Virchows Arch. pathol. Anat. u. Physiol.* **151**, 381.

Rich, A. R. (1942). Additional evidence of the role of hypersensitivity in the etiology of periarteritis nodosa; another case associated with a sulfonamide reaction. *Bull. Johns Hopkins Hosp.* **71**, 375.

Rich, A. R., and Gregory, J. E. (1943). Experimental demonstration that periarteritis nodosa is a manifestation of hypersensitivity. *Bull. Johns Hopkins Hosp.* **72**, 65.

Riveros, M., and Pack, G. T. (1951). The glomus tumor. Report of twenty cases. *Ann. Surg.* **133**, 394.

Riveros, M., and Pack, G. T. (1953). Tumor glomico. *In* "Cancer, Problemas Clinico-Terapeuticos" (M. Riveros, ed.), p. 459. Paz Montalvo, Madrid.

Robbins, W. C., Holman, H. R., Deicher, H., and Kunkel, H. G. (1957). Complement fixation with cell nuclei and DNA in lupus erythematosus. *Proc. Soc. Exptl. Biol. Med.* **96**, 575.

Rypins, E. L. (1941). The roentgenologic aspects of subungual glomus tumor. *Am. J. Roentgenol. Radium Therapy* **46**, 667.

Sabin, F. R. (1917–1918). Origin and development of the primitive vessels of the chick and the pig. *Contribs. Embryol.* **6**, 61.

Shaw, J. J. M. (1928). Haemangioma. *Lancet* **214**, 69.

Smith, C. C., Zeek, P. M., and McGuire, J. (1944). Periarteritis nodosa in experimental hypertensive rats and dogs. *Am. J. Pathol.* **20**, 721.

Smith, D. J. (1952). Constriction of isolated arteries and their vasa vasorum produced by low temperatures. *Am. J. Physiol.* **171**, 528.

Smith, W. G., Beahrs, O. H., and McDonald, J. R. (1957). Congenital arteriovenous fistula involving both ischioanal fossae: Report of a case. *Ann. Surg.* **145**, 115.

Sokoloff, L. (1956). Some aspects of the pathology of collagen diseases. *Bull. N. Y. Acad. Med.* **32**, 760.

Stout, A. P. (1935). Tumors of neuromyo-arterial glomus. *Am. J. Cancer* **24**, 255.

Stout, A. P., and Murray, M. R. (1942). Hemangiopericytoma: a vascular tumor featuring Zimmermann's pericytes. *Ann. Surg.* **116**, 26.

Sucquet, J. P. (1862). Circulation du sang: d'une circulation derivative dans les membres et dans la tête chez l'homme. *In* "Anatomie et Physiologie." Delahaye, Paris.

Taber, R. E., and Ehrenhaft, J. L. (1956). Arteriovenous fistulae and arterial aneurysms of the pulmonary arterial tree. *A. M. A. Arch. Surg.* **73**, 567.

Talbott, J. H., and Ferrandis, R. M. (1956). "Collagen Diseases." Grune & Stratton, New York.

Theis, F. V., and Freeland, M. R. (1939). The blood in thromboangiitis obliterans. *A. M. A. Arch. Surg.* **38**, 191.

Thomas, A. (1933). Tumeurs comparables à des tumeurs glomiques développées dans les muscles de la cuisse à la suite d'un traumatisme. *Ann. anat. pathol.* **10**, 657.

Thompson, A. W., and Shafer, J. C. (1951). Congenital vascular anomalies. *J. Am. Med. Assoc.* **145**, 869.

Vazquez, J. J., and Dixon, F. J. (1957). Immunohistochemical study of lesions in rheumatic fever, systemic lupus erythematosus and rheumatoid arthritis. *Lab. Invest.* **6**, 205.

von Albertini, A. (1943). Nochmals zur Pathogenese der Coronarsklerose. *Cardiologia* **7**, 233.

Ward, C. E., and Horton, B. T. (1940). Congenital arteriovenous fistulas in children. *J. Pediat.* **16**, 746.

Wegner, F. (1936). Über generalisierte, septische Gefässerkrankungen. *Verhandl. deut. pathol. Ges.* **29**, 202.

Wessler, S., Ming, S.-C., Gurewich, J., and Frieman, D. G. (1960). A critical evaluation of thromboangiitis obliterans: The case against Buerger's disease. *New Engl. J. Med.* **262**, 1149.

Woolard, H. H. (1922). The development of the principal arterial stems in the forelimb of the pig. *Contribs. Embryol.* **14**, 139.

Worthington, J. W., Jr., Baggenstoss, A. H., and Hargraves, M. M. (1959). Significance of hematoxylin bodies in the necropsy diagnosis of systemic lupus erythematosus. *Am. J. Pathol.* **35**, 955.

Yakovlev, P. I., and Guthrie, R. H. (1931). Congenital ectodermoses (neurocutaneous syndromes) in epileptic patients. *A. M. A. Arch. Neurol. Psychiat.* **26**, 1145.

Zak, F. G., Helpern, M., and Adlersberg, D. (1952). Stenosing coronary arteritis. Its possible role in coronary artery disease. *Angiology* **3**, 289.

Zeek, P. M., Smith, C. C., and Weeter, J. C. (1948). Studies on periarteritis nodosa. III. The differentiation between the vascular lesions of periarteritis nodosa and of hypersensitivity. *Am. J. Pathol.* **24**, 889.

Zimmermann, K. W. (1923). Der feinere Bau der Blutcapillaren. *Z. Anat. u. Entwicklungsgeschichte* **68**, 29.

Chapter XXII. ABNORMALITIES OF THE VENOUS SYSTEM

A. STRUCTURAL ALTERATIONS

By G. C. McMillan

The venous system is not as exact in its architecture as the arterial system. It is more variable in its distribution and considerably richer in anastomotic connections. Variations in its architecture, so gross as to constitute congenital anomalies of clinical interest, are also found.

1. Congenital Abnormalities

It is not appropriate to detail many of the anomalies here, but mention can be made of some of them. There may be two superior caval veins, the left being a persistence of the left anterior cardinal vein and the left duct of Cuvier. A left caval vein is seen occasionally with the adult form of aortic coarctation. The pulmonary veins may enter the left atrium as one, two, or three, rather than four vessels; this anomaly is of little interest. However, rarely some or all of the pulmonary veins may fail to connect with the left atrium and drain instead into systemic veins, such as the superior vena cava or innominate vein, or into the right atrium. Anomalous drainage from one lung often is accompanied by an atrial septal defect, which usually permits survival into adult life, death arising because of right heart failure. Anomalous pulmonary vein drainage from both lungs is very rare. It must be associated with an atrial septal defect to be compatible with life. Abnormalities of the venous drainage of other organs may also be of interest, as for example, anomalous venous drainage associated with renal ectopia. The variations in the portal vein are numerous and have become of considerable surgical importance. This vessel is formed as usually described in anatomy texts in only about one half of patients. Persistence of a congenitally patent umbilical vein is present in Cruveilhier-Baumgarten syndrome, and its dilatation from a channel with a rudimentary lumen to a gross vessel associated with the caput medusae in liver cirrhosis may be seen in this condition. (For normal embryology of veins, see Section A., Chapter VII.)

2. Varicose Veins

a. Definition: Circumscribed dilatations of veins are named varicosities, in contrast to diffuse dilatations which are termed ectasias. Varicosities may occur in veins without valves, as for example, the hemorrhoidal

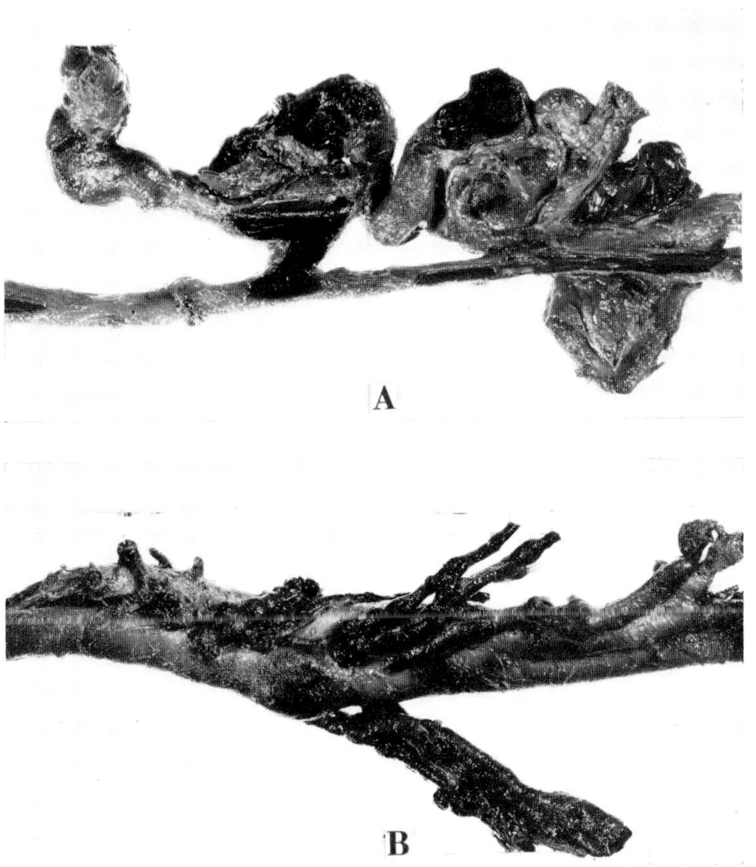

Fig. 138A. Varicose veins of leg illustrating dilatation, tortuosity and thrombosis in presence of normal deep veins.

B. Thrombosis and beginning varicose dilatation of leg veins, with maximal distension at the site of venous valves.

veins, in veins with occasional valves, as in the case of the pampinoform plexus, or in vessels with valves, such as those of the legs. Dilatation is usually accompanied by elongation, so that tortuosity develops (Fig. 138A).

Varicosities develop in veins subjected to increased hydrostatic pressure and poorly supported by tissue tension. The rate of flow in such vessels may be increased, as for example, in the esophageal varices arising in portal vein hypertension; or it may be decreased or retrograde, as in the varicosities of the superficial veins of the legs.

b. Etiology: The cause of varicose veins of the legs is not fully understood. It is apparent that hydrostatic pressure of unusual degree, due to venous obstruction, increased abdominal pressure, orthostatism, etc., is of great importance although not necessarily a sufficient cause of itself. It is held that such pressure must act either on veins that are congenitally weak or poorly supported because of obesity or wasting, or on veins in which the muscle tissue has been replaced by fibrosis. Varicose veins of the legs begin in early adult life and are progressive. They do not usually develop prominently for several years unless the insult is severe. The role of valvular defects in their development is not always clear. While valvular insufficiency occurs when varicosities develop, and while it undoubtedly contributes to the further extension of the ectasy, it has not been shown that defective valve function or valvular lesions precede the growth of varicose veins in the usual case. Indeed, when valved veins are overfilled by obstruction to their outflow, the valve sites distend as knots (Fig. 138B). Hence, it may be argued that the ectatic process begins with dilatation of the vessel wall immediately above the cusps of sufficient valves, leading in time to permanent weakening of the vessel and varicosity.

c. Pathologic changes: Varicose areas of the vein wall show varying degrees of muscle hypertrophy, attenuation of muscle, intimal and medial fibrosis, replication, attenuation and fragmentation of elastica, phlebosclerotic plaques, chronic inflammation, minimal focal lipid deposition, focal calcification and old or recent thrombus. The endothelial pattern, while abnormal, is not altered in any constant or specific way (Impallomeni, 1956). Valves may be impossible to identify. Adjacent saccular dilatations can press together and establish new communications at the point of contact. The vein wall distal to a varicosity is often under unusual pressure and will show ectasia, muscle hypertrophy and mild medial and intimal fibrosis, with some disruption of elastica. Embolism from thrombosis of varicose superficial leg veins is uncommon because the tortuous channels do not allow free transport of the emboli; moreover, the drainage of blood from varicose veins tends to be slow or even retrograde.

3. Phlebosclerosis

a. Definition: Phlebosclerosis is a pathologic process, somewhat analogous to arteriosclerosis, that leads to an intimal and medial fibrosis of the vein. It may be superimposed upon certain intimal thickenings that have been called endophlebohypertrophy and that seem to be a normal development of aging, or, it may, under the stimulus of inflammation, injury, or increased intraluminal pressure, develop in an unthickened area.

b. Endophlebohypertrophy: Lev and Saphir (1951) have observed a process of intimal thickening in veins that begins early in life, especially near the openings of tributaries, and that yields connective tissue plaques. There is gradual enlargement of such plaques with age by a proliferation of fibroblasts, smooth muscle cells, replication of elastic fibers and reticulum, and increase in ground substance. It is characteristic of the plaques that the cellular elements have an architectural arrangement and are oriented in the long axis of the vessel. In later life the plaques may hyalinize or even calcify. There is no evidence of inflammation or of lipid deposition. The process, which has been termed endophlebohypertrophy, is not held to be pathologic.

c. Pathologic changes: A less organized pattern of fibrosis of the intima occurs in phlebosclerosis with enlargement and confluence of intimal plaques. The growth and orientation of the cellular elements may be most variable, as may the pattern of the fibrillar elements that form in relation to them. The plaques may contain much or little ground substance and are prone to subsequent degenerative changes. Medial fibrosis may insert itself between muscle bundles and fibers and may occur in areas that do not show intimal phlebosclerosis. The medial and intimal lesions are not necessarily related, but in many cases they may be due to the same cause.

The lesions of phlebosclerosis are often extensive but do not cause stenosis of the lumen of the vein. Furthermore, they do not become grossly elevated or ulcerate. Their importance as a nidus for thrombus formation has not been determined, nor is their role in the development of varicosities certain. In fact, they do not appear in any consistent way to cause a disturbance of venous circulation. With the development of hyaline changes, intimal calcification may be seen, although it is usually slight. The intimal lesions not infrequently contain very small amounts of lipid, but it is rare to find any appreciable quantity. Occasionally at the point where the inferior vena cava is formed, there may be a plaque containing lipid. It is, perhaps, curious that vessels so prone to thrombosis should so rarely show lesions that resemble those of arteriosclerosis even slightly.

B. INFLAMMATION AND THROMBOSIS

By G. C. McMillan;[1] J. C. Paterson[2]

1. General Considerations

Much of the interest in phlebitis has in the past centered around two features, namely, the tendency for an inflamed vein to thrombose and the fact that thrombosis of a vein will result promptly in a mild, sterile phlebitis at the site of involvement.

a. Definition of terms: Thrombosis secondary or subsequent to inflammation of a vein is called thrombophlebitis. Thrombosis in a vein free from inflammation at the time of thrombus formation has been designated as phlebothrombosis. Since quite different mechanisms of etiology and pathogenesis are involved, and as these may have quite different therapeutic implications, it seems appropriate to emphasize the distinction. (For further discussion, see Subsection 3b, below.)

Not all phlebitis precipitates thrombosis. For example, the periphlebitis of small veins in chronic dermatitis is not accompanied by this process, while the phlebitis of allergic angiitis may or may not lead to it. In contrast, bacterial phlebitis commonly produces thrombosis, resulting in the development of a septic thrombus. Traumatic injury to endothelium may cause a sterile phlebitis which may also yield a thrombus. It would appear that even slight endothelial injury, with alteration or exposure of the intercellular substance between adjacent endothelial cells, is, in the presence of inflammation, a sufficient but not inevitable cause of thrombosis (Robertson *et al.*, 1959). The thrombus, itself, prob-

[1] Author of Subsections 1 and 2.
[2] Author of Subsection 3.

ably begins under such circumstances through the adherence of thrombocytes to the altered or exposed intercellular material (Samuels and Webster, 1952; Florey *et al.*, 1959).

At other times, thrombosis occurs in veins in the absence of inflammation or obvious endothelial injury. Under these conditions, there develops within 24 hours a mild, sterile acute phlebitis. This is analogous to the inflammation that may accompany a hematoma or a hemorrhage into a tissue space. The inflammatory exudate is relatively slight and does not extend into the thrombus. However, it may not be possible to distinguish such a lesion from a primary phlebitis.

A thrombus arising on the basis of phlebitis will be more adherent and may undergo more rapid organization than one developing in a normal vein, unless this process is inhibited by some injurious, lytic, or toxic agent. The adherence depends on extensive endothelial injury and the development of many points of fibrinous attachment. The more rapid organization probably occurs because of the larger area of endothelial injury and a greater area of contact with fibroblasts. Moreover, cells of the injured intima and endothelium may have already adopted a premitotic and migratory state or an antephase that allows them to initiate organization 24 to 48 hours earlier than in the uninflamed vein. Although the differences in adherence and in organization are not great, they are enough to influence the frequency of gross embolism, so that the latter process is less in primary phlebitis with thrombosis than in simple phlebothrombosis.

Where there has occurred appreciable inflammation or necrosis of the media and adventitia of the vein, fibrosis will result, and the vessel may show replacement of much muscle both by diffuse fibrosis insinuated between individual muscle fibers and bundles and by focal cicatrization. The intima may heal with the development of plaques of intimal fibrosis. The result is one of the forms of phlebosclerosis (see Section A.-3, above).

Organization of the thrombus may leave a dense fibrous plug in the vessel, with only minor recanalization, as, for example, in thromboangiitis obliterans. On the other hand, this process may fail, as in the small veins of the spleen, the periprostatic venous plexus or the perivesical plexus; under such circumstances the thrombus will be mineralized, subsequently to form a phlebolith. Or, commonly, it proceeds to yield a recanalized lumen and a fibrous intimal thickening. Experimental studies and clinical experience have shown that the valves of the veins may be seriously crippled in the process of organization (Edwards, 1937; Edwards and Edwards, 1937).

2. Phlebitis

a. Definition: An inflammatory reaction in a vein may be due to physical, chemical or allergic injury or to a living infective agent, or it may be idiopathic. Formerly the differences among these varieties of phlebitis were not appreciated, and their association with venous thrombosis or relationship to inflammation was not understood. So important did the clarification of the problem seem that Cruveilhier said that "La phlébite domine toute le pathologie" and Froriep advised the young Virchow to study the subject. Today there is little interest, possibly too little interest in phlebitis (Oertel, 1938; Long, 1961).

b. Role of acute purulent inflammation: Acute phlebitis of small veins is a very common occurrence in areas of acute bacterial inflammation. The vessel is part of the tissue involved, and the inflammatory reaction extends into its wall. Abscesses, meningitis, and pneumonia provide common examples of this type of lesion. If larger veins are affected, the adventitia is edematous and pus may be detected in it. The vasa of the vessel are engorged and red, and the intima may be swollen and discolored gray or yellow. A septic thrombus, adherent, but possibly partially liquefied and purulent, may be present. An ascending lymphangitis is commonly noted, because the adventitial coat of veins is particularly rich in lymphatics.

Rather uncommonly a vein may be infected by the implantation of bacteria from its lumen rather than from its adventitia or vasa. This may occur (1) due to the migration of bacteria in a septic thrombus that propagates along the vein; (2) as a result of the transport of viable bacteria within leucocytes through the intima of the vein; or (3) by phagocytosis of the organism by endothelial cells. It is uncertain how common the latter event is, but it is known that capillary endothelium is moderately phagocytic and that both venous and arterial endothelium may manifest such a response on occasion. The term, endophlebitis, may be used for inflammation which develops because of bacterial infection from the lumen of the vein (Benda, 1924).

c. Role of acute nonpurulent inflammation: Not all examples of acute inflammation are purulent. For example, the vasculitis of rheumatic fever, of drug sensitivity or of tuberculous meningitis may be classed as acute, but these varieties are essentially granulomatous in nature. Tuberculous endophlebitis is observed commonly in pulmonary veins adjacent to active caseous lesions of pulmonary tuberculosis.

Chronic phlebitis may result from an acute inflammation or it may represent a chronic process *ab initio.* The phlebitis of thromboangiitis obliterans is commonly acute at first but yields a chronic phlebitis in time. Tuberculous phlebitis is usually chronic except in the meninges. Numerous chronic skin diseases manifest a chronic periphlebitis of small vessels.

d. Unusual types of phlebitis: Among the unusual types of phlebitis are those due to protozoal parasites and also the chronic phlebitis and thrombosis attributed to senecio alkaloids. Several parasites may inhabit and lodge in veins, using them for habitation or transport. For example, Ascaris lumbricoides larvae pass from gut to liver in the portal veins and thence to the lung, where they rupture the capillaries to reach the alveolar spaces. In bilharziasis the ova of the schistosomes are carried by the veins to such organs as the liver, where they establish an endophlebitis. The consumption of senecio alkaloids with poorly winnowed wheat and cereals or in herbal and bush teas has been implicated as the cause of a toxic hepatocellular damage, with a mild phlebitis and widespread thrombosis of the radicals of the hepatic veins (Selzer and Parker, 1951; Bras *et al.*, 1954). This curious localization of the phlebitis gives rise to a Chiari syndrome. The disease attributed to the senecio alkaloids has been designated "hepatic veno-occlusive disease."

Saphir and Lev (1952) studying the femoral venous valve have described venous valvulitis and have noted that such valvulitis does not seem to be rare and that an old inflammatory lesion may predispose the valve to acute inflammation in instances of bacteremia.

e. Microscopic changes: In the acute phase the inflamed vein has its elements separated by edema. The vasa are congested. There is a prominent leucocytic infiltrate, comprising chiefly polymorphonuclear neutrophilic leucocytes. These are present in all layers but are most numerous in the outer coats. The intima is swollen, its endothelium disrupted, and tiny pus pockets may be found in its structure. If a thrombus has developed, bacterial colonies surrounded by leucocytes or pus may be present. The entire thickness of the vessel wall may be destroyed at some point, with fragments of elastica and muscle cells found in the pus.

In the late phase the healing of acute phlebitis is by organization of the thrombus and fibrosis of the vessel wall. If the inflammatory process has been relatively mild, organization is prompt and some degree of recanalization will occur. The periphery of the thrombus may begin to organize by fibroblastic and capillary invasion within 48–72 hours, and the thrombus will become firmly attached to the intima of the vein. At the same time, fibroblastic proliferation occurs in the intima, media, and adventitia. A scarred vessel, often with prominent vasa and a disrupted elastica, results. If valve cusps have been involved, they, too, may be fibrosed, distorted, or even destroyed. With a more severe inflammatory process, healing will be delayed, just as it is in an infected skin wound; the thrombus may fail to organize and may be softened and liquefied by bacterial activity and purulent reaction. Gross scarring and distortion of the vein and its lymphatics may result.

3. Deep Venous Thrombosis[3]

Deep venous thrombosis is still just about the same enigma it was in the days of Virchow. Its etiology remains obscure, its incidence is largely a matter of conjecture, and its prophylaxis and treatment seem to rest more on wishful thinking than on a solid basis of scientific knowledge. But the outlook for the eventual control of the disease is not as bleak as it was: fragments of factual information are now being reported that give some hope for the future, principally because they indicate the ways in which the problem can best be attacked.

The present section, although restricted to a discussion of thrombosis of the deep veins of the lower extremities and to emboli arising therefrom, is also applicable to venous thrombosis in other parts of the body.

a. General considerations: Contrary to the views expressed or implied in many recent reports, the incidence of venous thrombosis and pulmonary embolism (thromboembolic disease) cannot be determined by clinical means. Indeed, the fallacy that thromboembolism can be so diagnosed has probably done more to hold back progress on this disease than any other single factor. An attempt to justify these rather sweeping statements will be made in the succeeding paragraphs.

The literature is replete with surveys in which the diagnosis of venous thrombosis was made by the identification of certain clinical signs, which in many instances are now known to be misleading. For every case with presenting clinical signs there are many others in which they are absent, especially with so-called phlebothrombosis (Gage, 1953; Paterson and McLachlin, 1954; Gibbs, 1957; Coon and Coller, 1959). In the words of DeBakey (1954) "The diagnosis (of venous thrombosis) may be little more than an assumption based upon a high degree of suspicion and clinical criteria that are often vague and indefinite." It is indeed unfortunate that theories on pathogenesis and treatment should have been built on surveys with such weak foundations. But this is not to say that the clinical signs of venous thrombosis

[3] By J. C. Paterson.

may not be valid on occasion. The clinical picture of acute deep thrombophlebitis is so obvious that it need not be discussed here; but with phlebothrombosis, about all one has to go on are such indefinite findings as unilateral swelling of the foot, leg, or thigh; pain in the calf on forcible dorsiflexion of the foot; and tenderness in the calf or in the upper and inner portions of the thigh. The relative accuracy of these "diagnostic" signs has recently been assessed in a study in which they were searched for bi-weekly in seriously-ill patients and the findings compared with the results of venous dissection of the lower extremities at autopsy (Richards *et al.*, unpublished observations). In this unfortunately small series of cases, the index of unilateral swelling of the leg or lower thigh gave the best correlation, but even this was disappointing. In general, therefore, one must accept the dictum laid down by Coon and Coller (1959) that "Any series reporting the incidence of thromboembolic disease purely on the basis of clinical diagnosis is presenting highly inaccurate as well as misleading figures."

On the other hand, dissection of the veins of the lower extremities at autopsy gives an accurate picture of the incidence of venous thrombosis. Unfortunately, only a few surveys have been made in which this technique was followed, too few to establish the incidence of the condition in different age groups and with different disease states. But in general the incidence seems to vary from 34 per cent, in unselected male patients over the age of 40 years (McLachlin and Paterson, 1951), to the extremely high figure of 86 per cent, in older individuals of either sex with recent fractures of the femur (Sevitt and Gallagher, 1959). It is evident from these two surveys that the incidence of venous thrombosis of the lower extremities will depend largely upon the type of case examined. Many more such studies will have to be carried out before the matter is clarified, and these will probably have to be done elsewhere than in North America where the restrictions placed by most hospitals on complete leg vein dissections postmortem sharply limit the amount of reliable information on this phase of the problem (Coon and Coller, 1959).

The incidence of pulmonary embolism subsequent to venous thrombosis of the lower extremities is not quite as obscure as that of venous thrombosis *per se*. The clinical findings in patients with severe pulmonary embolism are said to be fairly characteristic, but it is significant that Coon and Coller (1959) found that only 27 per cent of fatal pathologically-proven emboli were diagnosed during life. Similar poor correlations have been reported by others (Belt, 1934; Barker *et al.*, 1941). Emboli of smaller size are even more difficult to diagnose clinically; they become evident only on careful dissection of the pulmonary arteries at autopsy, preferably by the "open-book" technique (Belt, 1936). In general, it seems fair to state that all suggestions for the prevention or treatment of pulmonary embolism should be based on carefully conducted pathologic studies. Fortunately there are quite a few of these, and they have been reviewed recently by DeBakey (1954) and by Coon and Coller (1959).

One other question about pulmonary embolism deserves consideration—how many venous thrombi of the lower extremities embolize and how many do not? McLachlin and Paterson (1951) found that only half of their cases with pathologically-proven venous thrombi had pulmonary embolism; but it is doubtful if this figure can be taken as the rule. It is probably fallacious to attempt to arrive at the incidence of venous thrombosis by simply determining the number of pathologically-proven cases of pulmonary embolism in a given population. The fact is that we are almost completely in the dark as to why some thrombi fracture and embolize while others do not (see Subsection 3d).

Finally, any discussion of the incidence of thromboembolism must take note of the fact that the disease is by no means peculiar to surgical patients. DeBakey (1954) analysed the data from the Charity Hospital, New Orleans, over a 7 year period and found that thromboembolism was equally common in surgical, medical, urological, obstetrical, and gynecological patients when the data were corrected for the admission rates to the various services.

b. Etiology: Surprisingly little has been added to our knowledge of the etiology of venous thrombosis in the past hundred years. Virchow's famous triad of causes are still in favor: local injury, increased blood coagulability, and venous stasis. Each of these etiologic factors has its proponents today, although most workers in the field seem to think that all of them may operate, in varying degrees, in individual cases. Before giving the evidence for each of these factors it is proper to decide whether a distinction should be made between thrombophlebitis and phlebothrombosis. The term "thrombophlebitis" implies that an inflammatory lesion in the vein wall precedes and causes thrombus deposition. "Phlebothrombosis," on the contrary, means a bland (noninflammatory) process the cause of which, in whole or in part, probably lies elsewhere than in the vein wall. Allen *et al.* (1946) did not attempt to separate the two types, and there is some justification for this point of view, since patients with thrombophlebitis are by no means as immune to the catastrophe of pulmonary embolism as was once thought (McLachlin and Paterson, 1951; Coon and Coller, 1959).

Inflammatory infiltrations in the vein wall can be demonstrated so easily in cases of clinically-obvious thrombophlebitis that they must be regarded as the major reason for thrombus deposition in these locations. Likewise, the thrombi that occur so often in association with fractures of the femur may well be the result of local injury to veins at the fracture site. But for the ordinary case of so-called phlebothrombosis (and these are in the great majority), no such local injury or inflammation can be found. Paterson and McLachlin (1954) performed complete microscopic serial sections on each of 21 tiny incipient thrombi found in the course of 165 venous dissections and failed to demonstrate any inflammatory infiltration, or other obvious histologic change, at the point of primary deposition of most of the thrombi. In a few instances a slight lymphocytic infiltration was noted in the adjacent vein wall but the authors were not impressed either with the severity of this inflammatory lesion or with its frequency. It must be stressed, however, that histologic studies like these do not eliminate the possibility that submicroscopic changes had occurred in the endothelial cells of the vein at the site of phlebothrombosis. Indeed, a number of workers have deduced from animal experiments that such minimal changes are present and that they are probably responsible for the primary depositions (O'Neill, 1947; Samuels and Webster, 1952; Robertson *et al.*, 1959). (For discussion of endothelial changes, see Subsection 3d.) It is evident that endothelial damage may be produced by trauma, by pressure, and even perhaps by anoxia. Whether such a change ever results in full-blown thrombosis is another matter, for it has been claimed that severe local trauma to the veins of animals does not regularly induce thrombi (Jaques, 1951). (For a discussion of thrombogenesis, see Section G.-3d and e, Chapter VI.)

Many substances have been suggested as causes of blood hypercoagulability and venous thrombosis. Among these factors is an increased blood concentration of antithrombin (Kay *et al.*, 1950), fibrinogen B (Cummine and Lyons, 1948), fibrinogen (Brinkhous, 1948), and total lipids (Keys, 1957). In addition, an elevated blood platelet count (Gage, 1953) and a decrease in the blood alpha tocopherol level (Kay *et al.*, 1950) have been implicated.

Paterson and McLachlin (1954) have put most of these suggestions to the test. Serial determinations of various blood constituents were made during life on seriously-ill patients and the results compared with the findings on venous dissection in those patients who died and on whom autopsies were performed. The results were consistently negative, no relationship being established between the presence or absence of venous thrombosis, on the one hand, and the blood levels of antithrombin, fibrinogen B, fibrinogen, and alpha tocopherol and the number of blood platelets, on the other (see also Paul *et al.*, 1954a).

No reliable data are yet available concerning the hypothetical relationship between alimentary lipemia and venous thrombosis. Many claims have been made along these lines but they have been based exclusively either on clinical evidence or on the demonstration of emboli in the lungs, not of thrombi in the veins (Keys, 1957; Keys *et al.*, 1957; Thomas *et al.*, 1960). Nor is any definite evidence at hand that hyperlipemia increases blood coagulability (Ratnoff, 1959). Indeed, Quick (1960) has stated: "It is still a moot point whether hypercoagulability (in general) has ever been unequivocally demonstrated. Progress in understanding thrombosis . . . has been hampered more by the myth that the clotting time is a measure of physiologic coagulability than by any other theory of coagulation." Until these matters are clarified, and until a definite correlation between high serum lipids and the presence of venous thrombi at autopsy is demonstrated (and vice versa), the numerous statements in the literature that a low fat diet goes hand in hand with a low incidence of thromboembolic disease must be accepted with reserve.

Impressive evidence has been aligned in support of the factor of venous stasis but all of it is indirect. It is well known that thromboembolic disease is restricted almost entirely to patients who are confined to bed and in whom venous stasis is therefore most apt to occur. Wright and Osborn (1952) have shown by tracer techniques that posture plays an important role in the venous return from the legs; elevation of the lower extremities by as little as 10° from the horizontal doubles the venous flow rate. McLachlin *et al.* (1960) have elaborated on this matter. Using radio-opaque materials injected into the ankle veins of normal subjects, they have shown (1) that although the leg veins empty fairly quickly when the subject lies horizontal in the supine position, the valve pockets do not; (2) that rapid emptying of valve pockets can be achieved by elevating the legs 15° above the horizontal; and (3) that valve pocket emptying is faster in young than in elderly subjects. Pathologically, Paterson and McLachlin (1954) showed that incipient venous thrombi in the deep leg veins form almost invariably at the apices of valve pockets—at points where stasis of blood should be most extreme. However, the illustrations in this latter report demonstrate clearly that sometimes only one of a pair of valve pockets was affected, thus suggesting that something other than stasis must have been involved in initiating the process. Along these same lines it has also been shown repeatedly that the clotting time of stagnant columns of blood in the ligated veins of animals is extremely slow, although it can be accelerated if homologous serum is injected into the blood stream (Wessler and Deykin, 1958).

c. Gross lesions: The exact location of the initial gross lesions of venous thrombosis is still debated. A large number of independent workers have reported that the primary site usually lies in the small veins of the calf (Rossle, 1937; Neumann, 1938; Hunter *et al.*, 1941; Bauer, 1946; Ochsner, 1948; Wilson, 1949; Eisendorff, 1949). However, some of these surveys were by no means complete. For example, Hunter *et al.* (1941) searched for thrombi only in the calf muscles which had been removed through a lateral incision in the leg; they did not make any other incisions in the lower extremities "for fear of losing the goodwill of the embalmers." Nevertheless, much has been made of this localization to the veins of the calf, both etiologically and therapeutically. Pressure on the calf veins while a patient is lying in bed results, it is claimed, in damage to the endothelium by pressure *per se* (Robertson *et al.*, 1959) or via the mechanism of endothelial cell anoxia (Frykholm, 1940). In either case thrombosis may result. From the therapeutic aspect, Homans (1934) suggested that surgical ligation of the superficial femoral veins should afford adequate protection against pulmonary embolism, since venous thrombi commonly originate below this point. McLachlin and Paterson (1951), however, reported that the usual site of initial thrombus deposition was not nearly so definite. In the course of 100 complete venous dissections, they found that more than half of the 76 individual thrombi encountered were located proximal to the upper end of the superficial femoral vein, i.e., above the usual site of venous ligation. These observations, which were so antagonistic to modern teaching, have been confirmed in a general way by Sevitt and Gallagher (1959). Frykholm (1940), while stressing the importance of the primary site in the small veins of the calf, found appreciable numbers of thrombi above the usual site of femoral vein ligation, just as many, in fact, as in the small veins of the calf. It is also of interest to note that Virchow (1846) suggested that thrombi which embolize tend to arise on the central side of the valves of the femoral and iliac veins.

The second major gross feature of the initial thrombus deposits is that they usually lie within valve pockets, not only in the pelvic veins (Virchow, 1846), but in the thigh and leg veins as well (Paterson and McLachlin, 1954). Paterson and McLachlin (1954) found that 80 per cent of 21 tiny incipient thrombi originated in this fashion. It is hard to understand why this peculiar localization has not been noted more often. Perhaps it is because propagation occurs both proximally and distally once the initial thrombus occludes the lumen, thus obscuring the primary site of deposition. In any event, it affords one more illustration of the benefits to be derived from studying early lesions rather than late ones.

The color and texture of venous thrombi vary considerably depending upon their age, the speed of their formation, their size and position, and upon whether they partially or completely occlude the venous lumens. Early incipient lesions within valve pockets are usually firm and white. As they grow by accretion and blossom out of the pocket, their surfaces are still pale and often show well marked lines of Zahn. Propagated thrombi, on the other hand, especially those which form in relatively stagnant columns of blood, tend to be reddish-black, shiny and smooth. Indeed, from their gross appearance, propagated thrombi are not unlike postmortem clots, but they can usually be distinguished from such masses by their friability. Furthermore, they are not elastic in the way postmortem clots are. In addition, most propagated red thrombi show gray streaks in the general background of reddish-black, shiny material.

d. Microscopic lesions: The earliest changes that occur in the walls of human veins at the sites of initial thrombosis have not yet been identified. Recognizable lesions of the endothelium have been produced experimentally by a variety of means, and special techniques have been developed to reveal the resulting lesions. These depend principally on silver impregnation *in situ,* followed by an examination of the endothelial surfaces of veins on their flat surfaces. The more important of the observations made using such procedures are given below.

O'Neill (1947) outlined the vasculature of the normal vein wall in dogs and found that the vasa venarum, both arterial and venous, were confined to the adventitial coat. Interference with these vessels resulted in desquamation of endothelial cells. However, platelet agglutination only occurred on the denuded surfaces when the blood flow through the vein was impeded. From these observations it might be deduced that two factors operate in venous thrombosis : poor nutrition to the vein wall (from either local or general causes) and venous stasis. Samuels and Webster (1952) obtained different results from those of O'Neill. They ascribed the endothelial cell desquamation, not to anoxia from cutting off the blood supply to the vein wall, but to a foreign-body reaction to materials used in the experiment. Furthermore, they considered the earliest site of fibrin and platelet agglutination to be at the intercellular cement line, i.e., the point of lowest surface tension. Robertson *et al.* (1959) were more interested in the experimental effects of pressure on the endothelium of veins, particularly of the same order as those that occur in people who are confined to bed or who sit in cramped positions for extended periods. They found that even the slightest touch to a vein produces endothelial damage, and that with greater external pressure, platelet agglutination occurs. No propagation of these platelet and fibrin agglutinations, nor of the ones described above in this paragraph, was observed, and it is therefore doubtful if they should be referred to as thrombi. However, they are reminiscent of the fibrin nubbins on the endothelial surfaces of arteries described as thrombi by Duguid (1949).

The earliest small thrombi that can be identified in human veins have been studied microscopically by Paterson and McLachlin (1954). Except for a very slight lymphocytic infiltration in a few cases, no obvious histologic lesion in the vein wall could be detected. The oldest parts of these small thrombi lay at the apices of valve pockets, and there they showed partial or complete organization. Springing from them in successive layers, lesser degrees of organized material were found. The most recent portions of the thrombi consisted of festoons of platelets and skeins of fibrin in which numerous red cells and leucocytes were enmeshed. The distinction between such recent thrombi and postmortem clots is not always easy to make, but if a recent thrombus is examined in its entirety, the presence of endothelialization or fibroblastic proliferation at some point proves its antemortem nature.

The microscopic structure of larger venous thrombi may provide clues to why some of them fracture and embolize while others do not. As mentioned above, recent thrombi are composed of festoons of platelets and skeins of fibrin, the latter varying considerably in quantity and architecture in different parts of the lesions. It has been suggested that the danger of thrombus fracture should be greater if the skeins of fibrin are poorly formed or unevenly distributed than if they are numerous and closely approximated. Carrying the argument farther, there are at least two reasons why the fibrin component of a thrombus may be defective: (1) a low blood fibrinogen level, and (2) a high concentration of fibrinolytic substances in the blood. Preliminary studies have been carried out which revealed that the blood fibrinogen level was lower in the three weeks period before death in 22 cases of thrombosis with embolism than in 18 cases of thrombosis without embolism, a difference that was statistically significant at approximately the 2 per cent level (Paterson *et al.*, unpublished observations). Blood fibrinolysins were determined in the same patients but by a method (Macfarlane, 1937) no longer considered accurate enough for research purposes. A significant relationship was not found in this latter comparison. Studies have also been carried out on other subjects to determine if intravenous saline and glucose infusions (which are often given to surgical patients) produce dangerously low levels of blood fibrinogen. Negative results were obtained with infusions of 1000 ml of these materials (Paul *et al.*, 1954b). However, such experiments should not be regarded as the only worthwhile approach to the mechanism of embolus production; physical factors, including those concerned with hemodynamics, should also be studied.

References

Allen, E. V., Barker, N. W., and Hines, E. A. (1946). "Peripheral Vascular Disease." Saunders, Philadelphia, Pennsylvania.

Barker, N. W., Nygaard, K., Walters, W., and Priestley, J. T. (1941). A statistical study of postoperative venous thrombosis and pulmonary embolism. *Proc. Staff Meetings Mayo Clinic* **16**, 17.

Bauer, G. (1946). Heparin therapy in acute deep venous thrombosis. *J. Am. Med. Assoc.* **131**, 196.

Belt, T. H. (1934). Thrombosis and pulmonary embolism. *Am. J. Pathol.* **10**, 129.

Belt, T. H. (1936). Demonstration of small pulmonary emboli at autopsy. *J. Tech. Methods* **15**, 39.

Benda, C. (1924). Venen. *In* "Handbuch der speziellen pathologischen Anatomie und Histologie" (O. Lubarsch *et al.*, eds.), Vol. II, Herz und Gefässe. Springer, Berlin.

Bras, G., Jelliffe, D. B., and Stuart, K. L. (1954). Veno-occlusive disease of liver with non-portal type of cirrhosis, occurring in Jamaica. *A. M. A. Arch. Pathol.* **57**, 285.

Brinkhous, K. M. (1948). *In* "Blood Clotting and Allied Problems," 1st Conference (J. E. Flynn, ed.), p. 39. Josiah Macy, Jr., Foundation, New York.

Coon, W. W., and Coller, F. A. (1959). Clinicopathologic correlation in thromboembolism. *Surg., Gynecol. Obstet.* **109**, 259.

Cummine, H., and Lyons, R. N. (1948). A study of intravascular thrombosis with some new conceptions of the mechanism of coagulation. *Brit. J. Surg.* **35**, 337.

DeBakey, M. E. (1954). A critical evaluation of the problem of thromboembolism. *Intern. Abstr. Surg.* **98**, 1.

Duguid, J. B. (1949). Pathogenesis of atherosclerosis. *Lancet* **ii**, 925.

Edwards, E. A. (1937). Observations on phlebitis. *Am. Heart J.* **14**, 428.

Edwards, E. A., and Edwards, J. E. (1937). The effect of thrombophlebitis on the venous valve. *Surg., Gynecol. Obstet.* **65**, 310–320.

Eisendorff, L. H. (1949). Phlebothrombosis of lower extremities. *Am. J. Surg.* **78**, 431.

Florey, H. W., Poole, J. C. F., and Meek, G. A. (1959). Endothelial cells and "cement" lines. *J. Pathol. Bacteriol.* **77**, 625.

Frykholm, R. (1940). The pathogenesis and mechanical prophylaxis of venous thrombosis. *Surg., Gynecol. Obstet.* **71**, 306.

Gage, M. (1953). Hidden thrombus in fatal pulmonary embolism. *J. Am. Med. Assoc.* **151**, 433.

Gibbs, W. M. (1957). Venous thrombosis of the lower limbs with particular reference to bed rest. *Brit. J. Surg.* **45**, 209.

Homans, J. (1934). Thrombosis of the deep veins of the lower leg causing pulmonary embolism. *New Engl. J. Med.* **221**, 993.

Hunter, W. C., Sneeden, V. D., Robertson, T. D., and Snyder, G. A. C. (1941). Thrombosis of deep veins of the leg. *A. M. A. Arch. Internal Med.* **68**, 1.

Impallomeni, G. (1956). The alteration and regeneration of the endothelium in venous thrombosis. *Angiology* **7**, 268.

Jaques, L. B. (1951). Discussion. *In* "Blood Clotting and Allied Problems," 4th Conference (J. E. Flynn, ed.), p. 42. Josiah Macy, Jr., Foundation, New York.

Kay, J. H., Hutton, S. B., Weiss, G. N., and Ochsner, A. (1950). Studies on an antithrombin. III. A plasma antithrombin test for the prediction of intravascular clotting. *Surgery* **28**, 24.

Keys, A. (1957). Diet and the epidemiology of coronary heart disease. *J. Am. Med. Assoc.* **164**, 1912.

Keys, A., Buzina, R., Grande, F., and Anderson, J. T. (1957). Effect of meals of different fats on blood coagulation. *Circulation* **15**, 274.

Lev, M., and Saphir, O. (1951). Endophlebohypertrophy and phlebosclerosis. 1. The popliteal vein. *A. M. A. Arch. Pathol.* **51**, 154.

Long, E. R. (1961). "Selected Readings in Pathology," 2nd ed., pp. 122 and 168. Charles C Thomas, Springfield, Illinois.

Macfarlane, R. G. (1937). Fibrinolysis following operation. *Lancet* **i**, 10.

McLachlin, A. D., McLachlin, J. A., Jory, T. A., and Rawling, E. G. (1960). Venous stasis in the lower extremities. *Ann. Surg.* **152**, 678.

McLachlin, J., and Paterson, J. C. (1951). Some basic observations on venous thrombosis and pulmonary embolism. *Surg., Gynecol. Obstet.* **93**, 1.

Neumann, R. (1938). Ursprungszentren und Entwicklungsformen der Beinthrombose. *Virchows Arch. pathol. Anat. u. Physiol.* **301**, 708.

Ochsner, A. (1948). Venous thrombosis. *Southern Surgeon* **14**, 477.

Oertel, H. (1938). "The Special Pathological Anatomy and Pathogenesis of the Circulatory, Respiratory, Renal and Digestive Systems." Renouf, Montreal.

O'Neill, J. F. (1947). The effects on venous endothelium of alterations in blood flow through the vessels in vein walls, and the possible relation to thrombosis. *Ann. Surg.* **126**, 270.

Paterson, J. C., and McLachlin, J. (1954). Precipitating factors in venous thrombosis. *Surg., Gynecol. Obstet.* **98**, 96.

Paul, R. M., Paterson, J. C., and McLachlin, J. A. (1954a). Blood coagulation factors in venous thrombosis. *Can. Med. Assoc. J.* **70**, 556.

Paul, R., Ramsay, A. G., and Paterson, J. C. (1954b). The effect of intravenous infusions on the blood fibrinogen level. *Canadian Med. Service J.* **10**, 62.

Quick, A. (1960). Prophylaxis in thromboembolism (Letter to the Editor). *Lancet* **i**, 169.

Ratnoff, O. D. (1959). Discussion. *In* "Connective Tissue, Thrombosis and Atherosclerosis" (I. H. Page, ed.), p. 31. Academic Press, New York.

Robertson, H. R., Moore, J. R., and Mersereau, W. A. (1959). Observations on thrombosis and endothelial repair following application of external pressure to a vein. *Can. J. Surg.* **3**, 5.

Rossle, R. (1937). Über die Bedeutung und die Entstehung der Wadenvenenthrombosen. *Virchows Arch. pathol. Anat. u. Physiol.* **300**, 180.

Samuels, P. B., and Webster, D. R. (1952). The role of venous endothelium in the inception of thrombosis. *Ann. Surg.* **136**, 422.

Saphir, O., and Lev, M. (1952). Venous valvulitis. *A. M. A. Arch. Pathol.* **53**, 456.

Selzer, G., and Parker, R. G. F. (1951). Senecio poisoning exhibiting as Chiari's syndrome. *Am. J. Pathol.* **27**, 885.

Sevitt, S., and Gallagher, N. G. (1959). Prevention of venous thrombosis and pulmonary embolism in injured patients. *Lancet* **ii**, 981.

Thomas, W. A., Davies, J. N. P., O'Neal, R. M., and Dimakulangan, A. A. (1960). Incidence of myocardial infarction correlated with venous and pulmonary thrombosis and embolism. *Am. J. Cardiol.* **5**, 41.

Virchow, R. (1846). Die Verstopfung der Lungenarterie und ihre Folgen. *Beitr. exptl. Pathol. u. Physiol.* (*Traube-Berlin*) **2**, 1.

Wessler, S., and Deykin, D. (1958). Theory and practice in acute venous thrombosis: a reappraisal. *Circulation* **18**, 1190.

Wilson, H. (1949). Surgery for the prevention of pulmonary embolism. *Am. J. Surg.* **78**, 421.

Wright, H. P., and Osborn, S. B. (1952). Effect of posture on venous velocity, measured with ^{24}NaCl. *Brit. Heart J.* **14**, 325.

PART FOUR

Lymphatic System

Chapter XXIII. LYMPHATIC VESSELS AND LYMPH

A. EMBRYOLOGY

By H. S. Mayerson

1. Growth and Development

It is now generally agreed that the lymphatic vessels are derived from veins. To quote Sabin (1911), "Lymphatics are modified veins. They are vessels lined by an endothelium which is derived from the veins. They invade the body as do blood vessels and grow into certain constant areas; their invasion of the body is, however, not complete, for there are certain structures which never receive them. The lymphatic capillaries have the same relation to tissue spaces as have blood capillaries. None of the cavities of the mesoderm, such as the peritoneal cavity, the various bursae and serous capillaries, forms any part of the lymphatic system. The lymphatic endothelium once formed is specific. Like blood vessels, the lymphatics are for the most part closed vessels."

a. Origin: In the embryo, the primary lymph sacs bud off from the veins. In the mammal two sets of paired sacs, jugular and iliac, and two unpaired sacs, retroperitoneal and cisterna chyli, are formed (Fig. 139). The jugular sacs bud off from the anterior cardinal veins and form large sacs in the neck. All the other lymphatic sacs bud off from the mesonephritic vein and the veins in the dorsomedial edge of the Wolffian bodies.

b. Further growth: From these primary lymphatic sacs, which are in communication with the veins, the peripheral lymphatic system grows by a process of endothelial sprouting. The jugular lymph sacs give rise to the lymphatics of the head and neck, forelegs, thorax, heart, and lungs. The iliac sacs are two long symmetrical sacs extending from the hilus of the Wolffian bodies to the level of the bifurcation of the aorta; from the caudal end grow the iliolumbar plexus, the femoral plexus and the plexus surrounding the umbilical artery. The cisterna chyli and the lower part of the thoracic duct arise in common with the iliac sacs from the mesonephritic veins. The thoracic duct is formed by the union of a duct, which grows downward from the left jugular sac, and a plexus arising from the cisterna chyli. The primary lymph sacs ultimately become primary lymph nodes, while the secondary nodes develop along the course of the lymph ducts.

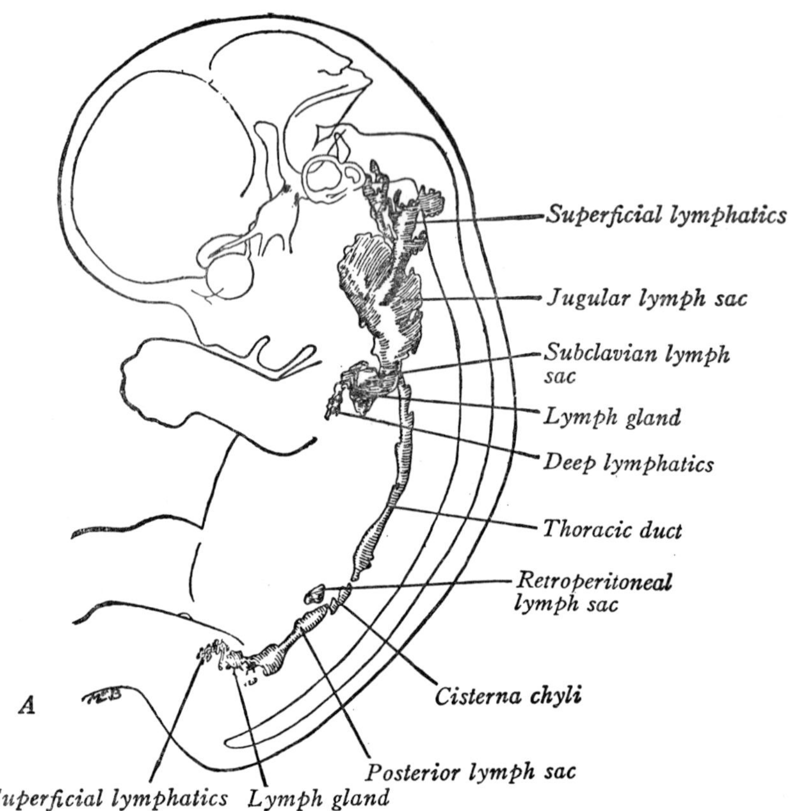

FIG. 139. Reconstruction of the primitive lymphatic system showing the primary lymph sacs in a human embryo of two months, after Sabin. (From Arey, 1954; reproduced with the permission of W. B. Saunders Co., Philadelphia.)

B. GROSS ANATOMY

By B. J. Anson

1. Main Lymphatic Channels

The terminal vessels in the lymphatic system are the thoracic duct and the right lymphatic duct (or other vessels which, in some instances, represent it). (For classical works on anatomy of the lymphatic system, see Mascagni, 1787, and Sappey, 1874–1875.)

a. Thoracic duct and cisterna chyli: The thoracic duct is by far the more capacious and the longer of the two. It is regarded typically as beginning in the abdomen as an elongated ovoid dilatation, the cisterna chyli, which actually rarely occurs (see Subsection 1e). When the latter is present, it ascends from the abdominal level, to pass through the aortic opening of the diaphragm and enter the posterior mediastinum as the thoracic duct. The thoracic duct ascends on the front of the vertebral column, to the right of the median plane, as high as the fifth thoracic vertebra (Fig. 140*a* and *b*). It then crosses from the right to the left of the midline, continuing an ascending course through the superior mediastinum to the root of the neck. Here it turns laterally, behind the carotid artery and internal jugular vein, and curves downward across the subclavian artery. The thoracic duct terminates at the medial border of the left scalenus anterior by joining the left innominate vein, in the angle of junction of the internal jugular and the left subclavian veins, or by ending in either of the tributaries of the innominate.

The thoracic duct is approximately 45 cm long and 2 mm in diameter. It may divide and reunite in its course through the thorax, and its terminal part is often broken up into a number of branches that end separately in the great veins. It is provided with numerous valves arranged in pairs, those most nearly perfect being near its termination.

b. Right lymphatic duct: This is an inconstant channel, since the three trunks that contribute to its formation, the right jugular, the right subclavian and the right mediastinal, usually open separately into the right internal jugular, subclavian, and innominate veins. When the right jugular and subclavian trunks (conveying lymph from the right side of the head and neck and from the right upper limb, respectively) unite before entering the veins at the point corresponding to that at which the thoracic duct enters on the left side, the vessel of confluence is called the right lymphatic duct. The right mediastinal trunk, which

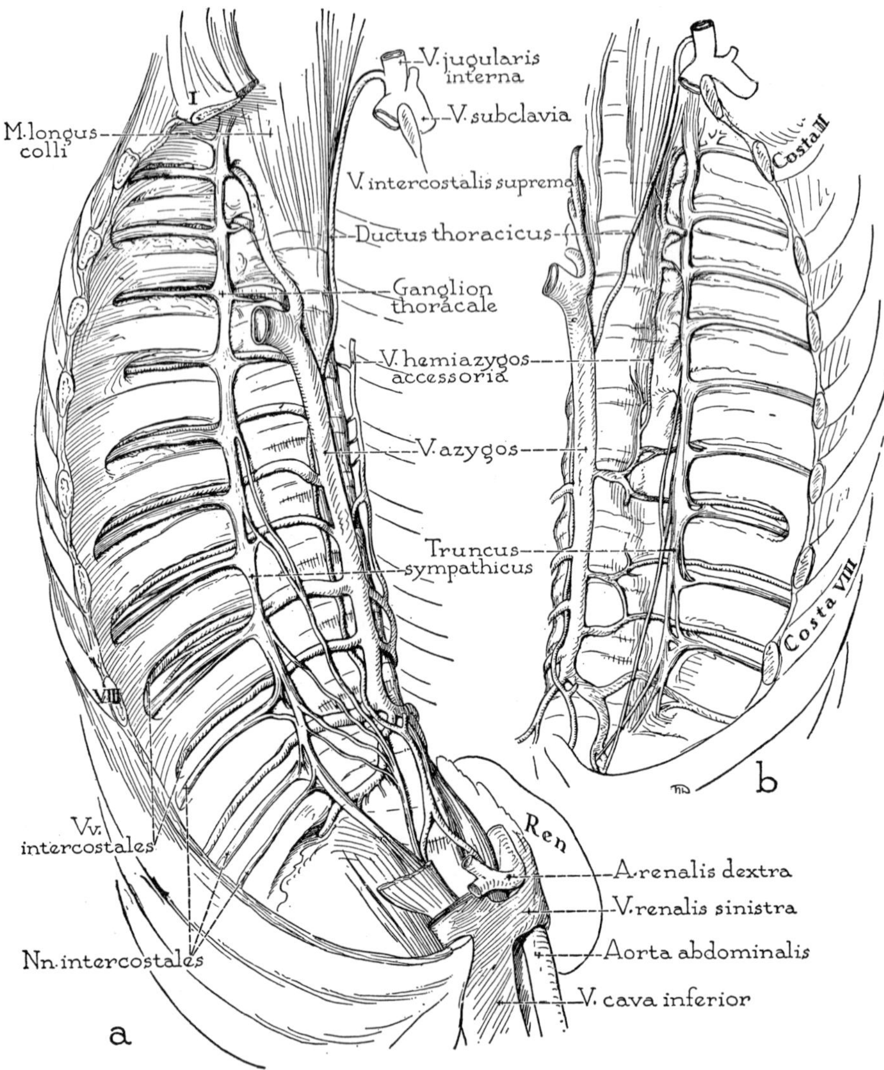

FIG. 140. The thoracic duct and related part of the venous and autonomic nervous systems. *a.* (Specimen's right side.) Relation of the duct to the azygos vein and to the latter's intercostal tributaries. *b.* (Opposite half of the body.) Course of the thoracic duct across to the vertebral column in its ascent to a termination in the internal jugular vein. (From Anson, 1950; reproduced with the permission of the W. B. Saunders Co., Philadelphia, Pa.)

collects lymph from an area corresponding to that drained by the left vessel (but including also the upper part of the right lobe of the liver), almost invariably enters the right innominate vein separately.

c. *Relations:* Typically, the cisterna chyli lies anterior to the first and second lumbar vertebrae and the corresponding right lumbar arteries, between the aorta on the left side and the azygos vein and the right crus of the diaphragm on the right side. In the posterior mediastinum, the thoracic duct is separated from the vertebral column and the anterior longitudinal ligament by the right posterior intercostal arteries and the transverse parts of the hemiazygos veins. In the lower part of its course, it is covered anteriorly by the diaphragm and the right pleura, and in the upper part, by the esophagus. The vena azygos is situated to its right, the descending aorta to its left. In the superior mediastinum the thoracic duct passes forward from the vertebral column across the left longus cervicis muscle (Fig. 140*b*). Emerging from behind, it is related to the esophagus anteriorly, and from below upward, to the aortic arch, the left subclavian artery and the pleura. At the root of the neck the duct arches laterally in front of the apex of the pleura and then downward across the first part of the left subclavian artery. It passes in front of the vertebral artery and vein, the roots of the inferior thyroid, the transverse cervical and suprascapular arteries, the medial border of the scalenus anterior and the left phrenic nerve, and behind the left carotid sheath and its contents.

d. *Tributaries:* The cisterna chyli is commonly described as receiving five tributaries: (1) the intestinal trunk, formed by the efferents of the superior mesenteric and celiac groups of lymph nodes, which convey lymph from the liver, spleen, pancreas, stomach, the small intestine and part of the large intestine; (2) paired lumbar trunks, formed by the efferents of the aortic glands, conveying lymph from the whole of the skin caudal to the level of the umbilicus, from the deep portions of the lower limbs, from the lower abdominal and the pelvic walls, from the remainder of the large intestine and the pelvic organs, and from the kidneys, suprarenal glands and genital glands; and (3) paired descending intercostal lymph trunks, formed by the efferent vessels from the lower intercostal glands, which descend to the cisterna through the aortic opening of the diaphragm.

The mediastinal segment of the thoracic duct receives efferents from the posterior mediastinal nodes and from those intercostal glands that do not send their efferents into the descending trunk. Through the posterior mediastinal nodes, the duct receives lymph from the esophagus and the dorsum of the pericardium and from the upper and posterior part of the liver.

In the superior mediastinum, the efferents of the upper intercostal nodes of both sides of the thorax open into the duct; additionally, the duct may receive communications from the mediastinal lymph trunks of both sides.

At the root of the neck, the thoracic duct, just proximal to its termination, may receive the following: (1) efferents from the nodes of the left upper limb (which may unite to form a subclavian trunk); and (2) the left jugular trunk, which carries lymph from the left side of the head and neck. Either or both of these vessels may end independently in the innominate or the subclavian vein (Fig. 140*a* and *b*). The left mediastinal trunk almost invariably has an independent entrance into the left innominate vein (not into the cervical part of the duct) and collects lymph from the following: deeper parts of the anterior thoracic wall; the upper part of the anterior abdominal wall, the anterior part of the diaphragm on the left side, the left half of the mediastinum, the left side of the heart and the left lung.

e. Variations: The cisterna chyli is commonly replaced by a plexus of vessels, formed by the intestinal and lumbar lymph trunks, from which the thoracic duct takes origin by several roots (Fig. 141*a* to *f*).

The thoracic duct, in its course through the thorax, frequently divides into two unequal branches which may run side-by-side for some distance before reuniting; or it may, like the so-called cisterna, be broken up into a plexiform arrangement of lymph vessels. It may end in a number of branches that enter the great veins separately; or it may terminate, as it sometimes begins, by dividing in the upper part of the thorax into two stems, of which the right joins the right jugular trunk to form one variety of the right lymphatic duct. The thoracic duct may end in either the internal jugular vein (Fig. 140*a* and *b*), the subclavian vein, or even the vertebral vein. In rare instances, it opens into the vena azygos.

The arrangement and mode of termination of the three cervical lymph trunks on each side of the neck (subclavian, jugular, mediastinal) are highly variable. They may either all open separately into one or other of the great veins, or unite in a variety of ways. Their number may be increased by division of one of them or by independent termination of their tributaries.

On the left side the jugular trunk ends in the thoracic duct, and the other two trunks may do the same. However, the subclavian trunk often terminates in the subclavian vein and the mediastinal trunk usually ends in the innominate vein.

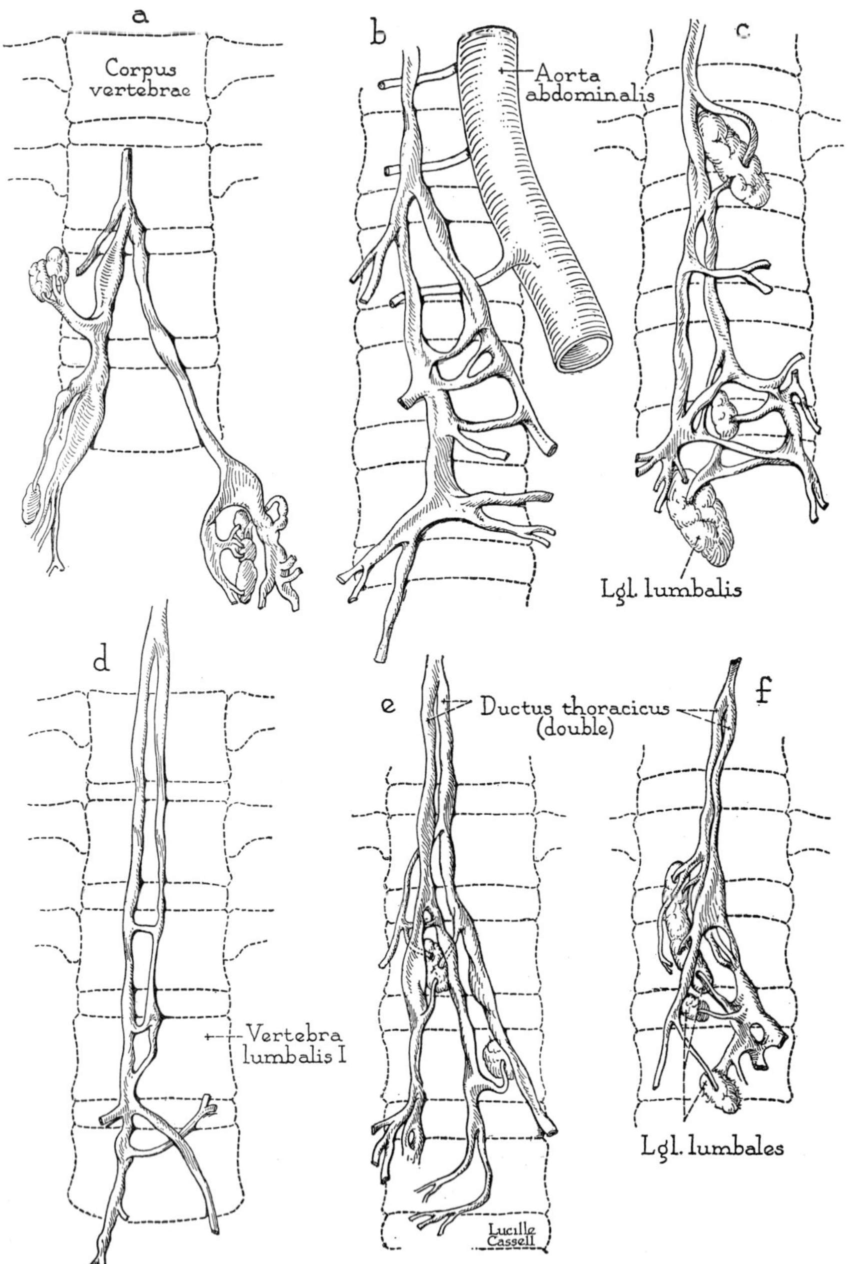

FIG. 141. Variation in pattern of thoracic duct. *a,* Duct of common type, with sacculations suggesting a cisterna. *b* and *c,* Tributaries of the thoracic duct, possessing numerous cross-anastomoses. *d,* Bilateral channels lacking sacculations. *e* and *f,* Trifid ducts.

(In a, c, e and f, prevertebral and paravertebral lymph nodes are included.) (From Anson, 1950; reproduced with the permission of the W. B. Saunders Co., Philadelphia, Pa., and Moen *et al.,* unpublished report.)

On the right side the subclavian, jugular, and mediastinal trunks usually have separate entrances into the subclavian, internal jugular, and innominate veins, respectively. But it is not unusual for the subclavian and jugular trunks to unite at the medial margin of the anterior scalene muscle, above the subclavian artery, the resulting vessel then being called the right lymphatic duct. In rare cases the right mediastinal trunk also joins the right lymphatic duct, and its internal mammary tributary sometimes enters the innominate vein separately.

C. MICROSCOPIC ANATOMY

By H. S. Mayerson

1. Structure of Lymphatic Capillaries

As indicated above (Section A.-1), the lymphatic capillaries are primarily endothelial tubes resembling blood capillaries but thinner. The medium-sized vessels (100–200 μ) have muscle fibers, whereas the larger lymphatics are composed of an endothelial layer covered by a diffuse connective tissue sheath in which elastic and muscular elements are irregularly scattered. Amyelinated nerves can be traced to the muscle fibers. Valves develop during intrauterine life in the large vessels and are always tricuspid. These structures determine the direction of flow when the vessels are compressed by muscular contraction.

2. Distribution of Lymphatic Capillaries in Various Vascular Beds[1]

Although the lymphatic capillaries spread into a tissue after the blood vessels, the density of the lymphatic plexus does not always run parallel with the richness of the blood supply. There is no invasion of lymphatics into the central nervous system, so that there are no lymphatic vessels in this tissue (Sabin, 1916). Likewise, no lymphatics have been described in bone marrow (see Section B.-5, Chapter XVIII). Voluntary muscle contains lymphatics only in fascial planes (for further

[1] For reference to lymphatics in the uterus, see Section A.-1a, Chapter XV; in the liver, see Section C.-7, Chapter XII.

discussion, see Section D.-2, Chapter XVII). It is generally believed that lymphatic capillaries do not actually reach the pulmonary alveoli, but that their distribution ceases at the beginning of the respiratory portion of the ultimate lung structure, the atrium leading into the alveolus. Likewise in the liver, the ultimate functional unit, the lobule, is not supplied with lymphatic capillaries. The fluid leaving the liver sinusoids passes through capillary endothelium and, in the lobule, lies between this endothelium and the liver cells. Lymphatic capillaries are found at the periphery of the lobule, and these carry the highly proteinized liver lymph to collecting trunks which join the thoracic and right lymph ducts. In the spleen, too, lymphatics are observed only in the capsule and the thickest trabeculae. Fluid which filters through the walls of the capillaries and sinuses must permeate the stroma before reaching the lymphatic vessels.

Lymphatic vessels in the kidney appear to begin blindly in two areas (Rawson, 1949). The first of these is near Bowman's capsule, and the second is beneath the mucosa of the papilla. Two networks of lymphatics then arise, accompanying venous and arterial blood vessels of the kidney. Those which originate in the medulla drain upward and outward toward the arcuate vessels, where they join with those beginning near Bowman's capsule and draining in the opposite direction. When the junction occurs, larger trunks then drain with the arcuate vessels toward the hilum of the kidney. They may be seen around the renal artery. There do not seem to be any demonstrable lymphatic channels in the glomeruli or about the afferent and efferent arterioles.

D. PHYSIOLOGY

By H. S. MAYERSON

1. FUNCTION

The blood capillaries are a closed system of vessels, with the blood only coming into actual contact with the cells of the tissue in the liver and perhaps in the spleen (see Section E., Chapter VI). The cells everywhere else are bathed by tissue fluid (interstitial fluid) which acts as an intermediary, supplying nutritive materials and receiving the products of metabolic activity. Phylogenetically, the need for the lymphatic system arose as soon as a closed cardiovascular system developed. In the mammal, the necessity of providing a high pressure system to insure an adequate supply of oxygen to the tissues created a situation favoring the transudation of fluid from the capillaries. The role of this mechanism was minimized by the increase in plasma proteins which exerted an oncotic pressure, but there still remained the problem of clearing the tissue spaces of substances which had leaked out or which were not absorbed by the blood.

Although, as Starling assumed, the large protein molecules do not readily pass through the capillary membrane, more recent studies (Courtice, 1951; Forker *et al.*, 1952; Courtice and Morris, 1955) indicate that in the course of a day some 50 per cent or more of the total circulating protein escapes from the blood vessels. This extravascular protein cannot be directly reabsorbed into the blood capillaries and hence must return by way of the lymphatics. The mechanisms involved are of fundamental importance and are an expression of the general function of the lymphatic system: the maintenance of a proper tissue environment. Not only do the lymphatics return protein, but they serve as the mechanism for clearing the tissue spaces of fluid, thus preventing undue increases in tissue fluid pressure. At the same time, the lymphatic return of protein and fluid is an important factor in plasma volume maintenance. Lymph transports vitamin K from the intestines to the blood stream (Mann *et al.*, 1949) and is the major, if not the exclusive, agent for the transport of absorbed long-chain fatty acids (Bloom *et al.*, 1951). There is also the suggestion that some enzymes may be transported to the blood stream via the lymphatics (Linder and Blomstrand, 1958). As part of its general function, the lymphatic system, with its nodes, is also involved in the response of the organism to infection and in the spread of disease from one part of the body to another.

2. Lymph Flow

In lower animals, particularly amphibia, the lymph flow is maintained by rhythmically beating lymph hearts. Mammals, however, do not have such structures, with the result that the flow of lymph is dependent chiefly on extrinsic mechanical factors in much the same manner as in the case of flow in veins.

a. Mechanisms involved: There is the suggestion from the literature that at least the larger lymphatics are contractile and that even the smaller ones show vasomotion under particular conditions (Smith, 1949;

Baez *et al.*, 1957). Lymphatics also are seen to be reactive to adrenalin and acetylcholine (Acevedo, 1943) and to sympathetic stimulation (Smith, 1949; Rusznyak, 1954). The role of these factors in the control of flow of lymph is still difficult to define. Likewise, although there is evidence that lymph flows by streaming in a quiescent area (McMaster, 1943), this mechanism is probably not of great importance since flow from such a location is very small. Of much more significance are the movements of muscles, particularly during activity, peristalsis, pulsating arteries or any factor which compresses lymphatic vessels. The rhythmic aspirating action of respiration is also important in lymph flow, as it is in venous flow. With every inspiration, the lacteals and abdominal part of the thoracic duct are subjected to a positive pressure and the intrathoracic part of the duct, to a negative pressure.

b. Thoracic duct: Flow through the thoracic duct has been studied more extensively than that in other lymphatics, since this vessel is larger and more accessible than others and has a high rate of flow even under resting conditions (for rate of lymph flow in different animals, see Yoffey and Courtice, 1956). Although in the experiments reported by different investigators there have been variations in the anesthetic used, the state of the animal's digestion and the duration of lymph collection, the average figures for groups of animals are of the same order, about 2 ml/kg/hr in nonruminants and somewhat more in ruminants.

In man, lymph has been collected from patients with thoracic duct fistulae and by catheterization of the thoracic duct. Crandall *et al.* (1943) noted that in their female patient of 40 kg with a traumatic thoracic duct fistula in the neck, the average control flow during 3 experimental days was 1.38 ml/kg/hr. This increased with drinking of water and eating, the highest flow of 5.8 ml/kg/hr occurring after eating a mixed meal. Courtice *et al.* (1951) found a flow of 1.0–1.6 ml/kg/hr in their case, a woman of 50 kg with a postoperative thoracic duct fistula in the neck. Bierman (1953) collected lymph by catheterization of the thoracic duct in patients with far advanced malignant disease, the volume obtained amounting to 475–2000 ml/24 hours or 0.4–1.2 ml/kg/hr.

The relatively large flow of lymph returning to the blood stream via the thoracic duct is striking and emphasizes the importance of the lymphatic system. In 24 hours it represents a turnover and return by the thoracic duct alone of a volume of fluid roughly equivalent to the plasma volume and from 50 to 100 per cent of the total circulating plasma protein.

c. Intestines: The flow of lymph from the intestines is variable, being dependent on the activities coincident with digestion and on the type of food being absorbed. The lymph itself is concerned, therefore, with the removal of the materials which have escaped from the blood stream, modified by the state of digestion at the time. Mention has already been made of its role in transporting vitamin K and long-chain fatty acids, but even in herbivora, whose diet is low in fat, intestinal lymph flow is high. It is increased during absorption of fluids and of foods, particularly fats.[2] Under normal circumstances, ingested fat, after passing through the intestinal mucosa, appears to enter the lacteals in the form of chylomicrons which consist mostly of triglycerides and some cholesterol and

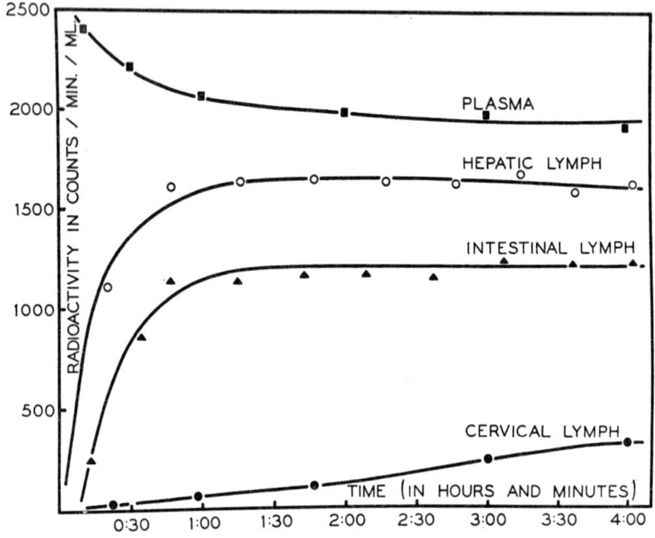

FIG. 142. Disappearance of albumin from plasma of dogs and its appearance in lymph of various areas. (From Mayerson *et al.*, 1960; reproduced with the permission of the American Journal of Physiology.)

phospholipid. The lipid, present as lipoproteins, does not alter appreciably during fat absorption. It comes mainly from the capillary filtrate and is small compared with the amount of fat in the form of chylomicrons.

d. Liver: It has been long thought that the liver furnished most of the thoracic duct lymph flow. The accumulated evidence suggests that the contribution is variable, being higher in the cat and dog than in the rat. Collection of lymph from various areas (Fig. 142) shows that, in the

[2] A considerable amount of work has been done on the absorption of fats and their uptake by intestinal lymphatics. The mechanism of their uptake is not clear. For a discussion of the evidence and different points of view, see Yoffey and Courtice (1956).

anesthetized dog, permeability of liver blood capillaries to protein is significantly greater than in the intestinal and cervical regions (Mayerson *et al.*, 1960); furthermore, there is no question but that the liver contributes a high proportion of the extravascular protein in thoracic duct lymph. Although there has been considerable discussion in the literature as to whether new protein enters the blood stream either directly or via the lymphatics, in the case of the liver the evidence indicates that only proteins which have filtered from the capillaries are found in lymph, while newly-formed protein enters directly into the permeable capillaries of the sinusoids (Morris, 1956; Wasserman and Mayerson, 1952).

The production of ascites in dogs by constriction of the vena cava results in an increased turnover of protein and, as in the case of any tissue in which venous congestion is present, lymph flow is considerably increased. In cirrhosis produced by carbon tetrachloride, lymph flow and its protein content are also increased (Nix *et al.*, 1951).

e. Lungs: The unique hemodynamic pattern in the lungs makes the role of the lymphatics a particularly important one. Pulmonary circulation is a low pressure circulation where there is a critical balance between the capillary hydrostatic pressure and the protein oncotic pressure. Increases in hydrostatic pressure which ordinarily would be relatively ineffective in increasing capillary exudation can, in the lungs, upset the usual direction of the flow of fluid into the blood stream and result in the abnormal leakage of protein. For example, Uhley *et al.* (1958) elevated the pulmonary venous pressure to 30 mm Hg by the introduction of a balloon into the left atrium of dogs and produced an increase in the flow of lymph containing a greater protein content. After maintenance of the elevated pressure for 30 minutes, critical pulmonary edema ensued. The balance may also be upset by relatively small decreases in plasma protein concentration and its consequent oncotic pressure (Paine *et al.*, 1949). Unfortunately, the study of pulmonary lymphatic drainage is complicated by the practical difficulty of obtaining lymph in pure form, since even collection from the right duct is a difficult procedure. In all types of pulmonary edema, the edema fluid contains a high concentration of protein. Experimental production of pulmonary edema by various procedures has shown that lymph flow increases considerably during the acute stage and more slowly in the recovery phase. (For more complete discussion of this topic and lymphatic flow from other areas, see Yoffey and Courtice, 1956.)

E. FORMATION OF LYMPH

By H. S. MAYERSON

1. MECHANISMS INVOLVED

Lymph is tissue fluid that has entered the lymphatic capillaries.[3] While it is clear that factors which increase the formation of tissue fluid will probably also have the same effect on lymph production, the mechanisms by which tissue fluid enters the lymphatic capillaries have not as yet been adequately demonstrated. McMaster (1947) investigated the relative pressures within the cutaneous lymphatic capillaries and in the surrounding tissues in the mouse's ear and found that the mean intralymphatic pressure was 1.2 cm of water and the interstitial pressure, 1.9 cm of water. This pressure gradient was, therefore, assumed to provide for a flow into the lymph capillary. However, while it is unquestionably of importance in lymph formation, many other factors must be concerned in the absorption of proteins and particulate matter into the lymph capillaries.

a. Permeability of lymphatic capillaries: Repeated demonstrations in the lymph of the proteins and other substances introduced intravenously, subcutaneously, or intraperitoneally have given rise to the assumption that the lymphatic capillaries are freely permeable to macromolecules (Yoffey and Courtice, 1956). Allen (1956) attempted to test the upper limit of lymphatic absorption by injecting intraperitoneally a variety of particles of various sizes up to 22.5μ in diameter. Since they all appeared in diaphragmatic lymph, he suggested that "As the diaphragm moves upward in expiration, the lymphatic plexus expands and a relative negative pressure is established in the lymphatic lumen. At the same time the triple-layered membrane which separates the peritoneal cavity from lymphatic lumen is stretched. On either side of the fenestrations of the basement membrane the peritoneal mesothelium and lymphatic endothelium open, sometimes form openings as great as 22.5μ in diameter. Through these openings suspensions are 'sucked' into the lymphatic lumen. As the diaphragm contracts, the tension on the lymphatic wall is released, the openings close, and are no longer demonstrable by the usual techniques, and compression of the plexus results in lymphatic flow." While this sequence of events may adequately explain absorption into diaphragmatic lymphatics, it is questionable whether such a view

[3] For discussion of factors concerned in the formation of tissue fluid, see Section E., Chapter VI.

is applicable to other locations. Much more data are needed to define the common mechanisms by which macromolecules enter the lymphatics.

b. Movement through lymphatic channels: Once the macromolecules enter the lymphatic vessels, they are carried back to the blood stream with minimal loss. In this regard, Patterson *et al.* (1958) injected radioactive iodinated albumin into a leg lymphatic of dogs and analyzed thoracic duct lymph and plasma. They found that less than 3 per cent of the infused albumin reached the circulation by routes other than the thoracic duct, except in unusual cases of right and left thoracic duct anastomoses. Of interest is the fact that lymph nodes do not phagocytize the serum albumin passing through them, and exchange of albumin between lymph and blood in lymph nodes is quantitatively insignificant. Further experiments with substances of various molecular weights (Mayerson *et al.*, unpublished) suggest that molecules at least as large as ±600 M.W. exchange from lymph to plasma with minimum or no hindrance, while those of ±6,000 M.W. are retained. The exact limits are still to be defined.

2. Composition[4]

a. Variations in different sites: It has long been known that there is a variable composition of lymph draining from different areas of the body, reflecting the variation in activity of the areas and differences in capillary permeabilities. The differences are chiefly in the content of macromolecules. All lymph contains proteins in lower concentration than plasma but of similar distribution, with the actual amounts varying. Thus cervical, leg, renal, and right duct lymph has from ⅓ to ½ the protein concentration of plasma, while intestinal, liver, and thoracic duct lymph contains approximately ⅔ as much protein as plasma. Liver lymph has the highest protein content, as would be expected in view of the high permeability of the capillaries of the liver sinusoids.

While the concentrations of fibrinogen and prothrombin in lymph are always less than in plasma and vary considerably in different regions, nevertheless all lymph clots. Brinkhous and Walker (1941) found that in 8 dogs the mean prothrombin level, expressed as a percentage of that in plasma, was 93.2 in liver lymph, 51.2 in thoracic duct lymph and 7.6 in leg lymph. The mean fibrinogen level was 211 mg per cent in thoracic duct lymph, compared with 410 mg per cent in the plasma.

[4] This discussion is based largely on the detailed and excellent presentation of data given by Yoffey and Courtice (1956). These authors have brought together and tabulated information available before 1956.

Mann *et al.* (1949) showed that when all of the intestinal lymph of the rat was drained externally, marked hypoprothrombinemia usually developed within 24 hours. If adequate amounts of vitamin K were administered parenterally, a normal level of prothrombin was maintained, despite loss of lymph and even of considerable amount of blood as well. Under the conditions of these experiments, it appeared that vitamin K was absorbed practically exclusively through the lymph and very little of it was stored.

Lymph also contains antibodies and enzymes which have leaked from blood capillaries. In some instances, the enzymes, such as alkaline phosphatase, are carried from their cells of origin to the blood stream via the lymphatics.

b. Ionic pattern: Data on electrolyte composition of lymph are very scanty, but the available information suggests that, for the most part, the ionic pattern of this fluid resembles that of plasma. An interesting exception is renal lymph, which has been shown to have a higher sodium and urea concentration and a lower glucose level than plasma (LeBrie and Mayerson, 1959; Kaplan *et al.*, 1943). This has been interpreted as indicating that renal lymph is a mixture of capillary filtrate and tubular reabsorbate and that tissue fluid and lymph from various parts of the kidney unquestionably vary.

c. Lipid and steroid patterns: Reference has already been made in a previous discussion of the permeability of blood capillaries to macromolecules (Section E., Chapter VI) that all the different lipoproteins present in plasma have been found in lymph and that the lipoprotein composition of lymph is different under varying conditions. In this regard, Vahouney and Treadwell (1957) infused rats with emulsions containing cholesterol, oleic acid and sodium taurocholate and studied the appearance of lipid fractions in thoracic duct lymph. The influence of added protein on the lipid levels was also determined. By itself, albumin added to administered saline was without effect on lymph flow or lipid fractions. However, in those animals receiving cholesterol, oleic acid and taurocholate, together with albumin, there was a rapid increase in total lymph lipid. Comparable changes were observed in the ester cholesterol, neutral fat and phospholipid fractions. Animals receiving no albumin showed less marked changes, except that the amount of lymph cholesterol for a 24 hour period was significantly greater than in the comparable group to which albumin was given. During cholesterol absorption, in both experimental groups the increase in total lymph cholesterol was attributable almost entirely to a greater esterified fraction, which comprised between 84 to 92 per cent of the absorbed sterol.

Further work (Vahouney and Treadwell, 1958) confirmed the finding that the absorbed cholesterol in lymph is esterified from 87 to 95 per cent and supported the concept that dietary cholesterol esters are hydrolyzed in the intestinal lumen prior to the absorption of the cholesterol moiety as free cholesterol. (For discussion of the role of cholesterol in the production of atherosclerosis, see (1) Atherosclerosis, Section B., Chapter XIX.)

References

Acevedo, D. (1943). Motor control of the thoracic duct. *Am. J. Physiol.* **139,** 600.

Allen, L. (1956). On the penetrability of the lymphatics of the diaphragm. *Anat. Record* **124,** 639.

Anson, B. J. (1950). "An Atlas of Human Anatomy," pp. 336, 337. Saunders, Philadelphia, Pennsylvania.

Arey, L. B. (1954). "Textbook of Developmental Anatomy," 6th ed. Saunders, Philadelphia, Pennsylvania.

Baez, S., Carleton, A., and Forbes, I. (1957). Mesenteric lymphatic adjustments during shock. *Federation Proc.* **16,** 5.

Bierman, H. R. (1953). The characteristics of thoracic duct lymph in man. *J. Clin. Invest.* **32,** 637.

Bloom, B., Chaikoff, I. L., and Reinhardt, W. O. (1951). Intestinal lymph as pathway of absorbed fatty acids of different chain lengths. *Am. J. Physiol.* **166,** 451.

Brinkhous, K. M., and Walker, S. A. (1941). Prothrombin and fibrinogen in lymph. *Am. J. Physiol.* **132,** 666.

Courtice, F. C. (1951). The effects of lymphatic obstruction and of posture on the absorption of protein from the peritoneal cavity. *Australian J. Exptl. Biol. Med. Sci.* **29,** 451.

Courtice, F. C., and Morris, B. (1955). The exchange of lipids between plasma and lymph of animals. *Quart. J. Exptl. Physiol.* **40,** 138.

Courtice, F. C., Simmonds, W. J., and Steinbeck, A. W. (1951). Some investigations on lymph from a thoracic duct fistula in man. *Australian J. Exptl. Biol. Med. Sci.* **29,** 201.

Crandall, L. A., Jr., Barker, S. B., and Graham, D. G. (1943). A study of the lymph flow from a patient with thoracic duct fistula. *Gastroenterology* **1,** 1040.

Forker, L. L., Chaikoff, I. L., and Reinhardt, W. O. (1952). Circulation of plasma proteins: their transport to lymph. *J. Biol. Chem.* **197,** 625.

Kaplan, A., Friedman, M., and Kruger, H. E. (1943). Observations concerning the origin of renal lymph. *Am. J. Physiol.* **138,** 553.

LeBrie, S. J., and Mayerson, H. S. (1959). Composition of renal lymph and its significance. *Proc. Soc. Exptl. Biol. Med.* **100,** 378.

Linder, E., and Blomstrand, R. (1958). Technic for collection of thoracic duct lymph of man. *Proc. Soc. Exptl. Biol. Med.* **97,** 653.

McMaster, P. D. (1943). Lymphatic participation in cutaneous phenomena. *Harvey Lectures, 1941–1942* **37,** 227–268.

McMaster, P. D. (1947). The relative pressures within cutaneous lymphatic capillaries and the tissues. *J. Exptl. Med.* **86,** 293.

Mann, J. D., Mann, F. D., and Bollman, J. L. (1949). Hypoprothrombinemia due to loss of intestinal lymph. *Am. J. Physiol.* **158,** 311.

Mascagni, P. (1787). "Vasorum lymphaticorum corporis humani. Historia et ichnographia." P. Carli, Senis, Italy.

Mayerson, H. S., Wolfram, C. G., Shirley, H. H., Jr., and Wasserman, K. (1960). Regional differences in capillary permeability. *Am. J. Physiol.* **198,** 155.

Morris, B. (1956). The exchange of protein between the plasma and the liver and intestinal lymph. *Quart. J. Exptl. Physiol.* **41,** 326.

Nix, J. T., Mann, F. C., Bollman, J. L., Grindlay, J. L., and Flock, E. V. (1951). Alterations of protein constituents of lymph by specific injury to the liver. *Am. J. Physiol.* **164,** 119.

Paine, R., Butcher, H. R., Howard, F. A., and Smith, J. R. (1949). Observations on mechanisms of edema formation in the lungs. *J. Lab. Clin. Med.* **34,** 1544.

Patterson, R. M., Ballard, C. L., Wasserman, K., and Mayerson, H. S. (1958). Lymphatic permeability to albumin. *Am. J. Physiol.* **194,** 120.

Rawson, A. J. (1949). Distribution of the lymphatics of the human kidney as shown in a case of carcinomatous premeation. *A.M.A. Arch. Pathol.* **47,** 283.

Rusznyak, I. (1954). Recent experiments on the physiology and pathology of the lymphatic circulation. *Minerva med.* **45,** 1468.

Sabin, F. (1911). A critical study of the evidence presented in several recent articles on the development of the lymphatic system. *Anat. Record* **5,** 417.

Sabin, F. (1916). The origin and development of the lymphatic system. *Johns Hopkins Hospital Repts.* **17,** 347.

Sappey, M. P. C. (1874–1875). "Anatomie, physiologie, pathologie des vaisseaux lymphatiques consideres chez l'homme et les vertebres." Delahaye, Paris.

Smith, R. O. (1949). Lymphatic contractility. A possible intrinsic mechanism of lymphatic vessels for transport of lymph. *J. Exptl. Med.* **90,** 497.

Uhley, H. N., Leeds, S. E., Sampson, J. J., and Friedman, M. (1958). *Proc. 31st Sci. Session Am. Heart Assoc.* **24,** 740.

Vahouney, G. V., and Treadwell, C. R. (1957). Changes in lipid composition of lymph during cholesterol absorption in the rat. *Am. J. Physiol.* **191,** 179.

Vahouney, G. V., and Treadwell, C. R. (1958). Absorption of cholesterol esters in the lymph-fistula rat. *Am. J. Physiol.* **195,** 516.

Wasserman, K., and Mayerson, H. S. (1952). Dynamics of lymph and plasma protein exchange. *Cardiologia* **21,** 296.

Yoffey, J. M., and Courtice, F. C. (1956). "Lymphatics, Lymph and Lymphoid Tissue." Harvard Univ. Press, Cambridge, Massachusetts.

Chapter XXIV. LYMPHATIC DISORDERS

A. INTRODUCTION

By A. Schirger and E. G. Harrison, Jr.

Disorders of the lymphatic system are not among the most common conditions encountered in medical practice. However, they often cause the afflicted individual much physical discomfort and, because of cosmetic disfigurement, much emotional suffering. At times, as in the case of cervicomediastinal cystic hygroma, they may endanger life. As with all medical rarities, timely recognition and proper treatment will be assured only by some familiarity with the entity in question. In the last few decades valuable information has been added to the previously established, classic clinical descriptions of these diseases.

This chapter, after a brief introductory discussion of lymphangiography and the classification of lymphedema, will be concerned mainly with four groups of abnormalities, namely, congenital lymphatic abnormalities, idiopathic lymphedema praecox, conditions associated with lymphatic obstruction and lymphatic tumors.

Lymph vessels are derived from mesenchymal cells and lined by a continuous layer of endothelial cells (see Section A., Chapter XXIII). They accompany major arteries and veins, and some are equipped with valves, not unlike those found in veins. (For detailed anatomy of these structures, see Sections B.-1 and C., Chapter XXIII.) Lymphatic disorders associated with obstruction are most commonly encountered in the extremities and, therefore, it is well to recall that both superficial and deep channels may be involved.

1. Lymphangiography

The introduction of a practical method of lymphangiography by Kinmonth *et al.* (1955a,b) greatly aided the study and understanding of congenital and other types of lymphedema. While various methods of visualizing the lymphatic system have been known to anatomists for years and attempted by clinicians from time to time, only in recent years through the wider application of the afore-mentioned procedure and of its modification has more extensive information about the lymphatics *in vivo* accumulated. Kinmonth *et al.* (1955a,b) suggested a two-step procedure: The most distal lymph channels of the extremity are made visible to the naked eye. A subsequent injection of a radiopaque solution makes radiographic visualization of the entire lymphatic tree of the extremity possible. After sterile preparation of the skin, 2.5 ml of patent diffusible blue dye is injected subcutaneously between the toes, approximately 0.5 ml into each web. Gentle massage aids diffusion of the dye into the tissues, and passive motion of the extremity for about 5 minutes helps in its distribution throughout the lymph channels of the lower extremity. The dye, eventually absorbed into the blood stream, gives the patient a blue appearance for about 24 hours. An incision is made across the dorsum of the foot, 5 cm proximal to the web of the toes, and the subcutaneous lymph channels are made visible by the patent blue dye. A radiopaque substance is injected through a needle inserted into one of the lymph vessels, and roentgenograms of the entire limb are taken as quickly as possible. The slow passage of lymph through the trunks of the extremity makes speed less imperative than in arteriography; however, diffusion of the contrast medium into the surrounding tissues in time may produce a hazy appearance of the vessels. The amount of radiopaque medium required for good visualization will be greater when large varicose lymph channels are present.

Since the original description, other investigators gained experience with lymphangiography (Kaindl, 1957a,b; Kaindl *et al.*, 1958; Gergely, 1958a,b; Arnulf, 1957, 1958; Málek *et al.*, 1956, 1959a,b; Marcozzi and Ricci, 1956; Colette, 1957; Tasatti, 1957; Leenhardt *et al.*, 1956), and attempts have been made at lymphadenography (Prokopec and Kolihová, 1958; Shanbrom and Zheutlin, 1959). Various modifications of the technique have been attempted. Laurentaci and Montinari (1956) found that preliminary infiltration with a local anesthetic did not affect lymphangiography in experimental animals. A method of visualizing the deep lymphatic system of the thigh through injection of the contrast medium into the posterior superficial lymphatics of the calf by means of an incision along the inferior aspect of the lateral malleolus has been suggested by workers of Málek's group (1959b). An extension of this method has been visualization of the lymphatics draining the testes by injecting into the spermatic cord (Málek *et al.*, 1956). The present status, indications and limitations of lymphangiography have been discussed in two excellent, recent reviews (Fuchs *et al.*, 1959; Málek and Belán, 1959).

2. Classification of Lymphedema (Figs. 143 to 154)

Much of the information concerning classification of lymphedema has been based on the work of Allen (1934), as shown in Table XX. Kinmonth *et al.* (1957) more recently has offered a modification of this classification.

TABLE XX
Classification of 300 Cases of Lymphedema[a]

	Per cent of cases
Noninflammatory lymphedema	
Primary	
1. Congenital lymphedema	
a. Milroy's disease	0
b. Simple	4
2. Lymphedema praecox	31
Secondary	
1. Malignant occlusion	11
2. Surgical removal of lymph nodes	20
3. Pressure	Less than 1
4. Radiation therapy	1
Inflammatory lymphedema	
Primary	14
Secondary	
Venous insufficiency	4
Trichophytosis	2
Systemic disease	2
Fibrosis	Less than 1
Local tissue injury or inflammation	11

[a] Modified from Allen *et al.*, 1955.

B. CONGENITAL LYMPHATIC ABNORMALITIES

1. Types

Three types of congenital lymphatic abnormalities affecting lymph flow may be encountered in human beings: congenital lymphedema, congenital spontaneous chylothorax, and idiopathic chylous ascites. The etiologic aspects of idiopathic chylous ascites and spontaneous chylothorax remain obscure. The former entity is characterized by the appearance of ascites early in life (Stirlacci, 1955); the latter is ushered in by dyspnea accompanying an often massive chylous pleural effusion which later may disappear (Kessel, 1952). Both conditions are mentioned here mainly because of their rare concomitant occurrence with congenital lymphedema (McKendry *et al.*, 1957; Warwick *et al.*, 1959).

2. Congenital Lymphedema

Congenital lymphedema is manifested by swelling of one or more extremities or another part of the body. On examination the characteristic finding is diffuse swelling of the affected portion of the body, commonly an extremity. Although usually present at birth, it may not become

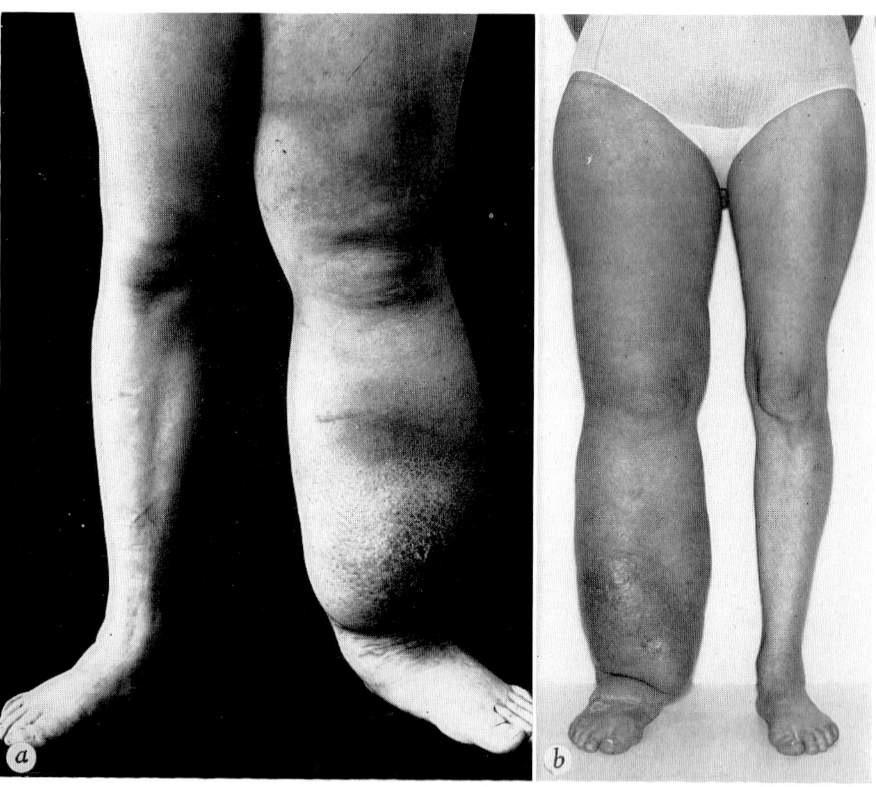

Fig. 143*a*. Congenital lymphedema in a woman, aged 34 years, with involvement of left lower extremity since birth, gradually increasing in size. Recurrent attacks of lymphangitis and cellulitis. (For histologic appearance, see Figure 144*a*.)

b. Lymphedema praecox in a woman, aged 42 years. Lymphedema of right leg followed severe contusion of right hip from a skating accident at age 33 years. Condition aggravated by pregnancy. Two attacks of diffuse lymphangitis. (For microscopic section see Figure 144*b*.)

noticeable to the parents until the child is several months old. The swelling is firm and initially will subside with elevation of the extremity. There is little discomfort, and secondary inflammation and ulceration are not common at first (Figs. 143*a*, 144*a* and 145*a*).

a. Etiology: The cause is unknown; however, considerable evidence indicates that hereditary and genetic factors play an important role in the production of this condition. Of interest in this regard is the recent

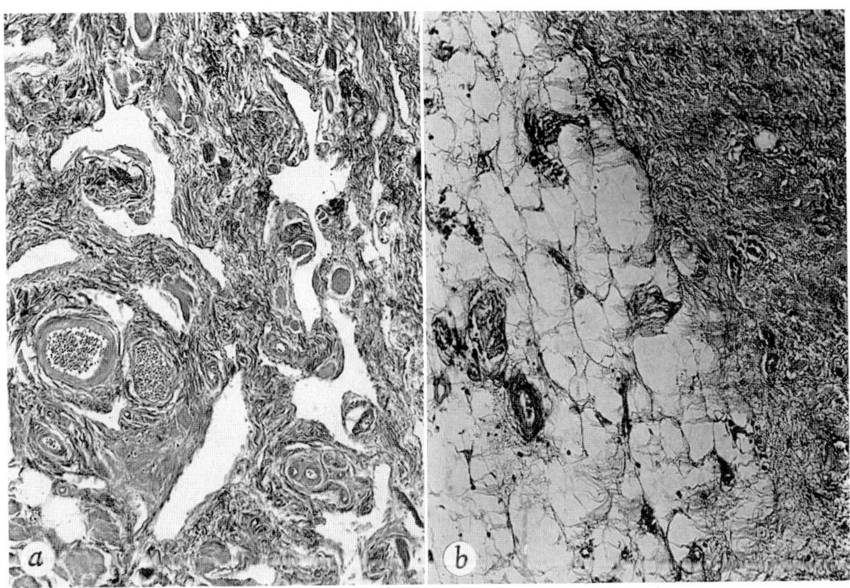

FIG. 144*a*. Same case as Figure 143*a*. Congenital lymphedema. Numerous dilated lymphatics (lymphangiectases) are seen in the thickened subcutaneous tissues (H and E; reduced from ×75).

b. Same case as Figure 143*b*. Lymphedema praecox. Edematous fibro-adipose tissue in subcutis with few lymphatics. Note fibrillar appearance of connective tissue (H and E; reduced from ×100).

description of congenital lymphedema in twin Ayrshire Australian calves by Morris *et al.* (1954). In man, congenital lymphedema may occur in an isolated form, affecting only one or a number of members of a family, often in several generations. The hereditofamilial form is known as Milroy's disease (1892, 1928) or Nonne-Milroy-Meige's disease, because of the description by Nonne (1891) and Meige (1898). A congenital and a late form are sometimes distinguished, the eponyms of Milroy-Nonne and Meige being affixed, respectively.

The condition is said to be transmitted as an autosomal dominant, with occasional incomplete penetration, accounting for a skipped generation (Verger and Blaquière, 1957). Transmission through linkage to the sexual chromosome has been postulated, and congenital lymphedema in the issue of a consanguineous marriage suggested a recessive mode of hereditary transmission (Klein and Doret, 1957). Thus, genetic factors seem important. That other mechanisms are operative is suggested by the fact that, as in other forms of lymphedema, females are affected more often than males. In the *forme tarde* of the hereditofamilial variety of

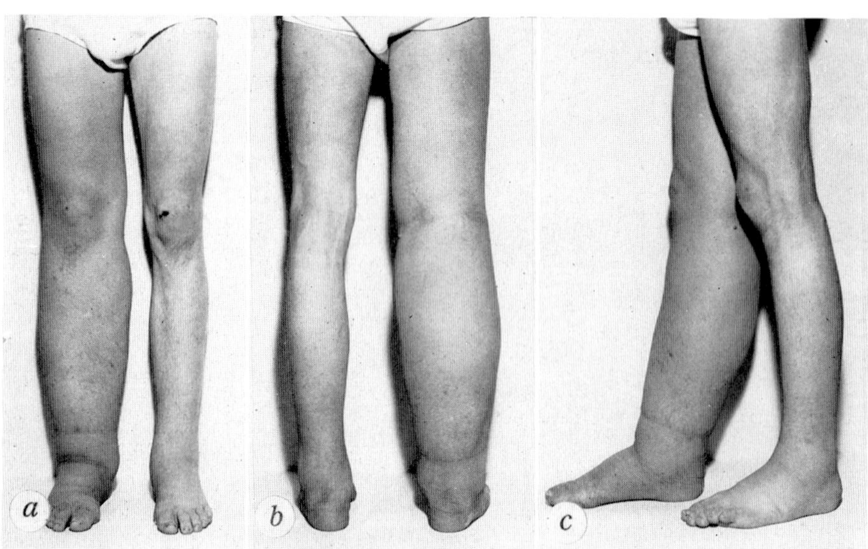

Fig. 145. Congenital lymphedema in a boy, 14 years old. Condition present in right leg since birth, with maximal involvement below knee but with some swelling also extending above it. Note mild swelling around left ankle and squaring of toes. During past 9 years, severe recurrent acute lymphangitis and cellulitis.

the disease, the symptoms, as in lymphedema praecox, often will appear during the second decade of life, pointing perhaps to a common precipitating factor rendering an underdeveloped lymphatic system overtly insufficient. It is likewise of interest that the *forme tarde* of the hereditofamilial variety may be associated with other developmental abnormalities, such as polydactyly, syndactyly and palatine fissures. While the familial occurrence is a *sine qua non* for the diagnosis of Milroy's disease, the occurrence in six generations of one family reported by Jennett (1956) is of interest.

Notwithstanding these observations, in the majority of patients with congenital lymphedema a positive family history cannot be elicited. In this regard, during a 5-year period 18 patients with congenital lymphedema were encountered by Schirger and Harrison (unpublished observations), and in none was the disease classified as hereditofamilial in nature. Thus, although genetic, hereditary, and familial influences are undoubtedly important in the production of congenital lymphedema, other poorly-understood factors likewise play an important role.

b. Pathology: Kinmonth *et al.* (1957) made lymphangiographic studies in 87 of 107 patients with primary lymphedema of the lower extremities, twelve patients in the entire group having congenital lymphedema. These authors concluded that in the latter condition there is an error in development of the lymphatic system, varying from aplasia to varicosity of the lymphatic trunks; since the draining pathways are malformed, fluid accumulates in the tissues and lymphedema results.

In a study of gross tissues in congenital lymphedema a characteristic finding described by Mason and Allen (1935) is a spongelike mass, seen particularly in the deeper subcutaneous fat, from which a watery albuminous fluid may be expressed. In some cases the spongelike character may not be apparent, although the thickened fibrous septa in the subcutaneous fat and the deep fascia may contain dilated lymphatics. Gross dermal thickening and epidermal roughening resembling "pigskin" may be observed; yet the bulk of the increased mass is in the subcutaneous tissue. Here fibrous septa may demarcate sharply rather pale, fat lobules, which bulge from the cut surfaces (Fig. 146*a*).

Numerous dilated endothelial-lined spaces in tissue affected by congenital lymphedema may be apparent microscopically; these are more evident along subcutaneous septa and near the deep fascia (Figs. 144*a* and 146*b*, *c* and *d*), and have been described as congenital lymphangiectasis and as varicose lymph trunks (Mason and Allen, 1935). When such dilated lymphatics are abundant, the basic pathologic features may be indistinguishable from those of diffuse congenital lymphangioma. In any event, a congenital background is evident and the malformation of the lymph vessels is associated with marked lymphedema. In other cases few superficial lymphatics may be present, and dilated lymph vessels may be seen only in the subcutaneous tissue or near the deep fascia.

Fibroblastic proliferation may occur in the course of chronic lymphedema of any cause. The fibrosis is often marked in the dermis, subcutaneous septa, and deep fascia. Dermal elastic fibers, although normal in amount, appear frayed and fragmented, probably due to tissue stretch from lymphedema. Separation of collagen fibers due to the excessive interstitial fluid is also seen readily. Focal deposition of lymphocytes and plasma cells may collar some lymph vessels, particularly when there

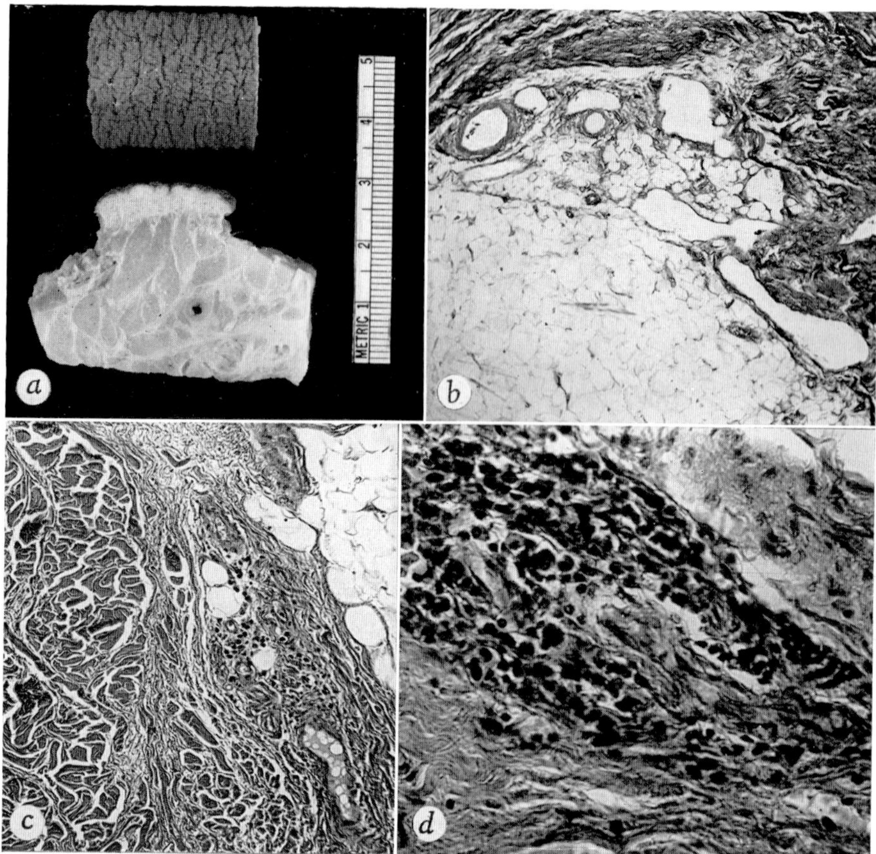

FIG. 146. Same case as Figure 145. Congenital lymphedema. *a.* Thickened, edematous skin and subcutaneous tissue in transection (below) which shows fibrosis in dermis, subcutaneous septa and deep fascia (lower right). Surface of skin (above) has a "pigskin" appearance. *b.* Dilated lymph vessels in subcutaneous tissue near junction with dermis (H and E; reduced from ×45). *c.* Dense fibrous tissue of subcutaneous trabecula. Perilymphatic lymphocytic infiltration at lower right (hematoxylin and eosin; reduced from ×100). *d.* Perilymphatic, lymphocytic and plasma-cell infiltration almost obliterating lumen of lymph vessel (center), apparently from recurrent lymphangitis (H and E; reduced from ×375).

is an associated recurrent lymphangitis. The walls of arteries and veins often appear slightly thickened and edematous but are otherwise not remarkable.

Either atrophic or hyperplastic changes may be found in the epidermis. There is often slight increase in basal pigmentation and the number of dermal appendages is decreased.

C. LYMPHEDEMA PRAECOX

Lymphedema praecox, originally described by Allen (1934), is a form of idiopathic lymphedema which commonly makes its appearance during the second or third decade of life, is more common in women than in men, and initially involves one lower extremity only but may slowly progress to affect both. Kinmonth *et al.* (1957) called attention to the fact that a *forme tarde* of idiopathic lymphedema also exists, and they put the dividing line at age 35 years.

1. Clinical Features

The patient, most commonly a woman in her early twenties or late teens, will become aware of slight puffiness about the ankle or in the foot. This may appear abruptly or insidiously. Initially, at least, it will subside during the night, only to become noticeable the next day. The swelling often will be worse during the hot weather and, in some patients, during menstrual periods. A small number of patients will volunteer the information that they first became aware of the swelling during pregnancy. The condition progresses slowly, and, in some patients, rapidly, over the next few months to years. The entire limb will become edematous (Fig. 143*b*), or the portion of the leg below the knee will be the only part affected. With time, the edema, pitting at first, will become resistant to gentle pressure and the overlying skin becomes roughened. The permanently enlarged extremity is a cause of discomfort to the patient and is also a great esthetic handicap. The incidence of acute lymphangitis and cellulitis in lymphedema praecox is given by Allen (1934) as 13 per cent.

2. Etiology and Pathogenesis

The etiology of lymphedema praecox is poorly understood. A preponderance of women is affected, and this fact, as well as the frequent appearance of the swelling during early adolescence and aggravation by menses, suggests a relationship to reproductive organs. Allen and Ghormley (1935) postulated that the female genital organs impose an additional load on the lymphatic structures, rendering them functionally incompetent or possibly facilitating the entrance of infection. In a series of 124 patients with lymphedema praecox, 13 gave a history of preceding minor trauma, often seemingly unrelated (Schirger and Harrison, unpublished observations). Kinmonth *et al.* (1957) also found a history of trauma in 17 of 107 patients with idiopathic lymphedema of the legs. The initial lymph stasis interferes with unhampered passage of tissue fluid into the vessels and serves as a stimulus for fibroblastic proliferation. The possible etiologic importance of developmental factors in congenital lymphedema has been mentioned, and it is probable that a similar factor may be important in the production of lymphedema praecox. Perhaps the malformed lymph system can meet normal physiologic demands, but stress periods, such as pregnancy, trauma or infection, may precipitate subtly obstructive changes in the regional lymph nodes or vessels themselves, with the result that overt edema makes its appearance.

Lymphangiographic studies performed in recent years have revealed hypoplastic lymphatic systems in many instances. Such underdeveloped lymph channels are unable to provide adequate drainage of lymph from the distal tissues, and lymph stasis ensues. This initially at least is proportional to gravitational factors, as it is more pronounced after prolonged periods of standing, walking, and sitting and diminishes or subsides after a period of rest in the horizontal position. As the disease progresses, nocturnal periods of rest no longer suffice to drain the excessive fluid accumulated in the dependent parts, and permanent lymphatic congestion results unless the periods of horizontal rest are prolonged. Eventually in some patients no amount of elevation will return the extremity to normal size. Chronic lymph stasis has produced a fibroblastic reaction, and tissue fibrosis poses a further hindrance to the free flow of lymph.

3. Pathology

Although congenital underdevelopment of the lymphatic system was suspected in lymphedema praecox (Allen *et al.*, 1955), the state of the lymphatics in lymphedema praecox remained obscure until the advent of lymphangiography. Kinmonth *et al.* (1957) demonstrated hypoplasia (deficiency in size or number) of lymphatic trunks in 55 per cent of 87 cases of primary lymphedema which they studied, twelve of which were congenital. Varicose lymph trunks were found in 24 per cent of the cases and dermal backflow (abnormal retrograde filling of dermal lymph channels through deep lymph channels with incompetent valves), in 22 per cent. Kinmonth *et al.* (1957) suggested that the difference between congenital lymphedema and lymphedema of delayed onset may be one of degree of lymphatic deformity.

Taylor *et al.* (1958) studied the protein content of the edematous fluid in patients with primary lymphedema and concluded that lymphatic insufficiency results in accumulation of protein molecules in the tissue spaces, with retention of water and production of lymphedema. They found no evidence of altered capillary permeability in uncomplicated lymphedema. Studies with radioactive plasma protein revealed a prolonged appearance time of the protein in the blood in patients with lymphedema as compared with normal persons (Taylor *et al.*, 1957), indicating a slower rate of removal of the labeled protein.

Gross inspection demonstrates that the edema and thickening of the tissue are due predominantly to increased thickness of the subcutaneous layer. Lymphangiectasia is never as extensive, nor is the tissue as spongy, as it is in some cases of congenital lymphedema. Gross changes may be otherwise similar to those in some cases of congenital lymphedema.

On microscopic examination fibrous tissue proliferation is usually extensive in the dermis, subcutaneous septa, and deep fascia (Fig. 144*b*). The collagenous fibers may be separated by edema fluid, and elastic fibers are fragmented. Superficial lymph vessels are usually few in number within the fibrous tissue, although a few dilated channels may be seen in the deep subcutaneous layer. Slight perilymphatic, lymphocytic, and plasma-cell infiltrations are readily observed. Arteries and veins are patent and do not show specific changes.

D. OBSTRUCTIVE LYMPHEDEMA

Obstructive lymphedema is secondary to some known disease process that causes obstruction in lymph flow.

1. Clinical Picture

This type of lymphedema is marked usually by rather sudden onset in the later years of life and often is characterized by rapid progression. The sudden appearance of lymphedema in an otherwise well person should always raise the suspicion of an underlying malignant lesion (Figs. 147*a* and 148*a*). Carcinoma of the prostate in men is particularly culpable. The swelling is usually marked, and the cutaneous changes seen in lymphedema of long standing may be absent.

2. Etiology and Pathogenesis

The causes of obstructive lymphedema are manifold. Metastatic lesions involving lymph vessels and lymph nodes are the commonest cause. Chylous ascites also often results from obstruction of the lymph channels by a retroperitoneal neoplasm, particularly a malignant lymphoma. Surgical removal of lymph nodes and deep lymph channels may eventuate in obstructive lymphedema. This is commonly seen after radical mastectomy, a form of lymphedema originally described by Halsted (1921). Obstructive lymphedema after radiation therapy

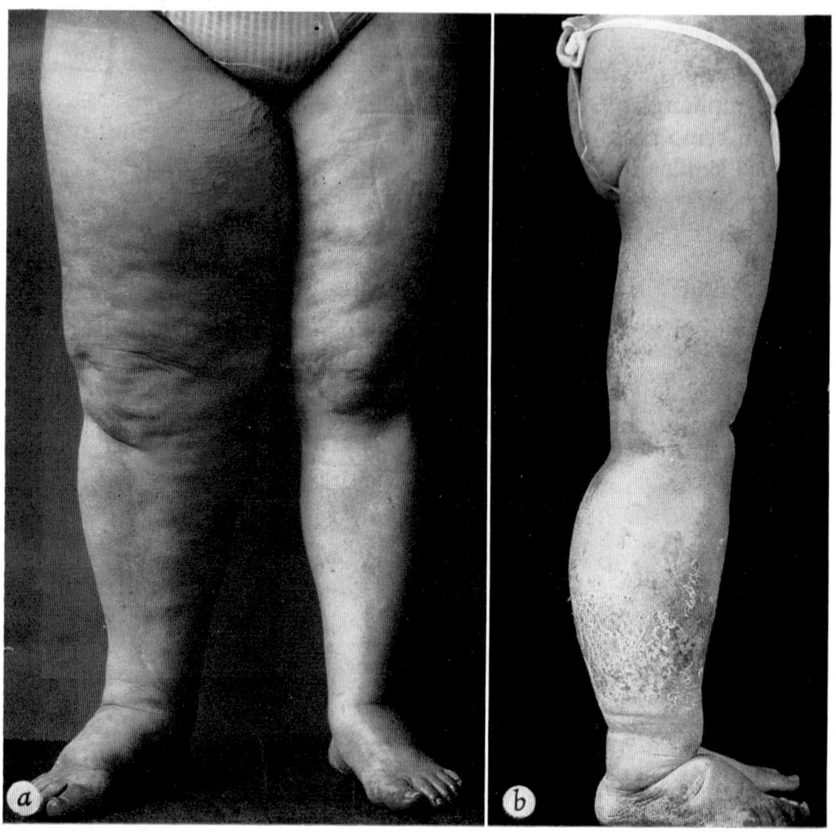

FIG. 147*a*. Obstructive lymphedema in a woman, aged 44 years, with obstructive lymphedema of right lower extremity which developed 9 months after hysterectomy and roentgen therapy for carcinoma of the uterus.

b. Primary inflammatory lymphedema in the right leg of a man, aged 61 years, due to many recurrent episodes of lymphangitis since age 34 years.

often poses diagnostic problems as to whether the lymphedema heralds recurrence of the original tumor or metastasis or whether it is due to scarring secondary to irradiation.

An unusual cause of obstructive lymphedema is so-called sclerosing retroperitonitis (Chisholm *et al.*, 1954). However, that swelling of the extremities is not common in this condition is apparent from the lack of this finding in a recent report of five cases (Hawk and Hazard, 1959).

The mode of production of this type of lymphedema is best explained through interruption of normal pathways of lymph flow by a malignant process, by disruption of the lymph channels through surgical procedure, or by fibrous tissue scarring and obliteration, as seen in

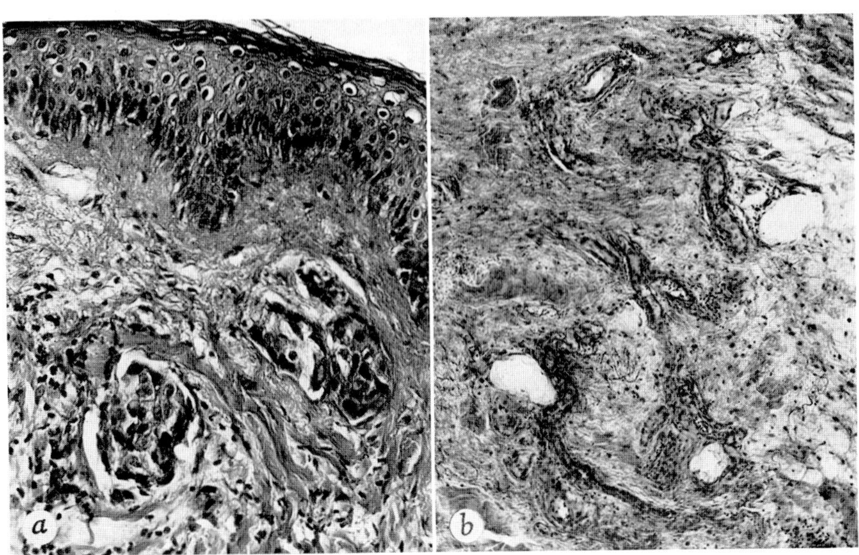

FIG. 148*a*. Skin from breast of a woman, aged 58 years. Retrograde infiltration of dermal lymph vessels by adenocarcinoma of breast usually signifies extensive axillary and mediastinal nodal involvement. Enlarged appearance of the breast, when present, is in part due to the resultant obstructive lymphedema (H and E; reduced from ×300).

b. Same case as Figure 147*b*. Primary inflammatory lymphedema. Moderate perilymphatic, lymphocytic and plasmacytic infiltration in subcutaneous tissue (H and E; reduced from ×75).

postirradiation lymphedema. A special type of obstructive lymphedema follows trauma, particularly fractures, in the region of the thigh.

3. Postmastectomy Lymphedema

Postmastectomy lymphedema is frequent enough to deserve special emphasis. Three types of swelling of the arm are known to occur after radical mastectomy (Stillwell, 1959). In some patients acute transient edema appears during the immediate postoperative period and is related directly to the surgical procedure. In others acute painful edema of the arm may be noted during the third or fourth week after operation, which disappears promptly on administration of adrenal steroids or phenylbutazone. In about 50 per cent of the patients some degree of painless swelling of the arm occurs weeks, months, or years after the operation, and in about 15 per cent it is a significant complication. Treves (1951) found that a swollen arm occurred in 41 per cent of 768 patients after radical mastectomy. He stated that a combination of several factors may operate in any given case of postmastectomy lymphedema, particularly extirpation or removal of the deep lymphatic vessels and nodes.

Other factors contributing to the production of this syndrome are undoubtedly subsequent obliteration of lymph channels by infection or radiation therapy, in addition to angulation or reflex spasm of the axillary vein and obesity. Britton (1959) stated that infection is the most important factor in production of postmastectomy lymphedema. If radiation therapy follows the surgical procedure, chance of postmastectomy lymphedema is enhanced. Thus, Haberlin *et al.* (1959), using an increase in circumference of the limb of 2 cm or more as a criterion, found lymphedema present in the arm in 38 per cent and in the forearm, in 32 per cent of 100 patients with carcinoma of the breast treated by radical mastectomy and postoperative irradiation.

Smedal and Evans (1960) have recently suggested, on the basis of clinical observations and venograms, that thrombophlebitis with secondary obstruction of the lymphatics in the vascular sheaths is the principal cause of postmastectomy lymphedema.

4. Pathology

Butcher and Hoover (1955) found that the superficial cutaneous lymph vessels were dilated and possessed extensive flow patterns in seven instances of "soft" pitting lymphedema, in three of which there was neoplastic involvement of regional lymph nodes. Vítek *et al.* (1958), studying two patients with post-traumatic lymphedema of the lower extremities by means of lymphangiography, found interruption of the lymph channels in both of these patients at the site of the fracture. This type of lymphedema is often not permanent, and the transient nature can best be explained by regeneration of the lymph channels, a phenomenon for which there is experimental support (De Lutio *et al.*, 1954; Málek and Kolc, 1959).

A few lymphangiographic studies are available in obstructive lymphedema. Vítek and members of his group (1954) found only slight dilatation of lymph channels in the severe degree and none in the milder type. They did not observe the so-called dermal reflux mentioned by others in this type of lymphedema. An interesting observation was made in the case of localized lymphedema due to contracture of the knee joint in a patient with rheumatoid arthritis (Csik *et al.*, 1958). The protein content of tissue fluid from the lymphedematous area was similar to that of the serum. Biopsy of lymphedematous tissue revealed fibrinoid, precipitation of calcium and severely depolymerized mucopolysaccharides. Further light may be shed on the etiology and pathogenesis of obstructive lymphedema by introduction of a method for detection of involvement of inguinopelvic nodes by lymphadenography (Colette, 1958).

The morphologic findings in noninflammatory obstructive lymphedema depend, of course, on the basic disease. Diffuse carcinomatous involvement of lymph nodes may produce obstruction to lymph flow, resulting in dilatation of lymphatic sinusoids and lymph vessels and, in turn, in lymphedema. For example, in some patients with carcinoma of the breast, there may be massive involvement of axillary and mediastinal nodes and retrograde metastasis even into cutaneous lymphatics (Fig. 148*a*). Swelling of the breast, if present, may be due not only to diffuse carcinoma, but also to obstructive lymphedema. Inguinopelvic or retroperitoneal neoplasms may cause lymphedema of one or both of the lower extremities. Radiation therapy at times will induce lymphedema through marked fibrosis and obliteration of lymph nodes and remaining lymph vessels, possibly already compromised by metastasis or surgical procedures. The lymphatics rarely are obliterated in marked retroperitoneal fibrosis of unknown cause (sclerosing retroperitonitis), to the extent that lymphedema results, although ureteral obstruction may be present.

E. INFLAMMATORY LYMPHEDEMA

In inflammatory lymphedema the inflammatory components play a prominent role in the production of symptoms and in the progression of the disease.

1. Clinical Picture

The hallmark of this variety of lymphedema is the acute episode of lymphangitis and cellulitis. The patient becomes ill suddenly and abruptly, the temperature becomes high, and shaking chills often follow. The affected extremity is hot, tender, and swollen, and the regional lymph nodes are tender. Symptoms usually subside promptly on admin-

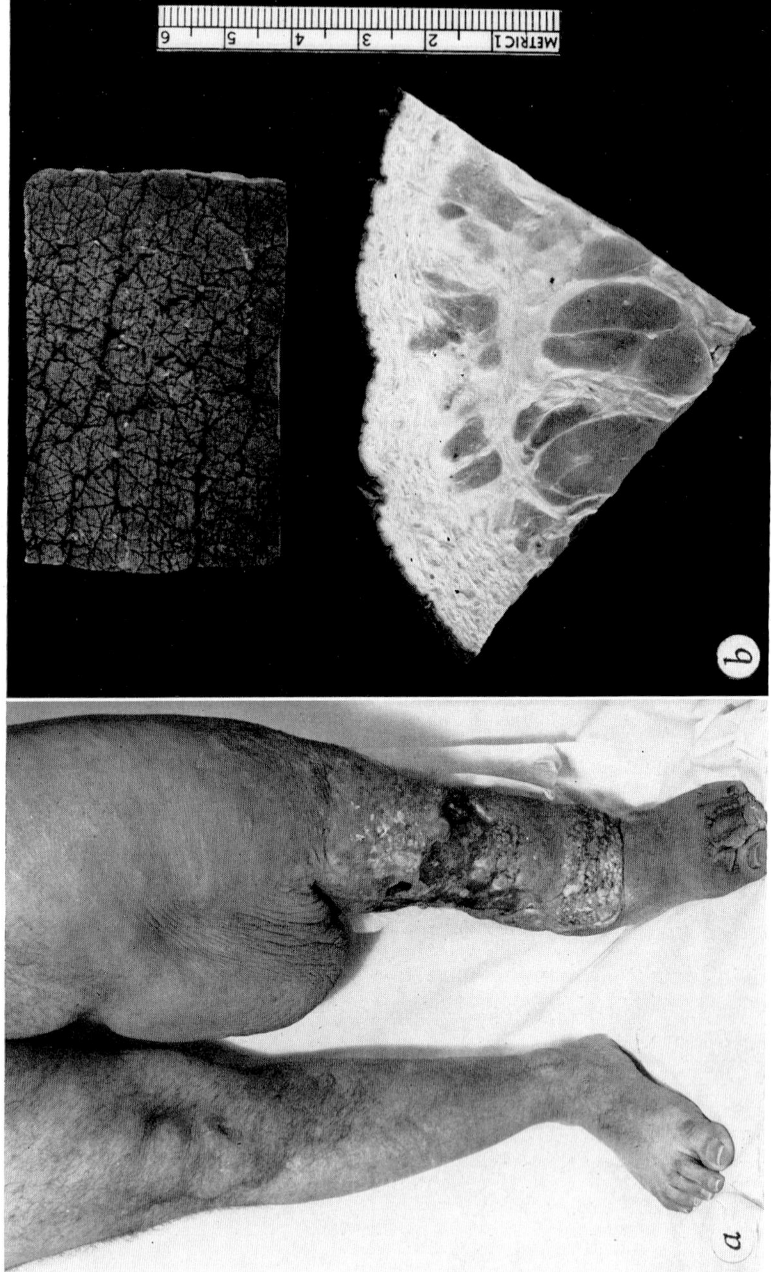
METRIC 1 2 3 4 5 6
b
a

istration of antibiotics, but the affected extremity may remain swollen for a time (Figs. 147*b* and 149).

2. Etiology and Pathogenesis

Inflammatory lymphedema is associated with recurrent cellulitis or lymphangitis of either known or unknown cause. The acute inflammatory episodes occur spontaneously in some patients and seem to be related to trichophytosis in others. It is often difficult to determine whether the swelling of the limb preceded or followed a first attack of lymphangitis and cellulitis and whether the swollen extremity presents a favorable environment for initiation of recurrent attacks of cellulitis and lymphangitis (Figs. 149 and 150). Trichophytes have been suspected as noxious agents in some patients either because of the possibility of direct invasion of the lymph channels or because their presence facilitated the entrance of secondary bacterial infection. Of the bacteria, streptococci seem to be the chief invaders even though Kinmonth *et al.* (1957) reported isolation of a staphylococcus in one of their patients with lymphedema. Young and DeWolfe (1960) have reported on 25 consecutive cases of recurrent lymphangitis associated with dermatophytosis. Fourteen of the patients had lymphedema. These authors pointed out that each episode of lymphangitis leads to thrombosis of lymph vessels that results in lymphedema which further disposes to infection. In filariasis, also, even though there is obstructive lymphedema from the adult worm associated with the reactive fibrosis in lymph nodes, the secondary bacterial infection has been suggested as responsible for the late manifestations of elephantiasis and recurrent lymphangitis and cellulitis so often seen in this disease. Each inflammatory episode leaves the patient with a

---

FIG. 149*a*. Complicated lymphedema (type?) in a man, aged 37 years, with a history of recurrent cellulitis and lymphangitis of the left lower extremity since age 4 years. Lymphedema was discovered at age 8 years and a Kondoleon operation was performed at age 13. Swelling increased at age 22 following open reduction and fixation of fractured left femur. Spontaneous ulceration in operative scar of left leg since age 35, with persistent drainage of lymph. This case illustrates the difficulty in classification of some cases of lymphedema. Although lymphedema began early in life, it followed recurrent lymphangitis and cellulitis and was later accentuated by trauma. The early onset suggests a basic developmental defect in the lymphatic system and that lymphedema may follow various precipitating factors.

b. Marked thickening of lymphedematous skin and subcutaneous tissue seen in wedge section (below). Extensive fibrous thickening of dermis with broad septal bands which extend into thickened deep fascia and outline the fat lobules. "Pigskin" appearance of surface (above).

somewhat greater degree of swelling, thus predisposing him to further inflammatory episodes.

Ochsner *et al.* (1940) stated that the increase in tissue fluid predisposes to bacterial infection, which, in turn, may be precipitated by previous trauma. Patients frequently will state that in the days or weeks preceding the acute episode they suffered a minor, seemingly unrelated injury. In the light of Hudack and MacMaster's observation (1933) that

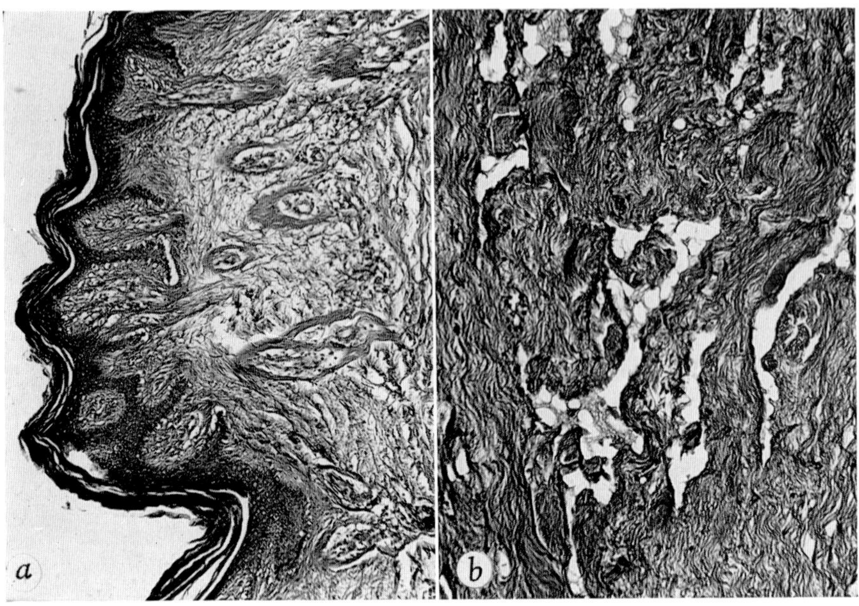

FIG. 150. Same case as Figure 149. *a*. Undulating atrophic epidermis and edema of dermis imparting a fibrillar appearance to the connective tissue. A few small lymph vessels and slight lymphocytic infiltration are present (H and E; reduced from ×90). *b*. A few dilated lymph vessels noted in subcutaneous tissue near deep fascia (H and E; reduced from ×100).

even slight injuries to the skin can tear open the intradermal lymph vessels, it is easy to understand how infection can gain entrance into the tissue. However, the abrupt onset of the symptoms still needs elucidation. Some interesting light has recently been shed on the role of lymphatics and infection. During the acute attack of lymphangitis, the small lymph channels become occluded by thrombosis and stasis of lymph. This, in turn, stimulates the fibroblastic reaction which further impedes the flow of lymph. Miles and Miles (1958) concluded on the basis of their studies in experimental animals that lymphatic blockade is not an effective defense mechanism against the spread of infection.

3. Pathology

Butcher and Hoover (1955) found soft lymphedema adjacent to areas previously afflicted by erysipelatous infection in four cases. No superficial cutaneous lymph vessels could be injected in the hard, non-pitting edema of the lower extremities of eight patients, six of whom had had many episodes of localized erysipelatous infection.

The amount of collagenous fibrotic thickening and chronic inflammation in the dermis increased as the number and extent of injectible dermal lymph vessels decreased. Butcher and Hoover also stated that obliteration of the cutis lymphatics is important in the genesis and chronicity of stasis ulceration.

Battezzati *et al.* (1959) described tortuosity and irregularity in caliber in two cases of postphlebitic syndrome and also described the appearance of tortuous lymph channels in a case of thrombosis of the popliteal vein. Jacobsson and Johansson (1959) studying 44 limbs of 25 patients with varicose veins found abnormalities in the lymphangiograms of 8 of 11 patients who had varices complicated by ulceration. In contrast, lymphangiograms of patients recently recovered from thrombophlebitis were normal. In patients with swelling of the leg accompanying venous thrombosis, Kaindl (1957b) observed dilated lymph vessels and concluded that the lymph vessels also were involved in the pathologic process. Sabaino *et al.* (1955), using lymphangiography, demonstrated slowing of lymph flow in lymphangitis and in cellulitis which was said to be due to slow passage of lymph through the diseased lymph nodes.

Inflammatory lymphedema may be pathologically almost indistinguishable from the late stages of lymphedema praecox complicated by lymphangitis. The thickened watery subcutaneous layer with densely thickened, fibrous septa and deep fascia is similarly readily seen in the gross surgical specimen.

Microscopic evidence of recurrent lymphangitis may be found in the numerous lymphocytes and plasma cells cuffing lymph channels (Fig. 148*b*) which may show obliterative changes, and marked fibrosis, like that often seen in lymphedema of various types, is present.

F. DIFFERENTIAL DIAGNOSIS

In the differential diagnosis of lymphedema, the common orthostatic edema and the likewise common lipedema must be considered. These conditions are readily distinguished from lymphedema by the characteristic softness of the tissue, by the relative mildness of the swelling, by the consistent decrease of the edema with rest, and by the increase in hot weather and on prolonged standing. Both lipedema and chronic orthostatic edema usually are seen in obese middle-aged women. Congenital or acquired arteriovenous fistulas are readily differentiated from lymphedema by the increase in length and girth of the affected extremity,

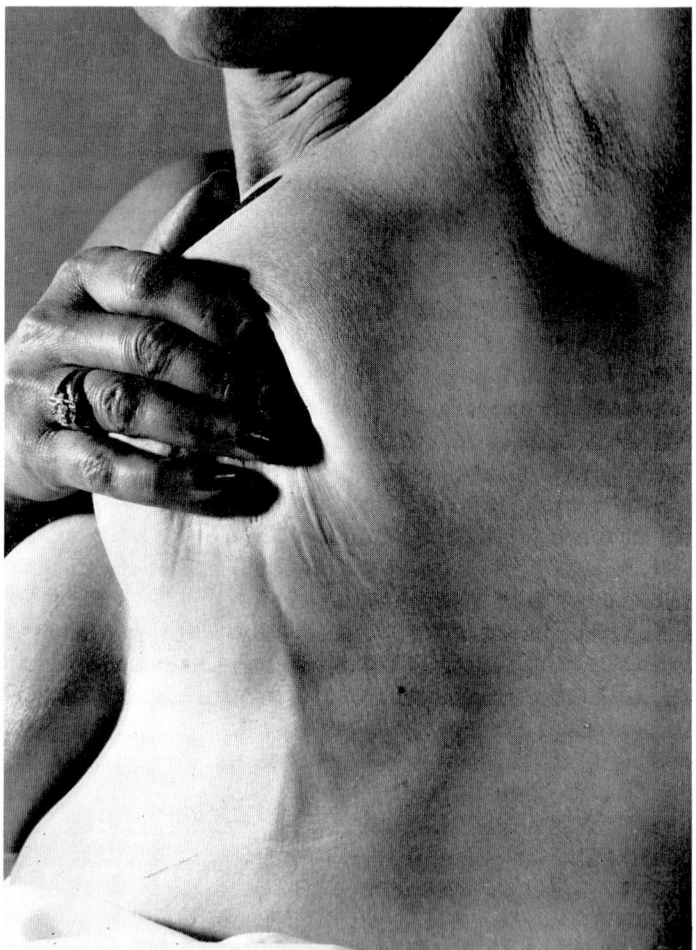

FIG. 151. Mondor's disease in a woman, aged 44 years, with tenderness along the fibrous bands in the inferior portion of the left breast and chest wall, which later disappeared spontaneously.

cyanosis, increased warmth, the presence of a bruit, evidence of unilateral venous insufficiency and the occasional finding of a positive Branham-Nicoladoni sign. Although patients with deep venous insufficiency also may have secondary edema, marked stasis pigmentation, ulceration and superficial varices that are frequently present may serve to distinguish this condition from lymphedema.

Mondor's disease, a rare entity, occasionally is seen and is mistaken for carcinomatous lymphangitis or superficial thrombophlebitis over the thoracic wall (Jönsson *et al.*, 1955). This disease is characterized by firm cordlike indurations in the subcutis commonly seen on the wall of the thorax or abdomen, rarely in the extremities (Fig. 151). The patient often will describe tension over the involved area and, on careful questioning, may mention previous trauma or a cutaneous or generalized infection immediately preceding the onset. Such an individual may be suspected of having superficial migratory thrombophlebitis or indeed a carcinomatous lymphangitis. Apparently this condition is the result of lymphangitis with a pronounced tendency to sclerosis.

G. LYMPHANGIOMAS

Lymphangiomas are benign overgrowths of lymph vessels and are classified microscopically as (1) simple or capillary, (2) cavernous, and (3) cystic (cystic hygroma). Whether these are congenital malformations, hamartomas or benign neoplasms cannot be categorically stated (Landing and Farber, 1956). However, they are often present at birth (Regenbrecht, 1959), usually lack the growth potential of a true neoplasm and apparently do not undergo malignant transformation. Sabin (1909) pointed out that the normal development of the lymphatic system proceeds through the formation of isolated lymph sacs, derived from veins that later are united by the thoracic duct, to peripheral growth of lymph vessels from endothelial sprouts that spread over body (see Section A.-1, Chapter XXIII). Similarly abnormal development may determine the type and distribution of a tumefactive lymphatic proliferation (lymphangioma). Proliferations of lymph vessels are far less common than those from blood vessels. Soule *et al.* (1955), in a study of 500 consecutive primary tumors of the soft tissues of the extremities, found that of the 13 per cent which were angiomatous only 5 per cent of these were lymphangiomas. (For discussion of angiomas, see (1) Vascular Anomalies Due to Developmental Defects, Section C., Chapter XXI.)

1. Lymphangiomas in the Extremity

Lymphangiomas in the extremity may range from small cutaneous lesions, lymphangioma cutis circumscriptum, which appear as small compressible nodules resembling deep-seated vesicles, through large cystic forms (hygroma), to diffuse involvement by lymphatic proliferation (lymphangiectasis; elephantiasis lymphangiectatica). The relationship of this latter form to congenital lymphedema has been mentioned (see Section B.-2b, above). Simple and cavernous lymphangiomas often are found in a mixed form, in that endothelial-lined spaces vary from capillary-like size to 1–2 mm in diameter, and they may be associated with hemangiomas (hemolymphangioma). Although more commonly seen in the skin or subcutaneous tissues, a lymphangioma occasionally may be encountered in almost any organ.

2. Cystic Lymphangioma (Hygroma) of Neck, Mediastinum and Related Structures (Fig. 151)

Cystic hygromas are loculated, fluid-filled, lymphatic tumors of developmental origin that usually occur in infancy, although they may appear later in life. They are seen more often in the neck and axilla than elsewhere and may involve the mediastinum. Embryologically they are thought to arise from sequestrated endothelial sprouts of the jugular sacs which later undergo independent proliferation. According to Goetsch (1938), extension of a cystic hygroma occurs through the proliferation of endothelial sprouts of the tumor. These then enlarge and develop additional locules and may infiltrate various tissues of the body.

Cystic hygromas may involve the face and mouth, may produce facial hypertrophy and asymmetry, and when the tongue is affected, distressing macroglossia (Robinson, 1959). Secondary infection of these abnormal lymph channels may produce increased swelling and complications of bacteremia, pneumonia and atelectasis. Wilson and Clarke (1955) reported death of a newborn infant due to macroglossia and macrocheilia from cystic hygroma.

Cervical hygromas, the most common form (Sariñana *et al.*, 1958), more often affect females usually before 2 years of age, although they may appear in middle age or in the elderly person (Fuller and Conway, 1959). Vaughn (1934) cited 155 cases from the literature to 1934, and Gross (1953) reported 112 cases of his own. Fuller and Conway recently reported 25 cystic hygromas seen in cervical, axillary, and cervicomediastinal locations and found that cosmetic complaints were due to the tumor in 80 per cent of their cases, although dysphagia and dyspnea were also present in some cases of mediastinal involvement. A gigantic cervical cystic hygroma in an infant necessitated a cesarean section due to dystocia; the hygroma was later successfully excised (Junco and Martin, 1956). However, an axillary hygroma resulted in death of an infant due to an arrest of delivery (Schubert, 1956).

Cystic hygromas are initially small; they often enlarge slowly, but may regress temporarily. Although frequently situated in the posterior triangle of the neck, they also may involve the axilla or the floor of the mouth and encircle the neck. Secondary infection of a cystic hygroma occasionally follows an upper respiratory infection; fatal toxemia may result. In cervicomediastinal hygromas symptoms may be due to encroachment and pressure of the tumor on vital structures in the neck and mediastinum (Lemon, 1931; Lim *et al.*, 1961). Dyspnea, wheezing, stridor, and cyanosis are caused by pressure on the respiratory passages and may require emergency surgical intervention to decompress or remove the cyst. Dysphagia, superior vena caval obstruction and paralysis also may occur but are less common. Crying, grunting or suddenly increased intrathoracic pressure may produce an increase in the cervical swelling. Chylothorax and chylopericardium are rare complications (Swift and Neuhof, 1946; Stratton and Grant, 1958). Roentgenographic examination of the thorax may demonstrate a mediastinal mass, often with a soft tissue shadow in the neck.

Cystic hygroma of the mediastinum alone usually presents as an asymptomatic mediastinal mass, the nature of which is not apparent until thoracotomy is undertaken for diagnosis and treatment (Childress *et al.*, 1956; Divertie *et al.*, 1960). Since it possesses no cervical or axillary counterpart and infrequently compresses vital structures, it may not be found until adult life. This tumor is even rarer than the cervicomediastinal variety (Hall and Blades, 1957; Divertie *et al.*, 1960). It usually lies anteriorly in the mediastinum and in the right side of the thorax. In a review of 58 primary tumors of the mediastinum in infants and children, Ellis and DuShane (1956) found that 6.9 per cent were cystic hygromas. Childress *et al.* (1956) found reports of 17 cases of mediastinal cystic hygroma treated surgically and added one case of their own. Divertie *et al.* (1960) reported two cases of mediastinal cystic hygroma and pointed out that these tumors may reach a large size, may be asymptomatic and rarely produce symptoms of mediastinal compression, due to the lack of a cervical component and a superior mediastinal communication.

a. Pathologic findings: Microscopically, these thin-walled cystic masses are lined by endothelium and in their walls there may be connective tissue, fat, smooth muscle, nerves, small blood vessels and lymphoid tissue (Fig. 152*b*, *c* and *d*). The finding of irregular, thin bands of smooth muscle and lymphoid aggregates aids in the microscopic distinction of cystic hygroma from other cystic tumors. The septa between locules may be complete or incomplete and on this depends the

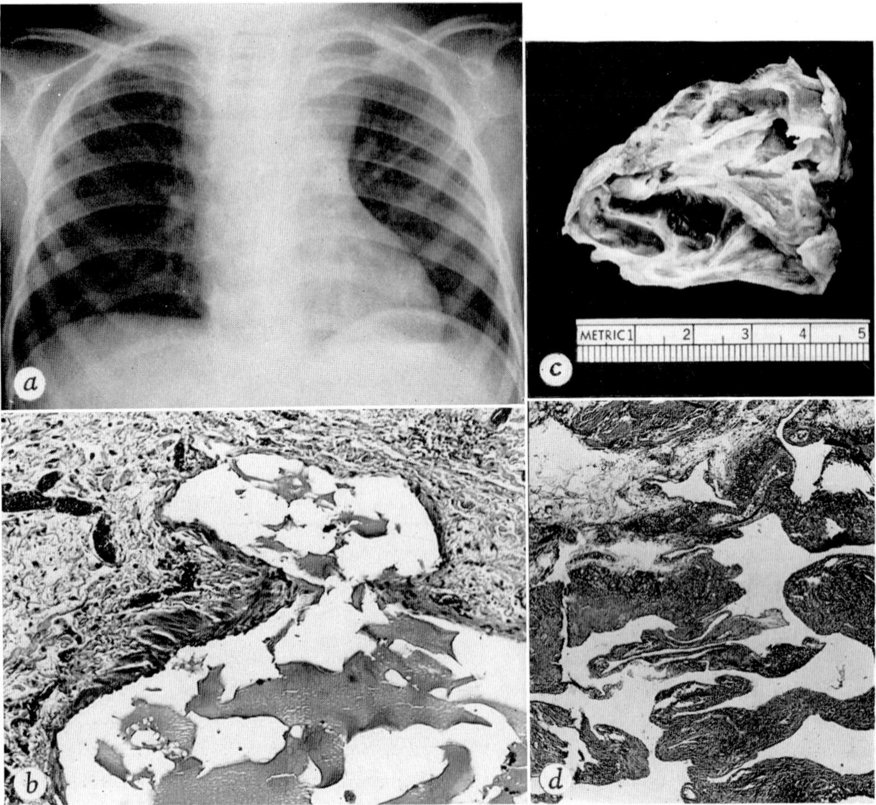

FIG. 152. Cystic hygroma. *a.* Chest roentgenogram showing a cervicomediastinal cystic hygroma in a 2½-year-old girl whose symptoms included exertional dyspnea, episodic cyanosis and slight dysphagia. *b.* Same case as *a.* Dilated endothelium-lined spaces of cystic hygroma from above which are filled with lymph and have connective tissue and variable smooth muscle in their walls (H and E; reduced from ×100).

c. Opened gross specimen of cervical cystic hygroma which suddenly appeared in the right posterior aspect of the neck of a 50-year-old man. Note the numerous locules and the thin walls. *d.* Same case as *c.* Communicating spaces of cystic hygroma above the endothelium-lined walls which contain thin bands of smooth muscle and focal lymphocytic infiltration (H and E; reduced from ×30).

communication of one portion of the tumor with another. The contained fluid is clear or straw-colored, although it is practically free of albumin and globulin and thus does not coagulate. White blood cells, phagocytes, nuclear fragments and other cellular debris may be centrifuged with the fluid. Cholesterol crystals may be present when there has been previous hemorrhage within the cyst. Rupture of small blood vessels in the wall of a cystic hygroma may produce a frankly bloody fluid content.

3. Visceral Lymphangiomas

Although lymphangiomas may occur in various isolated visceral sites, their occurrence in certain locations has received special attention in the medical literature.

a. Pulmonary lymphangiomas: Stephens *et al.* (1958) reported the third case of peripheral lymphangioma of the lung and suggested that this lesion may be due to mechanical blocking of the lymph channels by focal peripheral dilatation, or it may be on a congenital basis. Other pulmonary lymphatic disorders include pulmonary lymphatic cysts, which may be part of a generalized process, and congenital pulmonary lymphangiectasis. Frank and Piper (1959) found 21 cases of congenital pulmonary lymphangiectasis reported in the literature to 1959 and added two of their own. This developmental anomaly probably is due to the persistence of the lymphatics in fetal proportions while growth of pulmonary parenchyma continues (Laurence, 1959). It is usually found at necropsy on stillborn or neonatal infants. Progressive cyanosis is usually noted in the afflicted newborn, which may be due either to the relative inelasticity of the lung and immature parenchyma or to an associated congenital heart disease, rather than from poor aeration because of dilated lymphatics. In the lung, diffuse, irregular dilatation of the lymph vessels is seen.

b. Abdominal and retroperitoneal lymphangiomas: Abdominal lymphangiomas are extremely rare. Raiford (1932) found one in 11,500 necropsies. Davis *et al.* in 1959 collected 10 cases of lymphangioma of the gastrointestinal tract in the literature, in six of which the lesion was in the jejunum or ileum, in three in the rectum, and in one in the colon. They also reported the first lymphangioma in the duodenum of a 65-year-old man which was removed surgically. They stated that it arose from the lymphatic plexus of the submucosa into which the lacteals empty. Theories of origin include a development from misplaced embryonic tissue or a result of dilatation of lymphatics, due to back pressure from blockage of mesenteric glands.

Rauch (1959) found 20 cases of retroperitoneal lymphangioma in the literature and added two of his own. These lesions usually occur in cystic masses, either unilocular or multilocular, and they too are thought to be derived from congenitally misplaced lymphatic tissue, rather than from obstruction of existing lymph channels. Underhill (1959) suggested that the rapid development of a mesenteric lymphangioma may be due to dilatation of congenital lymphatic anomalies following torsion, inflammation, or changes in pressure. Symptoms usually are related to abdominal pain or a mass. Chyluria may be associated with retroperitoneal cystic hygroma. Such a case was reported by Morse *et al.* (1958) in which removal of 8 by 10 cm of tissue from the lumbar region and subsequent stripping of lymph channels of the renal pelvis resulted in cure.

An unusual lymphangioma occurs in the external genitalia or groin of children (Gueukdjian, 1958). It appears as a swelling or cyst and is mistaken clinically for hernia or hydrocele. The real nature of the lesion is discovered at operation.

Parsons in 1936 mentioned 500 cases of mesenteric cysts in the literature, only 10 of which were cystic lymphangioma. Chylous cysts of the mesentery are among the rarest of abdominal tumors (Simon and Williamson, 1956). Chylous cyst of lymphatic origin also may occur as a benign vascular neoplasm (chylangioma) or as a result of obstruction. Chyle, fibrin, inspissated fat, and blood may be found in the cyst. Symptoms include pain, a palpable abdominal mass and partial intestinal obstruction. To be distinguished from chylous cysts of lymphatic origin are enteric and dermoid cysts. Beahrs *et al.* (1950) presented 9 cases of chylous cyst of the abdomen found among 174 abdominal, mesenteric, retroperitoneal and omental cysts diagnosed in more than 1,000,000 patients seen at the Mayo Clinic. The five women and four men affected ranged in age from 24 to 62 years. The sizes of the cysts varied from 8 to 17 cm in diameter, and one filled the abdomen. Most of them were in the mesentery of the small bowel. Five had endothelial linings, and all contained milky fluid. Beahrs *et al.* also pointed out the difficulty in classification of chylous cysts which may be (1) embryologic or developmental, (2) traumatic or acquired, (3) neoplastic, and (4) infective or degenerative. Two patients were considered to have malignant lymphangioendothelioma and died after operation.

Raszkowski *et al.* (1959) pointed out two types of intra-abdominal lymphangiomas: one, in which the vast majority fall, consists of lymphatic cysts of the mesentery, which occur in infants and children, are unilocular or of dumbbell shape and are cured by simple excision; the other type, which is less frequent, is a multicystic lesion which invades by endothelial sprouting and has a high rate of recurrence. This rare lymphangiomatous lesion of the mesentery has been described as lymphangiectasis, since it is an extensive tumor; endothelial-lined spaces honeycomb the mesentery and make complete excision difficult (Amos, 1959). Clinical symptoms include (1) painless enlargement of the abdomen, (2) acute abdominal symptoms with volvulus, rupture or intestinal obstruction and (3) transient abdominal pains and vomiting.

4. Lymphangioma of Bone

Erosion of bone and cystic destruction have been reported in a few cases of lymphangioma of soft tissues secondarily involving bone. Three cases of cystic lymphangioma of bone are cited by Falkmer and Tilling (1956); in two of these there also were lymphangiomas in other organs. Cohen and Craig (1955) reported progressive destruction of multiple bones by lymphangiectatic lesions which resulted in compression of the cervical portion of the spinal cord and death of the child.

H. LYMPHANGIOSARCOMA IN CHRONIC LYMPHEDEMA

A rare primary lymphangiosarcoma may arise in chronic lymphedema of various types (Figs. 153 and 154). Stewart and Treves (1948) first described the development of a highly malignant lymphangiosarcoma of the upper extremity as a complication of long-standing postmastectomy lymphedema. According to McConnell and Haslam (1959), 28 cases of this syndrome, including 5 of their own, are reported in the literature. The incidence of this syndrome was estimated at 0.45 per cent in 894 patients with carcinoma of the breast who survived at least 5 years following radical mastectomy with or without radiotherapy.

At first the possibility of a circulating carcinogen was considered a factor in the development of this syndrome. Now importance is placed on the role of chronic lymphedema, and thus the mechanisms which produce it are etiologically important factors.

The clinical histories of patients with lymphangiosarcoma are similar, in that persistent lymphedema of the arm usually developed shortly after radical mastectomy for breast carcinoma (Ferraro, 1950; Fry *et al.*, 1959; Froio and Kirkland, 1952; Hilfinger and Eberle, 1953; Ogilvy *et al.*, 1959; Taswell *et al.*, 1961). After a cancer-free latent period, varying from 5 to 24 years (average, 8 years), lymphangiosarcoma develops in the lymphedematous extremity. The left arm has been involved almost twice as frequently as the right in reported cases. The patient first notes a solitary, purple-red macular or nodular lesion on the arm or antecubital region which she considers to be a "bruise." This is rapidly followed by multiple, satellite tumors on the arm, forearm, dorsum of hand, and even adjacent skin of the thorax. Individual lesions may

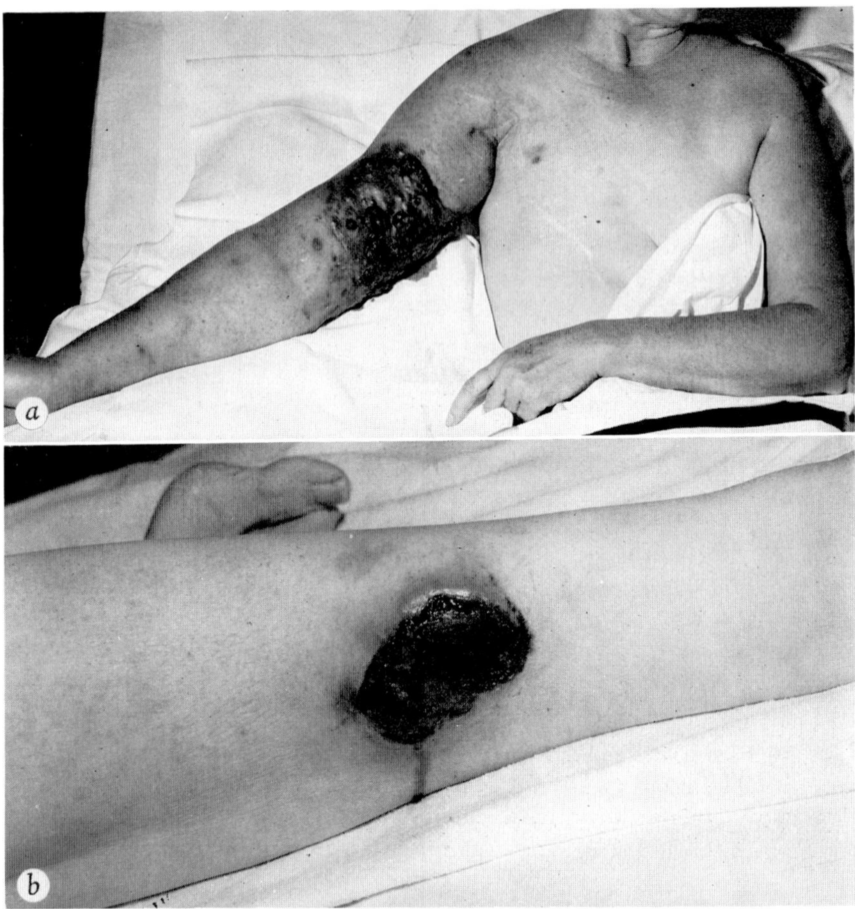

Fig. 153*a*. Lymphangiosarcoma in chronic lymphedema. Postmastectomy lymphangiosarcoma (Stewart-Treves syndrome) in a woman, aged 48 years. Radical mastectomy at age 40 years was followed by roentgen therapy for carcinoma of right breast. Lymphedema of right upper extremity developed thereafter. A bluish discoloration over right elbow noted 4 months prior to admission was followed by a nodular bluish ulcerating mass of lymphangiosarcoma. Additional lesions are seen on extremity and chest wall. Interscapulothoracic amputation was done and x-ray therapy was given, and patient died 3½ years later from acute coronary occlusion without known recurrence of lymphangiosarcoma. (See Figure 154*a* for histologic section.) *b*. Lymphangiosarcoma following primary lymphedema (Kettle's syndrome) in a 65-year-old woman with lymphedema praecox of 49 years' duration. In 3 months discoloration and then an ulcerating bluish mass of lymphangiosarcoma developed on medial side of left leg. Patient is alive 3½ years following an above-knee amputation, but recurrence has been excised. (See Figure 154*b* for histologic section.)

have a flattened surface with a grayish-red central zone and a purplish-red perimeter. When larger, they may become ulcerated and crusted or semi-necrotic and cystic; however, they usually remain confined to the dermis and subcutaneous tissue.

Microscopically, these tumors are composed of malignant endothelial cells, which may vary from almost solid sheets or nests of anaplastic

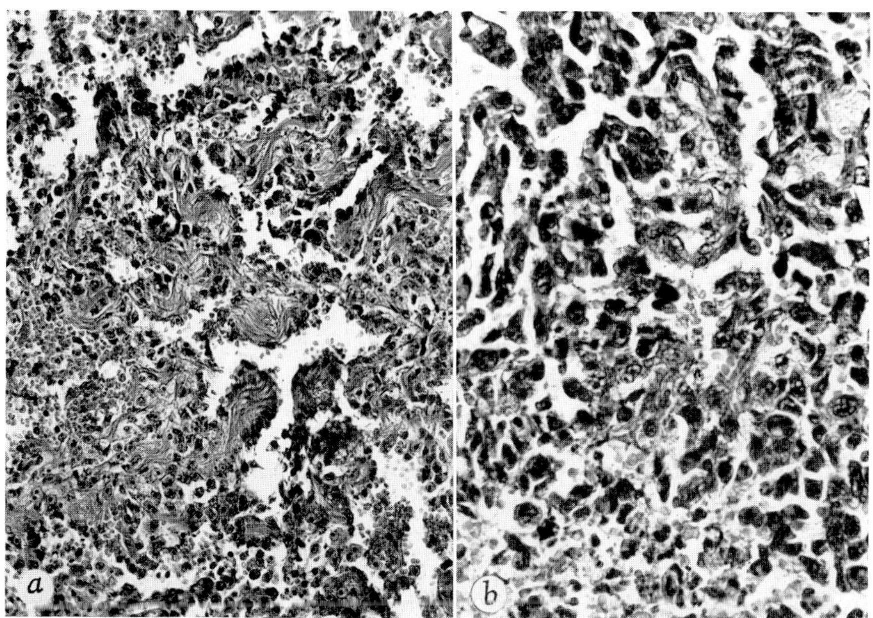

FIG. 154*a*. Same case as Figure 153*a*. Lymphangiosarcoma in chronic lymphedema. Angiosarcoma showing anastomosing spaces lined by malignant endothelial cells. Note papillary tuft formation and presence of some erythrocytes in spaces which suggest mixed blood vessel and lymphatic origin (H and E; reduced from ×200). *b*. Same case as Figure 153*b*. Lymphangiosarcoma following primary lymphedema. The malignant endothelial cells have a spindling pattern and some line anastomosing spaces (H and E; reduced from ×350).

spindle cells, at times mistaken for metastatic carcinoma, through papillary endothelium-lined spaces to formations mimicking lymph channels. Although these lymphatic-like formations are usually lined by a layer of frankly malignant endothelium, almost imperceptible gradations may be found, to lymphatics which are benign in appearance, although increased in number. McConnell and Haslam (1959) suggested that multifocal proliferation of endothelial cells in the lower part of the dermis and subcutaneous areas may be a "premalignant angiomatosis" of this syndrome and may correspond to the early evidence of bruising. Since some spaces may contain blood, a possible dual origin from lymphatic and blood vessels has been suggested (Hilfinger and Eberle, 1953; Ogilvy *et al.*, 1959). Venous infiltration is readily demonstrated in these tumors.

In several reported cases lymphangiosarcoma followed primary lymphedema (Hermann and Gruhn, 1957; McConnell and Haslam, 1959; Taswell *et al.*, 1961). The condition was first described in 1918 by Kettle in a case of congenital elephantiasis of the right leg complicated at the age of 44 years by angiosarcoma. The development and rapid progression of the malignant lesion are similar to those in postmastectomy lymphangiosarcoma and quite different from the slow evolution of Kaposi's sarcoma. Other features of distinction from Kaposi's sarcoma have been pointed out (Fry *et al.*, 1959; Froio and Kirkland, 1952; McConnell and Haslam, 1959).

References

Allen, E. V. (1934). Lymphedema of the extremities; classification, etiology and differential diagnoses: A study of three-hundred cases. *A. M. A. Arch. Internal Med.* **54,** 606.

Allen, E. V., and Ghormley, R. K. (1935). Lymphedema of the extremities; Etiology, classification and treatment: Report of 300 cases. *Ann. Internal Med.* **9,** 516.

Allen, E. V., Barker, N. W., and Hines, E. A. (1955). "Peripheral Vascular Diseases," 2nd ed. Saunders, Philadelphia, Pennsylvania.

Amos, J. A. S. (1959). Multiple lymphatic cysts of the mesentery. *Brit. J. Surg.* **46,** 588.

Arnulf, G. (1957). Précisions techniques sur la lymphographie des membres. *Lyon chir.* **53,** 772.

Arnulf, G. (1958). Practical value of lymphography of the extremities. *Angiology* **9,** 1.

Battezzati, M., Donini, I., Tagliaferro, A., Rossi, L., and Bocchialini, C. (1959). Modificazioni morfo-funzionali dei collettori linfatici dell'arto inferiore nella sindrome post-tromboflebitica: Studio linfografico. *Minerva chir.* **14,** 1209.

Beahrs, O. H., Judd, E. S., Jr., and Dockerty, M. B. (1950). Chylous cysts of the abdomen. *Surg. Clin. North Am.* **30,** 1081.

Britton, R. C. (1959). Management of peripheral edema, including lymphedema of the arm after radical mastectomy. *Cleveland Clinic Quart.* **26,** 53.

Butcher, H. R., Jr., and Hoover, A. L. (1955). Abnormalities of human superficial cutaneous lymphatics associated with stasis ulcers, lymphedema, scars and cutaneous autografts. *Ann. Surg.* **142,** 633.

Childress, M. E., Baker, C. P., and Samson, P. C. (1956). Lymphangioma of the mediastinum: Report of a case with review of the literature. *J. Thoracic Surg.* **31,** 338.

Chisholm, E. R., Hutch, J. A., and Bolomey, A. A. (1954). Bilateral ureteral obstruction due to chronic inflammation of the fascia around the ureters. *J. Urol.* **72,** 812.

Cohen, J., and Craig, J. M. (1955). Multiple lymphangiectases of bone. *J. Bone and Joint Surg.* **37A,** 585.

Colette, J. M. (1957). Étude radiologique de la circulation lymphatique superficielle et des relais ganglionnaires correspondants: Considérations expérimentales et cliniques. Les diagnostics lymphographiques. *Bruxelles Med.* **37,** 1869.

Colette, J. M. (1958). Envanissements ganglionnaires inguino-ilio-pelviens par lymphographie. *Acta Radiol.* **49,** 154.

Csik, L., Fodor, I., and Riesz, E. (1958). Experimentalis adatok juvenilis rheumatoid arthritisben kifejlödött elephantiasis vizsgálata kapcsan. *Orvosi Hetilap* **99,** 857.

Davis, J. G., Peck, H., and Gray, B. L. (1959). Lymphangioma of the duodenum. *Am. J. Roentgenol., Radium Therapy Nuclear Med.* **81,** 613.

De Lutio, O., Sabaino, D., and Fonda, G. (1954). Rigenerazione dei linfatici dopo la loro interruzione totale. *Arch. ital. chir.* **77,** 369.

Divertie, M. B., Lim, R. A., Harrison, E. G., Jr., Bernatz, P. E., and Burger, T. C. (1960). Mediastinal cystic hygromas: Report of 2 cases. *Proc. Staff Meetings, Mayo Clinic* **35,** 460.

Ellis, F. H., Jr., and DuShane, J. W. (1956). Primary mediastinal cysts and neoplasms in infants and children. *Am. Rev. Tuberc.* **74,** 940.

Falkmer, S., and Tilling, G. (1956). Primary lymphangioma of bone. *Acta Orthopaed. Scand.* **26,** 99.

Ferraro, L. R. (1950). Lymphangiosarcoma in postmastectomy lymphedema: a case report. *Cancer* **3,** 511.

Frank, J., and Piper, P. G. (1959). Congenital pulmonary cystic lymphangiectasis. *J. Am. Med. Assoc.* **171,** 1094.

Froio, J. F., and Kirkland, W. G. (1952). Lymphangiosarcoma in postmastectomy lymphedema. *Ann. Surg.* **135,** 421.

Fry, W. J., Campbell, D. A., and Coller, F. A. (1959). Lymphangiosarcoma in postmastectomy lymphedematous arm. *A. M. A. Arch. Surg.* **79,** 440.

Fuchs, W. A., Rüttimann, A., and del Buono, M. S. (1959). Klinische Indikationen zur Lymphographie. *Schweiz. med. Wochschr.* **89,** 755.

Fuller, F. W., and Conway, H. (1959). Cystic hygroma. *Surg., Gynecol. Obstet.* **108,** 457.

Gergely, R. (1958a). The roentgen examination of lymphatics in man. *Radiology* **71,** 59.

Gergely, R. (1958b). Die Bedeutung der Lymphangiographie in der Chirurgie. *Chirurg* **29,** 49.

Goetsch, E. (1938). Hygroma colli cysticum and hygroma axillare: pathologic and clinical study and report of 12 cases. *A. M. A. Arch. Surg.* **36,** 394.

Gross, R. E. (1953). "The Surgery of Infancy and Children: Its Principles and Techniques." Saunders, Philadelphia, Pennsylvania.

Gueukdjian, S. A. (1958). Lymphangioma of genitalia in children. *Pediatrics* **22,** 247.

Haberlin, J. P., Milone, F. P., and Copeland, M. M. (1959). A further evaluation of lymphedema of the arm following radical mastectomy and postoperative x-ray therapy. *Am. Surgeon* **25,** 285.

Hall, E. R., Jr., and Blades, B. (1957). Lymphangioma of the mediastinum: Report of two cases. *Diseases of Chest* **32,** 207.

Halsted, W. S. (1921). Swelling of arm after operation for cancer of breast—elephantiasis chirurgica—its cause and prevention. *Bull. Johns Hopkins Hosp.* **32,** 309.

Hawk, W. A., and Hazard, J. B. (1959). Sclerosing retroperitonitis and sclerosing mediastinitis. *Am. J. Clin. Pathol.* **32,** 321.

Hermann, J. B., and Gruhn, J. G. (1957). Lymphangiosarcoma secondary to chronic lymphedema. *Surg., Gynecol. Obstet.* **105,** 665.

Hilfinger, M. F., and Eberle, R. D. (1953). Lymphangiosarcoma in postmastectomy lymphedema. *Cancer* **6,** 1192.

Hudack, S. S., and MacMaster, P. D. (1933). The lymphatic participation in human cutaneous phenomena: Study of minute lymphatics of living skin. *J. Exptl. Med.* **57,** 751.

Jacobsson, S., and Johansson, S. (1959). Lymphangiographic changes in lower limbs with varicose veins. *Acta Chir. Scand.* **117,** 346.

Jennett, J. H. (1956). Persistent hereditary edema of the legs—Milroy's disease. *Clin. Orthopaed.* **8,** 122.

Jönsson, G., Linell, F., and Sandblom, P. (1955). Subcutaneous cords on the trunk. *Acta Chir. Scand.* **108,** 351.

Junco, T., and Martin, M. H. (1956). Successful surgical extirpation of gigantic congenital cystic hygroma: Case report. *Surgery* **39,** 315.

Kaindl, F. (1957a). Das Lymphgefäßsystem in menschlichen Extremitäten. *Wien. med. Wochschr.* **107,** 200.

Kaindl, F. (1957b). Zur Pathologie der Lymphbahnen in menschlichen Extremitäten. *Deut. med. J.* **8,** 209.

Kaindl, F., Mannheimer, E., and Thurnher, B. (1958). Lymphangiographie und Lymphadenographie am Menschen. *Fortschr. Geb. Röntgenstrahlen* **89,** 1.

Kessel, I. (1952). Chylous ascites in infancy. *Arch. Disease Childhood* **27,** 79.

Kettle, E. H. (1918). Tumors arising from endothelium. *Proc. Roy. Soc. Med.* **11,** 19.

Kinmonth, J. B., Harper, R. A. K., and Taylor, G. W. (1955a). Lymphangiography by radiological methods. *J. Fac. Radiologists* **6,** 217.

Kinmonth, J. B., Taylor, G. W., and Harper, R. K. (1955b). Lymphangiography: a technique for its clinical use in the lower limb. *Brit. Med. J.* **i,** 940.

Kinmonth, J. B., Taylor, G. W., Tracy, G. D., and Marsh, J. D. (1957). Primary lymphoedema: Clinical and lymphangiographic studies of a series of 107 patients in which the lower limbs were affected. *Brit. J. Surg.* **45,** 1.

Klein, D., and Doret, M. (1957). Lymphoédéme chronique héréditaire (maladie de Nonne-Milroy-Meige) du type récessif. *Bibliotheca Ophthalmol.* **1,** 576.

Landing, B. H., and Farber, S. (1956). Tumors of the cardiovascular system. *In* "Atlas of Tumor Pathology" (Armed Forces Inst. Pathol.), Section III, Fascicle 7, p. 124. National Research Council, Washington, D. C.

Laurence, K. N. (1959). Congenital pulmonary lymphangiectasis. *J. Clin. Pathol.* **12,** 62.

Laurentaci, G., and Montinari, M. (1956). Linfografie degli arti ed anestesia locale (Ricerche sperimentali). *Boll. soc. ital. biol. sper.* **35,** 1054.

Leenhardt, P., Colin, R., and Pélissier, M. (1956). Conception fonctionnelle du système lymphatique. *J. radiol. et électrol.* **40,** 574.

Lemon, W. S. (1931). Rare intrathoracic tumors. *Med. Clin. North Am.* **15,** 17.

Lim, R. A., Divertie, M. B., Harrison, E. G., Jr., and Bernatz, P. E. (1961). Cervicomediastinal cystic hygroma. *Diseases of Chest* **40,** 265.

McConnell, E. M., and Haslam, P. (1959). Angiosarcoma in postmastectomy lymphoedema: A report of 5 cases and a review of the literature. *Brit. J. Surg.* **46**, 322.

McKendry, J. B., Lindsay, W. K., and Gerstein, M. C. (1957). Congenital defects of the lymphatics in infancy. *Pediatrics* **19**, 21.

Málek, P., and Belán, A. (1959). Dnešní stav rentgenolymfografie, její cíle a perspektiva. *Českoslov. rentgenol.* **13**, 407.

Málek, P., and Kolc, J. (1959). Otázky obnovování toku lymfy. *Rozhledy chir.* **38**, 441.

Málek, P., Belán, A., and Kolc, J. (1956). Metoda znázornění hlubokého lymfatického systému lumbální krajiny. *Českoslov. rentgenol.* **13**, 343.

Málek, P., Kolc, J., and Belán, A. (1959a). Lymphangiography of the deep lymphatic system of the thigh. *Acta Radiol.* **51**, 422.

Málek, P., Kolc, J., and Belán, A. (1959b). Problematika rentgenolymfografie hlubokého lymfatického systému pánve a dolní končetiny. *Českoslov. rentgenol.* **13**, 54.

Marcozzi, G., and Ricci, G. A. (1956). Linfografia diretta transcutanea. *Ann. ital. chir.* **33**, 401.

Mason, P. B., and Allen, E. V. (1935). Congenital lymphangiectasis (lymphedema). *Am. J. Diseases Children* **50**, 945.

Meige, H. (1898). Dystrophie oedemateuse héréditaire. *Presse méd.* **102**, 341.

Miles, A. A., and Miles, E. M. (1958). The state of lymphatic capillaries in acute inflammatory lesions. *J. Pathol. Bacteriol.* **76**, 21.

Milroy, W. F. (1892). An undescribed variety of hereditary oedema. *N. Y. Med. J.* **56**, 505.

Milroy, W. F. (1928). Chronic hereditary edema: Milroy's disease. *J. Am. Med. Assoc.* **91**, 1172.

Morris, B., Blood, D. C., Sidman, W. R., Steele, J. D., and Whittem, J. H. (1954). Congenital lymphatic oedema in Ayrshire calves. *Australian J. Exptl. Biol. Med. Sci.* **32**, 265.

Morse, W. H., Diggs, L. W., and Raines, S. L. (1958). Unilateral chyluria associated with hereditary spherocytosis and retroperitoneal cystic hygroma: Case report. *J. Urol.* **79**, 153.

Nonne, M. (1891). Vier Fälle von Elephantiasis congenita hereditaria. *Virchows Arch. pathol. Anat. u. Physiol.* **125**, 189.

Ochsner, A., Longacre, A. B., and Murray, S. D. (1940). Progressive lymphedema associated with recurrent erysipeloid infection. *Surgery* **8**, 383.

Ogilvy, W. L., Franklin, R. H., and Aird, J. (1959). Angioblastic sarcoma in postmastectomy lymphoedema. *Can. J. Surg.* **2**, 195.

Parsons, E. O. (1936). True proliferating cystic lymphangioma of the mesentery. *Ann. Surg.* **103**, 595.

Prokopec, J., and Kolihová, E. (1958). Die Lymphadenographie in der klinischen Praxis. *Fortschr. Geb. Röntgenstrahlen* **89**, 417.

Raiford, T. S. (1932). Tumors of small intestine. *A. M. A. Arch. Surg.* **25**, 122.

Raszkowski, H. J., Rehbock, D. J., and Cooper, F. G. (1959). Mesenteric and retroperitoneal lymphangioma. *Am. J. Surg.* **97**, 363.

Rauch, R. F. (1959). Retroperitoneal lymphangioma. *A. M. A. Arch. Surg.* **78**, 45.

Regenbrecht, J. (1959). Das Lymphangiom. *Münch. med. Wochschr.* **101**, 2197.

Robinson, D. W. (1959). Lymphangioma of the lower face and neck involving mandible. *Plast. & Reconstruct. Surg.* **23**, 187.

Sabaino, D., De Lutio, O., and Dazzani, D. (1955). Sulle modificazioni della circolazione linfatica nelle linfoadeniti e nelle linfangiti. *Ann. ital. chir.* **32,** 984.

Sabin, F. R. (1909). The lymphatic system in human embryo; with a consideration of the morphology of the system as a whole. *Am. J. Anat.* **9,** 43.

Sariñana, C., Alamillo, J., Delgado, J. L., and Esparza, H. (1958). Linfangioma quístico del cuello. *Bol. méd. hosp. infantil (Mex.)* **15,** 657.

Schubert, E. (1956). Lymphangiom der Achselhöhle als Geburtshindernis Dekapitation, Wendung. *Geburtshilfe u. Frauenheilk.* **16,** 706.

Shanbrom, E., and Zheutlin, N. (1959). Radiographic studies of the lymphatic system. *A. M. A. Arch. Internal Med.* **104,** 589.

Simon, H. E., and Williamson, B. (1956). Retroperitoneal chylous cysts. *Am. J. Surg.* **91,** 372.

Smedal, M. I., and Evans, J. A. (1960). The cause and treatment of edema of the arm following radical mastectomy. *Surg., Gynecol. Obstet.* **111,** 29.

Soule, E. H., Ghormley, R. K., and Bulbulian, A. H. (1955). Primary tumors of the soft tissues of the extremities exclusive of epithelial tumors: an analysis of five-hundred consecutive cases. *A. M. A. Arch. Surg.* **70,** 462.

Stephens, F. G., Roberts, S. M., and Wolcott, M. W. (1958). Peripheral lymphangioma of the lung. *J. Thoracic Surg.* **36,** 182.

Stewart, F. W., and Treves, N. (1948). Lymphangiosarcoma in postmastectomy lymphedema: a report of six cases in elephantiasis chirurgica. *Cancer* **1,** 64.

Stillwell, G. K. (1959). Physical medicine in management of patients with post-mastectomy lymphedema. *J. Am. Med. Assoc.* **171,** 2285.

Stirlacci, J. R. (1955). Spontaneous chylothorax in a newborn infant. *J. Pediat.* **46,** 581.

Stratton, V. C., and Grant, R. N. (1958). Cervicomediastinal cystic hygroma associated with chylopericardium. *A. M. A. Arch. Surg.* **77,** 887.

Swift, E. A., and Neuhof, H. (1946). Cervicomediastinal lymphangioma with chylothorax. *J. Thoracic Surg.* **15,** 173.

Tasatti, E. (1957). La linfangiografia per lo studio dei linfatici e linfedemi dell arto inferiore. *Rass. clin.-sci.* **33,** 161.

Taswell, H. F., Soule, E. H., and Coventry, M. B. (1961). Lymphangiosarcoma arising in chronic lymphedematous extremities: Report of thirteen cases and a review of literature. Read at the meetings of the American Orthopaedic Association, Yosemite, California, May 22 to 25.

Taylor, G. W., Kinmonth, J. B., Rollinson, E., Rotblat, J., and Francis, G. E. (1957). Lymphatic circulation studied with radioactive plasma protein. *Brit. Med. J.* **i,** 133.

Taylor, G. W., Kinmonth, J. B., and Dangerfield, W. G. (1958). Protein content of oedema fluid in lymphoedema. *Brit. Med. J.* **i,** 1159.

Treves, N. (1951). An evaluation of the etiological factors of lymphedema following radical mastectomy: An analysis of 1,007 cases. *Cancer* **10,** 444.

Underhill, B. M. L. (1959). Acute intestinal obstruction due to mesenteric lymphangioma. *Arch. Disease Childhood* **34,** 442.

Vaughn, A. M. (1934). Cystic hygroma of the neck. *Am. J. Diseases Children* **48,** 149.

Verger, P., and Blaquière, R. (1957). Maladie de Meige-Milroy-Nonne ou trophoedème héréditaire chronique. *Pédiatrie* **12,** 37.

Vítek, J., Zejda, V., and Kašpar, M. Z. (1954). Naše zkušenosti s přímou lymforadiografií se zaměřením na lymforadiografii dolních končetin. *Českoslov. rentgenol.* 13, 349.

Vítek, J., Zejda, V., and Kašpar, M. Z. (1958). Naše zatímní zkušenosti s lymfografií. *Rozhledy chir.* **37**, 94.

Warwick, W. J., Holman, R. T., Quie, R. G., and Good, R. A. (1959). Chylous ascites and lymphedema. *Am. J. Diseases Children* **98**, 317.

Wilson, J. W., and Clarke, J. C. (1955). Neonatal respiratory obstruction due to hygroma colli cysticum. *Ulster Med. J.* **24**, 135.

Young, J. R., and DeWolfe, V. G. (1960). Recurrent lymphangiitis of the leg associated with dermatophytosis: Report of 25 consecutive cases. *Cleveland Clin. Quart.* **27**, 19.

AUTHOR INDEX

Numbers in italics indicate the pages on which the references are listed.

C

D

E

F

G

H

I

J

K

L

M

P

S

SUBJECT INDEX

A

B

C

D

E

F

T

U

W

X